NATURAL RESOURCES
Ecology, Economics, and Policy

SECOND EDITION

NATURAL RESOURCES
Ecology, Economics, and Policy

JERRY L. HOLECHEK

RICHARD A. COLE

JAMES T. FISHER

RAUL VALDEZ

New Mexico State University

Upper Saddle River, New Jersey 07458

Library of Congress Cataloging-in-Publication Data

Natural resources : ecology, economics, and policy / Jerry L. Holecheck ... [et al.].—2nd ed.
      p. cm.
   Includes bibliographical references and index.
   ISBN 0-13-093388-0
     1. Natural resources. 2. Natural resources—Government policy. 3. Ecology. I.
Holechek, Jerry.

HC85 .N37 2003
33.7—dc21   2002022043

Editor-in-Chief: Stephen Helba
Executive Editor: Debbie Yarnell
Editoral Assitant: Sam Goffinet
Production Editor: Lori Dalberg, Carlisle Publishers Services
Director of Manufacturing and Production: Bruce Johnson
Managing Editor: Mary Carnis
Marketing Manager: Jimmy Stephens
Manufacturing Buyer: Cathleen Petersen
Design Director: Cheryl Asherman
Senior Design Coordinator: Miguel Ortiz
Cover Designer: Marianne Frasco
Composition: Carlisle Communications, Ltd.
Printing and Binding: Courier Westford

Pearson Education LTD.
Pearson Education Australia PTY, Limited
Pearson Education Singapore, Pte. Ltd.
Pearson Education North Asia Ltd.
Pearson Education Canada, Ltd.
Pearson Educación de Mexico, S.A. de C.V.
Pearson Education—Japan
Pearson Education Malaysia, Pte. Ltd.

10 9 8 7 6 5 4 3 2 1
ISBN 0-13-093388-0

Contents

Chapter 3 Basic Ecology   63

Chapter 4 Conservation Economics   104

SECTION 2   AIR, WATER, AND LAND RESOURCES   157

Chapter 6   Atmospheric Resources and Climate   159

SECTION 3   THE LAND-BASED RENEWABLE RESOURCES 311

Chapter 10   Forests and Forestry 313

Chapter 12 Farmland and Food Production 391

Preface

The purpose of this book is to introduce students to the science of natural resource management by coupling the latest concepts and technology with proven traditional approaches. We hope our intended audience includes managers on public and private lands, foresters, wildlife biologists, marine biologists, earth scientists, farmers, ranchers, hydrologists, urban planners, environmental scientists, conservation biologists, economists, politicians, and the growing segment of the public interested in natural resource management. We have tried to provide a comprehensive text for those concerned with natural resource management, not only in the United States but in other parts of the world as well.

Our approach has involved coupling fundamental topics such as plant ecology, soil science, climatology, economics, and policy with the most recent research. Some traditional concepts and viewpoints on natural resource management have been substantially altered as a result of new findings. This is particularly true in areas such as forestry, range management, plant ecology, energy conservation, urban planning, and wildlife management.

The management of natural resources has become more integrated through its 100-year history. Our textbook on natural resource management attempts to integrate ecological, economic, and policy factors into a functional, applied framework. We believe sound natural resource management requires an understanding of the interactions between natural and social processes. While the natural processes are predominantly ecological, the relevant social processes are primarily economic and political. We consider economics, the measurement of relative value, a critical part of natural resource management decision making. Therefore, we have made economics an important component of our book, in contrast to many of the other books on natural resource management. Political policy affects the will of society in regard to choices concerning how natural resources will be conserved, produced, managed, and allocated.

We recognize that many natural resources cannot be readily valued monetarily, such as scenic beauty, open space, natural wonders, and rare plant and animal species, but they contribute greatly to human quality of life. It is the intent of our book to help managers make more rational choices between material and spiritual well-being. At the same time, the future welfare of society must be considered as well as the present.

Because a diversity of benefits is usually the object of most natural resource management, today's resource managers need a broad understanding of various natural resource categories as well as specialization in certain specific areas. Foresters, wildlife biologists, geologists, range managers, environmental scientists, and others need a basic understanding of all the other disciplines. Most comprehensive natural resource management is done through teamwork and requires strong communication and interpersonal skills. Therefore, we have made planning an important component of our book.

While many books on the environment and natural resources have been somewhat pessimistic about the future on planet earth, our book presents a view of guarded optimism. Although the human population of the world continues to increase, major breakthroughs in technology show great promise for alleviating resource scarcity, pollution, and degradation. Improved political economic systems are resulting in more equitable distribution of resources to human populations in many parts of the world. At the same time, we readily acknowledge that elimination of rain forests, reductions in biodiversity, global warming, loss of open space, loss of the ozone layer, and the human population increase are all important natural resource challenges. However, we believe that through science, technology, universal education, and application of proven socioeconomic principles humanity can overcome these daunting problems.

We freely acknowledge that this text will not convey everything there is to know about natural resource management. However, we believe it will provide the reader with a powerful understanding of how basic principles of ecology and economics can be integrated to successfully solve natural resource problems.

We have drawn heavily from many other great books on various aspects of natural resource management to develop our subject. We express deep gratitude to Raymond Dasman, G. Tyler Miller, Jr., Bradley Schiller, Oliver S. Owen, Daniel D. Chiras, Eric G. Bolen, William L. Robinson, Bernard J. Nebel, Richard T. Wright, and Ronald D. Knutson whose textbooks provided the basis for coverage of many of our subjects. Appreciation is given to Dr. Frank A. Ward for his contribution to Chapter 4, Conservation Economics.

We received both encouragement and helpful criticism from many of our colleagues. Special thanks are given to John Davis, William H. Fleming, Sam Fuhlenborf, James Teer, and James D. Yoakum for their valuable suggestions on our manuscript.

JERRY L. HOLECHEK
RICHARD A. COLE
JAMES T. FISHER
RAUL VALDEZ

CHAPTER 1

Natural Resources: An Overview

We live in an era of increasing social and ecological awareness. Profound questions exist about whether or not healthy societies can be sustained without destruction of the natural resource bases they depend on. Answering these questions requires understanding of the interactions between natural and social processes. While the natural processes are predominantly ecological, the relevant social processes are primarily economic and political. Thus the content of this book goes beyond the typical inventory of natural resources encountered in many introductory texts to a larger view of natural resource ecology, economics, and policy. This treatment introduces students to the problems and opportunities, and some of the potential approaches taken as professionals and informed citizens in the field of natural resource management.

NATURAL RESOURCES DEFINED

In this book we apply an inclusive definition to natural resources. They are divided into basic categories: air, water, soil, natural ecosystems, forest land, rangeland, farmland, parks, urban lands, wildlife, fisheries, biodiversity, energy, and minerals. Under our definition, natural resources are the raw materials used to satisfy human needs. Humanity has grown to more than 5 billion because of the mobilization of natural resources required to meet basic needs for food, water, habitation, infrastructure, and other material well-being. Natural resources include nonliving and living things that interact and affect human welfare. Natural resources are differentiated from goods manufactured using human ingenuity and labor. Trees, fish, and food crops are natural resources in contrast to furniture and prepared meals, which incorporate substantial human inputs. Solar radiation, wind, coal, and water are natural resources, but the electricity produced from them is not. Sand and lime are natural resources, but the concrete made from them is not. Similarly, the plastics, fibers, glass, and various refined metals that compose many of our manufactured goods are not natural resources even though they require natural resources for their production.

As in all things, the boundaries between natural and artificial (an artifact of human modification) are often difficult to define. Whereas wild crops are definitely natural resources, cultivated crops are less so. Yet cultivated crops make an important contribution to local landscapes, support wildlife, and influence other natural processes. Similarly, while land is the most fundamental of natural resources, the buildings, highways, dams, railroads, and other engineered structures set on the land are not. Also among artificial structures are lawns, gardens, parks, vacant lots, waterways, and urbanized landscapes that grade into more wild environments. We include land used for farm and urban development among our natural resources because it forms important resource and environmental links between humanity and the natural world.

Abiotic and Biotic Resources

Abiotic resources include minerals, air, water, solar radiation, and atomic radiation. Some abiotic resources (nonliving), such as oil and coal, were formed from once-living matter exposed to anaerobic environments under pressure for millions of years. Biotic resources are living organisms such as fish, wildlife, range grasses, and trees. Soil resources include a mix of biotic and abiotic properties that have unique attributes such as arability, fertility, and drainage.

Ecosystems Defined

We define ecosystem as a physical area with unifying ecological characteristics on which humans have placed boundaries for management purposes. An ecosystem is the interactive complex of biotic (plants, animals, microbes) and abiotic (e.g., soils, topography, climate) components occupying the defined area. For water resources, the watershed is used as the common management unit. For range and forest resources, vegetation in various topographical settings forms convenient boundaries for management. Particular plant and animal communities are associated with various ecosystems (Figure 1.1). Whatever boundaries are placed on ecosystems, the interactions among the biotic and abiotic components sustain a wholeness that is greater than the sum of the component parts. (Figure 1.2 shows the components of a grassland ecosystem and the products it produces that are usable to humans.

An important concept in natural resource management is that ecosystems differ greatly in their resistance to overuse. For example, heavily grazed rangelands in the tallgrass prairie of Kansas can return to pristine or climax conditions within 5–10 years through natural processes if soil erosion has not occurred. However, in the Chihuahuan Desert of New Mexico, 20 years or more may be needed to overcome the effects of poorly controlled livestock grazing. Generally, the resistance of terrestrial ecosystems to abuse increases as precipitation and temperature increase. Water and heat regulate biochemical activity and ecological rates of adjustment and recovery.

Resource Consumption, Use, and Renewal

Resources may be consumed during use or remain to be used again. When coal is burned and grass is eaten by livestock they are consumed and cannot be used again in the original form. A wilderness park may be visited, endangered species preserved, or a fish caught and returned without consumption. They could be used over and over with human benefit from each use.

This flow of materials often leads to recycling and renewal of resources through ecological processes. Burning of wood and coal, for example, generates carbon dioxide, which influences heat balance in the atmosphere and provides a resource for more photosynthesis and more wood production. The carbon dioxide may cycle into phytoplankton, forming a basis for fishery renewal. Fish and trees are renewable resources, as long as reproductive populations are maintained. Coal is a nonrenewable resource—at least within time frames relevant to resource managers. Similarly, genetic diversity is a renewable resource as long as the ecosystem functions needed to sustain various life forms are retained. While unmined iron and copper ore are not renewable, their manufactured products can be recycled, reducing the demand on ore reserves. Even the nonrecyclable resources such as fuel oil are not likely to be totally depleted because some substitute will be found or developed that is cheaper and as effective. The renewable resources include soils, range, farm products, forests,

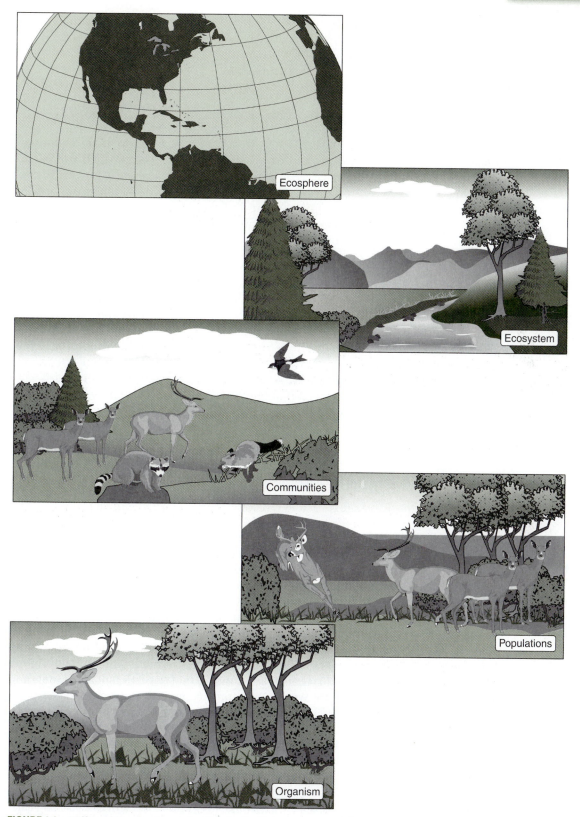

FIGURE 1.1 Different levels of organization in the living systems of the world (adapted from Miller 1990 by John N. Smith).

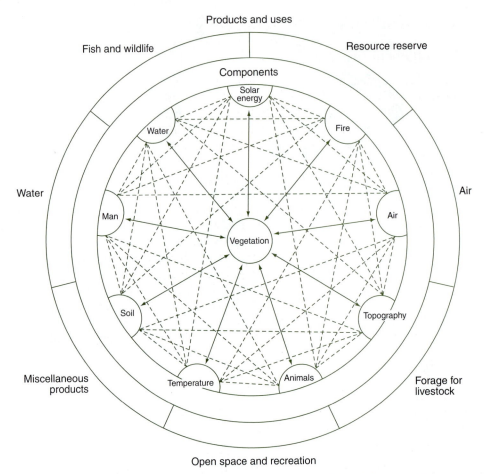

FIGURE 1.2 Grassland ecosystem components and products (from Blaisdell et al. 1970).

wildlife, fisheries, water, air, genetic information, and wildlands. These resources can be managed to meet basic human needs on a sustainable basis. However, their overuse by humans can also cause depletion or degradation beyond repair. Overgrazing or improper farming practices can destroy within a few years an inch of fertile topsoil that took 1,000 years to form.

Nonrenewable resources occur in fixed quantity, and so from a practical standpoint cannot be replaced by technology, although some can be recycled for new use. These resources include a number of energy sources (coal, oil, natural gas, uranium), nonmetallic minerals (magnesium, phosphates, potash), and metallic minerals (gold, silver, copper). During the 1970s there was a concern that depletion of fossil fuels and some minerals such as copper would constrain economic development and improvement in the human condition. These concerns were somewhat alleviated during the 1980s and 1990s with substitution, recycling, miniaturization, improved detection, improved extraction, and higher mechanical efficiency. An example of substitution involves modern telecommunications equipment. Fiber optics that use silica as the basic structural component have largely replaced copper wiring. In the next 10–30 years, electrical cars that use batteries rechargeable from renewable energy sources will probably replace cars that use fossil fuels. Within the last 20 years, the per capita consumption of fossil fuels was greatly reduced by improved vehicle fuel efficiency. Between 1973 and the late 1990s average fuel efficiency of new American cars doubled, increasing from 13 miles per gallon to 28 miles per gallon. These are just a few of the many examples of how technology has reduced problems of nonrenewable resource scarcity. Historically, the power of technology to solve natural resource problems has been both underestimated and overestimated.

NATURAL RESOURCE MANAGEMENT

Management Objectives

Natural resource management is the planning and active manipulation of ecosystem components and processes for human benefit. Management philosophy was once dominated by resource exploitation: take all that you can in a cost-effective manner and move on. Exploitation philosophy is shortsighted and tends to ignore future human needs. It still remains a predominant philosophy in some parts of the world. Conservation philosophy, which now predominates in the developed world, emphasizes the efficient use of natural resources while protecting their capability for renewal (when they are renewable).

As a rule, the object of natural resource management is to obtain the optimum combination of goods and services for society on a sustained basis. Goods include all material resources. Services include energy resources and other functions. For example, intact wetland ecosystems can function to provide a number of pollution, water supply, and flood control services while also sustaining genetic diversity, and providing various goods (a service), such as fish and wildlife and livestock forage. Natural resource management as we define it has two basic aspects: (1) sustaining or restoring the integrity of the energy, air, water, microbial, plant, and animal complex (the ecosystem complex), and (2) supplying products such as wood, food, water, minerals, energy, and wildlife consistent with demand and sustained production far into the future. In theory and practice renewable resources can be used while the integrity of the system that sustains their renewal is preserved. Even when the resources being managed are not renewable, the ecosystem they are extracted from is renewable and needs to be sustained.

Natural resource management is fundamentally concerned with ecosystem interactions—that is the interactions of energy, air, water, mineral, microbial, plant, and animal components. A key concept in natural resource management is that any human activity influencing one component will also affect the other components. Through research, many ecosystem interactions have been found to be predictable enough for reliable management results. Other processes remain to be discovered and clarified for improved management.

Private and Public Resource Management

Natural resource ecology, economics, and policy are closely linked with concepts of public and private resource ownership. While air, water, and wildlife have mobility, land is stationary. That simple reality allows land resources to be easily claimed and developed for either private or public gain, whereas water, air, and wildlife are more likely to be part of a more public domain. It has been only recently in the history of civilization that common individuals had the right to own land and other property independently of their ruler or their tribal chief.

Most anthropological evidence indicates tribal organization dominated human group behavior in prehistoric societies before the Agricultural Age began about 10,000 years ago. Before recorded history and written titles existed, occupancy and use of land were determined by tribal decision. Tribes decided "policy" in councils of various form, which represented various family or clan interests. Although ownership and trade of personal clothes, utensils, and art commonly occurred, there was no well-defined concept of land or resource ownership nor system of written title and laws protecting ownership of personal items. Settlement was usually temporary and possessions were few enough to allow easy movement elsewhere. There was little need for more than oral claim to personal property.

Once settlement became permanent and possessions accumulated, a more effective means evolved for protecting possessions and recording trade. As populations grew around reliable sources of cultivated food, division of labor and hierarchal organization evolved. During most of recorded history, resources were claimed by the "State," usually by kings, emperors, high priests, or other rulers at the top of the hierarchy who equated themselves with the "State." They developed the rules by which society lived and usually accorded themselves ultimate power backed by a warrior class. Although each member of the public used property, it typically was under the totalitarian control of the privileged few.

Land is the primary setting for most changes in the use of natural resources. History shows that informed people have been most willing to invest in sustaining the value of their own property, but they have tended to take for granted or to avoid responsibility for the perpetuation of environmental resources shared in common—the air, water, soil, and wildlife. Even if a farmer wanted to recover soil eroded from his land by wind or water, how could he ever reclaim it? On the other hand, why would he want it known that the eroded soil was his? He might be forced to compensate for the pollution his activities caused. Perhaps the pond he built filled with sediment because of some neighbor's "thoughtless" land management. Did he contribute to similar property degradation on other neighboring lands? This conundrum exists because ownership fails to completely constrain natural movements of air, water, and other material.

Our contemporary sense of environmental quality remains focused on air, water, and wildlife qualities more than in the quality of privately owned land. Air and water movements are the primary means by which ecological impacts such as soil displacement and contamination occur. Because these processes are not constrained by boundaries, environmental management is usually a shared public responsibility that is poorly served by independent private interests. The environment is composed of natural resources and its management is part of natural resource management. Land development and use is not only the basic source of wealth—to individuals and to nations—but also the origin of most air, water, and other resource degradation. Because of its ecological, economic, and policy implications, the history of land use control is central in the history of natural resource management.

An Integrated Approach to Management

The management of natural resources has become more integrated throughout history. This textbook attempts to integrate historical, ecological, economic, and policy factors into a functional framework. We consider economics—the measurement of relative value—a critical part of natural resource management decision making. Economics provides a means for comparing values assigned by people in order to best allocate scarce resources among competing uses for the greatest human benefit. Without a basic knowledge of economic principles, any selection of natural resource management alternatives will likely be far from ideal. However, not all valuations can be readily converted to a monetary standard. The monetary value for protecting genetic resources, for example, is not accurately estimable at this time. Although protective laws remain popular among most U.S. citizens, an inability to directly compare values of alternative uses of natural resources is the basis for substantial controversy in decision making. An important goal is the inclusion of all resources under a single monetary standard. However, politics and policy are likely to continue to play roles in the decision process for a long time to come, especially where there is a demand for a redistribution of wealth other than through private enterprise.

Most federal lands in the United States are managed to optimize public welfare. The harmonious use of public land for more than one purpose—such as water, timber, wildlife, and livestock production—is defined as multiple use. Multiple use philosophy also is becoming more common for resources managed privately because it usually leads to the greatest sustained benefit to the owners as well as to the greatest benefit for "the customers." Because a diversity of benefits is usually an object of most natural resource management systems, contemporary resource managers need a broad understanding of the various natural resource categories, as well as specialization in certain areas. Thus foresters, fisheries biologists, geologists, range managers, and environmental scientists need a basic understanding of all the other disciplines. Most comprehensive resource management is done through teamwork and requires strong communication and interpersonal skills.

CAREERS AND EDUCATION IN NATURAL RESOURCE MANAGEMENT

The twenty-first century will be a boon for natural resource careers and job opportunities. Hundreds of thousands of satisfying and challenging environmental careers have already been generated by the public and private sectors, nonprofit organizations, and academia. This is not surprising considering that the global market for environmental products and services is estimated at nearly $500 billion. The U.S. federal government alone employs more than 250,000 people (Table 1.1) and the many industrial companies employ more than 1 million people in a multitude of environmental jobs and through entrepreneurial consulting companies (Table 1.2). The range of career opportunities includes planners, economists, educators, air and water quality and solid waste and hazardous waste managers, energy specialists, and soil, water, fishery, wildlife, conservation biologists, and outdoor recreation professionals.

The environmental concerns and needs of present and future societies will require highly trained specialists and technicians including biologists, chemists, physicists, geologists, meteorologists, statisticians, economists, mathematicians, engineers, computer software professionals, and multidiscipline-trained specialists. Computers are the most influential new technological development. Their application to natural resource planning and management is ever increasing such as in analyzing large databases and predicting responses to environmental impacts. Students and professionals must keep abreast of advances in computer technology.

An excellent source of environmental careers information is the Environmental Careers Organization (*www.eco.org*) and their publication *The Complete Guide to Environmental Careers in the 21st Century* (Doyle et al. 1999). This publication devotes individual chapters to job descriptions and opportunities in various major natural resource career categories with salary ranges. It also lists numerous Web sites for environmental job seekers, print directories and books on environmental careers, addresses and Web sites of professional associations, and job and educational resources for minorities. Another important source, the *Conservation Directory*, lists thousands of state, federal, private, and nonprofit organizations and educational institutions that employ natural resources professionals. There are also numerous Web sites for job seekers such as *www.usajob.opm.gov*, which lists current federal government job openings. Several newsletters list job opportunities, such as *The Job Seeker*, which specializes in environmental work offerings.

TABLE 1.1 Federal Government Employment at Environmental and Conservation Agencies (1997)

Agency	Permanent	Temp/Seasonal	Total
National Park Service	15,729	6,000	21,729
Bureau of Land Management	8,760	1,900	10,660
Fish and Wildlife Service	7,000	500	7,500
U.S. Geological Survey	10,395	—	10,395
Other interior agencies	21,843	—	21,843
Environmental Protection Agency	18,165	—	18,165
USDA Forest Service	30,000	15,000	45,000
Other agriculture (enviro-related)	11,000	—	11,000
Department of Energy	16,983	—	16,983
Transportation (enviro-related)	12,610	—	12,610
National Oceanic and Atmospheric	7,500	—	7,500
Army Corps of Engineers	27,454	—	27,454
Other agencies	21,000	—	21,000
TOTAL	208,439	23,400	231,839

Source: Environmental Careers Organization.

TABLE 1.2   Estimated Environmental Industry Employment

Area	Entities	Jobs	Billions ($)
Services			
Analytical services	1,400	20,000	1.2
Wastewater treatment	27,000	113,700	25.9
Solid waste management	5,900	229,600	35.8
Hazardous waste management	2,500	53,300	5.6
Remediation/Ind. services	3,800	100,000	8.9
Consulting and engineering	5,800	162,800	15.6
Equipment			
Water (includes chemicals)	2,600	100,700	19.0
Instrument/Information system	500	24,800	3.4
Air pollution	800	83,300	16.2
Waste management	2,000	88,200	12.2
Process/Prevention technology	300	15,200	1.0
Resources			
Water utility	58,000	118,000	25.3
Resource recovery	2,000	128,300	16.9
Environmental energy	600	24,400	2.3
TOTAL	113,200	1,262,300	189.3

Primary source: Environmental Business International.

The future natural resource professional will need an educational background with breadth and depth of knowledge and communication skills (Figure 1.3). The highly complex and synergistic nature of environmental problems requires that professionals work in teams and be able to communicate with specialists in areas of expertise other than their own. The ecosystem approach to natural resource management also requires that professionals think and plan in terms of integrating biological, social economic, and political implications in formulating management strategies (Figure 1.3). Resource

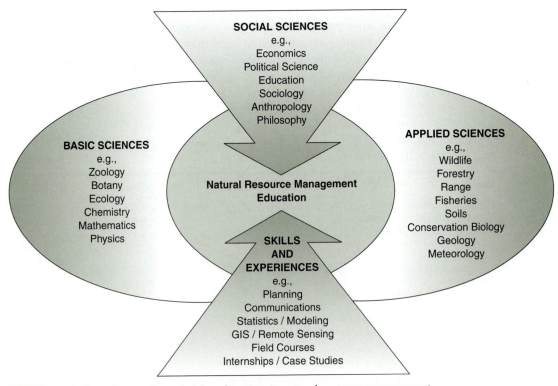

FIGURE 1.3   College classwork needed for education in natural resource management.

professionals are increasingly required to view problems from a broad perspective that crosses several disciplines. Many of the jobs in natural resource management require specialized knowledge of applied sciences such as wildlife, watershed, and forest and range management but with a background in ecology. A graduate degree is becoming increasingly necessary to prepare professionals for even entry-level jobs. Almost every professional organization publishes technical journals that provide students and professionals with the latest research findings. Students should refer to professional societies and their journals to gain technical knowledge required for specific careers and to develop their educational backgrounds. Many nonprofit organizations publish magazines that synthesize technical research with an applied framework.

THE NEED FOR NATURAL RESOURCE MANAGEMENT

The Human Population Increase

Many scientists consider the most pressing problem confronting humankind to be the tremendous increase in the human population expected in the next 50 years. In 2000 the world's human population reached 6.1 billion, which is twice as many as in 1960 (United Nations Development Programme 2000). Presently the world human population is growing at a rate of 80 million per year. In terms of numbers this is the most rapid rate of growth in world history (Table 1.3). Although fertility rates have declined faster than expected, the world population growth rate is 1.3% per year compared with 1.5% in 1990 (Table 1.4). By 2050 it is expected that 10 billion people will inhabit the planet (Figure 1.4). Over the past 300 years, world population growth rates per year increased from 0.3% in 1700 to 0.6% in 1900 to the current rate of 1.3% (McQueen 2000). This rate of increase has actually exceeded the growth rate calculated by Thomas Malthus (1768–1834) who predicted the world human population would exceed food supplies sometime between 1850–1900 (Kleiber 1961). Although birth rates have fallen substantially from six children in 1900 to 2.2 children per female in 2000, human longevity has increased from 40 years (1900) to 60–80 years (2000) depending on country. Improved health care, rapid increases in food production, and various other technological advances explain how the world today can support nearly four times more people than in 1900 at a much higher standard of living (McQueen 2000). Although various scenarios are possible, most experts believe the world's human population will level off between 10–14 billion by 2100. This will lead to unprecedented demands on the earth's natural resources.

   In the United States there are currently about 280 million people. The United States has been growing at nearly 1.0% per year for the last 30 years. The United States is projected to have 420–450 million people by 2050 and may have between 500 and 600 million people by 2100. This, of course, depends on birth rates, immigration rates, and

TABLE 1.3   World Human Population Milestones

Year	Population (billions)
1000	0.31
1500	0.50
1800	0.98
1900	1.65
1930	2.07
1950	2.52
1960	3.02
1974	4.00
1987	5.00
1999	6.00

Source: McQueen 2000; United Nations 1994; United Nations 1998.

TABLE 1.4   Growth and Density of the Human Population in Various Parts of the World in 2000

	Percent Net Growth per Year	Doubling Time (Years)	Population per Square Mile	Population per Square Kilometer
World	1.3	55	115	46
Africa	2.7	27	66	26
Kenya	2.4	30	128	50
Sudan	2.9	25	34	13
China	1.0	72	336	131
India	1.7	42	829	325
Europe	0.3	240	270	106
North America	1.2	60	40	16
Canada	1.2	60	8	3
Mexico	1.9	38	129	51
U.S.A.	1.0	72	75	29
Oceania	1.5	48	9	4
Australia	1.1	65	6	2
New Zealand	1.1	65	34	13
South America	1.7	42	48	19
Argentina	1.1	65	33	13
Brazil	1.2	60	50	20
Former U.S.S.R.	0.3	—	22	9

Source: USDC 2001 and FAO 2000.

life expectancy levels. Many ecologists and environmental planners have serious doubts about the earth's capability to support 12 billion people. The United States can probably support 500 million, but the quality of life is uncertain.

During our careers in natural resource management the authors have been influenced on human population issues primarily by two learned scholars who might be referred to as Dr. Doomsday (Paul Ehrlich) and Dr. Cornucopia (Julian Simon). In 1968, Dr. Ehrlich's book *The Population Bomb* made a convincing doomsday case that human population growth would overwhelm food production capability and cause severe pollution in the last 30 years (1970–2000) of the twentieth century. This supposedly would cause widespread human misery from famine, disease, and war leading ultimately to a major reduction in both human population and quality of environmental support systems (land, air, water, energy, genetic material). A contrast-

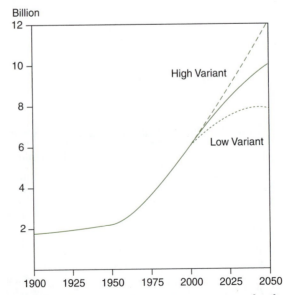

FIGURE 1.4   Projected world population growth from 1900 to 2050 under three assumptions (from U.N. Population Division).

ing cornucopian view espoused by Dr. Simon in his 1981 book *The Ultimate Resource* became popular under the Reagan administration of the 1980s. The essence of this view was that environmental problems were greatly exaggerated and people (brain power) are the most vital resource in solving the world's problems. In order to unleash this brain power, democratic, market-oriented social/political systems are required. Basically the idea here is that technology involving substitution, miniaturization, recycling, molecular engineering, and other innovations can overcome any kind of scarcity or environmental problem brought about by more people (Miller 1990). Generally, the doomsdayers have had their training in biological or ecological sciences while the cornucopians have an economic or financial background. The 1970s gave credibility to the doomsdayers as food and energy prices rose sharply, famines occurred in China and India, and the United States experienced protracted stagflation (Schiller 2000). However, the Reagan administration implemented a set of supply-side economic policies in the 1980s that were associated with improved economic growth and stabilized food and energy prices (Schiller 2000). The fall of Communism in the Soviet Union further reinforced the cornucopian view that human hardship was primarily a function of unsound social/political policies.

Presently great concern about human population growth remains among some of the conservation community, but the public in general is more apathetic than in 1980. Dr. Ehrlich's dire projections in 1968 have been somewhat discredited by various reports that show improving living conditions in a number of underdeveloped countries in the past 20 years (Schiller 2000). On the other hand, in several underdeveloped African and Asian countries little or no development has occurred to improve their lot. Many reports suggest the human carrying capacity of the world has been exceeded, and air, water, land, and genetic resources are being rapidly degraded (Wackernagel and Rees 1996). Who was more right in the long run (Dr. Ehrlich versus Dr. Simon) remains a serious question yet to be answered. It is generally accepted by ecologists that at some point physical resources will ultimately determine the limit of human population growth (McQueen 2000), and that limit may depend on what quality of life is tolerated. The newest technology and most efficient economic system will not be enough to overcome scarcity of water, arable land, and energy needed to provide basic necessities such as food, shelter, clothing and heating. Most estimates of world human carrying capacity fall between 4 to 16 billion people (McQueen 2000). These estimates do not take into account quality of life and they assume equity in distribution of resources. Presently 18% of the world's population (1.1 billion people) live in severe poverty and life is hardly comfortable (United Nations Development Programme 2000). The richest seven nations (United States, Canada, Great Britain, France, Germany, Japan, Italy) have only 10% of the world's population but annually account for 40% of the consumption of fossil fuels, forest products, and a variety of other commodities. The world's consumption rate of natural resources has grown even faster than the human population since 1950. However, climatic change and creation of pollution in excess of the world's decontamination capabilities are of more immediate concern than resource exhaustion (United Nations 1998).

The major constraints that will confront mankind worldwide over the next 50 years include food scarcity, water scarcity, water contamination, climatic change, air contamination, energy scarcity, farmland depletion, and loss of biodiversity (United Nations Development Programme 2000). The destruction of the rain forests is a particular concern because of the critical value of these areas in preventing climatic change and extinction of plant and animal species. About 35% of the world's forests have been removed since 1900 (Abramovitz 1997, United Nations Development Programme 2000). Extinctions of plant and animal species are now occurring at 50 to 100 times faster than their natural rate. Unsound farming practices and overgrazing have degraded 16% of the world's land area. World fisheries are a particular problem with 25% now depleted and another 44% being overharvested. The primary fear of the doomsdayers is that all these problems will converge at once causing devastation of the planet and elimination of 75% or more of the human population.

Trends in human population growth and economic development will have a considerable influence on how various countries will use natural resources in coming years. Although the emphasis may shift among natural resource products, rapid expansion in the human population will undoubtedly make natural resource management more important than ever before, and with different emphases than in the past. Recent international conferences about atmospheric change, decreasing biodiversity, declining ocean fisheries, and other resource issues would have been unheard of at the beginning of this century. International trade already is being influenced by the relative value different countries place on these resources.

Although a big part of the solution to emerging resource problems rests with decreased population growth and stabilization, more efficient and effective resource management is also important. History shows that the most effective way to encourage reduced population growth is to improve the quality of life through education, democratic process, and technology. Technology is a double-edged sword, however, which has contributed to many existing problems while providing solutions for many others. Although resource depletion and degradation are challenging problems, recent progress shows they are not entirely overwhelming. Improved natural resource management is one strategy that can contribute to problem solution.

LAND AREA AND LAND USES

About 11% of the land in the world is farmed; 24% is in permanent pasture; 31% is forest or woodland; and deserts, glaciated areas, high mountain peaks, and urbanized/industrialized land comprise the remaining 34% (Table 1.5). The total land covered by buildings, parking lots, highways, and other infrastructure is between 3 and 4%.

The total land area in the United States is about 1.91 billion acres (2.26 billion with Alaska). About 20% or 382 million acres is usable crop production. Cities, homes, factories, airports, highways, and other human structures account for 97 million acres or 5% of the total. Around 65% remains as forest and rangeland for wood, meat, fiber, mineral, and energy production (Figure 1.5).

TABLE 1.5 **Percentages of Farmland, Grassland, and Forest Land for Selected Countries and Regions in 1999**

	Farmland	Permanent Pastureland (Grassland)	Forest and Woodland
World	11	25	29
Africa	6	30	23
Kenya	4	7	4
Sudan	5	24	20
China	10	42	13
India	52	4	20
Europe	28	17	32
North America	13	17	32
Canada	5	3	35
Mexico	13	38	21
United States	20	26	31
Oceania	5	56	18
Australia	6	60	14
New Zealand	2	53	26
South America	6	28	46
Argentina	10	51	21
Brazil	7	22	58
Former U.S.S.R.	10	17	41

Source: Based on FAO (2000) data.

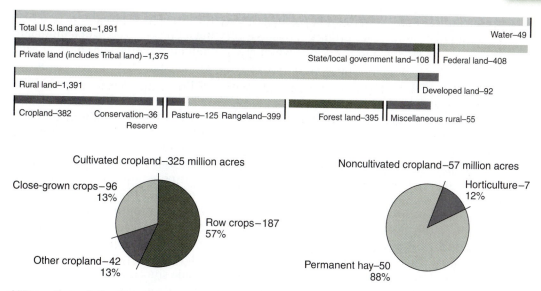

Cultivated cropland–325 million acres

Close-grown crops–96
13%

Row crops–187
57%

Other cropland–42
13%

Noncultivated cropland–57 million acres

Horticulture–7
12%

Permanent hay–50
88%

Millions of acres in the 48 contiguous states, Hawaii, Puerto Rico, and the U.S. Virgin Islands.

FIGURE 1.5   Total land area and land in the contiguous United States (from A Geography of Hope 1997. USDA-Natural Resources Conservation Service).

In the United States a shear lack of land will probably not compromise human welfare over the next 50–100 years. However, improper planning of land use over the past 50 years has unnecessarily increased congestion, pollution, and infrastructure costs and caused a severe decline in the open space that most Americans value. In our opinion much better land-use planning and landscape design will be critical to maintaining and improving human quality of life in the twenty-first century.

Natural Resource Issues

Many issues have developed out of imbalances in resource supply and demand. Resource sustainability requires understanding of the underlying mechanisms determining continued supply and the effect of human demand on those mechanisms. For renewable resources, much of that understanding comes out of ecological and economic science. Ecology is the science that provides understanding of resource renewal rates. Economics is the science that examines the relationship between resource supply and demand. Policy is the means by which societies organize guidance in balancing supply and demand based on ecological and economic information. At the policy level, there is an emerging sense that resource supply and demand questions need to be addressed in an ecosystems context.

Natural resource management has a long history with its modern roots established during the last century when a period of unchecked resource exploitation in the industrialized world resulted in widespread polluted air, contaminated water, declining wildlife and fisheries, depleted forests, degraded rangelands, and blighted urban areas. By the turn of the century natural resource management centered on conservation of water, forest, fishery, wildlife, and outdoor recreational resources. Conservation philosophy was vindicated by the "dust bowl" era when the social costs of farmland mismanagement in terms of human health and property destruction became undeniable. The role of mass communication in shaping public opinion was beginning to have its impact through movie theater documentaries and radio.

Following World War II (WWII), the development and rapid acceptance of television amplified the spread of information, including growing concern about environmental contamination by organic wastes, oil, toxic metals, synthetic chemical compounds, and atomic radiation. A period of unprecedented economic growth and optimism followed WWII as the world recovered from mass destruction. A "green revolution" in agriculture promised to solve growing food shortages. But a more pessimistic view emerged in the 1960s when environmental scientists began to present their growing concerns in popular books and through the mass media (television). In this period they were able to greatly influence the public toward greater environmental regulation. Other scientists observed that population growth was spurred to highs by the green revolution and that fertilizers and pesticides contributed to a growing environmental threat. Numerous pieces of regulatory legislation highlighted by the National Environmental Policy Act (1969) and the Endangered Species Act (1973) were passed by the United States Congress to halt environmental decline. Food, mineral, and energy costs rose while the world population continued to increase. The book *The Limits to Growth* from the Club of Rome (1972) used a computer model to show how human population increase and natural resource depletion could lead to widespread hunger and great environmental decline if trends remained the same. People responded with continued support for enforcing new environmental regulations, which soon had positive impacts on water and air quality.

During the 1980s, the economist Julian Simon made the case (*The Ultimate Resource, The Resourceful Earth*) that technology could potentially solve most problems related to resource scarcity if socioeconomic systems rewarded innovation and efficiency. The fact that the 1980s and 1990s have been characterized by declining real (inflation adjusted) prices for food, minerals, and energy resources and lower pollution levels in developed countries has given credibility to many of Simon's arguments, at least in the short run. However, environmental degradation and declining human conditions in parts of Africa, Asia, Latin America, and eastern Europe have been used as counterarguments against Simon's more positive outlook based on continued technological advances.

At the beginning of the twenty-first century there is general agreement that developed countries have made considerable progress in solving problems that relate to natural resource scarcity and pollution of water and land resources. However, depletion of ozone in the outer atmosphere, the "greenhouse effect" (atmospheric warming from increased CO_2 levels), and rapid depletion of the world's genetic diversity remain issues of controversy and serious concern. Whereas point sources of pollution appear under control, there are many problems related to water resources, such as groundwater contamination and depletion, that remain unresolved.

In developing countries, the destruction of the rain forest, loss of endangered species, loss of open space, and desertification have become major natural resource issues. In some of these countries, slowing the rate of population growth and increasing the rate of economic development have shown the potential to improve the human condition and slow or halt environmental decline. However, flawed socioeconomic systems and exploding human populations remain as major impediments to progress in many developing countries.

In the remainder of this section we briefly highlight the key issues for the various categories of natural resources.

Soils. Current primary problems are erosion and soil contamination from pesticides and herbicides. However, new tillage and grazing practices can greatly reduce erosion. Biological engineering and changes in cropping systems are reducing the use of herbicides and pesticides. The most serious soil problems are in developing countries and center around erosion and salinization.

Water. This resource in the future may be the biggest constraint on human population growth and economic development. Considerable progress has been made in cleaning up our rivers and lakes but groundwater contamination is still a serious problem. Groundwater depletion is rapidly becoming a factor limiting economic growth in many parts of the world.

Air. Generally there have been improvements in most U.S. cities, and the problem of acid rain appears to have been exaggerated. Global warming and ozone depletion are serious global problems that have not been resolved.

Biological Diversity. The loss of biological diversity is accelerating. Much of this problem centers around destruction of the rain forest in tropical developing countries because of rapid human population growth and flawed socioeconomic systems. One potential solution is to couple economic compensation for conservation to sustainable rain forest management programs. While rain forests are especially problematic, biodiversity loss is a concern in many other ecosystems.

Minerals. Generally the scarcity of minerals does not appear to be an immediate or long-term problem. The former Soviet Union has vast reserves of most metallic minerals that are now becoming available in the world marketplace as result of change from a communist to a market economy. Substitution with other materials is reducing the need for many scarce metals such as copper and silver. Recycling of metals such as iron and aluminum has many advantages beyond extending supplies. These include energy conservation, pollution reduction, and reduction in disposal problems. There will be an increased emphasis on the repair of environmental damages after mineral extraction in the United States and other developed countries.

Energy. New technology and globalization of the world economy (free trade) have reduced the concern over energy scarcity that existed in the 1970s. Solar energy, nuclear energy, and wind-driven energy can become desirable substitutes for fossil fuels. Since the early 1970s, increases have occurred in the efficiency of fossil fuel use in automobiles, industrial use, and home heating. The major challenges now center around minimization of air pollution and finding better methods for disposal of hazardous nuclear waste. However, the scarcity problem could reemerge sometime during this century.

Farmland. The concern over lack of food in the 1970s was replaced by a surplus of food in world markets in the 1980s. However, starvation remains a problem in many developing countries because of high human population growth rates and inefficient agricultural systems. In developed countries, the big challenges are preserving prime farmland and minimizing the use of herbicides and pesticides. In developing countries, flawed socioeconomic systems and rapid population growth are the root cause of soil erosion and food scarcity problems.

Rangeland. The historic problem on rangelands has been overuse of forage resources. Generally, economic benefits of sustainable grazing practices have not been well understood in either developed or developing countries. Government subsidies and public ownership of grazing lands have often undermined sustainable grazing practices. Improved knowledge of grazing management outcomes and better transfer of knowledge to range users has much potential to improve this natural resource both in the United States and other countries. We believe preservation of open space will be the biggest rangeland challenge in this century.

Forest Land. In the United States, forest land problems center around maintaining mature (old growth) woodlands and minimizing environmental damage from road construction and other logging operations. In developing countries, deforestation is a serious problem, particularly in tropical areas. Linking economic aid to changes in natural resource management in developing countries could have many benefits. Rain forest preservation is thought to be crucial in slowing global warming from the greenhouse effect.

Urban Land Use. Many of the resource management problems associated with cities have improved. Most notably the Clean Air Act and Clean Water Act have reduced point source emissions. However many problems remain, especially from nonpoint source emissions, urban sprawl, and city-center decay. These are complex issues that contribute greatly to climate change, loss of biodiversity in the developed world, and increased human stress. Many fundamental problems remain in the underdeveloped nations.

Recreation.   Tourism has become the world-class means of recreation. Ecotourism is an especially attractive joint solution to protecting ecosystem integrity while sustaining economic development. However, much remains to be done to create and maintain market-based ecotourism. Many unique wilderness areas are threatened with extinction. In the United States, which is rich with recreational land, the biggest challenge is to provide affordable access to uncrowded recreational land close to where people live.

Wildlife.   In the United States, many wildlife success stories have occurred over the past 20 years. Changes in government policies in the mid-1980s have reversed the decline in farmland wildlife. The net loss of wetlands has been nearly halted. Improved wetland management is causing a comeback in waterfowl and other wetland animals. Urban wildlife programs have been successful in creating favorable conditions for many species of birds and mammals in towns and cities, but careful animal damage control is an increasing challenge. Although some species remain endangered, such as the northern spotted owl and desert tortoise, market incentives show potential to alleviate some problems in the United States. In developing countries, the future of many wildlife species is less than optimistic. However, even here market forces could solve many wildlife problems when allowed to operate.

Fisheries.   Overexploitation continues to be the major problem confronting ocean fishery resources. Aquaculture has been successfully used for production of salmon, trout, shrimps, oysters, and so on, but it is not practical for large ocean fish such as tuna. Better international agreements are needed to ensure sustainability of large ocean fishes and mammals (whales). Bans on ocean dumping of waste and better control of oil spills would help many species of marine fishes and mammals.

Natural Parks and Wildlands.   Throughout the 1900s large areas have been set aside in the United States as wilderness and natural parks. The debate will continue on how much wilderness should be set aside. These areas preserve places with great scenic beauty and/or high natural biological diversity. Generally these areas do not contribute much to the national timber, energy, and mineral resources. Private groups have shown increasing willingness to buy up unique areas both in the developed and in the developing countries. We expect that the private sector will play an even greater role in natural resource preservation in the twenty-first century. However, for the large tracts of federal lands in the United States with no unique esthetic or biological value, there could be an opposite trend toward privatization.

AN OPTIMISTIC VIEW OF THE FUTURE

While many studies of the world natural resources in the decades since World War II have emphasized the limits to world growth, such as Club of Rome (1972), our position is more optimistic. Recent global trends in technology, education, acceptance of free market economies, economic productivity, and global concern for sustainability show promise for alleviating resource scarcity and degradation while providing improved quality of life. Over the past 20 years, many measures of environmental health on our planet show improvement, particularly in developed countries. Improvements in energy efficiency, reductions in air pollution, reductions in water pollution, increases in farm yields, reduction in pesticide and herbicide use, increases in farmland and urban wildlife, and a reversal in wetlands loss are some positive natural resource changes in the United States. The elimination of rain forests, desertification, reductions in biodiversity, watershed pollution, the loss of the ozone layer, global warming, and human population explosion are all important natural resource problems that remain to be solved. However, through science, technology, universal education, and the application of proven socioeconomic principles, we believe humanity can overcome these daunting problems.

OTHER SOURCES OF INFORMATION

We acknowledge this text will not convey everything there is to know about natural resource management. However, we believe it will provide the reader with a powerful understanding of how basic principles of ecology and economics can be integrated to solve natural resource problems. At the same time we encourage students and practicing natural resource managers to read the many other excellent books that exist on this subject. Some of the books that have influenced us and have been repeatedly cited in this work include Dasmann (1976), Miller (1990), Schiller (1994, 2000), Owen and Chiras (1995), Owen et al. (1998), and Nebel and Wright (1996, 1998).

ORGANIZATION OF THE BOOK

This book is organized into six sections. Each section forms a basis for more effective learning in the sections that follow. Summaries of the foundational knowledge areas underlying natural resource management are followed by sections that describe resource management in more detail. A final section expands the national focus of the book to a more global perspective and a more integrative emphasis. A glossary of terms is provided.

Section 1: Management Foundations

The first section introduces basic concepts in ecology, economics, policy, and planning process as they pertain to natural resource management. We emphasize those concepts that underlay the development of the chapters that follow. The lead chapter in the section is a brief history of natural resource ecology, economics, and policy in the United States. Understanding of history leading up to the present state is a prerequisite for understanding contemporary natural resource management. The second chapter, on ecology, summarizes important concepts in population, community, and ecosystems ecology, with an emphasis on principles that underlay management of renewable living resources. The next chapter focuses on the role of economics in aiding management decisions and policy development directed at the goal of sustained benefit from natural resources. The management operations addressed in later chapters are preceded by management planning and policy actions described in the last chapter of this section. As the chapters in this section of the book will show, most contemporary natural resource managers need a broad educational background in both natural and social sciences in addition to specialization in certain aspects.

Section 2: Air, Water, Land, and Living Resources

The second section introduces the basic resources associated with the atmosphere, hydrosphere, lithosphere, and biosphere. The chapter on climate discusses basic climatic processes and the various climatic types of the world. It includes a brief section on climatic history. Climatic problems that confront natural resource managers in the twenty-first century, such as the greenhouse effect, are addressed. Water is the basic resource addressed in Chapter 7, which describes its characteristics and shows how it is important for maintaining life processes, transporting natural materials, and navigating. It illustrates the hydrologic cycle, which does much to connect the atmosphere, hydrosphere, and lithosphere into geophysical processes that sustain the biosphere. The last two chapters discuss the lithosphere and the biosphere. The lithosphere provides both raw minerals and soils. Soil not only is the source of physical support and nutrition for most terrestrial and much aquatic production, it also provides the foundation for most terrestrial engineering, such as roads and buildings. Because of the diverse services provided by soils, urban development often competes with farming, forestry, and range use. The process of soil development, soil erosion, and management of soil resources are treated here. The last chapter in this section describes predominant ecosystem types defined mostly by the combination of topography (especially the distributions of surface water and dry land) and vegetation. Ecosystem attributes determine to great extent the types of resource development, use, and management concerns. At this juncture, the reader may continue on to Sections 3 or skip to Sections 5 before returning to Sections 3 and 4.

Section 3: The Land-Based Renewable Resources

The third section summarizes the development of resources we most frequently associate with living processes in the biosphere, including the urban habitats of humanity. A chapter on forest resources leads this section, followed by another on range resources. These two chapters go a long way toward covering the management of resources derived from noncultivated vegetation. An important chapter on farmland summarizes the state of cultivated plants and other farm resources with an emphasis on farm policy. Another chapter on rural parks, wilderness, and other reservations in which the vegetation is managed for purposes of human recreation completes most of the rural land survey of resources and their management. A chapter on urban landscapes addresses the use of land for human industry, residence, infrastructure, and recreation and discusses the extent to which this use competes with other uses. Certain urban land uses fall undeniably into renewable resource categories, such as parks, parkways, private lawns, and roadside vegetation. In addition, the land itself is a renewable source of space in which old structures may be removed and replaced with either natural or artificial structure. Development of all of the land-based resources has environmental effects addressed in each of the chapters.

Section 4: The Wild Living Resources

Chapters 15, 16, and 17 discuss those wild resources that are held in trust by the public. The chapter topics include wildlife, fisheries, and biodiversity. Terrestrial wildlife and biodiversity are most obviously dependent on the way the land-based resources are managed, but fishery and aquatic biodiversity also are impacted through the watershed processes that link land and water. The wildlife chapter discusses commercial and recreational resources as well as animal damage control. The fishery chapter summarizes the state of food and recreational fisheries, and management directed at declining aquatic biodiversity. The biodiversity chapter summarizes the dilemma of declining biodiversity and management directed at preserving genetic diversity and ecosystem services associated with diversity.

Section 5: The Mineral and Energy Resources

The first chapter in this section discusses the minerals of the lithosphere, which supplies many of the metals and other elements that have become so important in contemporary technology and architecture. Also supplied are the sands, gravels, clays, and other earth materials so important for sustaining the human habitat of concrete, tile, ceramics, and glass. The lithosphere also provides the inorganic nutrients used in fertilizers, such as phosphorus and potassium. The second chapter in this section is about nonrenewable energy resources and describes the state of the fossil and atomic fuels, which are the sources of energy most used in industrial and municipal processes. The final chapter in this section treats the renewable forms of energy including solar energy, hydropower, and wind energy. This introduction also defines the role of solar energy as the primary driving force in the world's ecosystems.

Section 6: Integration of Natural Resources Management

Trends in technology development; global, social, and natural processes; and the incorporation of more of a systems process in management all contribute to a growing emphasis on integrated natural resource management. This section contains three chapters that extend concepts of natural resource management in the United States to a global perspective. A common theme emerges for global natural resource management: the need for more integrated management. The first chapter discusses those attributes that appear to be needed to establish sustainable societies based on wise development of natural resources. The next chapter in this section reviews the performance of economic systems in the global perspective, emphasizing the trends in the world today. The last chapter summarizes the most recent trends in integrated resource management in support of sustainable development and the global issues that are likely to dominate natural resource management during the next few decades.

LITERATURE CITED

Abramovitz, J. N. 1997. *Valuing nature's services.* State of World Report 1997. New York: W.W. Norton & Company.

Blaisdell, J. P., V. L. Duvall, R. W. Harris, R. D. Lloyd, and E. H. Reid. 1970. Range research to meet new challenges and goals. *Journal of Range Management* 22:227–234.

Club of Rome. 1972. *The limits to growth.* New York: Universe Books.

Dasmann, R. F. 1976. *Environmental science.* 4th ed. New York: John Wiley & Sons.

Doyle, K., S. Heizmann, and T. Stubbs. 1999. *The complete guide to environmental careers in the 21st century.* Covelo, CA: Island Press.

Ehrlich, P. R. 1968. *The population bomb.* New York: Ballantine.

Food and Agriculture Organization (FAO). 2000. *Production yearbook.* United Nations FAO Statistics Series, No. 53.

Kleiber, M. 1961. *The fire of life.* New York: John Wiley & Sons.

McQueen, R. E. 2000. World population growth, distribution and demographics and their implications on food production. *Canadian Journal of Animal Science.* 80:229–234.

Miller, G. T. 1990. *Resource conservation and management.* Belmont, CA: Wadsworth.

Nebel, B. J. and R. T. Wright. 1996. *Environmental science: The way the world works.* 4th ed. Upper Saddle River, NJ: Prentice-Hall.

Nebel, B. J. and R. T. Wright. 1998. *Environmental science: The way the world works.* 5th ed. Upper Saddle River, NJ: Prentice-Hall.

Owen, O. S. and D. D. Chiras. 1995. *Natural resource conservation.* 6th ed. Upper Saddle River, NJ: Prentice-Hall.

Owen, O. S., D. D. Chiras, and J. P. Reganold. 1998. *Natural resource conservation.* 7th ed. Upper Saddle River, NJ: Prentice-Hall.

Schiller, B. R. 1994. *The economy today.* 6th ed. New York: McGraw-Hill.

Schiller, B. R. 2000. *The economy today.* 8th ed. New York: McGraw-Hill.

Simon, J. L. 1981. *The ultimate resource.* Princeton, NJ: Princeton University Press.

Simon, J. L. and A. Kahn, eds. 1984. *The resourceful earth.* New York: Basil Blackwell.

United Nations. 1994. *World population growth from year 0 to stabilization.* Population Division of the Department of Social Reformation. Document, 6/17/94. United Nations Secretariat, New York.

United Nations. 1998. *World population projections to 2150.* Population Division of the Department of Economic and Social Affairs. United Nations Secretariat, New York.

United Nations Department Programme. 2000. *World resources 2000–2001.* New York: Elsevier Science.

U.S. Department of Agriculture-Natural Resources Conservation Service. 1992. *National resources inventory.* U.S. Washington DC: Government Printing Office.

U.S. Department of Commerce (USDC). 2001. *Statistical Abstract of the United States.* 119th ed. U.S. Department of Commerce, Bureau of Census.

Wackernagel, M. and W. Rees. 1996. *Our ecological footprint.* Gabriola Island, British Columbia, Canada: New Society Publishers.

Management Foundations

Prerequisite knowledge for a full introduction to natural resource management must include basic principles of ecology, economics, and organizational management. Chapter 2 starts this section with a historical overview of natural resource development in the United States, pointing out the importance of ecology, economics, and policy in natural resource management. Chapter 3 is an introduction to ecology, which is the study of interactions among living organisms and their environment. Ecology is an integrative science guiding sustainable natural resource development. It addresses the natural and managed regulation of processes determining rates of living resource production and the impacts of resource development on the many services ecological systems provide to humanity. Principles of natural resource conservation economics are introduced in Chapter 4 to show how economic study helps reveal how best to distribute natural resources for more total benefit to people in present and future generations. The organizational planning process, described in Chapter 5, includes a variety of approaches used to identify the most effective management policies in advance of making policy decisions. The most effective natural resource planning integrates ecological and economic analyses to determine those management policies and strategies that protect those ecological processes sustaining the greatest benefit over present and future generations. This section establishes a foundation for the specialty chapters that follow.

CHAPTER 2

The Historical Perspective

"One of the few good descriptions of *Homo sapiens* is that he is, above all, a change making animal. The evidence of that lies in what he has done—his history" (Roberts 1993). History also shows us that the human condition is a product of previous change wrought by predecessors and a source of insight about future legacies. The history of natural resource development and management is a study of change in human interactions with their natural resources. One of the most valuable benefits of understanding history is that it can prevent repeating past mistakes. In this chapter, we review the history of natural resource management in the United States from the colonial era to the present. We also consider challenges that are confronting natural resource managers in the twenty-first century.

COLONIZATION (UP TO 1776)

Ecology and Landscape

More than 10,000 years ago, the Northern Hemisphere was in an ecologically dynamic time. Ice, fire, flood, drought, and other natural events caused by widespread climate change widely altered landscapes. Continental glaciers were in retreat and the land bridge from Asia to Alaska had submerged. This isolated the earliest human colonists of the Americas until European discovery thousands of years later. In this same period agriculture began in Asia (Roberts 1993). Livestock husbandry originated more than 11,000 years ago in Asia Minor. Inventions of axes, adzes, and other long-handled implements were critical in advancing the rate of clearing and cultivation. Agriculture appears to have advanced most rapidly in arid ecosystems along fertile river bottoms with reliable sources of water for irrigation.

Agriculture did not become established in North America until about 2,500 years ago (Butzer 1990, Roberts 1993). Irrigated agriculture developed in the arid Southwest along the easily worked floodplains of desert rivers. Woodland agriculture also developed throughout much of the warmer eastern deciduous forests at lower elevations where soils were relatively easily worked (Figure 2.1). There is less indication of extensive agriculture in the open prairies, where the thick grass roots resisted cultivation and bison and other game were plentiful alternatives. Native Americans depended on hunting, fishing, and foraging for calorie-rich fruits and seeds. For many tribes, mobility was critical for survival. Their ultimate adaptation to ecological and social stress was to move elsewhere. This mobility was accomplished with minimum intertribal conflict at population densities less than 1% of current levels. However, when tribes met intertribal conflicts often occurred over resources.

FIGURE 2.1 This early engraving graphically portrayed an Algonquian Indian village as it appeared to European colonists. Farming was widely practiced by eastern Native Americans (courtesy of National Park Service).

During the time of European colonization in the late sixteenth century, North America held an abundance of natural resources unmatched in western Europe, where many resources were depleted. The eastern coastal plain south of New England and river floodplains were ideal for farming once cleared of forest. The deciduous hardwoods made effective fuel, durable furniture, lumber, and tools while straighter softwoods made unexcelled ship masts, fences, and housing material. Under the forested Midwest lay rich glacial tills, which as farmland would be exceeded only by tallgrass prairie. Farther west the grasslands grew more arid, shorter, and more bunched, but served to nourish huge numbers of bison and other grazing animals.

The fisheries of the North Atlantic and North Pacific were among the richest anywhere (Waterman 1975). Early European explorers described "countless numbers" of large animals roaming the North American continent (Kline 1997, Bolen and Robinson 1995), which were recorded in art as well as journals. Bison, elk, turkey, deer, waterfowl, wild pigeons, and other game animals were widely distributed and commonly encountered. An estimated 60 million bison, 10 million turkey, and 2 billion passenger pigeons inhabited the continent. Stoddard et al. (1975) calculated the equivalent of more than 67 million animal units (cow and calf equivalent) of wild grazing animals on the central Great Plains grasslands. The great abundance of fur-bearing animals provided incentives for the earliest exploration by European colonists. Beaver were especially widely distributed and eventually became the most popular of furs. By the seventeenth century the monarchies of Spain, France, and England had established American colonies in anticipation of wealth from resource extraction.

The conditions discovered by European colonists had not been stable, but responded to dynamic ecological processes. Weather was variable, especially in the

FIGURE 2.2 J. M. Stanley was an early artist-natural historian who painted this scene of a herd of bison near Lake Jessie in eastern North Dakota (courtesy of U.S. Dept. of Interior).

heartland where droughts were frequent and seasons were extreme. Lightning-caused fire was a widespread natural phenomenon, which sustained a dynamic and diverse pattern of vegetation types and associated animal communities. Animal populations sometimes fluctuated dramatically because of disease, severe winters, and drought. In the larger river valleys, floods periodically reshaped river habitats and sustained a dynamic mix of river channels, wetlands, riparian forests, and small lakes and ponds. Fire was frequent in prairies and drier forests where it created a mosaic of fire-adapted grasslands, savannahs, and forests. Accounts by early European explorers indicate that bison (Figure 2.2) at least temporally overgrazed local areas (England and De Vos 1969). Prairie grouse and waterfowl populations may have been suppressed by lack of vegetation cover (Kirsch and Kruse 1972). Unlike most present-day livestock, however, bison moved freely over the range leaving vast areas ungrazed and able to recover until they returned years later (England and De Vos 1969).

European colonists found a vast forest with many openings and parklike savannahs caused by fires and aboriginal settlements. The mix of forest and upland openings supported an abundance of wild game including deer, turkey, grouse, doves, pigeons, squirrels, and rabbits (Clepper 1966). Extensive wetlands, lakes, and streams provided habitat for large flocks of waterfowl, various fish species, and fur-bearing beaver, muskrat, raccoon, otter, and mink. Because these clearings attracted the densest game populations and were easiest to farm, they were the first areas settled by European colonists on the East Coast.

The first cattle, sheep, and horses were brought into the arid Southwest from Mexico by Coronado in 1540. By the seventeenth century the Spanish had established numerous settlements near perennial sources of water. Escaped cattle and horses from Spanish settlements multiplied rapidly in the mild climate and abundant forage. Feral horses dramatically changed the lives of many tribes, who came to depend on them for their survival. Spanish colonial life remained agrarian long after the industrial age was established on the East Coast in the eighteenth century.

Unlike Native Americans, the first impulse of European colonists was to remain on the land to defend and "improve" their claim by further clearing, planting more crops, and grazing more livestock. The European concept of private claim, based in written law, encouraged permanent settlement. One reason certain game animals were rapidly depleted following settlement was their concentration near forest openings where early farming was most feasible. Wild game grew more scarce around farm margins as farms proliferated and farmers continued to hunt to augment farm produce. As the Europeans advanced across the continent, the progressive reduction of fish and game resources contributed to a wave of conflict with Native Americans.

Permanent establishment provided reproductive advantages that allowed the population of European colonists to expand while those of Native Americans declined. Introduction of Eurasian diseases decimated Native Americans. Learning from Native Americans, early European immigrants produced a variety of native crops such as corn, squash, and tobacco. However, while Native Americans tilled little and moved frequently to new areas, European tilling exposed soil to erosion and continuous cultivation that eventually depleted fertility. Typically, settlers moved to virgin land only after the soil productivity declined beyond that required for family subsistence. In the Northeast, many "hill farms" were located on infertile and stony ground. Cotton and tobacco grown by the southern colonists were especially exploitive of soil resources. Decreased agricultural productivity was a major motivator for western expansion.

Significant Events in Colonial America

~10,000 B.C.	Last of American colonization over land bridge from Asia.
~1,000 A.D.	Norse in Newfoundland formed the first European settlement in North America. It failed to persist.
~1440	Portuguese probably discovered fishing grounds at Grand Banks off Newfoundland.
1497	Cabot was the first to seek Northwest Passage to Asia in a 350-year quest that opened northern fur trade and fisheries.
1540	Seeking gold, Spanish introduced cattle, goats, and horses to the Southwest.
1609	The first successful planting and harvest of tobacco by European colonists followed the advice of American Indians. This established the colonies' most important cash crop.
1620	Following John Smith's discovery of fisheries, the Plymouth Company formed to take advantage of New England fisheries. The Mayflower Compact established a basis for participatory local government.
1623	The first saltworks, sawmill, and cattle were established in New England.
1647	The first public education was required of towns by the Massachusetts colony.
1664	Transfer of lands from the Crown to the Duke of York initiated land speculation, sale, and transfer to private ownership.
1705	British initiated laws that restricted colonial trade only to England and levied taxes on goods including rice, molasses, pitch, tar, rosin, turpentine, and hemp and, later, copper, furs, tobacco, sugar, textiles, coffee, indigo, and wine.
1709	The first copper mine was established in Connecticut followed by the first ironworks in 1714 in Virginia.
1715	Nantucket had become a whaling center for offshore sperm whales and established a resource for a large candle industry started in 1750.
1724	The first rice irrigation greatly expanded production in South Carolina.
1730	New York established pottery and stoneware industry followed in 1739 by glass factories in New Jersey and Pennsylvania.
1750	England passed the Iron Act, which limited colonial steel industry to protect England's industry from competition.
1760	One-third of all English ships were being built in the colonies with cheap timber and other maritime supplies.
1764	England passed the Sugar Act, which was one of a series of tax laws on resources and goods that led to the American Revolution.

Economics

Independence and Wealth. The discovery of bountiful natural resources by Europeans encouraged colonization and land claim in the New World despite great odds. Abundant and favorable climate and soils at the time of colonization formed the foundation for the great prosperity the nation experiences today. Much of present-day American character and culture is based on the development of a vast and diverse supply of natural resources (Clepper 1966). The early establishment of market economies, democratic governance, and individual right to titled land ownership established a strong philosophical foundation for the new nation once independence was declared.

East Coast European colonists were favorably positioned to adopt a strong market economy and democratic local government even before independence from Britain was declared in 1776. Private property was made widely available by the British Crown to entice stable settlement. Local and international trade soon became the basis for thriving business and the establishment of industry along the East Coast. The early colonists developed valued trade with Great Britain for North American resources in short supply, such as ship timber and furs.

The Origins of the Market Economy. The foundations for complex market economies extend back to the earliest cities, where labor diversified into specialized services and commerce flourished. The specialization of services required ways to exchange goods and services in kind. The first forms of money used for exchange originated more than 4,500 years ago while the first metal coins appeared about 2,700 years ago (Roberts 1993). The first true cities of diverse industrial, religious, commercial, and residential structure arose in the Fertile Crescent of western Asia more than 5,000 years ago. An apparent precursor to urban life was development of agriculture at locations with dependable water and conducive climates. Settlements grew into cities around fortified centers of administration and commerce, which became the first political "states." Professional militaries were needed to protect cities and to mount offensives elsewhere, which stimulated innovation. Metal refining advanced with discovery of ore smelting in Eurasia and mines became strategically important for equipping armies. Writing and records keeping appeared soon after the first cities. Specialized artisan services evolved from simple wood, clay, metalworking, and textile weaving.

Highly organized bulk trade came about after permanent settlement was secured around stable agriculture. Trade generated wealth and wealth generated taxes for the rulers, who encouraged more trade. Overland bulk trade was difficult because of the rudimentary animal drawn carts that were used. Waterborne trade was much more efficient and had much to do with the development of the Phoenician, Greek, and Roman Empires along the shores of the Mediterranean.

Although marketplaces were common in ancient Mesopotamia, Greece, Rome, and other civilizations, the important attributes of a capitalist free-market system had not fully evolved in any of them. Heilbroner and Thurow (1994) identify several attributes of a capitalistic society, which the ancient empires failed to exhibit. None of them left evidence that *all* people had a legally protected right to private property. The privileged classes alone were allotted that right. Also, there was no central market system independent of the whims of nobility or priesthood and no organized means for property exchange, money lending, or employment. The economies were traditional barter or command economies, dictated by the privileged few. Most peasants had to work when it was demanded of them.

European Feudalism. The redistribution of titled land from privileged class to commoner appears to have first occurred in Europe after the feudal system disintegrated following the Renaissance (fourteenth to sixteenth centuries). From A.D. 700 to A.D. 1200, much of Europe was split into small feudal states, organized around warlords who sustained a stable balance of power among themselves, monarchies, and the Christian Church. This feudal system exploited the serfs, farmers, and artisans who had to pay land-use tribute to the lords, churches, and kings (Heilbroner and Thurow 1994). Serfs shared access to the lords' estates in areas known as "commons." Serfs often were prohibited from hunting, fishing, or cutting wood on much of the lords' estates, forcing overuse of the commons. This practice left a strong cultural distaste for the private ownership of fish and wildlife. It also was to later influence the way in which many local communities in the North American colonies managed their common-use areas.

Market economies began a slow emergence in Europe following the Christian Crusades to Palestine (Roberts 1993). Western Europeans were introduced to Middle Eastern goods such as silks, spices, and perfumes. Some of the crusaders remained at ports on the eastern Mediterranean where they bought and sold goods for shipping to Venice and ports farther west. A new merchant class in Europe began to accumulate wealth independent from the land. In the sixteenth century the bubonic plague and trade from the New World and Asia destabilized the feudal economy and began replacing it with a market economy (Heilbroner and Thurow 1994). The plague caused a shortage of labor, which favored more liberal treatment of serfs. Eventually, reduced productivity and lower revenues from serfs drove the feudal landlords into debt. Merchants often took land title from the impoverished landlords in lieu of money or barter.

European Market Economy and Industrial Revolution.   The merchants split the land into smaller parcels and sold it for profit to a growing number of middle class artisans and merchants. They also invested part of their profit in more innovative and larger operations, hiring displaced serfs with valuable skills. Literacy increased rapidly after invention of the printing press in 1440. As the size of businesses grew, competition grew among merchants, forcing more investment in size, innovation, and efficiency. Once personal income was tied to innovation, invention accelerated into an Industrial Revolution. The market economy in England had become established long before Adam Smith first described its functioning in his classic book, *The Wealth of Nations,* published in 1776.

The Colonial Economy.   The economy of the New World was elemental at first; based mostly in raw natural resources such as game meat, furs and hides, agriculture, fisheries, and timber. In this era, people were few and natural resource supplies seemed unlimited. Much of the economy was based on direct barter of goods produced from local natural resources. They used game animals, birds, and fish for food and clothing (fur and leather). Although the forests of eastern North America had to be cleared for farming, they provided an abundance of wood for homes, furniture, utensils, charcoal, and boats (Clepper 1966). In addition to wood, the forests provided nuts, maple syrup, fruit, soap ingredients, tar, turpentine, rosin, and other staples. These pursuits demanded only hand tools, foot travel, and work animals.

Economies within the colonies were local and often quite isolated when situated any distance from water transport. Roads were little more than wide trails, except for a few freight routes between major cities (Oliver 1956). Trade in the north developed around whaling and fisheries, saltworks for fish preservation, ship building (various forest products), rudimentary textile mills, granaries, and glass and metal industries. Development of mineral resources was slow, with some mining of copper and iron in New England. In the South, slavery became the basis for a labor intensive agriculture emphasizing exportable tobacco, rice, indigo (a black clothing dye), cotton, and sugar.

The main sources of power in the colonial New World were wind power for boats, animal power for tillage, and water power for millwork (Figure 2.3). Fire for kilns, furnaces, stoves, and fireplaces was fueled by ubiquitous wood, including charcoal. Pursuit of wealth often translated into ownership of lands suitable for logging, farming, and settlements. Men like Benjamin Franklin grew wealthy from speculative land purchase and profitable resale (Freidenberg 1992).

FIGURE 2.3 Gristmills for water-powered grinding of grain into flour, similar to this one in nineteenth-century Georgia, were common in late colonial America before independence (courtesy of National Park Service).

Numerous issues precipitated the American Revolution in 1776. Most important was a desire by the colonists to be the masters of their own resources at home and in international market places. The British government was bureaucratic and slow to respond to citizen needs in the colonies. It resisted demands for sale of inland private lands to replace "worn out" lands. Colonists resented "taxation without representation" on goods entering and leaving the colonies (Jenkins 1997). The English aristocracy viewed the colonies as a huge source of natural resources and trade with outpost civilizations too fragile for independent governance.

Policy

Democracy, Land Ownership, and Resource Regulation. The earliest European settlers of the New World confronted a vast wilderness that provided them with both opportunities and challenges. Successful settlement, survival, and development of sustained civilization were driving concerns. A puritanical skepticism of secular pleasure and aesthetics was widespread. Except to avoid local depletion, resource protection was not a prominent concern (Kline 1997). Regulation of resource use was a controversial issue because preservation in Europe had favored the privileged few who had inherited their wealth. This wariness extended to lands and waters used in common, such as local fisheries, hunting grounds, sources of fuelwood, and, in the arid Southwest, irrigation water. Elected town leadership and democratic town meetings became the basis for deciding how local natural resources would be used.

In response to local resource depletion, early in the colonial period local townships established ordinances restricting fuelwood, fish, and game harvest. Other restrictions were ordered by the English Crown to reserve the best of certain resources. As early as 1691, the charter establishing the colonial province of Massachusetts contained a provision for reserving certain "royal trees" for the crown (Clepper 1966). These trees, the largest white pines, were valuable for shipbuilding. Such laws made sense to the colonies because the ships built from the pines carried trade between North America and Europe.

Other attempts to conserve the future timber supply in colonial America involved municipal penalties for unauthorized tree cutting, enactment of laws against setting forest fires, and stipulation that for every five acres cleared, one acre must be left in forest. Wild fish and game were assigned to common ownership early in colonial history, in contrast to European consignment to private control. This public behavior came in response to the widely despised aristocratic privilege in Europe. Most ordinances remained a local discretion. Except for a few edicts from England, there was no regional regulation of resource use until long after national Independence in 1783.

Town governments were responsible for trails, roads, navigation channels, public education, law enforcement, court proceedings, common defense, and common response to disasters from fire, storm, flood, and disease. Common-use areas, or "commons," were established by consistent public use of navigable waters, footpaths, highways, grazing areas, fuelwood, recreational areas, and meeting areas. These local government actions were to become the foundation for the public natural resource policy now serving the United States.

WESTWARD EXPANSION (1776–1860)

Ecology and Landscape

The Nation in 1776 was mostly agrarian with industry centered in coastal river cities and towns where water power and boat transport were available. Farming required a large amount of cultivated land. The colonial farmers had to be nearly self-sufficient (Clepper 1966). They raised a wide variety of crops that might include two or more grains, vegetables, fruits, and nuts. Large numbers of livestock were kept for pork, wool, poultry, eggs, and leather. Several horses and oxen were needed for transportation and to pull farm implements. Lands too poor for cultivation usually were maintained as pasture. By 1800 only small islands of virgin timber remained east of the Appalachians. These virgin lands typically were on mountaintops and in swamps. By 1825 nearly all of the Great Smokey Mountains had been cleared for grazing or farming. Cultivation was even more intensive in the South under cotton and tobacco culture. Many farmers gave up because the land they cleared could not sustain a family. They left for better lands farther west or industrial employment along the coast. These abandoned lands were typically in rugged terrain or areas prone to drought, flooding, fire, or erosion. Abandoned lands began ecological succession into second-growth forests, some of which are now more than two centuries old.

Significant Events During Western Expansion

1768	A series of treaties with American Indian tribes opened lands west of the Appalachians for settlement and land speculation.
1769	Steam engine was invented.
1776	Adam Smith published *The Wealth of Nations* and initiated the theory of market economics.
1784	Russians established permanent settlement in Alaska and Spain closed lower Mississippi to American navigation.
1785	Land ordinance legislation established township basis for public domain allocation.
1787	The U.S. Constitution was ratified and the first steamboats were operated on the Delaware and Potomac Rivers.

1796	Federal land disposal policy was established emphasizing privatization of public domain at low prices.
1803	The Louisiana Purchase opened the Mississippi to commerce and was followed by the Lewis and Clark expedition to the West Coast.
1806	The national road from Maryland to Illinois was initiated and completed 22 years later.
1824	Engineers of the War Department were authorized to maintain major waterways for navigation establishing barge canals.
1825	The Erie Canal was finished, linking the Atlantic with the Great Lakes. Steam locomotive was invented.
1830	The railroads rapidly extended westward from eastern states.
1833	Economic depression encouraged westward movement.
1836	The first wagon trains crossed what will become the Oregon and Santa Fe Trails.
1844	Telegraph was invented.
1845	Iron bridges were invented and greatly advanced highways and railroads.
1846	Sewing machine was invented.
1848	After the Mexican War, acquisitions from Spain in 1848 and 1853 completed what became the 48 contiguous states of the United States.
1849	The California gold rush was first of many western mining "rushes."
1850	Herbert organized the New York Game Protection Association.
1859	First major oil extraction and development of the Bessemer process for steel production.

Overland travel was arduous and long, requiring several days for what now takes a few hours on interstate highways. Waterways were the most reliable and cheapest travel routes. People concentrated along them as they moved across the Appalachians and down the Ohio River valley.

In 1800 wilderness dominated landscapes west of the Appalachian Mountains. Various accounts in the 1700s and early 1800s marveled at the abundant wildlife, fishery, grassland, and woodland resources in the present states of Ohio, Kentucky, Tennessee, Indiana, Illinois, and Missouri (Clepper 1966). Small outposts of European civilization occurred in the Ohio and Mississippi river valleys and in what is now Florida, Louisiana, New Mexico, and California.

Economics

The invention of the steam engine by James Watt in 1769 was a critical aspect of the industrial age. It drastically improved water and land transportation (steamboats, trains) and industrial output starting in the 1820s (Oliver 1956). Labor saving innovations resulted in widespread job loss in eastern industrial centers from 1830 through 1850 (Hibbard 1924). In addition, deplorable working conditions caused many laborers to quit. Many of the jobless returned to the land once they learned that fertile Midwestern farmland could be acquired from the federal government at a cheap price. The discovery of gold in California in 1849 was the first of many important mineral discoveries that encouraged people with poor labor and farm prospects to go westward.

The fur trade was a tremendous incentive for early western expansion into the vast forests of the upper Midwest and the mountainous Far West. By the Civil War (1861), beaver and many other furbearers were growing scarce.

FIGURE 2.4   During the first few decades of the nineteenth century, many miles of canal were constructed for mule-towed freight barges. As railroads developed, interest in developing more canals waned (courtesy of Railroads Memorial Museum, Altoona, PA).

Barge canals (Figure 2.4) were dug to link the Great Lakes, Mississippi tributaries, and the Atlantic Ocean. They made Buffalo and Pittsburgh pivotally important trade centers, followed by Chicago, Cleveland, Detroit, and Saint Louis. Most canals were dug between 1817 and 1840, when further development was eclipsed by railroad development.

Following the expedition of Lewis and Clark in 1803–1805, a chronological pattern of settlement occurred during westward expansion (Clepper 1966). In advance of transportation improvements, fur trappers entered the wilderness either before it was acquired from other governments or soon thereafter. Trading posts were established, linked by improved trails and then roads. Once transportation links were established, the U.S. government sent soldiers to protect commerce. Next came the farmers, who created an agricultural bonanza first in the forest openings and then in the flat prairie soils of the central Great Plains. Next physicians, merchants, blacksmiths, and other service providers moved to the posts and forts, forming nuclei for small communities.

Flat boats, poled upstream and floated downstream, opened limited commerce into the wilderness along the main rivers. By the late 1820s, paddle-wheel steamboats churned the Mississippi, Ohio, and Missouri Rivers (Figure 2.5) (Oliver 1956). After 1830, railroads were rapidly established through much of the East up to the edge of the Great Plains. Chicago and Saint Louis quickly grew to be large cities based on resource processing and commerce. Highways remained poor, however, except for a few good toll roads in the East. Deep rivers remained major barriers to cross-country travel. Even so, many wagon train expeditions were organized to follow the Oregon and Santa Fe Trails westward to new settlements during the 1840s and 1850s (Figure 2.6).

During the 1820 to 1860 period, numerous new inventions improved manufacturing and home life. The process of steel production was greatly improved, which was critical to expansion of the railroads. Invention of the suspension bridge in 1848 greatly enhanced railroad and highway construction and maintenance. Sewing machines, photography, vulcanized rubber, improved looms, and canned goods were invented. The petroleum industry got its start in 1859 at Titusville, Pennsylvania, replacing whale oil for lamps and other uses (Figure 2.7).

FIGURE 2.5 By the Civil War many steamboats used inland waters; especially the Mississippi and Ohio Rivers (courtesy of U.S. National Park Service).

FIGURE 2.6 A bone-weary family rests on the prairie along a wagon trail (courtesy of U.S. National Park Service).

FIGURE 2.7   An oil field developed in Pennsylvania during the Civil War (courtesy of Drake Well Museum).

Policy

Challenges for the New Nation.   After the War of Independence, the new federal government was preoccupied with national security, credibility, and solvency. Claims to the continent beyond the U.S. borders were split among several nations, often at war. There was concern that British and Spanish territories might attract trade from Midwestern settlers if the states were not quickly linked to Midwest territories. Public land was plentiful as the nation grew in size (Figure 2.8). The primary motivation for the Louisiana Purchase from France in 1803 was control of the Mississippi and its resources (Jenkins 1997). Subsequent exploration by Lewis and Clark and others revealed immense resources as an added bonus. Additional land purchases and treaties with Spain, Great Britain, and Russia continued until the nation's geography was completed with the purchase of Alaska in 1867. Most of the land acquired by the United States was owned and administered by the federal government until there was enough settlement to grant statehood. Because of the high demand for private land, sale of public domain was an attractive alternative to taxes for raising federal revenues (Hibbard 1926).

The land acquisition and disposal policy of the early federal government was critical in determining how natural resources were developed and managed. Federal land policy was guided by a strategy that (1) improved national security along U.S. borders, (2) gained revenue to pay back loans, and (3) responded to public demands for more private land. Congress passed land disposal legislation in 1796 and 1800 in accordance with that strategy. The resulting policy, charging a low price for public domain land, was a compromise between eastern establishment demand for federal revenue and western settler demand for free land (Hibbard 1924). Because of limited federal funds, management basically involved military exploration and general survey of new lands as they were acquired.

The Early Environmental Movement.   There was growing awareness about soil degradation in Europe by the eighteenth century, but few farmers knew anything about conservation practices. George Washington was well aware of the problem (Oliver 1956). When Thomas Jefferson returned to his Virginia farm in 1794 after several years in Europe, he was appalled by the soil erosion. He immediately applied a program of crop rotation, use of legumes, use of manure fertilizer, and contour plowing to rebuild his fields. Jefferson was an amateur scientist and a strong

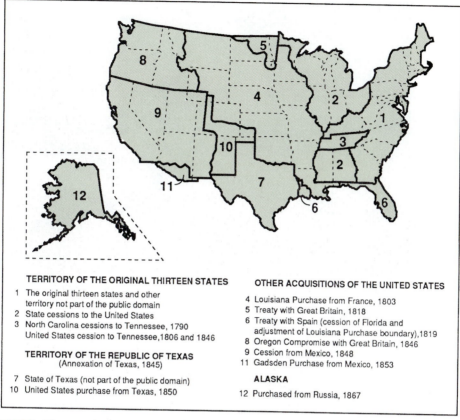

TERRITORY OF THE ORIGINAL THIRTEEN STATES

1   The original thirteen states and other
    territory not part of the public domain
2   State cessions to the United States
3   North Carolina cessions to Tennessee, 1790
    United States cession to Tennessee,1806 and 1846

TERRITORY OF THE REPUBLIC OF TEXAS
(Annexation of Texas, 1845)

7   State of Texas (not part of the public domain)
10 United States purchase from Texas, 1850

OTHER ACQUISITIONS OF THE UNITED STATES

4   Louisiana Purchase from France, 1803
5   Treaty with Great Britain, 1818
6   Treaty with Spain (cession of Florida and
    adjustment of Louisiana Purchase boundary),1819
8   Oregon Compromise with Great Britain, 1846
9   Cession from Mexico, 1848
11 Gadsden Purchase from Mexico, 1853

ALASKA

12 Purchased from Russia, 1867

FIGURE 2.8   The most basic of natural resources—land—was added to the public domain of the United States over a period of 77 years as the nation expanded westward (courtesy of U.S. Dept. of Interior-Bureau of Land Management).

advocate for agricultural education in colleges. However, his ideas on agricultural colleges and educational extension did not become a reality until after the Civil War.

As the Industrial Revolution progressed in Europe and the eastern United States, a new attitude toward wild nature began to emerge among the artists and intellectuals of cultured society (Kline 1997). It was closely aligned with the ideals behind democratic government. In this intellectual Age of Enlightenment, the scientific, aesthetic, and spiritual appeal of wild nature grew as it became tamed and less threatening. Led by George Catlin, Ralph Waldo Emerson, Henry Thoreau, and Thomas Cole, romantic art and transcendentalist essays became very popular among East Coast intellectuals. It contributed to the growth of early tourism, sport hunting and fishing, and suburban development in areas bordering East Coast cities. A growing network of railroads, horse-drawn trolleys, and steamboats provided the means for tourism. Catlin and Thoreau made the early public cases for national nature preserves (Reiger 1975), which were to first materialize in the establishment of Yellowstone National Park.

Early complaints about diminished hunting and fishing sport were voiced well before the Civil War (Reiger 1975). Henry William Herbert, a spokesman for sportsmen in New York, complained about unlimited market hunting of wildlife. He expressed the view that game animals had greater value for sport than meat. Just before the Civil War, he organized the first of numerous private conservation organizations, the New York State Game Protective Association. Their primary goals were to curtail market hunting, enforce protective fish and game laws, eliminate barriers to spawning fish movement, and to replenish depleted fisheries by stocking. Many of them became fish culturists who would later establish the American Fisheries Society.

FIGURE 2.9   Erosion became a chronic and widespread problem. After the Civil War much of the land in the South was farmed by tenants who had little incentive for soil conservation (courtesy of U.S. National Park Service).

THE GILDED AGE (1861–1899)

Ecology and Landscape

Ninety years of population expansion and settlement following the War of Independence had greatly modified the U.S. landscape by the onset of the Civil War in 1861. A growing network of highways, railroads, canals, and natural waterways connected much of the continent outside the mountainous West, the high plains, and "North-woods." People were most concentrated in the river valleys and coastal plains because they provided rich farmland and water for power, transportation, and waste removal. Settlement remained sparse over most of the interior. However, trappers and hunters had reached far into the wilderness and taken a high toll of the wildlife resources near waterways, roads, trails, and settlements.

Much of the deciduous forest lands of the western Appalachian and Great Lakes states had been cleared and converted to farmland. Fuelwood supplies were becoming scarce enough to encourage transition to coal use. By 1870 coal mining had begun to influence local landscapes in Pennsylvania. The pine forests of New England were nearly logged out, but large tracts remained in upstate Maine, New York, Pennsylvania, and the upper Great Lakes region. Many of the cotton and tobacco lands of the Southeast had been abandoned because of nutrient depletion and erosion (Figure 2.9). The forests and arid desert of the mountainous West remained mostly wilderness. Only small areas of timber around coastal cities had been cut on the West Coast, but logging accelerated following the discovery of gold in California. Damage from natural flooding aggravated by watershed deterioration began to increase as human populations concentrated in flood-prone areas.

White-tailed deer and turkey had become scarce in settled areas and several species of anadromous fish were becoming rare. Beaver had been trapped out of most of its range south of Canada (Bolen and Robinson 1995). With the demise of beaver, millions of beaver ponds disappeared, reducing the number of wetlands and altering the hydrology of watersheds.

Significant Events in the Gilded Age

1861	Onset of the Civil War and massive mobilization and destruction of U.S. resources.
1862	The Morrill Act established Land Grant Colleges that led in natural resources education.
1862	The Homestead Act gave 160 acres with stipulations it must be farmed.
1862	The Transcontinental Railroad Act was passed. It established mass transit across the United States.
1864	George P. Marsh published first major soil and water conservation work *Man in Nature.*
1865	The Civil War came to a close.
1867	Alaska was purchased from Russia virtually completing U.S. domain.
1868	Completion of the Transcontinental Railroad.
1870	The American Fish Culturists Association became the first conservation-oriented professional organization and predecessor of the American Fisheries Society.
1871	The U.S. Bureau of Fisheries was formed establishing a federal conservation mission.
1872	Yellowstone National Park was established by an Act of Congress.
1872	The General Mining Law of 1872 encouraged public domain mine development.
1876	Telephone was invented.
1879	The internal combustion engine and the electric light were invented.
1883	The Civil Service Act established technical competency requirement for government employees.
1886	The National Audubon Society was founded under leadership of sportsman-editor, George Bird Grinnell, two years before the Sierra Club was formed by John Muir.
1886	New York was the first city electrified from a central power plant.
1890	The Sherman Antitrust Act became the basis for later trust-busting under Theodore Roosevelt.
1891	The Forest Reserves Act authorized National Forests.
1898	Smith published first scientific account of widespread overgrazing on western rangelands.

Economics

Success in the Civil War clearly depended on technological advantage. Railroads, iron-clad ships, telegraph, photography, various weapons, preserved foods, and other inventions were further developed to win the war. The Civil War laid the foundation for the most inventive period in history (1870s). Rail and highway transportation difficulties initiated a larger role for the federal government in regulating commerce after the war.

The Gilded Age of the 1870s grew out of technological innovation and aggressive resource development. Many of the technologies so important to us today were first developed including electric dynamos, internal combustion engines, refrigeration, automated looms, automated food canning, and telephone communication. Kerosene lighting was soon replaced by electric lighting generated at central power stations and distributed by electric transmission lines. This occurred first in New York City in 1882, but in a few years areas as remote as Montana, Idaho, and Utah had electricity (Oliver 1956).

FIGURE 2.10   During the Gilded Age, railroad magnates held great wealth and political influence as illustrated here by the cartoonist Thomas Nast in *Harper's Weekly* (July 10, 1886) (courtesy of Library of Congress).

Big business propelled the nation to world economic leadership and unprecedented personal wealth by 1900, but at an alarming cost to natural and human resources. Powerful business leaders gained notoriety as "the robber barons" because of their ruthless tactics used to gain business advantage. Government regulation designed to promote fair competition was virtually nonexistent. The press was emerging as an important force in American politics. Cartoonists, such as Thomas Nast, depicted the hold the railroad industry had over the U.S. Senate (Figure 2.10). The most successful businessmen ruthlessly eliminated their competition, particularly in the steel, oil, railroad, and finance industries. They formed monopolizing "trusts" that avoided state regulations and often controlled resource extraction, transportation, and refinement (Jenkins 1997). They bribed and otherwise influenced many government officials. Although economic development substantially increased total national wealth, it was distributed disproportionately in favor of the few business empires.

As the frontier "closed" in the 1890s, the prevailing national sense of unlimited resources was reexamined. The better homestead lands had been claimed, and new westward migrants were generally disappointed by the marginal productivity of the semiarid lands they homesteaded. The closing of the frontier in the 1890s restricted one major recourse (homesteading) to unfavorable industrial conditions. Left with poor homestead alternatives, the closing of the frontier helped stimulate political movements for improved employment conditions in industry.

Both in Europe and the United States, employees were almost completely at the mercy of employers (Jenkins 1997). Children often started work at age 10. Pay was meager. Ten-hour days (or more) and six days a week, without break or vacation, was the typical regimen. There were no work benefits, such as unemployment insurance or compensation for injury on the job, and working conditions often were hazardous. In this environment it was not surprising that a labor movement would grow and the downside of unregulated capitalism would be questioned. Karl Marx wrote about the numerous social problems caused by unregulated capitalism (see Chapter 21). He advocated much more government regulation in favor of the "working class" (Heilbroner and Thurow 1994). Growing "threats" of social revolution were influential in determining the "progressive" course of the U.S. federal government during the early twentieth century. It also set the stage for the rise of communism in Eurasia and the associated international stress that dominated much of twentieth-century politics.

By the end of the nineteenth century, the entire country was served by a network of linked highways, waterways, and railways designed to move mineral and agricultural resources from privately claimed lands to markets all over the country and to international ports. Railroads throughout the nation's interior were used extensively to transport minerals, grain, livestock, timber, game, and manufactured products. Steam power had almost totally replaced sail on the Great Lakes, major rivers, coasts, and oceans. Urban electrification was escalating. The first important conversion of hydropower to electricity was completed in 1895 at Niagara Falls (Oliver 1956).

Logging in the Great Lakes and northwestern states accelerated as railroads made transport of building material from sawmill to destinations across the country much cheaper. Wildfires devastated many areas because combustible timber wastes were routinely left to catch fire. Destructive flooding increased downstream from exploited watersheds. The Johnstown flood in western Pennsylvania was a particularly deadly rallying event for the growing forest conservation movement.

Following the Civil War until the late 1880s, livestock use of western rangeland boomed in response to eastern market demands for meat and improved rail transport. The livestock industry rapidly developed in the 1870s as a result of East Coast and European investment (Holechek et al. 1998). The severe winter of 1885–1886 killed many cattle in the northern Great Plains. Severe drought during 1891–1892 resulted in heavy losses from starvation in the Southwest. These events depressed investment interest, devastated livestock producers, and substantially degraded rangeland productivity (Holechek et al. 2001).

Demand for game meat was high and market hunting flourished despite growing criticism (Clepper 1966). The railroads made it possible to rapidly ship large quantities of game animals, hides, and even bison bones to industrial centers. Market hunters almost completely eliminated bison between 1870 and 1876. Prairie chickens, ducks, geese, cranes, plovers, and pigeons were shipped by the trainload from the prairies to processing centers in Chicago and Saint Louis (Figures 2.11 and 2.12). Despite new protective state laws, poaching continued to contribute to reduced densities of whitetail deer, turkey, waterfowl, plumed birds, passenger pigeons, trout, and other fish in eastern forests, wetlands, and streams. The exploitation continued mostly because of insufficient education and law enforcement. By 1900, even the farmland game species of the southern and Midwestern states were beginning to decline from overhunting. Elk, bighorn sheep, and pronghorn antelope were scarce or eliminated throughout much of their range. In many locations, household consumption contributed more to decline than market hunting.

Policy

Public Land Disposal. Territorial lands and unclaimed federal lands remaining within state boundaries were opened to free settlement with the passage of the Homestead Act of 1862. This allowed a settler to obtain 160 acres after farming the land for five years or for $1.25 an acre after six months of residence. It resulted in rapid transfer of public lands in the Great Plains to private ownership.

Many early settlers tried to farm dry lands better suited to grazing. The land grants were too small to provide the minimum grazing land needed for a family ra

FIGURE 2.11 Sale of game animals at the Fulton Market, New York City, 1978 (courtesy of Johnson 1964 and North Dakota Game and Fish Department).

FIGURE 2.12 Mixed bag of small game taken by two hunters near Stump Lake, North Dakota, in the 1890s (courtesy of Johnson 1964 and North Dakota Game and Fish Department). By the late nineteenth century, most states had enacted protective game laws or were considering them to prevent overhunting.

ppropriate land grants led to many conflicts between farmers and ranchers, which ked in the 1890s. There was much fraudulent accumulation of grazing land and ent violence. By 1900 only public domain lands of low productivity remained ttled. Much land in the western states had been degraded by overgrazing and un- ssful cultivation.

The federal government adopted the agricultural education philosophy of Thomas Jefferson with passage of the Morrill Act in 1862. It established a system of Land Grant Colleges on public domain lands provided to each state or territory. The colleges were to have a technical emphasis in agriculture and engineering. These institutions were to play a successful role in training extension specialists, research scientists, and future farmers and ranchers.

Influential Early Environmentalists.   As exploitation increased, a small group of far-sighted individuals emerged to eventually change American attitudes and government policies toward renewable resources. They were a mix of artists, literary figures, magazine editors, wealthy sportsmen, scientists, and concerned politicians who built on the philosophy of earlier romantics and concerned sportsmen. Their interests in improved management of natural environments were based on utilitarian, recreational, spiritual, and esthetic values (Clepper 1966). George Perkins Marsh (Figure 2.13) published a book that called attention to degradation of soil, vegetation, and watersheds in many locations of the world (Marsh 1864). He established many of the basic principles of soil and watershed conservation (Clepper 1966). Marsh also described how managed cutting, followed by replanting, would save soil, water, and timber resources. The predecessor of the American Fisheries Society, the American Fish Culturists Association, was formed in 1870 with a strong conservation platform. In 1871, the U.S. Bureau of Fisheries was formed to assess food fisheries status and to develop conservation approaches. They emphasized fry stocking, building fish passages at dams, and outlawing market fishing in certain waters (Scott 1875).

Artists were very influential in shaping public attitudes (Kline 1997). Albert Bierstadt, Frederick Church, Thomas Moran, and numerous others captured the wilderness beauty in their art (Novak 1995). Exhibits of these paintings attracted thousands of influential viewers and sold for high prices. Early photographers, such as William Jackson, also began to chronicle wilderness aesthetics in black and white (Figure 2.14). These artists saw great beauty in wild nature and communicated directly to the influential educated class (Novak 1995).

The writers who influenced change in renewable natural resource management were of two philosophical schools: those who espoused the conservation of resources for sustained use and those who advocated preservation of nature. Outdoor-sports writers, led by George Bird Grinnell, educated the public on the need for sound conservation policies (Reiger 1975). Grinnell also was largely instrumental in establishing the National Audubon Society in 1886. In the West, John Muir led an effort to protect wilderness in National Parks from logging and dam building. Muir's followers promoted a zoned approach to resource use with liberal preservation of wilderness in parks and other reserves. Muir was instrumental in the transfer of Yosemite Park from state to national status in 1890 (Figure 2.15) and he was a founder of the Sierra Club (Kline 1997).

Forest Preservation.   Following the Civil War the rapid depletion of timber began to raise serious concerns (Reiger 1975, Williams 1989, Kline 1997). In 1873 the American Association for the Advancement of Science formally asked the U.S. Congress to pass laws for forest protection. In 1875, the American Forestry Association was formed. Congress created the Federal Division of Forestry in 1881 in the Department of Interior, but it was later transferred to the Department of Agriculture. In 1882, Massachusetts authorized acquisition of forests near major cities to protect water supplies. Similar actions soon followed in other states along the Atlantic seaboard. Watershed protection grew in the wake of increasingly disastrous floods in logged areas. New York set aside the Adirondack and Catskill Reserves in 1888. Pennsylvania began to set aside lands in the heavily logged upper Susquehanna watershed. The first serious commitment of the federal government to forest resource conservation was shown with passage of the Forest Reserves Act of 1891. This act allowed the president to set aside specific areas of public land, which were to become the National Forests. The act was strengthened in 1897 when specific provisions relating to watershed protection were attached to it.

GEORGE P. MARSH
1801–1882
Interpreter of Man and Nature

JOHN MUIR
1828–1914
Preservationist of Natural Resources

PRESIDENT THEODORE ROOSEVELT
1858–1919
"The Conservation President"

ALDO LEOPOLD
1886–1948
Scientific Wildlife Management

FIGURE 2.13   Important men in the early natural resource conservation movement (courtesy of U.S. Department of Agriculture and University of Wisconsin).

Mineral and Water Resources.   The General Mining Law of 1872 first established a federal policy supporting development of mineral resources on public lands. It made suitable public lands available for mineral exploration and purchase ($2.50 to $5.00 per acre), if mined. It permitted development without any royalty or other compensation to the United States. The law continues to the present time without significant modification. It is controversial because it subsidizes development that is often environmentally damaging.

FIGURE 2.14 Photograph of Yellowstone Falls taken by William Jackson, a famous nineteenth-century photographer of western landmarks who, along with painters such as Thomas Moran and Albert Bierstadt, influenced public land policy in favor of preservation (courtesy of Library of Congress).

FIGURE 2.15 Early photograph of Yosemite Valley, which gained much public support for the National Park Service. The efforts of John Muir, who founded the Sierra Club and initiated a public land preservation movement, played a key role in the establishment of National Parks and the National Park Service (courtesy of U.S. National Park Service).

Increasing demand for inland and coastal navigational improvements resulted in establishment of the River and Harbor Act of 1875. This act, and subsequent updates, authorized the Corps of Engineers to initiate and maintain many projects aiding navigation, including channel dredging and various kinds of engineered structures. Following exploration of southwestern water resources by John Wesley Powell, the U.S. Geologic Survey was established in the Department of Interior in 1879. Its primary responsibilities were to map public lands, classify land for appropriate use, and monitor water resources. This established a basis for federal water supply management. In 1899, the Corps of Engineers was authorized to regulate modification of navigable waters, including dumping of wastes into them.

Parks and Recreation. As America grew more affluent, wilderness and natural wonders acquired significant recreational value. Tourism boomed on the wheels of a rapidly growing railroad network. Untouched wilderness remained only in the most rugged, remote, and harsh terrain. Some of the first successful government preservation efforts were directed toward the more awe-inspiring areas, including Yosemite Valley in California, the upper Yellowstone in Wyoming, the Hot Springs of Arkansas, Niagara Falls in New York, Mackinac Island in Michigan, and the Adirondack Preserve in New York. The national parks were first established with the Yellowstone Act of 1872, which set aside 2.25 million acres (Yellowstone National Park) to be kept in natural condition as "a pleasuring ground" for the people of the United States.

THE PROGRESSIVE PERIOD (1900–1945)

Ecology and Landscape

Tremendous ecological transformation occurred in the United States during the nineteenth century. By the end of the century, two-thirds of the continent was settled and substantially developed wherever water and other necessities were consistently available. A network of water and rail-based transportation linked agriculture, logging, and mining across the nation to manufacturing centers on the nation's large rivers and coasts. Native Americans were allocated reservation lands amounting to a small fraction of what they once roamed.

By 1900, most of the natural landscape of the eastern and central United States had been altered and often degraded by logging and farming. The vast forests and prairies of the eastern and central United States had been placed under cultivation (Williams 1989). Large areas of northern pine forests had burned following wasteful logging practices. Overgrazing was rampant in the high plains and western mountain ranges. Inappropriate farming in the semiarid plains led to massive soil erosion and farm community failure during the "dust bowl" drought of the early 1930s. The Northwestern forest had been extensively cut at lower elevations and much of it converted to farms. Much of the cotton and tobacco farmland of the South was badly eroded and had been abandoned because of degraded fertility and gully erosion.

The remaining undeveloped natural areas in the United States were either too rugged, too dry, or too wet to develop. Much fertile arid land and wetland near river floodplains remained undeveloped because of the inability of the states to muster the resources to provide irrigation or drainage. Flooding was becoming more of a problem as clearing and agriculture altered the watershed capacity for slowing runoff rates and as more people settled the floodplains. Levees had been built on some of the major river systems to improve navigation and to help manage flooding.

The dramatic change in vegetation, human numbers, and hunting and fishing pressure was associated with widespread decline of certain wildlife and fish species. Elk, pronghorn, and turkey had been decimated and bison were nearly extirpated. Passenger pigeons were about to go extinct. Whereas natural forest and grassland species declined, species adapted to farmland conditions flourished, including bobwhite quail, cottontail rabbits, redwing blackbirds, fox squirrels, and various rodents.

Exotic species were becoming common by accident and design. Urban areas and farmland were especially altered by planting of exotic crops, gardens, and trees. Many European plant species became naturalized. Some were to become pest species, such as dandelions, purple loosestrife, and water hyacinths. Salt cedar, Russian olive, Russian thistle (tumbleweed), and cheat grass invaded the arid West. Also from Europe, rock doves (common pigeons), house sparrows, starlings, house mice, and Norway rats became ubiquitous members of urban and farm wildlife. Foresters, first educated in Europe, had begun to import "more manageable" European species for development of large plantations in the Northeast. Fisheries professionals stocked Eurasian brown trout and carp using hatchery methods from Europe. The extensive development of canals in the nineteenth century allowed exchanges of previously isolated aquatic faunas. Construction of the Welland Canal around Niagara Falls contributed to overfishing and invasion of the upper Great Lakes by sea lamprey and alewives.

Economics

Technology. The progressive movement led the nation into a new era based on high-minded social goals, scientific innovation, and more regulation by federal and state governments. Human welfare continued to improve during this period and wealth was more equally distributed throughout society as labor organized and as the federal government increasingly regulated big business. With help from the Civil Service Act of 1883, which required appropriate technical expertise for government employees, the federal government became more actively involved in natural resource development and management, especially of forest, range, and water resources. The completion of the Panama Canal in 1909 was an unprecedented engineering accomplishment of immense benefit. It became a model for many other large government water resources projects and public works.

The Progressive Period

1901	Theodore Roosevelt became President of the United States and led the "progressive movement" toward the "greatest good for the greatest number" through strong government.
1902	The Reclamation Act established federal role in developing water resources in the arid West resulting in some of the largest reservoirs ever built.
1903	The Wright brothers invented motorized air travel.
1905	The first wildlife refuge was authorized.
1905	The U.S. Forest Service formed under the direction of Gifford Pinchot, who coined the word conservation and promoted "wise use" and sustainable development of resources.
1914	The first industrial refrigerators greatly improved food transport and storage and the first permanent federal income tax provided unprecedented policy leverage.
1916	Migratory Bird Treaty Act established coordinated management of waterfowl and other birds with Canada.
1916	The Federal Highway Act of 1916 ushered in national highway planning and transcontinental "motoring."
1924	The first wilderness was set aside by the Forest Service and Aldo Leopold and Robert Marshall worked toward founding of the Wilderness Society in 1935.
1929	Television was invented and the stock market crashed for complex reasons associated with liberal lending policies and overproduction of cars and appliances.

1933	Bad farming practices and the dust bowl drought hammered agricultural production and aggravated the nation's worst economic depression. The U.S. Soil Conservation Service was first authorized as the Soil Erosion Service.
1934	The Taylor Grazing Act created the first government control over grazing on public domain lands.
1935	The term ecosystem was coined by Tansley, emphasizing links between biotic communities and abiotic environment.
1936	The Omnibus Flood Control Act established a nationwide federal leadership role in flood control and onset of dam construction and improved watershed management.
1940	The U.S. Fish and Wildlife Service was created in the Department of the Interior.

Technology advanced at a tremendous rate. Many improvements were made in metallurgy, mechanics, building materials, and farm practices leading to larger, stronger, more efficient, and safer ways to manage, extract, move, refine, and otherwise develop natural resources. Invention of the radio greatly expanded interstate communication. Advances based on the internal combustion engine, motorized flight, and electric generators lead to public pressure on government to assure that appropriate infrastructure was developed to facilitate interstate commerce.

Automobile manufacturers, led by Henry Ford, developed mass production. In 1908, the Model T Ford became the first car affordable to the middle class. Automobile ownership became as much a part of American aspiration as owning a home. Highways improved markedly with the paving of the U.S. highway system. The first hard-surface road crossed the continent in 1913. People began to "motor" across the country and fresh farm goods became widely available as a trucking industry burgeoned. Many new products were now ordered cost-effectively through national catalogs and shipped via rail and truck to local retailers or directly to the home.

Thanks to Thomas Edison, one of America's greatest inventors, electrification progressed rapidly at the turn of the century (Figure 2.16). Electrification powered the first household radios, electric stoves, washing machines, and industrial refrigerators just before World War I (1914–1918). Air conditioning was introduced to movie theaters in the late 1920s. Pop-up toasters, televisions, household refrigerators, stereo systems, garbage disposals, electric razors, blenders, tape recorders, and fluorescent lighting soon followed. The cities benefited from electrification first and the poorest rural regions were last. During the depression years (1930s), the federal government established the Tennessee Valley Authority and the Bonneville Power Authority to provide electricity by hydropower in depressed rural areas.

Numerous medical advances led to much improved understanding of diseases and how to treat them (the development of penicillin and insulin). In the absence of understanding the many positive functions, wetlands were drained to reduce mosquito-borne diseases such as malaria. Public health organizations became more influential and improved waste disposal practices reduced typhoid, cholera, and other deadly diseases.

Depression and Conservation.   The Progressive period was a time of great economic change and stress. The key event was the 1930s' depression, which started with the stock market crash in 1929. The depression was brought on by a complex assortment of factors, but its root cause was overinvestment in production capacity and excessive use of credit (debt) (see Chapter 22). The depression coincided with severe drought in the Great Plains, which brought about massive farm failures. Economic growth came to a standstill in the early 1930s. Depression dominated national policy until declaration of war in 1941.

FIGURE 2.16 The decades around 1900 were among the most inventive in history. Thomas Edison was influential in establishing central electrical supply and incandescent lighting. Rapid expansion of electrification increased development of coal and oil fuel sources and hydropower (courtesy of U.S. National Park Service).

After the stock market crash in 1929, the economy spiraled downward as personal income shrank, trade declined, and international markets sank into economic depression. Based in part on theories of John Maynard Keynes, the federal government led by Franklin Roosevelt used government tax and spend policies for the joint solution of environmental and economic problems (Heilbroner and Thurow 1994). Disastrous floods and soil erosion focused legislative attention on improved water and soil management. President Roosevelt promoted projects for irrigation, flood control, and hydropower development through reservoir construction. These projects provided much needed work (Figure 2.17). Roosevelt's administration believed careful construction could conserve living resources, such as migratory fishes, by proper engineering. He encouraged establishment of the Civilian Conservation Service and other federally sponsored work programs with joint social and environmental improvement goals.

Although the effectiveness of Roosevelt's "New Deal" prior to World War II has been questioned, the federal government would never step back to the passive role it had taken before 1900. World War II reinforced the role of federal government in the U.S. economy when it mobilized and guided the largest conversion of natural resources into war technology experienced up to that time. An economic triangle of big labor, big business, and big government was firmly established and interacted to determine the future course of natural resource development.

Policy

Progressive Emphases.   The main goal of the federal government during the Progressive movement before World War I was to encourage sustained use of natural resources and to assure that the benefits would be distributed equitably among the people. The conservation ideal was natural resource management for the greatest good of the greatest number over the long run. Because public lands were much easier to influence than private lands, the federal government shifted away from a policy of public land disposal to public land management.

UPPER MISS. RIVER-LOCK NO. 26
PWA-CONTRACT-NO.W-1103 ENG. 1706
VIEW UPSTREAM

FIGURE 2.17 Construction of a lock and dam structure on the upper Mississippi River in 1935 served both to stimulate work during the Great Depression and to improve commercial navigation (courtesy of U.S. Dept. of Interior).

The conservation movement of the Progressive Era was highly utilitarian, which fit with prevailing public sentiment. Progressive policy emphasized resource development that generated jobs, affordable housing, food, safety, and health. Outdoor recreation was considered among the important uses of public lands and waters. This was due to the growing economic value of tourism and general recognition that outdoor recreation was good for public health.

The movement promoting natural resources management had split into several camps by the time Theodore Roosevelt became president in 1901. One camp, led by John Muir, favored preservation. He advocated zoned public land use, in which a large proportion of public lands would be preserved in a pristine state for low-intensity recreational use, epitomized in the National Parks. In contrast, Gifford Pinchot favored multiple resource uses on public lands, and believed all resources should be developed to their full potential. Pinchot's philosophy gained the greatest following within the progressive movement. However, Muir's philosophy was very influential in establishing a large national park system and, later, a network of wilderness areas on other public lands.

In the progressive view of conservation promoted by Pinchot, nature was the servant of human needs and resources were "wasted" if not used. Therefore, the concept of conservation included management of natural threats to the development of land and water resource use. An important weakness in this utilitarian conservation philosophy was poor understanding of the value of ecosystem biodiversity. This ultimately led to the endangerment of numerous species.

While the federal government took control of natural resource management on public lands, it also acquired a powerful tool for influencing private behavior—the income tax. Although temporarily enacted during the Civil War, the federal income tax did not become permanently established until 1914. Through tax law, Congress could redistribute wealth and encourage desired behaviors in the private sector. Income taxes provided a dependable source of revenue for financing a federal government of expanding size.

Water and Mineral Resources. Transfer of public domain to State control under the Swamp Acts and Desert Land Acts had resulted in little development of those lands as required. The Reclamation Act of 1902 authorized federal construction of large dams and water delivery systems to irrigate arid land mostly for agricultural development. The Bureau of Reclamation originated under this act and continues to operate and maintain many irrigation projects in the western United States.

Maintenance and development of improved navigational resources remained an important emphasis of the Corps of Engineers. Amendments to the Refuse Act of 1899 in 1905 authorized the Corps of Engineers to enforce prohibition of garbage and other solid waste dumping into navigable waters. This was the first significant federal pollution-control legislation. The Rivers and Harbors Act of 1917 provided the Corps with the authority to conduct studies on flood control and related problems. It authorized flood control projects along the Mississippi and Sacramento-San Joaquin Rivers. The federal government assumed primary responsibility for flood control along those same rivers via the Flood Control Act of 1928.

Flood control authority was extended to all areas of the nation by the Omnibus Flood Control Act of 1936. It gave the secretary of agriculture authority to study watershed process and develop programs to retard runoff and soil erosion through the newly created "Soil Conservation Service." The Corps was authorized to build many large dams and levees for flood control and navigation purposes. This law was the first to require gross benefits of government projects to exceed costs.

Wetland drainage and filling accelerated early in the century in the name of progressive land conservation. Wetland mismanagement was symptomatic of the general ecological ignorance and expediency that pervaded government decision process. Except for waterfowl and certain fisheries, wetlands were thought of as wastelands because they impeded land resource development. There was little appreciation of the human benefits they provided, such as habitat for waterfowl, groundwater maintenance, and water quality maintenance. Their filling rate accelerated when mosquitoes were discovered to cause malaria and yellow fever. Many wetlands were channelized and filled with dredge material as the Corps of Engineers pursued its mission of improved navigation in coastal and riverine environments. Numerous wetlands were drained under government programs that served agricultural land users. Federal wetland management policy grew more controversial as more was learned about the beneficial roles of wetlands in supporting fish and wildlife habitat.

The Mineral Leasing Act of 1920 initiated the first significant change in Federal mineral policy since the General Mining Law of 1872. It placed fossil fuels and fertilizer minerals under a leasing system that provided monetary compensation to the United States. It required the land to be returned to U.S. control after mining was completed. This law provided a basis for the later expansion of government and enforced reclamation of mined lands.

Forest and Park Resources. By the early 1900s the value of forests for wood, watershed protection, and outdoor recreation was well recognized. Part of the solution to flooding was linked to reforestation and other watershed improvement. Federal forest management was strengthened in 1905 by formation of the U.S. Forest Service headed by Gifford Pinchot. He was an articulate politician who had a considerable influence on Theodore Roosevelt and the conservation movement. The Weeks Act of 1911 authorized federal purchase of deforested lands in the eastern United States to reforest them for watershed protection.

Most of the large national parks were established during the Progressive movement. They were formally brought together in one coordinated management system with formation of the National Park Service in 1916. The National Park Service was also authorized to manage the national monuments, which had been established under the Antiquities Act of 1906. It permitted preservation of lands with unique scientific or historic value through presidential proclamation. President Theodore Roosevelt, an ardent outdoor recreationist and conservationist, aggressively set aside lands for national forests, parks, and monuments.

Personnel within the Forest Service, led by Arthur Carhart, were among the first advocates of wilderness protection, which began in 1924 in the national forests. Aldo Leopold, a Forest Service employee, was instrumental in establishing the first Forest Service wilderness in 1924. He and another Forest Service employee, Robert Marshall, were founding members of the Wilderness Society in 1935. Marshall was a forester, explorer, and writer who played a critical role in strengthening the Sierra Club and National Parks Association as political forces in the management of public lands.

Range and Soil Resources.   Government intervention into grazing problems on the western range began in 1898 when the Department of Interior granted grazing permits to limit the number of livestock on federal lands. Even so, much of the public rangelands continued to degrade (Smith 1898, Stoddard et al. 1975). After 1905, the U.S. Forest Service established a process of forage allotment on national forests. Because of high demand for beef during World War I, severe overgrazing took place between 1915 and 1920.

During the drought in the early 1930s, the United States reaped the consequences of 60 years of range exploitation. Vast areas of the Great Plains had been plowed to take advantage of temporarily favorable climatic conditions and high grain prices. After the drought, millions of acres of plowed land returned to grazing use. The drought also precipitated passage of The Taylor Grazing Act of 1934. This act placed administration of remaining public domain land under the Grazing Service, which later became the Bureau of Land Management. It resulted in allocation of grazing privileges on the public domain lands that had not been transferred to private holdings or to agency management authority. Prior to the Taylor Grazing Act, there was no control over how public domain lands unassigned to federal agency management were to be used. If one person did not exploit the forage in an area, his or her neighbor probably would. This provides an excellent example of what Hardin (1968) later called "the tragedy of the commons" (Figure 2.18).

Until the 1930s, government policies continued to be oriented toward converting the remaining public domain land into private ownership under pressure from western states. The failure to recognize that large areas of the West could not sustain farming resulted in great land damage, social upheaval, and economic loss. The Enlarged Homestead Act of 1909 was passed to encourage settlement of the westernmost states. It increased homestead size to 320 acres, with the requirement that one-fourth be cultivated. This amount of land was insufficient for ranching and encouraged plowing of rangeland that would not sustain cultivation. The Stockraising Homestead Act of 1916 granted 640 acres for raising 50 cows. In most areas, this act caused rangeland destruction because 640 acres would not provide sufficient forage for 50 cows.

In 1928 Dr. Hugh Hammond Bennet wrote the first U.S. Department of Agriculture Bulletin on soil erosion entitled *Soil Erosion, a National Menace*. The severe drought on the Great Plains confirmed Bennet's warnings about impending disaster from soil erosion. Dust was blown thousands of miles, damaging buildings, human health, and economic activity. In 1933 Congress created the Soil Erosion Service in the Department of Interior, which in 1935 became the Soil Conservation Service in the Department of Agriculture.

Fish and Wildlife.   In 1900, unregulated market hunting was brought to closure after decades of massive slaughter of shore and wading birds to provide feathers for women's hats. Public outrage resulted in passage of the Lacey Act of 1900, which made interstate shipment of game killed in violation of state law also a federal offense (Trefethen 1966). In the same year, the Protection of Migratory Game and Insectivorous Birds Act was passed. It was followed in 1916 by the Migratory Bird Treaty Act. The two laws placed all migratory birds under federal protection, ended spring waterfowl hunting, and ended hunting for scarce shorebirds.

The first of many national wildlife refuges was established in 1905. Some species responded quite rapidly to the increased protection. Deer populations exploded in

FIGURE 2.18 "Tragedy of the commons" (from Holechek et al. 2001) (drawing by John N. Smith).

many areas after 1910. By the 1920s, major die-offs of deer occurred in various states where they had increased to such an extent that they exceeded their food supply (Trefethen 1966). Such incidents and growing public interest demonstrated a need for improved wildlife management. In the 1920s, Aldo Leopold published initial studies and theories on game management. His classic textbook *Game Management* was published in 1933. These and other works formed the basis for the new profession of wildlife management and led to the establishment of The Wildlife Society in 1939.

During the 1930s, several wildlife laws were enacted to aid farmers through tax breaks and conservation incentives, which improved waterfowl and upland game habitats. The Predatory Mammal Control Program of 1931 promoted development of methods of "eradication, suppression, or bringing under control" various mammals viewed as pests, including mountain lions, wolves, and other species now protected.

The Duck Stamp Act of 1934 required all waterfowl hunters under age 16 to purchase a $1.00 stamp, with the proceeds to be used to fund creation and restoration of wetland habitat. The Wildlife Restoration Projects Act of 1937 returned to the states revenues collected from federal excise tax on sporting arms and ammunition. To qualify for revenues the states agreed to use the funds for wildlife management and to apply state hunting license fees to fish and wildlife management. Prior to this law, it was common for state legislatures to use license fees for unrelated programs (Clepper 1966).

Today's U.S. Fish and Wildlife Service was first established in 1940 in the Department of Interior by consolidating the Bureau of Biological Survey and the Bureau of Fisheries.

Federal fisheries legislation was less active and targeted specific fishery resources. Growing competition for fishery resources on the open seas led to establishment of the International Halibut Commission in 1909 and the North Pacific Salmon Commission in 1937. These international agreements between Canada and the U.S. established a pattern for future ocean fishery management. However, they failed to halt the decline of many valued fish species.

World War II Impacts on Natural Resource Management. The Progressive period was brought to closure by the ending of World War II in 1945. During the war, changes in natural resource policy were few. The war split the twentieth century in two. Before the war, the United States looked mostly inward to solve problems at home. Even though the United States was becoming an important part of the global economy, national interests were primarily self-centered. The war forced Americans to deal with global problems. Rapid advances in communication further prevented a return to isolationist self-concern. World War II was the culmination of the industrial age and the harbinger for the information age. Inventions before and during the war were rapidly transformed into the diverse computing, communications, and knowledge-based industries of today.

NEOPROGRESSIVE PERIOD (1945–PRESENT)

Ecology and Landscape

The Western Hemisphere was spared the destruction of World War II. By the end of the war, the amount of land in public and private ownership had stabilized with most of the public land existing west of the Great Plains (Tables 2.1 and 2.2). Much marginal farmland was converted back to grassland or woodland with remaining farmland being more intensively cultivated. After the exodus of dryland farmers from the dust bowl in the high plains, range use became more compatible with sparse and uncertain rainfall. Groundwater resources were developed to irrigate many areas in the 11 western states. Groundwater depletion became a serious problem in several regions. To accommodate larger equipment and more efficient land use, farm fields also were enlarged and were more consistently planted to the same crops. Wild areas left between farmed lands were gradually eliminated, leaving the farmed landscape more monotonous.

Before WWII, the U.S. was fairly distinctly split into localized urban-industrial landscapes and widespread rural landscapes. Cities were compact and built around business and market centers. Even in large cities like New York, a person could drive

TABLE 2.1   Present Custodianship of Federal Lands in the United States

	Acres (Millions)	Percent of Total
Total land in the United States	2,271	100
Not owned by federal government	1,621	71
Owned by federal government	650	29
Bureau of Land Management	268	12
Forest Service	191	6
Fish and Wildlife Service	86	4
National Park Service	73	3
Department of Defense (except the Corps of Engineers)	16	1
Corps of Engineers	12	<1
Bureau of Reclamation	4	<1
Bureau of Indian Affairs	3	<1
Other Agencies	4	<1

Source: Holechek et al. 1998 (from USDI reports).

TABLE 2.2   Percent of Land in the 18 Western States Presently in Federal Ownership[a]

State	Percentage	State	Percentage
Alaska	68	New Mexico	31
Arizona	47	North Dakota	4
California	44	Oklahoma	2
Colorado	36	Oregon	52
Idaho	62	South Dakota	6
Kansas	<1	Texas	1
Montana	28	Utah	64
Nebraska	1	Washington	24
Nevada	83	Wyoming	48

[a]28% of the United States in federal ownership.
Source: Holechek et al. 1998 (from USDI reports).

to woodland, farms, and small towns in a short time. After the war, a new phenomenon of suburban growth began as people desired larger backyards, closer proximity to schools, and escape from urban blight and crime. Migration to the suburbs was facilitated by the development of interstate highways starting in 1956. During the 1970s, a small countercurrent of young and idealistic professionals began to move back into the inner city to buy cheap housing for renovation. Conditions have slowly improved in many blighted urban areas while suburban sprawl continues unabated.

Over the course of the last five decades, the Nation's landscape has been drastically altered by urbanization. Much of the farmland adjacent to larger cities has been converted to urban and suburban landscape. Natural ecosystems have become increasingly fragmented by roads and developed areas. In general, people in the United States have continued to move westward and southward. This trend has been driven by the relatively cheap land, cheap labor, mild winters, and appealing environment. The widespread development of water resources also has been important.

Water and air quality declined after WWII until the 1970s when environmental legislation slowed that trend and to some extent reversed it. However, air quality remains in violation of standards in many cities. Diffuse sources of water pollution are still a problem in many parts of the nation. Numerous mines, waste dumps, and abandoned industrial sites continue to leak contaminants. As the industrial age gave way to the information age, many industrial areas were abandoned leaving large expanses of barren land and crumbling structures that became known as "brownfields."

In recent years, there has been much interest in dedicating land to biodiversity preservation. Nongovernment organizations, led by The Nature Conservancy, have been instrumental in purchasing lands with exceptional natural value to add to public lands or to retain under nongovernment control. The leading federal land management agencies include the U.S. Forest Service, Bureau of Land Management, Fish and Wildlife Service, and National Park Service (Table 2.1). The U.S. Fish and Wildlife Service has initiated a nationwide inventory and analysis of land use to identify areas where biodiversity protection is needed.

Landscape changes in the undeveloped world have accelerated tremendously since WWII as human populations and development have expanded. Trends there mirror changes in the United States during the nineteenth century. Large areas of tropical forests have been cut and converted to other land uses, often with disastrous results. Desertification is occurring in semiarid nations as a consequence of overgrazing and overharvesting of fuel woods. Soil erosion has accelerated in many locations, decreasing the life of reservoirs built for flood control, electrification, and irrigation. Numerous cities have become huge, congested, and badly polluted as people have left the degraded land for employment. Air and water quality deteriorated tremendously in parts of industrialized Eastern Europe under communism.

National and international goals increasingly consider ecosystem protection and restoration as important elements for future human welfare. However, the continuing trend toward fragmentation of land ownership into many small parcels makes large-scale ecosystem restoration a challenging problem. Although most people in the United States favor ecosystem restoration, few will readily sell their personal land, even when fairly compensated. Therefore, where restoration is sometimes most needed, the costs are tremendous and often prohibitive.

Economics

After World War II, U.S. policy goals broadened to enhance worldwide welfare through democracy and free enterprise in an international market economy. The federal government played a more active role in almost every aspect of American life and in international politics. During the two decades following WWII, the United States had unprecedented economic growth because of its advantages in world trade. This period also was a time of rapid population growth. The baby boom following the war was to have disproportionate influence on later policy and economics. People were better educated than ever before, in part because of government investment in education, but also because of the escalating availability of information. In planning for their children's happiness and success, parents emphasized education, fun, beauty, health, safety, security, and material welfare.

The Neoprogressive Period

1941	Onset of U.S. involvement in World War II and massive commitment of natural and human resources.
1945	The atomic bomb was exploded and WWII ended with U.S. infrastructure and natural resources basically intact and a Cold War continuing for the next half century.
1946	The U.S. baby boom began along with a two-decade period of exceptional economic growth.
1948	The first nationwide TV broadcast and Vogt (1948) published a seminal study of world population problems.
1949	Xerox developed first photocopier and Aldo Leopold published a land-use ethic in *A Sand County Almanac*.
1951	The first commercial computer took up a large room.
1952	The first hydrogen bomb was exploded in the Pacific.
1953	Eugene Odum published *The Fundamentals of Ecology*, the first systems-oriented ecology textbook.
1955	The transistor was invented.
1956	The Interstate Highway Act was justified for nuclear attack evacuation, wartime mobility, and economic development.
1957	*Sputnik*, the first space satellite was launched by Russia, pressing the race for dominance in space.
1960	The Multiple-Use and Sustained Yield Act authorized the U.S. Forest Service to manage for sustained yield and multiple uses of resources.
1962	Rachel Carson published *Silent Spring*, an indictment of pesticide abuse that galvanized the environmental movement.
1964	The Wilderness Act authorized a public wilderness system.
1965	The Water Resources Planning Act established benefits-based planning in a riverbasin context and the Land and Water Conservation Fund was established using oil extraction taxes.
1968	Paul Ehrlich published *The Population Bomb*.
1969	The National Environmental Policy Act ushered in the "green" decade for laws pertaining to clean air, clean water, endangered species, fisheries conservation, and hazardous material cleanup.

1977	The first personal computer was invented.
1993	Web browser software opened up "information superhighway."
1999	The first dam was breached in Maine under federal order to restore natural flow for anadromous fisheries.

Television ushered in the information age and much improved awareness of national and world events. Photocopying, invented in 1949, resulted in a large leap in extending information. The first commercial computer was built in 1951. UNIVAC, as it was called, filled a large room but had less computing power than today's laptop portable. The Internet, developed first for security and research purposes, became the "information superhighway" starting in 1993. Modern computers ushered in the era of database management. They made it practical to use more complex analyses and mathematical models for resource management. Connected with computers, many resources are now inventoried and monitored using geographic information systems (GIS).

Soon after the war, scientists began lobbying for peaceful applications of atomic power for electric-power generation. Atomic power plants first showed up in the landscape in the 1950s (Figure 2.19). Atomic power, along with hydropower and fossil-fuel power, collectively provided cheap electricity for running an increasing array of labor-saving devices. Atomic power was promoted because it produced no air pollution, assuming accidental releases of radioactive emissions did not occur. A narrowly avoided disaster at Three-Mile Island nuclear power plant in 1979 and other influences led to greatly depressed investment in further nuclear power development.

By the early 1960s people were better off in the United States than any previous time. Construction of the federally supported interstate highway system in the late 1950s made access to the nation's public lands and historic monuments much easier. Visitation rates skyrocketed at national parks, national forests, federally built reservoirs, and lands managed by the bureau of land management (Table 2.3). Participation in sightseeing and outdoor recreation increased rapidly and the tourist industry mushroomed.

FIGURE 2.19   Atomic power plants began development soon after WWII and were touted for their low environmental impact. However, accidental radioactive releases, costly safeguards, and difficulty in developing socially acceptable radioactive waste storage have altered the once rosy prospects for fission-produced power (courtesy of U.S. Dept. of Interior).

TABLE 2.3 Estimated Total Visitor Days (1,000) and Hunter Days (1,000) of Recreational Use on Public Lands Under the Jurisdiction of the Bureau of Land Management in 1965 and 1980

	Visitor Days		Hunter Days	
	1965	1980	1965	1980
Arizona	808	7,430	120	150
California	3,162	16,969	290	1,237
Colorado	1,382	4,572	249	601
Idaho	700	9,421	130	402
New Mexico	659	6,978	114	306
Oregon	3,699	4,266	361	994
Utah	558	2,811	240	335
Wyoming	119	4,097	55	361
	11,087	56,544	1,559	4,386

Source: USDI reports.

Policy

Environmental Policy.   Agricultural and ecological research in the U.S. accelerated after World War II to help address world food shortages and potential environmental threats from atomic radiation and other contaminants. The awesome power of the atomic and hydrogen bombs both inspired and frightened the scientific community. It lobbied for improved understanding of atomic potential and risks. Ecological studies sponsored by the Atomic Energy Commission quickly revealed how contaminated materials were mobilized and moved through ecosystems, often to undesirable locations in human and wildlife food webs (Golley 1993). Agricultural research over the next two decades generated the green revolution based on a combination of improved plant breeding, tilling, pesticide application, and fertilization techniques. Use of pesticides and fertilizers escalated without much concern over side effects except in research laboratories. However, health scientists soon verified previously suspected threats of chemical toxicity.

Among numerous publications warning of environmental problems, a few in particular stood out. Soon after the war William Vogt (1948) published *Road to Survival,* a seminal work on human ecology and population pressures. Aldo Leopold's philosophy of a land ethic was published in his book, *A Sand County Almanac* (Leopold 1949). He was among the first to call public attention to the need for sustainable land use. Although Leopold greatly influenced other thinkers, he had little direct impact on natural resource and environmental policy at that time. The press, however, was beginning to identify oil, radiation, and chemical pollution problems. Air pollution was of increasing concern especially in the industrial cities (Figure 2.20). Scientific information about environmental contamination gradually accumulated. In 1962, Rachel Carson published *Silent Spring*, which was a pivotal work on threats from environmental contamination. Secretary of Interior Steward Udall attracted official attention with his 1963 book *The Quiet Crisis*. It called for comprehensive environmental planning and environmental regulation by government. In 1968, Paul Ehrlich published *The Population Bomb,* which was widely read. This book focused on how human population growth was jeopardizing ecological support systems.

By the 1960s, numerous federal agencies were responsible for various monitoring, research, and regulatory programs pertaining to the environment. The National Environmental Policy Act (NEPA) of 1969 led in establishing the present approach to environmental regulation. It was followed by authorization of the Environmental Protection Agency (EPA) in 1970, which pulled together the environmental functions of numerous agencies under a single administration. NEPA requires any public or private organization using federal funds to draft environmental impact statements for proposed actions affecting the environment. It mandated that the federal government provide an acceptable environment for future generations.

FIGURE 2.20 Air pollution was a serious health problem in many industrial cities of the United States before and following World War II. Pittsburgh was among the worst examples (courtesy of U.S. Dept. of Interior).

Earth Day was celebrated for the first time on April 22, 1970. This was a grassroots event organized outside of the mainstream environmental organizations of the time. Its success showed the extent of environmental concern in the baby boom generation.

Amendments to the Clean Air Act of 1970 established standards for a much larger number of air contaminants than those first identified in the Clean Air Act of 1963. It required industry to install scrubbers to remove noxious emissions and automobiles to be fitted with catalytic converters. The Federal Water Pollution Control Act of 1972 mandated restoration of water quality to standards that supported fishing, swimming, and other recreational uses. It created a public works program for funding the development of domestic sewage treatment plants. The Clean Water Act of 1972 authorized regulation of wetland filling by the U.S. Army Corps of Engineers. The intent was to decrease the rate of wetland loss through avoiding negative impact when possible and compensating through creation, restoration, enhancement, and preservation of wetlands when significant loss was unavoidable. These laws applied to private lands as well as public lands. The Ocean Dumping Act of 1972 and the Safe Water Drinking Act of 1974 added to the improvement of water resources.

Better management of toxic materials was addressed by the Toxic Substance Control Act of 1976 and the Resource Conservation and Recovery Act of 1976. These laws established safe handling procedures for wastes and encouraged recycling. The Environmental Response, Compensation, and Liability Act of 1980 established a trust fund (the "superfund") to recover waste storage areas associated with mines, industry, and military installations.

Water and Mineral Resources. The Water Resources Research Act of 1964 established a cooperative state and federal water research program. The Water Resources Planning Act of 1965 facilitated comprehensive river basin planning. It resulted in improved methodology for evaluating the benefits and costs of federally funded water projects. Both economic and environmental benefits were to be considered. The Water Resources Development Act of 1986 elevated ecosystem protection and restoration to higher consideration in the Army Corps of Engineers mission. The Soil Conservation Service broadened its mission and became the Natural Resources Conservation Service in 1994.

After years of strip mining that left parts of Appalachia with barren lands, acid streams, and acid lakes, Congress passed the Surface Mining Control and Reclamation

FIGURE 2.21   Much of northern Appalachia was scarred by abandoned coal mines, a practice that was stopped by the Reclamation Act of 1977. The photo on the top is a typical abandoned mine and on the bottom is a reclaimed coal mine (courtesy of Office of Surface Mining and Reclamation and Enforcement).

Act in 1977. It requires reclamation of surface mines to productive use after mining is completed (Figure 2.21).

Forest and Range Management.   Livestock grazing on federal rangelands was reduced 25% between 1960 and 1992 (Table 2.4). Range improvement on private lands accelerated during the 1980s and 1990s because of improved information and education programs by state and federal agencies and improved education of modern ranchers. Future range management on public lands is likely to place greater emphasis on water, wildlife, recreation, and biodiversity and less emphasis on livestock production. Because of the natural role of fire in range ecosystems, it is likely to receive greater use in manipulating rangeland vegetation.

Since the Taylor Grazing Act of 1934, there have been several movements by private interest groups to transfer remaining federal lands into private ownership. The most recent attempt, known as the "sagebrush" rebellion, occurred in the late 1970s and early 1980s. These movements have been unsuccessful because of heavy public opposition. The majority of citizens in the United States believe public lands are part of their natural heritage and should be retained in federal ownership. The Federal

TABLE 2.4 Trend (1960–1993) in Actual Animal Unit Months (thousands) of Livestock Grazing on Bureau of Land Management and Forest Service Lands

Type of Livestock	Bureau of Land Management				
	1960	1970	1980	1990	1993
Cattle	10,277	13,368	8,984	9,327	8,486
Horses & burros	198	400	72	72	62
Sheep & goats	4,790	2,741	1,351	1,445	1,211
Total AUMs	15,265	16,209	10,407	10,844	9,759
	Forest Service				
Cattle	8,084	—	7,594	—	6,784
Sheep	1,083	—	893	—	805
Horses & burros	210	—	176	—	130
Total AUMs	9,377	—	8,663	—	7,719

Sources: U.S. Department of Interior annual reports and U.S. Forest Service annual reports (from Holechek et al. 1998).

Land Policy and Management Act of 1976 made government retention of existing public land official policy.

The early 1960s were characterized by considerable change in the philosophy of public land management. The Multiple Use Act of 1960 mandated that Forest Service lands be managed for grazing, wildlife, timber, recreation, and mineral development, rather than for a single use. Previously, range research and management had been geared toward producing forage for livestock. This change in mandate had considerable influence on range and forest management practices. Multiple use was defined as management for the combination of uses that "best meet the needs of the American people."

The Forest Management Act of 1976 mandated forest management plans emphasizing multiple use. It specified that plans should provide for diversity of plant and animal communities. Wilderness was recognized as an important use. The act set many limits on timber harvest, addressing concerns about water quality, soil erosion, and fish and wildlife habitat.

Parks and Recreation. The Wilderness Act of 1964 strengthened the existing Forest Service policy of protecting lands with exceptional wilderness attributes. It extended consideration to a larger range of federal lands. In 1965, the Land and Water Conservation Fund was authorized, which used offshore oil-well revenues to purchase new public recreation areas. The Wild and Scenic Rivers Act of 1968 was passed to protect some of the free-flowing rivers from dam building and other modification. In the same year, the National Trails System Act was passed to maintain the scenic quality of major trails within the National Forests and in other locations managed by the National Park Service and the Bureau of Land Management.

Fish and Wildlife. The Fish and Wildlife Coordination Act in 1946, as amended in 1958, provided that fish and wildlife conservation receive consideration in water resource development. The Fish Restoration and Management Projects Act of 1950 established a federal tax on fishing equipment. The U.S. Fish and Wildlife Service redistributes these funds to state agencies for sportfishery development. It was expanded in 1985 in response to a tremendous increase in sportfishing.

The Fishery Conservation and Management Act of 1976 recognized growing problems with depleted ocean stocks of commercial fish. It established a 200-mile offshore authority for managing coastal fisheries more effectively. However, massive closures of fisheries in the early 1990s demonstrate that this law has failed to prevent overharvest.

The Endangered Species Act of 1973 and the Marine Mammals Protection Act of 1976 were designed to comprehensively protect all wild organisms from human-caused extinction. The Endangered Species Act authorized the U.S. Fish and Wildlife

Service and National marine Fisheries Service to identify threatened and endangered species, regulate their take, and recover their populations to unthreatened levels. Its power is most concentrated in government authority to protect critical habitat needed for species recovery regardless of whether the land is publicly or privately owned. The act was later amended to allow development of habitat conservation plans by landowners as a way to provide some flexibility. The Endangered Species Act also established in federal law an ecosystem perspective for management.

Present Status. While attempts to weaken environmental laws typically have met public disfavor, the public has resisted more costly environmental initiatives during the past two decades. Few federal environmental laws have originated since 1980, although many laws have been amended, usually to strengthen or make them more workable.

Amendments and agreements have added a dimension of ecosystem restoration and management to the activities of federal government, such as in the Water Resources Development Act of 1986 and 1996 and the Food and Agriculture Conservation and Trade Act of 1990. These actions are oriented toward protecting biodiversity from further deterioration. Wetlands have received exceptional restoration attention. Agencies like the Natural Resources Conservation Service (Soil Conservation Service before 1994) and the Army Corps of Engineers now facilitate restoration of wetlands that they once degraded or eliminated (Figure 2.22). Much of the future of natural resource management is likely to be associated with restoration ecosystem management and the emerging concept of sustainable development.

Since the 1960s the roles of nongovernment agencies have increased tremendously in influencing government environmental policy (Coggins 1993). They have acted through the courts to protect existing law and to encourage agencies to follow through as authorized. The National Wildlife Federation, Environmental Defense, Sierra Club, and Wilderness Society have been especially influential in protecting environmental interests through lobbying and litigation. This has precipitated policy recommendations to review environmental law enacted during what has been called the Environmental Overreach Period (Martz 1993). The main objective is to

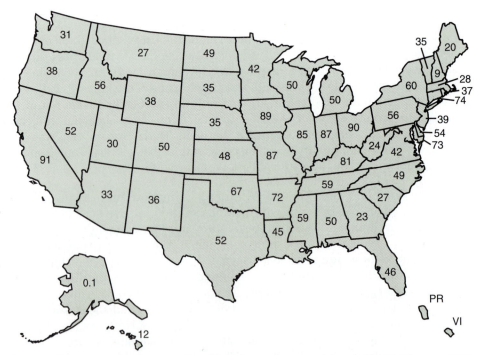

FIGURE 2.22 Percentage of Wetlands Acreage Lost, 1780s–1980s (after Dahl and Johnson 1991). About half of the U.S. wetlands have been lost from filling and draining. Federal administrative policy has sought to prevent any net wetland loss since 1989.

move management of resources away from polarizing litigation toward cooperative management approaches.

Coggins (1993) believes U.S. natural resource policy has too many land zoning categories, agencies, allocation mechanism, statutes, and regulations and "not enough consistency, clarity, uniformity, efficiency, or comprehensibility." In summary, according to Coggins (1993): "The present legal structure is not the product of conscious design; rather, it is simply the residue of more or less ad hoc developments over two centuries."

The twentieth century ended with growing awareness of natural resources problems and opportunities. Evidence increasingly indicates that humanity is causing costly changes in the biosphere, although the implications are not always clear. The major challenge to natural resource managers in the twenty-first century will be how to sustain economic development in a world where many resources are in rapid decline. Improving education, science, technology, and social ethics provides many opportunities for improved management of natural resources, including some restoration of what formerly existed.

LITERATURE CITED

Bolen, E. G. and W. L. Robinson. 1995. *Wildlife ecology and management.* 3rd ed. Englewood Cliffs, NJ: Prentice-Hall.

Boorstin, Daniel J. 1987. *Hidden history.* New York: Harper and Row.

Boyer, P. S. 1995. *The enduring vision.* Lexington, MA: D.C. Heath.

Bramwell, A. 1989. *Ecology in the 20th century.* New Haven, CT: Yale University Press.

Butzer, K. W. 1990. The Indian legacy in the American landscape (pp.27–50). In *The making of the American landscape,* edited by M. P. Conzen. New York: Routledge.

Carson, R. 1962. *Silent spring.* Boston, MA: Houghton Mifflin.

Clepper, H. 1966. *Origins of American conservation.* New York: The Ronald Press.

Coggins, George C. 1993. Trends in public land law (a title the inaccuracy of which should become manifest) (pp. 49–65). In *Natural resources policy and law: Trends and directions,* edited by Lawrence J. MacDonald and Sarah F. Bates. Washington, DC: Island Press.

Conlin, J. R. 1993. *The American past.* Fort Worth, TX: Harcourt Brace.

Conzen, M. P., ed. 1990. *The making of the American landscape.* New York: Routledge.

Cronan, W. 1991. *Nature's metropolis: Chicago and the great West.* New York: W. W. Norton.

Crosby, A. W. 1986. *Ecological imperialism: The biological expansion of Europe, 900–1900.* New York: Cambridge University Press.

Dahl, T. E. and C. E. Johnson. 1991. *Status and trends of wetlands in the conterminous United States, mid-1970s to mid-1980s.* Washington, DC: U.S. Department of the Interior, Fish and Wildlife Service.

Davis, S. M. and J. C. Ogden, eds. 1994. *Everglades: The ecosystem and its restoration.* Delray Beach, FL: St. Lucie Press.

England, R. E. and A. De Vos. 1969. Influence of animals on pristine conditions of the Canadian grasslands. *Journal of Range Management* 22:87–94.

Ehrlich, P. R. 1968. *The population bomb.* New York: Ballantine.

Feldman, David L. 1991. *Water resources management in search of an environmental ethic.* Baltimore, MD: Johns Hopkins University Press.

Freidenberg, D. M. 1992. *Life, liberty and the pursuit of land.* New York: Prometheus Books.

Golley, F. B. 1993. *A history of the ecosystem concept in ecology.* New Haven, CT: Yale University Press.

Graham, F. 1971. *Man's dominion: The story of conservation in America.* New York: M. Evans and Company.

Hardin, G. 1968. Tragedy of the commons. *Science* 162:1243–1248.

Heilbroner, R. and L. Thurow. 1994. *Economics explained.* 3rd ed. New York: Simon & Schuster.

Hibbard, H. B. 1924. *A history of the public land policies.* New York: Macmillan. (Re-copyrighted and published by the University of Wisconsin Press, Madison, in 1965)

Holechek, J. L., R. Peiper, and C. Herbel. 2001. *Range management principles and practices.* 4th ed. Upper Saddle River, NJ: Prentice-Hall.

Jenkins, P. 1997. *A history of the United States.* New York: St. Martin's Press.

Johnson, M. D. 1964. *Feathers from the prairie.* Bismark, ND: North Dakota Game and Fish Department.

Kirsch, L. J. and A. D. Kruse. 1972. *Prairie fires and wildlife.* Bismark, ND: North Dakota Game and Fish Department.

Kline, B. J. 1997. *First along the river: A brief history of the U.S. environmental movement.* San Francisco, CA: Acada Books.

Knight, Richard L. and Sarah F. Bates. 1995. *A new century for natural resources management.* Washington, DC: Island Press.

Leopold, A. 1933. *Game management.* New York: Scribner.

Leopold, A. 1949. *A Sand County almanac and sketches here and there.* New York: Oxford University Press.

Marsh, G. P. 1864. *Man and nature; or, physical geography as modified by human nature.* New York: Charles Scribner.

Martin, P. S. and R. G. Klein, eds. 1984. *Quarternary extinctions: A prehistoric revolution.* Tucson, AZ: University of Arizona Press.

Martz, Clyde, O. 1993. Natural resources lawman historical perspective (pp. 22–48). In *Natural resources policy and law: Trends and directions,* edited by L. J. Mac-Donnell and S. F. Bates. Washington, DC: Island Press.

Nester, W. R. 1997. *The war for America's natural resources.* New York: St. Martin's Press.

Novak, B. 1995. *Nature and culture: American landscape and painting 1825–1875.* New York: Oxford University Press.

Oliver, J. W. 1956. *History of American technology.* New York: The Ronald Press.

Reiger, J. F. 1975. *American sportsmen and the origins of conservation.* New York: Winchester Press.

Roberts, J. M. 1993. *A short history of the world.* New York: Oxford University Press.

Scheffer, V. B. 1991. *The shaping of environmentalism in America.* Seattle, WA: University of Washington Press.

Schubert, F. N. 1980. *Vanguard of expansion: Army engineers in the trans-Mississippi West, 1819–1879.* Office of the Chief of Engineers. Washington, DC: Superintendent of Documents, U.S. Government Printing Office.

Scott, G. 1875. *Fishing in American waters.* New York: Harper.

Smith, J. G. 1898. *Grazing problems in the Southwest and how to meet them.* U.S. Dep. Agric. Div. Agrost. Bull. 16:1–47.

Stoddard, L. A., A. D. Smith, and T. W. Box. 1975. *Range management.* 3rd ed. New York: McGraw-Hill.

Thompson, P. E. 1970. The first fifty years—the exciting ones (pp. 1–11). In *A century of fisheries in North America,* edited by N. G. Benson. Bethesda, MD: The American Fisheries Society.

Trefethen, J. B. 1966. Wildlife regulation and restoration. In *Origins of American conservation,* edited by H. Clepper. New York: The Ronald Press.

Trefethen, J. B. 1975. *An American crusade for wildlife.* New York: Winchester Press.

Turner, F. J. 1963. *The significance of the frontier in American history,* edited by H. P. Simonson. New York, Frederick Unger Publications.

Udall, S. L. 1963. *The quiet crisis.* New York: Holt, Rinehart and Winston.

Vogt, W. 1948. *Road to survival.* New York: William Sloan Associates.

Waterman, C. F. 1975. *Fishing in America.* New York: Holt, Rinehart and Winston.

White, L., Jr. 1967. The historical roots of our ecological crisis. *Science* 125:1205.

Williams, M. 1989. *Americans and their forests: A historical geography.* New York: Cambridge University Press.

CHAPTER 3

Basic Ecology

INTRODUCTION

Natural resource management depends on ecological science, or ecology as it is commonly called. Ecology is the knowledge and study of interrelationships among organisms and their environment (Colinvaux 1993, Ricklefs 1996). Ecology became identified as a scientific discipline with a definite body of accumulated knowledge at the beginning of the twentieth century. The foundations were laid in the 1700s with the development of biogeography and a classification system for living things. Darwin in 1859 developed his theory of evolution—a discourse on how interactions between organisms and their environment could result in adaptive change. An early pioneer in applied ecology, George Perkins Marsh, wrote an astute assessment of watershed processes and their relationship to vegetation conditions and soil erosion (Marsh 1865). The first major integrative work in aquatic ecology was published by Forbes in 1888, who conceived of "the lake as a microcosm." He established a basis for what would eventually become ecosystem science.

Frederick Clements stood out among early terrestrial biologists because his 1907 book, *Plant Physiology and Ecology,* was the first to formally organize ecology into a discrete subdiscipline (see Clements 1907). Allee et al. (1949) summarized the rapidly evolving field of animal ecology. Andrewartha and Birch (1954) added substantially to the science of population ecology. Odum (1959) organized the best synthesis and conceptual treatment of ecosystems science available up to that time. Numerous other general ecology books have been completed since then, but Colinvaux (1993) and Ricklefs (1996) are recent standouts.

Ecology is usually approached from three standpoints: the population (*autecology*), the community (*synecology*), and the ecosystem. Each examines organismic interaction with environment at a different scale. While early plant ecologists emphasized the community approach, animal ecologists emphasized the population approach. Ecosystems ecology emerged as an approach for unifying plant, animal, and environmental ecology. This chapter starts with an introduction to population and community ecology, and then culminates with a summary of important ecosystem processes and concepts.

POPULATIONS

Population Identity

Organisms are identified by how well their morphological, physiological, behavioral, and genetic attributes match other organisms. Those individuals that commonly exchange genetic material under natural conditions are assigned to the same species.

FIGURE 3.1   A population of bison on mixed prairie grassland in South Dakota (courtesy of Holechek et al. 2001).

Even within species, however, subspecies are often recognized. A *population* is defined as a group of individuals with similar traits and potential for reproductive interaction. They typically share the same locality and resources (Figure 3.1).

Distribution and Dispersal

Populations are distributed in the three dimensions of space and the fourth dimension of time in ways that reflect their life requirements. Arboreal species, adapted to life in trees, differ in distribution from fossorial species adapted to burrowing life. Similarly, coastal oceanic species rarely overlap in vertical distribution with the species adapted to the darkness and pressure of the oceanic abyss. At a still finer scale, fossorial species may be limited only to that volume of earth that has the proper moisture and particle sizes suitable for burrowing. In terms of time, different species are active at different times of the day and in different seasons. Many species hibernate, estivate, or otherwise exist through dormant periods.

Species that are broadly adapted to the environment tend to be more widely distributed than those that are narrowly adapted. However, the sharpness of environmental change also is important in determining distributions. For the most specialized species, even subtle differences in the environment may limit distribution. Changes caused by fire, flooding, grazing, and so on create or reduce connections between isolated islands of suitable environment. Many species depend on a variety of environments for activities such as feeding, resting, and rearing of young. These must be close enough together to provide needs within the time limits required for their life processes.

Population dispersal is defined as the movement away from place of birth. It is the means by which populations successfully reach isolated areas suitable for colonization. Populations within species often are isolated from other populations for variable periods of time depending on the environmental dynamics in the area. Population abundance also changes as distributions expand and contract with change in the environment. However, the extent of change depends on the population *density*, or numbers per unit area or volume. Density of large organisms is most frequently estimated as the total number of individuals per unit area (e.g., number/hectare). For small aquatic and soil organisms, numbers in a volume of water or soil are typically measured. Populations may spread out and contract without change in abundance, causing their density to change. Therefore, the total number of organisms in a population is the product of their average density within the occupied range times the area of the range. Density can vary greatly within the range. A herding population, for example, can have high density within the herd and low density throughout the range.

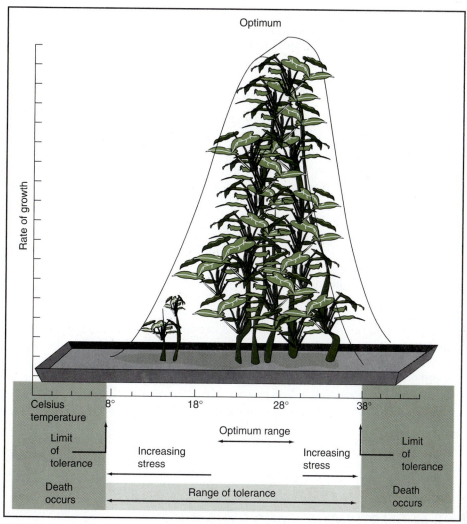

FIGURE 3.2 For every factor influencing growth, reproduction, and survival, there is an optimum condition. Above and below the optimum, stress increases until survival becomes impossible at the limits of tolerance (from Nebel and Wright 1996).

Factors Affecting Population Distribution

Understanding of population distributions requires knowledge of environmental regulating factors. Two basic concepts underlie most explanations of observed distributions. During the nineteenth century, Justus Liebig discovered that crops planted in a field conducive to their growth did not grow if a single nutrient was missing or in short supply. From these observations he concluded that a group of similar organisms is limited by the required resource that is in shortest supply. More recently it has been discovered that factors sometime interact in combination to establish a limiting minimum. The concept was later expanded by Shelford to include maximum as well as minimum environmental tolerances of the population (Colinvaux 1993).

From these concepts it was deduced that a population is able to exist and reproduce successfully only within a definite range of environmental conditions (Figure 3.2). The sum of the limiting conditions is referred to as the *environmental resistance.* Odum (1959) summarized ways in which organism tolerances to environmental conditions were most likely to be expressed. An organism may have a wide range of tolerance for one factor and a narrow range for another. Organisms with a wide tolerance for all factors are likely to be widely distributed. When conditions are not optimal for a species with respect to one ecological factor, the limit of tolerance may be reached for all factors. Tolerance is usually least broad during the reproductive period. Common variables limiting population distributions are temperature, water, light, substrate condition (e.g., soil, rock, sediment), nutrients, toxins, salinity, fire, and oxygen (in aquatic environments).

Population Habitat and Physical Niche

In population ecology, habitat and niche are two widely used words that can be conceptually confusing. *Habitat* usually is thought of as the site or place occupied by a population. It is characterized by obvious terrain and vegetation features (forest, desert scrub, marsh, or stream habitat). Most terrestrial and wetland habitats are identified more specifically by predominant vegetation (oak forest, cypress swamp) whereas aquatic habitat is more often identified by physical and chemical form (e.g., freshwater lake, coastal ocean).

The species *niche* is defined as all the environmental factors that a species needs to survive, stay healthy, and reproduce in a ecosystem. Numerous models of habitat suitability for various populations have been developed for use by the U.S. Fish and Wildlife Service and other agencies. They guide habitat protection, creation, and restoration based on the physical niche of the desired species. These models typically include limiting variables. Habitat determines the potential distribution of a population, but often not all suitable habitat is occupied. This is because of past extirpation of the population or because the habitat is new and never previously inhabited.

Habitat is more dynamic than the concept of place or site might imply. Environmental variables in the habitat often change daily (e.g., light, temperature), seasonally (moisture, temperature), or less predictably (fire, extreme flooding). Many populations are adapted to variation, which often triggers life-cycle events, such as reproduction, dormancy, migration, and feeding. Variability of habitat is required for life-cycle completion of many species. The reproduction of many aquatic species is triggered by light, temperature, or flooding. Numerous terrestrial plants reproduce only after exposure to fire. Reduction of environmental variation frequently leads to lower species diversity.

POPULATION STRUCTURE AND DYNAMICS

Population Structure

Population structure is defined as the distribution of the members among age categories and sexes. Unstable populations are indicated by bulges in the age distribution (Figure 3.3). A bulge among older ages indicates a population that has failed to reproduce enough to replace loss and is decreasing in number. A bulge in the youngest ages indicates a population poised for population increase (Figure 3.3, Mexico). Each species has a characteristic sex ratio for a stable population and variation from that ratio also indicates an unstable population. A higher ratio of females to males may signal a higher potential reproduction rate. Fecundity is gaged by the number of embryos per adult or per female in the population. A complete set of data describing the numbers of population members in age and sex categories can tell managers much about the history and future of the population.

Population Dynamics

Population structure depends on birth or germination rate (increase in number), death rate (decrease in number), immigration (invasion) rate (increase in number), and emigration (dispersal) rate (decrease in number). Birth and germination rates, or natality, are rates at which new members are added to the population. The death rate, or mortality, is the rate that members are subtracted from the living population. When natality exceeds mortality, the population increases. In a stable population natality about equals mortality.

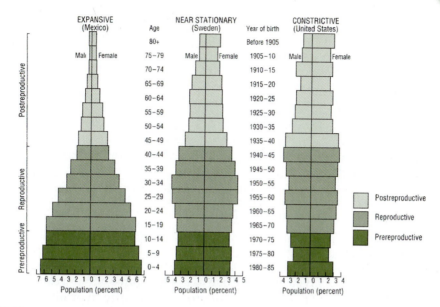

FIGURE 3.3 Age structures of human populations in Mexico, Sweden, and the United States (from Owen and Chiras 1995).

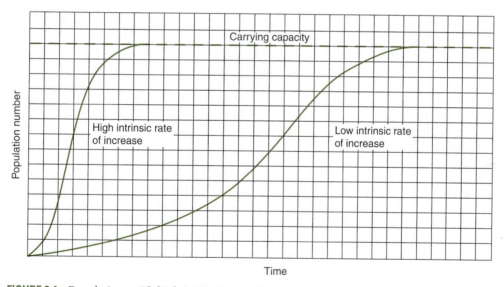

FIGURE 3.4 Populations with high intrinsic rate of increase more quickly reach the environmental carrying capacity when first introduced to an area than species with low intrinsic rate of increase.

Populations have an *intrinsic rate of increase* determined by individual growth rates as well as numbers (Figure 3.4). Members of populations grow by adding body mass (typically measured by weight and less commonly by volume). The intrinsic rate of population increase depends on fecundity (number of offspring produced) per adult and growth rate to maturity. For populations with a high potential number of progeny per adult, rapid growth rate, and early maturation, the intrinsic rate of population increase is high. The realized population increase in numbers also will be high when the death rate is low. Stable populations with a high intrinsic rate of increase must have a compensating high mortality rate. Populations with high intrinsic rate of increase have high *biotic potential*.

A very general relationship exists between the maximum size obtained by individuals in the population and the intrinsic rate of population increase. Organisms of large adult size typically have slower growth, lower fecundity, and a longer maturation time than smaller organisms. They also tend to have lower death rates. Healthy

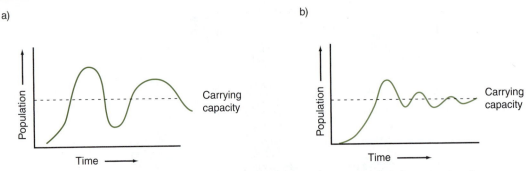

a)   b)

FIGURE 3.5 Two of the many possible types of population fluctuations around carrying capacity. Depending on species and circumstances, populations can vary widely and irregularly in density (a) or vary within narrow limits and quite regularly (b).

female elephants, for example, take years to grow to maturity and may live as adults for decades, but they typically have one baby every three or four years. Healthy female mice, in contrast, take only weeks to grow to maturity, typically survive only for months, have several young in each litter, and have a litter every few weeks. In general, smaller organisms live shorter lives and have a higher population turnover rate than larger organisms.

Populations in optimum habitat typically increase rapidly from low initial number. This type of situation often exists in changeable environments that suddenly become optimal after a period of difficult conditions, such as extended drought. It also exists when a population first gets established in a new and optimum habitat. After an initial lag, a newly established population typically reaches a maximum rate of increase, then the rate of increase slows, and finally reaches a somewhat stable number. At that point, the population has reached the carrying capacity of the habitat (Figures 3.4 and 3.5). As the population becomes more dense with respect to the resources available, competition for resources creates a "feedback" that slows the population increase to zero. Depending on conditions, the population number may remain stable at the carrying capacity or collapse and fluctuate (Figure 3.5).

Carrying capacity often varies because environmental factors such as climate vary and because the population can depress its own resources (Figure 3.6). Terrestrial plants, for example, often deplete the soil of nutrients as they accumulate in their tissues. Plant-eating animals often become so abundant they depress the plant regeneration. Ultimately the animal population declines from lack of food, the plants regenerate, and the cycle repeats. Climatic changes can alter soil water availability, which can limit plant growth. Because of these dynamics, population numbers are rarely constant. Typically they are in continuous adjustment to changing habitat conditions.

POPULATION ENERGETICS, BIOMASS, AND PRODUCTION

Energetics

Energy typically is expressed in calories or joules (J) received per unit area per unit time (e.g., $Cal/m^2/day$). Green plants use light and soil nutrients to create living matter, or *biomass* (also sometimes referred to as *standing crop*). Otherwise the energy used for metabolism comes from the chemical bonds in organic compounds.

Not all energy used by organisms within a population, through feeding or otherwise, is actually *assimilated* or retained by the organism (Figure 3.7). For animals, some unassimilated matter is egested as feces. Efficiencies of assimilation vary among organisms depending on the quality of the energy source available to them. For plants, assimilation of light by chlorophyll depends on the wavelengths of light

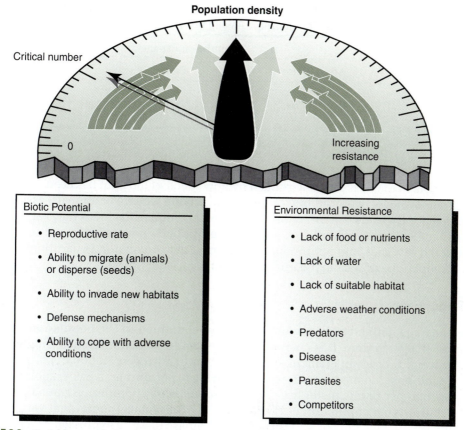

FIGURE 3.6 A stable population in nature is the result of a balance between factors tending to increase population and factors tending to decrease population (environmental resistance) (from Nebel and Wright 1996).

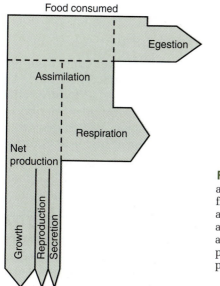

FIGURE 3.7 Energy partitioning among metabolic pathways of animals in a population. Of the food energy consumed, a fraction passes through to be egested and a fraction is assimilated into each member of the population. The fraction assimilated is partitioned between respiratory metabolism and population net production. The net production is the portion of the population that serves as a food base for predators or decomposers.

TABLE 3.1   Common Measures of Biomass and Productivity Expressed for a Hypothetical Organism, Showing General Relationships Among Dry Weight, Calories, and Carbon

	Population Parameter		
Measure	Mean Biomass	Productivity	P/B
dry weight	1000 g/m^2	5,500 g/m^2/year	5.5
carbon	400 g C/m^2	2,200 g C/m^2/year	5.5
calories	4,000 Cal/m^2	22.0 KCal/m^2/year	5.5

available to them. This is determined by how much of the light has been absorbed by the atmosphere, overhead tree canopy, or overlying water. For animals, assimilation efficiency depends mostly on the organic composition of the foods they eat. Once nutritional energy is assimilated, some is used in body maintenance and respiration, and the rest goes into net production. In turn, net production is partitioned among reproduction, growth, and various secretions necessary for life.

Population Biomass

Biomass is defined as the amount of matter contained in the population of organisms at a given time. It is commonly expressed in total weight including water, or dry weight, or as an element such as carbon or nitrogen (Table 3.1). To compare populations, biomass is expressed on a weight per unit area basis, such as grams of carbon per square meter (g C/m^2). The biomass of a population can be estimated by multiplying population numerical density times mean individual weight or by summing the weights of all individuals. Biomass often is expressed as a mean for some determined period, most typically one year. Because water content is highly variable among organisms, weights of different organisms are usually compared on a dry matter basis. To obtain dry matter, samples are dried to eliminate the water content.

Biomass often is expressed as carbon weight because carbon is the "backbone" or primary element in organic matter. It can be readily measured, and is linked closely to calories of energy contained in the biomass. Carbon is usually measured by burning weighed samples and measuring the carbon dioxide gas emitted using gas spectrophotometry. Calories are determined from the heat released upon combustion in a bomb calorimeter. The choice of biomass expression depends on the objectives of study or management.

Production, Productivity, and P/B Ratio

Organic production is the total generation of organic matter over some time period. Productivity is the rate of biomass generation. Thus, a production of 2,000 kg over 100 days has an average productivity of 20 kg/day. (Table 3.1). P/B ratio is the productivity divided by the mean biomass of a population for a given period of time. The P/B indicates the rate of population tissue turnover. P/B is related to mean size (biomass/individual). Small species typically have higher P/B ratios than large species.

POPULATION STRATEGIES

Populations have adapted to their habitats through two main strategies. One strategy is to become generally adapted to wide variation in the environment and give up some efficiency in the use of specific food and cover. Species that adopt this strategy are called *generalists*. The other strategy is to become highly adapted to the efficient

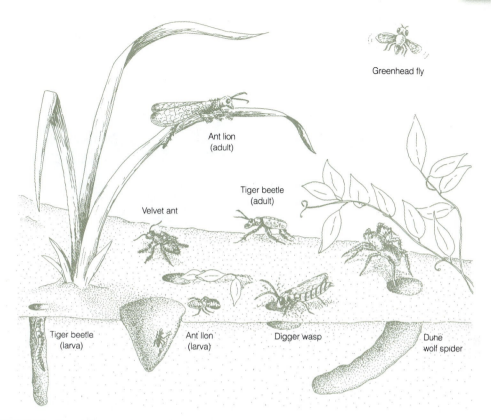

Greenhead fly

Ant lion
(adult)

Tiger beetle
(adult)

Velvet ant

Tiger beetle
(larva)

Ant lion
(larva)

Digger wasp

Dune
wolf spider

FIGURE 3.8 Specialization in insect use of sand dune habitat (from Lerman 1986). Specialists often are adapted to specific strata in an ecosystem. Sand dune specialists, for example, are narrowly adapted to subterranean, ground surface, dune vegetation, and aerial conditions.

use of specific resources in the ecosystem (Figure 3.8). These species are called *specialists*. Specialists are more likely to do poorly in unstable ecosystems. Widely distributed species are more likely to be generalists than specialists. Many species exhibit characteristics of both extremes.

Resilience refers to the capability of a population to "bounce back" quickly from stress. Many resilient populations are also quite tolerant of environmental variation and can disperse and colonize new areas rapidly. Generalists typically have greater resilience than specialists. Specialists tend to be better adapted than generalists to habitats with stable carrying capacity.

Populations with similar feeding behavior are sometimes lumped together, and are referred to as *guilds* after the professional unions of the Middle Ages. The concept of guilds has been most applied to birds, fish, and aquatic invertebrates. Organisms in the same guild are not necessarily related taxonomically. Because they are related through function, guild members typically compete for the same resources. However, they have often evolved subtle but effective means for sharing these resources.

Certain *keystone species* have outsized impacts and influence on the diversity of populations that can be supported in an ecosystem. Animals that shape their physical environment to exceptional degree can greatly influence the physical niches available for species evolution. Prominent species in this category include the dominant plants (aspen, cattail) and animals, such as muskrat (burrows and eats cattails), beaver (creates ponds and eats aspen), and of course humans (deconstruct, construct, reconstruct, and otherwise manage resources). It is often unclear how much effect a species exerts on other populations until after the species is greatly reduced or eliminated.

FIGURE 3.9   A community often is depicted by the dominant plants and animals, as indicated here for a marsh-pond community; however, most communities are composed of many more less obvious species (from Owen and Chiras 1995).

COMMUNITIES

Definition

A biotic community is an association of interacting populations occupying a given locality for a particular time period (Figure 3.9). Particular biotic communities are associated with prairies, woodlands, marshes, lakes, croplands, oceans, and so on. Communities usually include plants, animals, fungi, bacteria, and other major groups of organisms. Some communities, however, are reduced to fewer groups because of harsh environments, such as hot springs and caves.

COMMUNITY STRUCTURE AND FUNCTIONS

Community Habitat

Community habitat refers to the physical environment (climate, terrain, soil, water, and so on) that interacts with the community to provide needed resources. Many ecologists have come to accept the plant-oriented definition for community habitat provided by Daubenmire (1968). Under his definition, habitat is "an area characterized by a specific set of climatic, edaphic (substrate resources), topographic, and biotic factors." It implies potential for occupancy by a specific type of vegetation (the climax), if undisturbed by humans, fire, or weather. Because plants are so important in the distribution of most animals, the definition of community habitat used for plants can also serve for the animals in the community.

Communities interact to form larger geographical groupings, or *biomes*, indicated by similar vegetation type. A shortgrass prairie biome, for example, includes numerous communities associated with the edaphic, topographic, and microclimatic variation. Soil moisture is an important prairie variable, for example. Plants adapted to very moist soils occur at one extreme and many upland plants at the other extreme. Yet the biome vegetation is characteristically short and herbaceous throughout. The shortgrass prairie biome is characteristic habitat for adapted animal species, such as pronghorn antelope, prairie dogs, and bison (see Chapter 9).

A *stand* is a homogenous unit of vegetation that differs from surrounding vegetation occurring within a biome (Figure 3.10). Stands are dominated by one or a few species of plants. Stands result from differences in the environmental requirements of the various plant species and from environmental diversity. Tree planta-

FIGURE 3.10 A stand of longleaf pine trees in North Carolina (courtesy of U.S. Forest Service).

tions and crop land are examples of stands created by humans. Particular animal species are associated with specific stands of vegetation. Beaver, for example, are most abundant where stands of preferred food species, such as willows and aspen, are common.

Biodiversity

Biodiversity refers to the variety of species, or *species richness,* within the community. In more elaborate measures of diversity the relative *evenness of abundance* also contributes to the index of diversity. A community dominated by one or two species is less diverse than a community where the species are all relatively uncommon and more equal in abundance. Often species richness and evenness of abundance are correlated. In areas known for their high numbers of species, such as rain forests, no single species stands out in abundance. In such situations there is a low probability that any single species will dominate community functions. Regardless of what may happen to one species, community functions will be sustained. Species redundancy is central to the idea that diverse communities are functionally more stable.

The least diverse community, in theory, is composed of one species (monocultures). Communities are rarely completely monotonous, although many species in what appear to be single-stand communities (e.g., cattail marsh, aspen grove) are obscure and

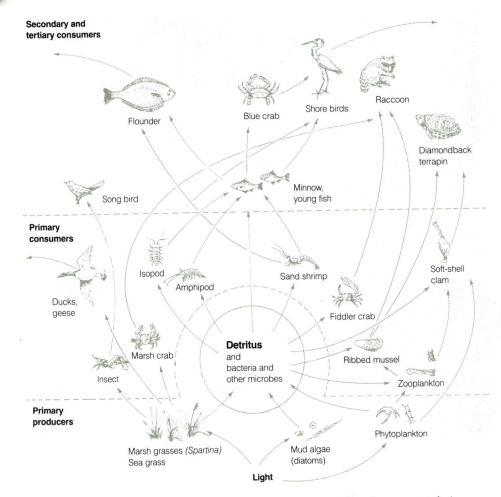

FIGURE 3.11   The various components of an estuarine community (from Lerman 1986). A community is composed of all of the primary producer, consumer, and decomposer organisms interacting in a given locality and at a given time.

rare. A community may be relatively rich in species and yet exhibit low diversity. Even in cultivated fields of corn or wheat, the diversity of invertebrates, microbes and vertebrates can be high, although the crop comprises most of the biomass.

Communities dominated by monotypic stands encourage species that will consume or parasitize them. Muskrats, for example, will consume cattails and tip the balance more in favor of other plants in the marsh. Humans have learned to exploit the most usable species in ecosystems. Cropland monocultures are the extreme in this direction. One price paid for large continuous crop monocultures is costly competition with pest species and depletion of soil nutrients (see Chapters 8 and 12).

The number of different communities present in a region and their arrangement also adds an important dimension to diversity. In some regions, species composition may change little throughout the entire area. In other areas of similar size, there may be many discrete communities.

Trophic Functions

Community functions are roles played by the different components of the community. Community biotic components include producers, consumers, and decomposers (Figure 3.11). The most basic community functions have to do with energy and material transfers among producer, consumer, and decomposer components. This will be discussed along with some management implications.

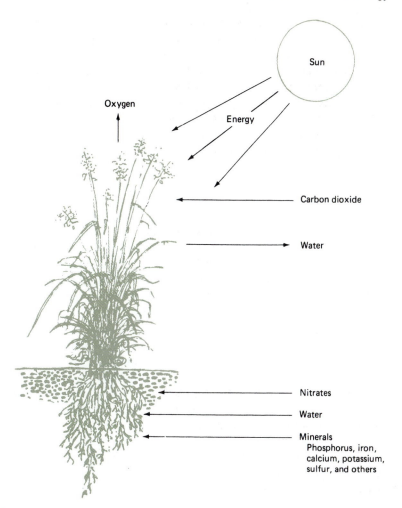

FIGURE 3.12   Materials used by grass plant for photosynthesis (drawing by John N. Smith) (from Holechek et al. 1998).

Producers.   *Primary producers* are plants, certain bacteria, and algae containing the pigment chlorophyll, which can convert solar energy to chemical energy through *photosynthesis.* The primary producers support, directly or indirectly, all other groups of organisms. *Primary productivity* has been defined as "the rate of organic matter storage by photosynthetic and chemosynthetic activity of producer organisms in the form of organic substances which can be used as food materials" (Odum 1959).

Photosynthesis is the process by which primary producers convert energy from the sun, carbon dioxide from the atmosphere, and water and minerals from the soil into nutritional resources for maintenance and growth (Figure 3.12). Many smaller primary producers have no specialized tissues and take up nutrients across their simple-cell surfaces. Oceanic and lake phytoplankton are small primary producers in this category responsible for most of the earth's primary production. Large terrestrial plants have leaves, roots, and vascular transport tissues (e.g., wood) that perform specialized life functions. Photosynthesis occurs primarily in the leaves (and sometimes the stems) where the pigment chlorophyll captures light that drives the formation of sugars, starches, fats, proteins, and other food compounds. Carbon dioxide from the air is taken up through leaf openings, called stomata. Plant roots obtain nitrogen, phosphorus, potassium, iron, and other nutrients from soil and aquatic sediments.

Many factors determine the rate of photosynthesis (primary productivity). These include the area of leaf or cell surface, intensity and quality of light, amount of carbon dioxide in the air or water, physiological efficiency of the plant, soil or water nutrient concentrations, water supply, and temperature. The amount of food manufactured is proportional to the amount of chlorophyll exposed to light. Light is often the limiting factor for primary production in ocean and lake depths, and under dense canopies of vegetation. Even where light is plentiful, photosynthesis is below potential because some other resource, such as nitrogen or phosphorus, is limiting. Under optimal conditions plant communities can use about 2% of the light that reaches them. However, under most conditions solar energy conversion efficiency is much lower. In many terrestrial communities water is the most frequent limiting factor. In most terrestrial and aquatic communities, inorganic nutrients, such as nitrogen and/or phosphorus, are likely to be most limiting.

Carbon dioxide in the air rarely limits terrestrial primary production. Carbon dioxide is more likely to be limiting in aquatic environments where it is less available as a gas dissolved in water than in air. More often, however, other nutrients limit photosynthesis. Two of the most common limiting nutrients are nitrogen and phosphorus, which often are the basis for commercial fertilizers. In the open ocean, iron is most likely to limit production. A variety of other inorganic nutrients can be limiting in certain circumstances and for particular groups of primary producers. For example, the availability of silica can determine the relative abundance of diatoms (a form of algae with a silica shell) in fresh waters.

Water is a necessary chemical constituent of photosynthesis. It provides hydrogen and, in terrestrial plants, it serves to keep the stomata open and the plant turgid. When the stomata close from water stress, carbon dioxide is denied entry. Water is also important in transporting dissolved minerals from the soil into and through the plant. For truly aquatic primary producers, water provides the environment for nutrient supply and light transmission as well as physical support. Water depth above and below the ground surface is key to the distribution of many aquatic, wetland, riparian (in river floodplains), and upland primary producers. Also, variation in the seasonal supply of water is important. This is because plant need for water varies with life stage and environmental conditions. Certain types of wetlands, for example, only exist where there is regular seasonal flooding and drying.

Physiological rates differ among species and are controlled by temperature. When low temperature forms ice, water is no longer available for uptake. Freezing within cells and tissues can, upon expansion into ice, disrupt cell walls and badly damage or kill plants. Concentrations of salts in plants can protect them against freezing.

Numerous other factors are influential in determining the rates of primary production for the entire community and for different species within the community. Even the casual observer will notice changes in plants as they walk a path from higher and dryer environments to moister lowlands, wetlands, and open water. Gradients of light, moisture, nutrients, toxic materials, acidity, salinity, sediment particle size, and so on combine to determine the relative productivity and survival of different species. The primary producer components change along these gradients, influencing the consumers and decomposers that depend upon them for food, cover, and other needs.

There can be many subgroupings within the primary producer component of any particular community. In a forest community, for example, the tallest trees function in the upper most stratum where they dominate the light. Understory trees and shrubs typically use light with greater efficiency than those in the overstory. In the soil, root systems of different plants occupy varying depths, which permits more efficient use of stored water and nutrients. Where communities are more structurally uniform, as in the phytoplankton community of oceans, the spatial differentiation into functional groups is less obvious than in forests or grasslands.

Decomposers.   *Decomposers* break down and absorb dead organic matter (detritus). The decomposer organisms include mostly bacteria and fungi, which possess enzyme systems necessary to break down resistant organic materials. In the process, organic matter is reduced to its inorganic components, which then become available once again for primary producer uptake. Without decomposers, elements would eventually be completely tied up in undecomposed organic material, and community function would stop.

Certain bacterial decomposers are among the most widely distributed and adapted life-forms. They are found in any environment where organic production occurs. Different species of decomposers vary in their tolerance to environmental conditions. There are significant differences among the bacteria and fungi, which are the most important of the decomposers. These diverse groups complement one another in their different approaches to decomposition. Bacteria group into three types with respect to their oxygen requirements: aerobic, facultative aerobic-anaerobic, and anaerobic. Whereas aerobic bacteria cannot tolerate an environment without oxygen, anaerobic bacteria are killed by oxygen. Bacteria of some kind occur in nearly all environments. Unlike fungi, many bacteria can function in anaerobic as well as aerobic environments, although the rate of decomposition falls sharply as oxygen concentration approaches zero.

Fungi have an advantage over bacteria when the organic matter is in larger masses, such as woody plant matter and soil humus, and is resistant to weathering into smaller particles. Fungi have threadlike hyphae, which can penetrate deep into organic tissue. In contrast, bacteria use the strategy of small spherical and rod shapes to develop a high contact surface area for the decomposition process.

Until raw sewage treatment became nearly universal in the United States, many urban rivers were anoxic (without oxygen) because of overloading with raw human wastes (Perry and Vanderklein 1996). Oxygen, an important rate regulator of organic decay, was depleted by the waste overload. Deep soils and sediments of wetland and aquatic ecosystems typically remain continuously anaerobic and accumulate organic matter. Through geologic time this organic matter may eventually become coal, oil, and natural gas. The decomposition rate also slows in acidic environments. Calcium appears to be an especially important limiting nutrient under acidic conditions. Where decomposition is slow, bogs, bog waters, and other "brown-water" lakes, streams, and swamps develop and create unique ecosystems with many life-forms (Horne and Goldman 1994). The brown stain in the water is caused by an accumulation of large organic molecules that decompose very slowly in the acidic and calcium-poor environment.

Consumers. Consumers are mostly animals, that ingest other living things and convert the energy into biomass (Figure 3.13). They are categorized into functional subcategories (primary, secondary, tertiary) based on the number of feeding transformations (or *trophic levels*) that have occurred (Figure 3.13). *Herbivores* feed directly on primary producers. They include cattle, grain-feeding birds, various insects, and molluscs. *Carnivores* are those animals that eat other animals including herbivores and other carnivores. Cats, snakes, adult frogs, many fish, and numerous invertebrates are strictly carnivorous. Many animals are *omnivores,* which feed on both plants and animals. Humans, chickens, turtles, bears, tadpoles, carp, and crayfish are all examples of omnivores.

Detritivores are consumers that ingest dead organic matter and decomposers (Figure 3.14). Some of the most obvious detritivores are ants, termites, and nematodes, which are critical, but often overlooked, components of terrestrial ecosystems (Paris 1969). Aquatic detrital food chains are particularly important in woodland streams, ponds, and bogs with rich supplies of organic detritus. Many aquatic insects, molluscs, worms, and crustaceans are detritivores. Consumers and decomposers compete for organic foods while complementing each other in converting organic matter back to inorganic elements.

Competitive Interactions Within Communities

Competition has been defined in many ways by different ecologists (Risser 1969, Ricklefs 1996). We define competition as two or more organisms trying to gain one or more scarce, required resources from the same ecosystem.

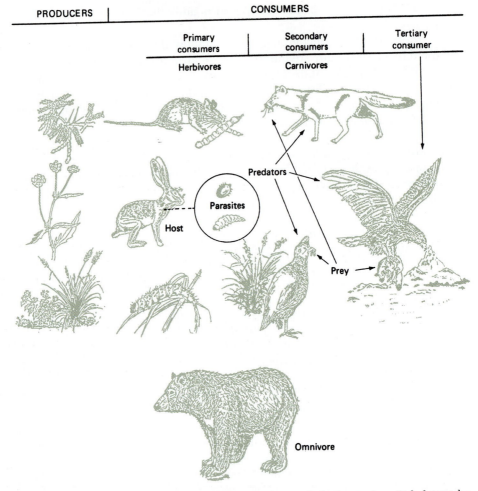

FIGURE 3.13   Feeding relationships among plants and animals (redrawn from Nebel 1981 by John N. Smith).

FIGURE 3.14   Various detritivores make use of the organic matter in a fallen tree (adapted from Miller 1990 and Nebel and Wright 1996).

FIGURE 3.15 American robins (left) and American woodcocks (right) both feed largely on earthworms, but find their prey in different ways and in different locations, thereby avoiding direct competition for the same food resource (from Bolen and Robinson 1995, design by R. Miller).

Generally, the factors for which competition occurs among plants are water, nutrients, light, oxygen, and carbon dioxide (Risser 1969). In some cases, plant competition may occur for pollinating agents or for suitable germination sites. For animals, competition typically occurs for food and habitat. Competition may occur among species (interspecific) or among individuals of the same species (intraspecific). The results of competition may manifest at the individual or at the population level. Individuals may show the effects of competition by reduced growth rate, reduced vigor, and death. At the population level, competition may be demonstrated by reduced density or by changing patterns of distribution (Risser 1969). Organisms often avoid competition by using the same resources at different spatial locations or seasons during the year (Figure 3.15).

Interactions Among Biotic Communities

Migratory Interactions. Communities are interactive in a variety of ways. Through wind, flight, and other movement organisms disperse from one community to another. Many mobile organisms are only temporary community members. Migratory animals are obvious examples. Many larger species of vertebrates, such as ducks and geese, require a different habitat for reproduction and early life stages than for other life stages. They move seasonally among communities as their need for resources changes. They can become very influential forces, as evidenced by the seasonal impact migratory grazing herds can have on grassland vegetation. Similarly, huge schools of planktivorous fish can have a local short-term impact on plankton populations as they move through lakes and oceans. Wild geese and duck populations have adapted to and affect biotic communities thousands of miles apart in order for them to use different summer and winter habitat. These migratory species link communities that are otherwise separated by large distances.

Ecotones. Communities are bridged by less mobile species at their *ecotones,* or areas where two communities come together (Figure 3.16). Ecotones are of particular importance to terrestrial vertebrates, which typically require more than one community to meet their daily needs. The three main types of ecotones include simple, gradual, and mosaic ecotones. The simple ecotone involves an abrupt change from one community to another (Figure 3.16). The gradual ecotone refers to an extended area where species from two or more communities gradually blend together. The mosaic ecotone involves a blend of two or more communities (Figure 3.16).

Simple ecotones are most common in desert and forested areas. They are usually caused by a sharp change of either soils or topography. Wetlands often form simple ecotones between terrestrial and aquatic communities. Gradual ecotones are characteristic of prairie areas where soils and topography change slowly across the landscape. Ecotones are especially gradual in the open oceans where temperature variation is the predominant environmental gradient. Mosaic ecotones are common in

FIGURE 3.16   A mosaic ecotone in central New Mexico where a woodland and grassland community come together. Another simple ecotone occurs along the fenced area in the foreground as a result of differences in methods of cattle grazing.

forest and grasslands where storm, fire, and groundwater close to the surface cause patchy distributions of vegetation. Whittaker (1975) discovered that many communities do not exhibit truly discrete identities at all, especially where they occur along continuous environmental gradients, such as mountain slopes.

Dispersal and Isolation.   When a community is partially or entirely eliminated by catastrophe, or newly created, the colonization of the area can occur either through population dispersal or evolutionary process. Many ocean islands formed from volcanic activity have unique flora and fauna as a consequence of very limited colonization and extended isolation. Hawaii and the Galápagos Islands are excellent examples. Recently formed islands are colonized by organisms that are present in the vicinity or somehow (transport by humans) make it to island shores. Much has been learned for example from Krakatoa, an island near Sumatra, which was entirely reformed by a volcano in the late eighteenth century. It has since been colonized by many species found on nearby lands.

Some basic principles of *island biogeography* are important in *landscape ecology* and *conservation biology*. Organisms dispersing from continents reach large islands faster than they reach small islands the same distance offshore. Organisms colonize islands closer to shore faster than ones farther offshore. These principles pertain even when islands are something other than land in water, such as lakes surrounded by land or patches of different vegetation in a "sea," or *matrix,* of another vegetation type. There are many examples of vegetational *patchiness*, such as swamps surrounded by upland forests, marshes and pothole lakes surrounded by upland prairies, and mountain forests surrounded by desert lowlands.

Keystone Species.   Keystone species have disproportionate impact on communities. They are often used to manage communities. Manipulation of animal grazers, such as domestic livestock and grass carp, can control the form and composition of terrestrial and aquatic plant communities (Figure 3.16). Introductions of plants and animals alters the value of living resources, sometimes positively and often negatively. Invasive plants and animals often are keystone species that can displace native species and greatly alter community structure and function. Because invasive species are usually widespread and displace rarer species of limited distribution, they can reduce genetic diversity.

ECOSYSTEMS

Ecosystems Defined

Definitions of ecosystem usually emphasize interactions among the living and non-living components occupying some defined space (Colinvaux 1993, Ricklefs 1996) (Figure 3.17). Forbes (1888) expressed this concept in terms of the lake "micro-cosm"—a miniature world among many such worlds on earth. Ecosystems can be very small, such as the intestinal community living in a flea, or as large as the biosphere that forms an envelope around the entire earth. Most practitioners of ecosystem management think of each ecosystem as a "functional unit consisting of the organisms (including man) and environmental variables of a specific area" (Van Dyne 1966). Ecosystem functions are time-dependent processes, such as organic matter production, mineral cycling, and succession following disturbance. Structure is the form of the ecosystems at any point in time.

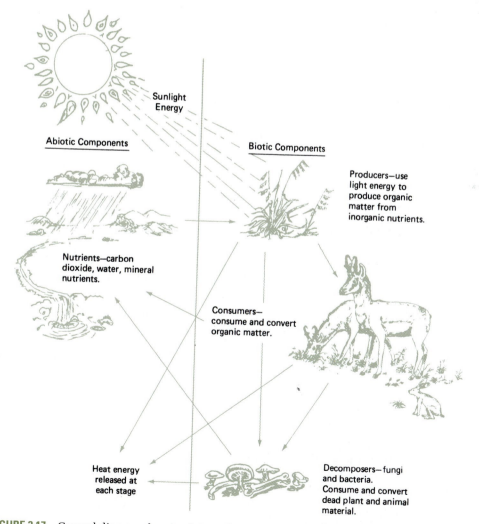

FIGURE 3.17 General diagram showing interactions among ecosystem components (from Holechek et al. 1998). An ecosystem is defined by the interaction of abiotic components in the habitat with biotic components in the community (drawing by John N. Smith).

TABLE 3.2   Examples of Ecosystem Processes, Services, and Goods

Functions
Production of Organic Matter
Decomposition of Organic Matter
Nutrient Cycling
Grazing Regime
Fire Regime
Hydrologic Regime
Infiltration
Runoff
Evapotranspiration
Soil Erosion Regime
Adaptations and the Evolutionary Process

Services
Maintenance of Atmospheric Quality
Control and Amelioration of Climate
Regulation of Freshwater Supplies
Origin and Maintenance of Soils (and their buffering capacity)
Detoxification and Degradation of Wastes (pollution dilution)
Natural Control of Pathogenic and Parasitic Organisms (pest control)
Pollination of Cultivated and Wild Plants
Purification of Air and Water
Renewal of Soil and Water Fertility
Retention and Delivery of Nutrients to Plants by Soils and Water
Genetic Resources (improve existing and developing new domestic plants and animals)
Aesthetic, Cultural, Spiritual Renewal
Recreational Services

Goods (valued structure)
Foods (mammals, birds, fish, shellfish and other invertebrates, plants, fruits, nuts, spices)
Fibers (cotton, flax, hemp, wool, cashmere, silk)
Fuels (botanochemicals)
Pharmaceuticals and Medicines (psychoactive drugs, Codeine, diuretics, painkillers, antibiotics)
Building Materials (lumber and other woody materials, resins, glues, shellac)
Industrial Products (waxes, rubber, dyes, vegetable fats and oils)
Cooking Oils (plant and animal fats and oils)

Sources: Modified from Daily (1997) and West (1995).

Ecosystem Services

Ecosystems play a critical role in providing societies with various services including the provision of goods (to serve means to provide) (Table 3.2). In most countries these services have generally been taken for granted or undervalued because of the vastness of land and water resources relative to human populations. However, during the twenty-first century the authors believe the value of ecosystems services will become better appreciated because of rapid shrinkage of natural landscapes and contamination of oceans coupled with an unprecedented increase in human numbers.

The functional integrity of the earth's processes determines the quantity and quality of natural resource goods and services provided in support of human welfare. The shallow layer of air, land, and water in which we live has serviced humanity for many thousands of years in ways that ecologists and economists are just beginning to appreciate.

While technology advances have done much to diffuse concerns about nonrenewable resources, the greatest concerns are now more broadly based in sustaining renewable resources for the services they provide. Technology has yet to cost-effectively replace many basic ecosystem services, such as clean air, productive soil clean water, uncontaminated living space, environmental stability, and the genetic information held in rare and endangered species.

We hope that the concept of natural goods and services will help societies worldwide recognize the value of intact ecosystems and take major initiatives to conserve natural ecosystems before human welfare becomes seriously jeopardized. In later chapters of the book we will discuss various ecosystems services and goods.

ECOSYSTEM STRUCTURE, FUNCTION, AND SERVICE

We define ecosystem structure as the spatial arrangement of the living and nonliving environment in a particular area. Structure results from the relative abundance, spatial arrangement, and form of the ecosystem's many component parts. The abiotic structural components consist mainly of matter in the rock, soil, water, and air. The structure of soil, for example, determines how much water, oxygen, and nutrients can reach plant roots and soil organisms. The hydrogen ion content of soil and water determines its acidity and alkalinity. It is important in determining the distribution and abundance of many species and the availability of some nutrients. Availability of inorganic nutrients such as nitrogen, potassium, and sulfur determines distribution, production, and abundance of primary producers.

Major categories of ecosystem function include energy flows resulting in production, consumption, and decomposition; material storage, flow, and cycling; genetic controls on life processes (genetic diversity); and successional processes. We will discuss each of these categories of function.

Energy Flow

The sun is the basic source of energy for all life on planet earth. Solar energy is received by grasses, forbs, and shrubs and is transferred by the process of photosynthesis into stored chemical energy in plant tissues. When herbivores eat plant tissue, they gain energy stored in plant tissues through the process of digestion (Figure 3.18). Carnivores, in turn, eat other animals and derive their energy requirements from their food. However, energy is dissipated at each step in the food chain through respiration.

Energy flow throughout the ecosystem operates under the first law of thermodynamics: Energy can be neither created nor destroyed, only changed in form. The organisms at each step in the food chain are not completely efficient at harvesting all available food resources, so energy retention diminishes considerably at each trophic level (Figure 3.19). Once energy dissipates to heat and is radiated back to space, it can never be recovered and reused. Thus energy flow is a one-way path and must be continually refueled by energy from the sun.

Trophic-Dynamics. Trophic-dynamics remains one of the earliest and most promising ways to understand the diverse interactions within ecosystems. Lindeman (1942) was the first to describe the concept of feeding (trophic) levels in communities based on the number of times food energy had been transformed from one life-form to another. Primary producers make up the first level, herbivores the second, first-level carnivores the third, and so on until the top carnivore level is reached (Figure 3.18).

Most ecosystems have 4 to 6 trophic levels. For example, lake phytoplankton (primary producers) support zooplankton (herbivores), which are eaten by small minnows (first-level carnivores), which feed yellow perch (second-level carnivores), which are consumed by northern pike (third-level carnivores), which are eaten by people (top-level carnivores), which most often decompose without being consumed. A species rarely occupies just one level. Many species are omnivores that feed from different levels. Other species go through life stage changes in which one life stage feeds in one level and another life stage feeds in another level. Many frogs, which are carnivorous as adults, have herbivorous or omnivorous larvae (tadpoles). Many top-level carnivores progressively feed at higher levels as they grow to a larger size.

Although it is far from a constant rule, the size of consumers tends to increase with position among trophic levels. The primary exception is in grassland and savannah

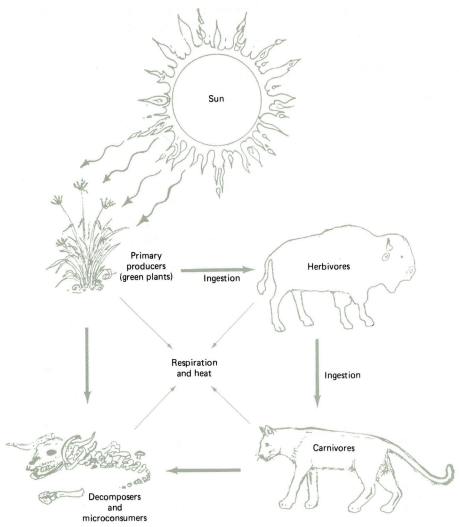

FIGURE 3.18   Generalized diagram showing energy flow through a grassland ecosystem (drawing by John N. Smith) (from Holechek et al. 2001).

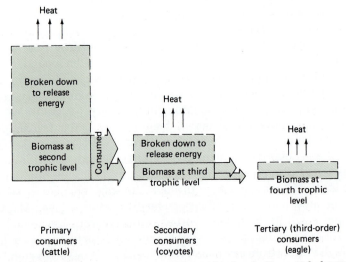

FIGURE 3.19   The loss of energy as it moves through trophic levels (from Nebel 1981).

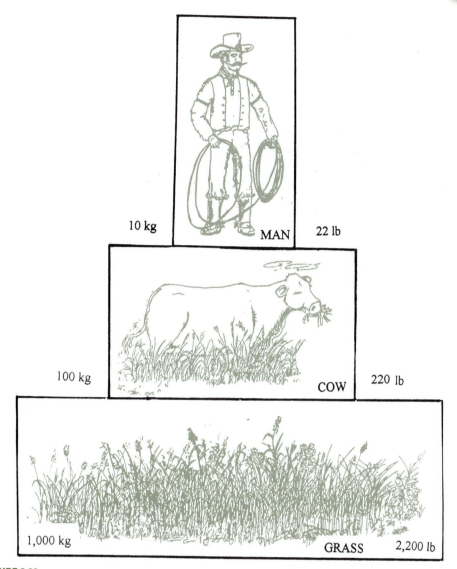

FIGURE 3.20 Biomass pyramid. It takes 1,000 kilograms of grass to produce 100 kilograms of beef needed to produce 10 kilograms of human biomass (adapted from Owen and Chiras 1995 by John N. Smith).

communities, which have exceptionally large herbivores (i.e., bison, elephants, rhinoceros). However, the largest biomass of terrestrial herbivores, even in grasslands, is found in the ants, termites, worms, and other invertebrate species.

Trophic Efficiency. The conversion of production in one consumer level to another consumer level is inefficient. A few studies have been conducted to estimate magnitudes of energy flow in ecosystems. These studies show that less than 2% of the solar radiation received yearly by primary producers is converted to production through photosynthesis, and less than 1% is the rule. Only a relatively small portion of the aboveground primary productivity (about 10%) is harvested by herbivores, including livestock (Lewis 1971, Pieper 1983).

An average of about 10% of the energy of each previous trophic level is retained by the next trophic level. By the time it reaches the top carnivore level, very little is left. For example, if 1,000 kg/m^2 is produced by plants, only 10 g/m^2 is retained at the first carnivore level. From a practical standpoint, this means that humans retain only 1.0% of the energy if they eat only pigs or chickens fed grain compared with 10.0% conversion if they eat the wheat or corn directly (Figure 3.20). The point here is that the world can feed far more humans if they eat only plant material than if they eat a large amount of meat.

Species tend to become progressively more specialized as they become larger. However, the biomass present is not necessarily indicative of the capability of a species to recover from stress. In world fisheries, rapid decimation by harvest has occurred for what appeared to be very abundant populations. Because they have low intrinsic rates of increase, the relatively high biomass in large fish species is a misleading indicator of production and sustainable yield. The list of the endangered species of the world has a remarkable number of large carnivorous species. Large compared to small species, require much more area of food producing habitat to support viable populations. Habitat fragmentation tends to affect the viability of larger species more than smaller species.

Consumers have developed many adaptations to take advantage of foods of all sizes and types. Some of the most productive consumers are detritivores—among the most impressive being earthworms in moist organic soils. One of the most important ways of reducing particle size is by consuming it and egesting fecal matter. Feces provide a rich substrate for decomposers. Thus, many detritivores specialize in consuming fecal matter (copraphagy).

Humans, of course, occupy dominant positions in food webs. Some ecologists believe that as the world becomes more densely populated people will be forced to eat more plant material and less meat. This belief is based on the idea that food demand will outpace technological developments. However, crop production so far has increased faster than the human population. Past advances have depended heavily on high amounts of fossil fuel, fertilizers, and pesticides (see Chapter 12). This approach is contributing to environmental damage, such as global warming and eutrophication of rivers and marshes. Thus, associated with any technological advance are trade-offs that need to be considered as societies determine their priorities. Natural resource managers increasingly will be involved in quantifying the benefits and costs associated with new technologies and providing that information to the public for policy decisions.

Resource Partitioning. Within trophic levels, populations compete for food while avoiding predation. They also complement each other through unique specializations that give them advantages under certain conditions. Because most ecosystems are dynamic and environments vary, a large number of niches are available. As new species evolve, other species evolve to respond to them as a food resource—competitor or predator. In this manner, the energy resources are partitioned among many specialist species, each with higher ecological efficiencies than their more generalized predecessors. The energy resources of the habitat are used more efficiently as diversification occurs, and the number of trophic levels increases.

Material Flow, Storage, and Cycling

The second basic functional process of ecosystems is material flow, storage, and cycling. This function includes physical, chemical, and biological processes, which are discussed below.

Material Flow. Materials flow both actively and passively into, through, and out of ecosystems (Figure 3.21). Water, wind, and living organisms are important transport agents. Water and wind act to reshape the earth through erosion. They redistribute the eroded material into deposits which form valley alluvium, outwash plains, floodplains, and sediments. Water also dissolves minerals and transports them through the soil, wetlands, streams, oceans, and in the water vapor of the atmosphere. These suspended and dissolved materials create the nonliving foundation for supporting biotic communities but also include inhibiting contaminants. Individual elements are absorbed by primary producers and are synthesized into living tissues. Herbivores consume plant tissue and the elements contained there. Some plant material is not

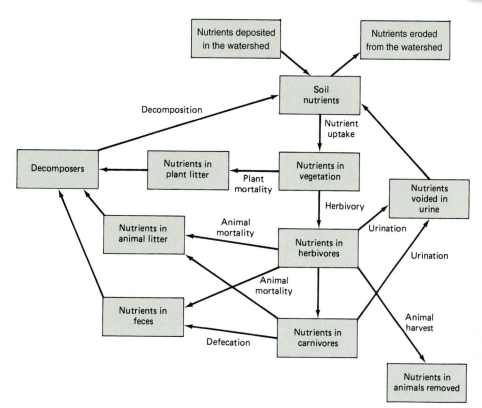

FIGURE 3.21 Generalized nutrient cycle diagram for a terrestrial grassland ecosystem (modified from Pieper 1977).

consumed by herbivores and is instead broken down by decomposers and microconsumers. The chemical elements are then returned to the soil. For herbaceous plants these turnover rates can be rather rapid, but, for woody plants, elements may be tied up in plant material for many years. The feces and urine of all consumers are eventually returned to the soil, where they may be taken up by the plants.

Salmon, for example, are important transport pathways for returning nutrients to the infertile streams where their young are born. By dying after reproduction their decayed carcasses enrich the system that supports the fry and fingerlings.

Unlike energy, which degrades and becomes unusable as it flows through ecosystems, chemical elements are reusable and cycle through the various compartments. Many cycles are completed in hours or days. Other cycles take millions of years to complete. Whereas many material cycles include only solid forms, others involve both solid and gaseous forms. Few materials become liquid or remain that way for long. Water is an important exception. Among the more important material cycles are those for carbon, oxygen, nitrogen, phosphorus, and iron, which we will briefly discuss.

Carbon and Oxygen. The primary element in organic matter is carbon. The original source of atmospheric carbon dioxide is from volcanic emissions from the earth's interior. However, some carbon dioxide comes from carbonate rocks such as limestone and dolomite. Carbon in fossil organic matter, such as coal and oil, can be returned to the atmosphere through natural combustion or burning by humans (Figure 3.22). Much of the earth's carbon is fixed in plant communities, especially in the wood of forests. Carbon dioxide builds up in the atmosphere when forests are cut and oxidized (burned). In tropical rain forests cutting and burning is now occurring at a more rapid rate than photosynthesis is locking up carbon in new organic matter. The accumulation of carbon dioxide in the atmosphere from forest cutting and fossil fuel burning is believed to be causing atmospheric warming, or the green house effect (Freedman 1995) (see Chapter 6).

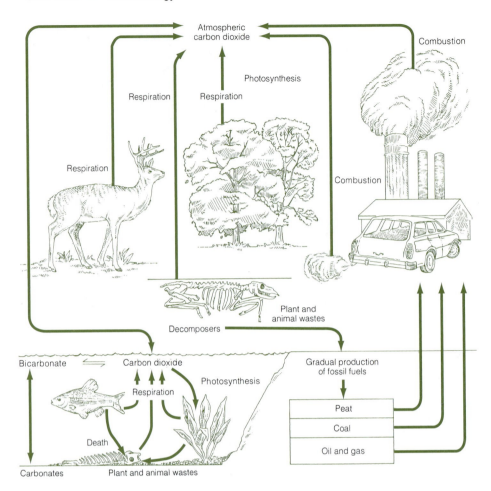

FIGURE 3.22   The carbon cycle (from Moran et al. 1980).

Nitrogen and Sulfur.   Nitrogen and sulfur are two nutrients of exceptional importance in ecosystem dynamics (Figure 3.23). Nitrogen is the most abundant gas in the atmosphere, but it is very scarce in the lithosphere (rock) and hydrosphere. Only certain bacterial forms can fix atmospheric nitrogen into organic nitrogen. Some of them, such as Cyanobacteria (commonly called blue-green algae), carry out photosynthesis in moist soils and in water. Other bacteria live in a symbiotic relationship with root nodes of leguminous plants. Nitrogen occurs in nitrate and ammonia forms.

Atmospheric nitrogen is also oxidized to nitrate-nitrogen by lightning. Nitrate-nitrogen is taken up by rooted plants, phytoplankton, and other primary producers. Ammonia nitrogen is a by-product of protein consumption (in waste products) and decomposition. Both ammonia and nitrate-nitrogen are very soluble in water, rapidly leached from soils, and are the main forms of nitrogen transported by streams and rivers. All nutrients, including nitrogen, cycle repeatedly through primary production and decomposition. In river channel and floodplain sediments this process is called *nutrient spiraling*. As it spirals through streams, rivers, wetlands, lakes, and estuaries, much of the nitrogen carried downstream returns to the atmosphere. Where decomposing matter gets trapped in anaerobic environments, such as deep sediments and wet soils, ammonia is reduced to gaseous nitrogen. It leaks back into the atmosphere in the form of small bubbles.

Nitrogen can be toxic in high concentrations. The fixation of atmospheric nitrogen by certain planktonic bacteria (Cyanobacteria) causes accumulation of toxic by-products. Nitrogen emitted with combustion products of petroleum contributes to the formation of acids in water vapor, causing acid rain.

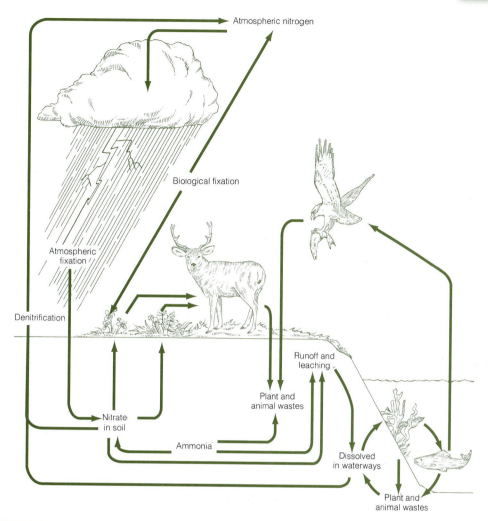

FIGURE 3.23 The nitrogen cycle (from Moran et al. 1980).

Sulfur is unlikely to be a limiting nutrient, but can reach toxic levels in certain aquatic environments. Wetlands are especially important in storing sulfur where it is reduced to hydrogen sulfide gas, which is toxic in high concentrations. The gas is oxidized rapidly when slowly released to the atmosphere in wetland gassing. There it dissolves in rain and is washed back to the land where it contributes to primary production. Human natural resource management has increased sulfur mobilization and redistributed it from the lithosphere to the atmosphere. Sulfur is artificially increased in the atmosphere by man's combustion of coal and oil. Under these conditions it can contribute to acid rain (see Chapter 6). Exposing sulfur-bearing minerals to the air during surface coal mining results in acid mine drainage, a major water pollution problem.

Phosphorus and Iron. Many nutrients are metals and related elements that have no gaseous phase. Among them, phosphorus and iron are most often limiting nutrients. Their compounds remain locked within the lithosphere and hydrosphere. They are part of the *sedimentary cycle,* which is the long geological process of continental erosion, sedimentation into depressions (usually the ocean), and uplift in new continents. Phosphorus is especially likely to limit primary production at high elevations in areas of slowly weathering rock such as granites and schists, where both

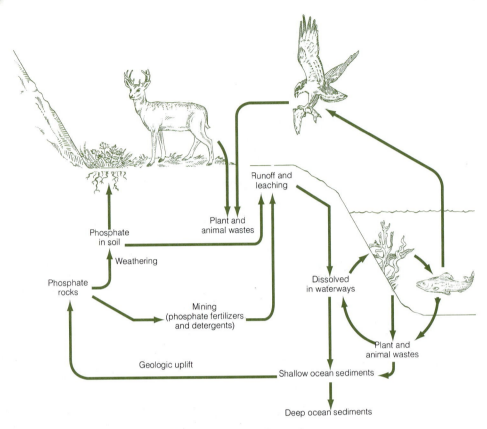

FIGURE 3.24   The phosphorus cycle (from Moran et al. 1980).

iron and nitrogen are more available (Figure 3.24). However, iron and phosphorus typically have closely associated geochemical transformations. In an oxidized state, phosphorus tends to form complexes with iron. Both elements precipitate and fall to the bottom of deep lakes and oceans where turbulence is low.

Because of erosion, rivers and streams transport phosphorus and iron toward lower elevations through nutrient spiraling. Stream turbulence keeps oxidized iron and phosphorus in suspension where they are available for biotic uptake. Once the rivers reach estuaries and oceans, much of the phosphorus and iron fall to the bottom either through chemical change or through biological uptake and settling of dead organisms. Iron is typically more trapped than phosphorus. It is usually the more limiting nutrient in the open ocean.

Where phosphorus and iron reach lakes, they gradually precipitate and fall to the bottom wherever the water is well aerated. Precipitation may be biological, through uptake by plankton which die and settle out, or through chemical and physical processes. Iron oxides and phosphates form complexes that gradually build mass and settle to the bottom. As long as water remains oxygenated, the precipitate will remain on the bottom unless it is disturbed by the wind. In some lakes, wind disturbance is an important means for recycling phosphorus to the surface where it can be taken up once again. Many lakes are deep enough to resist full mixing to the bottom for part or all of the year. They form a stratum of more dense water over the bottom—the hypolimnion—which is easily depleted of oxygen by decomposition when enough dead matter settles.

Relatively nutrient rich (*eutrophic*) lakes often produce enough organic matter to overwhelm all of the oxygen resources in the bottom stratum. Once this bottom layer becomes anoxic, changes in the chemical environment result in oxygen being stripped from phosphates and iron oxides and the two elements going back into so-

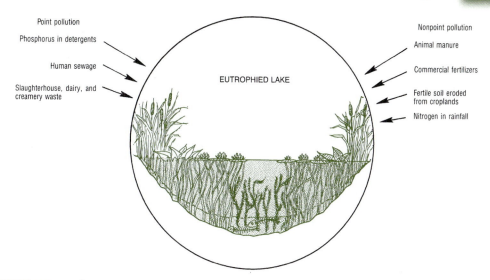

Point pollution
Phosphorus in detergents
Human sewage
Slaughterhouse, dairy, and creamery waste

EUTROPHIED LAKE

Nonpoint pollution
Animal manure
Commercial fertilizers
Fertile soil eroded from croplands
Nitrogen in rainfall

FIGURE 3.25 Anthropogenic sources of nutrients causing harmful eutrophication (from Owen et al. 1998).

lution. If the stratification is permanent, the two nutrients remain well below the surface. If the stratification is temporary, as it is in most U.S. lakes, the nutrients are returned to the surface upon remixing where they are available for more primary production. Most large lakes are efficient traps for nitrogen and phosphorus. *Cultural eutrophication* is a form of pollution that results from excess release of nutrients to natural waters (Figure 3.25). Intense production causes imbalances between consumers and decomposers, and results in severe oxygen depletion. This reduces the value of aquatic resources. Incomplete sewage treatment and overfertilized farmland are two main sources of phosphorus pollution (Perry and Vanderklein 1996).

Succession

Succession is defined as the replacement of one biotic community by another. Clements (1916) provided the first detailed description of plant succession. In recent years, some of Clement's theories have been challenged, but his ideas still remain the foundation for modern plant ecology. In the classical sense, plant succession in both terrestrial and aquatic ecosystems involves the replacement of one plant community by another until the final community is reached. This final, somewhat stable community is often called the *climax*. Although a given habitat type may support several successional plant communities, it will support only one climax plant community.

A climax community is considered to exist on an area if significant changes in species composition do not occur within 500 years (Barbour et al. 1987). Climax plant communities are not static, but the changes that do occur involve small random shifts in numbers and/or relative biomass that fluctuate around a long-term mean. Directional, cumulative, and generally nonrandom vegetative changes that occur within a period of 500 years are referred to as *seral stages* (Barbour et al. 1987). The term *sere* is used to describe an entire progression of seral stages from the first community that occupies bare ground to the final climax vegetation.

Ideas concerning successional changes and stability have been quite controversial, and the classical view of succession has been subject to numerous interpretations. Under the classical view of succession, plants are the overwhelming biological force driving successional rate and direction. The bacteria, fungi, microbial algae, and animals play interdependent roles. In recent years, the importance of animal pollinators, grazers, soil makers, and decomposers has become more appreciated.

CLIMAX STAGE
Dense stand of tall deep-
rooted bunch and sod grasses
not many forbs or other vegetation.

SUBCLIMAX STAGE
Short grasses with some
perennial forbs, a few shrubs.

FIRST HERB STAGE
Early maturing annual
forbs. Invading annual grasses,
few shallow-rooted perennial
forbs.

PIONEER STAGE
Lichens, algae, mosses.

PLANT DEVELOPMENT

HIGHLY FERTILE SOIL
Loam soil, few fine gravel,
rich in organic matter. Available moisture for plant
growth high.

FERTILE SOIL
Loam soil, slightly gravelly
with moderate organic matter.
Available moisture for plant
growth moderate to high.

POOR SOIL
Gravelly loam, little organic
matter. Water holding capacity
moderate to low.

FIRST STAGE OF
DECOMPOSITION
Bare rock formation.

SOIL DEVELOPMENT

FIGURE 3.26   Primary succession on grassland rangelands (from Gay 1965).

Terrestrial Primary Succession.   Succession may start from any point along the sere. Primary succession starts from bare rock and open water (primary areas). On bare ground suitable microhabitats are first colonized by lichens, algae, moss, bacteria, fungi, and animals (Figure 3.26). A rudimentary soil forms as rock is weathered, and water and organic matter are added from the partially decomposed organisms. Seeds from nearby plants are carried by wind, water, or animals to the newly formed soil where some germinate and establish the first vascular plants. Gradually more advanced soil animals and decomposers colonize the area, accelerating organic conversion, nutrient cycling, soil aeration, and water retention.

The earliest plant colonizers are often annual plants that can survive under rather harsh conditions (generalists), but die at the end of the growing season. With further weathering and soil formation perennial plants become established. Herbaceous plants typically precede woody plants. Each assemblage of plants influences soil and microclimate, making it more suitable for plants that have higher requirements for water, nutrients, and other resources.

Early colonist species alter the environment to an extent that it no longer is suitable for them and they are replaced by better adapted (more specialized) plants. Total biomass of all life-forms, total energy storage in organic matter, structural diversity, and rate of mineral cycling increase as succession proceeds (Lewis 1969). The processes of terrestrial primary succession are summarized as follows:

1. The development of soil from parent materials
2. Increasing longevity with successional advance
3. Replacement of species with broad ecological requirements by those occupying narrow niches complementary with other species
4. Greater accumulation of living tissue and litter per unit area
5. Modification of microenvironmental extremes
6. Change in size of plants from small to large
7. Increase in the number of pathways of energy flow

8. More nutrients tied up in living and dead organic matter

9. Greater resistance to fluctuation in the controlling factors

Periodicity and *plant stratification* tend to be highly developed as the climax community is approached. Periodicity involves different species making peak nutritional demands at different times during the year. Stratification is the development of layers of plants with different heights and rooting depths. These changes allow for a larger number of species, more pathways for energy flow, greater stability of total energy flow, and greater total productivity. As the habitat becomes more diverse, species become more specialized in habitat use and there is more efficient use of light, water, space, and nutrition. The microhabitat becomes less extreme as temperature, moisture, air movement, and other environmental factors become less variable and the community becomes more resistant to whatever fluctuations occur.

Whereas productivity generally increases as succession advances, it commonly declines as the climax is approached. Under climax conditions, nutrients are tied up in wood and other organic matter. Soil moisture is often quickly reduced by uptake from the larger plants. This, in turn, reduces the length of the growing season. Dense growth can reduce the efficiency of light use per unit of leaf area. In grasslands, excessive mulch accumulation can chemically and physically retard productivity. These observations suggest that terrestrial successional stages just preceding the climax support the highest production of goods such as timber and red meat. Genetic diversity is likely to be greatest in the late seral stages, but total regional diversity will be maximized where a mix of different seral stages exist.

Aquatic Primary Succession. Aquatic succession depends on the initial depth and nutrient concentration of the lake (Figure 3.27). In deep infertile waters, succession is extremely slow if it occurs at all. Studies of such lakes indicate that the total community production is respired away over an annual cycle and inconsequential amounts of organic matter are stored. The lake will never fill to a point where rooted plants can get established. Even more extreme conditions exist in the open ocean. However, production in fertile lakes usually exceeds respiration and dead organic matter accumulates on the bottom. This can eventually allow rooted vascular plants to become established. Once this occurs, the rate of filling may accelerate and the lake will become a marsh or swamp with emergent plants. Because emergent plants transpire water, they can contribute to lowering the water level. Most commonly, soil erosion from the watershed surrounding a lake contributes to filling. It accelerates the establishment of conditions favoring rooted plant communities and causes hydric succession toward a terrestrial climax. Thus, a close link exists between the successional stages in the terrestrial watershed and succession in ponds and lakes. If erosion accelerates in the watershed, succession in the receiving waters will accelerate. In another form of terrestrial encroachment, vegetation extends into the pond or lake from the sides where it dies and accumulates. Eventually the pond is filled with peaty decaying organic matter, and becomes a terrestrial environment.

Secondary Succession. Secondary succession is defined as vegetational change that occurs after some type of disturbance such as fire, grazing, windstorms, flooding, or logging. Secondary succession generally occurs much faster than primary succession, and generally in a more predictable fashion. Although primary succession is of educational interest, it typically plays a much smaller role in natural resource management than secondary succession. The variability in secondary succession is reduced as the climax is approached (Huschle and

FIGURE 3.27 Primary succession in ponds and lakes involves a process in which they are gradually filled and invaded by surrounding land vegetation (from Nebel and Wright 1996).

Hironaka 1980). Figure 3.28 shows the stages of secondary succession that occur in the abandoned cotton fields in the Piedmont region of Georgia.

Figure 3.29 depicts secondary successional stages following fire in piñon-juniper woodlands in three different areas. Succession is somewhat predictable because the changes are very similar. A skeleton of dead trees remains after a fire, but these areas are soon occupied by annuals, and then by some perennial grasses and forbs. The herbaceous vegetation is invaded by some shrubs and eventually the climax piñon-juniper woodland. Two of the diagrams show what happens when fire occurs during one of the seral stages. The intensity of the fire, characteristics of the fuel, and time of year all influence the damage the fire does. Rates of terrestrial succession are driven mostly by moisture and temperature. In humid ecosystems, such as the tallgrass prairie in the

FIGURE 3.28 Secondary ecological succession of plant communities on abandoned cotton fields in Georgia over a 100-year period (from Owen and Chiras 1995).

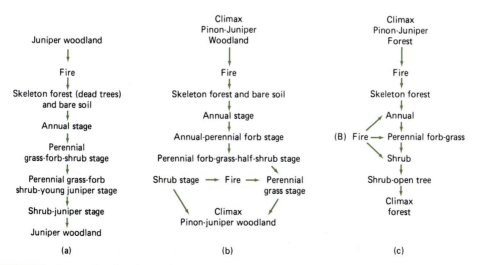

FIGURE 3.29 Successional pathways in piñon-juniper woodlands from different areas: (a) Central Utah (Barney and Frischknecht 1974), (b) Arizona (Arnold et al. 1964), and (c) Colorado (Erdman 1970).

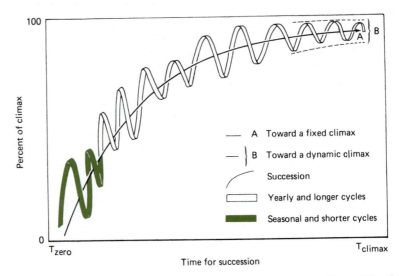

FIGURE 3.30   A model of succession asymptotically approaching a dynamic climax (B), which includes climatic variations associated with daily, seasonal, and yearly phenomena. The absolute climax might be defined as the midpoint of these variations at (A) (from Heady 1973).

eastern Great Plains, recovery to climax following disturbance can occur within five to ten years if severe erosion and soil nutrient depletion has not occurred. In the more dry Chihuahuan Desert, recovery from disturbance may require 20 or more years.

Climax Theory.   The final stage of succession has been viewed in different ways by different authors. Some have viewed the climax as "stable" while others have viewed it in "dynamic equilibrium" with the environment (Figure 3.30). Clements (1916) viewed the climax as ultimately controlled primarily by climate—mostly precipitation and temperature intensities, amounts, and temporal patterns. This he referred to as the "climate climax." He considered large areas of landscape as having the same climax to be governed by the same climate. Development of climax vegetation was considered a very slow process on the same time scale as geologic changes. He also recognized that periodic disturbance such as fire could sustain conditions deviating from the climate climax, which he referred to as *dis*climax stages.

Retrogression away from climax can occur from disturbances outside the community such as fire, storms, grazing, disease, and cultivation. However, some retrogressions to earlier stages can be caused by the community itself. In the tallgrass prairie of the central Great Plains, excessive mulch accumulation has caused the productivity of the climax grasses to be reduced and the composition to shift to earlier stages (Holechek et al. 2001). High levels of mulch tie up excessive amounts of nutrients, physically inhibits plant growth, delays plant growth in the spring by depressing soil temperatures, and provides habitat for pathogens and insects that can be harmful to the climax plants. There is evidence of similar community-controlled retrogression in the Chihuahuan Desert and in various forest ecosystems.

Although fire and grazing are often considered forces that cause retrogression, there is also much evidence that they play a crucial role in maintaining the climax vegetation in many forest and grassland ecosystems. The timing, intensity, and frequency of both fire and grazing are critical factors in determining whether they result in retrogression or progression (Holechek et al. 1998).

Other authors considered other types of climax in addition to climatic climax, such as edaphic climax, where soil characteristics may have an influential role in the type of vegetation that develops. In this view, each climatic region is characterized by a mosaic of climax types, sometimes called the "polyclimax" viewpoint (Whittaker 1953).Others considered vegetational development more a matter of chance, with discrete communities difficult to discern. This "individualistic" concept later gave way to the idea of the "continuum," which stated that the distribution of each species was independent of that of other species and that these distributions overlapped with each other (Curtis 1959). Whittaker (1967, 1975) found that species were often distributed along environmental gradients and that individual communities were only delineated somewhat arbitrarily along the gradient. In other words, ecotones were very gradual and extended in many locations. Therefore, discrete communities were hard to define. It is also obvious, however, that some ecotones are quite sharp between adjacent plant communities and discrete communities are easily recognized by the dominant plants in the community. The difficulty of identifying discrete communities increases as more rare species are included in the community.

Fire and Succession. Fire has been a dominant force in the evolution of most of the world's grasslands and forests, as well as some desert shrublands. Grasses and forbs are favored by periodic fires (every 3–10 years), but shrubs are suppressed (Figure 3.31).

(a)

(b)

(c)

FIGURE 3.31 Before fire control and livestock grazing, desert grasslands of the southwestern United States were maintained by fires that periodically reduced mesquite and other shrubs. The buildup in vegetation (a) creates a fuel load that will carry fire (b). After the fire (c), the perennial grasses gain a competitive advantage over the mesquite because they have dense fibrous root systems that allow them to regenerate more quickly than the shrubs (from Holechek et al. 1998).

Huge areas of continuous, flammable grassland vegetation occurred in the central Great Plains before the advent of settlement in the mid-1800s (Barbour et al. 1987). Early accounts describe how summer thunderstorms moving across the prairie would rapidly ignite a series of fires in their path. Until the advent of fences, roads, farming, and livestock grazing these fires could spread long distances until they burned out because of the lack of fuel or a weather change. Native Americans burned the grasslands of the Great Plains to improve conditions for game animals. The nutritious regrowth that occurs shortly after a fire attracts bison, elk, deer, pronghorn, and a wide variety of other life.

Most grasslands appear dependent on fire for recycling nutrients from vegetation to soil to new biomass (Mutch 1970). Ponderosa pine, the primary timber tree in the western United States, has heavy, fire-resistant bark. Periodic fires cleared out the small trees and shrubs, maintaining most ponderosa pine areas as an open parkland. In the southeastern United States, longleaf pine, another important timber tree, depends on fire for its maintenance. Fire is necessary to reduce understory competition, prepare a mineral seedbed, reduce fungal disease, and open the forest canopy (Barbour et al. 1987).

Grazing and Succession.   Many grassland ecosystems in the world evolved under heavy grazing pressure from large animals. During this process, plants developed defense mechanisms and animals have adapted to varying degrees to these defenses. Some of the mechanisms plants use to protect themselves from herbivory include low stature, delayed elevation of growing points, high amounts of seed production, synthesis of poisonous compounds, synthesis of compounds that reduce palatability, and growth of appendages such as thorns and spines that protect the plant (Holechek et al. 2001). Generally all these mechanisms require a reallocation of energy from growth to defense. Therefore they tend to make the grazing-resistant plants less competitive with plants not having these defense mechanisms when grazing does not occur, or occurs at light to moderate levels (Figure 3.32). Holechek et al. (2001) reviewed evidence showing that light to

FIGURE 3.32   Rangelands in Colorado showing the long-term effect of heavy and moderate cattle grazing on vegetation composition. The heavily grazed area in the foreground is dominated by big sagebrush while the moderately grazed area in the background is dominated by western wheatgrass. Big sagebrush has volatile oils that are toxic to cattle.

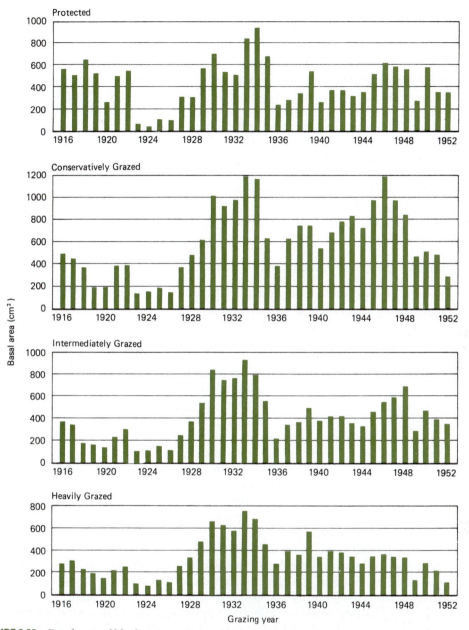

FIGURE 3.33 Basal area of black grama on areas protected from grazing and at three intensities of grazing on the Jornada Experimental Range, southern New Mexico, 1916–1953 (from Paulsen and Ares 1962).

moderate grazing tends to cause progression toward the climax and is important along with fire in maintaining high seral or climax conditions in mesic grasslands and many semiarid communities. Progressive effects of controlled grazing are less evident in more arid grassland communities.

Even under protection from large herbivores, vegetation is dynamic and fluctuates in response to other controlling factors, especially climate. Figure 3.33 shows fluctuations in black grama basal cover over time on the Jornada Experimental Range in southern New Mexico for nearly a 40-year period, under protection from grazing, and three intensities of grazing. In this study, heavy grazing represented utilization of more than 55% for black grama; intermediate grazing, 40% to 55% utilization; and conservative grazing, less than 40% utilization. Even without grazing, black grama cover was extremely low during the mid-1920s. Black grama cover was highest when the plants were grazed conservatively.

Management Applications.   Knowledge of plant succession processes is essential to forest, rangeland, wildlife, and national park managers. Many private natural areas are now being purchased throughout the world by conservancy groups, such as The Nature Conservancy, who will need a strong knowledge of plant succession in order to obtain their goals. Knowing which seral stages and what arrangement of seral stages are needed and how to obtain them is a critical part of forest, range, and wildlife management. In a few cases, a passive approach of merely letting the forces of nature cause progression to a higher seral stage or perpetuate the climax may be satisfactory. However, in most situations, active management—such as use of grazing, burning, logging, and/or cultivation—will be required to obtain the desired mix of plant communities within a reasonable time frame.

Ecosystem Manipulators: The Human Component

Humans are a keystone species in a class of their own. They are the only species that consciously plans and manipulates ecosystem components for their own benefit. Unfortunately, we humans are not perfect and have lost many benefits because of past mistakes. Human activities affect all trophic levels, and have had both positive and negative influences on natural processes and functions. In the past, intentional manipulations of ecosystems have been largely directed towards increasing those populations that contributed to amount of wood, meat, plant foods, fiber, minerals, energy resources, and recreational resources with little regard for the diversity of other organisms and their interactions with the environment (Figure 3.34).

More recently, managers have recognized that all organisms are intertwined in the pyramid of life. Therefore, any management practice affecting one species will, in turn, affect all others. This has lead to the present policy of multiple-use management on public lands in the United States.

Multiple use involves managing ecosystems for livestock, lumber, wildlife, minerals, water, fisheries, recreation, biodiversity, and other ecosystem services on a sustained basis. Obviously, no acre of land can be used for maximum production of all natural and enhanced goods and services. However, most land units will provide two or more products and regionally most if not all can be produced. The product or products emphasized will depend primarily on the characteristics of the ecosystem involved and demands by society.

Lately, the need for more effective management of all natural resources has led to a holistic approach called ecosystem management (e.g., Vogt et al. 1997). Although the need for specialization will continue, there will also be a need for awareness of how each specialty fits into an ecosystem context. Because ecosystem structure and function are so complex, natural resource researchers in the past have taken a mostly "reductionist" approach to ecosystem management. This involves using a one-factor approach, such as studying the influence of grazing on individual plants, the influence of fertilization on vegetation, and the influence of fishing on fish abundance. However, any management practice influences many components rather than just the intended one (Crisp 1964).

In the future, the demands placed on the earth's ecosystems for all products will accelerate because of a rapidly expanding human population on a shrinking land base. This means management must be much more intensive than in the past. Each ecosystem responds differently to management. However, the same basic ecological principles apply to all ecosystems.

FIGURE 3.34 In the 1970s agricultural systems in the United States typically involved planting large areas of the same crop year after year (top photo). In the 1990s there was a strong trend toward more emphasis on diversified farming, soil and water conservation, and wildlife habitat improvement (bottom photo) [photos courtesy of U.S. Dept. of Agriculture (*top*) and U.S. Dept. of Interior, Bureau of Reclamation (*bottom*)].

LITERATURE CITED

Allan, J. D. 1995. *Stream ecology.* New York: Chapman & Hall.

Allee, W. C., A. S. Emerson, O. Park, T. Park, and K. P. Schmidt. 1949. *Principles of animal ecology.* Philadelphia, PA: Saunders.

Andrewartha, H. G. and C. L. Birch. 1954. *The distribution and abundance of animals.* Chicago, IL: University of Chicago Press.

Arnold, J. F., P. A. Jameson, and E. H. Reid. 1964. *The piñon-juniper type of Arizona: Effects of grazing, fire, and tree control.* U.S. Dept. Agric. Rept. Res. Rep. 84.

Barbour, M. G., J. H. Burk, and W. D. Pitts. 1987. *Terrestrial plant ecology.* 2nd ed. Menlo Park, CA: Benjamin-Cummings.

Barney, M. and N. C. Frischknecht. 1974. Vegetation, changes following fire in the piñon-juniper type of west-central Utah. *Journal of Range Management* 27:91–96.

Bebee, A. and A. Brennan. 1997. *First ecology.* New York: Chapman & Hall.

Bolen, E. G. and W. L. Robinson. 1995. *Wildlife ecology and management.* 3rd ed. Englewood Cliffs, NJ: Prentice-Hall.

Carpenter, S. R. and J. F. Kitchell. 1993. *The trophic cascade in lakes.* New York: Cambridge University Press.

Clements, F. E. 1907. *Plant physiology and ecology.* New York: Henry Holt & Company.

Clements, F. E. 1916. *Plant succession.* Carnegie Inst. Wahs., Publ. 242.

Cole, G. A. 1994. *Textbook of limnology.* 4th ed. Prospect Heights, IL: Waveland Press.

Colinvaux, P. A. 1993. *Ecology 2.* New York: John Wiley & Sons.

Crisp, D. J., ed. 1964. *Grazing in terrestrial and marine environments.* British Ecological Society Symposium Number 4. Blackwell Sci. Public. Ltd., Oxford.

Curtis, J. T. 1959. *The vegetation of Wisconsin.* Madison, WI: Univ. of Wisconsin Press.

Daily, G. C., ed. 1997. *Nature's services: Societal dependence on natural ecosystems.* Washington, DC: Island Press.

Daubenmire, R. 1968. *Plant communities: A textbook of plant synecology.* New York: Harper & Row.

Day, J. W., C. A. S. Hall, W. M. Kemp, and A. Yanez-Arancibia. 1988. *Estuarine ecology.* New York: John Wiley & Sons.

Erdman, J. A. 1970. *Piñon-juniper succession after natural fires on residual soils of Mesa Vedge, Colorado.* Brigham Young Univ. Sci. Bull. Biol. Ser. 11:1–24.

Forman, R. T. T. and M. Godron. 1986. *Landscape ecology.* New York: John Wiley & Sons.

Freedman, B. 1995. *Environmental ecology: The ecological effects of pollution, disturbance, and other stresses.* 2nd ed. New York: Academic Press.

Gay, C. 1965. *Range management: How and why.* N. Mex. State. Univ. Coop. Ext. Circ. 376.

Heady, H. F. 1973. Structure and function of climax. In *Arid shrublands,* Proc. 3rd Workshop, U.S./Australia Rangelands Panel. Denver, CO: Society for Range Management.

Holechek, J. L., R. D. Pieper, and C. H. Herbel. 2001. *Range management principles and practices.* 4th ed. Upper Saddle River, NJ: Prentice-Hall.

Horne, A. J. and C. R. Goldman. 1994. *Limnology.* 2nd ed. New York: McGraw-Hill.

Huschle, G. and M. Hironaka. 1980. Classification and ordination of plant communities. *Journal of Range Management* 33:179–182.

Jenny, H. 1980. *The soil resource origin and behavior.* New York: Springer-Verlag.

Johnston, C. A. 1998. *Geographic information systems in ecology.* Malden, MA: Blackwell Science.

Lerman, M. 1986. *Marine biology: Environment, diversity, and ecology.* Menlo Park, CA: Benjamin-Cummings.

Lewis, J. K. 1969. Range management viewed in the ecosystem framework. In *The ecosystem concept in natural resource management,* edited by G. M. VanDyne. New York: Academic Press.

Lewis, J. L. 1971. The grassland biome: A synthesis of structure and function. In *Preliminary analysis of structure and function in grasslands,* edited by N. R. French. Range Science Series No. 10, Colorado State University, Ft. Collins.

Lindeman, R. L. 1942. The trophic-dynamic aspect of ecology. *Ecology* 23:399–418.

MacArthur, R. H. and E. O. Wilson. 1967. *The theory of island biogeography.* Princeton, NJ: Princeton University Press.

Marsh, G. P. 1865. *Man and nature; or, physical geography as modified by human nature.* New York: Charles Scribner.

Miller, E. C. 1938. *Plant physiology.* 2nd ed. New York: McGraw-Hill.

Moran, J. M., M. D. Morgan, and J. H. Wiersma. 1980. *Introduction to environmental science.* San Francisco, CA: W. H. Freeman.

Mutch, R. E. 1970. Wildland fires and ecosystems—a hypothesis. *Ecology* 51:1046–1051.

Nebel, B. J. 1981. *Environmental science: The way the world works.* Englewood Cliffs, NJ: Prentice-Hall.

Nebel, B. J. and R. T. Wright. 1996. *Environmental science: The way the world works.* 5th ed. Upper Saddle River, NJ: Prentice-Hall.

Nebel, B. J. and R. T. Wright. 1998. *Environmental science: The way the world works.* 6th ed. Upper Saddle River, NJ: Prentice-Hall.

Odum, E. P. 1959. *Fundamentals of ecology.* 3rd ed. Philadelphia, PA: W. B. Saunders.

Owen, O. S. and D. D. Chiras. 1995. *Natural resource conservation.* 8th ed. Englewood Cliffs, NJ: Prentice-Hall.

Owen, O. S. D. D. Chiras, and J. D. Reganold. 1998. *Natural resource conservation.* 7th ed. Upper Saddle River, NJ: Prentice-Hall.

Paris, O. H. 1969. The function of soil fauna in grassland ecosystems. In *The grassland ecosystems: A pro synthesis,* edited by R. L. Dix and R. G. Beidleman. Range Science Series No. 2, Colorado State University, Ft. Collins.

Paulsen, H. A. and F. N. Ares. 1962. *Grazing values and management of black grama and tobosa grasslands and associated shrub ranges of the Southwest.* U.S. Dept. Agric. Tech. Bull. 1270.

Perry, J. and E. Vanderklein. 1996. *Water quality: Management of a natural resource.* Cambridge, MA: Blackwell Science.

Pieper, R. D. 1977. *Effects of herbivores on nutrient cycling and distribution.* Proc. 2nd U.S./Australia Rangeland Panel. Australia Rangeland Society, Western Australia.

Pieper, R. D. 1983. *Consumption rates of desert grassland herbivores.* Proc. 14th International Grassland Congress, edited by J. A. Smith and V. W. Hays. Boulder, CO: Westview Press.

Primack, R. B. 1993. *Essentials of conservation biology.* Sunderland, MA: Sinauer Associates.

Ricklefs, R. E. 1996. *The economy of nature: A textbook in basic ecology.* 4th ed. New York: W. H. Freeman.

Risser, P. G. 1969. Competitive relationships among herbaceous grassland plants. *Botanical Review* 35:251–284.

Rosenzweig, M. L. 1995. *Species diversity in space and time.* New York: Cambridge University Press.

Van Dyne, G. M. 1966. *Ecosystems, systems ecology, and systems ecologists.* ORNL-3957, Oak Ridge, TN: Oak Ridge National Laboratory.

Vogt, K. A., J. C. Gordon, J. P. Wargo, D. J. Vogt, H. Asbjornsen, P. A. Palmiotto, H. J. Clark, J. L. O'Hara, W. S. Keeton, T. Petel-Weynand, and E. Witten. 1997. *Ecosystems: Balancing science with management.* New York: Springer.

West, N. E., ed. 1995. *Biodiversity on rangelands.* Logan, UT: College of Natural Resources, Utah State University.

Wetzel, R. G. 1983. *Limnology.* 2nd ed. New York: Saunders College Publishing.

Whittaker, R. H. 1953. A consideration of climax theory: The climax as a population and pattern. *Ecological monographs* 23: 41–78.

Whittaker, R. H. 1967. Gradient analysis of vegetation. *Biological Review* 42:207–264.

Whittaker, R. H. 1975. *Communities and ecosystems.* 2nd ed. New York: Macmillan.

Wilson, E. O., ed. 1988. *Biodiversity.* Washington, DC: National Academy Press.

CHAPTER 4

Conservation Economics

THE PROBLEM

Natural resources including land, water, wildlife, and minerals, can be preserved for future generations or developed into commercial products for current generations. These resources typically have many competing uses in any one time period, and there are many different periods in which they can be used. For example, ancient redwoods in northern California can be cut down to produce lumber for houses, they can be left standing for later timber harvest, or they can be designated as wilderness. Conservation economics deals with the problem of how to best distribute natural resources to provide more total benefit to people in current and future generations.

HARD CHOICES

Natural resource economics is about making hard choices when no single choice is best for all people in all generations. Since the beginning of recorded time, governments have tried to design institutions such as laws, policies, and regulations to help make better choices in meeting society's objectives. Government natural resource agencies in the United States are expected to improve the general welfare by choosing wisely among natural resource conservation proposals.

Adam Smith (1776), in his book *The Wealth of Nations,* was the first to demonstrate that *markets* are efficient mechanisms for allocating resources to maximize human benefits for current and future generations. This theme has been an important one in economic science ever since. For example, Howe et al. (1986) proposed that water markets be used for conserving water in the western United States. Under his approach people who need water could buy it from somebody who was willing to sell.

In this chapter, we apply basic economic principles to natural resource conservation. Underlying the chapter are two basic assumptions: First, environmental and natural resource management policies should produce the greatest possible total human benefits for present and future generations; and second, these policies should produce a fair distribution of those benefits. We will begin by considering hard management choices.

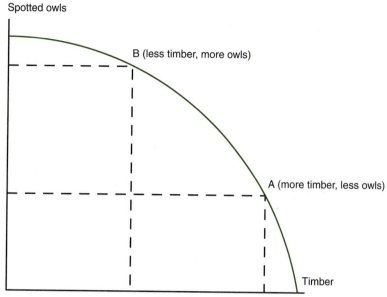

FIGURE 4.1 The trade-off between logs and spotted owls.

Timber versus Spotted Owls (Pacific NW)

In the Pacific Northwest there are only so many acres of land with ancient forests, called old growth trees. These trees are exceptionally valuable because they provide timber and high-paying jobs for loggers. However, the same trees also provide a critical habitat component for northern spotted owls. This species is so rare that in 1990 it was listed as an endangered species. Because the total supply of old growth forest land is scarce, it is essential that we make the right decisions in allocating it to these and other uses. Underlying the issue of logs versus owls is the more fundamental one of measuring the relative value associated with each use (Figure 4.1).

We are now at point A in Figure 4.1, producing lots of logs and few spotted owls. Montgomery and Brown (1992) have found that ensuring owl survival requires that high-quality forest not be logged, in other words that we move to point B, producing more owls and less timber. Choosing correctly in this situation depends on whether or not the benefits gained from moving to point B from point A outweigh the costs. Nobody knows the answer for certain, but a good part of it will be determined by attempts to measure those benefits and costs. Many hard choices like this are faced by natural resource managers.

Cheap Energy versus Functioning Ecosystems

How much more are you willing to pay in the price of gasoline to ensure that ecosystems are being protected? The U.S. Congress has considered passing laws to require oil companies to build double-hulled tankers so that accidents do not cause an ecological catastrophe such as the *Exxon Valdez* in Prince William Sound, Alaska. But that policy would raise the price of gasoline, because the double-hulled tankers are ultimately paid for by people who buy gasoline.

Clean Air versus Cheap Steel

People in many of America's biggest cities live with air pollution every day. This pollution can be cleaned up, but the costs are high. The Clean Air Act of 1970 and later amendments are strong laws restricting air emissions in many parts of our economy. Achieving air pollution control objectives set up by Congress in the 1990 amendments to the Clean Air Act will require an estimated $25 to $50 billion each year to be spent by both the public and private sectors. Before the 1990 amendments, there were fewer than 20 regulated pollutants. The Clean Air Act currently regulates more than 380 pollutants at a considerable cost to industry. The steel industry is particularly hard hit by the new laws. They will require the steel industry to spend additional billions of dollars over the next 10 years. This will add substantially to the steel industry's operating costs and to the cost of steel products we all buy, like new cars. So Americans must face the question: Do we want cheaper steel or cleaner air?

Endangered Wolves versus Livestock and Wildlife

How much are you willing to pay in higher beef and lamb prices and decreased abundance of deer and elk to restore wolves to native habitat? Reintroductions of wolves in Yellowstone Park and in Arizona may result in lower sustainable abundances of large wildlife. We all may have to pay a price in fewer deer, elk, and other animals watched and hunted to restore wolves. Wolves also eat cattle and sheep. So reintroduced wolves might reduce the U.S. beef and lamb supplies in our grocery stores, which may increase the price of beef and lamb.

Wetlands versus Housing and Transportation

The executive branch of the U.S. federal government has followed a policy of allowing no-net-loss of U.S. wetland acreage and function since 1989. For each acre of functional wetland converted to another use, such as housing development and road construction, an acre of similar wetland must be created or restored. This added cost is passed on to the public through increased home purchasing prices and transportation taxes. In continuing support of this policy, the public has made the hard choice between sustaining wetlands and reducing housing costs and taxes.

Hydropower versus Recreation

Glen Canyon Dam, which backs up the Colorado River on the Arizona-Utah border, was completed in the early 1960s by the U.S. Bureau of Reclamation, mostly to produce cheap electricity. The best time to let water go over the dam's turbines for power production is on hot summer afternoons, when power demand is high because people need it to cool down their sweltering Phoenix and Las Vegas homes. However, the best time to let high flows go past the dam for environmental benefits is in March, when the value of hydropower is nearly zero. Water flow in March reproduces natural spring runoff needed by rare fish and natural riparian vegetation downstream.

Here is the hard choice. By sending water past the dam in March for environmental benefits, Phoenix and Las Vegas lose some high-valued summer electricity that must be made up with expensive coal burning. However, by holding the high flows back until summer, all the March environmental benefits are lost. As federal taxpayers, we and the Bureau of Reclamation must try to make the optimum choice.

Electric Power versus Productive Lakes

Much of the western United States is endowed with abundant and high-quality deposits of low-sulfur coal. An important use of that coal is burning it to produce electricity. An unfortunate by-product of burning that coal is sulfur dioxide, which travels in the air thousands of miles to the north and east. When it falls back to earth as acid rain, the precipitation combines with the chemical to produce sulfur dioxide, which is thought to damage recreational fisheries in many lakes in Canada and the northeastern United States. How much are we prepared to pay in the value of lost lake recreation to ensure a continued cheap supply of coal-fired electric power?

Mining versus Recreation on Public Lands

People need coal and mineral products for their daily lives. However, mining activities of various kinds can conflict with outdoor recreation. One widely publicized conflict emerged in the late 1970s in southern Utah between coal development and nearby Bryce Canyon National Park. One of the proposed coal fields was suspected of damaging outdoor recreation by impairing scenic views at the southern end of the park. On the basis of an economic study that estimated recreational values displaced by the proposed mine (Haspel and Johnson 1982) by then Secretary of the Interior, Cecil Andrus, the coal mining project was shut down.

Riparian Ecosystems versus Livestock

Most of the southwestern United States is arid, which means that water for agriculture, cities, and wildlife is scarce and unreliable. Livestock in the deserts of Arizona and New Mexico congregate near the streams because that is where water and grass are most available. Livestock that are allowed to graze these areas affect wildlife and their habitat, such as the endangered southwestern willow flycatcher. Ensuring the survival of the flycatcher may require taking cattle off a large percentage of these streamside areas. So we are faced with the question of how many cattle we are willing to give up and how much more we are willing to pay for beef to guarantee survival of the flycatcher and other species that require dense riparian (streamside) habitats.

Hydro-Dam Relicensing

The Federal Energy Regulatory Commission (FERC) licenses all nonfederal power plants in the United States. Many of these licenses will soon come up for renewal. Changes in the amount and timing of water used for power generation can affect other uses of water downstream such as recreation, irrigation, and endangered species. The relicensing process gives environmental groups the opportunity to seek downstream flows more compatible with needs of outdoor recreation and endangered species. New licenses may contain terms and conditions relating to wildlife protection and enhancement (or may be denied in extreme cases). Typically, minimum flows will be required downstream from dams. In developing these terms and conditions, FERC must, by law, give equal consideration to power and nonpower values. So as a nation we are faced with the hard choice: Do we want cheaper electricity or more suitable flows for downstream recreation, fish, and endangered species?

CHOICES DISPLACE OPPORTUNITIES

Few decisions are easy to make because most entail costs as well as gains. Socrates as well as other historical figures taught us that people live happier lives when they continually reevaluate impacts of their choices, learn from their mistakes, and make better future choices. Socrates' famous quote is "the unexamined life is not worth living." Nowhere is this more true than in natural resource management. The common thread among all of the hard choices presented above is that every public or private choice involving natural resources creates certain opportunities while displacing others. The opportunity gained is the benefit. The opportunity displaced is the cost. Economics helps us think about which choices we should make to receive the maximum benefits with the least cost.

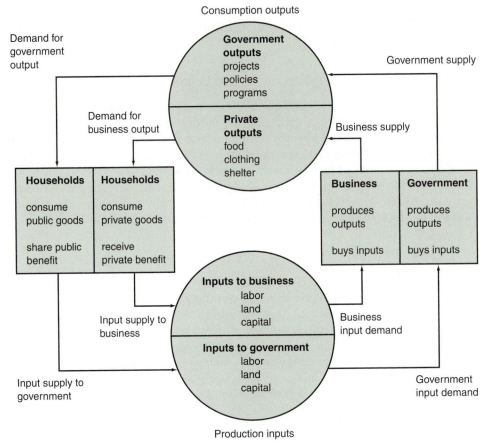

FIGURE 4.2   A general circular-flow diagram that illustrates what, how, and for whom things are produced in an economic system.

Economics Is About Choices

All civil societies invest a large amount of time, energy, and money in the *design of institutions* for assuring that good decisions are continually made, carried out, and judged. By institutions, we mean practices or organizations for governing peoples' behavior. The two main institutions used for solving a society's economic problems are *markets* and *government controls*. The way society goes about solving its economic problems tells us a lot about the way they approach their natural resource conservation problems. All societies face three basic economic problems that are about making choices (Figure 4.2).

First, societies need to determine **what** commodities will be produced and in what quantities. For example, will we use our public and private rangeland and forest land to produce only large quantities of timber and livestock with no outdoor recreation, water, and habitat (living area) for endangered species? Or will we use those lands to produce smaller quantities of livestock and timber, but more outdoor recreation, water, and endangered species habitat? Will we allocate our rivers to produce electric power and water for industry with nothing for recreation and endangered species, or will we use those rivers to produce smaller amounts of a greater number of products? Finally, should we let our rivers run wild to the ocean and take the inevitable floods and droughts that result, or will we take taxpayer resources (money, land, and water) and allocate them to building dams to reduce flood and drought damages?

Next, each society needs to determine **how** it will produce all these goods. By this we mean who, what resources, and what technologies will be used. For example, will businesses use environmentally friendly technologies for producing paper that keep pulp from being dumped in blue-ribbon trout streams, but make a ream of paper cost $15, or will businesses use environmentally damaging technology that reduces the price of a ream of paper to $3, but dumps paper wastes into streams and rivers? Will houses be produced at low cost with lumber harvested cheaply from big trees that displace spotted owls, or will they be produced with more expensive bricks, stone, and adobe while preserving spotted owl habitat?

Last, all societies must decide for **whom** the goods shall be produced. In other words, who will enjoy the benefits of the nation's products? To put it differently, this asks who gets the income needed to buy those produced commodities. Will forest product workers receive high wages and live in expensive houses in unpolluted environments, or will sawmill owners pay low wages to the workers, reserving high profits and a clean environment for themselves? Will laws be passed that do not protect the environment, thus reduce wages to environmental science students and increase incomes earned by lumberjacks and ranchers, or will we pass laws that require environmentally friendly technologies that increase salaries of biologists and reduce incomes for farmers and miners?

Importance of Good Decisions

Where natural resources are unusually scarce or where their demand is high, it is essential we make the right decisions because the cost of making mistakes is so high. For example, in desert climates like Albuquerque, Phoenix, or Las Vegas, water is scarce. Therefore, it is essential to design and apply sound plans for finding, storing, saving, and using water. Similarly, the cost of a failed effort to save a species from extinction is high—the species is forever lost. Finally, in nations where protein is scarce, such as much of Africa, or in situations of war or famine, policies that trade off cattle ranching with other activities, such as tourism, must be carefully thought out because the wrong policy may cause many people to starve.

What the Market Mechanism Does

The market is more than a place where goods are traded like the New York Stock Exchange or the local livestock or automobile auction. It is a process by which buyers and sellers of any good or service interact to determine its availability (quantity) and value (price). The driving force of the market is the notion that everything has its price: each commodity, each service, and each natural resource.

How does the market accommodate peoples' desire for more of something, which economists call increased demand? If people want to buy more of something, such as thick, juicy, beef steaks, a surge of new orders will be placed for them (Figure 4.3). As buyers scramble to buy more steaks, sellers raise the price because they can do it without losing customers. The higher price sends a signal for all to see. This signal lures ranchers to raise more beef cattle, packers to pack more steaks, and grocery stores to buy and stock more steaks on their scarce shelf space. The market process, through the price and profit signal, rewards people who respond to buyers' desire for more steaks by producing more steaks.

If people want less, the market process does just the opposite. If people want less chicken, they place fewer orders for chicken. Chicken sellers, who are anxious to get rid of their unsold chickens rather than have them spoil on the shelves, lower their price. They then produce and sell less chicken. Instead they use those scarce resources like chicken feed, chicken pens, chicken packinghouse space, and chicken shelf space for something else, such as beef production.

How the Market Solves the Three Problems

What Things Are Produced.
What things are produced is determined by consumers' dollar votes. If consumers want more housing, they vote with their dollars by buying more housing and less of something else. In the world of politics our votes lead

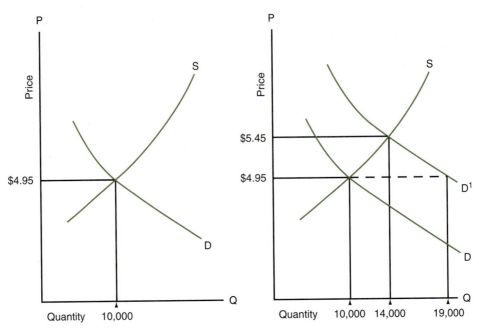

FIGURE 4.3   Market equilibrium for steaks before and after an increase in demand. As the demand for steaks increases (D to D¹), the price increases and stimulates an increased supply from 10,000 to 14,000. If the demand levels and the supply continues to increase to 19,000 steaks, the price will decrease to $4.95, the original price.

to action—they signal who will represent us and, indirectly, what actions we want our representatives to take. Dollar votes are in some ways similar to political votes. The size of dollar votes attracts added resources into producing more of what people desire.

How do those dollar votes attract resources to produce what people want? Business firms are enticed into producing goods in high demand by the higher potential to earn profit. Profit is what remains after a business subtracts its total cost from its total revenue. The opposite of profit is loss. Profit is the main reason businesses exist. It is the main measure of their success and lack of it is the main measure of their failure.

How Things Are Produced.   The way things are produced is determined by the competition among different producers who must all minimize costs to stay in business. All businesses compete for a limited number of customers and their dollar votes. Any business that fails to minimize its production costs will need to charge their customers higher prices than a competing business that successfully minimizes its costs. Therefore, businesses that fail to minimize costs lose business to competitors who do minimize costs.

Producers maximize profits by minimizing their costs. To keep costs at a minimum, they minimize the use of expensive inputs and maximize use of cheap inputs. Agriculture provides a good example of how this works in different parts of the world. In the United States, land is relatively plentiful and cheap and labor is scarce, so land prices are relatively low and wages are high. For American farmers, the low price of land and high labor price signal the use of laborsaving technology for cotton, wheat, and livestock production. Where labor is abundant and land is scarce, as in Asia, farmers find they can minimize their costs by producing cotton, wheat, and livestock with labor-intensive technology.

For Whom Things Are Produced.   In a market economy, goods are produced for people with income to buy them. The distribution of income in a market economy is determined by who owns productive resources (land, natural resources, labor, and capital) and how well the production meets the wants of consumers.

Each person's income in a market economy equals the price of the product or service he provides times the quantity he chooses to sell to the market. The distribution of income among people depends on the quantities of all goods or services they sell, and the prices of each unit. Ranchers who sell the most beef with the least cost will accumulate the most wealth. This is why ranchers oppose hikes in grazing fees and reductions in numbers of livestock that they can graze on public lands. It is also why cattle ranchers want you to eat as much beef as you can buy rather than chicken, pork, or fish.

Who Controls the Market?

No one person controls the market. The three bosses of the marketplace are consumers, technology, and profits. Consumers are directed by their natural or learned preferences. They vote with dollars for those preferences. TV marketers try to change your preferences so you want to buy what they have to sell. Jeans producers will try to convince you that if you buy their jeans, you will be sophisticated, better looking, and well-loved. They will be delighted if you alter your preferences and spend your dollar votes on their jeans.

Technology sets limits on what can be produced and at what cost. About 150 years ago, lonely, tired, cold, wet California gold miners wanted to talk to their families in the eastern United States. The lucky ones were willing to pay lots of gold dust for those intimate conversations. But until the telephone was invented and the lines stretched, they could not do it, no matter how high the demand. Demand must be backed by the technology to carry it out.

Profits reward private businesses who make and deliver the goods in highest demand to consumers at the least cost. Profits lure firms into areas where consumers want to buy more goods and to exit enterprises where consumers want less. They also reward businesses who use the cheapest methods of production.

The Invisible Hand of the Market

Adam Smith wrote a classic book in 1776 titled *The Wealth of Nations*. In it he emphasized the importance of self-interest as a human motivator. This fact has led some people to call economics the dismal science. Smith tells us that every individual uses his resources to produce the greatest value for himself. Smith's idea was that in selfishly pursuing only our own personal interests, we all are led as if by an invisible hand to accomplish the best good for everyone.

Translated, this means that producers who attempt to maximize their own profits, consumers who buy where prices are the lowest, and workers who go where wages are highest do a great service to society. Further, they do it more efficiently than government bureaucracies or social planners who would try to do the same thing. Profit has been criticized because it is associated with greed and self-interest. Still, profit incentives are in some ways like a well-oiled machine that can perform some amazing tasks for us.

For society, profit encourages businesses to produce what people want to buy, motivates them to produce using the least costly methods, and rewards them if they conserve on scarce resources. The profit signal tells businesses how to get rich and tells them when they are rich. In short, the profit motive, combined with free-market competition for customers, ensures that products are made at minimum costs and with maximum choices for us as consumers.

Can Profit Incentives Protect Natural Resources?

Thousands of farmers, oil companies, foresters, ranchers, wildlife managers, and others ask daily whether profit can protect natural resources. The answer depends on whether or not there are property rights (i.e., rights to own, use, buy, and trade natural resources) (see Anderson and Leal 1991). Where there are well-defined property rights to natural resources, the profit motive helps to protect private forests, private grazing lands, and private water. The following examples help explain why.

High timber prices encourage forest owners to plant more trees so they have something to harvest for the future. In places where wildlife can be privately owned (e.g., Texas), high tourist and hunter expenditures encourage landowners to invest in conservation of desirable wildlife.

Where water rights are secure and are marketable, high water prices encourage water right owners to use their scarce water carefully and/or to save it for possible sale to somebody else. Establishing an emergency drought water bank helped California get through a severe drought in the early 1990s with minimum hardship. California established high prices for farmers' water, if the farmers agreed to deposit it into a bank for resale to big cities. The California Water Banking Law included legal assurance that depositing water into the bank would not change farmers' permanent water right. This legislation was cheap, easy, and effective.

How Markets Settle Conservation Questions

By and large in the United States we accept free markets for allocating natural resources between the present and the future. There is little likelihood that this practice will be abandoned (Herfindahl 1961). We do not, for example, impose a tax on copper production in order to save copper for use by future generations. We do it to raise revenue to run the government. The question of the total quantity of minerals extracted over time is left to the mining company, the smelter, and the refiner. Likewise, the production and distribution between present and future generations of food buyers is largely in the hands of private farmers, grain distributors, food processors, and retail grocery stores.

The market settles conservation questions by discovering, renewing, producing, and selling natural resources over time to maximize their owners' current and future incomes. A private forest is a good example. The opportunity for current and future profit tells the private forester when to plant new trees, when to harvest old trees, and how long to wait for little trees to grow into big ones. Will these profit-maximizing decisions lead to depletion of their forests? If the private forest owner has secure property rights, the answer is a resounding "no." Private forest owners have a strong financial incentive to take care of their trees. Cutting down all their trees and selling them today without replanting means they will have nothing to sell in future years.

The profit system can also help provide for future national timber needs. If a private forest owner expects population to be growing and people to keep buying houses with wood frames, they can plan on high future demands for timber and, as a result, high future prices. So they have every incentive to plant more trees now, wait for more tree growth, and harvest fewer trees now and save more for the future when they can get a higher price. Their personal self-interest acts like an invisible hand in providing for future generations' timber needs.

When Are Markets Best at Promoting Conservation?

Markets are best at settling conservation questions in those situations when all potential resource users can effectively voice their demands in a monetary form. Take forests for example. If they were auctioned so that private bidders could effectively voice their demands for all forest products, including spotted owls, timber, livestock, water, and recreation, then the market system would make it profitable for businesses to produce the right mix of timber, spotted owls, and other forest products.

The following is a good general principle: When all resource users are able to voice their demands monetarily, prices of all resources are set by current and anticipated future demands for and supplies of those resources. When that happens, anybody who decides to use a particular resource at a particular time is faced with a price equal to the value of other uses displaced by that use. Then natural resources will be priced at their real cost, and private decisions will be made to use a resource now only when the public benefit of the use exceeds all public costs (Ward 1998). In that very special case, the market institution can be trusted to conserve natural resources.

Limits of the Market

There is one major limit to relying on the market system to settle conservation questions. Markets only account for private income produced by privately owned resources; they sometimes fail to account for all human benefits produced by natural resources. Many publicly owned resources, such as air and wildlife, are unpriced. The market system has little incentive to conserve them because it is unprofitable to do so. So the market system fails to conserve and therefore depletes or wastes any natural resource that is priced at less than its real cost.

Underpriced Resources.   One good example of the market system failing to conserve an underpriced resource is reflected in the price of coal used to produce electric power in the western United States. In an unregulated market, burning coal to produce electric power causes acid rain, which damages lake resources in the northeastern United States and in Canada (see Chapter 6). Damages to these resources are not included in the cost of the coal-fired electricity charged to customers in Phoenix and Las Vegas. So electricity buyers in those cities are not facing the full cost of their power. Where there are underpriced resources, unregulated markets fail to signal people with the real cost of their actions. This discourages conservation.

Spillover Benefits.   Spillover benefits are derived by society as a direct by-product of some economic trade between a business and its customer. An example of a spillover benefit is the pulp mill owner who considers spending money on controlling pollution from his pulp mill to reduce discharges into a blue-ribbon trout stream. However, he does not enjoy trout fishing and none of the anglers are likely to boycott his business. Even if the benefits to trout anglers and to society at large (e.g., people who provide angler services) exceed the costs to the mill owner, he will probably decide against the investment because he cannot make any money selling the cleaner water to anglers. Therefore, the mill fails to invest in pollution controls and the trout stream remains polluted.

Spillover Costs.   Spillover costs are a third example of where market incentives discourage conservation. Spillover costs are costs sustained by society as a direct by-product of some economic trade between a business and its customer. A good example is provided by the logging company that clear-cuts a patch of forest and is not charged for silting up the downstream river. The effect of the muddy stream is of no concern to the logger if he does not have to pay for the downstream damages. But these costs are of prime importance to downstream water users because they have to live with the siltier water or clean it up. The presence of spillover costs dull the potential power of market incentives to promote conservation.

In the case of either spillover benefits or spillover costs, these externalities do not enter into the market-pricing calculations of the parties undertaking the activity. So letting the market system settle conservation questions blocks conservation whenever natural resources are priced lower than their real cost. Put more simply, underpriced resources are inadequately conserved by the market system.

THE FUTURE

In future years, three problems will magnify the limitations of the unregulated market to conserve natural resources in the United States. One is the increase in population with its dramatic effect on the supply of water and open space. The other factor is the increasing per capita demand for many forms of outdoor recreation such as camping, hiking, and fishing. The third is the increased value people assign to endangered species.

HARNESSING THE POWER OF THE MARKET

Despite their limits, market forces have considerable power that can be harnessed to promote natural resource conservation (Anderson and Leal 1991). The idea is simple—business responds to profit opportunity. So making conservation profitable to business is a sound approach to natural resource management.

Concepts for Incentive-Based Pollution Control

By the 1960s, pollution problems in the United States were severe enough to warrant interest in finding ways to deal with them. One way of dealing with the pollution problem was environmental legislation, which reached a peak in the "green" decade of the 1970s. Politicians generally preferred to address environmental problems by passing laws that set limits on acceptable pollution. Their preferred method was "command and control" rules and regulations, such as design standards, which require firms to use prescribed technologies. Another example is performance standards, which specify the maximum amount of pollution that individual firms can dump into the environment.

Economists proposed that market incentives were a better way to control pollution than government regulations for two reasons. First, market forces could be harnessed to make pollution expensive to polluters. The effect of this is to both reduce pollution and make it profitable for businesses to search for production technologies that reduce pollution. Second, market forces would be used to lower the costs of pollution control by eliminating regulations and leaving decisions about the details of reducing pollution up to individual firms. The two most popular ways for harnessing the power of the market are *pollution taxes* and *marketable permit systems*.

Pollution Taxes

A pollution tax is levied on firms that discharge harmful waste such as raw sewage into environments such as streams, lakes, the atmosphere, or the soil. The tax is set equal to the damages the firm does. The power of the market is harnessed by setting the tax level according to the damage that the firm's pollution causes to society when dumped into the environment. A firm that emits a pollutant with low damages gets a low tax, but pollution sources that impose high costs to society are taxed more heavily. This is basically a resource use charge—just like paying to use the facilities in a national park.

An important advantage of pollution tax compared with rules and regulations is that the tax raises the costs of making the polluting product. Therefore, market forces make it profitable for the firm to invest in pollution control measures as long as the waste-dumping charge is more than the cost of controlling pollution. By investing in measures to reduce pollution, the firm reduces its costs of production because the pollution tax charge falls by more than the cost of installing pollution control.

Another advantage of pollution taxes is that they encourage firms to find the least-cost measure to reduce pollution. If there are several technologies for controlling emissions of raw sewage into a river, all of which reduce the pollution tax by X dollars, the firm has a profit incentive to find the cheapest pollution control technology. By contrast, regulations on pollution emissions specifying what kind of pollution control technologies must be used often are considerably more expensive because they do not give firms the freedom to search for ways to minimize the cost of compliance (Field 1994). With the market incentive, firms are encouraged to minimize their total pollution taxes and the cost of technologies required to reduce the taxes.

Pollution taxes also have advantages over fines commensurate with damages done. This is because fines typically involve court battles and presume some measure of guilt. Pollution taxes are treated like any other cost of production, and the firm will take steps to invest in pollution control that minimizes the tax.

Some of the better-known pollution taxes in the United States include the excise taxes on hazardous chemicals enacted in 1980 to fund the Environmental Protection Agency's superfund. A related reason for the superfund concept was that pollution problems often could not be assigned to a specific polluter. Similarly, after the 1989 *Exxon Valdez* oil spill, Congress imposed an added tax on petroleum products to pay for the Oil Spill Liability Trust Fund. One could argue that this added tax reflected the expected cost of ecological damages from future oil spills. More recently Congress has considered legislation to impose a carbon tax on fossil fuels like gasoline which, when burned, release carbon dioxide into the atmosphere, possibly increasing the greenhouse effect and contributing to global warming. However this approach has been rejected by President Bush and his administration.

Pollution taxes are not perfect. To make them an effective way for dealing with pollution, policymakers need to know the cost of the damages caused by the pollutant before setting the level of the tax. Otherwise, it would be easy to set the tax too low or too high. In fact, for several reasons it is difficult to know the cost of damages caused by the pollutant. First, it is hard to determine which activities (e.g., oil, chemical production) cause how much pollution. It is also hard to determine how much of the pollutant stays in the environment for how long, and how harmful the pollutant is to people who are exposed to it. It is also most difficult to determine the monetary cost of damage to health and premature death (Turner, Pearce, and Bateman 1993). Should those costs be based on lost lifetime income, lost days of work, or people's willingness to pay to avoid the pollution? This makes it difficult to determine what pollution taxes should be. Another problem with pollution taxes is that if one country imposes them on its own economy, they put their own firms at a disadvantage compared with foreign competitors.

In concept, taxes also could be levied to create market incentives for protecting endangered species and ecosystems. Once a tax is levied on further reduction of a species or ecosystem, a developer will be encouraged to restore habitat elsewhere, if the option exists and is cheaper. The basic problem remains, however—determining how much an endangered species or ecosystem is worth and what the taxes should be.

Marketable Waste Emission Permit Systems

The idea of marketable waste emission permits is simple: The government decides on how much waste the atmosphere or water body can assimilate safely (i.e., is nonpolluting) then issues waste emission permits (quotas) for no more than that amount (Figure 4.4).

This method of preventing pollution harnesses market forces by allowing the emission permits to be tradeable—that is, they can be bought and sold in the open market. Emission permits are especially attractive for many reasons (Pearce and Turner 1990): They are cheaper than command and control regulations, they allow for firms entering or leaving the area, they can be bought by environmental groups, they adapt to inflation or other changes in values, they adapt to geography, and they have low administrative costs.

Lower Cost.   Marketable permits let those who emit wastes (emittors) trade permits for cash. So emittors with low costs for abatement will find it cheaper to abate their emissions rather than buying permits. However, emittors whose technology is locked in and who find abatement very expensive will find it cheaper to buy the marketable permits from somebody. By giving emittors a chance to trade cash for permits on the open market, the total cost of preventing pollution is reduced compared with the regulatory approach of setting standards.

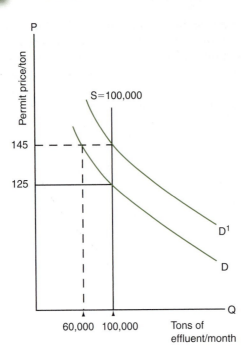

FIGURE 4.4 Supply and demand for marketable effluent permits, before and after new firms move into an area.

Entering or Leaving the Area.   Suppose we are dealing with a high-growth area like southern California, and new emittors move in. The effect of the new emittors coming to town is to increase the demand for permits. If the total capacity of the environment to assimilate wastes is unchanged, the authorities may wish to keep the total number of permits fixed to allow 100,000 tons per month (Figure 4.4). If new emittors want to come to town and do business, they will invest in the cheapest combination of buying new permits and waste emission abatement. Similarly, if existing emittors decide to leave the area, their waste emission permits can be marketed to others. If, for example, the government discovered that the pollution standard needed tightening up to 60,000 tons per month (Figure 4.4) they could buy some of the permits themselves, thereby taking permits out of the market. So the permit system makes it easy for the government and for individual waste emittors to adapt to changes in scientific knowledge or economic need.

Environmentally Friendly.   An open market in waste emission permits allows anybody to buy them including groups like environmentalists, sportsmen, or departments of game and fish. This option is most useful when a difference of opinion exists about the appropriateness of the standards. By buying permits a group could use the market to apply its own standards. If an environmental group such as the Sierra Club wanted to lower waste emission in the area, they could enter the market and buy some of the permits, keeping them out of the market for as long as they wished. If the group was a private fishing club and the permit was to limit waste emission that impaired a water body's capacity to support the most sensitive fish species, they could allocate some of their annual operating budget to buy permits, thus saving the need to stock large numbers of fish in poor-quality waters.

Adapts to Economic Changes.   In periods of high inflation or rapid changes in technology in the polluting industries, the initial allocation of permits may change in value or need to be changed to adapt to new technology. Because the price of the permits automatically adjusts to changes in supply or demand, these outside changes can be handled efficiently and quickly.

Geographical and Spatial Flexibility. The region in question may have complex dispersions of wastes, which may vary considerably with weather patterns, time of year, and the kind of firms that buy the permits. In principle, these conditions can all be handled under a permit system. Different quantities of permits could be issued for different seasons of the year, different weather patterns, different types of wastes, and more, and each kind of permit could operate in a fairly independent market. As long as the markets were open, flexible prices would clear the market and the costs of abating waste emission could be minimized.

Low Administrative Costs. Under a pollution tax system, the authority must estimate the cost of pollution damages and the costs of various abatement technologies. However, the authority can easily make mistakes that have high economic and political costs. Under a permit system, the authority does not need to estimate the cost of complying, but they need to know how many permits to initially issue. The waste emitting industries themselves are the only ones who need to know the details about the cost of abatement. They can buy the permits at the going price if abatement is expensive, and invest in abatement if the permits are more expensive.

APPLICATIONS OF
INCENTIVE-BASED REGULATIONS

Pollution Taxes

Solid Waste. The main U.S. experience with real-world pollution taxes has been in the arena of ordinary garbage disposal. In Highbridge, New Jersey, the city altered its flat fee for garbage hauling of $280 per year with a price that increased with the amount of trash picked up by haulers (Hanley 1988). Each 30-gallon can set out at a curbside each week had to have a sticker before it would be hauled away. Each household could buy 52 stickers for $140, about $2.75 per sticker. The effect of the experiment in raising the price of garbage hauling was to reduce the amount of trash put outside people's houses by 25%. Because the cost of disposing of large recyclable items like used furniture could use four or more stickers, more trash was recycled than before.

Water Pollution. Europe has more experience with pollution taxes than the United States According to Hahn (1989), Germany, France, and Holland have all set charges for various water pollutants to compensate for their social costs. The effect of the taxes has been twofold—to encourage pollution reduction and to help pay for the cost of cleaning up pollution. France (1960s) and Germany (1970s) have both enacted emission charges on water pollution. These charges seem to be accepted by the French people as a way of doing business to improve water quality. In Germany, the charges seem to be improving water quality (Hahn 1989). In Holland, the effect of water pollution taxes (administered on both volume and concentration of pollutants) has been to cause a slow but steady increase in water quality (Bressers 1983).

Environmental Adders for Electric Power. Electrical power generation makes use of unpriced environmental resources. However, it is difficult to attach monetary values to these damages. So some have suggested adding taxes to the price of electric power, called "adders." However, these taxes are used only for selected electric-power capacity expansion projects and are not actually charged to the consumer. The advantage of using adders is to promote cleaner electric-power sources through design and selection. Utility regulators in more than half the states are either using adders or are considering using them. However, because the monetary value selected for the adder can be arbitrary and because the consumer never actually pays for it, the debate on the use of adders is far from over and is likely to continue (U.S. Department of Energy 1995).

Marketable Permit Systems

The United States has had some experience with marketable permit systems since the mid-1970s. Amendments to the 1990 Clean Air Act have many features designed to clean up air pollution as efficiently and cheaply as possible. They allow businesses to make choices on the best way to reach pollution cleanup goals. Because businesses must make a profit, they have an incentive to clean up their pollution at the lowest cost possible. These new flexible programs are called market or market-based approaches. For example, the acid-rain cleanup program offers businesses choices as to how they reach their waste reduction goals and includes waste allowances that can be traded, bought, and sold. Consider the following examples of marketable permit systems.

Lead Banking. Lead causes developmental problems in children and heart disease and strokes in adults. Lead in gasoline is dispersed into the atmosphere when burned by cars. In 1985, the Environmental Protection Agency (EPA) set up a program that required lower lead standards for leaded gasoline. To reduce the very high cost to industry of complying with these standards, EPA set up a lead banking program (Hahn 1989). Under the banking program, gasoline refiners who were able to reduce the lead content in their gasoline by more than the legally required standard for a given period of time were allowed to bank the difference in the form of "lead credits." These credits could then be used or sold in any following time period up to 1987, at which time all refiners had to meet the new standard. This proved to be a successful program. More than half of refiners participated in the banking program. Hahn found that the cost savings from the program may have exceeded $228 million. The costs were saved by allowing old refineries, for whom complying was expensive, to trade banked credits with more modern refineries who could comply at a lower cost. Requiring all refineries to meet the standards, without allowance for the differential cost of compliance, would have been considerably more expensive.

Emissions Trading. The EPA has been experimenting with emissions trading since 1976 for five air pollutants. These include volatile organic compounds (VOCs), sulfur dioxide, carbon monoxide, nitrogen oxides, and particulates. The EPA allows credits to be earned whenever a polluting source reduces emissions by more than the legally required amount.

Sulfur Dioxide Allowance Trading. The 1990 Clean Air Amendments set up an ambitious system of tradeable permits for emissions of sulfur dioxide (SO_2), a chemical that can cause acid rain. Acid rain is formed when SO_2 and nitrogen oxide, released when fuel is burned, are transformed into sulfuric and nitric acids. These acids ultimately return to the earth dissolved in rain. Scientific studies over the last 20 years have shown that acid rain causes increased levels of acid in lakes and streams. Scientists suspect, but do not know for sure, that these acids may also damage forests. The acids can also erode buildings, bridges, and statues.

In establishing the Clean Air Amendments, Congress wanted to reduce SO_2 emissions by 10 million tons from 1980 levels. The EPA harnessed the power of the market by allowing sulfur dioxide allowances to be traded. Namely, the holders of allowances for SO_2 emissions—mostly electric utilities east of the Mississippi River—were allowed to transfer their permits among one another. Therefore, utilities who could reduce their emissions at the least cost had an incentive to do so. They were also able to sell their unused allowances to utilities who would save money by purchasing the allowance rather than reducing emissions. Allowances can also be banked for later use. The result of this banking plan was that the incremental cost of reducing SO_2 emissions was made equal across all sources. Making these incremental costs equal across all pollution sources has the effect of achieving total SO_2 abatement at minimum total cost. Utilities can also offer SO_2 emission allowances for sale at an annual government-sponsored auction.

This SO_2 allowance trading program has turned out to be quite successful. Desired reductions in emissions have been accomplished and even exceeded. More important, total SO_2 abatement costs have been much less than what they would have been without the trading provisions. Stavins (1998) estimates that tradeable permits for SO_2 have resulted in a cost savings of nearly $1 billion per year over the regulatory alternatives that Congress had considered in earlier years.

GOVERNMENTS AND CONSERVATION

How Government Settles Conservation Questions

When government gets involved in conservation, the instruments for resolving natural resource conflicts enter the political arena. Participants in this arena include elected officials, interested voters, government agency employees, lobbyists, the press, environmentalists, agriculture, mining, and more. Obviously some sort of resolution comes out of this process, but it is not always clear if the results promote better conservation and human welfare.

Limits of Government In Settling Conservation Questions

Despite the good intentions behind attempts to set up government conservation policies and plans, four kinds of "government failures" stand in the way of conservation. These failures include rent seeking, taxpayer ignorance, empire building, and abuses of economic analysis.

Rent Seeking

Rent seeking occurs when special interests use their private resources in lobbying and other activities aimed at getting government to pass laws that economically benefit the group. Successful rent seeking (i.e., getting the group's favorite laws passed) increases net benefits for the special interest group, typically at a cost to the larger society. An example with a long history is the iron triangle, a group who sponsored and completed Western water projects. This triangle consisted of Western politicians, irrigated agriculture in connection with banks and real estate, and the federal water management agencies.

Taxpayer Ignorance

Taxpayer ignorance occurs because we find it more bothersome than beneficial to stay informed about issues that do not affect us directly. For example, New York taxpayers have little interest in investing the time and money to become informed on proposed legislation that would develop a new irrigation reservoir on the Colorado River. This is because the reservoir costs the New York taxpayer only pennies per person. It simply is not worth their time to scrutinize public actions that have little direct bearing on them. Many federal water projects were built that had little chance of promoting the national interest simply because the local beneficiaries wanted the legislation that paid for the project and the rest of the nation simply did not care enough.

Empire Building

Empire building involves a government agency purposely increasing its own power and influence for its own sake. Empire building, especially when combined with a bureaucratic desire for job security, can impede programs aimed at conservation. Few government agencies or employees can resist the temptation to increase their own influence or power. They often do too much of it and for too long.

Government gets away with building empires because there is no market check to stop them. Private business loses money if people are not willing to pay for the products produced by their expansion. The owner of a bicycle shop will not open new stores unless there is enough customer demand for bikes to support them. However, if the government builds too many dams or drains too many swamps, there are no negative profit signals that require them to stop building and draining. The only support needed for government projects is a legislative majority, which can be bought by the small minority by financing political campaigns of key legislators.

Once initiated, government programs and agencies can take on a life of their own. The same agencies that made mistakes based on scientific ignorance were "rewarded" with a chance to correct those mistakes. After World War II, for example, the U.S. Congress gave the Army Corps of Engineers and Natural Resources Conservation Service (formerly the Soil Conservation Service) large budgets to drain and fill wetlands. Beginning in the 1980s, these agencies received budgets from Congress to create or restore wetlands. This problem typically starts in social impatience with careful research and planning, which often is looked upon by special interests as costly "foot dragging."

Abuses of Economic Analysis

The final government failure comes from agencies that ignore the economic consequences of their decisions. Sometimes a government agency chooses to ignore important costs or benefits because it is in the business of promoting one type of resource use (Herfindahl 1961). In fact, stimulating use of the resource may be implied in the agency's mission statement. Beginning in the early twentieth century, hundreds of dams were built in the western United States by the Bureau of Reclamation and Corps of Engineers whose missions were, respectively, irrigation development and flood protection.

Certain costs or benefits were ignored in major decisions because they could not be quantified in monetary terms or could be described only in vague nonquantitative terms. Several dams built in the 1950s and 1960s were subjected to poor benefit-cost analyses that assigned high values to lake recreation behind a dam while presuming that riverine recreation and wildlife habitat displaced by the dam had zero value, simply because those values were hard to measure.

Government programs promote conservation only if benefits and costs are properly measured and used. Only then is the public protected against rent seeking by special interests, taxpayer ignorance, and empire building. Nevertheless, economic analysis is expensive and often difficult to understand. It is seen by some as unreliable and often is unpopular with politicians. In summary, government regulations and public resource management have the potential to promote more inferior conservation plans than the imperfect market system whenever proposed public actions are not based on objective economic analysis.

What Government Can Do to Protect Natural Resources

Benefit-Cost Analysis (BCA) is an economic tool for comparing in monetary terms the desirable and undesirable impacts of proposed natural resource policies where desirability is measured in human welfare. It is a method for ranking the economic performance of natural resource projects, policies, and programs in which impacts are measured in nontechnical terms and estimated by scientific methods. It is a way to compare, in common units, all the gains and losses to people resulting from some action. BCA is required by a number of laws, but far from all.

BCA organizes information in a way that promotes the conduct of rational policy analysis. Rational policy analysis considers all the relevant alternatives and identifies and evaluates all the consequences that would follow from the adoption of each alternative. It then selects the alternative that would be preferable in terms of maximizing society's overall welfare.

In the Flood Control Act of 1936, Congress required that the gross benefits of water development projects, "to whomsoever they may accrue," should exceed all costs. Since that time, BCA has been used widely to decide whether any federal water project should be started. In more recent years BCA has been used to help make three kinds of public natural resource decisions: (1) a simple ranking of the comparative benefits of several possible actions, (2) the optimal size or scale of a project produced by a decision, and (3) the optimal timing or sequencing of several elements of a decision.

BCA uses a simple decision rule. If, for some proposed plan, the sum of its human benefits exceeds the sum of the human costs by a larger amount than any other proposed action with the same aim, the plan with the greatest gross benefit for the cost should be adopted. Otherwise it should not. This assumes that all people are treated equally, so that an additional dollar of benefit accruing to a rich person is valued the same as a dollar of cost paid by a poor person. An advantage of BCA is that the dollar, as a unit of measure for ratings, is easily understood by everyone.

For the concept of revenue to the private firm, BCA substitutes the notion of benefit to society. For the cost of the private firm, BCA substitutes the concept of opportunity cost—or the value of benefits displaced when resources are taken from other economic activities to support the proposed plan in question. For the profit of the business firm, BCA substitutes the concept of net benefit, which means the difference between gross benefit and cost.

BCA values all natural resource services in the common denominator—money. Consider the huge range of services produced by natural resource decisions: electric power, water quantity, water quality, critical habitat for endangered species, timber, forage for cattle and wildlife, recreation, fish, minerals, and others. By expressing all these outputs in a common denominator, it allows all natural resource policy decisions to be compared on a level playing field. Although most natural resource benefits can be valued in monetary terms, some values remain resistant to monetization.

Whenever choices can be made among project outputs of equal perceived value, the least-cost alternative is preferred. For example, if the fate of an endangered species can be improved the investment that returns the greatest improvement for the cost (e.g., number of individuals in the population) is preferred.

Why Measure Benefits and Costs?

Why spend all the effort and money necessary to conduct a benefit-cost analysis? Why not use plain honest-to-goodness profit and loss accounting used by any business?

One answer is that what counts as a benefit or a loss to one part of the nation—to one or more persons or groups—does not necessarily count as a benefit or loss to the whole nation. Therefore, with BCA, we are concerned with the economic welfare of the nation as a whole rather than some small part of it. BCA for natural resource decisions asks the same question that private companies ask of their accountants. However, instead of asking whether the owners of a private enterprise will be made better off by the company's proposed decision, BCA asks whether society as a whole will be made better off by undertaking the action.

BCA helps make better management decisions by using a time-tested economic framework for organizing economic data. In natural resource management, there are plenty of ways governments can make bad decisions. Management by tradition is a widely applied method for making poor decisions. Resource agencies pursue activities, such as stocking X pounds of trout at a Y water or leasing Z board feet of timber on some national forest, simply because that is what they have done for years. Without information on benefits and costs, resource agencies often have no choice but to manage by tradition.

Another way to make bad decisions is through fear of change. In his small but mighty book entitled *The Prince,* the fifteenth-century Italian writer Nicolo Machiavelli warned us there is nothing more difficult to take in hand, more perilous to conduct, or more uncertain of its success than to take the lead in the introduction of a new order of things. This is because the innovator faces as enemies all who have done well under the old conditions, and lukewarm defenders in those who may do well under the new. While change is resisted by virtually everybody, BCA gives a quantitative way to decide which changes are worthwhile by government and which ones should be left to the private sector or screened out altogether.

Poor conservation decisions are promoted by the necessity of government resource agencies to appease special interests. These special interests abound. In the western United States, for example, Congress sets prices for federal timber, water, energy, and livestock forage below market levels because powerful special interests (rent seekers) benefit economically, not because these low prices are good economic decisions for the nation as a whole.

Finally, unsound decisions happen when public resource managers' personal biases are allowed to influence decisions. America is "a nation of laws and not men." In the realm of natural resource policy this means we do not want some local Forest Service official deciding the mix of grazing and recreation on federal lands based on personal biases or political connections. The decision should be made through the consistent and objective application of time-tested economic and environmental principles. These principles should be set up at the national level and applied to resources and resource issues at the local level. Because not all environmental resources are readily expressed in monetary terms, economic principles have to be combined with environmental standards based on sound science and clear public choices.

HOW TO MEASURE BENEFITS AND COSTS

Basic Principles

Willingness to Pay. In principle, *the benefit of any natural resource conservation decision is measured by what people are willing to pay for it. Costs are benefits displaced by the decision.* These concepts apply to both goods and services that are sold in markets and those that are not.

Consider the case of market goods. Suppose the U.S. Secretary of Interior proposes limiting logging in the Southwest to produce a more critical habitat for the Mexican spotted owl. The benefit lost (cost) from such a policy is the timber left unharvested. Timber is sold in markets. So the total cost is the sum of prices times quantities of timber displaced by the decision.

As another example, consider the example of a nonmarket good in which Congress considers passing a law that requires shipbuilders to build double-hulled oil tankers in order to reduce future damages from oil spills. The benefit is the value of fish not killed, beaches not fouled by spilled oil, marine wildlife not killed, and ecosystems not damaged. While the concept applies to both market and nonmarket goods, for the latter, prices typically must be estimated by some indirect method. This might involve contingent valuation surveys, in which people are asked directly what the resource is worth to them.

Scarcity, Demand, and Benefits. Both market and nonmarket goods share an important characteristic. When they are scarce and in high demand, they are in great value, and the benefits of policies that supply more of them are quite high. For example, a policy that brings water (which is scarce) to growing cities in the desert to produce high-quality drinking water (which is in high demand) is likely to produce great benefits. Similarly, the policy that brings water to a humid, cool region like Minnesota (where water is abundant) to produce habitat for an obscure species (which is in low demand) is likely to produce negative human benefits.

How Benefit and Cost Information Should Be Used. Benefits and costs are measured and reported to facilitate sound conservation decisions. By sound, we mean the decisions are worth more to a nation than they cost. There are many wrong ways to use information about benefits and costs, but only one right way.

Implementation Principles

An Almost Right Way. Requiring the ratio of total benefits to total costs to exceed 1.0 is almost right. According to that rule, resources should be committed to support a conservation decision if the ratio of the present value of benefits to the present value of costs exceeds 1.0. This method is almost right but not quite. It is not quite right because it is too conservative. Only having the benefit-cost ratio equal 1.0 lets one implement decisions for which benefits are no greater than costs. A private business cannot stay in business if its revenues do not exceed its costs by a safe margin. For public natural resource decisions, benefits should exceed costs by the greatest amount possible.

The term *benefit-cost ratio* is widely heard in policy circles such as in Washington, D.C. It is easily understood by Congress and the press because it is expressed as dollars of benefit per dollar of cost. When Congress and other decision makers enact laws or make appropriations for their projects, they want to know if the benefitcost ratio exceeds 1.0. Once the ratio exceeds 1.0, these decision makers may consider the economics to be acceptable and then use its political attractiveness to make the final decision.

The Right Way. The *maximum net present value* (MNPV) gauge is the one right way to use information on benefits and costs. According to this rule, natural resource conservation should be carried out to maximize the present value of net benefits. If carried out, the MNPV rule will produce the decision that maximizes the economic performance of the available resources. With the MNPV decision rule, only plans with net present value greater than zero are accepted as economically performing. If a management decision has a negative net present value, society gives up more than it gets over the life of the action, or society is paying more for the plan than it is worth.

Implementation Practices, Incremental Benefits, and Costs

The incremental (marginal) benefits of a proposed decision should always be compared to its incremental cost. How does marginal benefit translate into the natural resource management decision process? Depending on restrictions imposed by law or custom, managers and policy makers with a goal of economic efficiency emphasize those products generating the greatest additional public benefit for the added cost of management.

If additional logging resulted in less additional benefit than the added cost of implementing the logging, efficient actions would make other choices, such as outdoor recreation, livestock grazing, flood reduction, or improved functioning of ecosystems to support clean water, clean air, and biodiversity. As long as the gain in benefits exceeds the added costs compared with logging, other actions should be selected over more logging. Managers can increase total economic benefits by choosing policies for which the marginal benefits exceed marginal costs.

Measurement Problems

Unfortunately, all of the benefits and costs generated by management are not equally easy to measure. Some benefits cannot be accurately estimated given the present state of economic knowledge. For example, the benefits from restoring ecosystems and recovering most endangered species cannot yet be accurately estimated. Also, the real cost of managing for timber, grazing, recreation, and flood reduction may not be estimated because the negative side effects on other ecosystem services are not fully known. That does not mean ecosystems and endangered species are worth zero or worth whatever they cost to produce. The continuation of the Endangered Species

Act, along with a majority of public support, suggests its value to society is at least as great as its costs. Thus, in the political arena, the willingness to pay for the benefits of nonpriced goods and services is negotiated through the political process and legislation. The legislative decision about how much to regulate use or to allocate to non-monetized benefits is harder than it might be with some reliably monetized benefit.

SCOPE AND LIMITS OF BCA

The Bad News

BCA is difficult for most people to understand. Most politicians and other decision makers do not really understand it, nor does the average voter. It is technical and requires a good deal of economic theory. So people who read a BCA report either must accept the numbers on faith or ignore them.

Measurement problems plague BCA. Dollar values of things like endangered species habitats are hard to measure because one cannot buy them at the downtown grocery store. So the accuracy of monetary values of natural resource programs is questionable when the natural resource is not traded in commercial markets. Many of those nonmarket benefits require a contingent valuation analysis. Here people are asked what they would be willing to pay for the service even though such payment is rarely required and people have no previous payment experience. For that reason many decision makers do not believe some contingent valuation numbers. Therefore, BCA often ignores most important parts of government projects, such as saving wildlife habitat or wilderness land, because these resources are not traded in markets. So, unless a lot is spent on estimating benefits of the decision, the most important values are not necessarily included in the BCA.

Ethical problems also plague BCA. It may be morally wrong to use BCA to evaluate policies such as the Clean Air Act, which restricts technologies used by coal companies. How can the added length of miners' lives be valued in cold money terms? When a politician or other decision maker sees dollar values placed on such a proposed piece of legislation, it may confuse rather than enlighten. Being fair to future generations is another ethical issue poorly handled by BCA. Valuing future benefits requires using a discount rate to deflate future benefits so they can be expressed in a common denominator with the current generations' benefits. Any discount rate greater than zero may be unfair because it assigns a lower value to future generations than to current generations.

Some say that BCA is unfair because it ignores the distribution of benefits and costs. It only measures total benefits and total costs of an action. A proposed policy that does well on a BCA may be consistent with an economic arrangement that makes the rich richer and the poor poorer.

Finally, government agencies who do their own BCAs are like foxes guarding the hen house. These agencies have every incentive to cheat by inflating the benefits of their programs and ignoring costs. By inflating their own program's benefits, they may get a bigger budget in the future. Federal water construction agencies such as the Army Corps of Engineers and Bureau of Reclamation have been accused of this practice in past years.

The Good News

BCA is an objective way to judge proposed policies by connecting economic principles to policy actions. It is based on an established body of economic theory that has been scrutinized, debated, and improved by many economists over the past 100 years. It attempts to measure the values of all the people who benefit from and who pay for a government policy decision, and not just a select few. Therefore, it is democratic and not elitist. By valuing impacts in dollars, it measures impacts in units that are understandable to both decision makers and the average adult. Other policy evaluation measures use some sort of indices, which typically are arbitrary. There is a vast body of literature on applications of BCA to various natural resource conservation issues.

Regardless of how BCA information is abused, the information allows the public, government employees, and elected officials to separate truly inferior alternatives from ones that have some merit. It can also be used to help isolate economically weak features of proposed plans that should be dropped so that the stronger ones can be carried out.

Summary of BCA

The use of BCA permits government to use economic information to design better natural resource conservation proposals, defined as those for which the economic benefits exceed the costs. It can play an important role in legislative and regulatory policy debates on how natural resources should be managed. Although formal BCA should not be viewed as either required or sufficient by itself for designing good public policy, it can provide a powerful framework for consistently organizing diverse information. In this way it can greatly improve the process and therefore, the outcome of natural resource policy analysis (Arrow et al. 1996). Moreover, even if the data available for a BCA are poor, the steps required to carry it out force analysts to raise the right questions. It is more important to measure the right thing in a crude sort of way (doing the right thing) than to measure the wrong thing with impressive refinement (doing the thing right).

CONCLUSION

Economics gives us insight about and better understanding of human behavior. Through that understanding we can design more efficient and equitable private and government programs for natural resource management. By influencing that behavior, particularly with economic incentives, we can better meet the needs of current and future generations of people. Economics also informs us about how we can better allocate scarce resources among people competing for them. If you think a career in natural resource conservation may be for you, we believe the investment you make in economics courses will produce a high return (high BCA). For further reading on natural resource economics, the authors recommend Anderson and Leal (1991), Field (1994), Tietenberg (1996), and Schiller (2000).

LITERATURE CITED

Anderson, J. L. and D. L. Leal. 1991. *Free market environmentalism*. Boulder, CO: Westview Press.

Arrow, Kenneth J., Maureen L. Cropper, George C. Eads, Robert W. Hahn, Lester B. Lave, Roger G. Noll, Paul R. Portney, Milton Russell, Richard Schmalensee, V. Kerry Smith, and Robert N. Stavins. 1996. Is there a role for benefit-cost analysis in environmental, health, and safety regulation? *Science* 272 (April 12):221–222.

Bressers, J. 1983. *The effectiveness of Dutch water quality policy*. Universiteit Twente, The Netherlands.

Field, Barry C. 1994. *Environmental economics: An introduction*. New York: McGraw-Hill.

Forbes, S. A. 1888. *The lake as a microcosm*. Bulletin Peoria Scientific Association 1888:77–87.

Goodstein, Eban S. 1998. *Economics and the environment*. 2nd ed. Upper Saddle River, NJ: Prentice-Hall.

Hahn, R. W. 1989. Economic prescriptions for environmental problems: How the patient followed the doctor's orders. *Journal of Economic Perspectives* 3(2):95–114.

Hall, D. C., 1998. *Public choice and water rate design.* Department of Economics, California State University, Long Beach, prepared for the workshop on the political economy of water pricing implementation. The World Bank, November 3–5.

Hanley, R. 1988. Pay by bag trash disposal really pays, town learns. *New York Times,* November 24, p. B1.

Haspel, A. E. and F. R. Johnson. 1982. Multiple destination trip bias in recreation benefit estimation. *Land Economics* 58:361–372.

Herfindahl, O. 1961. What is conservation? (pp. 1–12). In *Three studies in mineral economics, resources for the future.* Washington, DC: U.S. Government Printing Office.

Howe, C. W., D. R. Schurmeire, and W. D. Shaw. 1986. Innovative approaches to water allocation: The potential for water markets. *Water Resources Research* 22(4): 439–445.

Kahn, James R. 1997. *The economic approach to environmental and natural resources.* 2nd ed. Orlando, FL: The Dryden Press.

Krutilla, John. 1967. Conservation reconsidered. *American Economic Review* (September):777–786.

Machiavelli, Nicolo. 1505. *The prince.* Florence, Italy.

Montgomery, C. and G. M. Brown. 1992. Economics of species preservation: The spotted owl case. *Contemporary Policy Issues* 10:1–12.

Pearce, David W. and R. Kerry Turner. 1990. *Economics of natural resources and the environment.* Baltimore, MD: The Johns Hopkins University Press.

Pinchot, G. 1947. *Breaking new ground.* New York: Harcourt, Brace and Co.

Prato, T. 1998. *Natural resource and environmental economics.* Ames, IA: Iowa State University Press.

Samuelson, P. and W. D. Nordhaus. 1985. *Economics.* 12th ed. New York: McGraw-Hill.

Schiller, B. S. 2000. *The economy today.* 8th ed. New York: Random House.

Smith, A. 1776. *The wealth of nations.* London, England.

Stavins, R. N. 1998. What can we learn from the grand policy experiment? Lessons from SO_2 allowance trading. *Journal of Economic Perspectives* 12(3):69–88.

Tietenberg, T. 1996. *Environmental and natural resource economics.* 4th ed. New York: HarperCollins.

Turner, R. Kerry, D. Pearce, and I. Bateman. 1993. *Environmental economics: An elementary introduction.* Baltimore, MD: The Johns Hopkins University Press.

U.S. Department of Energy. 1995. *Electricity generation and environmental externalities: Case studies.* Energy Information Administration, Office of Coal, Nuclear, Electric and Alternate Fuels, Coal and Electric Analysis Branch, DOE/EIA-0598, Distribution Category UC-950, September.

Van Kooten, G. Cornelis. 1993. *Land resource economics and sustainable development: Economic policies and the common good.* Vancouver, BC: University of British Columbia Press.

Wantrup, S. V. C. 1968. *Conservation: Economics and policies.* Berkeley, CA: University of California Agricultural Experiment Station.

Ward, Frank A. 1998. Economics of water conservation. *New Mexico Journal of Science* 38:127–139.

CHAPTER 5

Planning, Policy, and Administration

ELEMENTS OF ORGANIZATIONAL FUNCTION

Managing Problems and Opportunities

Chances are you have already encountered a problem today that you somehow managed through planning and action. In our daily routines we encounter many small problems that we either solve quickly, set aside, or ignore. Sleeping too late, for example, requires problem identification (I could be late!), searching for solutions (What choices do I have?), choosing a solution (skipping breakfast or making up an excuse), applying the solution (going hungry), and evaluating the result (Was I on time?). These are the basic elements in all planning and implementation of solutions to problems (Figure 5.1).

In identifying the problem and taking stock, we often forecast conditions and consequences (What if the traffic is bad?). In determining a course of action, we usually weigh alternatives available to us and pick the best one for our circumstances (skip shower, breakfast, or dog walking). A thorough evaluation may lead to rethinking how we wake ourselves up. This could lead to a new personal policy—a procedural rule—for a new wake-up routine (set two alarm clocks). Permanent adjustments of our personal affairs will probably require careful administration (i.e., facilitating more effective activities).

Organizational Integration

Planning, policy, and administration are complementary elements in problem identification and management and are essential for integrating organization functions to achieve desired results (Figure 5.2). Natural resource management organizations have to identify problems and the means for solving them before acting and evaluating the results. This is true for any natural resource organization, whether local, state, or federal agency; for-profit business; environmental advocacy group; educational or research institution; or other type of nongovernment organization (NGO). The care taken in this integrated process of planning and plan implementation can greatly improve management effectiveness. Management can be defined simply as the effective

FIGURE 5.1   The basic elements of management: problem and opportunity identification, analyzing for resolution, implementation, choosing an alternative, and evaluation.

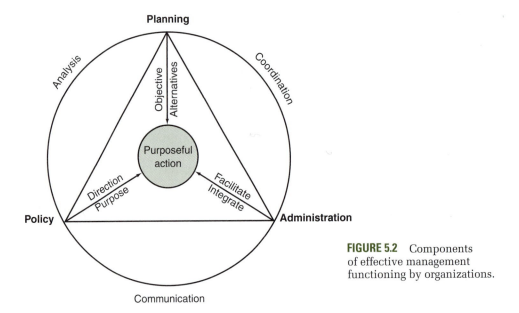

FIGURE 5.2   Components of effective management functioning by organizations.

implementation of plans. Planning is the process of identifying problems or issues and their possible solutions, then mapping out actions and evaluating results. Policy is guidance developed by an organization to carry out plans. Administration facilitates coordination of actions into an organizational integrity that serves a purpose held in common by all members of the organization.

Planning, policy, and administration determine organizational effectiveness in accomplishing purpose from the natural, human, and technical resources on hand. All members of organizations have roles and stakes in the outcome from organizational effort. Organizational achievement follows from the coordinated achievement of individuals. While styles of management are diverse, all organizations need to coordinate the planning and actions of individual members into an effective organizational integrity if they are to efficiently solve their problems and take advantage of important opportunities. An important function of administration is to identify co-ordination needs for management planning and plan implementation.

Contemporary natural resources management concepts have roots in science and its applications. Management starts with identification of problems, establishes scientific alternatives for problem solution, and analysis for the most cost-effective solution. Planning and policy analyses parallel the scientific method, which emphasizes testing of alternatives (hypotheses), evaluating results, drawing conclusions, and integrating them into scientific knowledge. *Adaptive management* is an approach to natural resource management that promotes continuous scientific evaluation of management effectiveness and continual adaptation of management to conditions as new knowledge is gained (Walters 1986).

Excellent communication skills are a prerequisite to linking personal planning with others in an organizational planning process. The need for strong communication skills grows as the complexity of organizational tasks grows and involves more people. In such situations, the planning process often becomes a more public and formal organizational effort to keep management from dissolving into disarray or chaos. Effective organizational planning integrates personal planning into a coherent institutional process directed toward accomplishing organizational purpose.

MANAGEMENT SYSTEMS

Organizations as Management Systems

Each organization forms a planning and implementation system that interacts with other management systems in environments held in common and around purposes of mutual interest (Lorange and Vancil 1977). Management systems also overlap and interact with ecosystems and economic systems to redirect their processes toward planned outputs and outcomes (Figure 5.3). While many specific approaches are taken to management, the same four basic elements of a planning and implementation cycle are common to all effective management systems (Figure 5.4): (1) inventorying the planning environment; (2) focusing management intents through mission, goals, and objectives; (3) allocating organizational resources to appropriate operations (implementation); and (4) evaluating outcomes (Lorange and Vancil 1977, Crowe 1983, Koteen 1989). Planning requires an *inventory* of existing conditions in the management environment, including the resources at hand to do the job and those stakeholders who will be affected by it. Inventory often includes *forecasting* future events, based on trend analysis or other information, and developing different

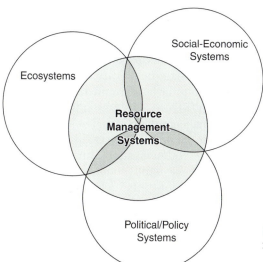

FIGURE 5.3 The basic components of natural resource management systems.

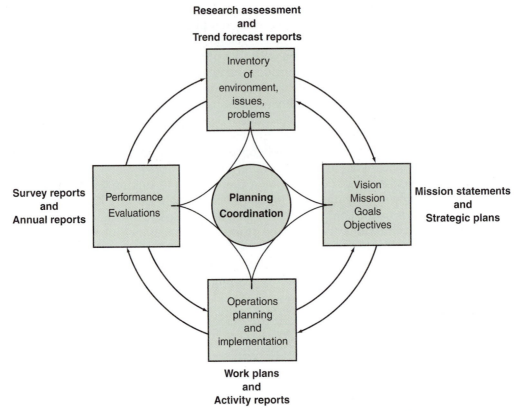

FIGURE 5.4 Basic planning elements of effective management and their associated communication products. While the process of planning and implementation progresses forward through inventory, objective setting, operations, and evaluation, there is learning feedback at each step in the process. Each step generates communication designed to facilitate the management process.

plans for a series of possible future conditions. Planning focuses on transforming a vision of future desirable conditions through statements of mission and goals into specific objectives and the strategies and tactics used for their accomplishment. *Operations* are designed to accomplish the objectives that will realize the vision. Performance *evaluation* is needed to judge progress made toward realization of the desired future condition. Performance evaluation provides new information for modifying the inventory and the planning process that follows. In this way, the planning process cycles refined and improved information for organization adaption to a changing planning environment.

Management System Boundaries. Unique natural resource management systems exist for diverse organizations including private businesses, advocacy groups, government, and any other type or combination of organizations. Fully stated organizational authorities and purposes usually establish boundaries around the unique internal management environment of the organization (Figure 5.5), which forms the management domain within each organization's management system. The external management environment lies outside the organizational boundaries and is the source of resource materials, information, financing, and other system inputs. It also is the destination for management influences on ecosystems, social systems, and other management systems. These management-system outputs include all goods and services affecting the general welfare outside the organization. Basic boundary-setting authorities and responsibilities are determined by law, such as the purposes defined for agencies and the right to private property so important to business and other private organizations. The management mandates, mission, and goals usually define the resources of concern to the organization, the actions to be taken regarding the re-

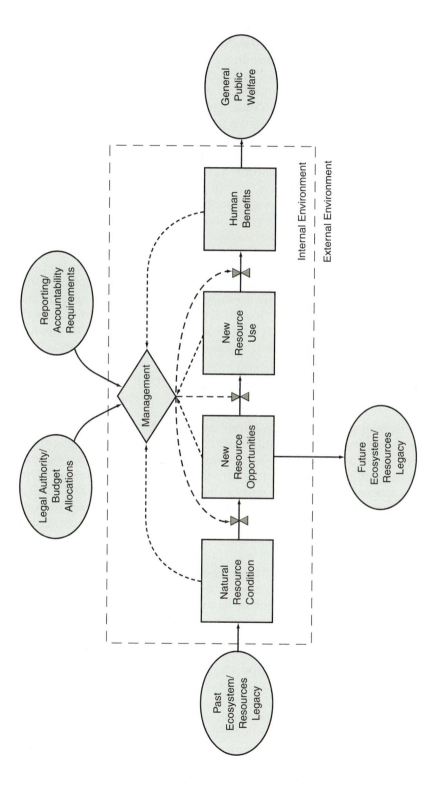

FIGURE 5.5 The management system. Management "drives" a process of assessing resource condition and developing new resource opportunities based on expected resource use and human benefits. Factors in the external environment provide ecological and social inputs that enable management system function. System outputs affect the future condition of the ecosystem and resources as well as general public welfare.

sources, and the geographic area to be developed, regulated, or otherwise influenced by the organization.

The U.S. Forest Service, for example, is bounded by its legal authority to manage all national forest natural resources according to legally authorized policies. Within the geographical limits of its authority, the Forest Service has a multipurpose management authority over grazing, timber, recreation, biodiversity, watershed, and other resources. In contrast, the U.S. Environmental Protection Agency (EPA) is primarily authorized to prevent environmental degradation through enforcement of regulatory policy applied throughout all of the United States. Whereas it is a single-purpose organization, limited to environmental protection, the authority of the EPA is geographically unlimited within the United States. The U.S. Fish and Wildlife Service has a more complex authority including management jurisdiction over an extensive refuge system, regulatory enforcement of numerous federal fish and wildlife laws, and administration of funding to states gained through federal taxation on fishing and hunting equipment and supplies. State, county, and municipal agencies have no jurisdiction beyond their political boundaries, but otherwise are similar to federal agencies in the way their management systems are bounded.

A private organization holding land is authorized through recognition of private property rights to manage the ecosystem associated with that property within legal constraints imposed by government. In the United States, for example, a private land holder cannot, without a state permit, manage or harm the wildlife on the land, because the wildlife are a public resource. Compared to government agencies, the boundaries of private organizations usually are determined more by the mission and goals circumscribed by the organization than by law. In contrast, the flow of goods and services across government agency boundaries, and human access to them, is governed by law, starting with the U.S. Constitution. The difference between private and public service flow is well illustrated by recreational beaches. While private owners have a right to close their beaches to public access, a recent Connecticut Supreme Court decision confirmed that a city-owned beach cannot be legally closed to nonresidents.

Natural resource management is often based in geographically identified units, regardless of whether an organization is authorized control over land through private ownership or through the public trust. Organizational functions vary with geographical area as resources and human preferences vary from one region to another. Many natural resource management systems classify resources through some kind of geographically based reference system. Management units are often split up on a regional basis, and management plans often define the boundaries of each separate area to be affected by management and to what degree.

The Internal Management Environment.   Management planning as well as plan implementation is the driving force (symbolized by the diamond in Figure 5.5) in the system, which operates on the natural resource condition of the relevant ecosystems. The organization mobilizes its human resources and facilities to enhance the condition of natural resources and provide improved opportunities for use and human benefit. Whether private or public, managers operate with the intent of serving the interests of people who use the provided opportunities, either to profit from them if they are private businesses or to benefit them according to legal mandate if they are government agencies.

Private businesses may manage natural resources to sell to customers or to use themselves in manufacture of refined products that are then sold. Many owners of timber, minerals, fossil-fuels, and other resources open them to sale and extraction, much as the Forest Service, Bureau of Land Management, and other agencies do on public lands. Other private land owners extract resources, such as timber, and then use these resources to produce refined products, such as lumber and furniture. Private businesses compete in the market for customers and in so doing are driven to benefit the customers while they too benefit through continued employment and profit.

Opportunities may be provided in a diversity of ways other than increasing the production level and sustainability of goods and services from the natural resources. Providing more access often is a critical element in providing opportunity to use resources such as by way of roads, boat ramps, improved navigation channels, trails, railroads, and airports. Provision of facilitative services can be important, such as providing power, fuel, food services, housing alternatives, and a diverse array of other wants and needs. Management effectiveness in facilitating resource use and in benefiting customers often depends on how well a mix of opportunities is provided.

Government agencies are mandated through their authorizing laws to serve resource-user "customers." A state game and fish agency, for example, provides opportunities for recreational anglers, hunters, and wildlife watchers, all of whom are customers of the agency's services. However, government agencies often find it more difficult than private businesses to determine exactly who the customer is because the resource is "owned" by all of the public, not just those who actively use it. When the Forest Service sells timber the purchasing logging company may momentarily stand out as a customer, but all of the taxpaying public are also customers affected by the government income generated from the sale (reducing fees and taxes), the cheaper price of wood resulting from greater supply, and any environmental costs associated with the sale. Because of their public mandates, government agencies must serve all of the public well while private businesses can specialize or diversify more as they choose within broader regulatory limits.

The External Management Environment.

No management system is totally independent of other management systems and thus is influenced by and influences ecosystems and socioeconomic systems in the external environment. Each management system depends on inputs from outside the system and, in turn, each management system's outputs contribute to the inputs of other systems beyond their boundaries. For natural resource management systems, most of the more influential inputs and outputs are ecological and economic.

The ecosystem condition "inherited" by an agency or firm determines the initial resource condition, economic demand for resource use, and cost of management. Past natural and management history and the location of the ecosystem with respect to human demands and impacts are critical factors. Ecological influences outside their management domain often influence their management system. No where is this more evident than in water resource management. An agency or firm situated high in the watershed has a much different ecosystem perspective than management systems located lower in the watershed in large rivers or estuaries. Upper-watershed agencies are more concerned about output impacts than about changes in inputs from external sources. Lower-watershed management organizations need to be concerned with all of the waterborne inputs that can have a major impact on the manageability of their authorized systems. Similar kinds of connecting flows among the jurisdictions of management systems occur along plant and animal dispersal routes, animal migration routes, coastal and oceanic currents, and prevailing winds.

Thus, some of the more vexing resource problems in recent years pertain to large animal migration and dispersal from public to private lands, downwind acid rain fallout across state and international borders, international incongruities in managing migratory fish and birds, and progressive water quality deterioration from upper- to lower-watershed ecosystems (Figure 5.6). Simply documenting the characteristics of material flows (e.g., water, contaminants, migrating animals) between management systems is basic to solving some of the related management problems. Poor land and water management often results in the system-output of materials that become system-input problems somewhere else, such as sediments in streams and lakes and atmospheric pollutants. The net effect is lowered total public welfare requiring government correction. The import-export connections between management systems are often the focus of mediating regulation administered by an overlying management system. The Environmental Protection Agency, U.S. Fish and Wildlife Service,

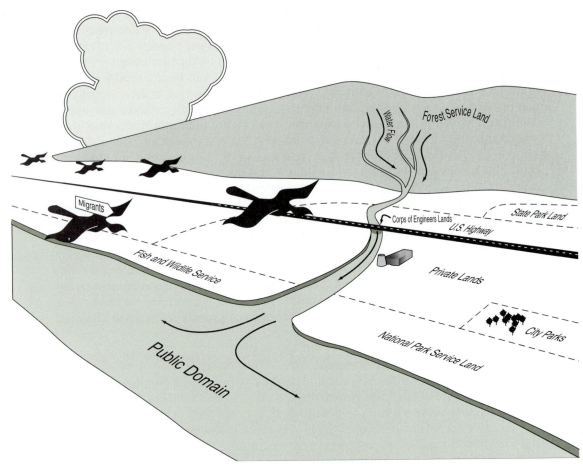

FIGURE 5.6   Political authorities often cut across natural processes, such as weather, water flow, and animal migration. When these ecosystem processes are ignored within political planning boundaries, problems usually result. Integrated planning is needed to contend with such problems.

and National Marine Fisheries Service are good examples of nationwide regulatory management systems superimposed over the public and private land-based management systems.

Among economic resources, all organizations need continuous investments of labor and capital in addition to the land-based resources. Most fundamental is the human resource, which is basically prepared by an educational system outside the management-system boundaries. The ability of the human resource to seek outside information for application inside the management system often makes the difference between success and failure. Also critical is the operation budget needed for salary, housing, equipment, supplies, travel, and services. Money is to a management system as energy is to an ecosystem. Without a budget, the organization fails to function (even voluntary organizations require budgets). Money links the economics of the management system to the larger economic system of the Nation and the world.

The outputs from management systems include all of the goods and services contributing to the public welfare, which determines public attitudes and future willingness to pay the price or the taxes for continued management-system products. Whenever little useful opportunity is provided, benefits usually are insufficient to justify the management cost. For private firms, this results in rapid adjustment to the marketplace or business failure. For government, it leads to escalated complaints, tax revolts, and the demand for smaller government. Environmental exports also may be important.

Management organizations often interact in ways that require coordination to develop and sustain maximum benefit. Reservoir recreational management is a good example. Many agencies and private firms provide boat ramps and related access and services among different reservoirs, often without considering the effects of uncoordinated efforts on use and benefit. This frequently leads to doubling up of unneeded opportunity and changes use patterns in ways that otherwise complicate efficient management. Reservoirs are just one of a large number of resources managed by a mix of public and private organizations, often as if the other management organizations were incidental or irrelevant. Such isolating activity, especially among government agencies, has spurred a variety of planning and coordination laws. Even so, true coordination is more-rare among government agencies than the ideal.

Managing the System for Results

Simply providing opportunity does not guarantee use where it is desired. Too many public resource management organizations have emphasized provision of opportunities over customer benefits, using the premise, "If you built it, they will come." As a consequence numerous underused roads, campgrounds, harbors, boat ramps, reservoirs, stocked fisheries, and the like were provided in place of more useful provision of opportunities elsewhere. Resources developed for a use that never materializes benefits no one. In fact total benefits can conceivably be negative because the management money used is diverted from development of beneficial opportunities elsewhere. Osborne and Gaebler (1992) wrote an influential book entitled *Reinventing Government,* which emphasizes outcomes assessment in the form of public "customer satisfaction" as the way to gauge government effectiveness. For private corporations, as long as competition is sufficient, profitability is an unexcelled goad to "customer satisfaction."

Public, Private, and Advocacy Systems

Public Organizations. Numerous federal, state, and local government agencies make up an interactive complex of natural resource management systems (Table 5.1). Some agencies regulate the activities of other agencies, for example, as the EPA does under the authority of the National Environmental Policy Act of 1969. Most of these federal agencies form management subsystems within cabinet departments in the executive branch of federal government. The departments of Agriculture, Energy, and Interior are most consistently organized around natural resource management, but with some exceptions. Other departments play smaller roles in natural resource management, but few play no role at all. The numerous federal management systems can interact through complementary and overlapping jurisdictions, oversight, coordinative legislation, executive orders, various directives, and networking through professional organizations, workshops, publications, Internet services, and other communication. Because of the diverse purposes authorized for different agencies, the natural resource management subsystems do not integrate as well as they might into a single federal natural resource organization.

There has been periodic discussion about possible rearrangement of government agencies. The last major change in federal government pertaining to natural resource management was to integrate most of the diverse environmental protection functions into the Environmental Protection Agency. Insufficient integration of a natural resource planning process at state and local agencies also has impeded coordination among local, state, and federal management systems, interfering with the development of an effective planning process.

Business Organizations. Private business management systems are similar to government systems in their basic attributes, the main difference being the importance of competitive forces in their function. The nature of market-driven business systems is described in more detail in Chapter 4. Unlike government agencies, profit-motivated businesses have investors who expect to realize substantial profit from

TABLE 5.1   Major Federal Agencies Responsible for Natural Resources

U.S. Agency	Department	Natural Resource Authorities
Army Corps of Engineers (Civil Works)	Defense	Water resources management for navigation maintenance, flood damage reduction, recreation, environmental improvement, hydroelectric, and other water use; also enforces wetland protection law and administers public lands
Bureau of Land Management	Interior	BLM public lands management for range, forest, watershed, recreation, mineral, fish and wildlife, and other authorized use
Bureau of Reclamation	Interior	Water resources management for irrigation supply, recreation, environmental improvement, hydroelectric, and other water use; also administers public lands associated with water resource projects
Departments of Army, Navy, and Air Force	Defense	Authority over resources on military installations
Environmental Protection Agency	Stands alone	Enforces U.S. environmental protection and restoration law and administers public water treatment facilities programs
Federal Highway Administration	Transportation	Oversees development and maintenance of U.S. highway system and associated lands
Federal Power Authority	Energy	Oversees development and coordination of U.S. energy production
Fish and Wildlife Service	Interior	FWS land management of the fish and wildlife refuge system, enforcement of U.S. fish and wildlife law including Endangered Species Act, and monitors fish and wildlife resources
Forest Service	Argiculture	USFS public lands management for forest, range, watershed, recreation, mineral, fish and wildlife, and other authorized use
Geological Survey	Interior	Federal natural resources research and mapping agency including hydrological, geological, and biological processes
National Marine Fishery Service	Commerce	Monitors marine resources and administers international marine fisheries and marine mammal laws including Endangered Species Act
National Park Service	Interior	NPS public lands management including national parks, monuments, recreational shore, wild rivers, and other reservations of natural and cultural heritage
Natural Resources Conservation Service	Agriculture	Agricultural resources services associated with water supply and drainage, soil maintenance, erosion control, and farmland, fish, and wildlife habitat management
Office of Surface Mining, Reclamation and Enforcement	Interior	Oversees and enforces federal legislation pertaining to environmental protection from surface mining and reclamation of lands for productive use following mining
Tennessee Valley Authority	Energy	Water resources management for hydroelectric, flood damage reduction, recreation, fish and wildlife, and other uses in the Tennessee River Basin

company activity. Private businesses are quite diverse in size and complexity of goods and services they sell in the marketplace. Most businesses are small and usually specialize in a narrow range of goods and services. Common types of small businesses oriented around natural resources management include ranches, farms, engineering and construction firms, landscape and gardening services, travel services (restaurants, shops, gas stations/garages, hotels/motels), bus and other transportation companies, logging companies and sawmills, guiding and tour services, equipment and retail services, outdoor recreational services (ski slopes, marinas, golf courses, recreational vehicle rentals), mineral extraction companies, and local power and water-supply companies.

Larger companies may be "vertically assembled" based on ownership, extraction, transportation, delivery, and refinement of resources. Large mineral-resource industries often have an extensive network of landholdings, extraction equipment, refineries, and product delivery services. Most land-based businesses focus on development of specific resources, but some have diversified. Interest in developing recreational services on privately held rangeland and forest land has increased, for example.

Advocacy Organizations. Advocacy groups are extremely diverse and represent a wide variety of public and private land-use concerns. Advocacy groups are non-government organizations (NGOs) that are typically nonprofit organizations. Usually, they are motivated to shape public or private policies through various strategies including marketing their positions through communications media, business boycotts, demonstrations, court actions, lobbying legislation, strategic property purchases, and, rarely, unlawful sabotage. Over the past several decades, they have become the primary avenue for representation of stakeholder interests in outcomes pertaining to public resources management. Environmental advocacy groups have been especially effective in promoting their special interests.

The goals of advocacy groups are organized with special interest objectives in mind, but sophisticated organizations realize they can be more successful in a situation where none of the stakeholders are perceived to be total losers. Public management systems are increasingly involving advocacy groups early in the planning process to help work through difficult issues, especially those pertaining to private land use and development. Numerous advocacy groups are concerned about natural resources and the environment. Among the better known are The Nature Conservancy, Sierra Club, Wilderness Society, World Wildlife Fund, Audubon Society, and National Wildlife Federation. Professional societies also have moved into advocacy roles. Examples include the American Society of Foresters, the Range Society, the Wildlife Society, the American Fisheries Society, the Society for Conservation Biology, and the Ecological Society of America.

PLANNING PROCESS

Inventorying Planning Environments

Information Gathering. Inventory is the process of gathering and analyzing information. It sets the stage for development of vision and its translation into reality (Figure 5.7). Inventory activities go by numerous titles including survey, monitoring, research, intelligence, environmental scanning, cost accounting, forecasting, trend analysis, systems analysis, issue and problem identification, SWOT analysis, stakeholder analysis, and information resources and management. Problems and opportunities are identified early through continuous and astute inventory methods. For natural resource organizations, inventory centers on the state of natural resources, their development and use, and the profits or benefits derived from them. The most basic intelligence pertains to changes in the "balance" between resource supply and resource demand, changes in the influential planning environment, such as laws and other mandates, organizational condition, and stakeholder needs. Perhaps most importantly, inventory helps to clarify visions of desired future resource conditions. Inventory examines the state of things both inside and outside the management system.

Reviewing Mandates. Planning inventory often starts with a review of the laws or the mandate authorizing actions or regulating action (Bryson 1995). For public organizations, laws provide authorities to work toward certain ends. But authorization laws often are quite broad, leaving latitude for the organization to create a sense of purpose or mission that extends beyond existing law. Planning often inventories the legal mandate, organizational authorization, and policies before proceeding into new initiatives.

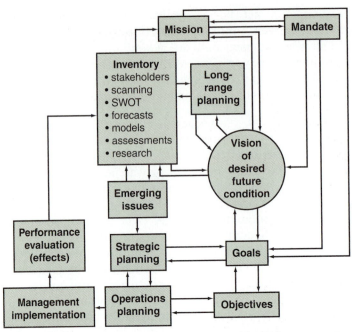

FIGURE 5.7   The general planning process by which mission and mandates are accomplished. Planning starts with a vision of a desired future condition and identification of goals and objectives that will realize the vision through operations planning and implementation. An inventory of existing conditions, including emerging issues, is completed to develop effective strategy.

Private organizations also are bound by laws that limit their operations to socially acceptable practices. These laws often originate from government action designed to promote constitutional ideals, which strive to protect individual rights while assuring improvement in the general public welfare. Most organizational legal conflicts arise over private rights versus the general public welfare. Whereas legal mandates both authorize and regulate the operations of public-service organizations, they only regulate the operations of private organizations.

Some extralegal public expectations from government can approach the force of legal mandate. Public opinion may at times be strong enough to approach the force of law. Many state fish and wildlife agencies, for example, are legally mandated to serve the needs of all state citizens. Although management of fish and wildlife resources benefit many nonconsumptive users, a select group of hunters and anglers typically pays most of the management costs in many states. As a consequence, the desires of hunters and anglers can act as an extralegal "mandate" whenever they differ collectively from those of other citizens. Public service favoring special interests continues to be common whether the agencies regulate or otherwise manage water, food, range, forest, real estate, fish, wildlife, biodiversity, mineral oil, or other resources. Government agencies vary in their success at optimizing the demands of special interests consistent with serving the local, state, and national interests.

Reviewing Mission.   The mission is the organizational rudder—the fundamental means for communicating the essence of organizational purpose among members and all others affected by its performance. The mission usually identifies who is to be served by the organization and how and where that service is to be provided. An effective mission statement conveys an accurate sense of organizational attitude toward its employees and the world it intends to influence. Inspirational mission statements are motivational and impressive, contributing to morale of the organization membership and the respect of stakeholders, if the organization follows through. For agencies, a good mission gives focus to legal authorities and frames values in such a way that goals may flow naturally from it. The mission statement is the main communication of intent within the organization and between the organization and outsiders. It needs to be reviewed to sustain or change direction as new information is received through inventory (Figure 5.7).

Environmental Scanning. Environmental scanning is a sweeping approach to reviewing planning environment trends relevant to organization management. Trends and emerging issues that have significance as management opportunities or problems are brought to the planning process for identification of the most effective alternative. This process usually requires perusal of trade magazines, newsletters, various digests, legal briefs, news media, press releases, Internet home pages, and other communications from within and among diverse stakeholder groups. The focus is on finding emerging issues—potential problems, solutions to problems, or other opportunities in their earliest stages. These are tracked and incorporated into the planning process as they more fully emerge or are discarded as they fade in importance.

Stakeholders. Stakeholders are individuals and groups affected by actions of the organization. Stakeholders are best assessed and brought into the planning process as early as possible (Bryson 1995). Certain stakeholder groups often are the object of organizational performance as service recipients, promoters, or detractors.

SWOT Analysis. Effective planning often requires SWOT analysis of the strengths (S) and weaknesses (W) of the organization and the opportunities (O) and threats (T) originating outside the organization (Bryson 1995). Forecasts of change are an important part of effective SWOT analysis. In a dynamic planning environment SWOT analysis, or some variation of it, can become a frequent part of the planning process.

Forecasting. Forecasting is the process of predicting future trends and/or future conditions. The future usually is dependent on too many events outside management influence for perfect forecasting. The most useful approach is to plan for several feasible alternative futures. Planning for alternative futures introduces an element of flexibility and proactive process in anticipation of change. The effects of resource demand on continued supply, for example, depend on trends in both resource supply and demand. Forecasting analysis allows prediction of future supply based on estimated demand. From this information, stakeholder benefits can often be tracked. Inventory also includes monitoring of all organizational costs incurred by project activity, allowing benefits to be compared to costs.

Mathematical models often are used to integrate information into analytical tools for forecasting the effects of different management policies. Typically the models represent relationships among variables determining resource supply and demand. Development of precise predictive models can greatly reduce inventory costs, but often with substantial investment in the developmental research. Predictive models exist, for example, to characterize water resources, fish and wildlife habitat, timber production, range production, viability of rare species, and nonrenewable resource depletion. Many organizations use qualitative approaches for analyzing future change. Some models are comprehensive enough to represent whole systems and are used for systems analysis. Sometimes physical models, including natural ecosystems, are used as a reference for the planning process.

Focusing Management Intent

Mission. The process of focusing management intent into implementation of a plan typically starts with the statement of organizational mission and values. When well stated, mission and value statements guide the formation of planning vision, goals, and objectives for organizational accomplishment. Basic elements of all U.S. government missions follow from the mandates of the U.S. Constitution, which emphasizes improvement of the general welfare, including the interests of future generations. For those agencies responsible for managing renewable resources, the mission often identifies resource conservation or sustainability as a guiding philosophy. In response to political influences, most agency missions historically have focused their service on the most active special interests identified in their authorities: timber and recreation in the Forest Service, agricultural irrigation in the Bureau of Reclamation, navigation improvement and flood-damage reduction in the Army Corps of Engineers, livestock forage in the Bureau of Land Management, environmental improvement in the Environmental Protection Agency, hunting and fishing improvement in the Fish and Wildlife Service, and many others.

Planning Vision.   At the center of systematic planning (Figure 5.7) is the focusing of a vision of desired future condition based on achievable tasks (Bryson 1995). The establishment of an acceptable vision of desired future condition, or outcome, may be the most critical part of an effective natural resource planning process (Bryson 1995). If the vision is vague or unrealistic, much time and money may be spent ineffectively. Without visions of what needs to be accomplished, an organization can drift through the management system environment without having much effect.

A planning challenge is building consensus among diverse stakeholders in the development of a *shared vision* of desired future condition of resources. The approach to building consensus is to provide an environment that favors the open exchange of information among the resource experts knowledgeable about trends and stakeholder's needs. A variety of approaches have been developed to facilitate an orderly discussion of issues. A critical part of the discussion hinges on the knowledge of how benefits will be distributed among stakeholders as a consequence of the policies selected. Visions can be developed at different organization levels ranging from small everyday projects to the vision behind organizational mission.

Goals.   Goals are more general descriptions of elements in the vision to be accomplished and the general means to be used in accomplishment. They do not specify the results precisely. Where organizational visions of successful management are reasonably clear, goals are relatively easily developed in pursuit of vision realization. The goals of public resource agencies usually emphasize generations of public benefit while sustaining resources. Private enterprise emphasizes a larger market share, a greater profit, and better employee income and benefits. Goals also identify the approaches that are to be used in pursuing the mission, including education, regulation, research, and management. Goals typically are not scheduled for a specific completion date and sometimes remain unchanged for many years. Goals are most easily developed in stable planning environments where visions of desirable future conditions are clear (Bryson 1995).

Issues.   Even in the most stable planning environments, organizational threats and opportunities emerge. Issues usually can be disassembled into problems that need solution if the organization is to most efficiently pursue its purpose. A special area of the goal-driven planning process deals with issue management (Bryson 1995); threats are mitigated and opportunities are exploited.

In organizations with unstable planning environments and vague visions of organizational success, most planning is dominated by disruptive threats. Many resource organizations have become less stable in recent decades as conflicting demands of diverse stakeholder groups have grown in intensity. In those situations, visions of success often become muddled and issues drive the planning process. Strategies and tactics are developed to mitigate or exploit each issue. As greater control is developed over issues, they may become the focus of new goals and new visions of success (Bryson 1995). But until then, the issues replace the goals as targets for planning accomplishment.

Translating Strategy Into Operations

Strategic Planning. Strategy is a proactive, as opposed to reactive, and adaptive, as opposed to standard, approach to identifying and solving problems and taking advantage of opportunities (Bryson 1995). Strategic planning typically addresses time horizons beyond the present budget planning period, which usually extends less than three years forward. These shifts may derive from past natural resource practices, new knowledge, changes in law, changes in customer and public preferences, new technologies, natural catastrophes, and new trends identified during inventory of planning environments. Strategic planning is most effective where long-range planning has provided sound information about major resource trends.

Long-range planning is a category of strategic planning that involves analyzing trends and then setting far-off goals, and/or avoiding strategies that are ultimately incompatible with desired long-range outcomes. The goals of long-range planning typically are very general. Since most natural resource management is intent on improving human quality of life, long-range targets typically focus on sustaining natural resource options for future use. For natural resource management, knowledge of resource abundance, renewal, extraction rates, and environmental regulation are critical elements in estimating long-range status. Long-range analysis and planning addresses generic issues about trends in resource supply and demand, availability of substitute resources, and the need for regulating demand to sustain resource supply. It also examines the recovery times needed to sustain renewable resources at levels compatible with social demand. The U.S. Forest Service, for example, uses 50-year planning horizons based on timber cutting cycles. Even 50 years is far short of the centuries needed to reestablish old-growth forests. Similarly, recent legislation has authorized ecosystem restoration by the Corps of Engineers, which also requires long-range visions of desired future conditions and strategies for their realization. Strategic planning needs to be flexible, adaptable to changes in the planning environment, and continuous. Strategic planning is focused more on relative program emphasis and new program needs than it is on project planning. For example, developing criteria for ranking project priority would fall into strategic planning.

Operations Planning. Operations planning links the organization budget to plan implementation. In operations planning, project objectives are clearly defined, projects are ranked in priority, and budgets are assigned to each area of management directed at project completion. The typical target of the most detailed operations planning is the budget year—which may be based on the calendar year or some other initial and ending date. When done well, a number of alternative approaches to objective accomplishment are compared and the most cost-effective alternative is used to accomplish each project.

Objectives are developed to focus tasks on measurable achievement. In contrast with the open-ended time frames typical of goals and issues, objectives have deadlines for specific accomplishments. The specific procedures used to pursue objectives compose the operational tactics—or *operations*. Typically, a suite of specific objectives is used based on indicators of the desired future condition. For example, a restored wetland may be the planning objective. A completely functional wetland may be anticipated if the objectives detail the important attributes of the landscape, the water supply, nutrient and sediment supply, and the species that are envisioned in the structure and function of that wetland.

Operations planning focuses on objectives formulated for completion according to schedule. An important part of operations planning is the assignment of personnel time to scheduled tasks. Within the Forest Service, budgeted projects for recreational service might include numerous outputs like road improvement, installation of new campgrounds, development of nature trails, improvement of user-law compliance, development of brochures, and research reports. Operations planning usually follows a predictable annual cycle in both public and private management systems and is linked to strategic planning.

Organizational Performance Evaluation.   Without evaluating accomplishment, there is no way to confirm management success. Thorough evaluation usually requires a well-organized monitoring process, which minimally monitors management costs and benefits. Performance evaluation may also monitor some of the intermediate management outputs designed to provide customer opportunities. The main reason for this monitoring is to determine inefficiencies in providing benefits and how to improve upon them. Performance can be measured in terms of process outputs and in terms of customer satisfaction or benefits. Job performance usually is based on completion of assigned tasks, but ultimate organizational performance is gauged by the benefits provided.

Private businesses have no question about the utility of performance evaluation. Without it, they would have no knowledge of how much income exceeded costs and the proper base for taxes and distributing profits. Government agencies have had more problems evaluating performance in benefit terms, tending to count opportunities provided (e.g., roads built, wildlife counted) or process rates (publications produced, law violations cited) rather than benefits generated. Although some agencies are required by their authorities to generate more benefit than cost, techniques for estimating benefits were not available for certain activities until very recently (e.g., recreational use). The benefits of other activities, such as much environmental protection and improvement, cannot be entirely monetized using widely accepted techniques to estimate benefit in a way that can be compared directly to cost. Instead, the relative significance of the results needs to be gauged as a proxy for benefit. The Government Performance and Results Act of 1994 institutionalized more consistency in federal government with respect to performance assessment.

PLACING BOUNDARIES ON PLANNING PROCESS

Organizational Planning Boundaries

Practicality dictates that planning be manageable in scope. Boundaries may be set on planning process in a number of diverse ways. Planning is frequently delineated by organizational activities including project, program, and comprehensive organizational activities. Planning also is frequently organized geographically or regionally. Integrative management planning process facilitates organizational interaction.

Most organizations divide their labor according to program and project boundaries. Program and project planning often are done separately, but involve interactive processes. Program and project management can be viewed as subsystems in organizational management systems. Management subsystem boundaries often are based on the dimensions and relationships of work activities organized to assemble materials, energy, and information into a new or improved output or "product." The output often is a good or a service of some kind, or a combination of goods and services.

Project Planning.   Project planning typically focuses on organizing natural, human, and fiscal resources into the desired kind and amount of output in which some form of accomplishment can be measured. The most obvious types of projects are physical constructs of some kind—for example, dams, levees, boat ramps, toilets, campgrounds, roads, city parks, trails, and habitats. Projects also may include new methods for identifying and extracting natural resources, ways to refine raw materials, and development of improved information services in the form of books, videos, and advertising in various media. Conceptually at least, there are no activities in organizations that could not be assigned to a project.

Projects form a basic unit for planning and operations in natural resource management. They typically set up objectives for accomplishment with deadlines. Project boundaries often cross administrative lines, and require careful coordination usually through a project manager.

Many organizations separate "routine operations" from projects. Because routine operations typically are not expected to have clear outputs and outcomes, they usually are not considered projects. However, some management organizations attempt to integrate most, if not all, activities into projects. Even operations as general as administration contribute to different projects. That contribution must be integrated into project cost if the real cost of the project is to be estimated. Certain administrative initiatives may be categorized into projects of their own. Most, if not all, operations can be linked to accomplishable objectives. This approach provides targets for improvement or, at very least, sustained performance. Otherwise the routine operations can steadily degrade in performance, diverting important resources away from innovative accomplishment.

Many government projects include a mix of agencies and private contractors in project development. A federal water resource project, for example, typically involves local and federal agency "partnering" and contracting with private services, all of whom need to be coordinated through the project planning process. Most projects also have a variety of stakeholders who usually need to be involved in the planning process, especially early in the development of a shared vision.

Program Planning. Programs typically are organized around customer service and associated benefits. A state conservation organization, for example, may have sportfishery, commercial fishery, wildlife, forest, range, and mineral-resource programs, each with numerous projects. Programs often are linked to specific sources of funding, but may be funded out of a larger pool. In either case, program managers need to argue for funding based on the benefits that will be produced from the program. Commonly, programs form around the special interests of certain stakeholder categories, which makes for easier analysis of program effectiveness than when they cut across linkages between resources, opportunities, uses, and benefits. However, many agencies are organized that way, with planning in one program and implementation in other programs.

Programs envelop groups of projects with similar attributes and purposes into a larger organizational decision process. A program-project matrix may be used to organize project activities into program areas. Each program includes projects pertaining to outcomes desired of each program and program managers typically seek those projects that contribute most to program performance in generating outcomes as planned. When left up to the program management, an important function is to decide how to cost-effectively distribute budgets among projects so as to generate the greatest public benefit or profit. Program management also provides centralized support services to all projects, thereby cutting project costs.

Comprehensive Organizational Planning. Comprehensive management planning coordinates across groups of programs within an organization (Crowe 1983). It can be viewed as systemwide planning for organizational management systems. Comprehensive planning addresses decisions about future program emphasis, the development of new programs, and the retirement of others. Comprehensive planning seeks integration among all programs under the organizational umbrella. It may seek to develop new and/or terminate old programs. Much of the demand for government programs is expressed in new laws. Over recent decades, for example, natural resource agencies have responded to new laws by directing more management resources to programs that emphasize protection of resource options (e.g. endangered species) and ecological services (e.g., flood-damage reduction, erosion control, and fish and wildlife habitat).

In most planning processes, long-range, strategic, and operations planning grade together. Strategic planning uses outputs from long-range planning to help identify appropriate goals, program emphases, and strategies. Operations planning uses outputs from strategic planning to formulate project objectives and tasks for their accomplishments. The most effective planning involves continuously adding output information to the planning inventory and adapting plans to new conditions. This is called adaptive management.

Regional (Geographic) Planning Boundaries

A common way to place boundaries on planning and management systems is through geographic or regional boundaries. The boundaries are most commonly defined by political boundaries, but are increasingly being defined by natural boundaries.

Politically Bounded Planning. Political boundaries traditionally have been emphasized in the identification of planning regions such as city, county, state, and national boundaries; the boundaries associated with different government agency authorities; and private land boundaries. Urban planning often ends at the city limits, but increasingly it is extended to a metropolitan area and beyond (Kaiser et al. 1995, Cullingworth 1997). Regional planning usually is chosen when the object of planning is land and water resource use that requires linkages across political boundaries. Before the 1960s, most regional planning was economically motivated, and often focused on one or two forms of development, as was the Tennessee River basin. It was developed for electric power generation by the federal government based on economic improvement of an especially depressed area. Transportation planning also is regionally based. Regional planning for utilities and transportation provision were among the first ways business and governments integrated activities through the planning process; therefore, they were also very influential in determining the way in which the nation developed during the twentieth century. However, more recent regional planning has incorporated environmental quality considerations in response to a rapidly degrading environment and corrective laws passed mostly in the 1960s and 1970s.

Naturally Bounded Planning. Regional planning is increasingly adopting principles of landscape ecology into the planning process (Steiner 1991, Dramstad et al. 1996). When this is done, it emphasizes natural connections among ecosystems and more natural bounding within and among planning regions. Management systems also may be naturally bounded at least in part such as by watersheds, coastal zones, geological formations, soil types, and vegetation types. Coastal zone planning has been encouraged by the Coastal Zone Management Act of 1972 and is done by most coastal states (Kaiser et al. 1995). Watershed planning became useful for water resources management when the Water Resources Planning Act of 1965 established river basins to delineate national water resources planning. Within the past several years, the EPA resurrected a watershed approach, which is gaining greater attention among many natural resources agencies. Many resource management agencies apply natural boundaries, such as watersheds and vegetation types, within their authorized political boundaries.

Integrated Resource Management Planning

Integrated resource management planning seeks the development of a more holistic approach to planning across organizational planning boundaries. Many of the natural resource agencies in the past interacted relatively little except through policy mandates. They have been encouraged toward more integration by laws that require greater cooperation and coordination. At the heart of integrated resource management is the development of a shared vision of desirable future condition. Policy development in the form of new laws and their interpretation into rules and guidelines is a critical part of this integrated management process. The problem has taken on an international perspective, most recently with concern over ocean fisheries, global warming, and biodiversity loss. The new concept of ecosystem management (see

Chapter 23) demands a more integrated effort than has been required previously. This will be especially true if a shared vision of future landscape conditions is to be realized for large watersheds, oceans, and other large ecosystems. In this view, the ecosystem becomes the common boundary for guiding each organization's system management toward a larger systems integration.

Several decades ago a form of urban regional planning called comprehensive planning was first initiated to integrate across local political jurisdictions. The process has had mixed success, depending greatly on the extent that states have developed planning guidelines (Kaiser et al. 1995). Attempts at integrated planning have been derailed when the participating organizations came to the table with their own mostly inflexible plans. By then they had invested too much in the planning process to modify it readily. Integrated planning often is a long process involving any number of stakeholders who can stymie a shared vision. Increasing emphasis on protection of personal property rights has been a major contributor to the slowdown in integrated land-use planning. Although difficult and often prolonged, integrated resource management planning is the most promising approach to many pervasive problems such as urban sprawl, habitat fragmentation, and water conflict resolutions.

EXAMPLES OF AGENCY PLANNING

Brief descriptions of planning in three very different federal agencies reveal consistency in the basic elements of planning process. They also show how quite different natural resource organizations integrate into a larger interorganizational planning and implementation process. In addition, they show examples of how federal agencies interact with state and private organizations to accomplish national goals and objectives as identified in federal law. Similar interactions between the public and private sector exist at state, county, and municipal levels of government.

The Environmental Protection Agency

When formed in 1970, the EPA was directed to administer a host of already existing environmental laws as well as new laws. These are the basis of a predominantly programmatic planning process. The EPA is authorized to set environmental standards, seek congressional funding for and allocate that funding for corrective measures, conduct research to develop more effective standards and means for meeting them, enforce the rules and regulations developed to meet the standards, and evaluate progress in meeting standards. Establishing and enforcing the meeting of standards is the means for carrying out the EPA mission, which is "to protect human health and to safeguard the natural environment—air, water, and land—upon which life depends." The administrative authority and authorized funding associated with each law established a programmatic basis for agency planning, which is reflected in the administrative divisions of the agency. The EPA does relatively little project planning and development itself, but does fund other organizations for that purpose, as authorized by Congress. While the corrective technology, planning, and implementation of projects are funded mostly by the regulated organizations, the EPA administers grants to facilitate compliance with the law. For example, federal funding is passed to municipalities for their waste treatment projects and to commissions for watershed-based planning designed to comply with water quality standards.

EPA strategic planning, like that of other federal agencies, is designed to respond to the Government Performance and Results Act of 1994 by identifying its main programmatic goals and objectives and establishing annual performance measures to evaluate agency performance in achieving objectives. For the EPA, the goals and objectives of the strategic plan are based on achieving the approved standards for air, water, and other environmental quality. The application of economic criteria, such as a net-benefit criterion, varies among programs, depending on the requirements imposed in individual laws. The risk associated with lost life expectancy is an important criterion for determining environmental standards and prioritizing program activities.

The EPA influences the planning of all other agencies in its role as the administrator of such nationally inclusive legislation as the National Environmental Policy Act, the Clean Air Act, the Clean Water Act, and Superfund legislation. It also develops guidance for the planning and implementation process, such as for watershed-based planning, then makes it available to individuals and institutions concerned about meeting environmental standards. Other federal agencies having similar regulatory influence are the Fish and Wildlife Service and the National Marine Fisheries Service, which together administer the Endangered Species Act. The effectiveness with which the EPA and other regulatory agencies accomplish their regulatory objectives as required under the Government Performance and Results Act of 1994 depends on how effective they are in getting all agencies and other organizations and individuals to comply with rules and regulations intended to meet environmental standards. In this way, the regulatory agencies work programmatically to encourage effective project planning and implementation in other agencies and in the private realm directed at preventing and correcting environmental damage.

The Corps of Engineers

The Corps of Engineers (Civil Works) is authorized to plan and implement justifiable water resource development projects primarily for navigation, flood-damage reduction, and ecosystem restoration services of benefit to the nation, and secondarily including storm-damage prevention, hydroelectric power, recreation, water supply, and other services. Similar to the other federal water resources agencies, the Corps planning process emphasizes project planning, starting with project development and continuing through operation and maintenance of projects built entirely with federal funds before 1986. Since 1986, all projects have required a local sponsor who bears part of the costs and assumes all operation and maintenance responsibilities. Federal funding is authorized through congressionally approved allocation of funds to each project.

The project development planning process for federal water resources agencies is defined by a president-approved set of principles and guidelines that established a national goal for water resources project planning, which is "to contribute to national economic development consistent with protecting the nation's environment." The national economic development goal is linked with the environmental protection goals identified in standards developed by EPA, FWS, NMFS, and other authorized agencies. The project is rejected if planning analysis reveals the benefit does not justify the cost or that environmental standards cannot be met. Ecosystem restoration projects provide a means for local agencies to meet national environmental objectives identified in the standards established by the EPA, FWS, and other regulatory agencies. In these ways, the authorities of diverse agencies are integrated into pursuit and accomplishment of national goals and objectives identified in congressional law.

A six-step process is defined for project planning that (1) specifies the problems and opportunities; (2) inventories, forecasts, and analyzes the water and related land resource conditions; (3) develops alternative plans; (4) evaluates the effects of each alternative plan; (5) compares the alternative plans; and (6) selects the recommended plan based on the comparison of the alternatives. The recommended plan must be that plan most consistent with the national economic development goal. The primary criterion for meeting the national economic development goal is value added to the national goods and services—that is, the benefits must exceed the costs. In budgetary planning, projects are ranked in order of greatest difference between benefit and cost (net benefit). Those projects not fitting within the congressionally authorized budget are postponed for consideration in the next budget.

Policy directs that a watershed-based regional approach be taken for analyzing project feasibility and alternatives effectiveness. A chronic problem with the Corps planning process is that all benefits cannot be readily measured in dollars, and ecosystem restoration projects, especially, must be justified based on criteria other than those used to assess national economic development. This presents a conundrum for prioritizing individually authorized projects in budgets—a problem encountered by all government agencies but most clearly faced by the water resource agencies.

The USDA Forest Service

The Forest Service has one of the more comprehensive natural resources planning authorities in the United States. Its planning authority is grounded in its land management authority for the national forest resources and in a number of other laws that link national forest planning to regional and national inventories of resource supply and demand. The Forest Service mission is "to sustain the health, diversity, and productivity of the Nation's forests and grasslands to meet the needs of present and future generations." To do that, the planning processes conducted at each of the 155 national forests and 20 national grasslands are interfaced with regional planning in each of the 9 regions, including resources on private lands, and with national planning coordinated through Washington headquarters.

At the national level, the Forest Service is authorized to conduct a resource planning inventory to periodically assess and report on the status of all of the Nation's public and private forest and rangeland resources including future needs, and to develop a fully integrated plan for management of National Forest resources within defined planning areas. The primary objectives of land and resource management planning done by the Forest Service are to assess the capability of a planning area to provide forest and rangeland goods and services in the most cost-efficient, beneficial, and environmentally sound way. Based on regional assessments of supply capability and demand, each of the National Forests is assigned resource supply targets consistent with long-term sustainability. Each Forest Service region develops a set of alternative resource management plans and determines the preferred plan based on how well the regional issues and management concerns are addressed. The Forest Service must comply with all laws administered by federal and state agencies. As with all natural resource agencies, the National Environmental Policy Act, the Endangered Species Act, and the Government Performance and Results Act in particular have had large impacts on the planning process and objectives formulation.

At the national forest level, a forest plan is developed that guides all management activities at program and project level according to multiple-use and resource sustainability principles as authorized by Congress under the Multiple-Use and Sustainability Act of 1960. The forest plan develops goals and objectives for land use and resource outputs consistent with environmental standards identified by the EPA, FWS, and other agencies. Forest plans are programmatic and identify projects for funding within the annual budget period. Each project undergoes its own planning process in which details and budget needs are identified and the forest plan refined as needed. Then the management measures proposed to meet objectives are analyzed to determine the most cost-effective combination of tactics for objective achievement. A set of alternative plans is developed for a 50-year planning period, with emphasis on the first 15 years. Each alternative plan must address all of the relevant issues and resource targets, as well as environmental constraints. All forest plans must be updated within 15 years. The effects of alternatives are analyzed to identify the plan most likely to meet the objectives cost-effectively while satisfying a set of agency-identified management principles. A monitoring program must be identified to evaluate implementation success and guide the plan modification through a process of adaptive management.

POLICY

What is Policy?

Policy is the means by which organizations guide behavior within and outside their membership. For many organizations, this begins with the writing of a charter, constitution, or other document defining why the organization exists and generally how it will act in pursuit of its purposes. Policy may be public or private. Most public policy is written to guide the interpretation of law into specific guidance, rules, standards, regulations, and other actions. In the United States, the Constitution is the foundational law guiding all public policy. Because planning guides pursuit of purpose, effective policy depends on effective planning. Because policy guides organizational behavior, including planning, effective planning depends on effective policy. The two are closely aligned.

Both private and public policies contribute to natural resource development and management in the United States and many other countries. Private policy is determined through a variety of organizational management styles, which must also respond to relevant public policies. Public policy is developed and administered by government through a system of laws, regulations, rules, and standards. An important function of the U.S. Supreme Court is to assure that all public and private policy is consistent with the U.S. Constitution. All state constitutions and laws must be consistent with the U.S. Constitution as well as any county, municipal, or other government law.

Most laws place authority for their enforcement in a specified branch of government. These laws often authorize an agency to develop regulations and management standards. Most laws pertaining to natural resource management authorize discretion to agencies in developing and enforcing regulations, but usually require public involvement in the process. For natural resource management, this process usually requires agency knowledge of resource availability, renewal rates, if any, demand for consumptive and nonconsumptive use of resources, and the environmental and social implications of resource use. Some authorities are linked closely to public lands assigned to agencies for management responsibility, such as the U.S. Forest Service. Other authorities are less directly tied to land ownership, such as water resource management by the U.S. Army Corps of Engineers, Bureau of Reclamation, and the Natural Resource Conservation Service, but usually involve land-use agreements if not public acquisition. The regulatory authorities of some agencies, such as the Environmental Protection Agency and the U.S. Fish and Wildlife Service, extend to all public and private lands and waters.

Roles of Federal, State, and Local Governments

The need for more policy consistency and effectiveness was one reason why federal policy eventually came to dominate natural resource management. Natural resource public policy was locally developed and applied before federal government became active in developing resource policy toward the end of the last century. A mix of local, regional, national, and international policy now exists, creating a complicated mix of jurisdictions and regulations. This mix requires coordination in regional management and complicates management across jurisdictions. Federal policy development often has benefited from earlier state legislation that attempted to deal with natural resource issues. Many state and local natural resource laws were enacted before the federal government addressed these issues. These regional laws often served as models for the development of national law.

During the twentieth century, the federal government assumed leadership over state and local governments in developing comprehensive natural resource policies. Many of the issues associated with natural resources required a large view of national trends such as universal resource depletion and environmental degradation. The federal government has taken the lead because many local governments had neither the broad perspective nor the resources to coordinate legal solutions to universal issues. Also, most public land ownership resides with the federal government. State and local governments control relatively little public land.

Budget Policy

Although legal authority guides agency function, budgetary authority drives it by paying salaries and other operation costs. All public policy is contingent upon appropriate allocation of budget among government functions. Many agency authorities are limited by the size of the budget allocated to them. Budget development and justification is a crucial aspect of all organizational activity. Budget justification begins with how effectively salary is used to attain authorized outcomes. Job description and performance assessment often are tied back to the anticipated performance of the agency.

Budget policy varies among different private and public organizations, often complicating the coordination of activities. Although federal government budgets extend from October 1 through September 30, the budget year for other organizations commonly start in January or July. Planning for budget allocation is diverse, even among agencies within governments. These complicating differences reflect past independence in the planning and management process, many of which have contributed to existing resource management problems.

Public and Private Policy

Public policy takes precedence over private policy. Public policy can effect desired human behavior through either punishment or reward. Prohibitive laws allow no personal choice and administer punishment when violated. In contrast, tax policy often provides incentives for desired behavior. Tax policies that mimic the private marketplace provide positive incentives in place of the negative incentives from threats of fine and imprisonment. Policy can be relaxed, or laissez faire, leaving much personal choice or can be quite restrictive, leaving little personal choice. With respect to natural resource management, public policy has become increasingly restrictive as resources in general have grown scarcer and user actions have had increasing impact on each other.

This is no clearer than with air quality laws. Clean air, a resource "owned" in common and supporting human health and happiness, has grown much scarcer because the atmosphere is used as a sink for industrial by-products. Population growth, industrial development, and growing use of carbon-based fuels have made clean air more scarce. The determination of resource value and scarcity depended mostly on government-sponsored research. Until recently, the government has emphasized regulation of emission amounts under threat of fines. More recently, as knowledge has grown, there is growing interest in selling permits to allow set amounts of waste emission at permit prices that would encourage reduction of waste emission (see Chapter 4). The permits could be sold to others. This approach allows more choice and relies more on market incentives.

MAKING POLICY

U.S. Constitution and Policy Development

The U.S. Constitution guides all government policy development in the United States. Its interpretation is the source of some of the more vexing social problems pertaining to resource management. Constitutionally compatible laws only provide equal opportunity to gain economic and other resources. They do not assure equal economic resource allocation among citizens. In this way, the Constitution supports a free enterprise system based on fair competition.

The Constitution was established to assure that the U.S. government served the public interest by improving the general welfare of all citizens. This imperative establishes guidance for government decisions that affect all aspects of welfare and has influenced the development of economic and other measures of welfare derived from government action. Although the intent of welfare improvement seems clear, there is much room for interpretation. General welfare can be promoted overall while harming individuals. Thus two important issues underlie many natural resource management decisions. First is the determination of how much benefit is generated from management with respect to the costs. Second is the redistribution of benefits needed to compensate those who lost benefits.

The Constitution also asserts protection of liberty for both present and future generations of citizens. A broad interpretation of this assertion precludes avoidable deprivation of the rights of future citizens to the use of public resources. Unnecessary destruction of resources reduces the freedom to choose among options for pursuing opportunity. Thus the government in the United States exists to improve the general welfare of its citizens while preserving the freedom of future generations to develop and use the Nation's resources. These constitutional mandates establish the goals of public natural resource management in the United States.

The Bill of Rights and many of the amendments that follow clarify personal privileges of citizenship. Through these amendments, the Constitution respects individual rights to pursue opportunities as they see fit as long as they do not threaten the rights of others. The government is obligated to consider the desires of all public sectors.

Whereas the Constitution grants the right to private property, it also grants the right of the government to purchase upon demand (condemnation authority) and at fair price, any private land deemed necessary for public functions. This provision protects options for developing beneficial public infrastructure such as transportation systems and other common-use areas. More recently this has been the basis for demanding compensation for decreased land values associated with enforcement of the Endangered Species Act. This controversial interpretation will probably remain a policy issue for some time to come. A 1984 amendment to the act allows a habitat conservation plan to be developed by the landowner as a way to mitigate the restrictive effect on the landowner, but problems remain to be settled.

Legislation

The legal system in the United States and other democracies often works imperfectly, but it has been an effective way to communicate and coordinate principles for group behavior. Although many people equate law with constraint of personal behavior, it is a means by which society resolves conflicts among individuals and groups, and fosters opportunity. Lawmaking is an incremental adaptive process. Many long-established natural resource laws have been amended numerous times and have had various sections repealed.

Most laws are developed through interactions among legislators, legislative aids, special interest groups, and agencies authorized to carry out the policies. The press plays an important role by communicating the issues. Special interest groups usually assign volunteers or pay lobbyists to represent their interests. The word lobbyist derives from a time when special interest representatives cornered legislators in the lobbies of state and federal capitols to influence legislation. Although one-on-one interaction remains, especially in state and local government, more often lobbyists bring their issues to legislative aids, where legislative concepts are first discussed, outlined, and drafted. Special interest groups also affect legislative action through financial support of campaign activities of legislators. Government agencies also influence legislation based on past experience in use of authority (Rourke 1984).

The special interest approach to legislation has advantages and disadvantages. The main advantage is that special interests are able to organize behind spokespersons who efficiently influence legislative change. As long as all interests are well represented by special interests, this often is the most efficient way to systematically sort through the pros and cons of each issue before legislative action. A pure democracy is vulnerable to apathy and ignorance about issues. For many issues, the majority of people are disinterested and not very well informed. Disadvantages arise when interests are low profile but widespread, without articulate representation, and their financial backing is overwhelmed by wealthier interests. They also exist when the agencies authorized to regulate excesses of special interests instead become advocates of the special interests.

In response to what has been seen as too much special interest manipulation of the policy process, an increasing amount of state and local policy is developed directly by public referenda. Setting up a vote by referendum typically requires a public petition of significant size and approval of the legislature. This strategy may be effective when the issue has general support and limited opposition from special interests. However, the ultimate social effectiveness of legislation by referendum remains to be decided. The federal government does not as yet present referendums to the public for vote.

Government agencies in the United States execute their legal authorities under the chief executive, the President. Executive orders have the force of law within the administration, at least until the next president takes office. Some policy is sustained through executive order. State governors have similar authorities. Laws also may be called into question with regard to their constitutionality. The judicial branch of government has the authority for interpreting whether law is constitutional. Many natural resource and environmental laws have undergone constitutional "tests." Court decisions may be appealed through a series of court levels culminating with the U.S. Supreme Court.

Common Elements of Law.   Natural resource law is voluminous and diverse. Many laws include a brief justification for the law. Numerous laws are reminders of the need to improve the general welfare and assure that public benefit exceeds management cost. Programs authorized by law often have a spending limit attached to them. While most laws typically authorize a specific agency to take leadership, many comprehensive laws direct coordination among agencies and the public.

Most authorization law directs development of specific codes; that is, the standards and regulations to be determined and obeyed. Such things as pollutant classification and concentration, wetland qualities, hunting and fishing regulations, engineering specifications, grazing rates, and timber harvest rates are recommended by technical employees in the field for approval at the highest levels. In contemporary proceedings regarding rules and regulations, the public frequently is invited to comment in advance. Technical recommendations are not always accepted by the public. The communication skills needed to extend information in a meaningful way to the public are in high demand. Even so, policy often is shaped by considerations other than technical merits.

Heads of agencies most typically are appointed by the president, governor, or other chief executive officer, but some state agency officials are elected. Often an oversight commission or board is appointed or elected to assure the public is well served. Although public representation is the idea behind directorships, commissions, boards, and other authorized governing bodies, public input is rarely polled directly using contemporary techniques that minimize bias. The system continues to function primarily through special interest persuasion.

ORGANIZATIONAL ADMINISTRATION _____

Integrating Organizational Activities

Organizational planning and natural resource management are integrative processes both impeded and facilitated by organizational structure. Organizational units are formed from the integration of tasks performed by individuals into a work product that is more than the sum of the individual tasks. The effectiveness of planning and management is influenced by organizational structure. Reorganization often is an unavoidable adjustment to new conditions and responsibilities or to more effectively meet established responsibilities. However, reorganization also is disruptive of established links among units resulting in trade-offs. Each reorganization must be weighed against the anticipated improvement.

In large organizations, units of increasing size envelop smaller units. Each unit functions to produce refined outputs from less refined inputs passed on to it by other units. Outputs may be material products like lumber produced from inputs of logs, sawmill facilities, payroll, and human resources. Many units in large organizations exist for the purpose of refining raw information into more refined or applicable reports, plans, and other outputs through data acquisition, analysis, and conceptual synthesis. Each functional unit is in turn integrated into larger units with similar functions to form the entire organization. Large organizations, like the executive branch of the U.S. government, have many organizational branches and many subunit supervisory levels. The uppermost level is occupied by the organizational leader, a Chief Executive Officer (CEO). At the next lower level, organizational units branch off into functional groupings.

Natural resource management is administered and planned by numerous private enterprises at national, regional, and local governments. The coordination of management activities to accomplish organizational goals is enhanced by an integrative and comprehensive organizational planning process. Plans are meaningless without adept administration, which leads the organization in developing and executing organizational plans. Administration facilitates the planning process and plan realization through coordination and communication within and among administrative divisions and units. The organization of these units usually derives from the diverse responsibilities assigned to people throughout the organization and from the geographic location of the activities.

The executive branch of the U.S. government, for example, is divided into departments, each of which has a central focus. Most of these departments are involved with natural resource issues in one way or another. However, natural resources are the primary focus of the Departments of Agriculture, Interior, and, to a lesser extend, Commerce. The Department of Interior, for example, has a wide variety of natural resource responsibilities mixed with certain social responsibilities. It has broad authority over fish, wildlife, certain water development, rangeland, and mineral development, especially with respect to public lands and waters. In the Department of the Interior, the Fish and Wildlife Service is administered through programs and geographic regions—each area of which has a similar organizational structure. Administrative regions are determined primarily by the boundaries of states composing a region.

Information Flow and Authority

Each position within an organization is assigned a certain amount of authority over its own function and the function of supervisees in a lower level. Authority is delegated downward from upper administrative levels to lower levels in supervisory lines (Figure 5.8). Authority delegation determines the decision discretion for each position and is usually described for the position. Supervisory authority is indicated by the subordinate positions assigned in the job description or in modifications made from positions of higher authority (Figure 5.9). The chief administrative officer of an organization has the most comprehensive position in the planning and operations process. Planning, the decision process, and operations become more focused as one progresses from the top of the organizational pyramid to the bottom. The implications of decisions and operations become increasingly broad with progress up each rung of the administrative ladder.

Staff positions stand in support of line authority, but are not in the direct chain of command. Planners, advisors, and other special service positions often reside outside the chain of command in a "side-bar" support role. They respond to the requests of line authority with analyses and recommendations. They may have their own su-

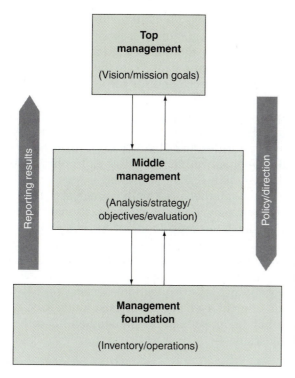

FIGURE 5.8 Information flow in a management organization. In general, policy directives and guidance flow from upper levels of management to lower levels. Information in the form of various written and oral reports with recommendations move upward from the foundational "field-level" of management.

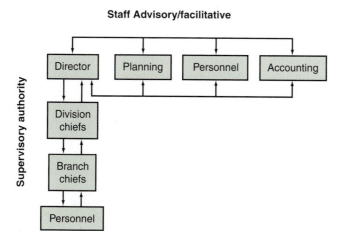

FIGURE 5.9 General lines of authority and information flow within a natural resource management organization. Levels below the Director also may have support staff with specialized skills.

pervisees outside the line of authority, but do not make command decisions that reach outside the staff unit.

Administrative structures exist to bring more cohesiveness to people operating toward the completion of a single or small set of outputs. An appropriate organization facilitates an efficient and an effective decision process leading to more desirable work outputs. The form or structure of a management organization has a lot to do with how efficiently organizations function.

Information Management

All natural resource organizations are faced with a tremendous information resource from which they need to distill relevant information for an effective decision process. Regardless of other functions, each subsystem must use, manage, and add to information needed for organizational function. The structure of a management organization facilitates an efficient communication process when form follows function. In efficient organizations, each administrative unit is a subsystem designed for information input, processing, and output. Input information travels "bottom-up" to the most comprehensive level of the decision process. Only information appropriate for organizational function is accepted as input at each level in the administrative ladder. The "noise" is filtered out. Reporting needs to be readable, comprehensive, to the point, and concise.

Linkages among administrative units serve as conduits for information flow upward with appropriate filtration at each level. In this way, an effective organization passes an optimum quality and quantity of information to the uppermost administrative levels. Obviously, to work well, each administrative unit must be appropriately informed about organizational priorities and about the resources and conditions that exist inside and outside the organization. Priorities and policy guidance flow top-down from the command level of the CEO.

Organizational flexibility must be counterbalanced with enough stability to sustain effective pursuit of purpose. Bureaucracy is a form of organizational arteriosclerosis that occurs when structure becomes too rigid to adapt to functional requirements. Like blood moving through stiff arteries and damaged kidneys, information flow is slowed and inappropriately filtered. In such organizations, priorities and policies change little despite dissatisfaction with organizational outputs. Bureaucracy is relatively rare in private profit-making organizations because they compete poorly and go out of business. Government agencies can persist much longer in an inflexible inefficient state—until they no longer can be tolerated and have budgets cut or are entirely eliminated or reorganized.

Networking

In the past, organizational structure controlled information flow much more than it does now. Horizontal information flow moved within the same administrative level, most often to people occupying the same common work space. Vertical information flow occurred between supervisors and supervisees within an administrative line of authority. Opportunities for information to flow obliquely and to jump levels was possible, but more limited. With the development of contemporary electronic networking, information has the potential for moving vertically, horizontally, and obliquely among workers within organizational structures. Information flow also can readily jump intermediate levels in either direction. It can also jump from organization to organization. This technically mediated freedom of information flow is a boon to integrated resource management. However, there is also more opportunity for misinformation to be passed. Whereas organizational vigilance over information flow was once more centralized, technological networking forces more individual responsibility on all personnel.

SUMMARY

Planning, policy, and administration control the functions of management systems operations. They determine the effectiveness and efficiency with which all organizations pursue their purposes. Whether business, government, or advocacy groups, organizations operate based on principles held in common. Most organizations function to benefit humanity in some way. They use a variety of strategies to determine the most appropriate distribution of benefits among members of society. Evaluation of result is necessary for determining the degree of success and for improving the efficiency and effectiveness of the management process.

In an era of increasing complexity, many organizations divide functions into program and project subsystems. They integrate activities across divisions within organizations and across organizations. Both private and public organizations have concentrated more on their internal environment in the past than will be necessary in the future. Advances in communications and computing technology have enhanced possibilities for more effective integration. Among the biggest challenges are determining and executing that level of organizational integration needed for maximum sustained benefit to society.

LITERATURE CITED

Braybrooke, D. and C. Lindblom. 1963. *A strategy of decision.* New York: Macmillan.

Bryson, J. M. 1995. *Strategic planning for public and nonprofit organizations.* San Francisco: Jossey-Bass.

Crowe, H. A. 1983. *Comprehensive planning for fish and wildlife.* Wyoming Department of Game and Fish.

Cullingworth, B. 1997. *Planning in the U.S.A.: Policies, issues, and processes.* New York: Routledge.

Dramstad, W. E., J. D. Olson, and R. T. Forman. 1996. *Landscape ecology principles in landscape architecture and land-use planning.* Washington, DC: Island Press.

Garvin, A. 1996. *The American city: What works, what doesn't.* New York: McGraw-Hill.

Kaiser, E. J., D. R. Godshalk, and F. Stuart Chapin, Jr. 1995. *Urban land use planning.* 4th ed. Urbana, IL: University of Illinois Press.

Koteen, J. 1989. *Strategic management in public and non-profit organizations.* New York: Praeger.

Lorange, P. and R. F. Vancil. 1977. *Strategic planning systems.* Englewood Cliffs, NJ: Prentice-Hall.

MacDonald, L. J. and S. F. Bates. 1993. *Natural resources policy and law: Trends and directions.* Washington, DC: Island Press.

Moreau, D. H. 1998. *A giant step backward.* Water resources update. Council on Water Resources. Issue No. 110.

Osborne, D. and T. Gaebler. 1992. *Reinventing government.* Reading, Mass: Addison-Wesley.

Rourke, F. E. 1984. *Bureaucracy, politics and public policy.* 3rd ed. Boston: Little, Brown and Company.

Steiner, F. 1991. *The living landscape: An ecological approach to landscape planning.* New York: McGraw-Hill.

Walters, C. 1986. *Adaptive management of natural resources.* New York: McGraw-Hill.

Air, Water, and Land Resources

Air, water, and land interact via earth systems to provide the necessities for terrestrial life, including the basic needs of humanity. In Chapter 6, the atmosphere is shown to be a reservoir for weather and the life-essential gasses, such as oxygen, carbon dioxide, and water vapor. Atmospheric water is linked to the land through the hydrosphere—the realm of water presented in Chapter 7. Water is the universal solvent in which the many materials essential to and affecting life processes are suspended, dissolved, and transported. Soils, discussed in Chapter 8, act with air, water, and the diverse living ecosystems of the earth, described in Chapter 9, to support forest, range, farm, park, urban, wildlife, fishery, and other renewable resources. Soil is the product of and the foundation for ecosystem process. Erosion at rates in equilibrium with undisturbed ecosystem process contributes to the flow of nutrients and sediments supporting life in wetland and aquatic environments. Accelerated erosion caused by careless use of the soil resource diminishes both the soil's fertility and the useful productivity of aquatic ecosystems. Human conversion of the carbon held in vegetation and fossil fuels to carbon dioxide in the atmosphere may be altering the heat content of the biosphere and changing the balance of liquid water, water vapor, and ice in the biosphere. The redistribution of materials from natural resource development and its consequences will be a central concern of natural resource management in future decades.

CHAPTER 6

Atmospheric Resources and Climate

THE IMPORTANCE OF ATMOSPHERE AND CLIMATE

The atmosphere holds many resources of direct value to humanity and influences the sustainability of several others. The atmosphere stores oxygen and nitrogen, essential elements for all life. It plays a critical role in the hydrologic cycle, which sustains fresh water supplies. It can assimilate a certain amount of gaseous waste products. The atmosphere maintains temperatures conducive to life processes. It is the source of climate and weather, which interact with water and earth to determine the attributes and functioning of the world's ecosystems. Climate here refers to the seasonal and annual patterns of atmospheric conditions over relatively large regions of the earth's surface. Weather is the state of the atmosphere at a particular time and place with regard to temperature, humidity, cloudiness, precipitation, and wind. Climatic variation and weather are important sources of uncertainty in management outcomes for many natural resources.

A sound understanding of atmospheric processes and climate is essential for management of living resources. In this chapter, we will discuss basic principles pertaining to the atmosphere, then use those principles to explain how they influence the attributes of living resources in different geographical areas of the United States and elsewhere. Human impacts on climatic stability will also be discussed. Our coverage of this subject is derived in part from Trewartha (1961), Rumney (1968), Ricklefs (1976), Roberts and Lansford (1979), Godish (1997), Holechek et al. (1998), McKnight (1999), and Dunn (2001).

GLOBAL WARMING AND OZONE PROBLEMS

During the past decade, world climate changes associated with global warming and ozone depletion have been elevated to the highest level of natural resource management concerns. In the 1980s, an "ozone hole" was discovered over Antarctica. Since then a similar hole was discovered over the Arctic with evidence of generally increased depletion over the entire globe. Global warming started to become a serious

159

concern about the same time, initiating increased international research. There is now enough evidence to convince many atmospheric scientists that global warming is occurring at least in part because of fossil-fuel use and other human activity (Houghton et al. 1996, OSTP 1997, Godish 1997, and Houghton et al. 2001). This evidence is derived from both empirical observations and predictions from mathematical models based on understanding of underlying processes. However, model predictions of change rates, eventual climatic outcomes, and consequences of different proposed management changes remain somewhat uncertain. They vary widely around an average predicted increase of 3° C over the next century compared with a 0.6° C increase during the last century (Houghton et al. 2001).

A recent report of the Intergovernmental Panel on Climate Change (Houghton et al. 2001), established by the World Meteorological Organization (WMO) and the United Nations Environment Programme (UNEP) in 1988, summarized other climate changes congruent with climatic warming. Temperature measurements indicate that 14 of the hottest years on record have occurred during the past two decades (Bright 1997, United Nations Development Programme 2000). Low temperatures have become less extreme while high temperatures have become more extreme as carbon dioxide and other greenhouse gasses have increased in concentration. Other changes that occurred during the twentieth century included a two-week decrease in ice cover in the Northern Hemisphere, widespread retreat of mountain glaciers outside the polar regions, a 10% decrease of sea-ice area and a 40% decrease in summer ice thickness in the Northern Hemisphere, a rise in sea level of about 0.1 to 0.2 m, an increase in global heat content, increased precipitation in the temperate zone and the Tropics, decreased precipitation in the subtropical regions (such as the sub-Saharan region), increased incidence of heavy precipitation events, increased cloud cover in middle to high latitudes, and increased drought frequency and intensity in parts of Africa and Asia. Since the 1960s, average snow cover has decreased about 10%. In addition, the ranges of many wildlife species have expanded toward the Poles. The process and implications of global warming and ozone changes are described in this chapter following a description of basic principles.

CLIMATIC FACTORS

Climate for a given area is primarily a function of the seasonal and annual patterns of atmospheric elements; particularly wind, temperature, and precipitation. The most important factors determining climate are atmospheric composition and pressure, solar radiation, earth shape, latitude, continental distribution, topography, marine currents, and the earth's orbital revolution around the sun. To a lesser extent, the vegetation in an area influences the climate. Humans have impacted local climates by cutting forests, building dams, and building factories that release large amounts of carbon dioxide (CO_2) and other compounds into the atmosphere. Volcanoes have been among the primary natural factors that have modified local and world climatic conditions for short time periods.

Atmospheric Composition and Pressure

About 99% of the earth's atmosphere is made up of nitrogen and oxygen (Figure 6.1). Among the remaining 1% are water vapor, carbon dioxide, ozone, methane, and various other gasses. It is also composed of colloidal aerosols, dust, and ions in aqueous solution. The atmosphere is most dense near the earth's surface. Density decreases at higher altitudes. Nitrogen, oxygen, and water vapor are concentrated close to the earth's surface and decrease in density with movement upward from earth. Natural ozone forms a layer in the outer atmosphere. Because gasses contract at cooler temperatures and expand at warmer temperatures, heating and cooling changes density and causes mixing as cool air sinks and warm air rises. This also causes areas of low and high barometric pressure and changing pressures at any one spot.

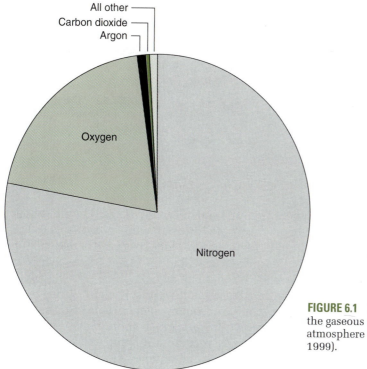

FIGURE 6.1 Proportional volume of the gaseous components of the atmosphere (from McKnight 1990, 1999).

Solar Radiation

Solar radiation is composed of electromagnetic waves of energy characterized by wave length. Only a small fraction of solar radiation is visible light with wavelengths between about 450 and 700 nanometers. The human eye recognizes different wavelengths of visible light as different colors. Wavelengths shorter than the visible range fall into the ultraviolet range, and longer wavelengths are infrared radiation. In the visible range, blue light is shorter than red light with greens and yellows in between. The sun radiates wavelengths in unequal amounts and more red is radiated than blue. However, blue light has about twice the energy of red light.

These attributes of solar radiation become important when they interact with gasses, aerosols, and particles in the atmosphere. Dust particles tend to reflect light uniformly back to outer space, where it is lost from the earth forever. Aerosols of sulfur and other compounds also reflect most visible wavelengths. Therefore, dust and aerosols cause atmospheric cooling. Other compounds reflect and absorb light more selectively. Ozone also intercepts, absorbs, and reflects high energy ultraviolet light and acts as a shield to this potentially harmful radiation. Infrared radiation is absorbed by water vapor, carbon dioxide, methane, and a variety of anthropogenic gasses called greenhouse gasses. The greenhouse gasses maintain a higher temperature at the surface of the earth than would occur without them.

Earth Latitude, Shape, and Rotation

Radiation reaches the earth with greatest intensity at the equator. With distance from the equator, measured by latitude, the slope of the earth's spherical surface increases the angle of incident radiation. As a consequence, the intensity of radiation per unit area of landscape decreases from the equator to the Poles. Because radiation is less intense at higher latitudes, less is absorbed and transformed to heat.

In general, the earth's mean annual climate is warmest at the equator and becomes progressively cooler toward the Poles. Day length becomes increasingly variable and

FIGURE 6.2   The distribution of land and water (from McKnight 1990, 1999).

light less intense with movement from the equator to the Poles. The sun's highest position in the Tropics is directly overhead, or close to it, while near the Poles it is closer to the horizon or entirely behind it. Major geographical patterns in temperature, rainfall, and wind are caused by the uneven distribution of the sun's energy over the earth's surface as a result of the earth's spherical shape.

The ability of air to hold water increases as it warms. Each 10°C rise in temperature nearly doubles the rate of evaporation from a water surface. This explains why warm tropical air around the equator can hold much more water and precipitate much more than arctic air. Tropical air is very humid compared to polar areas because water is cycled more rapidly through the atmosphere over the Tropics. The high temperatures cause greater evaporation from the oceans and evapotranspiration (a combination of plant transpiration and evaporation) from the continents. Warming air expands and rises until cooled to the temperature of surrounding air. The process of cooling with gain in altitude causes condensation and precipitation.

As tropical air rises and cools, water condenses and precipitates almost daily. The rotational spin of the earth around its axis moves faster than the atmosphere and causes easterly tradewinds away from the equator. As the rising air moves toward the middle latitudes it cools, condenses into clouds, releases precipitation, and dries out. Therefore, the middle latitudes of the continents are occupied mostly by arid lands. Farther from the equator, in the Temperate Zone, the atmosphere gains moisture mostly from oceanic sources and humidity rises, becoming greatest in summer heat, and precipitation increases. With greater cold and less ice-free exposure of water, both evapotranspiration and precipitation decrease markedly as the Poles are approached.

Distribution of Continents

The distribution of continents affects climate because water evaporates more rapidly from exposed surfaces of water than from land (Figure 6.2). Therefore, inland areas generally have much lower precipitation than coastal areas. The greater area of water surface also accounts for the greater rainfall in the Southern Hemisphere compared with the Northern Hemisphere. Oceans and lakes comprise 81% of the Southern Hemisphere compared with 61% of the Northern Hemisphere. In addition, the ability of the continents to absorb and store heat is less than that of the oceans. As a result, areas inland from the ocean vary much more in seasonal temperature than areas near the ocean. In addition, ocean heat storage causes a lag effect in coastal temperatures, causing warmer autumns and cooler springs than occur inland.

Topography

Altitude.   Topography influences climate because as air gains altitude it cools, contracts, and concentrates water vapor. This results in condensation and precipitation. As air descends the leeward slopes of the mountains it warms, expands, and dries out. Areas on the leeward side of mountains are usually quite dry. The Great Basin

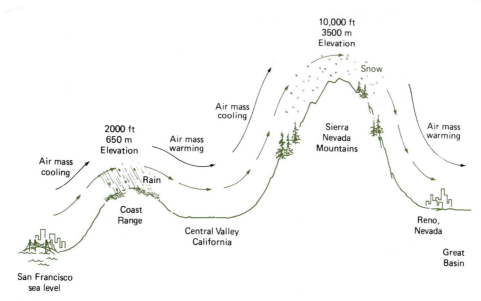

FIGURE 6.3   As air masses move inland from the Pacific Ocean, they rapidly lose their moisture due to cooling as a result of elevation increases from the Coast Range and Sierra Nevada Mountains. This is called the *orographic effect* (from Holechek et al. 1998) (drawing by John N. Smith).

area of the western United States is quite arid because it lies between the Rocky Mountains to the east and the Cascades to the west (Figure 6.3).

Topography interacts with temperature and precipitation to determine the climate for any area. Much of the world is characterized by large topographic variation. This is particularly true in North America. The wide difference in topography between the eastern and western United States largely explains the difference in climate and vegetation. Hopkins (1938) related altitude, latitude, and longitude to vegetation in a law known as *Hopkins's bioclimatic law*. The interpretation of Hopkins's law reveals that a 305 m increase in altitude will result in essentially the same phenological (relationship between climate and plant growth cycle) changes that would be encountered traveling 107 km north with no increase in elevation. This law applies to many parts of the western United States, where extremes in plant development during the spring may occur within a few kilometers because of elevation differences.

Aspect.   *Aspect* refers to the directional orientation of slopes. The amount of solar radiation and temperature on slopes in North America increases as aspect changes from the north to east or west, and from west to south. The aspect of the slopes influences the vegetation the slopes support. Where warmer south- and west-facing slopes support grasses and herbs, the cooler north- and east-facing slopes often support trees (Figure 6.4). Aspect is much less important in the Tropics where solar radiation is more evenly distributed over all slopes.

Degree of Slope.   Degree of slope is of considerable importance because it affects both plant productivity and human ability to harvest plant resources. In land surveys, slope is commonly expressed as percent. As slope increases, vegetation productivity declines per unit of precipitation because less water enters the soil and more runs off as overland flow.

Marine Currents

The primary role of marine currents in climate determination is through heat transfer. Because of the same rotational forces that effect wind movements, marine currents move clockwise north of the equator and counterclockwise south of the equator (Figure 6.5).

FIGURE 6.4   Grasslands occur on south-facing slopes and forest lands occur on north-facing slopes in northeastern Oregon.

Warm current

Cool current

FIGURE 6.5   The major ocean currents. Water movement generally proceeds clockwise in the Northern Hemisphere and counterclockwise in the Southern Hemisphere (from McKnight 1990, 1999).

Cold water moves along the western coast of the continents toward the Tropics while warm water from the Tropics moves along the eastern coasts of the continents away from the Tropics. Therefore, the Pacific coast of southern California and Mexico is drier than the Gulf and Atlantic coasts of the same latitude in the eastern United States.

Interactions between Climatic Factors

The climatic factors and elements interact to help determine the various world ecosystems. Rainfall during warm months generally promotes more plant growth than the same amount of precipitation during cold months. In the northern Mixed Prairie country of western North Dakota, 16 inches of precipitation is more effective than 25 inches in the southern Mixed Prairie of western Oklahoma, even though precipitation amounts are similar during the summer. This is because evapotranspiration is much less during the summer in North Dakota, and cold prevents plant use of winter precipitation in western Oklahoma. To help reflect on such interactions, Weaver and

Clements (1938) developed a system of classification of vegetation potential based on the ratio of precipitation to evaporation (P/E). Under their system, an area characterized by a P/E ratio of less than 0.2 would be a desert, 0.2–0.4 a shrubland, 0.4–0.6 a grassland dominated by short or midgrasses, 0.6–0.8 a grassland dominated by tall grasses, 0.8–0.10 a deciduous forest, and greater than 1.0 would be a coniferous forest.

Mean air temperatures drop with increases in altitude and latitude. Every one kilometer increase in elevation is characterized by about a 6°C decrease in temperature. An 800-kilometer (500-mile) increase in latitude results in a 6°C drop in temperature. However, high-elevation tropical areas are not characterized by the same climatic conditions as low-elevation temperate areas. This is because in the Tropics there is little seasonal variation in climate while large variations occur in temperate areas.

Vegetation as a Climatic Factor

Vegetation responds to climate and can also contribute to climate. Forest vegetation typically has a greater impact on climate than vegetation with less biomass and vertical structure. Vegetation modifies topography and, in doing so, alters wind patterns and changes the light transmission to the ground. Forest interiors are darker, more humid, more stable in temperature, and less windy than adjacent deforested areas. Precipitation intensity is also moderated. Transpiration from vegetation dries the soil and adds humidity to air above the forest. In addition, vegetation sequesters (holds) carbon in biomass, peat, and other stored organic matter, thereby influencing the pool of carbon dioxide in the atmosphere. Photosynthesis removes carbon dioxide from the atmosphere and generates oxygen as a by-product.

CLIMATIC ELEMENTS

Wind

Wind, like precipitation and temperature, is caused by the solar radiation interacting with the earth's rotation and gravity. Rising air in the Tropics moves both north and south in the upper layers of the atmosphere. The rotation of the earth on its axis results in a deflection of the rising air to the right in the Northern Hemisphere and to the left in the Southern Hemisphere (Figure 6.6). This deflective force is called the Coriolis force. The region of rising air at the equator is referred to as the doldrums

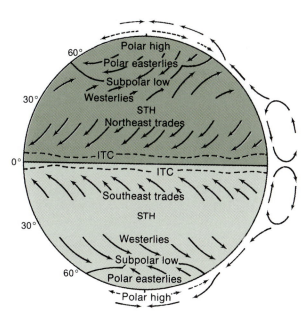

FIGURE 6.6 The general circulation of the atmosphere disregarding the major land masses (from McKnight 1990, 1999).

because there is almost no directional air movement. The early sailing ships were stranded for long time periods in this zone.

Wind can substantially reduce precipitation effectiveness by increasing soil evaporation losses and increasing plant transpiration. Wind has little influence on soil moisture below 20 to 30 cm (Veihmeyer 1938). The highest wind velocities occur in flat terrain with few trees. The Great Plains in the central United States are characterized by high wind velocities, particularly in most of Washington, Oregon, and California, where topographic barriers are at a maximum. Hot summer winds during dry years greatly magnify the effect of drought in the Great Plains by increasing both soil moisture losses and plant transpiration.

Temperature and Frost-Free Period

Temperature.   Temperatures in different seasons vary substantially in different parts of the world (Figure 6.7). In tropical areas, monthly and yearly temperatures vary little. Seasonal variation increases with latitude and distance inland from oceans and

(a)

(b)

FIGURE 6.7   Mean January (a) and July (b) temperatures (degrees Fahrenheit) for the world (from McKnight 1990, 1999).

large lakes. Extreme cold reduces the availability of moisture to vegetation by freezing the soil. Temperature also affects enzyme and other biochemical reaction rates.

For example, in the mountain areas of the northwestern United States, temperature can have as much influence as precipitation on annual variation in productivity of vegetation. Nearly all native plants in this region are cool-season plants. They grow to limited extent during intermittent warm periods in the winter, but most of their growth occurs in the spring. Years with above-average spring temperatures are usually characterized by above-average plant growth even though total precipitation may be below average. However, in years when below-freezing temperatures occur late in spring, plants often grow more slowly regardless of precipitation amount. This often occurs because much of the moisture in the soil is evaporated by wind before temperatures are warm enough for high rates of growth. When high wind velocities occur, snow is evaporated directly from the snowpack on high-elevation areas through a process called sublimation. In most years, summers are too dry for plant growth in the Northwest. In the northern portions of the Great Plains, particularly Montana and North Dakota, spring temperatures vary considerably from year to year and can have a substantial influence on productivity of cool-season prairie grasses.

Above-average temperatures usually occur concurrently with drought in the Great Plains. This increases evaporation of the limited soil moisture and greatly magnifies the effect of drought.

Frost-Free Period. On high mountainous areas (over 2,000 m elevation) in the western United States, the frost-free period at many locations is less than 100 days. This gives plants only a very short period to complete their growth cycle. The low temperatures on these lands are a much bigger constraint on vegetation productivity than is precipitation.

Precipitation

Importance. Precipitation is the most important single factor determining the type of vegetation and its productivity in arid to semiarid terrestrial areas. Vegetation productivity increases rapidly as precipitation increases up to about 500 mm per year. Above 500 mm of precipitation per year, soil characteristics can assume much greater importance than precipitation in determining vegetation productivity. The amount of water held in the soil is significantly important. Wetlands are among the most productive ecosystems. Critical characteristics of precipitation that affect ecosystems are the total amount, the distribution, the form, and the annual variability. The average annual precipitation for various parts of the world is shown in Figure 6.8.

Precipitation Variability. The growth-limiting effects of relatively low annual precipitation in the western United States is made more severe by relatively high year-to-year variability (McKnight 1990, Figure 6.9). The variability in precipitation increases rapidly as the annual total drops below 450 mm per year (Conrad and Pollak 1950). Even slight reductions from normal precipitation can reduce plant yields in areas with less than 300 mm of precipitation, while much greater reductions in precipitation may have no influence on plant yields in areas with more than 800 mm of precipitation (Klages 1942). Annual variability in timing of precipitation can be more important than variability in the total amount that occurs.

Drought. *Drought* is defined as prolonged dry weather, which is commonly indicated by annual precipitation less than 75% of the average (Holechek et al. 1998). Using this criterion, over the past 54 years (1944–1998) drought in the United States has occurred in 43% of the years in the Southwest, 13% of the years in the Northwest, 21% of the years in the northern Great Plains, and 27% of the years in the southern Great Plains. The most severe and widespread drought in the Great Plains region of the United States in recorded history occurred between 1933 and 1935. In the southwestern United

FIGURE 6.8 Average annual precipitation for different parts of the world (from McKnight 1990, 1999).

Centimeters	Inches
less than 25	less than 10
25–50	10–20
50–100	20–40
100–150	40–60
150–200	60–80
More than 200	More than 80

0 1000 2000 3000 mi
0 1000 2000 3000 4000 km

Percent

	Under 10
	10–15
	15–20
	20–25
	25–40
	Over 40

FIGURE 6.9   Precipitation variability in the world indicated by the yearly percentage departure from the average (from McKnight 1990, 1999).

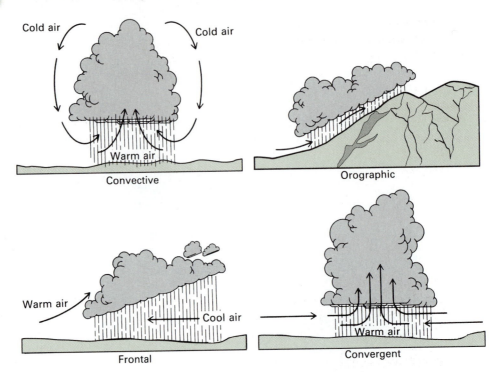

FIGURE 6.10   The four basic types of storms (from McKnight 1990, 1999).

States the most severe drought occurred between 1951 and 1956. These droughts all had tremendous influences on native vegetation.

Precipitation Systems

Four types of storm systems typify local precipitation conditions. These include orographic, frontal, convective, and convergent storms (McKnight 1990). Knowledge of these storm systems is useful for understanding the climatic causes of different ecosystems.

Frontal Storm Systems.   Frontal storms are caused by one air mass displacing another along the line of contact between air masses known as a front. When this happens, the warmer air mass is forced upward (McKnight 1990, Figure 6.10). Frontal types may be distinguished by the way they move relative to the air masses involved. If warm air is replacing cold air, it is termed a warm front. If cold air is replacing warm air, it is a cold front. A front that is not moving is termed a stationary front. The leading edge of contact between two air masses is the front itself. Warm fronts tend to result in mild, prolonged storms because the warm air moves gradually into the upper cold air. In contrast, cold fronts usually result in intense, short-duration storms because the dense cold air moves very rapidly into the overlying layer of warm air. Frontal storms are the primary source of precipitation in the prairie country of the central United States.

Orographic Storms.   Orographic storms are caused by the lifting of a moving air mass over a topographic barrier. This type of storm causes much of the precipitation in the Appalachian, Rocky, Cascade, Sierra Nevada, and Coastal Mountains in the United States. This type of storm results from adiabatic cooling, which is the cooling of an air mass with increased elevation. The adiabatic cooling rate of dry air is about 10°C with each kilometer of increase in elevation. The adiabatic cooling rate is less for wet air because it can hold more heat. With increasing temperature, the adiabatic cooling rate decreases while the opposite is true of decreasing temperature. Orographic storms can cause considerable climatic variation between mountainous areas in close proximity.

Convective Storms. Convective storms result from a strong differential heating of the earth's surface, which causes an upward lift of associated air. This type of storm is common during the summer in the southwestern United States, and is usually a warm-weather phenomena. Reflection is a primary factor causing affected air to become warmer than its surroundings. The affected air rises very quickly and rapid condensation releases heat preventing the air from cooling to surrounding air. The condensation continues until the raindrops are heavy enough to fall through the updraft. This results in a localized high-intensity storm sometimes called a "cloudburst."

Convergent Storms. Convergent storms are less common than the other three storm types. They occur when air converges and results in a general uplift because of crowding (McKnight 1990). Convergent storms are frequently associated with convection and are most common in low latitudes.

Humidity

Humidity refers to the amount of moisture in the air. It is usually expressed as *relative humidity,* which is the percentage of the maximum quantity of moisture that the air can hold at the prevailing temperature. Cold air can hold less moisture than warm air. Rates of water evaporation and plant transpiration increase as the relative humidity decreases. Therefore, areas with high humidity give greater plant growth per unit of precipitation than do areas where humidity is low. The lowest summer humidities in the United States occur at low elevations in Nevada, Arizona, New Mexico, and California. Relative humidities increase with movement eastward across the Great Plains toward the Atlantic Ocean. The highest humidities occur in the southeastern states, particularly in Florida.

WORLD CLIMATE TYPES

Several approaches have been used in classification of climate. The system described by Dansereau (1957) has been accepted by many ecologists in recent years. This system involves categorization of climate into six broad types based on annual regimens that produce definite types of vegetation.

Equable Climate

This climatic type is characterized by lack of seasonality, or in other words, little annual fluctuation in temperature and precipitation. The most clear-cut cases of the equable climate are in central Africa (Sudan, Chad, Niger) and northern South America (Brazil).

Desert Climate

A low ratio of precipitation to evaporation (< 0.2) characterizes desert climate. Temperatures may vary seasonally, but precipitation is low year-round. Variability in precipitation from year to year is usually quite high. Vegetation is typically small shrubs, succulents, and ephemerals. Examples of this climatic type occur in the Mojave Desert in California, the Great Basin in Nevada and Utah, the Sonoran Desert of southern Arizona, and the Chihuahuan Desert of southern New Mexico.

Polar Climate

Areas with a polar climate have a cool summer with a very cold winter. Here average temperatures of the warmest months are below 10°C. Vegetation is characterized by low-growing grasses, sedges, and lichens. High mountain alpine areas and the tundra in northern North America and Asia have this kind of climate.

Mediterranean Climate

The Mediterranean climate is manifested in a cool, wet winter with a moderately warm, dry summer. This climate is, in general, mild. It is found at midlatitudes on west coasts. This climate favors evergreen woody shrubs with broad leaves. Central and southwestern regions of California have this kind of climate, as do Spain and Morocco.

Continental Climate

The continental climate is characterized by the large difference in temperature between the warmest and coldest months, unlike the types previously discussed. Several temperature and precipitation variations occur within this climate type, which result in different kinds of vegetation. The interiors of North America and Eurasia have continental climates.

Tropical Wet and Dry Climate

The tropical wet and dry climate characterizes southeastern Asia, central Africa, India, northern Australia, and parts of South America. This pattern is described by Wallen and Gwynne (1978) and Heady and Heady (1982). It is characterized by high temperatures throughout the year, with sharply defined wet and dry periods. Considerable variability exists in precipitation between years in most areas receiving this pattern. Temperatures below freezing occur only on high mountain areas. Tropical wet and dry climates with 130 to 250 mm of precipitation typically support desert shrubs, grasses, and cactus. Where annual rainfall is between 250 and 500 mm an open savanna of grasses and trees occurs. A dry forest with large trees and scattered shrubs occurs in the 500-, to 1,140-mm precipitation range. Tropical forest dominates most landscapes that have more than over 1,140 mm of precipitation.

U.S. REGIONAL CLIMATE TYPES

The major patterns characterizing the climate of the United States were placed in six categories by Humphrey (1962). These include the Pacific pattern, the Great Basin pattern, the Southwestern pattern, the Plains pattern, the Eastern pattern, and the Florida pattern (Figure 6.11). These will be discussed individually following Humphrey (1962) and Holechek et al. (1998).

The Pacific Climate

The pacific climate characterizes the Pacific coastal areas of California, Oregon, Washington, and British Columbia. It extends inland to the Cascade and Sierra Nevada Mountains. Summers are quite dry and winters are wet (Figure 6.12). The percentage of precipitation occurring during the winter increases from north to south, although total precipitation decreases. Precipitation also decreases rapidly with movement inland. The percentage of the total precipitation that falls during summer ranges from about 10% in northern Washington to 1% in most of California. The temperatures associated with this pattern are mild near the coast but become progressively colder with movement inland.

   The vegetation of the southern part of this climate type, in California and southwestern Oregon, is primarily chaparral and grassland. In Washington, British Columbia, and northwestern Oregon, dense coniferous forest occurs as a result of the high total rainfall (900 to 2,500 mm). Because this entire area is characterized by little summer rainfall, the herbaceous species are almost entirely cool-season. Grasslands occur

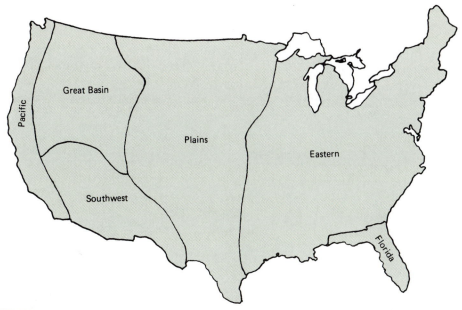

FIGURE 6.11 Climate types in the United States (from Holechek et al. 1998) (based on the classification of Humphrey 1962).

in low-elevation areas with 250 to 460 mm of annual precipitation. The warmer temperatures associated with these areas permit growth of the cool-season grasses in the winter and spring when moisture is available. Other parts of the world with this pattern of climate and similar vegetation include southern Europe around the Mediterranean Sea and the coastal areas of South Africa, Australia, and southwestern South America.

The Great Basin Climate

The Great Basin climate occurs largely in the states of Washington, Oregon, Idaho, Nevada, Utah, and portions of eastern California. In California, Oregon, and Washington, this pattern occurs east of the Sierra-Cascade Mountains. The Great Basin region has much less precipitation than the Pacific region, but the pattern of precipitation is similar. Like the Pacific climate, summers are the driest period. However, more precipitation falls in the spring, and summers are less droughty than in the Pacific region. Total precipitation ranges from 500 mm in northeastern Oregon and eastern Washington to 100 mm in Nevada. Low precipitation is largely in reflection of the rain-shadow effect of the Sierra-Cascade and Rocky Mountains. Winter temperatures over much of this region are cold, with the exception of palous prairie in Oregon and Washington.

Shrubs of the genera *Artemisia, Atriplex,* and *Juniperus* dominate most of this region because they can use moisture stored deep in the soil profile during the summer dry period. Grasses in the region are generally all of cool-season type because of the lack of summer moisture. Portions of the Middle East, former Soviet Union, South Africa, Australia, and western South America have this pattern of climate and similar associated vegetation.

The Southwestern Climate

The southwestern climate occurs in New Mexico, Arizona, southern Utah and Nevada, and southwestern Texas. This region has a bimodal precipitation pattern with winter precipitation, spring drought, summer precipitation, and fall drought. This bimodal pattern has maximum development in Arizona. Temperatures are warm

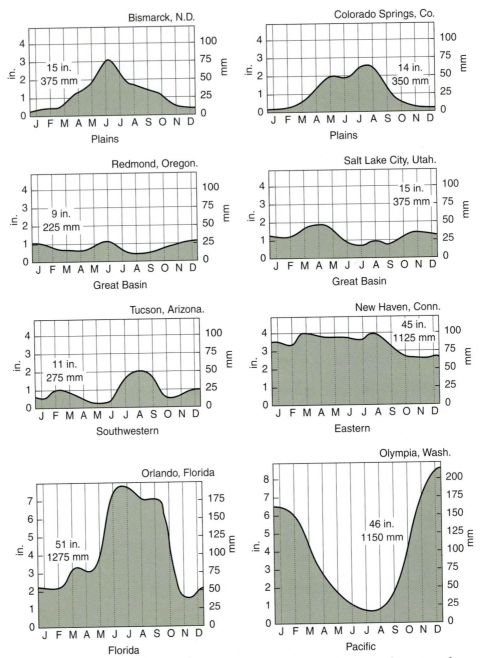

FIGURE 6.12   Precipitation patterns of the United States (adapted from Trewartha 1961 and Holechek et al. 1998).

throughout this region during the entire year. Total precipitation ranges from 150 to 460 mm. Pacific air moving in from the west causes the winter rains, which are primarily frontal storms. Summer rains are primarily convectional storms that result from air moving in from the Gulf of Mexico. These storms are local, usually of short duration, and very intense. Their effectiveness in generating forage growth is much lower than that of the more frequent, light-intensity storms in the central Great Plains.

Shrubs in the genera *Prosopis, Larrea, Opuntia,* and *Acacia* dominate many areas because they are well adapted to warm temperatures and long dry periods. Nearly all of the perennial grasses in this region are warm-season plants and grow as the result of summer rainfall. Winter precipitation favors cool-season annual forbs.

The Plains Climate

The plains precipitation pattern characterizes the central United States. Total precipitation ranges from 250 to 900 mm. Maximum precipitation delivery occurs in the late spring and summer, with a moderate to light amount in the fall. The least amount of precipitation occurs during the winter. It is significant that most of the precipitation in the central plains area comes from frequent, light-intensity rains during the spring-summer growing season.

This climatic pattern in large part explains why the central United States is a grassland. Because precipitation occurs during both the spring and summer growing season, this region is dominated by a mixture of warm- and cool-season grasses. Other parts of the world with this type of climate and similar vegetation include the lee side of the Andes in South America and continental, mid-latitude Europe and Asia.

The Eastern Climate

The eastern climate covers all of the eastern United States with the exception of the Florida peninsula. It is distinguished by uniformity in amount of precipitation during each month. Over most of the East the precipitation ranges from 900 to 1,400 mm, with greater amounts in the southern portion and the mountains. With movement northward, an increasing percentage of the precipitation is snow.

A large amount of precipitation with even distribution in a temperate climate provides favorable conditions for deciduous trees. This explains why much of the natural vegetation is deciduous forest. Like the eastern United States, northern Europe, including the British Isles, has this pattern of climate and similar vegetation.

The Florida Climate

The Florida climate is restricted to the Florida peninsula. This area receives very heavy rainfall (250 mm or more per month) during the summer period from June to October, but the period from November to May is quite dry. Annual total precipitation for this area ranges from 1,270 to 1,500 mm. Because of the heavy rainfall and warm climate, this climatic type is the most productive of all regions for forage. However, the rainfall also leaches the soil, and several minerals are deficient in the forage.

CLIMATIC INSTABILITY AND NATURAL RESOURCE MANAGEMENT

One of the questions asked by many natural resource managers, particularly those involved with farmlands and rangelands, is whether technology in the twenty-first century will make humans invulnerable to climatic instability. This question may have even greater significance now than in the past because the world's population of 5.8 billion people is four times that in 1900. It will likely double within the next 50 years to 12 billion. Any failure on the part of technology or unfavorable change in world climatic conditions could spell doom for literally billions of people because of severe reduction in food production.

World Climatic History

World and local climatic conditions can vary among decades, centuries, and millenniums. An interesting review of world climatic conditions over the last one billion years is provided by Roberts and Lansford (1979). They provide evidence the world has gone through many cycles of cold and warm periods (Figure 6.13). The world has been warming over the last million years, but is still in an ice age compared with the last 50–100 million years.

(a)

(b)

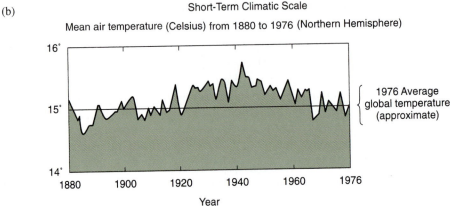

FIGURE 6.13   These two graphs of global temperature fluctuations show that the earth's climate is variable on both long (a) and comparatively short (b) time scales (National Center for Atmospheric Research, from National Academy of Sciences and other sources) (from Roberts and Lansford 1979).

The Last 25,000 Years.   Starting about 22,000 years ago, the world experienced extensive continental glaciation, which began to retreat about 12,000 years ago. By 7,000 to 8,000 years ago, the ice sheets in the Northern Hemisphere had retreated to their present extent. Evidence indicates that the period from 7,000 to 5,000 years ago was warmer than now with exceptionally cold intervals occurring about 2,800 and 3,500 years ago (Roberts and Lansford 1979).

The Last 1,000 Years.   According to Lamb (1972) the earth was in a dry, warm period about 1,000 years ago. During this time the Viking voyages to Iceland and Greenland occurred. Vineyards apparently flourished in England indicating higher summer temperatures than at present.

A 200-year cooling trend began around A.D. 1200 and was characterized by remarkably severe European winters. Vineyards in both continental Europe and England fell on hard times (Roberts and Lansford 1979). The Viking colonies in Iceland and Greenland declined because of an increase in ocean ice and colder, more stormy inland weather.

From 1400 to 1550, the climate became more favorable, but then a 300-year cold spell set in, known as the "Little Ice Age." During this period vineyards disappeared from England and farming became impossible in the mountainous portions of western Europe.

In much of Europe, the year 1816 was one of the very worst of the "Little Ice Age." It was characterized by a wet winter, a long cold spring, and a rainy summer. Widespread famine occurred across England and Europe because of failure of the wheat crop. This year was known as "The year without a summer in the United States." Heavy snow in June and frost in July and August occurred in the northeastern United States (Roberts and Lansford 1979). In northern New England, the corn crop was nearly wiped out. Widespread hunger was prevalent during the cold winter that followed.

The Last 100 Years.   Since the 1850s, a warming trend has occurred over the Northern Hemisphere. However, there was a 20-year period between the 1940s and 1960s when temperatures decreased slightly, but the upward trend resumed in the 1970s and appears to be still underway.

La Nina–El Nino Cycle.   At roughly 22-year intervals a climatic phenomenon known as the "La Nina–El Nino Cycle" affects the central Great Plains and southwestern United States. Basically it involves 11 wet years (El Nino) and 11 dry years (La Nina). It is responsible for the eight extended droughts in the region that occurred in the 1830s, 1850s, 1870s, 1890s, 1910s, 1930s, 1950s, 1970s, and may now be in effect. The explanation centers around spots on the sun that change their polarity every 11 years. This in turn causes a cooling in the South American Pacific Ocean, which results in high pressure systems that deflect air flows from the Gulf of Mexico.

During drought periods in the Great Plains when La Nina is in effect, other parts of the world such as China, India, and Russia also appear to experience unfavorable climatic conditions. This was particularly true in the early 1970s when world food production fell 2% (1972) after increasing every year since 1945. In the mid-1990s unfavorable climatic conditions due to La Nina effects in the United States, China, Russia, India, and Australia again depressed world grain production and increased world grain prices.

Climatic Lessons from the Past

Knowledge of the climatic history of various parts of the world can be invaluable in planning for future adversity. Although it cannot be known for sure that past patterns will be repeated, it is certain that the climate in a particular area will fluctuate and ultimately change if enough time passes.

Failure to heed climatic lessons from the past caused the "dust bowl" of the 1930s in the Great Plains of the United States. During the late 1910s and 1920s, millions of acres of native grasslands in the Great Plains were plowed for large wheat growing operations. Memories of the homesteaders who had been wiped out in past droughts, such as in 1893–1894 and 1910–1915, were forgotten. When the worst of all Great Plains droughts began in 1930, everyone was completely unprepared. Thousands of tons of topsoil blew away and, at the peak in 1934, the dust darkened the sky in places as far away as New York City. Starting in 1935, a wide array of government programs went into effect to repair the damage. Much farmland was returned to pasture, trees were planted to provide windbreaks, and farming techniques were developed to maintain crop residue for soil protection. However, by the mid-1950s, when drought struck again, many areas had been converted from grassland to cropland. After a 10- to 12-year period of application of conservation practices following the drought of the 1950s, there was another plow-out in response to world grain shortages and high prices in the early 1970s.

In the mid-1980s, low grain prices and the threat of another dust bowl led to a new wave of federal government programs aimed at retiring erodible lands to permanent grasslands and incentives for conservation tillage (see Chapters 8 and 12). It is encouraging that the mid-1990s run-up in grain prices did not trigger major expansion of crop culture in the Great Plains. However, only time will tell how well the lessons from the climatic history of the Great Plains have really been learned.

CLIMATE CHANGE AND HUMAN ACTIVITIES

During the past 20 years, there has been growing concern that the earth's climate will become warmer because of human activities. Acid precipitation and ozone depletion are other climate problems associated with human activities. Another concern is that certain areas are becoming more arid (desertification) because of destruction of the vegetation by grazing, woodcutting, and cultivation. We will discuss these concerns and their validity.

The Greenhouse Effect and Radiation Balance

The "greenhouse effect" refers to an increase in atmospheric temperature caused by increasing amounts of carbon dioxide and certain other gasses that absorb and trap heat radiation, which normally escapes from the earth (Godish 1997). Various activities by humans, such as burning fossil fuels and cutting forests, are thought to be increasing carbon dioxide levels in the atmosphere and, thus, contributing to the "greenhouse effect" (Figure 6.14).

The earth's atmosphere reflects, absorbs, and emits radiation. Through radiation balance it maintains surface temperature on the earth about 16°C higher than would otherwise occur. Without this environmental service, life on earth could not exist as we know it. Past fluctuations of mean atmospheric temperatures have occurred up to at least 5°C since world continental glaciation about 15,000 years ago (Houghton et al. 1996, OSTP 1997). The causes for past climate changes are unclear. However, dramatic changes can occur within centuries.

Climate change results from the interaction of incoming solar radiation and outgoing infrared radiation with changing concentrations of gasses and other material in the atmosphere. Unaltered albedo is one source of climate change. Albedo is the reflectivity of sunlight from the earth's surface. Ice, snow, bare soils, clouds, aerosols, and atmospheric dust have high albedos. As the climate changes, albedo changes further. Warming creates more water vapor, which forms more clouds and melts more snow and ice.

External forcing factors that drive the climate system include, most importantly, the sun's energy output and the earth's rotation. However, such things as volcanic events and human activities are also important. Human activities—such as logging,

FIGURE 6.14 Industrial production and automobiles are major contributors to the "greenhouse effect" because they burn large quantities of fossil fuels and release carbon dioxide into the atmosphere.

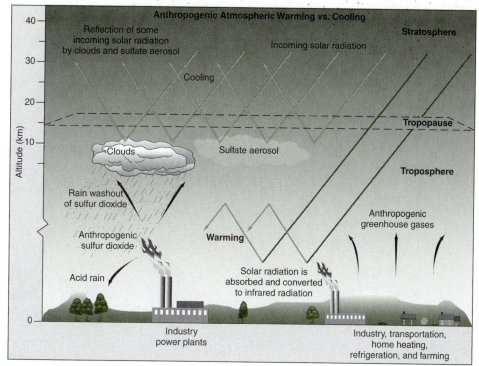

FIGURE 6.15 Human (anthropogenic) factors that are involved in atmospheric warming and cooling. Increases in some gasses, such as carbon dioxide, promote global warming while others, such as sulfur dioxide, cause cooling (from Nebel and Wright 1996).

farming, grazing, and urbanization—have altered the face of the earth, changing its albedo. Human activities generate certain gasses, aerosols, and particles that can influence climate by changing both the albedo and the greenhouse effect.

The radiation and energy balance of the earth is determined mostly by atmospheric composition and earth features. Nitrogen and oxygen, which make up 99% of atmospheric gas, reflect and absorb incoming light, but are transparent to infrared energy leaving the earth (Figure 6.15). In contrast, greenhouse gasses absorb departing infrared energy and reradiate it in all directions, returning some of it back to the earth. Heat builds up as the proportion of infrared radiation returned to the earth increases. The primary greenhouse gasses causing the greenhouse effect are water vapor, carbon dioxide, methane, nitrous oxide, and ozone. They are products of both natural and human activities.

Rice paddies, industry, fossil-fuel use, biomass burning, landfills, and agriculture all contribute greenhouse gasses. Carbon dioxide is released in largest quantities from combustion of fossil fuels and burning of forests. Methane is released from anaerobic locations, such as in landfills, rice paddies, natural wetlands, and feedlot wastes. Nitrous oxide and ozone are products of automobile and other internal combustion. Large sinks of carbon occur in the oceans, where carbon is mixed into and stored in the abyss with residence times of centuries or more. It is also stored in marsh sediments, peat bogs, and forests in huge quantities. If these stores of carbon were released, they would greatly increase infrared emission back to the earth and accelerate global warming.

Although water vapor also is a product of human activity, the overwhelming source is evapotranspiration from the earth's land and water masses. Water vapor increases following some type of change that results in greater temperature, such as increased concentration of carbon dioxide and methane in the atmosphere. Higher temperatures increase evaporation and raise the average humidity. Because water vapor

Global Average Temperature

FIGURE 6.16   The global average temperature has risen by about 0.6°C over the last century (from OSTP 1997).

is a highly effective greenhouse gas, its increased formation in the atmosphere further increases the earth's surface temperature.

Evidence from glacial ice cores indicates that the earth's climate has fluctuated drastically as the various factors that drive it have changed. Human activity is implicated in documented carbon dioxide increase and associated change in atmospheric temperature.

Not all scientists are in agreement that the "greenhouse effect" is due entirely or in part to human activities. Some believe it is more of a recovery process from the "little ice age" (1550–1850) brought about by natural forces. As recently as the 1980s some scientists were sounding an alarm about a dangerous trend toward global cooling. There continues to be uncertainty about the causes of the greenhouse effect, but a growing consensus asserts that human activity is a major cause. However, the global emissions of major greenhouse gasses have begun to decline (World Resources Institute 1994). This appears to have resulted from improved manufacturing processes, reduced carbon emissions from automobiles, and a lower rate of tropical deforestation. Even so, greenhouse gasses will continue to accumulate in the atmosphere maintaining the warming potential and risk of global climatic change.

Global Warming and Fossil Fuels

The mean surface temperature of the earth has increased about 0.6°C during the past century (Houghton et al. 1996, OSTP 1997, Houghton et al. 2001). The trend line suggests future increase at the same or greater rate (Figure 6.16). Projected changes in mean temperature over the next century range from 1.4°C to 5.8°C. This variation in estimated temperature increase is based on differences in the assumptions used in different mathematical models for temperature prediction. The main cause for this increased temperature is believed to be increasing concentrations of the greenhouse gasses previously discussed.

One of the most important greenhouse gasses, carbon dioxide, has been increasing since about 1860 (Figure 6.16). The data in Figure 6.17 are from direct measurements at Mauna Loa Observatory in Hawaii since 1960, and on carbon dioxide concentrations

Carbon Dioxide Concentrations

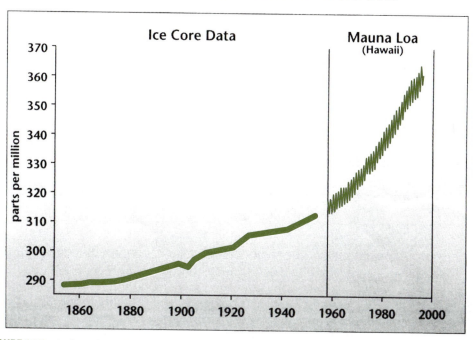

FIGURE 6.17 Carbon dioxide concentration measured in ice cores before 1960 and at the Mauna Loa Observatory since 1958 (from OSTP 1997).

trapped in glacial ice of known age before 1960. Close correlations between temperature and atmospheric carbon dioxide concentration have been derived from study of bubbles in glacial ice up to 160,000 years old. Because carbon dioxide absorbs infrared radiation leaving the earth and emits some of it back to the earth, it increases temperatures at the surface of the earth (Figure 6.17). Other greenhouse gases that could be contributing to temperature increase include water vapor, methane, and nitrous oxide.

Past increases in atmospheric carbon dioxide started about the time (1850s) that industry began using significant quantities of coal for production of steel and other material goods. Annual world emissions of carbon from combustion of oil, gasoline, coal, natural gas, and other fossil fuels now exceed six billion tons, of which the United States contributes 22%. An additional one to two billion tons of carbon are emitted from forest cutting and burning mostly outside the United States in the developing world. Future emissions are projected to nearly double over the next 35 years. Because much of the future increase will occur in the developing world, the United States percentage is projected to fall to about 15% by 2035.

The Implications of Global Warming

The resource management implications of global warming are staggering. There are many threats and some opportunities. One expected change is a rise in ocean water level of 15 to 90 cm with a most probable rise of 50 cm by the year 2100. The rise in water level will occur because water expands as it warms and more glacial ice will melt. Obviously, wherever the ocean inundates present human habitation, it will degrade private property value. Similarly, public property set aside for coastal wildlife refuges may no longer provide the necessary habitat. Estuarine and salt-marsh ecosystems would shift inland at rates depending on the coastal slope and existing engineering, such as sea walls. In coastal areas of low slope, presently dry property is projected to become shallow wetland.

A global temperature rise will increase evaporation from land and water surfaces. One net result is more water in the atmosphere and a greater greenhouse effect from the increased water vapor. This phenomenon is one reason why in some models the rate of warming increases at an increasing rate. Because land heats more than water per unit of radiation absorbed, heat is expected to become more concentrated over the central continents. The northern part of the Northern Hemisphere is expected to have the greatest warming because a high proportion of the continental land mass is located there. Inland freshwater runoff volume is expected to decrease while coastal freshwater runoff could increase substantially in certain regions. While some estuaries will become fresher, others will become more saline.

Continental heating is predicted to accentuate the rate of hydrologic cycles and cause greater seasonal extremes of temperature and water runoff. The incidence of intense storms already has increased about 20% over the past century and is expected to continue to increase as mean temperature rises. This will result in greater wind and flood damage if people continue to concentrate in coastal zones and river flood plains.

Because terrestrial plant and animal distributions are primarily determined by temperature and moisture, global climate change is expected to result in the shifting of whole ecosystems. Massive die-offs of forest and other vegetation are possible. This means that many lands now managed for certain forest, range, farm, recreation, wildlife, and biodiversity values will have to be managed for a new set of resource values. Places that formerly had optimum temperature and moisture for certain crops could become much drier and other locations may become much wetter. Public lands now set aside for specific wildlife, forest, grassland, or other management may no longer be suitable to the management purpose. On private lands, the rich wheat and corn belts of the central plains states could shift northward into Canada over the next century. This may increase agricultural productivity in Canada while decreasing it in the United States. The biggest concern for agriculture is that the climate will become so erratic that crop production will be much more unreliable. A similar shift northward could occur for the prairie pothole ponds important for waterfowl production.

The regional changes based on prediction models vary substantially at this time. There is a need for improved models and management strategies for different regions under different scenarios of global warming.

Possible Strategies for Managing Global Warming

Various strategies for dealing with global climate change are now under discussion. The three main strategies are to slow emissions of greenhouse gasses, manage ecosystems to favor greater sequestration of carbon in carbon "sinks," and manage ecosystems to allow species to adapt to changes as they occur. Even if emissions could be totally stopped at this time, the lags in heat absorption by the oceans probably will result in a continued rise in temperature for many decades. Adaptation appears to be necessary at least to the extent that thermal lags cannot be prevented.

Greenhouse gas emissions can be slowed through a combination of technology improvements and conservation. The use of noncarbon alternatives to energy production, such as solar and wind energy, is part of the first strategy. Preservation of carbon "sinks" in forests and peaty wetlands, as well as resource management to restore and create new carbon sinks, are important aspects of the second strategy. Adaptation includes improved understanding of climate change, improved forecasts of impacts, and changes in resource management that mitigate damage and take advantage of possible benefits. A rise in carbon dioxide concentration could increase productivity of many plants where other factors are not limiting. This might have beneficial impacts on the resources of some woodland, cropland, and rangeland ecosystems if other environmental factors remain favorable.

FIGURE 6.18 Regions receiving acid rain deposition. The numbers refer to the pH of the acid deposition (from Nebel and Wright 1996). A pH of 5.5 is considered natural for uncontaminated rainwater.

Acid Precipitation

Atmospheric precipitation has a natural weak acidity caused by the interaction of water and carbon dioxide which forms carbonic acid. Acidity is determined by the concentration of hydrogen ions in solution. About three decades ago, hydrogen ion concentrations greater than expected from natural sources were first reported for North American precipitation (Godish 1997). Acid rain was first discovered in Europe and later linked to industrial emissions into the atmosphere. Further study revealed that snow could be more acidic than expected from natural causes. Many lakes were found to be unnaturally acidic. Elevated hydrogen ion concentrations were found in clouds, dew, and atmospheric crust. By 1980, research scientists had mapped hydrogen ion concentrations in precipitation. They found the greatest concentrations over the eastern Great Lakes, upper Midwest, and mid-Atlantic states (Figure 6.18).

The major source of acid rain is combustion of fossil fuels used to generate electricity, power automobiles, and fire metal smelters and industrial boilers. The major source of acidity, sulfuric acid, is associated with sulfate aerosols. The sources of sulfate aerosols are mostly from high-sulfur coal. Nitric acid, a minor contributor, is produced mostly from oxides in automobile exhausts. The oxides of sulfur and nitrogen can be transported many hundreds of miles before they form acid precipitation. About half of the acidic precipitation from the United States ends up in Canada and some Canadian emissions end up in the United States.

Most of the ecosystems sensitive to acid rain exist in Canada. However, significant sensitive areas also exist in the northeastern and north central United States, and in isolated high mountain areas. As lakes become more acidic, their productivity and biodiversity decline. Losses of fish populations have been associated with elevated acidity at numerous sites. So far, forest and crop production do not appear to be significantly affected by acid rain, although gradual depletions of nutrients may in time reduce productivity of exposed terrestrial ecosystems. Poor air quality due in part to acidity contributes to respiratory problems in humans.

Sulfur emissions are controlled by regulations of the United States Environmental Protection Agency and government regulatory agencies in other nations. Sulfates are removed from coal both before and after combustion using a number of processes (see Godish 1997 for details). Less effective means are available for removing nitrogen oxides.

One ramification of removing sulfate aerosols to prevent acid precipitation is that their associated albedo is reduced. Early model projections of greenhouse warming failed to consider this artificial source of atmospheric cooling and overestimated the warming. However, recent information suggests there could be more removal of sulfate aerosols by emissions treatment and, therefore, less reduction of global warming than in the recent past.

Ozone Depletion

In the late 1960s there existed the possibility that ozone, O_3, at the outer edge of the stratosphere could be depleted by high-altitude flight and nuclear weapons testing (Godish 1997). This was disturbing because ozone shields the earth's surface from potentially harmful ultraviolet radiation. Because of these theories, United States policy steered away from both activities. In the 1970s, scientists proposed a theory that chlorofluorocarbons used in aerosols, refrigerants, and foam propellants could deplete ozone once chlorine, a strong oxidant, was released and the molecules decomposed. Later, they theorized that atmospheric accumulation of N_2O—a product of fertilizer breakdown—could react with and deplete the ozone.

In the late 1980s, a scientist working with data accumulated in Antarctica discovered what appeared to be depletion of ozone over the South Pole. Subsequent study revealed concentrations of chlorine in the area of the "ozone hole." Since discovery of the ozone hole over Antarctica its size has increased, and a hole was discovered over the Arctic. Recent measurements indicate that ozone depletion may be occurring over the entire globe. Ozone depletion is of great concern because of the role stratospheric ozone plays in radiation balance. It is especially effective in absorbing ultraviolet radiation that can cause cell destruction and related diseases—including skin cancer. Studies in Canada indicate that ultraviolet light has increased at the earth's surface, as would be expected with significant ozone depletion. Ultraviolet light regulates plant growth and has been implicated with depression of phytoplankton photosynthesis in oceans near Antarctica. These observations have caused concern over the sustainability of primary productivity in both land and water. Following the signing of an international treaty in Montreal, most of the developed nations have taken major steps to reduce ozone-destructive emissions.

Desertification

Desertification is the formation of desert-like conditions, largely through human actions, in areas that do not have desert climates. Biological productivity declines while the prevailing climatic conditions are thought to remain nearly constant. Human activities implicated as causes of desertification include uncontrolled livestock grazing, burning, woodcutting, and temporary cultivation and abandonment of semiarid to arid lands (Figure 6.19). Africa has been the focal point of concern over desertification during the past 20 years because of continued droughts in the Sahel region. These droughts have caused tremendous losses of livestock and human hardship. Compre-

FIGURE 6.19 Overgrazing by cattle and drought have contributed to desertification in the arid Sahel region of Africa.

FIGURE 6.20 Rainfall fluctuations in the Sahel and Sudan. Expressed as a percent departure from the long-term mean, 1901–1984 (from Brown and Wolf 1985).

hensive reviews on the subject of desertification are provided by the United Nations (1977), Glantz (1977), Postel (1989), and Mouat and Hutchinson (1995).

Past climatological data show that drought has been a recurring phenomenon in the Sahel (Wallen and Gwynne 1978, Winstanley 1983) (Figure 6.20). However, the effects of drought on the vegetation have been magnified in recent years because of rapidly increasing human and livestock populations. There is also some evidence that recent droughts have been more severe than those in the past (Winstanley 1983).

As human populations increase in semiarid to arid areas, desertification may become an important problem in other parts of the world as well as Africa (Mouat and Hutchinson 1995). Application of soil and range management practices have considerable potential to reduce or reverse the desertification problem in many areas (see Chapters 8 and 11).

LITERATURE CITED

Bright, C. 1997. *Tracking the ecology of climate change. State of the World Report 1997*. Washington, DC: World Watch Institute.

Brown, L. R. and W. C. Wolf. 1985. *Reversing Africa's decline*. World Watch Paper No. 65. Washington, DC: World Watch Institute.

Conrad, V. and L. W. Pollak. 1950. *Methods in climatology*. Cambridge: Harvard University Press.

Dansereau, P. 1957. *Biography: An ecological perspective*. New York: The Donald Press. 394 pp.

Dunn, S. 2001. *Decarbonizing the energy economy. State of the World Report 2001*. Washington, DC: World Watch Institute.

Finch, V. C. and G. T. Trewartha. 1942. *Elements of geography: Physical and cultural*. New York: McGraw-Hill. 823 pp.

Glantz, M.H., ed. 1977. *Desertification: Environmental degradation in and around arid lands*. Boulder, CO: Westview Press.

Godish, T. 1997. *Air quality*. 3rd ed. New York: Lewis Publishers.

Heady, H. F. and E. B. Heady. 1982. *Range and wildlife management in the Tropics*. New York: Longman.

Holechek, J. L., R. D. Pieper, and C. H. Herbel. 1998. *Range management principles and practices*. 3rd ed. Upper Saddle River, NJ: Prentice-Hall.

Hopkins, A. D. 1938. *Bioclimatics: A science of life and climate relations*. U.S. Dept. Agr. Misc. Pub. 280. 188 pp.

Houghton, J. T., L. G. Meira Filho, B. A. Callander, N. Harris, A. Kattenberg, and K. Maskell. 1996. *Climate change 1995: The science of climate change*. Published for the Intergovernmental Panel on Climate Change. New York: Cambridge University Press.

Houghton, J. T., Y. Ding, D. J. Griggs, M. Naguer, P. J. Van der Linden, and D. Xiaosu, eds. 2001. *Climate change 2001: The scientific basics*. London: Cambridge University Press.

Humphrey, R. R. 1962. *Range ecology*. New York: Ronald Press. 234 pp.

Klages, K. H. W. 1942. *Ecological crop geography*. New York: Macmillan.

Lamb, H. H. 1972. *Climate: Present, past, and future*. London: Methuen Publishing.

McKnight, T. L. 1990. *Physical geography: A landscape appreciated*. 3rd ed. Upper Saddle River, NJ: Prentice-Hall.

McKnight, T. L. 1999. *Physical geography: A landscape appreciation*. 6th ed. Upper Saddle River, NJ: Prentice-Hall.

Miller, G. T. 1990. *Resource conservation and management*. Belmont, CA: Wadsworth.

Mouat, D. A. and C. F. Hutchinson, eds. 1995. *Desertification in developed countries*. London: Kluwer Academic Publishers.

Nebel, B. J. and R. T. Wright. 1996. *Environmental science: The way the world works*. 5th ed. Upper Saddle River, NJ: Prentice-Hall.

Nebel, B. J. and R. T. Wright. 1998. *Environmental science: The way the world works*. 6th ed. Upper Saddle River, NJ: Prentice-Hall.

Office of Science and Technology Policy. 1997. *Climate change: State of knowledge*. Executive Office of the President. Washington, DC.

Postel, S. 1989. *Halting land degradation. State of the World 1989*. World Watch Institute Report. New York: W. W. Norton.

Ricklefs, R. E. 1976. *The economy of nature*. Portland, OR: Chiron Press.

Roberts, W. O. and H. Lansford. 1979. *The climate mandate*. San Francisco, CA: W. H. Freeman.

Rumney, G. R. 1968. *Climatology and the world's climates*. New York: Macmillan.

Trewartha, G. 1961. *The earth's problem climates*. Madison: University of Wisconsin Press.

United Nations. 1977. *Desertification: Its causes and consequences.* Oxford: Pergamon Press.

United Nations Development Programme. 2000. *World Resources 2000–2001.* New York: Elsevier Science.

U.S. Department of Agriculture (USDA). 1987. *The Jornada Experimental Range, Las Cruces, New Mexico.* U.S. Dept. Agric. Misc. Publ. 689–479.

Veihmeyer, F. J. 1938. Evaporation from soils and transportation. *American Geophysical Union Transaction* 1938:612–619.

Wallen, C. C. and M. D. Gwynne. 1978. *Drought—A challenge to rangeland management.* Proceedings International Rangeland Congress 1:21–32.

Weaver, J. E. and F. E. Clements. 1938. *Plant ecology.* 2nd ed. New York: McGraw-Hill.

Winstanley, D. 1983. Desertification: A climatological perspective (pp. 185–213). In *Origin and evolution of deserts,* edited by S. G. Wells and D. R. Naragan. Albuquerque, NM: University of New Mexico Press.

World Resources Institute. 1994. *World Resources 1994–1995.* New York: Oxford University Press.

CHAPTER 7

Water Resources

THE IMPORTANCE OF WATER

In the fall of 1997, the world watched with excitement as images of the surface of Mars were captured by robotic equipment. Such images inspired a sense of awe but also gave us deep appreciation for the earth's living landscapes. More specifically, such images make us more appreciative of water, a life-sustaining resource that has been conspicuously absent in space probe images of other planets. In contrast, the photographs of Earth taken by astronauts show a strikingly blue planet with water covering more than two-thirds of its surface. Water is a critical component of all life. It can be beautiful, powerful, and destructive (Figure 7.1).

Like many other renewable natural resources, supplies of water can be both scarce and abundant at the same time and place, depending on the required quality. An example is Los Angeles, where seawater is abundant but usable "fresh" water is limited. While water in the oceans is inexhaustible, without processing to remove salts it is unsuitable for most human uses. Lack of potable water is becoming a major obstacle to further urbanization of the southwestern United States. It is also the primary constraint on the development of several African, Latin American, and Asian countries. In many parts of the world, demand for water of usable quality is far greater than available supplies. Problems of loss in biodiversity are often associated with man's control of water (drainage of marshes, building dams, contamination of estuaries).

Modern man uses water for agriculture, transportation, industrial processes, waste disposal, power generation, and recreation as well as for consumption. The tremendous advance of civilization over the past 200 years has depended heavily on the human capability to control floods, develop irrigation systems, and use water as a source of power (steam engines). Without question, the destiny of humankind in the twenty-first century will depend heavily on being able to solve problems relating to water.

This chapter provides an overview of water uses, water resources management (including watershed management), water quality management, and water needs in the United States. We also discuss the hydrologic cycle and how it interacts with atmospheric and watershed processes. For more detailed discussions of watershed management, water quality problems, and other issues related to water resources, the authors refer the reader to Gleick (1993), Heathcoate (1988), Perry and Vanderklein (1996), Owen et al. (1998), and Nebel and Wright (2002).

(a)

(b)

(c)

FIGURE 7.1 Water is vital to life on planet Earth. It can be (a) beautiful, (b) powerful, and (c) destructive (courtesy of the United States Department of Agriculture).

WORLD WATER PROBLEMS

The availability of useful water will likely be the biggest constraint on world population growth in the near term. It is already a contentious issue in the western United States. Water management problems include depletion of underground aquifers, contamination of surface and groundwater, siltation of dams, salinization of irrigation water, prolonged drought, flooding, and loss of habitat for endangered species. By 2025, 40% of the world's population will likely experience water shortages (United Nations Development Programme 2000) because the demand for water has outpaced the economic capability to clean and distribute water as needed. Because of human population growth, global availability of water has dropped from 17,000 m^3 per person in 1950 to 7,000 m^3 in 1998. Currently 20 countries are experiencing water stress (less than 1,000 m^3 per person). Past famines in the African Sahel and other locations have been a consequence of inadequate water.

Over the past six thousand years irrigation has transformed human societies like no activity in previous history (Postel 2000). It created large food surpluses that led to centralized governments, and freed large numbers of people to pursue other activities. Human populations increased in size and density, which led to the first cities. Today about 40% of the world's food comes from 17% of the cropland that is irrigated. Between 1970 and 1994 the world's irrigated farmland grew by 38%. However, most of the potentially irrigatable land has now been brought under production. One out of five acres is damaged by salt. Large underground aquifers in China, India, Pakistan, North Africa, and the United States are being rapidly depleted by massive pumping for irrigation. There is growing concern that mounting water deficits and increasing human numbers will increase regional food shortages within the next 10 to 20 years.

Rapidly growing cities and industries are increasingly reducing water available for agriculture. By 2020 this alone could cause world grain production to drop by 16%. This problem is particularly acute in China, India, and Indonesia.

As surface supplies are stretched to their limits, people become more dependent on groundwater—wherever it can be developed cost-effectively. However, problems relating to groundwater pollution are now ubiquitous. Scientists are uncovering cases of pollution in aquifers near farms, factories, and cities on every continent (Sampat 2001).

Without question, providing adequate water quantity and quality to a human population that will likely increase from 6 billion to 10–12 billion will be one of the greatest natural resource challenges in the twenty-first century. Major changes in water supply and irrigation systems, domestic use, and control of groundwater contaminants provide a means to meet this challenge.

THE PROPERTIES OF WATER

Physical Attributes

Water supports life on our planet because of its unique physical and chemical properties. Several of its properties contribute to the massive transfers of energy that take place in the earth's atmosphere and oceans. In turn, these energy transfers lead to a more even heat balance away from the earth's equator. They maintain the continuous circulation of moisture that renews the freshwater supply throughout the globe. To perform its critical roles, water must remain in a fluid state over wide extremes of temperature. It can do this because it has a rather high boiling point (100°C, 212°F) and a low freezing point (0°C, 32°F). Because of its solvent properties, water contributes to biogeochemical cycling as it carries dissolved minerals from land to sea (Smil 1997). On a much smaller scale, water transports nutrients wastes and other minerals through individual ecosystems and through individual organisms. Water is essential in maintaining the circulatory systems and temperature equilibrium in higher plants and animals, and it supplies the hydrogen necessary for photosynthesis.

The ability of water to buffer the temperature of the earth's surface can be attributed to its high heat of vaporization and capacity for storing heat. Because liquid water can store a large amount of heat, large bodies of water do not warm or cool rapidly. Without the vast volume of ocean water, earth temperatures would vary more.

Water shows remarkable changes as it moves through its liquid, solid, and vapor states. Liquid water is the only common substance that expands rather than contracts when it freezes. As a result, ice is less dense than liquid water and floats above water when oceans, lakes, and rivers freeze. Consequently, water bodies do not freeze solid in cold climates and aquatic life is more abundant than would otherwise be the case. Water releases heat when it changes from liquid to ice and from steam to liquid. The extremely high surface tension of water is created by the forces that attract water molecules to one another. This property allows liquid water to adhere to soil and move through plants. Water's low viscosity makes it an outstanding medium for distributing nutrients through soils and plants as well as for distributing heat in ocean eddies and currents.

WATER FORMS AND DISTRIBUTION

The Planet's Reserves

Water is distributed throughout the planet, but most (97.4%) occurs as ocean water (Figure 7.2). The sea covers 71% of the total area to an average depth of 4 km (2.5 mi). Its massive size is maintained by return flows from rivers and melting ice. Seawater is highly saline and cannot be used as a source for drinking water or for crop irrigation without expensive treatment. Sodium chloride and magnesium sulfate are the most abundant dissolved salts found in ocean water. These salts, as well as a host of other substances, are carried to the sea by rivers. Salts are released from the landmasses through erosion and weathering processes. Across the continents the amount of salts and water that each river contributes to the ocean varies greatly, but constant mixing maintains a relatively stable concentration of salt in the oceans, amounting to about 3.5% based on weight.

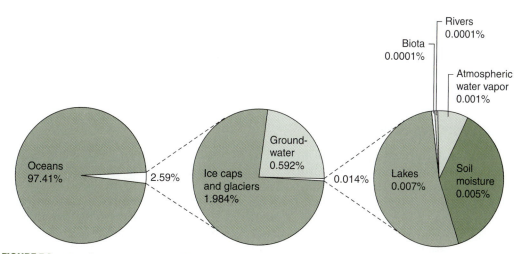

FIGURE 7.2   Distribution of the world's water (courtesy of United States Department of Agriculture). *Source:* David H. Speidel and Allen F. Agnew, "The World Water Budget," in *Perspectives on Water Uses and Abuses,* David H. Speidel, Lon C. Ruedisilli, and Allen F. Agnew, eds. (Oxford University Press, New York, 1988).

TABLE 7.1   The Distribution of the World's Ice by Geographic Region

Glacier Region	Area of Ice-Covered Surface (10^3 km^2)
North polar regions	2,000
Temperate countries of the Northern Hemisphere	190
Tropical countries	0.1
Temperate countries of the Southern Hemisphere	26
Antarctica	14,000
Total of the earth	16,000[a]

[a]Total given in the original table does not add to total of entries because of rounding.
Source: Gleick 1993.

Glacial Ice

About 10% of the earth's land surface is covered by glacial ice. Most (85%) glacial ice is found in Antarctica and Greenland (Table 7.1). Much of the remaining portion is scattered across temperate mountain valleys. As glacial ice has advanced and retreated over geologic time, the proportion of freshwater found on the planet in a frozen state has varied. At present, glacial ice comprises a rather small portion of the planet's total water (1.7%) (fresh and saline). However, it accounts for 68.7% of its fresh water. If all of this ice suddenly melted, the oceans would rise about 60 m (200 ft). Some of the world's largest cities—such as London, Tokyo, New York, Seattle, and Los Angeles—would be inundated.

Lakes

A variety of geologic events such as earthquakes (e.g., Lake Nyasa in East Africa), volcanic activity (e.g., Crater Lake, Oregon), and glaciation (e.g., the Great Lakes) has led to the creation of lakes (Figure 7.1). Freshwater lakes are found mostly in regions of ancient and recent glaciations, in regions of large tectonic fractures in the earth's crust, and along rivers and ocean coasts. Lakes do not persist indefinitely over geologic time. Increased aridity, caused by climatic shift, has caused numerous massive lakes to become spacious salt beds. Utah's Great Salt Lake is what remains of ancient Lake Bonneville. Lakes also disappear as they slowly fill with sediments and become colonized by plants adapted to a hydric environment.

In terms of total water volume, lakes appear insignificant alongside oceans. Nevertheless, various lakes throughout the world cover large areas (e.g., the Saint Lawrence Great Lakes, Aral Sea, Lake Victoria). Nearly half (48%) of the world's large lakes (i.e., greater than 500 km^2) are found in North America (Herdendorf 1990). In the United States alone there are 41 million acres of lakes and reservoirs. While the number of reservoirs built for recreation outnumber all other purposes, most large reservoirs have been built primarily for water supply and flood control. Most large reservoirs are multipurpose structures. See Chapter 9 for more information on aquatic ecosystems.

Streams and Rivers

Streams and rivers account for much less of Earth's total aquatic surface area than lakes. However, they are the primary arteries of water flow across continents, and they recharge and drain the lakes. Rivers such as the Amazon are massive in their width and depth and greatly influence the biology of some continents. Though less awesome in size, rivers traversing the world's arid regions provide water that can radically change ecosystems and settlements. Variation of flow is natural for most large rivers, which typically form floodplains that are periodically inundated. Rivers unconstrained by engineered controls (e.g., dams, levees) continuously cut new channels through their floodplains and create a mix of backwater lakes and wetlands, which support diverse natural communities. In the United States, some rivers, such as the Mississippi, remind us that their path is self-determined in spite of the best efforts at flood control. In the United States, rivers and streams have a total length of about 3.5 million miles. However, only about one-third of these flow continuously.

Wetlands

Wetlands exceed both lakes and flowing waters in total U.S. surface area, amounting to more than 200 million acres. About 5% of that occurs in estuaries along ocean coasts. Most wetlands are isolated from rivers and lakes, but because they are typically shallow, they are not as important as lakes for water storage. Even so, wetlands are important for erosion protection, groundwater recharge, flood reduction, trapping nutrients and sediment, and providing fish and wildlife habitat. Chapter 9 describes wetland ecosystems in more detail.

Oceans

The oceans occupy the largest area and volume of water among the different types on Earth. A salinity averaging about 35 parts per thousand (3.5%) makes ocean water useless for drinking and agriculture unless it is desalinated. Although ocean water desalinization is technically feasible, high cost makes it a last resort. The oceans are important for their fisheries, other biological resources, minerals, recreation, and transportation uses. The deep oceans remain among the least-explored areas on the earth.

Groundwater

A significant portion of the world's water is underground where it must be accessed through excavation and deep wells. The quality and quantity of groundwater is highly variable among the different continents and the world. Much groundwater is too saline for drinking water or irrigation. Some of the best-quality water resources, however, occur underground where they must be developed for use.

WATERSHED MANAGEMENT

The Hydrologic Cycle

The process by which the sun causes water from the land and oceans to vaporize into the atmosphere, condense as the result of cooling, return to the earth as precipitation, and flow to points where it vaporizes once again is termed the *hydrologic cycle* (Figure 7.3). Other important sources are from transpiration (moisture released from plant parts, mainly leaves) and evaporation from streams, lakes, and land surfaces. Man can influence the hydrologic cycle by modifying vegetation and soil. These modifications, which can have a major impact on both water yield and quality, are discussed later in the chapter. That part of the hydrologic cycle that occurs in the landscape is often referred to as the watershed process.

The Watershed

A *watershed* is the total area of land that contributes runoff water above a given point on a waterway. The areas drained by rivers and streams are called drainage basins, or watersheds. Watersheds are hierarchical; that is, small watersheds contribute to large watersheds in a progressive accumulation of watershed area as drainages join. Hydrologists sometimes refer to first-, second-, and third-order and larger stream channels as they describe watersheds. A second-order stream channel forms when two first-order channels join, and so on. Depending on management concern, one could choose a massive watershed of high order (e.g., tenth order), such as the area drained

FIGURE 7.3   The hydrologic cycle (courtesy of United States Department of Agriculture).

by the Mississippi (an area encompassing numerous states), or a watershed served by a minor tributary of low order.

Precipitation that falls to the watershed surface and enters the ground is called *infiltration*. The portion of the moisture that fails to infiltrate is surface runoff. After water infiltrates and water moves downward in the soil, it eventually reaches an impervious layer of rock or dense clay where it accumulates until it completely fills all the spaces above the impervious layer. This accumulated water is called *groundwater*, and its upper surface is the water table (see Figure 7.3). Where surface slope intersects the groundwater table, groundwater resurfaces as springs. Stream channels often cut into aquifers and gain flow from groundwater. Because water flowing through the ground is filtered of particulates, often enriched with dissolved nutrients, and has relatively constant flow rate and temperature, springs can have a large impact on water quality. Any activity that reduces infiltration and increases surface runoff diminishes the usually beneficial effects of springwater inputs. Where streams are perched above the groundwater table, they lose surface flow through bottom infiltration.

Groundwater flows through layers of porous material called *aquifers*. Sand and gravel layers provide good aquifers because of their permeability. In contrast, aquifers comprised of mostly clay or crystalline rocks are poor aquifers. Except for fissures that allow spring flow to the surface, artesian aquifers are overlain by some type of impervious matter such as clay or shale. Artesian aquifers often form fountain-like springs caused by the pressure from the weight of overlying mass. Many thousands of years ago—during periods of wetter climate—some parts of the world, such as Israel and the southern Great Plains, accumulated massive amounts of water deep beneath the earth's surface. These are called fossil aquifers, and water withdrawal from them is commonly referred to as "water mining." Groundwater also typically accompanies rivers with well-developed floodplains. There it often sustains riparian vegetation adjacent to the river and isolated ponds and wetlands during periods of low river flow.

Watershed Management Objectives

Watershed management involves the manipulation of soil and vegetation on a particular area (a watershed) to reduce erosion, improve water quality, stabilize discharge, and increase water supplies. Basic approaches to watershed management involve control of water infiltration, runoff, erosion, and pollution. Specific watershed management practices include conversion of vegetation type, construction of small reservoirs, water harvesting, water spreading, and control of livestock grazing, farming, logging, recreation, and urban land-use practices. Allen and Sharpe (1960) summarized the objectives of the water resources manager as follows:

A. Reduce floods and high water flows
   1. Maintain high watershed infiltration
   2. Encourage deep water percolation
   3. Reduce snow-melting rates
   4. Reduce surges by mechanical barriers

B. Control water quality
   1. Maintain or increase soil cover (e.g., vegetation, leaf litter)
   2. Improve watershed infiltration
   3. Protect stream banks and beds against erosion
   4. Reduce water temperatures by maintaining or increasing shade over streams (enhances fish habitat)

C. Increase water yields
   1. Reduce evaporation and transpiration losses
   2. Reduce losses to precipitation interception by vegetation
   3. Increase snow accumulation
   4. Decrease evaporation from streams and reservoirs

D. Alter timing of water yield
   1. Reduce snow melt in winter and prolong it in spring
   2. Spread water into subsurface storage areas (meadows) for later return to streams
   3. Raise lake outlets or create reservoirs for increased temporary storage

E. Control and prevent erosion
   1. Reduce exposure of bare soils
      a. Prevent unnaturally destructive fires
      b. Regulate grazing
      c. Discourage cultivation of eroded slopes
      d. Plant vegetation
   2. Retard runoff
      a. Construct check dams and contour trenches
      b. Restore beaver
      c. Cover eroded areas with brush, litter, or other debris

VEGETATION TYPE

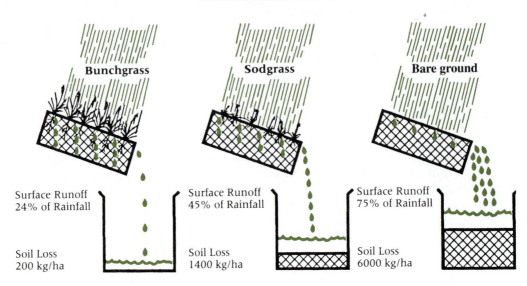

FIGURE 7.4   Influence of vegetation type on sediment loss, surface runoff, and rainfall infiltration from 10 cm of rain in 30 minutes (adapted from W. H. Blackburn et al. 1986 by Knight 1993).

More recently, watershed management includes the objective of restoring watersheds to a previous natural condition whenever that strategy is most cost-effective in providing the greatest beneficial use of watershed services. Wetland restoration, for example, has been emphasized in part because of wetland effects on flood reduction and water quality improvement.

Watershed Processes

Infiltration and Percolation.   Water moves into the soil by infiltration and through the soil by *percolation*. After a raindrop reaches the soil surface, it can infiltrate the soil, evaporate, or become a part of overland flow. The primary factors influencing infiltration rate are intensity of precipitation, amount and kind of vegetation cover, and soil surface properties (texture, structure, and organic matter) (Figure 7.4). Fine-textured soils (clays) generally have lower infiltration and percolation rates than do coarse-textured soils (sands). However, some fine-textured soils have infiltration rates comparable to those of coarse texture. This results from the cementing of fine particles into aggregates that act like larger particles.

When infiltration is high, much of the precipitation is stored in the soil for plant use, and part of it may percolate to groundwater where it can be recovered from springs or wells. Conversely, high rates of surface runoff contribute to soil loss and flooding. The most manageable factor influencing infiltration is vegetation cover. When raindrops fall on unprotected soil, they dislodge soil particles and erode away the soil surface. As plant cover declines, infiltration decreases (Table 7.2). This reduces soil moisture available for forage production and contributes to desertification of arid areas.

When percolation is high, as in sandy soils, little water is stored for long in the soil. Such soils dry quickly following rains and the vegetation they support is especially vulnerable to drought. Optimum water infiltration and storage typically occur on gentle slopes in soils with a mix of particle sizes and organic matter.

Water repellency has been found under a variety of climates and vegetation types, including many in the United States (DeBano 1969, DeBano and Rice 1973). It affects the hydrologic cycle by reducing infiltration rates (DeBano 1969). Water-repellent soils occur under a variety of conditions, but often are associated with a recent burn.

TABLE 7.2 **Standing Air-Dried Herbage, Mulch, and Rate of Water Intake on Heavily, Moderately, and Lightly Grazed Watersheds in South Dakota**

Grazing Intensity	Total Herbage (lb/acre)	Mulch (lb/acre)	Water Intake Rate (ln/hlr)
Heavy	900	456	1.05
Moderate	1,345	399	1.69
Light	1,869	1,100	2.95

Source: Data from Rauzi and Hanson 1966.

FIGURE 7.5   Relationship between percentage of bare soil and average annual runoff from 17 watersheds in western Colorado (from Branson and Owen 1970, cited in Branson et al. 1981). The r^2 of 0.74 indicates that 74% of the variation in annual runoff can be explained by the fraction of bare soil in these 17 watersheds.

On burned watersheds the water-repellent layer occurs just below the surface of the soil. The thickness of the layer depends on the intensity of the fire and the type and amount of mulch. Plants with high levels of volatile oils are most often associated with water repellency (De Bano 1989, Dennis 1989).

Runoff and Erosion.   Surface runoff is initiated when the amount of precipitation exceeds the infiltration and storage capacity of the soil. The primary factor influencing runoff is the amount of vegetation available to retard water movement over the soil surface (Figure 7.5). Runoff declines as soil cover increases.

Erosion and sediment (suspended mineral materials in water) deposition are major problems caused by excessive runoff. Maximum sediment yields occur at about 250 mm (10 in.) of annual precipitation. A decrease occurs below 250 mm precipitation because of insufficient runoff for sediment transport. Above 250 mm precipitation, erosion is less because increased vegetation reduces precipitation impact and roots the soil in place. Excessive sediment is economically significant because it is detrimentally deposited on land and plants, reduces depression storage capacities, causes increased flood hazard, and pollutes stored water supplies (Higgins et al. 1989, Walling and Webb 1996). On the other hand, some "normal" natural erosion is necessary to sustain river floodplains, deltas, and estuaries.

Geologic erosion is normal erosion for a natural environment undisturbed by humans. Human disturbances such as overgrazing, logging, farming, and road construction

FIGURE 7.6   Severe gullying and loss of soil from unsound management practices in Mississippi. This area was a cotton field 35 years prior to when the picture was taken (courtesy of U.S. Department of Agriculture, Natural Resources Conservation Service).

can cause accelerated erosion, which proceeds at a higher rate than geologic erosion. Differentiating the two types of erosion is a challenging problem to the watershed manager. Watersheds often have high rates of geologic erosion because of steep slopes, aridity, thin soils, and sparse vegetation cover.

Accelerated erosion occurs when human activities destroy the vegetation cover that retards soil loss from the forces of water and wind (Figure 7.6). The best protection against erosion is to establish and maintain a good vegetative cover. The inverse relationship between accelerated erosion and plant cover is well established (Osborn 1956, Marston 1958, Thurow et al. 1986). Accelerated erosion is the most severe consequence of watershed mismanagement because replenishment of lost soil is a slow process. Several hundred years are required to form an inch of soil; therefore, losses of soil result in nearly permanent reductions of rangeland, farmland, woodland, and watershed productivity. During the initial phases, changes in watershed management can often bring accelerated erosion under control. However, in the more advanced stages, costly measures are often necessary, such as mechanical structures and revegetation.

Managing Watershed Land-Use Practices

Farming.   Erosion and runoff control practices on farmlands include conservation tillage, terracing, strip-cropping, contour farming, gully reclamation, and windbreaks. These practices are discussed in considerable detail in Chapter 8. In addition, runoff of fertilizers and pesticide residues is a major source of water pollution. We also refer the reader to Brady and Weil (1996) for in-depth discussions on farmland watershed management.

Logging.   Properly designed forest cutting and log removal can increase downstream water yields while minimizing erosion. Selective cutting and shelter wood cutting, in contrast to clear-cutting (removal of all trees on the logged area), leave some of the trees to protect the site from soil erosion (see Chapter 10). Clear-cutting forests in small strips or patches, as opposed to large blocks, can minimize soil erosion problems. Because

FIGURE 7.7 The overgrazed range on the left shows severe water erosion compared to the moderately grazed range on the right. The soil surface is nearly a foot lower and dominated by rocks.

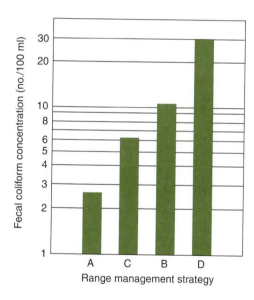

FIGURE 7.8 Fecal coliform bacteria loadings for four range management strategies for the period 1979–1984 in central Oregon (A = no grazing, B = grazing without management to improve livestock distribution, C = grazing with management to improve livestock distribution, D = grazing with cultural practices to increase forage and improve livestock distribution) (from Tiedemann et al. 1987).

forest logging roads are major sources of sediment, road closures and forest restoration are important means for erosion management.

Livestock Grazing. The key to maintaining healthy hydrological conditions on rangelands is through grazing practices that develop and maintain a good plant cover (Holechek et al. 1998 and Chapter 11). Perennial grasses, because of their high basal area and excellent soil-binding properties, play the critical role in watershed stability. Moderate stocking rates in conjunction with practices that promote even livestock distribution over the range are the best approaches to maintaining a good perennial grass cover (Figure 7.7). The success of any grazing program geared toward watershed maintenance and enhancement is best measured by the residue of living and dead vegetation (mulch) it maintains on the site throughout the year.

Fecal wastes from range livestock grazing can cause local pollution problems (Figure 7.8). More serious problems can occur at livestock feedlots. Fecal coliform bacteria counts in water have been used as an indicator of infectious bacterial contamination (Wadleigh 1968). However, the coliform bacteria themselves are not pathologically harmful. Poorly managed livestock operations have caused increased bacterial pollution in rangeland streams (see review by Holechek et al. 1998). The severity depends on livestock numbers, timing of grazing, frequency of grazing, and access to the stream. In Wyoming, livestock grazing, recreational activities, and wildlife all contributed to increased coliform bacterial loads in stream waters (Hussey et al. 1986). Practices that improve livestock distribution and attract livestock away from streamside areas are recommended based on Tiedemann et al. (1987) and Binkley and Brown (1993). The reader is referred to Holechek et al. (1998) for more detailed information on livestock grazing management in watersheds.

Outdoor Recreation.   Areas of intense recreational use are sources of accelerated erosion, especially on steep slopes and fine soils. For parks and other areas used intensively for sports, turf management not only provides a safe substrate but also holds the soil in place. Controlling access to hikers and recreational vehicles is also a key strategy.

Urban Land Use.   Urban development can create great disturbance of the watershed and accelerate erosion from building and road construction. The high fraction of impenetrable watershed surface in urban environments can aggravate downstream flooding and alter groundwater recharge. Intense use of fertilizers, pesticides, and other elements in cities causes major water quality problems from "nonpoint" sources.

Managing Riparian Areas

Riparian management is critical for controlling bank erosion, providing habitat for endangered and recreational species, and maintaining water quality (Thomas 1986). Riparian zones usually receive a disproportionate amount of the grazing, urban-development, farming, recreation, and forest cutting impact. Fish and wildlife populations have been severely affected by uncontrolled grazing, road construction, farming, and forest practices in riparian zones (Figure 7.9). A summary of 20 studies by Platts (1982) consistently (with one exception) showed that poorly controlled

FIGURE 7.9   Livestock tend to concentrate in riparian areas. The fenced area on the left has been excluded from grazing. Grazing practices that disperse rather than concentrate livestock are recommended.

FIGURE 7.10 Typical stream channel cross section under heavy grazing, after 2 to 3 years without grazing, and after 5 to 10 years without grazing (from Bowers et al. 1979).

grazing degraded the streamside environment and the local fishery (Figure 7.10). Reviews of how livestock grazing can be manipulated to reduce adverse impacts on riparian areas are provided by Heady and Child (1994) and Holechek et al. (1998).

Riparian zones often are the most productive sites in a region because floodplains frequently have rich soils with plentiful moisture. They have a greater diversity of plant and wildlife species than adjoining ecosystems. Healthy riparian systems purify water as it moves through the vegetation by retaining sediment and by retaining water in aquifers beneath the floodplain. Riparian zones often are a diverse mix of wetland and upland vegetation, all of which are linked closely with the floodplain groundwater. Maintaining proper amounts of vegetation is a critical part of increasing sediment deposition and enhancing channel restoration in small stream systems (Clary et al. 1996). Many wildlife species depend on the diverse habitat niches in riparian areas. They also serve as a focal point for many forms of recreation (camping, fishing, picnicking, bird-watching, and so on).

Special Watershed Management Techniques

Vegetation-Type Conversion. Several studies reviewed by Branson et al. (1981) show that conversion of shrubland or woodland to herbaceous vegetation can greatly increase

FIGURE 7.11   Piñon-juniper woodland in New Mexico where small areas have been converted to grassland to increase forage for livestock and wildlife and increase water yields.

water yields (Figure 7.11). This is because grasses and forbs generally transpire much less water than do woody plants. Water yield for rangelands has been improved with chemical and mechanical brush control (see reviews by Vallentine 1989 and Holechek et al. 1998). Methodologies for vegetation-type conversion are discussed by Vallentine (1989), Heady and Child (1994), and Holechek et al. (1998).

Water Harvesting.   Water harvesting is the process of collecting and storing precipitation for beneficial uses from land that has been treated to increase runoff. The process has been used in Israel to grow olives, apricots, and other crops. In the United States and Australia, water harvesting is used to provide drinking water for livestock and wildlife and to create wildlife habitat on arid ranges. Water harvesting structures in the United States are called rain traps, catchment basins, paved drainage basins, trick tanks, and guzzlers.

Water Spreading.   Water spreading is a technique that involves diversion of water from drainages onto the surrounding landscape through a system of dikes, dams, or ditches (Branson et al. 1981). The diversion of runoff water with earthen banks to areas favorable for cultivation dates back to the early inhabitants of the Middle East and South America. Water spreading has three main functions: (1) increasing crop or livestock forage production by spreading of floodwater, (2) reducing erosion in drainageways, and (3) reducing downstream flooding and sedimentation. Stream channels (dry most of the time, but flowing for short periods) generally provide the water supply for water spreading schemes. Water ponding has been used in Australia to reclaim bare areas.

MULTIPURPOSE WATER RESOURCE MANAGEMENT

The Water Resource Management Agencies

Five agencies are responsible for much of the federal management and research of water resources in the United States. The Army Corps of Engineers (COE) is the oldest of the resource management agencies, starting early in the nineteenth century. In addition to its original authority for improving navigation of rivers, harbors, and coastal waters, the COE has assumed the lead in flood-damage reduction. In recent years, it has assumed an environmental improvement mission equal in emphasis to navigation and flood-damage reduction. The Bureau of Reclamation (BOR) is mainly responsible for developing and managing surface water irrigation systems, mostly in the arid West. The Natural Resources Conservation Service (NRCS) is authorized to help manage waters on private agricultural lands, including building small reservoirs and drainage systems. It emphasizes erosion control and environmental protection and improvement. The Tennessee Valley Authority (TVA) was authorized to operate a system of reservoirs and connecting waters in the Tennessee River. It emphasizes hydropower and navigation development among numerous other purposes. In addition to the federal water management agencies, the U.S. Geological Survey (USGS) is the main research arm of the federal resource agencies. It operates a branch that concentrates on monitoring water supply and has numerous stations located throughout the United States. It recently consolidated biological research into a new Biological Resources Division.

These federal agencies coordinate with state water resource organizations, often in cooperatively funded projects. They also coordinate more loosely with the federal land management agencies that include watershed yields among their objectives, especially the U.S. Forest Service and Bureau of Land Mangement. Other water resources management is conducted by local governments and private utilities, mostly for power production, but also for irrigation and municipal water supply.

Integrated Management Objectives

The federal water resource agencies develop multiobjective projects for efficient provision of water resource benefit. That means that wherever feasible, the projects provide for all significant needs. Therefore, large reservoirs often are operated for an assortment of water storage (e.g., for irrigation and municipal supply), flood reduction, hydroelectric production, navigation, recreation, and environmental protection. It is not uncommon for two or more agencies to operate in the same river basins. After years of sometimes working at cross-purposes, they are now expected to integrate activities within a basin context. Careful monitoring by the USGS and development of water-routing models (mostly by the COE) have been particularly helpful in making integration possible. We will briefly discuss the primary objectives of water resource management.

Flood-Damage Reduction. Recurrent flooding of rivers has been both beneficial and detrimental to the development of civilization. Ancient societies flourished along the Nile River in Egypt and the Tigris and Euphrates Rivers in Iraq because the periodic floods sustained rich floodplain soils for agriculture. On the other hand, major floods throughout history have caused massive property destruction and loss of human life. This problem has been compounded with the advance of civilization and its associated activities. Cultivation of erodible lands, deforestation, overgrazing, mining, and urbanization all contribute to flooding problems because they remove vegetation and soil, thereby reducing the land's ability to infiltrate and retain water. In the United States, about $96 billion was spent on flood control during the twentieth century. Areas where major flooding problems occur are shown in Figure 7.12. Various watershed practices to control floods center around maintaining vegetation cover on the watershed, dam construction, levee construction, dredging, and stream channelization.

Irrigation and Water Supply. The largest reservoirs in the United States were built for irrigation purposes in the western United States by the BOR. Irrigation systems include dams for water storage and water diversion, channels and piping to deliver the water to use areas, and control gates and valves for allocating water to each user. Irrigation management also includes vegetation management to reduce transpiration loss and to increase water yield.

Navigation. A system of U.S. waterways is now maintained in most of the large rivers and coastal zones of the Gulf of Mexico and the Atlantic. In addition, harbors are maintained along the ocean coasts and in the Laurentian Great Lakes. Navigation

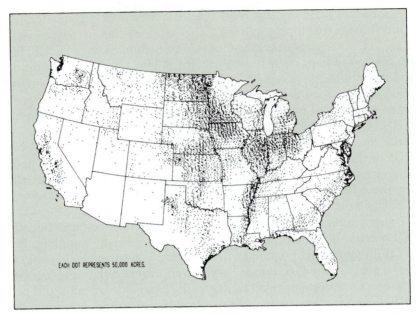

EACH DOT REPRESENTS 50,000 ACRES.

FIGURE 7.12   Areas in the United States with flooding problems (from USDA-Natural Resources Conservation Service).

is one of the cheapest and environmentally cleanest forms of bulk transport. Many locks have been built and are operated by the COE to facilitate navigation past rapids, shoals, and dams. In addition, thousands of miles of river, harbor, and coastal channels are dredged to sustain navigation.

Recreation.   Recreation is also a purpose of water resources management. It includes boating, swimming, fishing, and water-based wildlife observation and hunting. Marinas, docks, boat ramps, and numerous other facilities support recreational activity.

Environmental Protection and Improvement.   Since enactment of the National Environmental Policy Act of 1969, all federal agencies have been responsible for protecting the environment as approved by the U.S. Environmental Protection Agency (EPA). They are required to adopt the least environmentally damaging cost alternative within reason. In recent decades, the water resources agencies also have adopted missions for improving the environment, including restoration of impaired ecosystems. A difficult challenge, however, is balancing environmental protection and improvement with more tangible benefits that have quantifiable monetary values.

Water Management Engineering

Reservoir Construction.   Reservoirs are a valuable means of capturing water from rain and snowmelt that would otherwise go unused for irrigation, municipal supplies, hydropower, recreation, or navigation (Figure 7.13). They also permit control of river flooding and thereby allow people to farm and live on fertile flood plains downstream. Recreation, hydroelectric power, and irrigation water are other important products from reservoirs.

Dams and reservoirs often are controversial because they have significant costs as well as benefits. Construction is usually expensive; however, as long as the benefits exceed costs, the expense is justified. Most reservoirs will ultimately fill up with sediment after a 25- to 300-year time span (Figure 7.14). As long as that loss is included in the analysis the benefits still may justify the dam. Vast areas of scenic beauty, fish and wildlife habitat, and farmland have been altered by dams. On the other hand, beauty and productive habitat have been created by other dams. The reservoirs behind the world's 13,000 dams over 50 feet in height have displaced millions of people (Miller 1990).

FIGURE 7.13 The Hoover Dam on the Colorado River was one of the first large multipurpose dams built during the 1930s Depression Era (courtesy of U.S. Bureau of Reclamation).

Dams have greatly contributed to the 90% decline in salmon populations in the northwestern United States while providing inexpensive navigation and electricity. Some outdoor sports enthusiasts believe dams replace desirable forms of water recreation (white-water canoeing, kayaking, rafting, stream fishing) with less desirable, more "artificial" forms (motorboating, sailboating, lake fishing). However, many other people use and benefit from reservoir recreation nationwide, especially boaters and anglers. Unsafe dams in populated areas are a hazard to human safety. Dam failure in Rapid City, South Dakota, killed 237 people while another in Buffalo Creek, West Virginia, killed 125 (Miller 1990); however, thousands of lives have been protected by flood-control dams. The U.S. Army Corps of Engineers now espouses the use of nonstructural (i.e., without dams and levees) solutions where it is in the Nation's interest. Dams built and operated under federal auspices undergo a careful benefit-cost analysis to ensure wise public investments. These analyses include environmental costs and benefits. However, budget priorities are decided by Congress based on political as well as economic and environmental criteria.

Levee Construction. Levees are dikes usually of earth or concrete erected along the banks of a river to prevent flooding. Most major rivers in the United States have some levee development along them. They have been heavily used along the Mississippi River with mixed results. They were originally built on the Mississippi River to facilitate barge traffic by forcing floodwaters to cut deeper central channels. They have also permitted extensive development (agriculture, industrial, residential) of floodplains that would not otherwise have occurred. Levees are also criticized because they sometimes increase flood severity by concentrating river flow downstream, they can destroy floodplain wetlands that store floodwaters and provide habitat for wildlife, they often impair the aesthetic value of the river, they can encourage unwise floodplain urban development, and they are expensive to maintain (Owen and Chiras

(a)

(b)

FIGURE 7.14 The Mono Reservoir in Santa Barbara was built to stop silt flowing into the Gibraltar Reservoir. By 1938, it had completely silted in (photo a). In 1949, vegetation had invaded the silted area (photo b) and woodland was forming (courtesy of the U.S. Forest Service).

1995). As the 1993 Mississippi River flood and the 1996 Red River flood showed, levees also fail.

In recent years, government agencies (U.S. Army Corp of Engineers, the Bureau of Reclamation, Natural Resources Conservation Service) have increasingly adopted the view that restricting development in floodplains and relocating people away from floodplains is more practical than floodplain protection, at least in areas of relatively low population density. In other areas, such as around large cities, even larger and more effective levees may be needed in the future. Climate change predictions are especially relevant in this regard because of the connections between flooding, navigation needs, and other water resource management.

FIGURE 7.15   Long stretches of the Rio Grande in southern New Mexico have been channelized to increase farmland and to deliver water more quickly to irrigation uses. Channelization has also adversely impacted wildlife and fish populations and impaired the river's aesthetic quality.

Dredging.   Dredging involves widening and/or deepening river channels and coastal harbors to facilitate navigation and reduce flooding. Dredging also maintains the intracoastal canal, which parallels much of the Gulf and Atlantic Coasts. This practice prevents navigation channels from clogging with silt, but often impairs aesthetic values and can be damaging to wildlife habitat. It can contribute to degradation and destruction of wetlands bordering the river by reducing water inflow into these areas. The COE oversees most dredging activities for navigation improvement and the filling of waters feeding into navigable waters. The COE has developed programs for developing beneficial use of dredged material, including creation of wetlands and islands for fish and wildlife habitat.

Stream Drainage Channelization.   Stream channelization involves the straightening and deepening of streams to reduce flooding and make more land available to agriculture or other development (Figure 7.15). The United States Department of Agriculture Natural Resources Conservation Service has directed the stream channelization program. Basically, channelization converts meandering, vegetated streams into straight ditches. The NRCS has often constructed small reservoirs on channelized streams, which can benefit wildlife. However, channelization has many drawbacks that include loss of trees and shrubs associated with the streambank, increase in water temperatures because stream shading from trees and shrubs is reduced, loss in wildlife diversity, and lower nutrient levels in the stream (Owen and Chiras 1995). In many cases, downstream flooding increases because the dissipating effect of streamside vegetation and meandering is reduced or eliminated. There was considerable opposition to stream channelization by the NRCS in the 1990s. This caused the government to scale back this program as the evidence accumulated indicating that the benefits failed to exceed the costs.

WATER QUALITY PROBLEMS

Types of Water Quality Problems

Surface water quality degradation is widely publicized, but has not become a major limitation to water availability or use nationwide. A relative abundance of good-quality surface water still exists, although water quality problems have developed in some areas.

TABLE 7.3   Five Leading Causes of Water Quality Impairment

Rank	Rivers	Lakes	Estuaries
1	Bacteria	Nutrients	Nutrients
2	Siltation	Siltation	Bacteria
3	Nutrients	Oxygen-depleting substances	Oxygen-depleting substances
4	Oxygen-depleting substances	Metals	Habitat alterations
5	Metals	Suspended solids	Oil and grease

Source: EPA 1995.

TABLE 7.4   Five Leading Sources of Water Quality Impairment Related to Human Activities

Rank	Rivers	Lakes	Estuaries
1	Agriculture	Agriculture	Urban runoff/storm sewers
2	Municipal sewage-treatment plants	Municipal sewage-treatment plants	Municipal sewage-treatment plants
3	Hydrologic/habitat modification	Urban runoff/storm sewers	Agriculture
4	Urban runoff/storm sewers	Unspecified nonpoint sources	Industrial point sources
5	Resource extraction	Hydrologic/habitat modification	Petroleum activities

Source: EPA 1995.

Under the Clean Water Act and other legislation, the EPA monitors the conditions of the nation's lakes, streams, and estuaries with the aim of identifying and regulating pollutants at their sources. About one-third of our lakes, streams, and rivers show at least some impairment.

The five leading causes of surface water impairment include bacteria, siltation, nutrients, oxygen-depleting substances, and metals. Because rivers, lakes, and estuaries differ greatly in their physical attributes, specific categories of pollutants affect them differently (Table 7.3). Health-threatening bacteria, for example, are the leading cause of impairment in river waters but nutrients are the prime cause of lake and estuary impairment.

The EPA has also identified the five leading sources of surface water pollutants (Table 7.4). These include agriculture, municipal sewage-treatment plants, aquatic habitat modification, urban runoff/storm sewers, and resource extraction. They impact lakes, rivers, and estuaries with different intensities. Agriculture, because of its role in producing nutrient and sediment problems, is the leading cause of lake and river impairment. Estuaries suffer most from urban runoff and storm sewers that carry petroleum contaminants. Estuaries are also highly vulnerable to eutrophication caused by treated sewage waste. While secondary sewage treatment removes organic pollutants, much of the inorganic nutrient remains.

Surface Water Impairment

Water pollution problems are most severe along coastal areas, much of the Midwest, and in the eastern United States (Figure 7.16). In the United States, there are countless forms of pollutants. We have grouped them into eight major categories, which will be discussed.

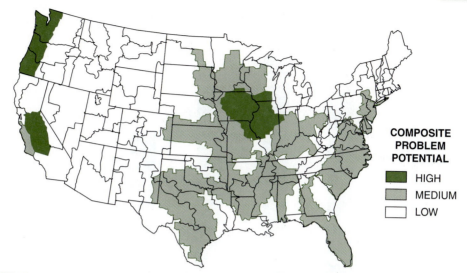

FIGURE 7.16 Composite potential surface water pollution associated with agriculture in the United States. Isolated locations may have greater composite problem potential than indicated at the scale used on the map (from USDA Natural Resources Conservation Service).

Disease-Causing Organisms

The pollutants of prime concern in this category are bacteria and other pathogens such as viruses and protozoa. Waterborne pathogens cause human illnesses that range from typhoid and dysentery to minor respiratory and skin diseases. Pathogens can enter waters through a number of routes including inadequately treated sewage, stormwater drains, septic systems, runoff from livestock pens, and sewage dumped overboard from recreational boats. Municipal waste can carry numerous pathogenic microorganisms (Nash 1993). Widespread decreases in fecal bacteria have occurred as municipal waste treatment has improved. In 1993 and 1994, for example, only 30 reported disease outbreaks were associated with drinking water, 23 were associated with public drinking water supplies, and seven were with private wells. This low incidence reflects the effectiveness of sewage treatment overseen by the EPA and local government agencies.

Under rare conditions, certain protozoan pathogens such as *Cryptosporidium* can pass through water treatment filtration and disinfection processes in sufficient numbers to cause health problems. *Cryptosporidium* causes cryptosporidiosis, a gastrointestinal disease. An outbreak of this pathogen occurred in Milwaukee, Wisconsin, in 1993. More than 400,000 persons contracted the disease, over 4,000 were hospitalized, and more than 50 died.

Nutrients

Nutrients can reach such high concentrations in water that its ability to support certain native species is altered. Nutrients can overstimulate the growth of algae and aquatic plants in a process known as cultural eutrophication (see also Chapter 3 and Chapter 9). In advanced cases, cultural eutrophication interferes with the recreational use of lakes, increases the cost of water treatment before it can be used, and reduces diversity in native fish, plant, and animal populations. It is one of the leading problems facing our nation's lakes and reservoirs. Phosphorus and nitrogen are among the common nutrients that cause eutrophication. They are carried along into streams and lakes by runoff and wastewater. The origin of nitrate in surface waters can also be traced to atmospheric deposition, which involves nitrogen oxide emissions from power plants and motor vehicles. Nutrient enrichment is caused by a wide variety of land-use practices including treated and untreated sewage wastewater, feedlots, concentrations of range livestock,

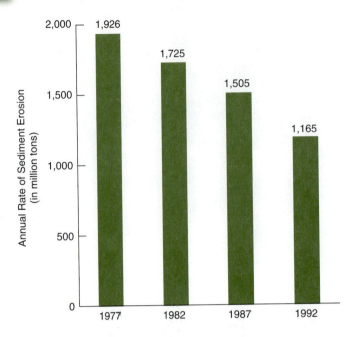

FIGURE 7.17   Sediment erosion from cropland has decreased over the period from 1977 through 1992 (EPA 1994, *Environmental Indicators of Water Quality in the United States*).

farm and yard fertilizers, overabundance of certain wildlife (e.g., geese), and any disturbance that accelerates watershed erosion.

Nitrate-contaminated drinking water creates a grave risk for humans. Nitrates get into groundwater from many anthropogenic causes. Often they are associated with septic tanks that release them into groundwater serving private wells. Contamination can also be caused by animal wastes from farm feedlots as well as heavy applications of nitrogen fertilizer applications to sandy soils. Infants are the group at greatest risk. In some susceptible babies under six months old, nitrates can be converted to nitrites by stomach bacteria. These nitrites can then bind with blood hemoglobin and prevent oxygen from circulating to the rest of the body. The disease can produce "blue baby symptoms" and may prove fatal. Nitrates are not usually harmful to adults or older children.

Silts and Suspended Solids

As fine silt and suspended solids accumulate in aquatic habitats, fish have difficulty respiring and their eggs are smothered. Turbidity reduces plant productivity to the extent that photosynthesis is impaired by reduced sunlight. Recreational opportunities can also be lost. In a water quality context, sediment usually refers to soil particles that enter the water column from eroding land. Sediment consists of particles of all sizes including fine clay particles, silt, sand, and gravel. Suspended sediments can often be traced to agricultural sources. However, progress has been made toward reducing sediment loading from cropland (Figure 7.17).

Biochemical Oxygen Demand (BOD)

The decomposition of organic materials reduces the availability of dissolved oxygen that is crucial to respiration of fish and aquatic invertebrates. Industrial sources currently contribute about one-third of this biochemical oxygen demand (BOD) load nationwide. Most industrial BOD load reduction occurred in the mid-1970s shortly after the Clean Water Act of 1972 was passed. Municipal reductions occurred later, in the early 1980s. Within a decade after the 1972 legislation was passed, municipal loads of BOD decreased 46% and industrial BOD loads decreased 71% nationally.

Salinity and Other Dissolved Solids

Salts typically make up most of the total dissolved solids in water. Salinity impairs the use of water for drinking and crop irrigation, and adversely affects aquatic ecosystems. In the West, surface waters have been contaminated with salts by irrigated agriculture. In coastal areas, saltwater intrusion from ocean sources also causes local contamination of groundwaters where too much is pumped for use.

Toxic Materials

Toxic materials can have adverse effects when present in extremely low concentrations. They can cause death, mutation, or reproductive failure in fish and wildlife, and may pose carcinogenic or other health threats to humans. They may be persistent or dissipate quickly. More than 60,000 commercial chemical substances are currently in use in the United States. They potentially threaten our water if handled irresponsibly. About 3.5 billion pounds of formulated pesticide products are used each year. Pesticides, including insecticides and herbicides, are applied extensively to crop-, pasture-, and forest land throughout the nation. In urban areas, they are used on lawns and gardens, and to exterminate pests in buildings and homes. The greatest release of pesticides has been on farms.

Mining operations contribute directly and indirectly to heavy metal toxicity. However, the mining industry has cleaned up abandoned mines and controlled discharges from active ones. The silver, lead, and copper mines found in the western United States can directly contribute metal-laden runoff through mine tailings piles and mine seepage. Many of the worst problems come from mines operated and abandoned many decades ago. Pollution from abandoned mines is addressed in the Federal Surface Mining Control and Reclamation Act of 1977. Discharges from active mines are regulated by discharge permits issued by the EPA and states. Many states also use best management practices (BMPs) to reduce the pollution threats posed by mines. Best management practices applied to mining, forestry, and other activities are designed to prevent rather than correct water quality problems.

Aerial sources of toxic metals also originate from fossil-fuel ignition and wind erosion. In addition to natural sources, coal-burning power plants appear to be a source of mercury contamination. Aerial sources come by way of both dry (dust) and wet precipitation. One of the worst cases of air pollution occurs in the Owens River Valley in southern California. There, the dried-out bed of an extinct lake is a source for tremendous storms of contaminated dust, which are transported long distances. Because the lake is extinct as a consequence of water transfer from the Owens River to Los Angeles, there is a movement to return at least some of the water for lake restoration. However, there is also resistance because Los Angeles now depends on the water.

Polychlorinated biphenyls (PCBs) are used to cool electrical transformers. They were discharged into water systems prior to regulatory controls in the 1970s. PCB contamination of stream sediments has led to moratoria in 15 states on the consumption of fish caught in streams below points of known PCB discharges.

Acid Mine Drainage

Acid mine drainage problems occur when sulfur-bearing minerals are exposed to water and atmospheric oxygen, which forms sulfuric acid. Acidic water draining or seeping from mines dissolves metals from geologic formations and carries them into waterways. On entering more alkaline streams, iron compounds dissolved in the acidic water precipitate and impede production of bottom-dwelling aquatic organisms.

Thermal Discharges

Power plants and industries release cooling water with elevated temperature. The discharged waters often average 10°C or greater. When ambient temperatures are warm, this heating can block river migrations of fish or trap fish in areas where they are exposed to pollutants or other harm. In winter, fish often are attracted to the warmed water. Although this can result in excellent fisheries, it also leaves fish vulnerable to thermal shock if the power plants have to shut down operation. For these reasons among many, very large cooling towers are often used to cool water before it is returned. In other instances, large cooling ponds are built and dedicated to cooling purposes and compatible uses.

APPROACHES TO WATER QUALITY MANAGEMENT

Classification of Water Pollution Sources

Anthropogenic pollution can be categorized as emanating from either municipal, industrial, or agricultural sources (Figure 7.18). Pollutants can be further classified as originating from either point or nonpoint sources (Table 7.5). Point sources discharge pollutants directly into surface waters from a conveyance. Point sources include municipal sewage-treatment plants, industrial facilities, and combined sewer overflows. Nonpoint sources deliver pollutants to surface waters from diffuse origins. Nonpoint sources include urban runoff, agricultural runoff, and atmospheric deposition of contaminants in air pollution. Habitat alterations such as construction, dredging, and streambank channelization can also degrade water quality.

In the mid-1970s, point-source pollution was believed to be more serious than nonpoint pollution. As the large point-source discharges were brought into compliance, it became more obvious that nonpoint sources were also an important cause of water quality problems. Therefore, more recent efforts to improve the quality of the Nation's water have given greater attention to nonpoint pollution.

The Watershed Approach

The EPA has begun applying a watershed approach to water quality problems. Nationwide, about 22% of all watersheds need capital investments to restore water quality. About one-half need improved management to attain long-term water resource goals. Watershed plans integrate programs for control of point and nonpoint sources. They provide decision-makers with the opportunity to protect and restore habitat for aquatic life. The watershed approach to pollution control fosters more comprehensive environmental solutions, more cost-effective programs, and improved public involvement.

Water Treatment

Water suppliers use a variety of treatment processes to remove contaminants from drinking water. Water typically passes through a series of treatment processes to remove common contaminants. Specific treatments may be needed at some locations but not others. The most commonly used processes include filtration, flocculation and sedimentation, and disinfection (Figure 7.19). Some treatment programs also include ion exchange and adsorption. Flocculation refers to water treatment processes that combine small particles into larger particles, which settle out of the

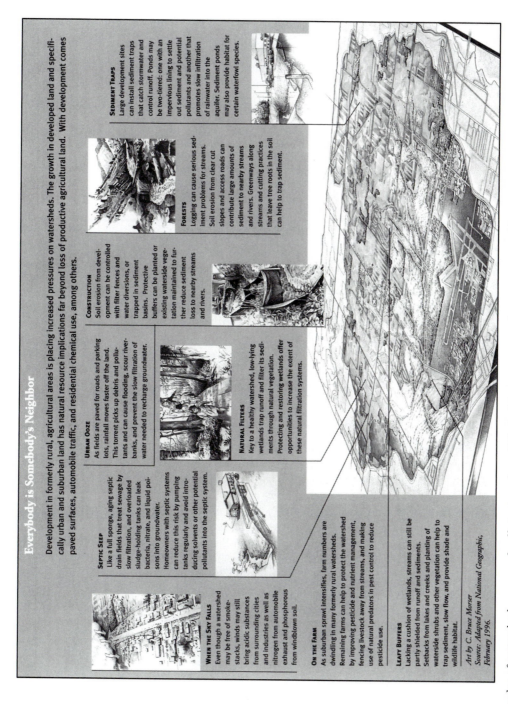

Everybody is Somebody's Neighbor

Development in formerly rural, agricultural areas is placing increased pressures on watersheds. The growth in developed land and specifically urban and suburban land has natural resource implications far beyond loss of productive agricultural land. With development comes paved surfaces, automobile traffic, and residential chemical use, among others.

WHEN THE SKY FALLS
Even though a watershed may be free of smokestacks, winds may still bring acidic substances from surrounding cities and industries as well as nitrogen from automobile exhaust and phosphorous from windblown soil.

ON THE FARM
As suburban sprawl intensifies, farm numbers are dwindling in many formerly rural watersheds. Remaining farms can help to protect the watershed by improving pesticide and nutrient management, fencing livestock away from streams, and making use of natural predators in pest control to reduce pesticide use.

LEAFY BUFFERS
Lacking a cushion of wetlands, streams can still be partly shielded from runoff and sediments. Setbacks from lakes and creeks and planting of waterside shrubs and other vegetation can help to trap sediment, slow flow, and provide shade and wildlife habitat.

SEPTIC SEEP
Like a full sponge, aging septic drain fields that treat sewage by slow filtration, and overloaded sludge-holding tanks can leak bacteria, nitrate, and liquid poisons into groundwater. Homeowners with septic systems can reduce this risk by pumping tanks regularly and avoid introducing solvents or other potential pollutants into the septic system.

URBAN OOZE
As fields are paved for roads and parking lots, rainfall moves faster off the land. This torrent picks up debris and pollutants and can cause flooding, scour riverbanks, and prevent the slow filtration of water needed to recharge groundwater.

NATURAL FILTERS
Key to a healthy watershed, low-lying wetlands trap runoff and filter its sediments through natural vegetation. Protecting and restoring wetlands offer opportunities to increase the extent of these natural filtration systems.

CONSTRUCTION
Soil erosion from development can be controlled with filter fences and water diversions, or trapped in sediment basins. Protective buffers can be planted or existing waterside vegetation maintained to further reduce sediment loss to nearby streams and rivers.

FORESTS
Logging can cause serious sediment problems for streams. Soil erosion from clear cut slopes and access roads can contribute large amounts of sediment to nearby streams and rivers. Greenways along streams and cutting practices that leave tree roots in the soil can help to trap sediment.

SEDIMENT TRAPS
Large development sites can install sediment traps that catch stormwater and control runoff. Ponds may be two-tiered: one with an impervious lining to settle out sediment and potential pollutants and another that promotes slow infiltration of rainwater into the aquifer. Sediment ponds may also provide habitat for certain waterfowl species.

Art by C. Bruce Morser
Source: Adapted from National Geographic, February 1996.

FIGURE 7.18 Examples of watershed sources of pollutants and conditions that prevent or reduce pollution loads (from *A Geography of Hope*, 1997; United States Department of Agriculture, Natural Resources Conservation Service).

TABLE 7.5   Examples of Point and Nonpoint Pollution Sources

Point Sources

Industrial	Pulp and paper mills, chemical manufacturers, steel plants, metal process and product manufacturers, textile manufacturers, food processing plants
Municipal	Publicly owned sewage treatment plants, which can receive discharges from industrial facilities or businesses as well as homes and public services
Combined sewers	Single facilities that treat both storm water and sanitary sewage, which may become overloaded during storm events and discharge untreated wastes into surface waters

Nonpoint Sources

Storm sewers/urban runoff	Runoff from impervious surfaces including streets, parking lots, buildings, lawns, and other paved areas
Agricultural	Crop production, pastures, rangeland, feedlots, other animal holding areas
Silvicultural	Forest management, tree harvesting, logging road construction
Construction	Land development, road construction
Resource extraction	Mining, petroleum drilling, runoff from mine tailing sites
Land disposal	Leachate or discharge from septic tanks, landfills, and hazardous waste sites
Hydrologic modification	Channelization, dredging, dam construction, streambank modification

Source: EPA 1995.

water as sediment. Filtration removes the remaining particles from the water supply. Ion exchange processes are used to remove inorganic constituents if they cannot be removed adequately by filtration or sedimentation.

Chlorine, chloramines, or chlorine dioxide are commonly used for disinfection. Residual concentrations can be maintained to guard against biological contamination in the water distribution system. Ozone is a powerful disinfectant, but it is not effective in controlling biological contaminants in distribution pipes.

Home Water Treatment

Some homeowners choose to add further protection to their drinking water. Numerous appliances are commercially available for this purpose. Home filtration units use activated carbon filters that adsorb organic contaminants and constituents, which can cause taste and odor problems. They also remove chlorination by-products, some cleaning solvents, and pesticides. They do not remove metals such as lead and copper. Ion exchange units can be used to remove minerals, particularly calcium and magnesium, and are sold for water softening. Some ion exchange softening units remove radium and barium from water. Ion exchange systems that employ activated alumina are used to remove fluoride and arsenate from water. Reverse osmosis treatment units generally remove a more diverse list of contaminants than other systems. They can remove nitrates, sodium, other dissolved inorganics, and organic compounds. Distillation units boil water and condense the resulting steam to create distilled water.

WATER USES IN THE UNITED STATES _____

Types of Water Use

Since 1900, the U.S. population has increased 200% and now numbers about 260 million. Over the same period, however, per capita water use increased about 650%. This reflects tremendous growth in all human uses for water. Water is used either in place in the stream (*in-stream use*) or it is withdrawn for *off-stream use*.

PRELIMINARY TREATMENT
remove debris and grit

Bar screen

Grit chamber

Velocity
slows, coarse
grit settles.

Raw sewage

Grit removal

Debris
removal

To landfill

PRIMARY TREATMENT
remove particulate organic material

Water enters at center and flows
out over weir at edge. Particulate
organic material settles. Constitutes
raw sludge.

Fat and oil float
to top and are
skimmed off.

Rotating "plow"

Raw sludge
removal

Composting

Sludge
treatment

Anaerobic digestion

Clarified
water

Fertilizer   Methane

BIOLOGICAL NUTRIENT REMOVAL Excess activated sludge
(BNR)/SECONDARY TREATMENT
To remove colloidal
and dissolved
organic material

Organisms settle and become
next batch of activated sludge.

Activated-sludge system

Organisms feed
on organic
material in
oxygen-rich
environment

Aeration tank

Activated sludge returns

Forced air

Disinfection,
release

FIGURE 7.19 Example of wastewater treatment (from Nebel and Wright 1998).

In-stream water uses do not require the diversion or withdrawal of water from its source. In-stream uses include navigation, hydropower generation, fish and wildlife propagation, recreational activities, and the maintenance of estuary salinity. The quantity of fresh water withdrawn in the United States for all uses increased steadily from 1950 to 1980 (Figure 7.20). Each year, considerably more water was obtained from surface than from ground sources. From 1980 to 1990, the total amount withdrawn from surface and groundwater sources actually decreased slightly.

Consumptive use of water precludes further use downstream. Virtually all water use involves some conversion to water vapor through transpiration, perspiration, and evaporation. Other water use diverts water from the original river flow to another watershed and thus is no longer available to downstream users. From the standpoint of downstream users, the conversions and transfer out of the stream are tantamount to water "consumption." It is no longer available to them for use. Many types of water use return most of the water withdrawn from streams, lakes, or groundwater. In cities, most water is returned through sewers and storm drainage systems after it is poured or flushed

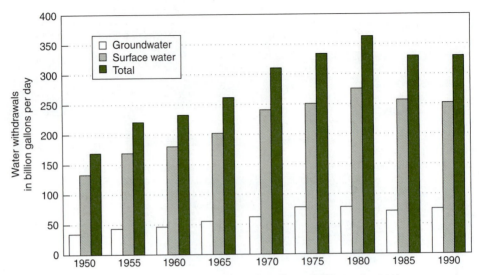

FIGURE 7.20   Freshwater withdrawals from 1950 to 1990 (from Solley et al. 1993).

down drains and runs off roofs, parking lots, and streets. The returned water may be altered in quality, but can be improved through treatment. In crop production, some water is consumed as it evaporates from the soil surface or is transpired. Water also is consumed by manufacturing and food processing industries. Water consumption typically is greatest in arid environments where evapotranspiration exceeds precipitation.

Average per capita use in 1990 was 1,340 gal/day of freshwater. Declines in withdrawals from 1980 to 1990 reflect numerous factors. Higher energy prices, more efficient application techniques, extent of irrigation development, regional droughts, increased competition for water, domestic water conservation, and water recycling all played a part in reversing the steady increase in water use that occurred in the preceding decades. About 17% of all water used in the United States is saline.

Off-Stream Uses

The United States Geological Survey (USGS) periodically reports water withdrawals by category. Especially useful are reports prepared by Solley et al. (1988) and USGS (1993). For simplicity, we will confine the discussion to thermoelectric, agricultural, industrial, and domestic categories.

Thermoelectric.   Water in this category is used mainly for condenser cooling in thermoelectric power plants, which rely on steam production. Steam electric generation requires water boiling by burning fossil fuels (e.g., coal, gas, oil) or by nuclear fission. The steam cycle is maintained by condensing the steam, once it has been used, and boiling it again for steam regeneration. Thermoelectric cooling can use saline as well as fresh water. In 1990, about 131 billion gallons of fresh water were used each day in thermoelectric power production (Figure 7.21). This is roughly equivalent to the flow of ten Mississippi Rivers. More than 30% of the water used for thermoelectric cooling is saline.

Agriculture.   Irrigation is the primary use of the nation's freshwater (Figure 7.21). It accounts for 84% of the freshwater consumed each day. Irrigation was used in the United States by Native Americans before European colonization. Irrigation water has been used to increase crop yield and to increase the number of plantings each year. Without irrigation, growers would have to confine crop production to seasons of high precipitation. Today, irrigation is used on 15% of the Nation's total cropland and it contributes 38% to the total value of all crops. For individual crops, irrigated

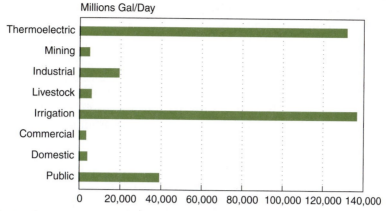

FIGURE 7.21 Freshwater withdrawals by use category for 1990 (from Solley et al. 1993).

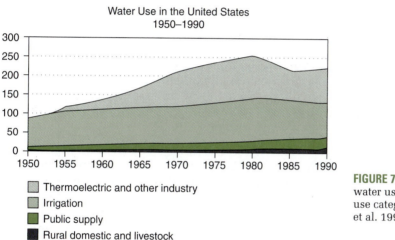

FIGURE 7.22 Trends in water use 1950–1990 by use category (from Solley et al. 1993).

acreage as a share of total acreage is most significant for rice (100%), orchards (81%), vegetables (64%), and cotton (36%).

Irrigation water can be supplied by farms themselves or by irrigation companies or districts. Water is applied to crops by two general methods. Flood irrigation includes applications made by flooding or by using furrows. Nonflood irrigation is done with center pivot, traveling gun, trickle, and drip systems. Irrigation water statistics include water used to irrigate public and private golf courses as well as water applied to farm and horticultural crops. They exclude irrigation of lawns and gardens, which is a substantial use of water in urban settings.

Total irrigation withdrawals increased progressively from 1965 to 1980 when use peaked (Figure 7.22). From 1980 to 1985 irrigation actually decreased 9% and in 1990 it had not returned to 1980 levels. Most of the decrease in total water use was due to decreased irrigation use as a consequence of more efficient techniques and retirement of some lands from irrigation agriculture.

The water used by livestock was included in the irrigation category but represented an insignificant amount. Livestock use includes water withdrawn for feedlots, dairies, and other farm needs.

Industry and Mining. Industries use water primarily for cooling and washing, and for creating all sorts of products. Water is used, for example, to produce paper, steel, chemicals, prepared foods, and computer chips, Mining uses water for extraction, sorting, and processing. In 1990, industries and mining accounted for 4% of the fresh water withdrawn. One-third of the water used by mining is saline as is 15% of the water used for industrial purposes.

Domestic and Commercial.   Domestic water use includes water used by households for drinking, food preparation, bathing, washing clothes and dishes, flushing toilets, and watering lawns and gardens. Commercial water use includes water for motels, hotels, restaurants, office buildings, other commercial facilities, and public institutions (e.g., government buildings and lands, schools, museums, military bases, auditoriums, stadiums). In 1990, domestic and commercial uses accounted for 7% of all water consumed.

Most water for domestic and commercial uses is provided by public-supply facilities, which divert, treat, and deliver the water. In 1990, domestic deliveries averaged 105 gallons per day for each person served, the same as during 1985. The per-capita use remained about the same for the last decade owing to conservation efforts including the installation of additional meters and water-conserving plumbing fixtures, and regulated irrigation of lawns and gardens.

In-Stream Uses

Important in-stream uses include hydropower production, recreational uses, and maintenance of ecosystem functions. While off-stream uses typically are more consumptive and environmentally problematic, not all in-stream uses are without environmental effects.

Hydropower.   In 1990, hydroelectric plants produced about 11% of the Nation's total power output (see Chapter 20). From 1985 to 1990, in-stream use increased about 8%. Nature largely determines how much energy can be produced from hydropower (dams) each year. When surface flow decreases during periods of drought, so does energy production. The Pacific Northwest uses by far the most water for hydroelectric power production. Hydropower production plants can have major impacts on natural ecosystems. Much of the damage done to migrating salmonid populations in the Northwest is believed to be caused by hydropower dams, which block upstream passage and kill a proportion of the young fish as they move downstream through the turbines. However, hydropower neither causes climate change nor thermal pollution. A major challenge is developing improved compatibility, if possible, with fish passage.

Navigation.   Historically, the most important in-stream use of water was navigation and it continues to be a critical use to this day. Thousands of miles of navigable waterways are maintained in the United States as well as other parts of the world. Without water transportation, the cost of many bulk goods would be significantly greater.

Recreation.   In-stream flows support fish and wildlife, boating, rafting, and other forms of water-based recreation. The values of such use can be substantially greater than the prevailing off-stream use values. However, in-stream uses for recreation have been historically belittled or ignored in many western states where water is owned and used mostly in off-stream settings. In recent years, most states have adopted laws to provide for in-stream flow in support of recreational use.

Ecosystem Support.   In addition to recreation, maintenance of in-stream flow is justified to maintain natural biodiversity and ecosystem function, especially in streams supporting threatened or endangered species. Most states have laws that provide in some way for minimum in-stream flow needs.

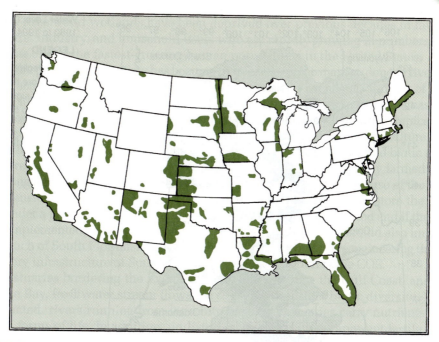

FIGURE 7.23   Regions where groundwater depletion is an important problem (United States Department of Agriculture).

Regional Trends in Water Use and Consumption

Groundwater depletion is an important problem in the southwestern United States (Figure 7.23). Domestic uses of surface freshwater are more concentrated in the eastern states, which are more heavily populated. Industrial uses are greater in the East and are especially concentrated in the northeastern states. Irrigation is the predominant use in the West. Here supplemental irrigation is needed to assure reliable crop production. Among all states, California leads the United States in water consumption, with 20% of total water consumed. This high use results from a combination of large population demand and large irrigation demand.

WATER USE PROBLEMS AND CONFLICTS _____

Water is a hot issue today owing to controversies and problems surrounding its use. Basically, the problem is too little water relative to various demands. More now than ever, the public and user groups are keenly interested in water issues. Space will not permit discussion of all of the important water issues confronting the Nation or even a specific region (see Postel 1996, EPA 1995, Perry and Vanderklein 1996). However, we will briefly examine some of the more prominent problems and issues.

Overpumping

One important problem in several regions is that water is being withdrawn faster than nature is restoring it. In such cases, it can be said that water is being mined. This is happening in geographic locations as remote as Lubbock, Texas, in the High Plains (Supalla et al. 1982), Phoenix, Arizona, and Florida. As water is pumped to meet the growing demands of irrigated agriculture and escalating urban growth, land actually subsides, often damaging buildings, roads, and other structures. East of Phoenix, water tables have dropped more than 120 m. If groundwater withdrawals continue at current rates, water tables in Albuquerque, New Mexico, will drop an additional 20 m by 2020 (Postel 1996).

Guldin, R. W. 1989. *An analysis of the water situation in the United States: 1889–2040.* U.S.D.A. Forest Service Gen. Tech. Report RM-177.

Heady, H. F. and R. D. Child. 1994. *Rangeland ecology and management.* San Francisco, CA: Westview Press.

Heathcoate, I. W. 1988. *Integrated watershed management: Principles and practices.* New York: John Wiley & Sons.

Herdendorf, C. E. 1990. Distribution of the world's large lakes (pp. 3–38). In *Large lakes: Ecological structure and function*, edited by M. M. Tilzer and C. Serruya. New York: Springer-Verlag.

Higgins, D. A., S. B. Maloney, A. R. Tiedmann, and T. M. Quigley. 1989. *Storm runoff characteristics of grazed watersheds in eastern Oregon.* Water Resour. Bull. 25:87–100.

Holechek, J. L., R. D. Pieper, and C. H. Herbel. 1998. *Range management principles and practices.* 3rd ed. Upper Saddle River, NJ: Prentice-Hall.

Hussey, M. R., Q. D. Skinner, and J. C. Adams. 1986. Changes in bacterial populations in Wyoming mountain streams after ten years. *Journal of Range Management* 39:369–370.

Jensen, M. E. and J. D. Bredehoeft. 1983. New efficiencies in water use vital for nation (pp. 18–27). In *Using our natural resources, 1983 yearbook of agriculture.* Washington, DC: U.S. Department of Agriculture.

Knight, R. W. 1993. Managing stocking rates to prevent adverse environmental impacts. In *Managing livestock stocking rates on rangeland*, (pp. 97–107). TX. Agric. Ext. Serv., College Station, TX.

MacKichan, K. A. 1951. *Estimated use of water in the United States, 1950.* Geological Survey Circular 115. Washington, DC: U.S. Department of the Interior, Geological Survey. 13 pp.

MacKichan, K. A. 1957. *Estimated use of water in the United States, 1955.* Geological Survey Circular 398. Washington, DC: U.S. Department of the Interior, Geological Survey. 18 pp.

Marston, R. B. 1958. Parrish Canyon, Utah: A lesson in flood sources. *Journal of Soil and Water Conservation* 13:165–167.

Miller, G. T. 1990. *Resource conservation and management.* Belmont, CA: Wadsworth.

Nash, L. 1993. Water quality and health (pp. 25–39). In *Water in crisis: A guide to the world's freshwater resources*, edited by P. H. Gleick. New York: Oxford University Press.

Nebel, B. I. and R. Wright. 2000. *Environmental science: The way the world works.* 8th ed. Upper Saddle River, NJ: Prentice-Hall.

Osborn, B. 1956. Cover requirements for the protection of range site and biota. *Journal of Range Management* 9:75–80.

Owen, O. S. and D. D. Chiras. 1995. *Natural resource conservation.* 6th ed. Upper Saddle River, NJ: Prentice-Hall.

Owen, O. S., D. D. Chiras, and J. P. Reganold. 1998. *Natural resource conservation.* 7th ed. Upper Saddle River, NJ: Prentice-Hall.

Perry, J. and E. Vanderklein. 1996. *Water quality: Management of a natural resource.* Cambridge, MA: Blackwell Science.

Platts, W. S. 1981. *Effects of sheep grazing on a riparian-stream environment.* U.S. Dept. Agric. For. Serv. Res. Note INT-307.

Platts, W. S. 1982. *Livestock and riparian-fishery interactions: What are the facts?* In Trans. 47th North American Wildlife and Natural Resources Conference, pp. 507–515. Washington, DC: Wildlife Management Institute.

Postel, S. 1996. *Dividing the waters: Food security, ecosystem health, and the new politics of scarcity.* Washington, DC: World Watch Institute.

Postel, S. 2000. *Redesigning irrigated agriculture.* In: State of the World 2000. Washington, DC: World Watch Institute.

Rauzi, F. C. and C. L. Hanson. 1966. Water intake and runoff as affected by intensity of grazing. Journal of Range Management 19:351–356.

Sampat, P. 2001. *Uncovering Water Pollution.* In: State of the World 2001. Washington, DC: World Watch Institute.

Smil, V. 1997. *Cycles of life: Civilization and the biosphere.* New York: Scientific American Library.

Smith, Z. A. 1989. *Groundwater in the West.* New York: Academic Press.

Solley, W. B., C. F. Merk, and R. P. Pierce. 1988. *Estimated use of water in the United States in 1985.* Geological Survey Circular 1004. Reston, VA: U.S. Department of the Interior. 82 pp.

Solley, W. B., R. R. Pierce, and H. A. Perlman. 1993. *Estimated use of water in the United States in 1990.* Geological Survey Circular 1081. Reston, VA: U.S. Department of the Interior.

Speidel, D. H. and A. F. Agnew. 1988. The world water budget. In *Perspectives on water uses and abuses*, edited by D. H. Speidel, L. Ruedisilli, and A. F. Agnew. New York: Oxford University Press.

Supalla, R. J., R. R. Lansford, and N. R. Gollehon. 1982. Is the Ogallala going dry? *Journal of Soil and Water Conservation* 37(6):310–314.

Thomas, A. E. 1986. Riparian protection/enhancement in Idaho. *Rangelands* 8:224–227.

Thurow, T. L., W. H. Blackburn, and C. A. Taylor, Jr. 1986. Hydrologic characteristics of vegetation types as affected by livestock grazing systems, Edwards Plateau, Texas. Journal of Range Management 39:505–508.

Tiedemann, A. R., D. A. Higgins, T. M. Quigley, H. R. Sanderson, and D. B. Marx, 1987. Responses of fecal coliform in streamwater to four grazing strategies. *Journal of Range Management* 40:322–329.

United Nations Development Programme. 2000. *World resources 2000–2001.* New York: Elsevier Science.

U.S. Environmental Protection Agency. 1984a. *The cost of clean air and water.* Report to Congress, 1984. Report No. 230/05-84-008. Washington, DC: U.S. Environmental Protection Agency.

U.S. Environmental Protection Agency. 1984b. *Nonpoint-source pollution in the U.S.* Report to Congress. Washington, DC: U.S. Environmental Protection Agency.

U.S. Environmental Protection Agency. 1996. *Environmental indicators of water quality in the United States.* EPA 841-R-96-002.

U.S. Environmental Protection Agency, Office of Water. 1995. *National water quality inventory.* Report to Congress, 1994.

U.S. Environmental Protection Agency, Office of Water. 1997. *Water on tap: A consumer's guide to the nation's drinking water.* EPA 815-K-97-002.

U.S. Geological Survey. 1993. *National water summary 1990–1991, Hydrologic events and stream water quality.* U.S. Geological Survey Water Supply Paper 2400. Data on trends of selected pollutants found in surface water.

Vallentine, J. F. 1989. *Range development and improvements.* 3rd ed. Provo, UT: Brigham Young University Press.

Wadleigh, C. H. 1968. *Wastes in relation to agriculture and forestry.* U.S. Dept. Agric. Misc. Publ. 1065.

Walling, D. E. and B. W. Webb. 1996. Erosion and sediment yield: A global overview. *International Association Hydrological Science* 236:3–19.

World Resources Institute. 1994. *World resources 1994–95.* New York: Oxford University Press.

CHAPTER 8

Soil: The Basic Land Resource

IMPORTANCE OF SOIL

The most basic of resources is the land and its primary component, the soil. Knowledge of soil characteristics, classification, and management is essential to the natural resource manager. Soil is the primary factor determining the potential for crop, timber, wildlife, and forage production of an area within a particular climate. Throughout history, productive soils have consistently been the cornerstone of the flourishing civilizations. On the other hand, soil destruction or mismanagement was often the root cause of those that declined. Excessive cutting of timber in watersheds, overgrazing by livestock, unsound cultivation practices, and farming of erodible lands are some of the costly mistakes societies have made in managing their soils. In this chapter we provide basic understanding of soil and soil problems. Excellent sources for the topic are provided by Miller (1990), Brady and Weil (2002), and Owen et al. (1998).

Soil is defined as the dynamic, natural body of the surface of the earth in which plants grow (Brady and Weil 1996). It is comprised of minerals, organic materials, and living forms. Soil is often distinguished from a developed *regolith* (unconsolidated rock and materials) by a higher organic matter content, more intense weathering, the presence of horizontal layers, and the presence of living organisms (Figure 8.1). Regolith could be considered potential soil. It performs some of the same resource functions as soils, such as foundations for road and building construction, but is typically less suitable for life-support functions.

Soil characteristics of importance to the natural resource manager include color, texture, structure, depth, pH, organic matter content, and mineral status. The interaction between these seven factors with climate and topography determines the type and quantity of vegetation that an area is capable of supporting. They also determine the nature of support and stability for road, housing, and other construction.

The Soil Profile

Most soils have distinct layers known as horizons. Variation in characteristics and appearance of these layers—referred to collectively as the soil profile—distinguishes

FIGURE 8.1 Soil profile showing the different layers (courtesy of McKnight 1990, p. 290, p. 348, R12, 36 USDA-NRCS).

the different soil types. In mature or developed soils, the different horizons are quite distinct. These horizons are caused by the interaction of precipitation, temperature, and biotic materials through time. Soils in wet, humid areas generally have much more horizon development than those in dry, cold areas.

The first horizon near the soil surface is often an accumulation of fresh or partly decomposed organic matter (Figure 8.1). This thin layer of plant litter is referred to as the O horizon. Next is the A horizon, which is a layer of humus (decomposed organic matter) and mineral matter. The more soluble minerals have been washed from this zone by rainwater. Soil organisms and biotic activity are at a maximum in this zone. Some soils have an E (also called A2) horizon which is a zone of maximum leaching of soluble minerals. The top soil ends and the subsoil begins where B horizon adjoins the A horizon. The B horizon is a zone of accumulation of minerals leached out of the topsoil. Here the original parent materials are greatly altered and clays have tended to accumulate. The true soil ends at the C horizon, which is a layer of unconsolidated mineral materials and rock fragments not greatly modified by living organisms. The final zone is an impenetrable layer of bedrock referred to as regolith.

Soil classification centers around the nature of these various horizons. The suitability of land for production of crops, timber, and other human needs primarily involves management of the O and A horizons where the nutrients essential to life are present and readily available.

Soil Formation

Soil formation is initiated by the process of *weathering*. Two types of weathering occur, which include mechanical and chemical weathering. Mechanical weathering is the breaking down of hard rocks by alternate wetting and drying, and by frost action. The process of chemical weathering causes the rocks to be dissolved and structurally weakened primarily by carbonic acid. This acid is formed from the carbon dioxide in the atmosphere and water as follows: $CO_2 + H_2O = H_2CO_3$.

Although this acid is chemically weak, it can dissolve minerals from rocks and bring them into solution as carbonates. Many of these carbonates are associated with important inorganic nutrients required by plants. As carbonates are dissolved the nutrients also are dissolved and become available for plant uptake. Much of the earth's surface is composed of carbonate rock such as limestone and dolomite.

Presently it is well established that climatic conditions and parent material physical and chemical characteristics have considerable influence on weathering. Climate has more effect than any other factor on the rate and kind of weathering that will take place. Weathering rates are generally most rapid in warm, wet, tropical areas. Dry areas are usually characterized by mechanical weathering, which causes only a reduction in particle size. Chemical weathering is the main type occurring in moist areas. This type of weathering may also bring a change in chemical composition.

Soil formation is a function of five factors, which include climate, topography, parent material, biological materials, and time. Climate is probably the most influential of these factors because temperature and precipitation determine the rate of physical and chemical reactions.

Arid and desert soils have little profile development compared with soils of the Tropics. Climate also has a major influence on the living organisms that are associated with a particular area. Living organisms contribute organic matter to the soil. Plant roots strongly influence soil structure. Soil animals such as earthworms play an important role in mixing soil materials. The chemical composition of plant parts affects soil chemical properties. Soils occupied by conifers are usually acid in pH while grassland soils are slightly basic (Figure 8.2). Parent material can control the

FIGURE 8.2   Profiles of grassland and forest soils. (From R. F. Dasmann, 1976, *Environmental Conservation*, 4th Edition, copyright © 1976, John Wiley & Sons, Inc., New York. Reprinted by permission of John Wiley & Sons, Inc.)

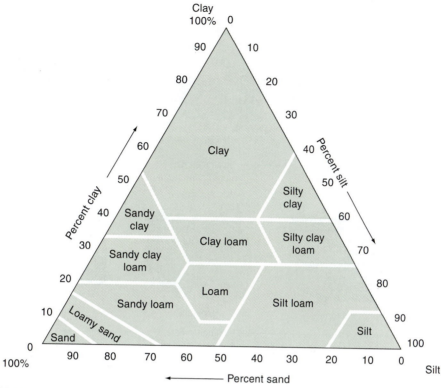

FIGURE 8.3 Soil texture designation based on percentages of sand, silt, and clay (USDA-Natural Resources Conservation Service).

rate of weathering and the natural vegetation that occupies the area. Limestone, for example, is generally high in bases and can delay acidification in high rainfall areas. Topography influences soil development by modifying climate. This is mainly through its effect on natural erosion. Rolling or steep topography is much more subject to erosion than smooth, flat country. Basins where water collects often have rapid soil accumulation and formation. It may be better to consider time as a dimension rather than a soil-forming factor. This is because time is the period in which the other four factors have to interact. Volcanic soils of the Pacific Northwest show lack of development compared with southeastern soils with the same precipitation because they are much younger.

Soil Characteristics

Texture. *Soil texture* refers to the size of the mineral particles comprising the soil. Soil particle sizes, from the smallest to largest, are clay (less than 0.002 mm), silt (0.002 to 0.02 mm), fine sand (0.02 to 0.2 mm), coarse sand (0.2 to 2.0 mm), fine gravel (2 to 5 mm), and coarse gravel (more than 5 mm) (Figure 8.3). Based on the proportions of these particle sizes, a soil can be classed as sand, loamy sand, sandy loam, silt loam, clay loam, or clay.

To a considerable degree, soil texture determines the fertility of the soil. Soils with a high clay content retain nutrients such as nitrogen, phosphorus, and potassium compounds much better than do sands. This is because the small particles have a much greater surface area for attracting and binding nutrients per unit volume than do large particles.

Soil texture also plays an important role in determining the soil moisture content. Water enters coarse, sandy soils much more rapidly than fine clay because there is more space between particles. On the other hand, clay soils retain water much better

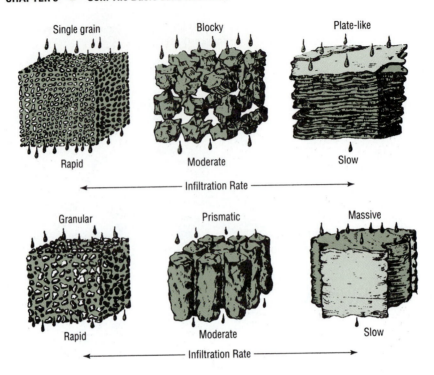

FIGURE 8.4 Note the variety of shapes occurring in soil aggregates and the relative rates at which water can move through them (from Owen and Chiras 1995).

than do sandy soils. The best balance between moisture infiltration and retention is obtained with loamy soils, which have a mixture of sand, silt, and clay.

In arid regions such as the western United States, sandy soils provide a more favorable habitat for plants, particularly grasses, than do clay soils. This is because the limited precipitation that occurs rapidly infiltrates the soil rather than moving overland. In the higher-rainfall areas (over 900 mm of annual average precipitation) east of the Mississippi River, moisture and nutrient retention by soils has more influence on vegetation productivity than rate of water infiltration. In these areas, soils high in clay are more productive than those high in sand.

Structure. *Soil structure* refers to how soil particles are arranged. Soil structure is quite important because it determines the rate at which water can enter the soil (Figure 8.4). Soil is classified in six basic types of soil structure: platelike, prismlike, blocklike, spheroidal, single-grained, and massive. A massive soil structure is tightly compacted with little or no space between soil particles for infiltration and percolation. This structure type often occurs on overgrazed areas or areas compacted by excessive vehicle traffic. Single-grained structure refers to a loose arrangement of individual soil particles that do not stick together, such as sands. Platelike structure refers to soil particles aggregated together into leafy plates. This type of structure can result naturally from parent material or soil-forming forces. Water infiltration is minimized with this type of structure. Prismlike structure refers to soil particles aggregated in vertical columns. This type of structure is most common in arid areas. When the tops are round, this type of structure is referred to as *columnar*. Columnar soils in the West are often associated with a high-sodium content and unfavorable soil-water relations. Soils with blocklike structure have a cubical aggregation. The blocks have sharp edges and are found in the subsoil. The spheroidal soils have the best characteristics for water infiltration and storage. The aggregates are found and easily shaken apart. Soils high in organic matter usually have a spheroidal structure. This type of structure is associated with grassland areas.

Depth. Soil depth influences rooted plant productivity because it determines how much moisture the soil can hold. A deep soil in an area with moderate precipitation will often produce more than a thin soil receiving heavy precipitation. Conversely, in some arid areas, grasses on thin, sandy soils underlain with a hard impermeable layer of calcium carbonate (caliche) are more productive than grasses on deeper sandy soils. The impermeable layer restricts moisture to the portion of the soil profile near the surface where it can readily be used by the fibrous roots of the grasses. Depth usually means distance from the soil surface to bedrock or the unconsolidated material. Soil depth, in conjunction with texture and structure, largely determines the potential of a site for cultivation.

pH. Soil pH indicates the status of the soil in regard to exchangeable mineral ions. The H in pH is for hydrogen and the pH value expresses whether a soil is basic or acidic. The proportional weight of the free hydrogen ions in the soil solution is given by the actual numerical value. The pH value is a logarithmic expression and is demonstrated as follows: pH 1 = one hydrogen ion per 10,000 ions, pH 2 = one hydrogen ion per 100,000 ions, pH 3 = one hydrogen ion per 1 million ions, and so on. The scale ranges from 1 to 14. A highly acid soil would have a pH of 4, while a highly basic soil would have a pH of 10. Minerals are most available to plants in soils with pH values between 6 and 8. A pH value near 7, which is the pH of pure water, is considered ideal for the growth of most plants. However, many species of plants grow best where soils are either highly acidic or basic.

Most soils in the western United States are slightly basic because they are derived from calcareous parent material and receive low precipitation. In contrast, most eastern soils are slightly acidic. The low rainfall in the West results in little leaching of soil minerals and the associated base losses. *Leaching* refers to the downward transport of dissolved soil minerals as water percolates through the soil. Soils of the Southeast are the most heavily leached in the United States. South America has large areas of soils called Oxisols, which are the most heavily leached soils in the world.

Organic Matter. Soil organic matter represents an accumulation of partially decayed and partially synthesized plant and animal residues. It usually represents a relatively small part of the soil (< 1 to 6%) and is found primarily in the upper 30 cm of soil. Partially decomposed organic matter that has been incorporated into the soil is called humus. Humus provides a constant, although small, supply of nutrients to plants. More important, it serves as a binding site for cations and prevents them from leaching out of the soil profile. Three other important functions of humus include binding of soil particles together, increasing soil moisture-holding capacity, and providing food for microorganisms. For these reasons, it is important to maintain as much humus in the soil as possible. Grassland soils are generally much higher in organic matter than are forest soils. Wetland soils are exceptionally high in organic matter because decay is slowed by anaerobic and toxic conditions.

Fertility. Next to water, fertility is the factor that most limits crop production. On high-rainfall ecosystems (over 1,000 mm of annual precipitation), soil fertility can be a more limiting factor to crop production than precipitation. From the standpoint of plant growth, nitrogen is the most deficient element in most soils. Farmland soils are generally deficient in phosphorus and some in potassium and sulfur. Soils in high-rainfall areas are heavily leached and have low levels of elements required by plant crops, such as copper, cobalt, magnesium, sodium, and zinc. Plants growing on these soils often have inadequate levels of these elements to meet crop requirements. Selenium and molybdenum are two elements that cause both toxicity and deficiency problems for livestock depending on their levels in the soil. Low soil fertility is a major problem limiting forage and livestock production in tropical areas of Africa, Asia, Central America, and South America. Forest soils are generally much more leached and less fertile than grassland soils.

Soil Classification

The soil classification system presently in use was developed by the Soil Survey Staff of the U.S. Department of Agriculture. It is called the "Comprehensive Soil Survey System." The major features of this system are based on the characteristics of soils as they are found in the field. It classifies soils rather than soil-forming processes. Soil names give the major physical characteristics. The levels of classification in this system are as follows: (a) Order, (b) Suborder, (c) Great Group, (d) Subgroup, (e) Family, and (f) Series. This system uses diagnostic surface and subsurface horizons as the basis of soil classification.

The 12 orders of the present soil classification system are discussed as follows.

Entisols—They occur in mountainous regions, floodplains, beaches, tundra, and other dynamic locations. These are soils found in the Rocky Mountain area of the United States. These are the youngest of soil orders and are lacking in horizon development.

Gelisols—These soils occur in alpine and tundra areas where the soil is frozen much of the year. The principle definite feature of these soils is the presence of a permafrost layer. Permafrost is a layer of material that remains at temperatures below 0°C for more than two consecutive years (Brady and Weil, 2002).

Inceptisols—These soils are slightly more advanced in profile development than Entisols. They are formed from volcanic ash and are found primarily in the Pacific Northwest and other areas of recent volcanic activity.

Andisols—These soils are formed on volcanic ash and cinders near volcanic sources (Brady and Weil 1996). They are not highly weathered. They are young soils with unique properties due to their mineral composition and high percentage of organic matter (see Brady and Weil 1996).

Aridisols—These are desert soils with little profile development. They are dry more than six months of the year and are common in the American Southwest.

Vertisols—These soils have a high content of swelling clay, which shrinks and cracks when wet. In the United States, they are found mostly in the southcentral states.

Mollisols—These are the natural grassland or prairie soils. They typically are deep, have a high percentage of organic matter, a high base supply, and moderate profile development. In the United States they are most widely developed on the Great Plains of the Midwest.

Alfisols—Alfisols are similar to mollisols but have a higher percentage of organic matter. These soils are most associated with the eastern deciduous forest in the United States.

Spodosols—These are leached mineral soils with a distinct organic horizon and a developed B horizon, but with less accumulation of clays than Ultisols. These soils are found primarily in boreal coniferous forest. In the United States they are common in Alaska, and the northeastern and northcentral states.

Ultisols—Ultisols are the most highly leached soils in the United States, where they are found primarily in the Southeast. These soils are moist much of the time and have a B horizon of clay accumulation, a leached A horizon, and a low base supply. They usually support forest or savannah woodland.

Oxisols—These are the most developed of all soils. They occur in tropical areas and are rare in the contiguous United States. Oxisols have the lowest fertility of all orders. They are least suited for farming.

Histosols—These hydric soils have a very high percentage of organic matter and are often referred to as peats and mucks. They are found in diverse locations where wetlands occur.

Different types of vegetation are associated with individual soil orders (Figure 8.5). A map of the major soil orders in North America is shown in Figure 8.6.

(Entisol)
Variable climate
(Grass and Forest)

(Aridisol)
Warm xeric climate
(Desert)

(Mollisol)
Temperate mesic climate
(Grassland)

(Alfisol)
Temperate hydric climate
(Deciduous forest)

(Spodosol)
Cold hydric climate
(Coniferous forest)

(Oxisol)
Hot hydric climate
(Tropical rain forest)

FIGURE 8.5 Soil profiles of the major soil orders typically found in six different biomes (adapted from Miller 1990).

We will review the six major soil orders in greater detail. Our discussion of the soil orders follows Brady (1974) and Brady and Weil (2002).

Mollisols.   These are the world's most important agricultural soils. They have moderate profile development and the surface horizons have a granular or crumb structure that does not harden when the soils are dry. This explains their name, which comes from the Latin word *mollis* meaning soft.

Generally mollisols develop under grassland or prairie climatic conditions. They dominate the central Great Plains of the central United States (North Dakota, South

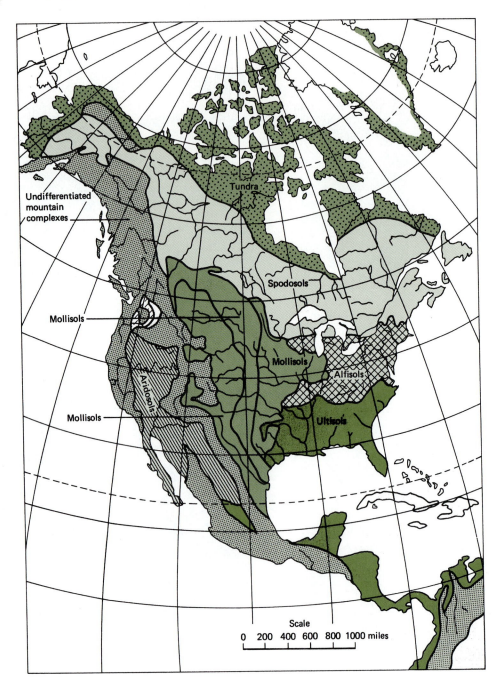

FIGURE 8.6 Major soil orders of North America (from Owen and Chiras 1995).

Dakota, Nebraska, Kansas, Oklahoma, Texas). However, sizable portions occur further west in the palouse region of Washington, Oregon, and Idaho. Much of the heartland of the former Soviet Union is characterized by mollisols, which also dominate parts of South America (northern Argentina, Paraguay, Uruguay) and Asia (northern China, Mongolia).

Mollisols are considered the very best soils in the world for grains (corn, wheat, sorghum, barley, oats) and many other crops because of their high natural fertility. These soils are characterized by generally moderate rainfall (15–30 inches) that arrives mostly in the spring and summer. Periodic drought (about 1 year in 10) can cause crop failure, but when properly managed these soils can produce bumper crops.

Alfisols.   Alfisols rank second only to mollisols in terms of suitability for crop production. In the United States, these soils occur east of the Great Plains in the Midwestern states (Iowa, Illinois, Indiana, Ohio, portions of Minnesota, Michigan, Pennsylvania, New York). In other parts of the world, Alfisols are prominent in central China, much of Europe, southern Africa, eastern Brazil, and southern Australia. Generally Alfisols are characterized by a temperate climate with 30–50 inches annual precipitation fairly evenly distributed throughout the year. Deciduous forests (such as beech, maple, oak, and hickory trees) are the natural vegetation associated with Alfisols.

Alfisols differ from Mollisols in being a bit more leached but less droughty. Early pioneer farming depended heavily on these soils. Corn, potatoes, tobacco, various fruits, various vegetables, and a wide range of forages grow well where Alfisols are properly managed.

Spodosols.   Spodosols are mineral soils differentiated from Alfisols by being colder and more leached. They naturally support coniferous forests (pines, firs). They have distinct A horizons depleted of soluble minerals, but with an accumulation of acidic organic matter at the soil surface. The subsurface B horizon has an accumulation of iron compounds and organic materials. Generally the leached A horizon is light gray or tan while the lower B horizon is a distinct brown. Spodosols characterize the northeastern United States and much of central and eastern Canada. Other large areas include northern Europe, Siberia, and the most southern parts of South America.

Spodosols are naturally infertile. In the early 1900s, many small farms throughout New England and northern Michigan, Wisconsin, and Minnesota were abandoned because of low fertility and a short growing season. Most Spodosols are best suited for timber production. However, when properly fertilized, some Spodosols can be quite productive. Examples are the potato growing region of northern Maine and vegetable producing area in Michigan and Wisconsin.

Ultisols.   Ultisols are highly weathered, acidic soils that developed under warm, consistently moist climatic conditions where annual precipitation is 40 to 70 inches per year. Ultisols are less acidic than Spodosols and less leached than Oxisols. The natural vegetation occurring on Ultisols is forest or savannah woodland. Ultisols dominate the southeastern United States, but are also found in Hawaii and western portions of California, Oregon, and Washington. They are prominent in eastern Australia, southern China, southeast Asia, Paraguay, and southern Brazil.

Although low in fertility, Ultisols can be productive for certain crops when properly managed. Some of these crops include cotton, corn, tobacco, peanuts, and various cultivated forages for livestock (Owen and Chiras 1995). Because of abundant rainfall, a long growing season, and good tillage characteristics, these soils rank third after Mollisols and Alfisols in terms of suitability for agriculture. In the southeastern United States, many areas of Ultisols are hilly and highly erodible. These areas are most suited for wood, forage, and wildlife production.

Oxisols.   Oxisols are the most weathered and wettest of the 11 soil orders. Annual precipitation is typically in excess of 1,200 mm per year. These are the soils of the tropical rain forest, and occur only in Florida in the contiguous United States. The largest areas of Oxisols occur in South America (northern Brazil and neighboring countries) and Africa.

One of the biggest controversies in natural resource management is the conversion of tropical rain forest into agricultural land. Although it would seem these areas could be productive farmland, most agriculture has been unsustainable because of erosion, infertility, and poor soil-tillage properties. Under improved management it may be possible to overcome these problems in certain areas. An alternative approach is to better use natural products from the rain forest such as rubber, cocoa, various tropical fruits, and tree extracts effective in treating human diseases.

The high iron content of Oxisols gives them unique properties. Initially after the trees are removed by burning, Oxisols are moderately productive because of the release of nutrients locked up in the vegetation. However, after a few years the organic matter and nutrients are depleted leaving a high proportion of iron oxides and aluminum. In the hot equatorial sun these materials quickly bake into a hard bricklike substance referred to as laterite. These Oxisol "bricks" make great construction materials for buildings because of their hardness and durability (Owen and Chiras 1995). On the other hand, once Oxisols turn into laterite they become almost useless for plant production. Management of the rain forest is one of the great dilemmas that confronts humanity in the twenty-first century.

Aridisols.   Aridisols, as their name implies, are associated with dry, desertlike climates. In the United States they dominate the Southwest (Mojave, Sonoran, and Chihuahuan Deserts). In other parts of the world they occur over large regions of China (Gobi and Taklamakan Deserts), the former Soviet Union (Turkestan Desert), Africa (Sahara Desert), the Middle East, and Australia. In their natural state these soils are dry for over six months of the year and are high in nutrients because they lack leaching. Generally plants are widely spaced in desert areas where wind and erosion are often severe. This results in a layer of stones called "desert pavement" on the surface of many Aridisols (Owen and Chiras 1995).

Aridisols are unsuited for cultivated crops without irrigation. Historically these lands in Africa and the Middle East have been grazed by livestock, particularly sheep and goats, under nomadic herding. Generally productivity is low and rapid human population increase has lead to severe degradation or "desertification" of many desert lands.

When irrigation water becomes available, Aridisols can be productive for high-value crops, such as citrus, dates, celery, and nuts, because of their high nutrient content (Owen and Chiras 1995). Over time soluble salts (particularly sodium) may accumulate on the surface of these soils. This process, known as salinization, can be mitigated through management practices involving leaching or flushing, improved drainage, conversion of sodium salts to other compounds, reducing evaporation through improved irrigation practices, and the use of salt-resistant crops.

Soil Erosion

Soil erosion is the detachment and movement of soil particles from one place to another (Miller 1990). Wind and water are the primary forces causing soil erosion. Some geologic or natural erosion always occurs because of these forces. With the exception of desert areas, natural plant vegetative cover and roots keep the loss of soil in balance with topsoil replacement (Figure 8.7).

Soil erosion is categorized into four types: splash, sheet, rill, and gully (see Brady 1974, Brady and Weil 2002). Splash erosion is the spattering of small soil particles caused by the impact of raindrops on very wet soils. The loosened particles may or may not be removed by surface runoff. Sheet erosion involves wide flow of water moving across a sloping field. This kind of erosion typically occurs when a recently tilled and leveled piece of farmland with no protective plant cover is subjected to heavy rainfall. Rill erosion results from water moving through small channels in the soil surface (Figure 8.8). This kind of erosion occurs mostly on farmland that is partially tilled with some vegetative residue or on moderately overgrazed rangeland where plants of low palatability remain to hold the soil in place. Gully erosion occurs where water is concentrated into a high-velocity flow through a large channel that lacks vegetation. This type of accelerated erosion commonly occurs in hilly areas that are heavily grazed or farmed. When human activities such as farming, mining, woodcutting, grazing, highway construction, and other disturbances remove natural plant

FIGURE 8.7 Vegetation and organic litter protect soil from all forms of erosion (drawing by John N. Smith, based on Nebel 1981).

FIGURE 8.8 Sheet and rill erosion on an unprotected slope. Serious gullying is imminent if protective measures are not instituted soon. Strip-cropping or even terracing should be employed if cultivated crops are to be grown. Perhaps better, the field could be seeded and put into the USDA conservation reserve program (courtesy U.S. Soil Conservation Service).

cover, soil erosion can accelerate to rates that threaten human health and destroy land productivity (Figure 8.9).

Accelerated erosion can severely reduce the productivity of croplands, woodland, and grazing land because the topsoil contains most of the plant nutrients and has the greatest water-holding capacity (Figure 8.10). Suspended sediment, which collectively refers to the dislodged soil particles, is the largest source of water pollution. It

FIGURE 8.9   Severe gully erosion in southern New Mexico as a result of overgrazing by livestock.

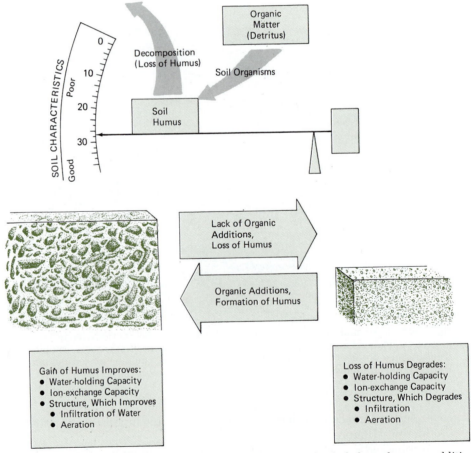

FIGURE 8.10   The amount of humus present in the soil is the result of a balance between addition of organic matter and oxidation of humus. Physical and chemical characteristics of soil improve or degrade according to amount of humus present (from Nebel 1981).

FIGURE 8.11 Average annual erosion rates on various types of land in the United States (courtesy of Natural Resources Conservation Service).

also causes considerable economic damage by clogging rivers, irrigation ditches, and reservoirs with sediment deposits.

Topsoil is the layer of soil typically moved in cultivation (Brady and Weil 1996). It is a renewable resource because it is constantly being formed by natural processes. It takes somewhere between 220 and 1,000 years to form an inch of topsoil, depending on precipitation and temperature (Miller 1990). Warm, wet, and flat areas have more rapid rates of soil formation than areas where the opposite conditions occur.

Soil erosion problems are greatest on farmlands (Figure 8.11). However, about half of the erosion in the United States is caused by other activities such as grazing, mining, logging, and urban development. Construction sites have greatly increased throughout the United States during the 1990s. The first thing developers and builders usually do on construction sites is to remove all vegetation. Frequently little is done to prevent erosion until the project is complete, which may take several years. Erosion rates on shopping center and highway construction sites can be 100 to 200 times the natural rate.

Soil Erosion in the United States

The United States has been through several alternating periods of both high and low rates of soil erosion on its croplands. These periods have been linked to socioeconomic conditions in the country. During periods of war, demand for agricultural products accelerates and farmers have strong incentives to maximize production. During these periods, soil conservation practices have been discarded and erosion rates have been excessive. World War I largely created the conditions that led to the infamous "dust bowl" era of the 1930s. Because agricultural production was greatly reduced in Europe during World War I (1914–1918), prices for most crops skyrocketed. Because the United States was not directly affected by the war, it became the major supplier. Much of the western Great Plains was plowed in areas unsuited for sustained cultivation. Although crop prices decreased somewhat during the 1920s, farmers continued to intensively crop their lands. Generally the land was plowed after each harvest and left bare without strip cropping, crop rotation, or the use of wind breaks. At the same time, heavy livestock grazing destroyed the grass on large areas of rangeland that could not be farmed because of rocky soils on sloping terrain.

Disaster finally struck in the form of concurrent drought and depression in the early 1930s. In 1934, crop prices had dropped over 30% from 1929 levels. It also had the distinction of being the driest year of the century. In the worst portions of the dust bowl (Oklahoma Panhandle, southwestern Colorado, and southeastern Kansas) (Figure 8.12) hot, dry windstorms caused dust clouds so thick that darkness occurred in midday. Several human fatalities resulted from breathing the dust-filled air. Dead birds and rabbits commonly were found in the wake of these storms. In the spring of 1934, topsoil from the Great Plains was carried across the eastern United States to far out into the Atlantic Ocean (Figures 8.13 and 8.14). New York City was literally covered with dust. In

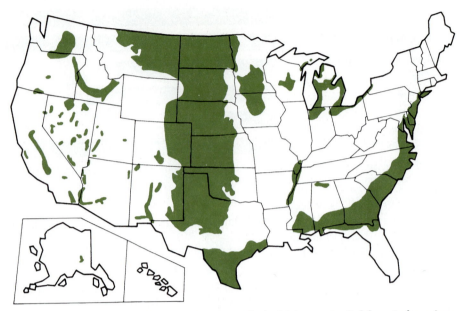

FIGURE 8.12 Cropland areas of the United States with the highest potential for wind erosion (courtesy of USDA-Natural Resources Conservation Service).

FIGURE 8.13 A dust storm during the early 1930s in southeastern Colorado. Such storms were the combined result of a natural drought cycle and inappropriate plowing of the western Great Plains. (courtesy of U.S. Government).

May 1934, congressional action was taken after dust from the Great Plains leaked into a Washington conference room where legislators were considering programs to deal with the problem (Miller 1990). The USDA Soil Conservation Service (now the Natural Resources Conservation Service) was established in 1935 to promote conservation practices on private lands in the Great Plains. This started an era of improved farmland and rangeland management that lasted until the late 1960s.

The era of improved soil conservation was followed by a period of severe soil erosion that began in the early 1970s. This resulted primarily from mismanagement of

FIGURE 8.14 Land in Oklahoma severely damaged by wind erosion during the dust bowl era of the early 1930s (courtesy of U.S. Government).

the U.S. economy and adverse climatic conditions in several parts of the world. In order to pay for the Vietnam War and finance its social programs, the U.S. government under Presidents Johnson and Nixon decided to print more money rather than borrow it or raise taxes. This devalued the dollar against most foreign currencies, making U.S. farm products relatively inexpensive. In 1972, the former Soviet Union started importing food from the United States rather than rationing food to make up for production shortfalls. These factors in combination greatly increased crop prices and encouraged plowing of erodible farmlands during the 1970s. However, the 1980s ushered in a new set of economic policies after Ronald Reagan was elected president. The Federal Reserve Board switched from a loose to a restrictive monetary policy, which brought inflation under control and raised the value of the dollar. During most of the 1980s, world climate conditions were favorable for crop production. At the same time, yields of many crops were greatly increased through new technology. By 1984 serious problems were associated with both huge agricultural surpluses and severe soil erosion in many parts of the Great Plains.

In 1985, the U.S. Congress passed a new farm bill (Food Security Act) that had several provisions to retire erodible lands and encourage soil conservation. Over the past 17 years soil erosion has again been contained in the United States. Another potential dust bowl was narrowly avoided. During the early 1980s there were increasing reports of severe dust storms on the western edge of the Great Plains. A severe drought occurred over most of the Great Plains in 1988, but by then the 1985 Food Security Act had returned most of the erodible lands to perennial grasses.

Although estimates vary, USDA-NRCS survey data indicate U.S. croplands have lost an average of about 30% of their original topsoil. This loss may cost the country as much as $30 billion per year. Until the 1985 Food Security Act, about one-fourth of the Nation's cropland accounted for about two-thirds of the annual soil loss (Figure 8.15). However, some 30 million acres of these erodible lands have been returned to permanent grasses and soil erosion losses have been cut by over half. Under present policies (1985, 1990, and 1996 Farm Acts), we do not consider soil erosion of cropland to be a serious threat to food security or human health in the United States, although local problems still exist. New farming technologies show potential for overcoming some of the topsoil loss that occurred in the past. We consider the

Soil Erosion as a Multiple of the Tolerable Rate (T), 1982

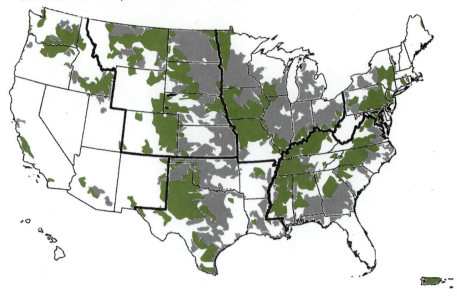

Soil Erosion as a Multiple of the Tolerable Rate (T), 1992

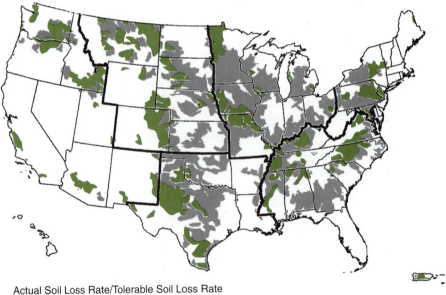

Actual Soil Loss Rate/Tolerable Soil Loss Rate

LEGEND

■ 2T or More
■ T-2T

NOTE:
Average annual soil erosion by wind and water where cultivated
cropland is greater than 5% of land area.

FIGURE 8.15   Soil erosion as a multiple of the tolerable rate in 1982 and 1992. The area designated in blue is eroding at more than two times the tolerable rate. The area in gray is eroding from one to two times the tolerable rate (courtesy of USDA-Natural Resources Conservation Service).

conversion of prime farmland into urban uses to be the most serious threat to sustaining high-quality food production at low prices.

SOIL CONSERVATION PRACTICES

Generally the key to controlling soil erosion and other land conservation practice is to maintain a good vegetative cover. Farming practices used to contain erosion include conservation tillage, strip-cropping, contour farming, terracing, gully reclamation, windbreak planting, and the retirement of highly erodible lands from cultivation. We will summarize each of these practices, but we refer the reader to Brady and Weil (2002) for in-depth coverage of soil conservation techniques.

Conservation Tillage

In order to develop this subject, we will first describe conventional tillage and alternative procedures. Our discussion follows Miller (1990) and Brady and Weil (1996). Generally one pass is made over the field with a moldboard plow that inverts crop residues in the upper 15 inches of soil. After a disk is then used to break up the large clods, the field is harrowed to break up the small clods and smooth the surface. After planting, a cultivator is sometimes used to remove weeds that compete for nutrients and moisture and can be a source of contamination when harvest occurs.

An alternative to this system is conservation tillage farming. It is also known as minimum tillage or no-till farming depending on the degree of soil disturbance (Miller 1990). Under this approach the new crop is planted into the residue of the previous crop with only partial disruption of the soil surface (Figure 8.16). With minimum tillage, a special machine is used to loosen but not turn over the soil surface and previous crop residues. With no-till farming, seeds, fertilizers, and herbicides are injected into slits in the unplowed soil.

Conservation tillage greatly reduces erosion, but it also has other benefits. These include decreased soil compaction, soil water loss, and fuel costs associated with tillage. Maintaining crop residues also improves habitat conditions for many types of desirable wildlife such as bobwhite quail, pheasants, various species of songbirds, water fowl, and cottontail rabbits. Yields have generally been similar to those under

FIGURE 8.16 An example of no-till farming where a specially designed machine applies seeds, fertilizer, and weed killers all at once (courtesy of USDA-Natural Resources Conservation Service).

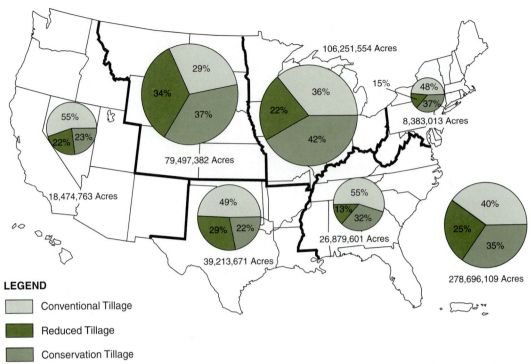

LEGEND

☐ Conventional Tillage

■ Reduced Tillage

■ Conservation Tillage

FIGURE 8.17 The percentage of farmland in different regions of the United States under conservation tillage in 1992 (courtesy of USDA-Natural Resources Conservation Service).

conventional tillage. Although it varies with soil type and crop, more extensive cultivation is needed about every 3–7 years. The main disadvantage is the increased need for herbicide to control weeds. However, this problem can be minimized with improved selection, application, and timing techniques.

In 2002 conservation tillage was used on about 55% of the Nation's farmland (Figure 8.17). The USDA estimates that this has reduced the soil erosion rate by about 20–25%. With no-till farming, around 50–100% of the previous crop residue remains on the soil surface compared to 30–45% with conservation tillage and 0–5% with conventional tillage. The primary factor limiting use of conservation tillage is the greater management skill required of farmers.

Strip-Cropping

Strip-cropping involves the planting of crops in alternating strips to reduce erosion and pesticide and fertilizer demands (Figure 8.18). It is often combined with crop rotation so that a soil-enriching legume crop (alfalfa or clover) is planted after a soil-depleting grain crop (corn, wheat, barley). The annual crop is often referred to as a row crop and the legume crop is referred to as a cover crop. The alternating pattern of row crops and cover crops reduces water erosion and creates barriers to the spread of pests and diseases. In the arid portions of the western Great Plains (Montana, Wyoming, Colorado), a modification is used where strips of wheat or other small grains are alternated with fallow. Every other year the strip is tilled and planted. Most of this land is marginal for crop production, but with alternate-year planting, enough moisture can be stored to ensure a crop. Maintaining the residue during the fallow period greatly reduces erosion and water evaporation losses from the soil. Herbicides are used to retard weeds during the fallow period. On sloping lands, contour strip-cropping in combination with terraces can reduce soil erosion by 50% or more.

FIGURE 8.18 A strip-cropping system designed to reduce wind erosion of soil in Montana (courtesy of USDA-Natural Resources Conservation Service).

FIGURE 8.19 Contour farming of sloping land to protect the soil and grass filter strips to protect water quality in the adjacent stream (courtesy of USDA-Natural Resources Conservation Service).

Contour Farming

In contour farming, crops are planted across rather than up and down sloping land (Figure 8.19). This forms a barrier that helps to hold soil and slow water runoff. Contour farming and strip-cropping are often used in combination.

Terracing

This practice has been used for centuries by societies in Europe, China, India, and South America (Owen and Chiras 1995). In these areas, high human populations relative to the amount of arable land often necessitated farming mountain sides in order to prevent starvation. When terraced, the slope is converted into an alternating

(a) Broad-Base Terrace
    Allows farming over the whole terrace
    May be built on slopes of up to 8%
    Water flow checked

(b) Grass Backslope Terrace
    Has a steep backslope that is not farmed
    May be built on slopes of up to 18%
    Water flow checked

FIGURE 8.20   Ridge terraces. A comparison of the broad-base and grass backslope terraces (from Owen and Chiras 1995).

series of broad, level benches and sharp or gradual slopes. On gently sloping land (under 10% slope), broad-base terraces are developed to allow farming of the slope (Figure 8.20). On steeper lands (10 to 20% slope), backslope terraces are planted with permanent grass on the slopes. Channel terraces differ from ridge terraces in that a diversion ditch (rather than a flat bench) is dug across the slope to halt and redirect water flow. Although terraces are effective in decreasing soil erosion by reducing the amount and speed of water runoff, they often are incompatible with modern farming methods involving large equipment. In the United States, the approach has been to encourage permanent retirement of most land where terraces would be required. In mountainous developing countries, such as Nepal and Peru, failure to apply terraces on newly cultivated lands is leading to permanent damage from soil erosion.

Gully Reclamation

Overland flow of water, if not retarded by vegetation, can quickly form gullies (large water-erosion channels) on sloping land (Miller 1990). Recently formed gullies are a major warning signal of a severe soil erosion problem. Generally, retarding erosion on surrounding uplands is more effective than restoring the gully itself. Once this is done, gullies can be reclaimed by seeding quick-growing annual grains such as wheat, oats, rye, and barley. Small temporary dams can be built with hay bales or old tires to check water flow and collect silt in the deeper, more quickly-forming gullies. Once a soil cover is established, perennial grasses and rapidly growing shrubs and trees should be planted to permanently stabilize the soil. Smaller diversion channels can be used to disperse water away from the gully.

Windbreaks

Windbreaks, or shelterbelts, are rows of trees planted perpendicular to the prevailing wind direction to protect soils, crops, homesteads, and more from wind and snow (Brady and Weil 2002). They disrupt the movement of wind and reduce the exposure of soil to its shocking force (Figure 8.21). The shading effect of windbreaks reduces the rate of soil moisture evaporation. Windbreaks are especially effective in reducing erosion in areas with sandy soils such as the Southern Great Plains of Kansas, Colorado, Oklahoma, and Texas. Farm residences in this area are often ringed with windbreaks. They protect the home from wind damage, and reduce the cost of heating in winter and cooling in summer. They provide high-quality habitat for various birds, pest-eating and pollinating insects, and other types of wildlife. Windbreaks add aesthetic diversity to plains landscapes that are otherwise uniform and monotonous (Figure 8.22).

FIGURE 8.21 The effect of a windbreak on wind velocity (drawing by John N. Smith).

Shelterbelt
30-40 Ft. High

FIGURE 8.22 Windbreaks, or shelterbelts, reduce erosion on this farm in Trail County, North Dakota. They also reduce wind damage, help hold soil moisture in place, supply some wood for fuel, and provide habitat for wildlife (courtesy of USDA-Natural Resources Conservation Service).

After the 1930s dust bowl, the USDA Soil Conservation Service implemented an aggressive windbreak planting program on farmlands of the Great Plains. Unfortunately, many of these windbreaks were removed during the 1970s when high crop prices encouraged farming all tillable land. Large farm equipment and irrigation systems became vogue. Following enactment of the 1985 Food Security Act, windbreak plantings have increased in a few areas.

Retirement of Erodible Lands

Land retirement from farm use is an approach to reducing soil erosion that goes back to the drought years of the middle 1930s when the federal government bought out farmers on erodible land in the Great Plains. The National Grasslands, administered by the USDA-Forest Service, were acquired from these buyouts. In the 1950s, a program called the "Soil Bank" was implemented by the USDA to control soil erosion and reduce food surpluses. It retired about 20 million acres of land from farm use, most of which went back into production in the 1970s when crop prices skyrocketed.

The most recent land retirement program was initiated in 1985 when Congress passed the Food Security Act. It called for removal of 45 million acres of marginal cropland into a conservation reserve for a 10-year period. This is about 10% of the Nation's cropland base. The primary intent was to reduce food surpluses, but soil erosion reduction and wildlife habitat provision were also objectives. Under this program, the farmer contracts with the USDA to withdraw erodible cropland for 10 years and establish a permanent vegetation cover (perennial grasses, shrubs, trees, legumes) that is not to be hayed or grazed unless a drought or some other unusual circumstance occurs. The average farmer has received about $50 per acre per year (Knutson et al. 1998). This program has cost taxpayers about $1.8 billion per year. Generally it is considered a success in terms of controlling erosion and improving habitat for wildlife. However, critics consider it too costly and contend it could have been implemented more efficiently. Under the 1996 Food Security Act the Conservation Reserve Program was retained with provisions to encourage further retirement of highly erodible lands and better reward farmers who seeded their lands with plants favoring wildlife.

Conservation Buffers

Conservation buffers are strips of vegetation installed where one type of land ends and another begins (Figure 8.23). They can be valuable lines of defense from the top to bottom of any watershed to protect surface and underground water quality. Generally, conservation buffers are applicable anywhere in a watershed where runoff water enters a pothole, stream, farm pond, reservoir, or another body of water.

We will briefly discuss the various categories of conservation buffers following NRCS descriptions.

Contour buffer strips—Strips of perennial vegetation alternated with wider cropland strips, which are farmed on the contour. They can reduce sheet erosion and reduce movement of sediment, nutrients, and pesticides.

Filter strips—Strips of grass or other vegetation used to intercept or trap sediment, organics, pesticides, and other pollutants before they reach a water body.

Riparian forest buffer—Streamside vegetation consisting of trees, shrubs, and grasses that can intercept pollutants from both surface and groundwaters before they reach a stream.

Field border—Strips of perennial vegetation planted at the edge of a field. They can be used for a turn area or travel lanes for farm machinery.

Windbreak/shelterbelt—Row or rows of trees, shrubs, or other vegetation used to reduce wind erosion, protect young crops, control blowing snow, reduce soil

FIGURE 8.23 An artist's image of conservation buffer strips (courtesy of USDA-Natural Resources Conservation Service).

moisture evaporation (especially on sandy soils), reduce heating and cooling costs for homes, and provide habitat for recreational wildlife, pollinators, and pest predators.

Grassed waterway—Strips of grass on areas where water concentrates as it runs off a field. While they are used primarily to prevent gully erosion, waterways can be designed or combined with filter strips to help filter contaminants.

Alley cropping—An agroforestry practice consisting of growing trees or shrubs in rows or corridors with alleys of agronomic crops or forage between them. Both the forestry and agronomic crops are harvested.

Other practices—considered as buffers or closely associated with them are hedgerow plantings, herbaceous wind barriers, crosswind trap strips, streambrank protection measures, and vegetative barriers.

The benefits of buffer and filter strips are great and the drawbacks are few. Properly installed and maintained buffer strips can do the following (Miller and Gardiner 2001):

- Trap fertilizers (50% or more)
- Trap pesticides (50% or more)
- Trap pathogens (60% or more of certain species)
- Trap heavy metals
- Trap sediments (75% or more)
- Trap snow
- Reduce noise and odor
- Provide turn rows for farm equipment
- Provide food, cover, and shelter for wildlife
- Improve the aesthetic appeal of the property

Strategically placed, conservation buffers such as windbreaks, shelterbelts, and living snow fences also screen some activities and operations from view. By slowing

FIGURE 8.24 Erodible cropland retired from farming in northeastern Oregon under the Food Security Act of 1985. It has been reseeded with perennial grasses.

down the wind they also help minimize drift from ground and aerial spraying. Field borders, grassed waterways, filter strips, crosswind trap strips, and herbaceous wind barriers of tall (4–6 ft) grasses, forbs, and legumes, and short (6–10 ft) woody shrubs are other examples of conservation buffers that help stop drift.

After the 1930s, the Soil Conservation Service implemented an aggressive conservation buffer program. Many conservation buffers were converted to farmland in the 1970s in response to high crop prices and use of large farm equipment and irrigation systems. Some plantings of conservation buffers increased after the 1985 Food Security Act (Figure 8.24).

Landowner incentives for conservation buffers were an important part of the 1996 Farm Bill passed by the U.S. Congress. Since the inception of the National Conservation Buffer Initiative in 1997 more than 170,000 miles of buffer strips have been installed. Readers interested in more information on conservation buffers are referred to Henry et al. (1999), Gilley et al. (2000), and Nakao and Sohngen (2000).

Soils and Carbon Sequestration

As discussed in Chapter 6, burning fossil fuels is the primary cause for the increased levels of CO_2 (global warming) in the atmosphere during the last 150 years. Agriculture has also contributed to the problem. Many forests and grasslands that were CO_2 consumers and storehouses for carbon were cleared for farming. However, agriculture can be modified to reduce the problem of excessive CO_2 in the atmosphere through carbon sequestration. This (carbon sequestration) involves locking up carbon in solid compounds through conservation tillage, maintenance of fertile topsoil, and use of cover crops. Conservation tillage further reduces carbon emissions by requiring less use of fossil fuel. We refer the reader to Allmaras et al. (2000) for a detailed discussion of carbon sequestration through conservation tillage.

Maintaining Soil Fertility

In the United States and other developed countries, commercial inorganic fertilizers are primary means for maintaining soil fertility. In developing countries, animal manures and human waste (sometimes referred to as night soil) are still commonly used to fertilize crops. The use of inorganic commercial fertilizers throughout the world has increased by 10- to 15-fold since 1950 and about 20% over the past 10 years (World Resources Institute 1994). They have had the advantages of being easy to transport and apply. It is estimated they have boosted annual food production by 35–40% over the last 50 years. Farmers commonly have their soil tested to determine the proper mix of nutrients. Nitrogen, phosphorus, potassium, and sulfur products are the common forms of inorganic fertilizers. However, other plant nutrients are often added based on soil tests.

The biggest drawback to inorganic fertilizers is that they do not add organic matter (humus) to the soil (Miller 1990). Without periodic additions of organic matter, the soil's ability to hold water and nutrients will decrease. This results from the lower cation exchange properties of soil minerals compared to organic matter, and decreased soil porosity when organic matter is depleted. Inorganic fertilizers also cause water pollution, particularly in areas with sloped land or a water table near the soil surface.

Periodic application of animal manures, green manures, and compost can reduce the problems associated with inorganic fertilizers. Animal manures increase organic nitrogen content, improve soil structure, and stimulate soil microbial growth.

Green manure is green vegetation that is grown to be plowed into the soil to increase organic matter (Miller 1990). Although it often involves legumes such as alfalfa and clover, it can be weeds and grasses in an uncultivated field.

In many areas of the United States, the application of animal manures is impractical today. This is because mixed crop-animal operations of the past have been replaced by more specialized enterprises. Often manure transportation costs between feedlots for livestock and crop culture are excessive. Therefore, some farmers have turned to compost (plant wastes such as cuttings and leaves) as an alternative to manure.

Crop rotation is another means of preventing soil nutrient depletion (Miller 1990). Crops such as cotton, wheat, corn, and tobacco can mine the soil of nutrients if grown year after year in the same field. However, if these crops are rotated with legumes and forage grasses, soil nutrients can be restored and erosion minimized.

CONSERVATION PLANNING

One basis for land conservation planning is a classification system. A primary accomplishment of the NRCS has been the development of a land capability classification system and detailed mapping of soils throughout the United States (Owen and Chiras 1995). Under the NRCS land capability classification system there are eight categories into which land can be placed (Figure 8.25). Class I land is flat, fertile, and well suited for crop production. Classes II and III are suitable for growing crops but require soil erosion control measures. Classes IV through VII are rangelands and woodlands suitable for grazing and tree production. Class VIII land is suitable only for wildlife habitat, wilderness, and/or recreation due to excessive slope, rocky soils, and/or low fertility. The intent of this system is to help farmers, ranchers, foresters, urban planners, and so on best use different types of lands. In the past, however, it helped relegate wildlands to the least productive category. Now significant examples of ecosystems that once occupied intensively farmed areas have become scarce.

The NRCS also helps farmers and ranchers develop conservation plans for their lands. Four steps are involved in development of this plan (Owen and Chiras 1995). They include first conducting an intensive survey of the farm or ranch in which each unit (acre) is mapped on the basis of its capability category. Second, a management plan is developed by the farmer/rancher and a NRCS technician. Under this

FIGURE 8.25   Classification of land according to capability using the USDA-Natural Resources Conservation Service approach (courtesy of USDA-Natural Resources Conservation Service).

plan, decisions are made for each land unit regarding use and soil protection measures. Alternative uses might include farming, livestock grazing, tree production, wildlife habitat, home sites, or some combination of these uses. Third, the planned uses and treatments are implemented. The NRCS technician may provide technical and/or monetary assistance (cost-sharing) to the farmer/rancher for certain conservation practices such as shelterbelt establishment, terracing, conservation reserve plantings, brush control, and water development. Finally, the farm/ranch plan is monitored for effectiveness from year to year, and updated to incorporate new technologies and government conservation initiatives.

SOIL POLICY

In 1985, the U.S. Congress passed the Food Security Act, which has several provisions that reversed the trend of accelerating erosion during the 1970s. As previously discussed, this act authorized the government to create a conservation reserve by paying farmers to retire their most erodible lands to perennial grasses, legumes, and trees. During the 10-year contract period, the land cannot be farmed, grazed, or hayed. Lands must have an erosion rate three times that of natural soil formation to be eligible. Farmers who violate those contracts are forced to return all government payments with interest.

By 1995, about 38 million acres or 9% of the Nation's farmland had been retired under the conservation reserve program (Figure 8.26). Under the 1996 Farm Act about 20 million acres will be retained in the conservation reserve for the next 10 years. It is generally agreed the conservation reserve has been effective in meeting its goals of reducing soil erosion, improving water quality, and enhancing wildlife habitat. Although estimates vary, the program probably has reduced soil losses by 50 to 60%.

Another part of the farm bill links conservation practices to eligibility for farm subsidies. All farmers must now have a soil conservation plan for their entire farm to receive government benefits. Farmers who agree not to farm highly erodible cropland or wetlands for 50 years can qualify for partial or complete reduction of debts owed to the Farmer's Home Administration. These lands must be converted back to grass and trees or wetland. Farmers who drain or destroy wetlands on their property are denied federal farm benefits.

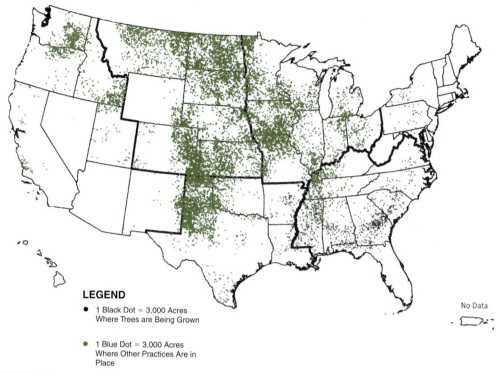

LEGEND

● 1 Black Dot = 3,000 Acres
Where Trees are Being Grown

● 1 Blue Dot = 3,000 Acres
Where Other Practices Are in
Place

No Data

FIGURE 8.26   Conservation Reserve Program (courtesy of USDA-Natural Resources Conservation Service).

In total, soil protection under the 1990 and 1996 Farm Acts has cost the federal government about $2 billion per year. Although there have been some complaints by farmers over excessive regulation and by conservationists over laxity in enforcement of certain provisions (particularly wetland protection provisions), overall it appears these acts have been effective. Even though concern remains over cost, it must be kept in mind that $2 billion represents less than two-tenths percent of the government's 1.5 trillion dollar 1997 federal budget.

Overall we consider the 1985 Food Security Act to be one of the better pieces of legislation that has been developed to protect farmlands from erosion. Modifications in the 1995 Food Security Act better separate highly erodible lands from those with only minor erosion. Only potential lands with definite erosion problems can qualify for renewal of land retirement contracts.

We generally believe that market forces can better solve conservation problems than government interventions. However, soil erosion on farmlands and urban land is among the exceptions to the common rule that market forces are better than government regulations in assuring resource conservation. Here, short-term profit opportunities can often outweigh long-term individual or social welfare from the standpoint of individual farmers or developers. A farmer whose soil gradually erodes away may not be directly impacted by these losses for several years or until after he or she retires or dies. The problems of silt and dust that result from poor management practices are born out by society as a whole rather than strictly by the farmer or developer. For these reasons, coupling conservation compliance to qualification for tax benefits and farm subsidies seems logical. At the individual state level, stiff fines and other penalties are increasingly being used to discourage practices such as leaving soil bare for extended periods by developers and farmers.

Although the United States has made definite progress in controlling soil erosion over the past 10 years, the outlook is somewhat pessimistic in many developing countries (World Resources Institute 1994). In these situations, developed countries could

The 1996 Farm Bill's Commitment to Conservation

The 1996 Farm Bill, passed by Congress and signed into law by the President on April 4, 1996, has been heralded as the most progressive environmental farm bill to date. Conservation provisions in the legislation will affect farmers well into the next century. The new provisions build on the conservation gains made by landowners over the past decade. They simplify existing programs and create new programs to address high-priority environmental protection goals. The key provisions:

■ Environmental Quality Incentives program consolidates four existing conservation programs (Great Plains Conservation Program, Agricultural Conservation Program, Water Quality Incentives Program, and Colorado River Basin Salinity Control Program) and directs cost-sharing and technical assistance to locally identified conservation priority areas. Half of EQIP funds are dedicated to livestock-related conservation problems.

■ Wetlands Reserve Program and Conservation Reserve Program are extended through 2002.

■ Farmland Protection Program provides assistance to states that have farmland protection programs to purchase conservation easements.

■ Swampbuster and wetlands provisions from the 1985 and 1990 Farm Bills are modified to provide farmers with more flexibility to meet wetland conservation requirements.

■ Wildlife Habitat Incentives Program helps landowners improve wildlife habitat on private land.

■ Flood Risk Reduction Program provides incentives to move farming operations off frequently flooded land.

■ Emergency Watershed Protection Program allows purchase of floodplain easements.

■ Conservation of Private Grazing Land Initiative offers landowners technical and educational assistance on private grazing land.

■ National Natural Resources Conservation Foundation is created as a nonprofit corporation to foster conservation research, education, and demonstration projects.

■ Conservation Farm Option allows farmers with market transition contracts to consolidate CRP, WRP, and EQIP payments annually, under a 10-year contract, in return for adoption of a conservation farm plan.

■ State Technical Committee membership is broadened to include agricultural producers and others with conservation expertise.

From: A Geography of Hope. 1997. USDA-Natural Resources Conservation Service.

link foreign aid and trade opportunities to application of sound conservation practices. At the same time, slowing population growth and increasing the rate of economic development appear crucial to halting or slowing soil erosion problems in these countries. We refer the reader to Knutson et al. (1998) for in-depth consideration of how government policy influences soil conditions in the United States and other parts of the world.

LITERATURE CITED

Allmaras, R. R., H. H. Schomberg, C. L. Douglas, and T. H. Dao. 2000. Soil organic carbon sequestration potential of adopting conservation tillage in U.S. croplands. *Journal of Soil and Water Conservation* 55(3):365–371.

Brady, N. C. 1974. *The nature and property of soils.* 8th ed. New York: Macmillan.

Brady, N. C. and R. R. Weil. 2002. *The nature and property of soils.* 13th ed. Upper Saddle River, NJ: Prentice-Hall.

Dasmann, R. F. 1976. *Environmental conservation.* New York: John Wiley & Sons.

Gilley, J. E., B. Eghball, L. A. Kramer, and T. B. Moorman. 2000. Narrow grass hedge effects on runoff and soil loss. *Journal of Soil and Water Conservation* 55(2):245–251.

Henry, A. C., D. A. Hosack, C. W. Johnson, D. Rol, and G. Bentrup. 1999. Conservation corridors in the United States: Benefits and planning guidelines. *Journal of Soil and Water Conservation* 54(4):645–650.

Knutson, R. D., J. B. Penn, and W. T. Boehm. 1995. *Agricultural and food policy.* 3rd ed. Upper Saddle River, NJ: Prentice-Hall.

Knutson, R. D., J. B. Penn, and B. L. Flinchbaugh. 1998. *Agricultural and food policy.* 4th ed. Upper Saddle River, NJ: Prentice-Hall.

Miller, G. T. 1990. *Resource conservation and management.* Belmont, CA: Wadsworth Publishing Co.

Miller, R. W. and D. T. Gardner. 2001. *Soils in our environment.* 6th ed. Upper Saddle River, NJ: Prentice-Hall.

Nakao, M. and B. Sohngen. 2000. The effect of site quality on the costs of reducing soil erosion with riparian buffers. *Journal of Soil and Water Conservation* 55(2):231–237.

Nebel, B. J. 1981. *Environmental science: The way the world works.* 2nd ed. Upper Saddle River, NJ: Prentice-Hall.

Owen, O. S. and D. D. Chiras. 1995. *Natural resource conservation.* 6th ed. Upper Saddle River, NJ: Prentice-Hall.

Owen, O. S., D. D. Chiras, and J. P. Reganold. 1998. *Natural resource conservation.* 7th ed. Upper Saddle River, NJ: Prentice-Hall.

U.S. Department of Agriculture. 1989. *The second RCA appraisal in soil, water, and related resources on federal land in the United States.* Washington, DC: U.S. Government Printing Office.

U.S. Department of Agriculture. 1994. *Soil erosion by wind.* Natural Resources Conservation Service Agr. Inf. Bull. No. 555.

U.S. Department of Agriculture. 1997. *A geography of hope.* Natural Resources Conservation Service. Washington, DC: U.S. Government Printing Office.

World Resources Institute. 1994. *World resources 1994–1995.* New York: Oxford University Press.

9

Ecosystems of the United States

INTRODUCTION

Anyone with even a casual interest in their environment who has done much traveling in the United States has observed both subtle and striking variation in forms of land, waters, and vegetation. Combinations of topography, geology, and climate determine the distribution of waters, soils, and vegetation and establish many of the characteristics of ecosystem form and function. Effective management of natural resources depends on understanding the factors that define the different ecosystems. Structural characteristics differentiate ecosystems into three major groups: terrestrial, aquatic, and wetland ecosystems. We will provide a description of the terrestrial, aquatic, and wetland ecosystems of the United States. The reader is referred to Chapter 10 for detailed coverage of forest ecosystems. For introductory coverage on the topic see Dasmann (1976), Lerman (1986), Miller (1990), Cole (1994), Barbour et al. (1999), Mitsch and Gosselink (2000), Wetzel (2000), and Holechek et al. (2001).

TERRESTRIAL ECOSYSTEMS OF THE WORLD

Biomes

Grasslands, desert shrublands, savanna woodlands, forests, and tundra are the basic natural land cover types of the world. Precipitation, temperature, and topography interact to determine the type of vegetation that occurs in the area (Figure 9.1). Each of these types is comprised of several plant associations that support a slightly different biota. Activities such as logging, cultivation, grazing, and industrialization have substantially altered the natural biota in all the land cover types. The major land cover types of the world are shown in Figures 9.2, through Figure 9.4.

The term *biome* refers to a set of natural plant and animal communities that exist together and have distinguishing structural characteristics. Broad vegetation categories such as prairie, desert, woodland, grassland, savanna, shrubland, and forest are usually used in naming biomes. All biomes are ecosystems, but not all ecosystems are biomes. While single communities and their habitats are ecosystems, biomes often include numerous communities and habitats. Thus biomes are larger landscape organizations of basic ecosystem units. The concept of biome is rarely applied to aquatic

FIGURE 9.1   Average precipitation and average temperature interact to determine the natural vegetation found in a particular area (from Nebel 1981).

ecosystems. In this chapter, we will discuss the physical factors that interact to cause the major land biomes of the world, and then provide a more detailed description of the biomes in the United States.

Grasslands of the World

Grasslands are the most productive lands in the world for farming and grazing. When properly managed they also are an important source of clean water. Grasslands are typically free of woody plants (shrubs and trees) and are dominated by plants in the family Gramineae (grasses). Grasslands occur from sea level to 5,000 m but are most common on relatively flat, inland areas at elevations from 1,000 to 2,000 m.

Grasses, Forbs, and Shrubs Defined.   Grasses are distinguished by having hollow, jointed stems; fine, narrow leaves with large parallel veins; and fibrous root systems (Figure 9.5). Many grasslands have a high fraction of sedge (*Carex* sp.) species and rushes (*Juncus* sp.). These grasslike plants have leaves and fibrous roots like true grasses but differ in having nonjointed, solid stems. Forbs are also an important component of many grasslands. Forbs are plants unlike grass and characterized by taproots, generally broad leaves with netlike veins, and solid nonjointed stems. Shrubs, which are a minor component of most grasslands, have woody stems that branch near the base and have long, coarse roots.

   In contrast with shrubs, trees have a definite trunk that branches well above ground. Shrubs and trees have stems that remain alive during dormant periods, but the aboveground parts of most grasses and forbs die back to the crown during winter in temperate areas or during the dry season in the Tropics.

Grassland Climate and Soils.   Grasslands generally occur in areas receiving between 250 and 900 mm of annual precipitation. This precipitation generally occurs as frequent light rains over an extended period (90 days or more). In temperate areas, extended light rains during the summer favor grassland over shrubland because the shallow, fibrous roots of grasses use moisture near the soil surface more efficiently than do the long, coarse roots of trees and shrubs. Winter rainfall with dry summers generally favors shrubland or woodland over grassland in temperate areas because moisture levels near the soil surface are low during the summer period when temperatures are

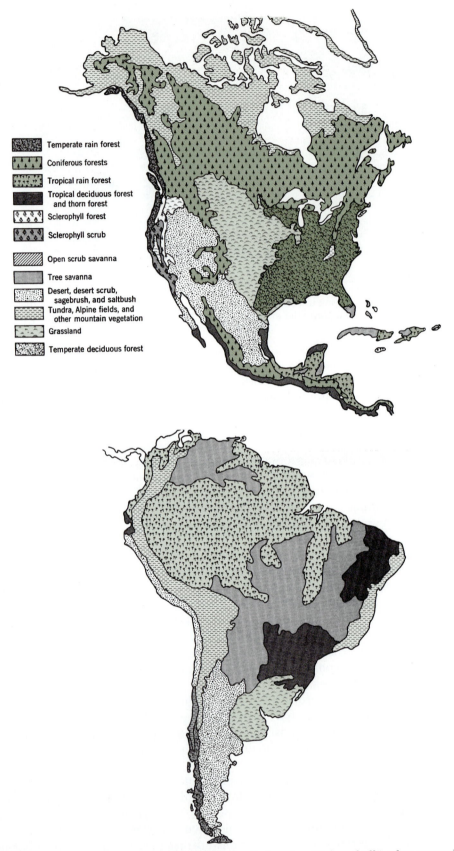

FIGURE 9.2 Major vegetation types of North and South America. "Sclerophyll" refers to woody plants with broad, hard leaves (from R. F. Dasmann, 1976, *Environmental Conservation,* 4th Edition, copyright © 1976, John Wiley & Sons, Inc., New York. Reprinted by permission of John Wiley & Sons, Inc.).

FIGURE 9.3 Major vegetation types of Europe, Asia, and Africa (from R. F. Dasmann, 1976, *Environmental Conservation*, 4th Edition, copyright © 1976, John Wiley & Sons, Inc., New York. Reprinted by permission of John Wiley & Sons, Inc.).

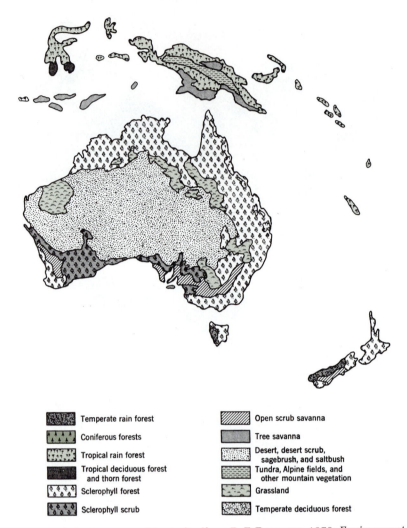

FIGURE 9.4   Major vegetation types of Australia (from R. F. Dasmann, 1976, *Environmental Conservation,* 4th Edition, copyright © 1976, John Wiley & Sons, Inc., New York. Reprinted by permission of John Wiley & Sons, Inc.).

high enough (at least 10°C) for plant growth. During the summer dry period, the long, coarse roots of shrubs and trees can use moisture stored deep (more than 1 m) in the soil profile more efficiently than can the shorter, fibrous grass roots. In the Central Valley of California and in the Mediterranean region of Europe and northern Africa, grasslands occur in a wet winter, dry summer type of climate. However, winter and spring day temperatures in these areas are above freezing during most days and are warm enough to permit growth by cool-season grasses (grasses that grow best between 20 and 25°C). Warm-season grasses (optional growth between 30 and 35°C) dominate tropical areas and temperate areas with dry winters and wet summers.

Soils associated with grasslands are usually deep (more than 2 m), loamy, textured, high in organic matter, and very fertile (Mollisols). These characteristics make them highly suitable for cultivation. In sandy arid areas (less than 300 mm of annual precipitation), soils (less than 600 mm deep) often support grassland, while deeper soils on the surrounding area support shrubland. The thin soils retain the limited moisture near the soil surface that can readily be used by the fibrous grass roots. In contrast, moisture quickly percolates through the deeper sandy soils and is more readily available to the longer taproots of the shrubs.

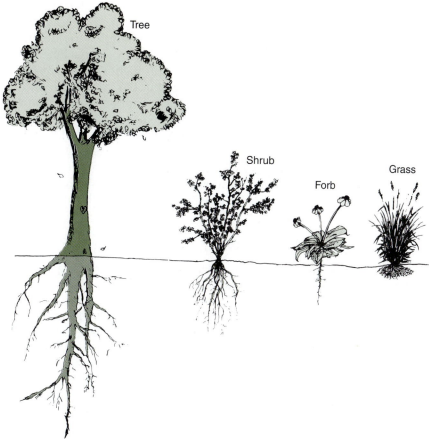

FIGURE 9.5 Rooting depths of grasses, forbs, shrubs, and trees (from Holechek et al. 2001) (drawing by John N. Smith).

In areas with more than 250 mm of annual precipitation, heavy clay soils often support grassland while surrounding loamy or sandy soils support forest. The fine clay soils permit less moisture infiltration and retain more moisture near the soil surface than do loams or sands. This results in more favorable conditions for fibrous-rooted than for tap-rooted species.

Desert Shrublands of the World

Desert shrublands are the driest of the world's terrestrial ecosystems and cover the largest area. Woody plants less than 3 m in height with a sparse herbaceous understory characterize vegetation of this biome. Of the biomes that are grazed by livestock, desert shrublands have been the most degraded by heavy grazing and are the slowest to recover. In some cases, desert shrublands have been created by degradation of arid grasslands by heavy livestock grazing. Desert shrublands provide habitat for numerous unique and rare species, especially where water and stream-side riparian vegetation occurs.

Desert shrublands generally receive less than 250 mm of annual precipitation. The amount of precipitation varies much more from year to year than in the other biomes. In hot desert shrubland areas, precipitation occurs as infrequent, high-intensity rains during a short period (less than 90 days) of the year. This usually results in long periods where the water content of the soil surface is below the permanent wilting point. This provides less favorable conditions for short, fibrous-rooted plants, such as grasses, than for coarse-rooted shrubs. Coarse-rooted plants can collect moisture from a much greater portion of the soil profile than can those with short, fibrous roots near the soil surface. Desert shrub roots extend considerable distances laterally as well as downward. The sparse spacing of desert shrubs permits individual plants to collect moisture over a large area. This explains why they can survive long, dry periods much better than can grasses. In temperate areas with high winter snowfall and dry summers, winter moisture remains available deep in the soil profile during the summer growing season. Shrubs can use this moisture much better than can grasses because of their longer roots.

Sandy to loamy textured soils of variable depth are typical of desert shrublands. Coarse-textured soils permit deep water infiltration and retain little moisture near the soil surface unless there is a restrictive layer. Grasses are more common in desert areas wherever heavy clay soils and sandy soils with a shallow restrictive layer occur. Desert shrubland soils are mainly Aridisols, although some are Entisols.

Savanna Woodlands of the World

Savanna woodlands are dominated by scattered, low-growing tress (less than 12 m tall). They have a productive herbaceous understory if not excessively grazed. Heavy grazing usually reduces the understory grasses and increases the density of trees and shrubs. Fire suppression also increases shrub and tree densities on many savanna woodlands. Savanna woodlands often occur as a transition zone between grassland and forest. Shifts toward grassland or forest take place continually in this biome, depending on grazing intensity, fire control, logging, drought, and climate change.

Potential exists for conversion of savanna woodlands to grassland when they occur on flat, nonrocky soils more than 1 m deep. Coarse-textured and/or thin, rocky soils often favor forest over grassland because they retain low amounts of moisture near the soil surface but store considerable moisture deep in the soil profile and/or rocky crevices. Furthermore, many woody species have long lateral roots that can absorb moisture over a large area of very thin, rocky soil. Without periodic fire, most of the wetter portion of the tallgrass type with loamy to sandy soils is quickly invaded by trees and shrubs because sufficient moisture reaches that portion of the soil profile below 2 m.

Forests of the World

Forests are distinguished from savanna woodlands by having closely spaced (less than 10 m apart) trees more than 12 m in height. In many areas, forests are managed primarily for timber production, recreation, and watershed protection and are too dense to have any grazing value. However, they can produce considerable forage for both livestock and wildlife when thinned by logging or fire or when in open stands. As concern grows about global warming, forests increasingly are valued as storage sinks for carbon.

Forests generally occur in high-rainfall areas (more than 500 mm). Under high rainfall, that portion of the soil profile below 3 m has a high water content during most of the year. Much larger quantities of moisture are needed to support the higher biomass of trees compared to grasses and shrubs. Forests occur under as little as 450 mm of annual precipitation in temperate areas of the western United States that have high snowfall in the winter and dry summers. In temperate areas with a nearly even distribution of precipitation during the year, such as the eastern United States, at least 800 mm of annual precipitation is required for forest growth. Forest can occur in tropical areas with as little as 500 mm of annual precipitation where soils are coarse textured and the wet season is short (less than 150 days).

Forest soils typically are less fertile than grassland soils because of greater leaching. Temperate, deciduous forests in most parts of the world are characterized by soils in the order Alfisol. Coniferous forests usually occur on either Spodosols or Ultisols. Oxisols, the most highly leached soils, support tropical forests.

Tundra of the World

Tundra refers to a level, treeless plain in arctic or high-elevation (cold) regions. Arctic tundra covers about 5% of the world's surface. Large areas of arctic tundra occur in North America, Greenland, northern Europe (the former Soviet Union), and northern Asia. Low-growing, tufted perennial plants and lichens dominate this type. Shrubs in the genus *Salix* are the main type of woody plants. The tundra surface is frozen for more than seven months of the year. A permanent layer of ice, the permafrost, exists below the surface and restricts trees growing on tundra. Precipitation is low (250 to 500 mm) over much of the area. Even so, much of the tundra remains moist to water saturated during the short growing season because evapotranspiration is low. Strong winds further increase the severity of the area for plant life. Most tundra remains wilderness with limited use by herding and hunting cultures.

Alpine tundra occurs at high elevations (above 2,800 m) above timberline in mountainous portions of the western United States and Canada, Europe, South America, and New Zealand. Precipitation tends to be greater than on the arctic tundra and most of it runs off to lower elevations. Because of its limited area, rough terrain, high aesthetic value, and short grazing season (less than 90 days), Alpine tundra is of little importance as a livestock grazing type. Exceptions are Peru and Tibet (China), where alpacas, llamas, and yaks make heavy use of this type.

TERRESTRIAL ECOSYSTEMS OF THE UNITED STATES

In the United States, 15 basic land biome types are economically important because of their natural resources and/or total area (Table 9.1). The arrangement of these biomes with movement across the United States is shown in Figure 9.6. Important physical and biological aspects of each type will be discussed using Branson (1985) and Holechek et al. (2001) as primary references.

TABLE 9.1   Present Extent of the 15 Major Biome Types in the United States

Item	Area (million ha)	Percent in Federal Ownership
Grasslands		
Tallgrass prairie	15[a]	1
Southern mixed prairie	20	5
Northern mixed prairie	30	25
Shortgrass prairie	20	5
California annual grassland	3	6
Palouse prairie	3	15
Desert shrublands		
Hot desert	26	55
Cold desert	73	77
Woodlands and Forests		
Piñon-juniper woodland	17	65
Mountain shrubland	13	70
Western coniferous forest	59	62
Southern pine forest	81	6
Eastern deciduous forest	100	7
Oak woodland	16	4
Tundra		
Alpine	4[b]	99

Source: USDA 1972, 1977.
[a]Includes sandhills and coastal prairie.
[b]Does not include Alaskan tundra.

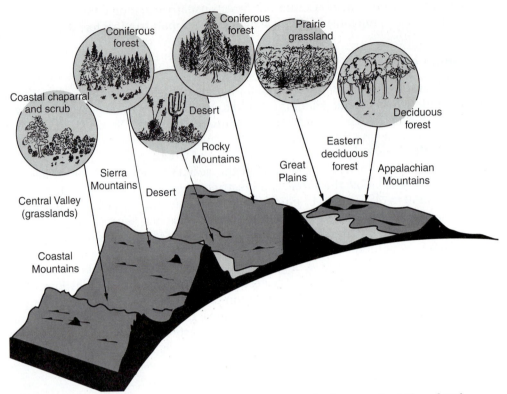

FIGURE 9.6 Gradual transition from one major biome to another between the 34th and 38th parallel crossing the United States. The major factors causing these transitions are changes in average temperature and precipitation (adapted from Miller, 1990 by John N. Smith).

GRASSLANDS

Tallgrass Prairie

The tallgrass prairie is located in the central United States east of the mixed and short-grass prairies and west of the deciduous forest (Figure 9.7). Precipitation falls mostly as rain during the warmer months and ranges from about 500 mm in the northern and western edges to 1,000 mm on the eastern and southern edges. The soils are Mollisols that are typically deep, very fertile, free of large rocks, and not leached enough to reduce fertility. These conditions favor cultivated grasses, such as corn (*Zea mays*) and wheat (*Triticum aestivum*), and most tallgrass prairie has been converted to that use. Native tallgrass prairie is now scarce. Other than scattered small patches, the major remnant occurs on thin rocky soils extending from the Flint Hills of eastern Kansas to the Osage Hills of Oklahoma. Conservation groups have been buying up relict areas of tallgrass prairie particularly in Iowa, Indiana, Minnesota, Wisconsin, and Illinois.

The climate of the tallgrass prairie is subject to periodic summer drought, which favors grasses over tree seedlings. Tallgrass prairie burned during dry spells every 3 to 4 years, which further favored grasses and fire-adapted forbs over trees. Some attempts have been made to restore the tallgrass prairie, and restoration success seems to depend on frequent burning (Jordan 1997). Dominant grasses (Figure 9.8) include little bluestem (*Schizachyrium scoparium*) in the drier uplands and big bluestem (*Andropogon gerardii*) in the wetter lowlands. Yellow indiangrass (*Sorghastrum nutans*) and switchgrass (*Panicum virgatum*) are the other two major grasses. Among the forbs, leadplant (*Amorpha canescens*) and scurfpea (*Psoralea* sp.) are important legumes with nitrogen-fixing ability. To the West, the tallgrass prairie grades into drier shortgrass prairie. On the eastern edge and along the waterways, tallgrass prairie grades into deciduous forest dominated by oaks (*Quercus* spp.), hickories

Palouse bunchgrass

California annual grassland

Desert grassland

Tallgrass prairie

Mixed prairie

Shortgrass prairie

FIGURE 9.7 The approximate major locations of the six basic grassland types in the United States (adapted from Branson 1985 and Barbour et al. 1999).

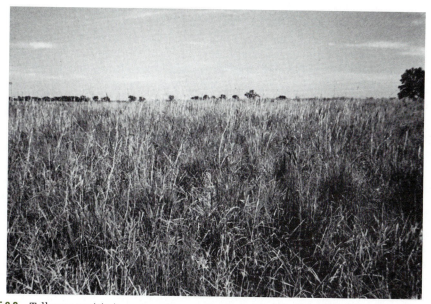

FIGURE 9.8 Tallgrass prairie in eastern Kansas dominated by big bluestem.

FIGURE 9.9   Oak savanna in central Texas. Note browse line on oaks from overuse by whitetail and goats.

(*Carya* spp.), elms (*Ulmus* spp.), and, especially in the river bottoms, roses (*Rosa* spp.), cottonwood (*Populus deltoides*), black willow (*Salix nigra*), green ash (*Fraxinus pennsylvanica*), and boxelder (*Acer negundo*). Most tallgrass species are widespread. Among the more indicative reptile species are the prairie skink (*Eumeces septentrionalis*), fox snake (*Elaphe vulpina*), and the plains gartersnake (*Thamnophis radix*).

Southern Mixed Prairie

The southern mixed prairie extends from eastern New Mexico to eastern Texas and from southern Oklahoma to northern Mexico. The climate is semiarid and precipitation varies from 300 mm in eastern New Mexico to 700 mm in central Texas. The soils are diverse, including Mollisols, Aridisols, and Entisols. Because of the wide range of soil and climate, grassland composition and productivity also vary greatly. However, the southern mixed prairie supports more free-range livestock production than any other western grassland. Most of the grasses associated with southern mixed prairie tolerate grazing because they evolved with dense concentrations of bison (*Bison bison*). The southern mixed prairie includes subtypes of true mixed prairie, desert prairie, high-plains bluestem, and oak savanna (Holechek et al. 2001). Much of the true mixed prairie and high-plains bluestem communities are now under cultivation, although sizable portions still remain as native grassland grazed by livestock.

Grasses that occur throughout the entire vegetation type include blue grama, buffalograss (*Buchloe dactyloides*), little bluestem, various three-awn species (*Aristida* spp.), silver bluestem (*Bothriochloa saccharoides*), vine mesquite (*Pannicum obtusum*), and sideoats grama (*Bouteloua curtipendula*). To the west, blue grama and ecotone tobosa (*Hilaria mutica*) and dominant. Moving from west to east in Texas, climatic conditions become more moist, silver bluestem and Texas wintergrass (*Stipa leucotricha*) become more common, and grassland grades into oak savanna (Figure 9.9). Taller grasses dominate on deep sandy soils throughout the southern mixed prairie, including primarily big bluestem, silver bluestem, and little bluestem. In southeastern New Mexico and adjacent Texas, these grasses grow in association with sand sagebrush

(*Artemisia filifolia*) and shinnery oak (*Quercus harvardii*). The vertebrates of the mixed prairie include a mix of shortgrass and tallgrass species.

Northern Mixed Prairie

The northern mixed prairie occupies much of the western half of North and South Dakota, the eastern two-thirds of Montana, the northeastern quarter of Wyoming, southeastern Alberta, and southern Saskatchewan. Transition to southern mixed prairie occurs in Nebraska and Kansas. It is adapted to semiarid climate (300 to 650 mm of precipitation); long, severe winters; and warm, moist summers. Most of the precipitation occurs in spring and summer, peaking in June. The soils are mostly Mollisols suitable for cultivation, but crop production is limited by severe winters, short growing seasons, periodic droughts, and low average precipitation. Much of the northern mixed prairie was farmed from 1900 to 1933, and wheat still is grown in isolated locations, but repeated drought has resulted in much conversion to range use.

The northern mixed prairie supports the highest diversity of grasses of all the western land types and a diverse composition of shrubs and forbs. Because of the great plant diversity, the northern mixed prairie is one of the most nutritious western grasslands for grazing wildlife and livestock. Cool-season grasses such as rough fescue (*Festuca seabrella*), bluebunch wheatgrass (*Agropyron spicatum*), and various bluegrasses (*Poa* spp.) provide good early spring feed for grazing animals. Green needlegrass (*Stipa viridula*), needleandthread (*Stipa comata*), western wheatgrass, and various forbs provide excellent late spring feed, and little bluestem, blue grama, and sideoats grama provide high-quality summer and fall forage. The wide variety of shrubs found in northern mixed prairie supports a diverse wildlife community. Important shrubs include silver sagebrush (*Artemisia cana*), big sagebrush (*Artemisia tridentata*), skunkbrush sumac (*Rhus trilobata*), western snowberry (*Symphoricarpos occidentalis*), various species of rose (*Rosa* spp.) fourwing saltbush (*Atriplex canescens*), and black greasewood (*Sarcobattus vermiculatus*). The vertebrates are a mix of those adapted to shortgrass and tallgrass prairie.

Shortgrass Prairie

The shortgrass prairie extends from northern New Mexico into northern Wyoming and in disjunct areas scattered throughout central Wyoming, western South Dakota, and southern Montana (Figure 9.7). The climate is semiarid (300 to 500 mm of annual precipitation), with cool winters and warm, moist summers. The climate favors warm-season grasses with very shallow but extensive root systems dominated by blue grama and buffalograss. Soil are largely Mollisols with locally high concentrations of sands (Entisols) and clays (Vertisols), which vary widely in water retention and soil aridity.

The widespread medium-textured loam soils support primarily blue grama and buffalograss and, secondarily, western wheatgrass. Little bluestem occupies the sandy soils while the heavy clay soils are dominated by western wheatgrass. Because these grasses evolved with bison grazing, they are well adapted for livestock grazing (Figure 9.10). Winterfat (*Ceratoides lanata*), an important shrub, and scarlet globemallow (*Sphaeralcea coccinea*), an abundant forb, are avidly consumed by both domestic and wild ungulates. Continuously intense grazing causes the grasses and preferred forbs to be replaced by cactus (*Opuntia* spp.), snakeweed (*Gutierrezia sarothrae*), Russian thistle (tumbleweed, *Salsola kali*), and fringed sagewort (*Artemisia frigida*). Indicator vertebrates of the shortgrass and midgrass prairies include bison (*Bison bison*), pronghorn (*Antilocapra americana*), swift fox (*Vulpes velox*), black-tailed prairie dog (*Cynomys ludovicianus*), Great Plains toad (*Bufo cognatus*), and Plains spadefoot (*Scaphius bombiforus*).

FIGURE 9.10   The shortgrass prairie is well suited for grazing by cattle.

FIGURE 9.11   California annual grassland (from Holechek et al. 2001).

California Annual Grassland

The California annual grassland is found primarily west of the Sierra Nevada Mountains (Figure 9.11) and west of the Cascade Mountains in Oregon. The climate is characterized by mild, wet winters and long, hot, dry summers. Precipitation varies from about 200 mm in the southern foothills to almost 1,000 mm in some areas near the coast. Most of the precipitation falls between October and May, with a peak in January. Many of the soils of the California annual grassland have prairie-like profiles (Mollisols) while others display desert attributes (Aridisols). Other soils are poorly developed (Endisols) from volcanic ash.

Much of the valley and coastal areas have been grazed by livestock since Spanish settlements during the seventeenth century. The original vegetation was predominantly comprised of cool-season bunchgrasses that did not evolve with intense grazing by large herbivores. Based on description and remnants, the native grassland was quite lush, but suffered severe degradation from livestock grazing during European settlement. The original vegetation was dominated by grasses of midheight, such as

FIGURE 9.12 This palouse bunchgrass prairie is dominated by bluebunch wheatgrass and Idaho fescue in northeastern Oregon. Most of the palouse is now farmed because of excellent soils and climate for wheat production (courtesy of USDA-Forest Service).

purple needlegrass (*Stipa pulchra*), nodding needlegrass (*Stipa cernua*), prairie june-grass (*Koeleria pyramidata*), pine bluegrass (*Pos scabrella*), California melicgrass (*Melica imperfecta*), creeping wildrye (*Elymus triticoides*), and California oatgrass (*Danthonia californica*).

Practically all the native perennial grasses have been replaced by introduced, cool-season, annual bromegrasses and by forbs. Numerous fires and overgrazing probably account for the change from perennial to annual grasses over a century ago. The moisture and temperature are ideal for growth and reproduction of cool-season annuals, and their seeds remain dormant during the dry summer period. Today almost 400 species of introduced plants from many lands are found in California and less than 5% of the herbaceous cover is composed of native perennials (Sampson et al. 1951). The vegetation now is dominated by slender oat (*Avena barbata*), wild oat (*Avena fatua*), soft brome (*bromus mollis*), ripgut brome (*Bromus rigidus*), foxtail brome (*Bromus rubens*), and little barley (*Hordeum pusillum*). Trees and shrubs become progressively more common as the grasslands grade into chaparral with increased elevation in the foothills. Unique indicator species are the San Joaquin pocket mouse (*Peronathus inornatus*), Heermann's kangaroo rat (*Dipodomys heermanni*), yellow-billed magpie (*Pica nuttalli*), bluntnose leopard lizard (*Gambelia silus*), California legless lizard (*Anniella pulchra*), and Gilbert's skink (*Eumeces gilberti*).

Palouse Prairie

Before much of it was converted to wheat, pea, and other crop culture, one of the most productive grasslands was the palouse prairie and similar grasslands in Washington, Oregon, Idaho, Utah, and Montana (Figure 9.12). The soils originated as loess (wind-blown) dunes that now have excellent textural, structural, and chemical properties for agriculture. Most of the remaining native grasslands are located in Utah and Montana and in the Blue Mountain region of northeastern Oregon and southeastern Washington. Rainfall in the area ranges from 300 to 640 mm annually with most falling during winter. Midsummer is driest. Palouse winters are relatively mild and summer temperatures are seldom over 35°C.

The climax vegetation of the remaining palouse prairie is dominated by blue-bunch wheatgrass, which is a codominant with either Idaho fescue or Sandberg blue-grass (*Pos sandbergii*) (Figure 9.12). Unlike the central prairie of North America, where the grasses form a sod, the palouse prairie grasses grow in bunches with open interspaces between plants. When intensively grazed, cheatgrass (*Bromus tectorum*) replaces most of the perennial grasses and Sandberg bluegrass increases. In some ar-eas the unpalatable medusahead rye (*Taeniatherum asperum*) has replaced downy brome after long-term heavy grazing. The palouse prairie supports a wide variety of forbs. Under climax conditions arrowleaf balsamroot (*Balsamorhiza sagittata*) is dominant. Western yarrow (*Achillea millefolium*) becomes more dominant when range condition declines. Several deciduous shrubs occur in the type. Big sagebrush occupies the drier sites and increases with heavy grazing. The vertebrates of the palouse prairie are not unique to the ecosystem. Most are common in adjacent cold desert and inland coniferous forest ecosystems.

DESERTS

Hot Desert

The hot desert is characterized by extremely little precipitation and high tempera-tures. Humidity is low and evaporation greatly exceeds precipitation. The hot desert occurs at the lowest elevations in southern California, Arizona, New Mexico, south-western Texas, Nevada, and northern Mexico from below sea level in Death Valley to about 1,400 m. Precipitation increases with elevation from 130 to 350 mm annually, and occurs most often during summer and, to a lesser extent, in winter. Summer rains occur in short, intense convection storms caused by solar heating. Much of the sum-mer precipitation quickly runs off the land and fails to penetrate the soils. Winter rains are less intense and more prolonged products of frontal movements that leave nearly as much effective soil moisture as the summer rains. Many areas may go 2 to 3 years without a killing frost and some areas rarely are exposed. The soils are mostly Aridisols. Some well-drained areas have established a nearly impermeable layer of calcium carbonate (caliche) at or just under the soil surface.

Three desert areas occur within the U.S. hot desert: the Mojave, Sonoran, and Chihuahuan deserts (Figure 9.13). The Mojave desert—found over a range of eleva-tions in southeastern California, southern Nevada, and northwestern Arizona—is the driest (Table 9.2) and higher elevations freeze. It has the least diversity of vegetation and is dominated by shrubs. The Sonoran desert occurs at intermediate elevations in southern Arizona, is virtually frost free, and is the wettest of the three areas. Warm, moist winters explain the wider diversity of species found in the Sonoran desert compared with the other hot deserts (Figure 9.14). The Chihuahuan desert of southwestern Texas and southcentral New Mexico has less species diversity because of freezing winter temperatures and drier winter weather.

Historic evidence from early European settlement indicates that parts of the Sono-ran and Chihuahuan deserts were more open grassland or grasslands with scattered shrubs (Buffington and Herbel 1965). Much of this area now supports a mixture of shrubs and grasses, with the shrubs more or less dominant. Mesquite (*Prosopis* sp.), catclaw (*Acacia* sp.) and creosote bush (*Larrea tridentata*) are common throughout the area. The spread of shrubs has been owed separately to overgrazing, fire control, climatic change, and seed dissemination by domestic animals, but a combination probably occurred.

Over much of the hot desert, creosote bush is a dominant species, but many other shrub species occur along sandy arroyos where more water is stored in the soil. Among the grasses, the climax dominants are almost completely of the genera *Bouteloua*, *Aristida*, and *Hilaria*. Intensive grazing results in a disclimax of low-pro-ducing perennials with annuals in the genera of *Bouteloua* and *Aristida*. In many

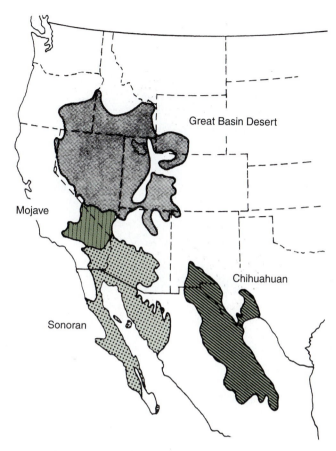

FIGURE 9.13 Deserts of North America. The Great Basin desert is the cold desert (after Shreve 1942 by Blaisdell and Holmgren 1984).

TABLE 9.2 Some Climatic Features of the Four Desert Regions of North America

	Great Basin (Cold Desert)	Mojave	Sonoran	Chihuahuan
Area (km²)	409,000	140,000	275,000	453,000
Annual precipitation (mm)	100–300	100–200	50–300	150–300
Precipitation in summer (% of total)	30	35	45	65
Snowfall (cm; 10 cm snow = 5 1 cm rain)	150–300	25–75	Trace	Trace
Winter mean max/min temperatures (°C)	+8/−8	+15/0	+18/+4	+16/0
Hours of frost (% of total)	5–20	2–5	0–1	2–5
Summer mean max/min temperatures (°C)	34/10	39/20	40/26	34/19
Elevation (m)	>1,000	Variable	<600	600–1,400

Source: Barbour et al. 1987.

areas the grasslands are reduced to almost pure stands of annual plants. Broom snakeweed is an unpalatable persistent plant that now dominates many heavily grazed desert grasslands.

Many vertebrates have adapted to the hot desert and adjacent grasslands by spending much time underground, adopting nocturnal behavior, and conserving water in various ways. Indicator species are collared peccary (*Dictoyles tajuca*), hog-nosed skunk (*Conepatus mesoleucus*), desert cottontail (*Sylvilagus audubonii*), several species of pocket mice (*Perognathus* spp.), several species of kangaroo rat (*Dipodomys* spp.), scaled quail (*Callipepla squamata*), cactus wren (*Campylorhynchus brunne-icapillus*), curve-billed thrasher (*Toxstoma curvirostre*), shovelnose snakes (*Chonactis* spp.), lyre snakes (*Trimophordon* spp.), several lowland species of rattlesnake (*Crotalus* spp.), and numerous lizard species.

FIGURE 9.14   Sonoran desert in southcentral Arizona supports a wide variety of cactus and shrubs because of warm winter temperatures and a biomodal pattern of precipitation (winter and summer peaks).

Cold Desert

The cold desert is composed of sagebrush desert steppe and salt desert, which together occupy so much of the Great Basin (Figure 9.13) that it is sometimes called the Great Basin desert. The sagebrush desert steppe and salt desert intermingle over the Great Basin and adjacent areas. The sagebrush desert steppe generally occurs at higher elevations than the salt desert. It is characterized by higher precipitation and less saline soils. Although the frost-free period for both ecosystems is typically less than 180 days, the salt desert has warmer temperatures because of its occurrence at low elevations.

Sagebrush desert steppe is one of the most extensive western ecosystems with about 39 million hectares at elevations over 1,200 m and in areas free of salts. Mountain ranges occur throughout the Great Basin and cause a rain-shadow effect that accounts for the aridity. Between 200 and 400 mm of precipitation occurs, with an average of 250 mm. About half of the precipitation comes as snow or rain in the cooler months and summers are dry. Temperatures in the area are extreme, dropping as low as −37°C in winter and climbing up to 38°C in summer. The soils are primarily volcanic with much sand and extensive areas of lava.

The sagebrush-desert vegetation is characterized by big sagebrush (Figure 9.15), bluebunch wheatgrass, bottlebrush squirreltail (*Sitanion hystrix*), Idaho fescue, western wheatgrass, Indian ricegrass (*Oryzopsis hymenoides*), needleandthread, and Great Basin wildrye (*Elymus cinereus*). Originally, native wheatgrasses composed most of the total aboveground vegetation. With overgrazing, big sagebrush becomes more prominent and the non-native cheatgrass replaces wheatgrass in the understory. Other important shrubs include black sagebrush (*Artemisia nova*) on rocky and shallow soils, low sagebrush (*Artemisia arbuscula*) on very dry soils at higher elevations, rabbitbrush (*Chrysothamnus nauseosus*) on sandy soils with a low salt content, antelope bitterbrush (*Purshia tridentata*) on sandy to rocky soils at higher elevations, and Mormon tea (*Ephedra* spp.) on coarse soils.

The salt desert occupies the same general area as the sagebrush desert steppe where much of it occurs in a mosaic with sagebrush desert. The salt desert occurs at the lower elevations, mostly at elevations below 1,200 m, where drainage is often restricted and the water table is high. The salt desert is extremely dry; it has the lowest precipitation of all desert ecosystems except the hot Mojave desert. Precipitation ranges from 80 to 250 mm,

FIGURE 9.15 The sagebrush shrub steppe is one of the largest types in the western United States. The high levels of volatile oils associated with the various sagebrushes make them unpalatable to livestock. Sagebrush does provide native big-game animals such as mule deer and pronghorn with important winter feed.

FIGURE 9.16 The salt desert is dominated by sparse, low-growing evergreen shrubs. Shadscale and winterfat dominate this salt desert rangeland in central Nevada.

with an average of 120 mm, of which nearly half is snow, nearly half is spring or fall rain, and very little occurs in summer. The soils are similar to those in sagebrush-desert steppe except that the higher evaporation rates concentrate more salts in the surface soil as evaporation draws soil water upward and causes the salts to accumulate at the soil surface. The highest salt concentrations occur where water collects following storms.

Salt desert vegetation is characterized by a few widely spaced, spiny, salt-tolerant shrub species with small grayish leaves (Figure 9.16). Shadscale saltbush (*Artiplex confertifolia*) is the shrub species dominating the largest area, with winterfat the second most common. Important understory species include inland saltgrass, Indian ricegrass, bottlebrush squirreltail, and galleta grass (*Hilaria jamesii*). Much of the salt desert has been severely overgrazed but can improve slowly under proper management (Yorks et al. 1992).

Vertebrates of the cold deserts are less diverse than the hot deserts and less likely to be uniquely found there. Unique indicator species at lower elevations include the great basin pocket mouse (*Perognathus parvus*), certain kangaroo mice (*Microdipdops spp.*), and the chisel-toothed kangaroo rat (*Dipodomys microps*). At higher elevations are the sagebrush vole (*Lagarus curtatus*), Townsend's ground squirrel (*Spermo-philus townsendii*), sage grouse (*Centrocerus arophasianus*), sage thrasher (*Oreos-coptes montanus*), and sage sparrow (*Amphispiza belli*).

WOODLANDS

Piñon-Juniper Woodland

Piñon-juniper woodland is one of the most widely distributed ecosystems in the western United States (Figure 9.17). Although it occurs on the slopes of the Cascades and Sierras, most piñon-juniper woodland is found in the states of Utah, Colorado, New Mexico, and Arizona (Figure 9.18). The climate of the piñon-juniper ecosystem is harsh for tree growth and characterized by low precipitation, hot summers, high wind, low relative humidity, very high evaporation rates, and intense sunlight. The annual precipitation varies from 300 to 500 mm. Soils typically are poorly developed

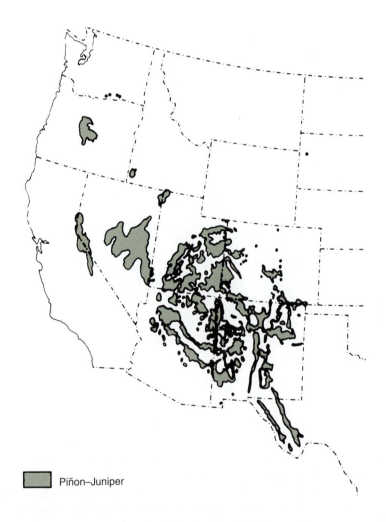

Piñon–Juniper

FIGURE 9.17   Distribution of the piñon-juniper type. This type in Oregon consists mainly of a single tree species, western juniper (*Juniperus occidentalis*) (from Franklin and Dyrness 1969; modified from Clary et al. 1974 and USDA 1936 by Branson 1985).

FIGURE 9.18 An open stand of piñon-juniper in New Mexico with a good grass understory.

Entisols and Aridisols. Widespread overgrazing of understory species has reduced protective soil cover and resulted in severe soil erosion in much of this ecosystem. Prior to European settlement, piñon-juniper stands were confined largely to rocky slopes or more level sites with shallow soils over fractured rock.

North of Utah this ecosystem consists entirely of juniper composed mostly of Rocky Mountain juniper (*Juniperus scopulorum*), which is the most widespread of several western juniper species. Rocky Mountain juniper is more cold resistant and can tolerate lower precipitation than the piñon pines. *Pinus monophyll* and *Pinus edulis* are the most widespread and abundant of a number of piñon pine species. This ecosystem occurs at elevations from 100 to 2,400 m and at its upper limits grades into Gambel oak (*Quercus gambelii*) and ponderosa pine (*Pinus ponderosa*), while the lower limits may grade into shortgrass prairie, sagebrush desert steppe, or hot desert.

Since European settlement, the juniper trees have spread into grasslands and become more dense because of overgrazing, fire control, seed dissemination by animals, and, possibly, climatic change (Johnsen 1962). Natural wildfires historically occurred at 10 to 20-year intervals, but grazing interests and proliferating rural homeowners have voiced strong support for fire suppression. As a consequence, fuel loads have accumulated, which make wildfires, once they do occur, far more devastating than those occurring more frequently under natural conditions. Vertebrate species of the piñon-juniper woodland include many of the species common at higher forested elevations and some from desert grasslands at lower elevations. The piñon jay (*Gymnorhinus cyanocephalus*), Scott's oriole (*Iterus parisorum*), and piñon mouse (*Peromyscus truei*) are good indictor species.

Mountain Shrub

The mountain shrub ecosystem occurs primarily in the Rocky Mountains and Sierra-Cascade Mountains in a narrow, intermittent strip from 1,200 to 2,800 m above sea level. It typically is sandwiched between grasslands and sagebrush steppe at lower elevations and coniferous forests at higher elevations. It occurs mostly at higher elevations and latitudes than piñon-juniper vegetation and is most prevalent in Colorado, Wyoming, Utah, Oregon, and Idaho. The climate of mountain-shrub is intermediate between that favoring grassland and forest. Annual precipitation averages 460 to 500 mm and the growing season is relatively short, ranging from 100 to 120 days. The soils are mostly undeveloped Entisols and Inceptisols.

FIGURE 9.19   Foothill ranges of the southwestern Rocky Mountains are dominated by stands of Gambel oak (*Quercus gambelii*) and other shrubs. Gambel oak is toxic to livestock in the spring, but provides important food and cover for deer.

The vegetation is dominated by shrubs and small trees 1 to 10 m tall (most 2 to 4 in tall). Almost any western shrub requiring mesic conditions may occur in this type, depending on location. Mountain-mahogany (*Cercocarpus* spp.) often dominates with scrub oaks (*Quercus* spp.) (Figure 9.19), chokecherry (*Prunus virginiana*), buckbrush (*Ceanothus* spp.), serviceberry (*Amelanchier* spp.), snowberry (*Symphoricarpos* spp.), bigtooth maple (*Acer grandidentatum*), bitter cherry (*P. emarginata*), and antelope bitterbrush (*Purshia tridentata*). In Arizona, mountain shrub includes manzanitas (*Arctostaphylos* spp.) and is sometimes referred to as interior chaparral. Mountain shrub provides habitat for a mix of vertebrate species that integrade from ecosystems at lower and higher elevations. It is especially good mule deer (*odocoileus hemionus*) habitat.

Oak Woodlands

This group of ecosystems is scattered about the western United States and is dominated by several different species of oak. These include the shrubby sand shinnery oak (*Quercus havardii*) in Texas, Oklahoma, eastern New Mexico and northern Mexico, and open oak savanna types dominated by tree oaks in California, Oregon, southern Arizona, and the Edwards Plateau in central Texas (Figure 9.9). The climate of the oak woodlands is variable depending on location. Oaks are sensitive to winter cold and do not occur much farther north than the Oregon white oak (*Quercus garryana*), which reaches southwestern British Columbia. Soils of the oak woodland ecosystems type vary, but usually are well drained. The vertebrates include a mix of grassland and forest species.

Chaparral

The chaparral ecosystem is dominated by broad-leaved evergreen shrubs and small trees located in lower to intermediate elevations in California along the Pacific Coast ranges mostly south of San Francisco Bay and around the Central Valley. The climate is Mediterranean with warm, dry summers, mild, moist winters, and a low percentage of time below freezing. Most precipitation occurs from November through April and precipitation ranges from about 300 to 1,000 mm. The chaparral is a droughty, fire-disclimax ecosystem, which burns naturally every 10 to 20 years, especially during the dry period of late summer and early fall. The soils vary in development but much is Entisol occurring on relatively steep slopes. The typically hot fires mineralize the surface soils and expose the watersheds to increased erosion, mudslides, and flooding. Much of this ecosystem is now densely settled and fire protection has resulted in exceptionally dense growth, which eventually burns with more devastating effect.

The chaparral includes a large variety of shrubs adapted to droughty conditions and periodic fire. The chaparral often is codominated by chamise (*Adenostoma fasciculatum*) with manzanitas and ceanothus (*Ceanothus* spp.). Common associates include the buckthorns (*Rhamnus* spp.), poison oak (*Rhus toxicodendron*), and several species of scrub oak (*Quercus* spp.). Vertebrate species intergrade between the California grasslands and hot deserts and the woodlands and forests at higher elevations. Some generally indicative bird species include California quail (*Lophortyx californicus*), wrentit (*Chemaea fasciata*), and California thrasher (*Toxostoma redivivium*).

FOREST ECOSYSTEMS

Western Coniferous Forest

The western coniferous forest occurs in two major types: the Western interior forest and the West Coast forest (Figure 9.20). Coniferous forest occurs above 1,100 m in the western interior, mostly on Endisols and Inceptisols, where it is dominated by ponderosa pine (*Pinus ponderosa*) at lower elevations (16 million hectares) and interior Douglas-fir (*Pseudotsuga menziesii*) at higher elevations (17 million hectares). The interior western coniferous forest occupies the east slopes of the Sierra Nevada and Cascade Mountains and extends from there to Nebraska and the Dakotas, and from southern Alberta and British Columbia to northern Mexico. The West Coast forest stretches from the Alaskan Panhandle to the San Francisco Bay area. There, coastal Douglas-fir descends to sea level where it often dominates numerous other coniferous species in numbers and height. It is second in height only to the redwood (*Sequoia sempervirons*), the world's tallest tree species.

1. Northern forests of white-red-jack pine, spruce-fir, aspen-birch, and maple-beech-birch groups.
2. Central forests of oak-hickory.
3. Southern forests of: Oak-pine, loblolly-shortleaf pine, and longleaf-slash pine groups.
4. Bottom land forests of oak-gum-cypress.
5. West coast forests of: Douglas-fir, hemlock-sitka spruce, redwood, and some western hardwood groups.
6. Western interior forests of: Ponderosa pine, lodgepole pine, Douglas-fir, white pine, western larch, fir-spruce, and some western hardwood groups.

FIGURE 9.20 Natural forest regions of the United States (from USDA-Forest Service, from Shirley 1973).

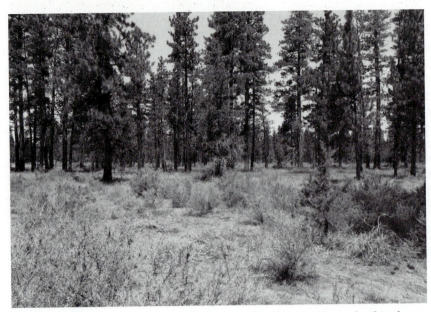

FIGURE 9.21   Ponderosa pine is the primary timber tree found on coniferous land in the western United States.

Ponderosa pine forests are the largest and driest of the true forest types in the western United States and commonly mix with Douglas-fir at the higher elevations of its range. Ponderosa pine ecosystems typically occur from 1,100 to 2,500 m above sea level (Figure 9.21), grading into piñon-juniper ecosystems at lower elevations. Precipitation there ranges from 450 to 650 mm with most as snow in the north and as rain in the south. Ponderosa pine forests are frequently exposed to extended drought and naturally recurrent fire, which sustains the dominant species in open stands when fires are frequent and not very intense. Fire protection has resulted in denser than natural growth in many locations and increased likelihood of intense, damaging fire. Ponderosa pine often forms nearly pure stands, especially in the southwestern Rocky Mountains. In the western-most mountain ranges, it is more likely to associate with other species including Jeffry pine (*P. jeffryei*), lodgepole pine (*P. contorta*), limber pine (*P. flexilis*), sugar pine (*P. lambertiana*), incense-cedar (*Libodedrus decurrens*), western larch (*Larix occidentalis*), grand fir (*Abies grandis*), white fir (*A. concolor*), and giant sequoia (*Sequoia gigantea*). In the Southwest, most of these species other than white fir drop out and are replaced with Apache pine (*P. engelmannii*). Chihuahua pine (*P. leiophylla*), and Arizona cypress (*Cupressus arizonica*).

Interior Douglas-fir forest occurs at elevations just above the ponderosa pine zone up to 3,100 m above sea level where it grades into alpine tundra. The Douglas-fir forest occurs mostly in Colorado, Idaho, Wyoming, Montana, Oregon, and Washington, and in isolated parts of Arizona and New Mexico. Consistent with higher elevation, this ecosystem is wetter than pine forest (640 to 900 mm per year) and it receives much more snow. The topography usually is quite steep and rocky; more so than the lower slopes occupied by pine forest. Unlike Ponderosa pine, Douglas-fir is shade tolerant and cannot survive fire. It often will be found in the understory of the wetter pine forests. Other frequent associates include lodgepole pine, western larch, white fir, and western white pine, the mix being influenced by frequency and intensity of fire as well as climate and soil

Where fire occurs in the interior Douglas-fir forest, it typically establishes persistent preclimax successional stages, with about 8 million hectares now dominated by lodgepole pine, 5 million hectares in quaking aspen (*Populus tremuloides*), 2 million hectares in western larch, and 1.2 million hectares in brush. Quaking aspen is favored following destructive fire because it establishes quickly following fire and grows rapidly in full light, outcompeting other species. Lodgepole pine has an advantage where fire is a common recurrent phenomenon, which sustains it for long periods. Original indicator species included grizzly bear (*Ursus arctos*), yellow-bellied marmot (*Marmota flaviventris*), mountain cottontail (*Sylvalagus nuttalli*), golden-mantled squirrel (*Citellus lateralis*), bushy-tailed woodrat (*Neotoma cinerea*), blue grouse (*Dendragapus obscurus*), band-tailed pigeon (*Columba fasciata*), Clark's nutcracker (*Nucifraga columbiana*), and Steller's jay (*Cyanocitta stelleri*).

Along the wet Pacific Coast, the coastal Douglas-fir forest extends from central British Columbia down the west side of the Cascades to sea level in the wet northwest coastal areas, where it is joined by a number of spruce, cedar, hemlock, and other coniferous species (Figure 9.20). The coastal climate north of San Francisco Bay is humid and rainy for all but the dry summer period and some locations are the wettest in the United States. It grades into the northern spruce-fir forests at higher and colder elevations in the Cascades. This lush, wet ecosystem is one of the most productive forests in the United States. Common forest associates of Douglas-fir include Sitka spruce (*Picea sitchensis*), western redcedar (*Thuja plicata*), western hemlock (*Tsuga heterophylla*), western white pine (*Pinus monitcola*), grand fir, Pacific silver fir (*Abies amabilis*), noble fir (*Abies procera*), and, to the south, tanoak (*Lithocarpus densiflorus*) and redwood. Most of the conspicuous interior forest vertebrates extend to the coastal forest. The exceptional high precipitation and mild winters along the coast explains why the mountain beaver (*Aplodontia rufa*) and numerous species of salamanders occur there and not farther inland.

Northern Forest

The northern forest is composed of several forest types, including a hardwood forest type described in the next section on eastern deciduous forests. If the northern forests can be characterized by a forest type, however, it would be the spruce-fir forests that dominate much of inland Alaska and link western coastal and interior coniferous forests through Canada to the northcentral and northeastern states (Figure 9.20). The spruce-fir forest creates an ecosystem link between East and West for numerous cold-tolerant animal species. These are the boreal forests of long, cold winters and short, cool, moist summers (see Chapter 10). Precipitation is 700 mm to 1,200 mm, much as snow.

In the uplands of the far northern tier red spruce (*Picea rubra*) and balsam (*Abies concolor*) dominate over lichen-, moss-, and shrub-covered openings on Spodisols and Histosols. The most common hardwood members of the boreal forests include quaking aspen and paper birch (*Betula papyrifera*), which seed in rapidly following fire, acting as nursery trees for spruce and fir. The shrubby speckled alder (*Alnus rugosa*) and green alder (*Alnus viridis*), and a number of shrubby willows (*Salix* spp.) often dominate in dense stands, especially along stream margins and in wet depressions. To the north, soils are sandy and acidic Spodosols with surface layers of slowly decaying organic matter. These grade into the Alfisols of northern deciduous forests. In the boggy and swampy lowlands, where peaty Histosol soils have accumulated, black spruce (*Picea mariana*), tamarack (*Larix laricina*), and northern white cedar (around the Great Lakes, *Thuja occidentalis*) dominate over sphagnum moss (*Sphagnum*), sedges (*Cyperaceae*), and a variety of shrubs mostly in the heath family (Ericaceae). Also throughout the northern states, and especially on sandy glacial soils, the northeastern pine forests of white (*Pinus strobus*), red (*Pinus resinosa*), and jack pine (*Pinus banksiana*) dominate in a climate that might ultimately support either spruce-fir or northern hardwoods on less droughty soils. These pine forests appear to be sustained on the drought-susceptible, sandy soils as fire-disclimax forests. The spruce-fir and pine forests grade southward into the northern deciduous forests described later. With few common exceptions, the vertebrate assemblage of the spruce-fir and northern pines are dominated by the same cold-tolerant mammals and birds that span the Canadian boreal forest from the western coniferous forests of the

FIGURE 9.22   Eastern deciduous forest in Ohio.

United States. Conspicuous examples include moose (*Alces alces*), lynx (*Lynx canadensis*), snowshoe hare (*Lepus americanus*), red squirrel (*Tamiasciurus hudsonicus*), and spruce grouse (*Canachites canadensis*).

Eastern Deciduous Forest

More than one-third of the land area in the contiguous United States would be covered by deciduous forest (Barbour et al. 1987) (Figure 9.22) in its climax successional state; most of it east of the Great Plains and south of the northern forest. Generally the Alfisol soils are enriched with humus formed from the annual leaf fall and are well suited for cultivation. Annual precipitation in the eastern deciduous forest varies from 625 mm in the northwestern part of its range, where it is commonly associated with reliable sources of water in river floodplains, to as much as 1,875 mm in the southeastern highlands. Precipitation is fairly evenly distributed throughout the year.

Because of leaf loss, the mature deciduous forest dramatically changes in seasonal appearance (Barbour et al. 1987). Under the leafless canopy, shade-intolerant herbaceous plants grow rapidly to early flowering maturity starting soon after the snow melts. As spring progresses, the trees leaf out closing the canopy and a new group of shade-tolerant species replaces the spring plants. In autumn, the tree leaves turn a variety of colors as leaf nutrition ceases and green chlorophyll degrades, leaving more persistent red and yellow pigments behind. The leaves are shed soon after and form an insulating layer of leaf litter over roots and the bulbs of ground-level plants. The period from full leaf formation to first shedding lasts about half of the year.

The species dominance in the eastern deciduous forest ecosystems varies along climatic, elevation, aspect, and soil gradients. The forest can be split into three main east-west bands of forest type adapted to different climatic conditions (Figure 9.20). In the northern forest, the birch (*Betula alleghaniensis*), beech (*Fagus grandifolia*), and maple (*Acer saccharum*) forests are distributed around the Great lakes, southern

New England, the Mid-Atlantic states, and at higher elevations to the south. Common associates include basswood (*Tilia americana*), white ash (*Fraxinus americana*), black cherry (*Prunus serotina*), and the coniferous eastern hemlock (*Tsuga canadensis*), white pine, and pitch pine (*Pinus rigida*). Snow makes up a significant fraction of the precipitation.

In the central forest, a very diverse species assemblage composed mostly of oaks and hickories including white (*Q. alba*), northern red (*Q. rubra*), chestnut (*Q. prinus*), and bur (*Q. macrocarpa*) oak; pignut (*Carya glabra*), shagbark (*C. ovata*), and bitternut (*C. cordiformis*) hickory; walnut and butternut (*Juglans nigra* and *J. cinerea*); yellow-poplar (*Liriodendron tulipifera*); and numerous other rarer tree species. Common understory trees and shrubs include dogwoods (*Cornus* spp.), rhododendrons and azaleas (*Rhododendron* spp.), viburnums (*Viburnum* spp.), hornbeam (*Carpinus caroliniana*), hop hornbeam (*Ostrya virginiana*), sassafras (*Sassafras albidum*), and witch hazel (*Hamamelis virginiana*). Before the chestnut blight early in the twentieth century, American chestnut (*Castanea dentata*) also was a dominant species. The blight is a good example of how quickly the structure of an ecosystem can shift as dominants come and go. This forest type extends up river valleys like fingers into the tallgrass prairie. The soils grade from Alfisols into Ultisols and snow contributes most significantly to precipitation in the north.

Where terrian is fairly flat, large areas of deciduous forest have been cleared (early through middle 1800s) and converted to some of the Nation's best farmland, particularly in the oak-hickory forests of Iowa, Illinois, and Indiana. Most of the eastern deciduous forest has been altered by farming, logging, and industralization. Much of it is in various successional stages as a consequence of past farm abandonment. The activities and succession have favored some wildlife species while discouraging others. While mammals and birds continue to dominate in the deciduous forest, warmer temperatures allow more reptiles and amphibians to survive the winters of the central deciduous forest compared to the cold northern deciduous forest.

Commonly encountered mammals include white-tailed deer (*Odocoilus virginianus*), opossum (*Didelphis virginia*), raccoon (*Procyon lotor*), gray fox (*Urocyon cineoargenteus*), gray squirrel (*Sciurus carolinensis*), cottontail rabbit (*Sylvilagus floridanus*), eastern chipmunk (*Tamias*), and white-footed mouse (*Peromyscus leucopus*), common crow (*Corvus brachyrhynchos*), blue jay (*Cyanocitta cristata*), red-bellied woodpecker (*Melanerpes carolinus*), tufted titmouse (*Parus bicolor*), eastern box turtle (*Terrapene carolina*), five-lined skink (*Eumeces faciatus*), timber rattlesnake (*Crotalus horridus*), the American toad (*Bufo americanus*), and woodland salamanders of the genus *Plethodon*.

Southern Forest

The southern forest is deciduous in its climax stage but is now dominated by southern pines (Figure 9.23). Precipitation averages 1,250 mm or more per year and autumn is the driest season. The Ultisol soils of this area are highly leached, typically acidic, moist much of the year, and underlain with a horizon of fine clay. Snow is incidental. The productivity is the highest of all vegetation types in the United States because of warm temperatures and generally moist soils. However, the area is exposed to periodic drought of intensities that can result in extensive natural burns. Because of their tolerance to fire, southern pine forests may be sustained wherever natural fires are common, creating a patchy mix of coniferous and deciduous forest. But the pine forests are exceptionally extensive now because of widespread abandonment of cotton farms in the last century and controlled burns set to sustain them for forest products.

Following fire, southern pines often dominate in a late stage of succession for long periods before the climax establishment of southern oaks such as cherrybark (*Q. falcate*), scarlet (*Q. coccinea*) and black oak (*Q. velutina*), and hickories such as

FIGURE 9.23   Southern pine ranges are highly productive for wood, wildlife, and livestock. Prescribed burning is used to eliminate excessive accumulation of vegetation and control hardwoods.

mockernut (*Carya tomentosa*) and pignut hickory (*C. glabra*). Most of these species occur in the understory now dominated by pines. Longleaf (*Pinus palustris*) and slash pine (*Pinus elliotii*) dominate at the lower elevations and latitudes. The higher elevations and latitudes are dominated by shortleaf (*Pinus echinata*) and loblolly pine (*Pinus taeda*). Common shrubs include evergreen hollies (*Ilex* spp.) and green-briars (*Smilax* spp.). Pine forests exposed to frequent light fire are generally free of woody understory and support extensive grass cover (Grelen 1978). The common grasses are in the genera *Andropogon, Panicum, Aristida, Paspalum, Sporobolus*, and *Cynodon*.

Many of the species are typical of eastern forests including the widespread white-tailed deer, grey and red fox, raccoon, and opossum. More indicative are the eastern woodrat (*Neotoma floridana*), golden mouse (*Ochrotomys nuttalli*), bobwhite, red-cockaded woodpecker (*Picoides borealis*), brown-headed nuthatch (*Sitta pusilla*), green anole (*Anolis caroliniensis*), eastern glass lizard (*Ophisaurus ventralis*), east-ern diamondback rattlesnake (*Crotalus adamanteus*), oak toad (*Bufo quericus*), and southern toad (*B. terrestris*).

Southern River Bottomland Forests

In the southern river floodplains and other periodically flooded lowlands of the eastern deciduous forests, a mix of "bottomland" species is dominated in the wettest locations by bald cypress (*Taxodium distichum*), a conifer, and the deciduous wa-ter tupelo (*Nyssa aquatic*). Regular river flooding and near-surface groundwater sup-plant the importance of direct precipitation in sustaining vegetation. The soils are a complex mix of Entisols, Histosols, and developing Ultisols. A diverse group of de-ciduous tree species occurs at slightly higher elevations in the floodplain bottom-lands including lowland oaks (e.g., *Q. lyrata, Q. phellos, Q. nigra, Q. laurifolia*), low-land hickories (e.g., *C. aquatica, C. myristiciformis*), cottonwood, sweetgum (*Liquidambar styraciflua*), American elm (*Ulmus americana*), red maple, black tu-pelo (*Nyssa sylvatica*), and other facultative species that tolerate temporary flood-ing, but do not require it. Few vertebrate species are uniquely adapted to southern bottomland forest. One of the most indicative, the ivory bill woodpecker (*Campe-philus principalis*), appears to be extinct.

TUNDRA ECOSYSTEMS

Tundra is a cold, treeless plain in arctic and alpine regions (Figure 9.24) typically underlain with a permanently frozen layer of permafrost. Arctic tundra covers about 5% of the world's surface mostly in North America, Greenland, northern Europe, and northern Asia. Outside of the north slope of Alaska, tundra in the United States occurs only at high elevations. The highest of the biomes, alpine tundra occurs above the timberline (usually above 2,800 m) in mountainous portions of the western United States and Canada, as well as high mountains elsewhere in the world. In the United States, alpine tundra is found primarily in Alaska, Colorado, Washington, Montana, and California, but Oregon, Idaho, Utah, New Mexico, Arizona, and Wyoming have small amounts on the highest peaks.

Alpine ecosystems occupy areas having short, cool growing seasons and long, intensely cold winters. Alpine tundra soils are shallow, and rocky Endisols and boggy Histisols are permeated by permafrost a short distance below the surface. Precipitation on arctic tundra is typically low (250 to 500 mm), but alpine tundra may receive substantially more, mostly as snow. Strong winds further increase the severity of the area for plant life. Even during the summer, plants are stressed by high winds, freezing cold at night, and high ultraviolet radiation during the day. Relatively few species have adapted to these extreme conditions.

Low-growing plants dominate the biotic communities of these ecosystems. The vegetation (Figure 9.24) includes both herbaceous and shrubby vascular plants less than 20 cm tall, extensive mats of mosses (e.g., *Selaginella* spp.), and lichens (*Cladonia* spp.). The alpine tundra flora typically is represented by no more than 200 to 300 species and most of these are widely distributed throughout alpine tundra ecosystems. Members of the bluegrass (*Poaceae*) and sedge (*Cyperaceae*) families occur throughout alpine tundra. Additional groups with wide alpine distribution are the saxifrage (*Saxifragaceae*), rose (*Rosaceae*), mustard (*Brassicaceae*), buckwheat (*Polygonaceae*), and pink (*Caryophyllaceae*) families. The few shrub species are predominantly members of the willow (*Salicaceae*) and heath (*Ericaceae*) families.

In the United States, arctic tundra supports caribou (*Rangifer articus*) in summer and musk oxen (*Ovibos moshatus*) year round. Other year-round residents include

FIGURE 9.24 The alpine tundra type in central Colorado.

arctic fox (*Alopex lagopus*), arctic hare (*Lepus articus*), arctic ground squirrel (*Citellus undulatus*), lemmings (*Discrostonyx* and *Lemmus*), tundra vole (*Clethrionomys rutilus*), willow ptarmigan (*Lagopus lagopus*), snowy owl (*Nyctea scandiaca*), and snow bunting (*Plectrophenax nivalis*). In the alpine tundra, pikas (*Ochotona princeps*), yellow-bellied marmots, and rock ptarmigan (*L. mutus*) are among the more conspicuous vertebrates.

VERTEBRATES IN TERRESTRIAL ECOSYSTEMS

Vertebrate Importance

We have already mentioned a few of the uniquely situated vertebrate specialists that are indicator species of specific ecosystems. Here we briefly describe how some of the more ecologically and socially important vertebrate groups are associated with terrestrial ecosystems. Many of the vertebrates have significant impact on ecosystem function and form, mostly through the interactions between trophic levels (see Chapter 3). Top carnivores control herbivores, which can sometimes overwhelm the plant community when carnivores are decimated. Many of the small but productive herbivores are a food resource of general importance such as the abundant species of rats, mice, rabbits, and hares. Other species impact the abiotic habitat, such as the effects of burrowing and pond-building species. When something exceptional impacts these groups—disease, overharvest, extreme weather—the whole system can be profoundly influenced.

Vertebrate Ranges and Ecological Limits

The ranges of terrestrial vertebrates usually follow ecological boundaries set by the availability of water, temperature, and the intensity and duration of freezing conditions—some of the same factors that determine the distribution of plant species and vegetation form. The range maps of vertebrates provide clues as to how generally or specially adapted they are. Range maps can be misleading, however. For example, some bat species and many bird species have wide-ranging distributions only because they migrate according to season and are not year-round residents facing the rigors of severe winter. No vertebrate species ranges throughout the entire North American continent all year round, but a few genera come close, such as the fox genus (*Vulpes*), the small weasels in the genus *Mustela*, the mouse genus (*Peromyscus*), and the cottontails (*Sylvalagus*). These are all endothermic ("warm-blooded") and relatively small mammalian species able to survive extreme cold by body heat and burrowing into snow, and able to persist through hot desert days deep in the moister recesses of burrows and fissures. Despite their endothermy, the bird genera are not as widely distributed year round, relying much more on migration to avoid extremes. A few highly conspicuous and common bird species span much of the continent south of the boreal forest, however, including the red-tailed hawk (*Buteo jamaicensis*), northern harrier (*Circus cyaneus*), turkey vulture (*Cathartes aura*), and red-winged blackbird (*Agelaius tricolor*).

The ectothermic ("cold-blooded") reptiles and amphibians are excluded from the arctic and alpine tundra by thorough freezing of their aquatic and soil refuges from the cold. The few widespread genera that penetrate the boreal forests, such as frogs (in the genus *Rana*) and gartersnakes (*Thamnophis*), are able to persist in ground refuges below the depth of freezing. Many genera of reptiles and amphibians are widely distributed continentally from East to West Coast—in forest, prairie, and desert—but are limited to more southern latitudes. Good examples include the spiny lizards (*Sceloporus*), skinks (*Eumeces*), whiptail lizards (*Cnemidophorus*), rattlesnakes (*Crotalus*), coachwhip snakes (*Masticophis*), king snakes (*Lampropeltis*), gopher snakes (*Pituophis*), true toads (*Bufo*), and mole salamanders (*Ambystoma*).

Moisture and Surface Water

Some genera appear to be limited to eastern or western conditions, which vary along elevation and moisture gradients. Other than birds, relatively few terrestrial genera are strictly eastern. The box turtles (*Terrapene*) and hognose snakes (*Heterodon*) are two common examples of reptiles that penetrate the semiarid shortgrass prairie but not the arid deserts or cold mountains. Many of the common western taxa do not get beyond the Great Plains including numerous bird genera, and the ground squirrels in the genus *Spermophilus*, the kangaroo rats (*Dipodomys*), pocket mice (*Perognathus*), magpies (*Pica*), western quails (*Callipepla*), earless lizards (*Cophosaurus*), collared lizards (*Crotaphytus*), horned lizards (*Phyrnosoma*), spiny lizards, patchnose snakes (*Salvadora*), blind snakes (*Leptotyphlops*), shovelnose snakes, and numerous other lizards and snakes. The greater diversity of small mammals, snakes, and lizards in the West appears related to the more diverse landscape and mountain barriers to movement. Despite the capability of flight, numerous bird genera sort out as eastern or western. This may have to do with the north-south orientation of most annual migration and the use of landmarks including mountain ranges, which in the United States tend to run north and south.

The distance from unfrozen surface water is another critical feature for many species, regardless of vegetation form. This is especially important for true frogs (*Rana*), tree frogs (*Hyla*), crocodilians, and numerous turtle species, which are often found on land but not very far from permanent water. The true toads and mole salamanders are more thoroughly terrestrial as adults, and certain species even occur in deserts, but they require access to temporary bodies of water to reproduce. The spadefoot toads (*Scaphiophus*) are best adapted among amphibians to arid conditions, but they do it only by burrowing deep in the ground for much of the time. Similarly, the beaver (*Castor canadensis*), muskrat (*Ondatra zibethicus*), mink (*Mustela vison*), otter (*Lutra canadensis*), belted kingfisher (*Megaceryle alcyon*), redwinged blackbird (*Agelaius phoeniceus*), and the gartersnake genus (*Thamnophis*) range widely, but always close to water where their food, cover, and other habitat needs are met. The ranges of other species extend along the riparian corridors into grasslands and even desert, but rarely far from woody vegetation. These species include whitetailed deer (*Odocoileus virginianus*), and mule deer, red fox, striped skunk (*Mephitus mephitus*), tree squirrels (*Sciurus*), rat snakes (*Elaphe*), woodpeckers (e.g., *Picoides*), ravens and crows (*Corvus*), warblers (e.g., *Dendroica* and *Vermivora*), titmice and chickadees (*Parus*), mourning dove (*Zenaida macroura*), and wild turkey (*Meleagris gallopavo*).

Food Distribution

The distribution of food is an important secondary determinant of species distributions. The foxes and weasels could not be active throughout the arctic winter without a good supply of their rodent prey. Similarly, the lynx and bobcat (*Lynx rufus*) could not survive without their major prey—the cottontails, hares, and jackrabbits—which collectively occupy most of the continent. And the prey could not survive without access to lichens (*Cladonia* spp.), dwarf willows (*Salix* spp.), sedges (*Carex* spp.), and other plants under the snow. While endothermic, no terrestrial predatory birds persist entirely on the arctic tundra throughout winter, probably because they hunt by sight and hearing and their food is under snow. And while caribou are the most abundant large mammal on the tundra in summer, they move back to the cover of boreal forest in winter, followed by their major predator, the gray wolf (*Canis lupus*).

Land and Vegetation Form

Also very important is the physical form of the habitat, especially the vegetation form. Basically, vertebrates split up into inhabitants of open grasslands and more closed forest and shrublands. For herbivores, the vegetation form also determines the food form. Species within genera often break out along vegetation boundaries. The foxes split into kit fox (*Vulpes macrotis*) and swift fox in dry grasslands and the red fox (*V. vulpes*) almost everywhere else except deep forest. The deer mouse genus (*Permomyscus*) has many locally distributed species specially adapted to specific land and vegetation form, but often overlapping with the widely distributed and generally adapted white-footed mouse (*P. maniculatus*). Similar examples are found in the widespread shrew genus (*Sorex*), the vole genus (*Microtus*), the cottontails, the warbler genus (*Dendroica*), and the gartersnake genus (*Thamnophis*).

Certain important genera are widespread in open grasslands including the bison, pronghorn, black-tailed jackrabbit (*Lepus californicus*), badger (*Taxidea taxus*), prairie dogs (*Cynomys* spp.), thirteen-lined ground squirrel, black-footed ferret (*Mustela frenata*), dickscissel, eastern meadowlark, and prairie chickens (*Tympanuchus*). In contrast are species that were once linked broadly to most North American forested conditions including the black bear (*Ursus americanus*), gray wolf, the porcupine (*Erethizon dorsatum*), hairy woodpecker (*Picoides villosus*), and great-horned owl (*Bubo virginianus*).

More subtle differences exist between the coniferous and deciduous forest species. Many of the same species of mammals and birds inhabit the Alaskan boreal forest, western, and eastern coniferous forests because their range links through the northern coniferous (boreal) forests. Examples include moose, lynx, fisher, marten (*Martes americana*), snowshoe hare, red squirrel, and red-backed vole (*Clethrionomys gapperi*), the goshawk (*Accipter gentilis*) and sharp-shinned hawk (*Accipter striatus*), spruce grouse, common loon (*Gavia immer*), and gray jay (*Perisoreus canadensis*). Fewer nonmigratory species seem most linked to deciduous forest, probably because food is very scarce in winter. Some species such as the wild turkey, raccoon (*Procyon lotor*), eastern chipmunk, Carolina chickadee (*Parus carolinensis*), and ruffed grouse (*Bonasa umbellus*) extend from deciduous forest into the margins of coniferous forest.

Species Interactions

Species interactions also appear to have an impact on species distributions. For example, the coyote (*Canis latrans*) appears to have been limited mostly to the prairie before wolves were extirpated from much of their original forest distribution. Since then coyotes have moved into much of the forested United States. And of course, humans have had an impact. Black-footed ferrets became endangered as their prairie dog prey were decimated. The elk (*Cervus elaphus*) and mountain lion (*Felis concolor*) are examples of western species all but eliminated from much wider distributions well into the eastern United States.

AQUATIC ECOSYSTEMS

Wetland Ecosystems

Overview. Wetlands form unique ecosystems along hydric gradients between terrestrial and aquatic ecosystems. They are among the most diverse of ecosystem categories (Mitsch and Gosselink 2000). Wetlands range in size from small isolated "pocket wetlands" at groundwater seeps and springs, to huge expanses over thousands of square kilometers, such as the original Everglades of south Florida and the coastal wetlands of Louisiana. Wetlands often are associated with shore zones of lakes, rivers, and estuaries where they form ecotones that bridge terrestrial and aquatic ecosystems. As a consequence, many wetlands are not easy to identify as discrete ecosystems. All wetlands, however, are associated with water at some point in

FIGURE 9.25 Bogs are unproductive wetlands with acidic, peaty soils formed from the slow decay of dominant moss and woody shrubs. In the photograph above, the bog is a floating mat extending into and over the pond. Northern spruce forest is in the background (courtesy of USDI-Fish and Wildlife Service).

the annual cycle, often have hydric soils that differ from nearby upland areas, and exclude the many plant species that cannot tolerate flooded soils.

Wetlands may be permanently covered by water, have permanently saturated soils with no surface water, or be intermittently flooded for only part of the year. Wetlands may be barren of plants (e.g., mudflats), only seasonally inhabited by wetland plants, or permanently and densely inhabited. Wetlands thrive in a wide range of salinity from high marine and arid inland salinity to little more than rainwater concentrations of salts. A great diversity of wetlands occurs in different classes determined mostly by their hydrology, setting in the landscape, water and soil quality, and climate. We refer readers to Cowardin et al. (1979) for detailed coverage of one popular classification scheme of aquatic habitats including wetlands. A common way to grossly categorize wetlands is by whether or not they are vegetated. Vegetated wetlands often are sorted into bog, marsh, and swamp classifications. Wetland plants may be dominated by mosses (*Bryophyta*), sedges (*Cyperaceae*), and woody shrubs in bogs (including muskegs and fens); herbaceous grasses (*Graminae*), rushes (*Juncaceae*), cattails (*Typhaceae*), and other nonwoody plants in marshes; or by larger trees and shrubs in swamps.

Bogs and Fens. Also called peatlands, bogs (Figure 9.25) occur in areas of poor drainage, typically with no drainage outlets and with low concentrations of nutrients in the surrounding watershed. Because of the absence of alkaline carbonates, the water supply to bogs usually is mildly acidic. Calcium deficiency, anoxic soils, and acidity limit the activity of decomposer bacteria causing an accumulation of peat and dissolved organic acids. A common bog plant, sphagnum moss (*Sphagnum* spp.), substantially reduces nutrient and calcium concentration and increases acidity.

Fens form on more calcareous and alkaline sites where groundwater typically is the main water source. The high concentrations of calcium chemically interact with phosphorus to cause both to precipitate from solution and limit production and decomposition. Fens typically support fewer woody plants than bogs and a higher herbaceous plant diversity than any other U.S. wetland. Because of slow decomposition and low

FIGURE 9.26 Salt marshes typically are dominated by cordgrasses (*Spartina* sp.) and occur in protected embayments along the coasts, in estuaries, and in arid inland areas of high salinity (courtesy of Virginia Carter and USGS).

nutrient concentration, bogs and fens are low productivity ecosystems with a high accumulation of biomass and organic matter.

The plants of bogs and fens include numerous members of the sedge, grass, and rush families, and sphagnum moss. Woody plants in northern bog-swamps include the blueberry genus (*Vaccinium*), Labrador tea (*Ledum groenlandicum*), sleep laurel (*Kalmia angustifolia*), black spruce, tamarack, and northern cedar. Adapting to low nitrogen availability, a number of bog-plant species are carnivorous including pitcher plants (*Sarracenia* spp. and *Darlingtonia* spp.), sundews (*Drosera* spp.), and bladderworts (*Utricularia* spp.).

Marshes.   Bogs often grade into marsh communities dominated by herbaceous plants standing in flooded soils or shallow water. Marshes are most characterized by the dominance of emergent and floating herbaceous vegetation rooted in soils that are flooded at least part of the year (Figures 9.26 and 9.27). Although marshes may take root in acidic, peaty sediments, they are commonly encountered in less organically enriched soils and moderately alkaline to alkaline environments. Marshes generally form where ionic concentrations are higher than those that favor bog formation. The diversity of marsh plants tolerating high salinity is rather low. However, they contribute to extensive salt marshes along the ocean coasts and at inland sites associated with salt formations. Coastal salt marshes often extend well up into estuaries where salinities vary with tidal flux.

Some warm-water marshes are among the most productive ecosystems. This is because of the ideal supply of water, aeration, and sedimentary cycle nutrients in rich soils and plentiful light for photosynthesis above water. Decomposition and mineral cycling is rapid. The vertical stem and leaf structure characteristic of marshes is highly efficient in using light at high stem densities. A diverse array of invertebrate and vertebrate species contributes to and depends on the consumer production. Many primary consumers in marshes are detritivores, which feed on marsh plants once they have died. In marine salt marshes, diverse worms, snails, clams, scallops, oysters, crabs, shrimps, amphipods, and isopods are among the more obvious invertebrate species

FIGURE 9.27 Freshwater marshes often are dominated by diverse herbaceous plant species in the grass, sedge, cattail, and rush families, and many other plant species are incidentally represented (courtesy of USDI-Fish and Wildlife Service).

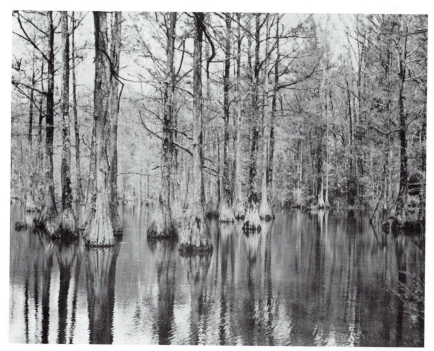

FIGURE 9.28 Swamps are dominated by woody plants such as the bald cypress (*Taxodium distichum*), which occurs throughout the southeastern United States. (courtesy of USDI-Fish and Wildlife Service).

encountered. These support numerous marine bird and fish species. Mammalian and reptile species adapted to feeding on marine invertebrates are more restricted.

Swamps. Swamps are shrub and forest wetlands characterized by emergent woody vegetation rooted in soils that are flooded at least part of the time (Figure 9.28). Swamps often grade into marshes and bogs. Few trees and shrubs have adapted to permanently flooded conditions because their roots require some aeration. A few species have developed areal extension of roots that function as "breathing tubes," such as the "knees" of bald cypress, which form extensive "blackwater" swamps in the southeastern United States. Other than the mangroves, most swamp plants have adapted to neutral or acidic soils inundated by freshwater. Fallen leaves typically contribute to maintenance of acidic sediments and the darkly stained water.

Coniferous swamp species often occur in association with bogs in the Great Lakes region and along the Atlantic coast. Bottomland hardwood swamps often occur adjacent to marshes in the seasonally flooded areas of riverine floodplains. The trees and shrubs in bottomland hardwood swamps survive flooding at certain times of the year, but typically do not tolerate persistent flooding. Although swamp plants are not particularly diverse, swamps often support abundant and diverse consumer populations. This diversity is associated with moderate productivity and the complex structural mix of water, islands, and adjacent uplands that typify swamp landscapes.

In semitropical to tropical areas, mangrove swamps (tidal woodlands) frequently dominate the shores (Figure 9.29). Among woody species, only the mangroves have

FIGURE 9.29 Mangrove swamps occur in semitropical and tropical estuaries with waters of variable salinity. They are common in south Florida (courtesy of Richard Frear and the National Park Service).

successfully adapted to saline waters. Because they cannot tolerate freezing, yet can survive for long periods at sea, mangroves are circumtropical in distribution along marine coasts and estuaries. Mangroves occur along protected coasts with low tidal flux; often behind barrier islands and in seawater somewhat diluted by freshwater inflows. Mangroves are composed of a number of diverse species from different plant families that have developed parallel adaptations to estuarine conditions. Some of these include waxy leaves to cut freshwater loss, salt pores for secreting salts through the leaves, and specialized roots that extend areal taps to the atmosphere above the anoxic muds. Mangrove species typically attain heights of no more than 10 to 15 m and support a few estuarine species in great abundance. Leaf fall from the mangroves is the basis for a detrivore community of oysters, crabs, shrimp, and a variety of fish species, some of which, such as snook and tarpon, are very popular sport species. A variety of large birds (e.g., storks, egrets, herons, spoonbills, ibises) rely on mangrove communities as sources of food and for roosting. Mangroves are the primary refuge for the American crocodile, an endangered species.

Nonvegetated Wetlands. Some important wetlands are not occupied by plants. Mudflats are extensive and important benthic habitats. Many species of worms and clams burrow in the mud of coastal mudflats and provide food resources for shore birds and other predators. Barren zones of high salinity above the tide mark where salts accumulate and create toxic conditions for plants. Numerous temporary shallow ponds that form after intense rains in arid regions support no plant growth. However, many hold the desicated eggs of adapted crustaceans, such as the fairy shrimps (*Anostraca*). They hatch upon inundation and complete their life cycle within a few weeks before the ponds dry up. Although temporary, these shallow ponds can be quite productive for the time they are inundated.

Wetland Functions. Wetlands form zones of habitat convergence for species with both terrestrial and aquatic attributes such as semiaquatic birds, mammals, and reptiles; some molluscs and crustaceans; and most amphibians. They also support, either directly or indirectly, a diverse array of fully terrestrial and aquatic species. Because of their position between land and sea, wetlands act as areas of exceptional material and energy flux. While they act to trap limiting nutrients, such as phosphorus, they also export carbon and other nutrients in the form of detrital organic matter. They are important in material cycling, especially sulfur and carbon. Peatlands are especially, important "sinks" for carbon and contribute to regulating greenhouse gasses. Wetlands also dampen hydraulic energies associated with waves and currents. Therefore, they decrease flood damage and erosion, trap suspended particles in sediment, and contribute to water clarity and offshore productivity. By acting like sponges for retaining water, some wetlands not only dampen floods, but also can contribute to groundwater recharge.

Lake Ecosystems

General Attributes. Lakes seem easy enough to classify and place boundaries on, yet they often grade into wetland, riverine, and oceanic environments in ways that make their definition more difficult than at first thought. Many people think of lakes as distinct from oceans because their water is fresh, but numerous lakes are saline, such as the Great Salt Lake in Utah. Waters like the Caspian Sea in Asia confound those who do not know whether to class it as the largest lake or the smallest ocean. If we define lakes as bodies of inland water disconnected from the oceans and with different salt compositions, then the Caspian Sea must be considered the largest lake. Lake Baikal, also in Asia, is considered the largest freshwater lake by volume. Lake Superior is considered the largest North American lake. At the other extreme of size, lakes include very small waters, often referred to as ponds (Figure 9.30). Wetzel (2000), Cole (1994), and Horne and Goldman

FIGURE 9.30   Lakes include a diverse array of aquatic environments typified by impoundment behind a natural or artificial barrier to water flow (courtesy of President's Council on Recreation and National Beauty).

(1994) provide excellent texts on limnology, the science of lakes and other inland waters.

Lakes are "flat water," formed behind a natural or artificial barrier to water flow with surfaces at the same elevation throughout. Wind is the main source of current and mixing. In contrast, river surfaces follow a slope in elevation and flow in a consistent direction because of the force of gravity. However, on closer inspection, lake surfaces vary in elevation as wind, rain, tributaries, and air pressure changes create currents and a somewhat uneven surface. These variations may be as great as several meters from one end to another in large lakes exposed to strong and consistent winds.

Lake Origin.   Lakes may be artificially built impoundments—reservoirs behind dams or lakes excavated by mechanical means. Natural lakes were formed in many ways, but most originated from the action of glacial ice. Glacial lakes dot the Northeastern United States, high mountains of the West, and much of the landscape of Canada and Alaska. Floodplain lakes (lakes created in isolated sections of old river channels) also are quite common as are lakes backed up behind beach dunes along large lake or ocean shores. Other lakes are caused by the gradual splitting of the earth's crust associated with plate tectonics. These are among the deepest lakes, such as Lake Baikal, and Lake Tahoe in the United States. Many temporary lakes form in wind-eroded depressions, or playas (pans in Africa), in arid and semiarid plains. They typically are seasonal and dry up before the rains or snowmelt returns. Volcanic craters sometimes form lakes, such as Crater Lake in Oregon. A number of other less-common origins have been documented and some origins remain a mystery.

Biotic Communities.   Except at the shallow edges, lake communities are dominated by phytoplankton and zooplankton, many of which are microscopic. Although most plankters have evolved ways to adjust position in the water column, they are, for the most part, at the mercy of wave turbulence. Phytoplankters vary in size; the larger ones being more common in nutrient-rich waters and when it is warm. A predictable succession of smaller to larger species occurs as the seasons warm and cool. A similar succession of zooplankters changes occur, with smaller species dominating in winter and larger ones in summer. However, plankton-feeding fish and invertebrates can markedly influence the sizes of the plankton present because they tend to select the larger plankters for food.

One group of phytoplanktonic bacteria, the Cyanobacteria (blue-green algae), include large species that fix atmospheric nitrogen and produce toxins. Where these nuisance algae dominate in warm nutrient-enriched water, the efficiency of herbivores in consuming phytoplankton is reduced. They add to the accumulation of organic matter that settles to the bottom and imparts a bad taste to drinking water. In part because phytoplankton production often is most controlled by phosphorus, water treatment focuses on phosphorus removal to reduce cultural eutrophication.

Exchange Rate and Drainage. Lakes exchange water at vastly different rates. Large lakes, like Lake Superior in northeast North America, would take over 140 years to drain. Many small reservoirs would drain in a few days. Lakes with an exceptionally slow exchange rate typically have low nutrient concentrations and low productivity. Deep lakes with high exchange rate, however, may have their plankton production and dependant fish production limited by high flushing rate. Therefore, lakes of intermediate exchange rate appear most productive for most species of sport and commercial fishes. All but a few natural lakes drain from the surface. Many reservoirs drain from depths below the surface. Reservoirs built for irrigation supply typically can be drained from the bottom.

Stratification and Trophic Status. Deep enough lakes stratify seasonally and cease to mix top (epilimnion) and bottom (hypolimnion) strata. Some lakes stratify permanently because of high salt concentrations in the bottom stratum. Lakes with a greater amount of settled organic matter than oxygen to decompose it will become anoxic if stratification lasts long enough. This causes numerous environmental changes that affect the aquatic community, often in undesirable ways from the standpoint of resource use. Oligotrophic lakes, with low nutrients and low production, sustain oxygen concentrations at saturation levels determined mostly by water temperature. Eutrophication results where production of organic matter exceeds the supply of oxygen needed for decomposition (see Chaper 4). If the lake is a reservoir with bottom release, hypolimnetic oxygen depletion can extend far downstream in the river below the dam, interfering with productive use.

Watersheds. Although lakes can be thought of as ecosystems separate from their tributaries and watersheds, aquatic ecologists have recognized for many years that the nutrients and sediments entering lakes come from the watershed. When watersheds are low in nutrients and fine-particle soils, as are some sandy regions, lakes remain oligotrophic. Other watersheds are rich in nutrients and are natural nutrient sources for eutrophic lakes. Human activity in watersheds has accelerated nutrient and sediment input through many different routes, but especially through crop culture, fertilization of yards, and domestic waste elimination. This cultural eutrophication often degrades lake resources by causing greater water treatment costs, less desirable fish species, and nuisance algae and plant growths.

Stream and River Ecosystems

General Stream Attributes. The key attribute of streams and rivers is consistent downslope flow caused by gravity. Streams vary in size from step-across origins to the largest rivers of the world. Streams flow in channels of their own making during times of high flow and at low or base-flow streams often occupy a small fraction of the channel or dry up entirely. Perennial streams occupy at least some portion of their channel all of the time (Figure 9.31). Intermittent streams leave a channel dry part of the time. As aridity increases, the extent of dry channels contributing to drainage systems increases, forming extensive networks of arroyos in the deserts, which flash flood temporarily during storms. Driven by gravity and the water mass, streams act as an erosive force on the channel. As the channel gradient (slope) becomes steeper and the

FIGURE 9.31 Streams interact intimately with the landscape they flow through. In forested areas, woody debris is as important as rock and fire sediment in defining the habitat (courtesy of USDA-Forest Service).

depth increases, streams become more turbulent and more capable of digging into the bottom substrate and moving the loosened material downstream.

Floods and Floodplains.   Most stream erosion and channel changes occur during floods. Floods are caused by extreme rain or snowmelt events, which increase the depth, turbulence, and erosion in the channel. As the channel is topped and flood-water moves over the floodplain, it loses some of its turbulence and capacity for carrying fine sediment, which settles on the floodplain. Floods also erode new channels and leave disconnected channel depressions, which form lakes or wetlands. These often are reconnected during seasonal flooding, sustaining reproductive sites and nurseries for many species of fish and other organisms.

Organic Detritus and Decomposers.   The stream community under dense riparian vegetation depends on the influx of organic matter and its colonization by bacterial and fungal decomposers. This organic detritus is consumed, decomposers included, by relatively large invertebrates such as stoneflies (Plecoptera), caddisflies (Trichoptera), and crane flies (Tipulidae; Diptera). They break the large leaf material down to smaller particles, mostly through generating undigested fecal pellets. These are carried downstream to be consumed by other invertebrates adapted to filtering or otherwise gathering up the fine organic matter. Much of the nutrition provided by detritus is in the form of the colonizing bacteria and fungi. Without the decomposers, the organic matter would be virtually useless calories lacking sufficient nitrogen and other nutrients to support the needs of the consumers.

Temperature.   Streams and rivers often are classified by their water temperature into cold-water and warm-water streams. In areas of high relief, cold-water streams at high elevations contribute to warm-water rivers at lower elevation. Of course temperature also changes along a continuum and there are intermediate "cool" streams as well, causing variation in the details of this simple temperature-based classification scheme. The variation in temperature also is important. Streams that gain much of their water from groundwater sustain more nearly constant temperatures because of the insulating properties of the earth. A few streams originate in hot springs, which derive their heat from volcanic sources. These typically support few species, but may have luxuriant growths of certain algae and bacteria that tolerate the heat.

Water Chemistry.   Streams also can be classified by their chemistry. Saline streams occur in regions of extreme aridity and geologies of soluble salt deposits. There are limestone streams high in bicarbonate alkalinity and carbonate formations known as travertine. These streams often are rich in inorganic nutrients because the soluble carbonate rocks often are associated with other nutrients. Streams flowing from regions with rock of low solubility, such as granites and schists, usually have low inorganic nutrient fertility. These streams may become acidic blackwater streams stained dark by dissolved organic matter much like bog ponds.

Geomorphology.   Streams also may be classified by physical shape and structure, including channel gradient, width, and meandering. Related to gradient, bottom types include bedrock, boulder, cobble, gravel, sand, silt, and clay. The shape of the valley is important, varying from steep-sided deep valleys to broad floodplains in shallow valleys. Streams also may be categorized by the degree of artificial channelization that has been imposed on them (e.g., navigation channels, irrigation ditches).

The River Continuum.   Many species of stream organisms have come to depend on the flow of materials carried down river along the river continuum. Other species migrate among the different river and floodplain habitats, and some, such as the West Coast salmon species (*Onchorynchus* spp.), move between headwaters and oceans to complete their life cycles. The river continuum concept of Vannote et al. (1980) integrates understanding of geophysical and biological process into a theory of stream and river changes as small tributaries converge into progessively larger rivers. Many stream species have adapted to the continuum of river flow. The interruption of material transport and migratory movement by dams, levees, pollutants, and other barriers has greatly altered stream ecosystems over little more than a century of water resource development. The extent to which this has contributed to the degradation of ecosystem services to humans and especially the endangerment of species, is an active area of scientific exploration.

Freshwater Aquatic Vertebrates.   In the United States, the greatest diversity of aquatic species is associated with streams and rivers. They are least diverse in fast-flowing cold-water streams and rivers, which are dominated by just a few genera in the salmon and whitefish family (*Salmonidae*), including the West Coast salmon and trout species (*Oncorynchus* spp.), chars (*Salvelinus* spp.), arctic grayling (*Thymallus articus*), Atlantic salmon (*Salmo salar*), and whitefish in the genus *Prosopium*. Two other families that include mostly cold-water stream species are the sculpins (*Cottus* spp.) and the smelts (*Osmerus* spp. and *Hypomesus* spp.). A few species of other families are mostly limited to cold-water riverine conditions including certain sturgeons (e.g., *Acipenser medirostris*), longnose sucker (e.g., *Catostomus catostomus*), lamprey (e.g., *Lampetra japonica*), and minnows such as longnose dace (*Rhinichthys cataractae*), pearl dace (*Margariscus margarita*), and lake chub (*Couesius plumbeus*). Few reptiles and amphibians are adapted to very cold flowing waters.

Fish become much more diverse in the warmer flowing waters of the United States. The two most diverse families are the minnows (Cyprinidae) and darters (Percidae), most of which are adapted to flowing waters. The minnows are predominantly small fish with a large fraction in the genus *Notropis*, but the pike-minnows

(*Ptychoceilus* spp.) attain large size in the Colorado, Sacramento, and Columbia Rivers. Other important riverine genera include the sturgeons (*Acipenser*), paddle-fish (*Polydon*), gars (*Lepisosteus*), freshwater eel (*Anguilla*), herrings (*Alosa* and *Dorosoma*), pikes (*Esox*), pike-perches (*Stizostedion*), temperate basses (*Morone*), black basses (*Micropterus*), rock basses (*Ambloplites*), suckers (*Catostomus* and *Moxostoma*), catfishes (*Ictalurus*), madtoms (*Noturus*), and sculpins. The softshell turtles (*Trionyx*), snapping turtles (*Chelydra*), and stinkpot (*Stenotherus*) are the turtles most adapted to warm rivers and streams with some current. Water snakes in the genus *Nerodia* are common riverine species. Many salamanders and a few frog species are adapted to rapidly flowing streams. The tailed-frog (*Ascaphus truei*) lives in cool streams of the Northwest. The hellbender (*Cryptobranchus alleganiensis*), which is the heaviest U.S. amphibian, is completely aquatic and lives in medium-size cool-water rivers. Numerous salamanders live as larvae in cool springs and streams including species in the genera *Pseudotriton*, *Eurecia*, and *Desmognathus*.

Lake and pond vertebrate species exhibit less diversity than in streams and rivers, from which most evolved. The cold-water glacial lakes of the United States in both the East and West are dominated by the lake trout (*Salvelinus naymacush*) and white-fishes in the genus *Coregonus*. The only freshwater member of the cod family, the ling cod (*Lota lota*) is adapted to cold lake waters. The lake sturgeon (*Acipenser flavescens*) and northern pike (*Esox lucius*) are also common in cold lakes. All of the above fish are common in the upper Laurentian Great Lakes. Reptiles are not well adapted to cold lakes although the common garter snake often enters them in summer. The mountain yellow-legged frog (*Rana mucosa*) inhabits alpine lakes in the Sierra Nevada Mountains. It is one of the many amphibians that appear to be in decline.

Before reservoirs were built, the warm lakes and ponds of the United States usually were formed from riverine and coastal processes. Therefore, most of the cool- to warm-water lake species derive from riverine groups. Among the best adapted to warm-water lakes and ponds are the gars (*Lepisosteus*), bowfin (*Amia calva*), shad (*Dorosoma*), chain pickerel (*Esox niger*), minnows in the genera *Notemigonus*, *Pimiphales*, and *Notropis*, suckers in the genera *Ictobius*, *Carpiodes*, *Xyrauchen*, and *Catostomus*, bullheads (*Ameiurus*), sunfishes (*Lepomis*), black basses, crappie (*Pomoxis*), pike-perches, and yellow perch (*Perca flavescens*). A wide variety of turtles inhabit warm-water lakes and ponds, but sliders (*Psuedemys*), painted turtles (*Chrysemys*), snapping turtles, mud turtles (*Kinosternon*), and map turtles (*Graptemys*) stand out. Among amphibians, many of the true frogs (*Rana*) are adapted to permanent warm ponds and small lakes. The mudpuppy (*Necturus maculosus*), a completely aquatic salamander, commonly occurs in lakes and ponds as well as streams in the eastern and central United States.

Freshwater marshes are attractive to muskrats, nutria (*Myocastor coypus*), especially among mammals, as well as rice rats (*Oryzomys palustris*), marsh rabbits (*Sylvilagus palustris*), meadow voles, and meadow jumping mice (*Zapus hudsonius*). Mink, otters, and raccoons commonly frequent swamps and marshes. Many species of ducks in the genus *Anas* reproduce in marshes as do redwing blackbirds, yellow-headed blackbirds (*Xanthocephalus xanthocephalus*), and marsh wrens (*Cistothorus* spp.). The rails, coots, and gallinules (Rallidae), and herons and bitterns (Ardeidae) typically inhabit marshes and swamps. Virtually all of the aquatic turtles of lakes and rivers also inhabit marshes or swamps. Especially in southern wetlands, the water-snakes, cottonmouth, ribbon snake, rainbow snake (*Abastor erythrogrammus*), mud snake (*Faranica abacura*), black swamp snake (*Seminatrix pygaea*), and crayfish snake (*Regina rigida*) commonly feed on amphibians and fish.

The frogs in the family Ranidae are typically associated with wetlands and most frogs and toads require wetlands for reproduction. The two-toed amphiuma (*Amphiuma means*), greater siren (*Siren lacertina*), and lesser siren (*S. intermedia*) are eel-like amphibians that lead entirely aquatic lives in southern swamps and marshes. Many of the warm-water lake fishes move into wetlands to feed and often to reproduce (see examples given earlier in this section), but relatively few fish species complete their life cycles in wetlands. The most common are small species including the top

FIGURE 9.32   Riparian vegetation grows in the floodplain adjacent to streams and rivers. It usually depends on groundwater and often on periodic flooding (courtesy of Bureau of Land Management).

minnows (*Fundulus* spp.), mosquitofishes (*Gambusia* spp.), mudminnows (*Umbra* spp.), the small sunfishes (*Enneacanthus* spp.), the pirate perch (*Apredoderus sayanus*), and the swampfish (*Chologaster cornutus*). The greatest diversity of all wetland fish and amphibians is in coastal freshwater marshes and swamps.

Few vertebrate species are linked closely to bog waters, many of which have closed drainages. Where associated with swamps, the mudminnows and small sunfishes enter bogs along the Atlantic Coast. Reptiles include the rare bog turtle (*Clemmys muhlenbergi*) and the spotted turtle (*Clemmys guttata*). The most commonly encountered amphibians are mink frogs (*Rana septentrionalis*) and wood frogs (*R. sylvatica*).

Riparian Communities.   Where floodplains occur and groundwater is close to the surface, a riparian plant community develops different from that on adjacent higher lands and wetlands (Figure 9.32). Although the soils of floodplains tend to be undeveloped Entisols with complex mixtures of cobble, sands, clays, and organic matter, nutrient-rich sediments typically are left behind following floods. The combination of permanent water below the surface and nutrient-rich soil makes them especially productive. Riparian communities are most obvious where forest communities extend into areas dominated by grasslands or desert shrub in the upland areas adjacent to the floodplain. These riparian forest communities, however, are not typical of the upland communities in mesic forests. Only certain species become particularly abundant in the floodplain. These species typically are not drought tolerant and tolerate some exposure to flooding, but are not quite as tolerant of wet conditions as the few swamp species described under wetlands. Typical members of riparian communities are box elder, sycamore (*Platinus* spp.), cottonwood (*Populus* spp.), willow (*Salix* spp.), green ash (*Fraxinus pennsylvanica*), red maple (*Acer rubrum*), alder (*Alnus* spp.), and lowland oaks and hickories. The Southern bottomland forests are also riparian. Riparian communities are typically quite dense with limited shade-tolerant growth of herbaceous flowering plants. Most grasses and other shade-intolerant species are common only in openings where flood events have torn out trees, and along the river margins. In some locations dense impenetrable growth of shrubs becomes established.

The introduced salt cedar (*Tamarix chinesis* spp.) has become dominant in many southwestern riparian communities following the failure of native species to reproduce after cutting and grazing. Salt cedar develops exceptionally thick stands. They have a high salt tolerance and concentrate salts in the surface soil as their leaves drop and decay. The salty soils inhibit growth of native plants. Native western riparian communities frequently had substantial herbaceous development of grasses and forbs. These areas can have high understory productivity before the invasion of salt cedar and, in association with shade and water, are often exceptionally attractive to range livestock. Many riparian areas have been greatly altered by grazing and a number of riparian restoration efforts, done mostly by fencing out cattle, have demonstrated that impact (Holechek 2001).

Many species of birds, amphibians, and mammals are associated with riparian communities with water in close proximity. Because they form edge communities with the adjacent uplands, the diversity of species typically is highest in riparian areas. Riparian communities in the semiarid and arid parts of the country contribute less than 5% to the total of all vegetation types and support several times that fraction of the diversity. Because riparian communities are more productive, wildlife are usually most abundant there. In many locations riparian areas formed corridors that numerous species travel through under protection of more dense growth than occurs in upland areas. With widespread grazing, crop culture, road development, building construction, dams, levees, and other development, much of the original riparian communities have been altered significantly and greatly fragmented into disconnected patches. The impacts are not entirely well understood, but there is increasing interest in restoring corridors, especially where endangered species might recover as a result.

Estuarine Ecosystems

General Attributes.   Estuaries form in the transition zone (ecotone) where rivers enter the ocean. They are characterized by salinity gradients, predictable tidal flux, and aquatic species that can tolerate sometimes dramatic changes in water elevation and salinity. Because saline ocean water is more dense than freshwater, rivers flow over the seawater forming a saltwater wedge with a relatively sharp boundary. In addition to salinity change, tidal flux often results in large expanses of estuary bottom being exposed to variation in temperature and nutrient (Figure 9.33).

The shapes and dynamics of estuaries are quite variable depending on the coastal landform. A rise in sea level inundated terrestrial landscapes and extended into river valleys following the melting of the last continental glaciers more than 10,000 years ago. Where rising seawater inundated flat coastal plains with broad river valleys, mixing became more graduated and formed more extensive esturine transition zones. Many estuaries have become more shallow from the accumulated deposits of river-transported sediments, which settle out on flat gradients. In many locations, adjacent estuaries merge and contribute to larger estuaries such as Chesapeake Bay, or form long lagoons behind barrier beaches. Where the shore steeply rises, as it does along much of the West Coast of North America, estuaries tend to be more confined. In those situations, river channels tend to be steep and river mixing with the ocean is rapid, limiting the extent of estuary development. Glacial scouring of river valleys adjacent to the sea also created deep fjord estuaries such as those in parts of Canada and Scandinavia.

Barrier Islands and Estuaries.   Formation of coastal barrier islands in shallow coastal environments contributes to constrained mixing and extension of estuarine characteristics. Coastal barrier islands are a prominent landform along the flat Mid-Atlantic coastal plain and the Gulf of Mexico, where they baffle the mixing of river and sea and extend estuarine environments along the coastline in lagoons behind the barrier islands. Where coasts are flat and rivers and barrier islands are common, the mix of ocean and freshwater can become quite complex. Because barrier islands are dy-

FIGURE 9.33 Tidal flux in estuaries often exposes large expanses of estuary bottom as shown here in this East Coast estuary where the water is fresh (courtesy of U.S. Fish & Wildlife Service).

namic and frequently breached and moved by storms, the circulation of coastal plain estuaries is quite variable. Extensive dredging to build and maintain an extensive system of canals and harbors has added to the dynamics. The barrier islands are quite unstable structures, which are continuously moving inland as wind and sea mobilize sand. Large storms often open channels through the islands and alter mixing. Behind the barrier islands, depending on tributary inflow and evaporation, salinities can vary from higher than ocean concentrations to much lower.

Estuarine Communities. Sea grasses of several genera (e.g., *Zostera, Thalassia, Phyllospadix*) have adapted to the protected lagoons behind the barrier islands and to less erosive environments along exposed coasts. The open-water lagoons often are bordered by salt marshes. The lagoons often are rich sources of molluscan (oysters, clams, scallops) and crustacean (shrimp and crabs) fisheries as well as productive recreational fisheries for a variety of fin fishes in the sea bass, flounder, and temperate bass families.

Shallow estuaries typically are very productive aquatic environments, but species diversity is limited for a number of reasons. Diversity is limited to species that have been able to adapt to the widely varying and physiological stressful environmental flux associated with the tides. However, the productivity of the successful species makes them particularly abundant and attractive as human resources, including various species of clams, oysters, shrimp, crabs, and fish. Shallow estuaries owe their productivity to plentiful light for photosynthesis, efficient trapping of fine organic sediments washed out of their source rivers, and a fresh supply of nutrients carried in on daily tides. They also are reliably flushed by waste-removing tides, which also sustain high oxygen concentrations.

Landform is critical to determining the productivity of estuaries. Deep fjord estuaries are less productive because they trap nutrient-rich sediments below the most illuminated waters where photosynthesis is least limited. Shallow estuaries trap sediments at depths where they can be successfully colonized by tolerant algae and vascular plants, such as submerged sea grasses (e.g., *Zostera marina*) and emergent salt marsh

FIGURE 9.34 Salt grass communities, shown here, and submerged sea grass communities are important components of shallow estuarine ecosystems, providing physical habitat and food for many estuarine species (courtesy of USDI-Fish and Wildlife Service).

grasses (*Spartina* spp.). The grasses are consumed directly by few organisms (Figure 9.34), but they become substrate for attached algae and both dead grass and algae contribute to huge exports of detrital organic matter, which supports many of the bottom invertebrates in estuary and coastal waters and indirectly supports many fish species and birds dependent on estuarine food webs. Crabs, isopods, amphipods, and snails feed directly on the decaying plants. Worms, clams, scallops, oysters, and shrimp filter fine organic particles suspended in the water.

Marine Ecosystems

The Oceans.   Although oceans cover nearly three-fourths of the earth's surface, the oceanic depths are among the least well defined of all ecosystems. Oceanic environments are distinguished from estuaries and inland saline lakes by a relatively uniform salt concentration varying near 35 g/liter and composed mostly of NaCl. The different communities and ecosystems are defined as zones, which exhibit gradual transition along gradients determined primarily by depth and proximity to bottom. The *benthic,* or bottom zone, is differentiated from the *pelagic,* or open-water zone. The pelagic waters are further divided into the fertile, nearshore waters, or *neritic* zone, and the typically infertile, illuminated waters offshore, the *oceanic* zone. The neritic zone and oceanic zone together compose the photic zone. Below the oceanic zone is the aphotic zone, which includes all water deeper than about 200 m. The benthic communities change along a depth gradient from the intertidal zone nearest shore, through the continental shelf, and along the continental slope into the deep abyssal zone (Figure 9.35).

The *benthos* compose all organisms that live in, on, or just above the bottom. The *plankton* and *nekton* inhabit the water column above the bottom. The benthos frequently use life strategies that sink or anchor them to bottom. In contrast, many planktonic and nektonic species have adapted to open water for completion of their entire life cycles based on the capacity of water density to suspend biomass. Whereas the

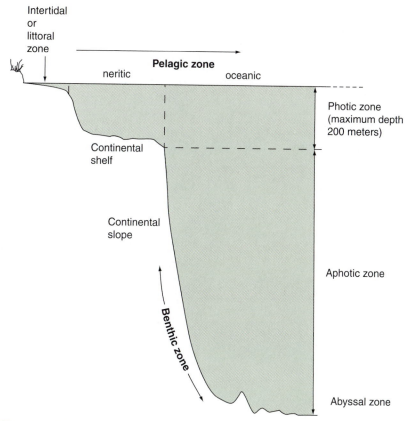

FIGURE 9.35 Major life zones in an ocean ecosystem (from Lerman 1986).

plankton depend greatly on currents for movement, the nekton are strong swimmers, including most of the fin fishes and all other marine vertebrates. Oceanic depth is an important indicator of physical variables determining the character of oceanic systems. Most importantly, the depth of water determines the amount of light available for photosynthesis and sight-dependent ecological processes, the amount of heat concentrated (temperature) from absorbed light, and the depth to which heat is mixed by the wind.

Many planktonic species can control density enough to control their position in the water column with little expenditure of energy. The bodies of most marine organisms are more than 90% water and their density usually is quite similar to seawater. Small variations in body density make the difference between floating upward in the water column or sinking. Many of the plankton are especially watery creatures, which can adjust body density through control over salt content, gas vacuoles, fat content, body shape, subtle motion, or other means. Many fish species have gas bladders that allow them to adjust their density. Most bottom fish, on the other hand, have no effective floatation. Many of the benthos take up dense minerals in carapaces and shells that weight them to the bottom. Others adapt behaviorally by wedging themselves into recesses and burrows much of the time. A third group secretes glue-like holdfasts to rock and other solid substrates.

Productivity typically is greatest in the illuminated neritic zone near shore where nutrient concentrations from continental runoff are higher than in the offshore oceanic zone (Figure 9.36). In some locations, deep onshore currents cause nutrient-rich upwellings at the shelf edge where they often support productive fisheries. A diverse array of phytoplanktonic species form the photosynthetic base for the food web. Copepods are common microcrustacean zooplankton that feed on the phytoplankton, but many other species are zooplanktonic, especially in early life stages. Krill is an important larger zooplankton crustacean, which sustains many fish, birds, and whales. Offshore, the oceanic zone is divided into several zones based on depth

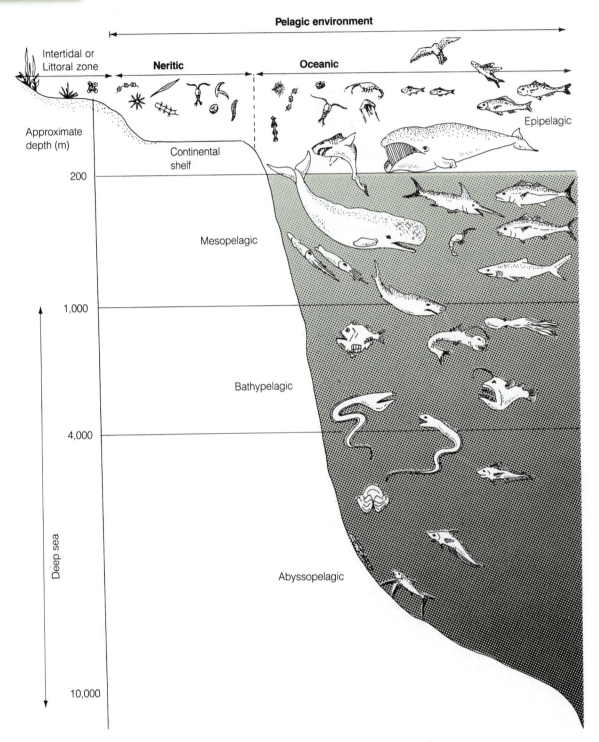

FIGURE 9.36   Stratification of the pelagic environment (from Lerman 1986).

and light penetration. The epipelagic zone offshore has similar planktonic diversity but the plankton typically are less dense than near shore because of lower offshore productivity. Many of the largest vertebrates (whales, whale sharks) live here; however, where they feed on the plankton by filtration. Just below is a dark aphotic zone where many nektonic fish, shrimp, squid, and other species retreat during the day to return to the epipelagic zone to feed under cover of darkness. In the perpetually dark, deep abyssal zone, life exists in continuous cold at very high pressure. Adaptation to the pressure restricts upward movement of the diverse deep sea fishes and invertebrates, which must maintain a pressure equilibrium.

The benthos of the deep oceans live below the disturbance of tides and winds in zones of consistently cool and dark water. The particle size of the bottom sediments tends to become progressively smaller with distance from shore, but is patchy. These are stable environments. The main limitation seems to be food, which comes from detrital fallout and decreases in amount farther offshore as the oceanic zone becomes less fertile. Although the density decreases with depth, the diversity of benthic species increases with depth, probably as a consequence of a long enough evolutionary history to become highly specialized in the stable environment. In specializing, they have evolved several feeding strategies: filtering suspensions from the water, feeding off deposits of particles, absorbing dissolved organic matter, scavenging, and preying on other species. A specialized benthic community inhabits areas around hot volcanic vents where water is warmed locally and bacteria production is high based on chemosynthesis using volcanic elements. A very productive community of tube worms, crustaceans, molluscs, and other species depends on the bacterial production.

Intertidal Ecosystems. The character of the intertidal community (Figure 9.37) in seawater depends firstly on whether the bottom is sandy beach, muds, or consolidated rock and other substrate. On rocky shores, the diversity of marine species increases with consistency of water coverage. In the upper tidal zone the common species are periwinkle snails (*Littorina*) and limpits, which scrape algae from the rocks. In the middle intertidal zone diversity increases, but barnacles and mussels often dominate the space by attaching in fixed positions on the rock surface. They are able to filter the water of detrital and planktonic food particles while submersed. Macroalgae, much of it rockweed *(Fucus* spp.), also share this space with a variety of crab, snail, starfish, and other predaceous species that feed on the mussels, barnacles, and other crustacean,

FIGURE 9.37 In the intertidal zone the diversity of living organisms increases with distance offshore as the amount of time water covers the bottom increases. Here, oysters (*Crassostiea* sp.) are the dominant surface organisms. Many other species are buried in sediment (courtesy of U.S. Fish and Wildlife Service).

molluscan, and echinoderm species. The lower intertidal zone is most diverse including a number of kelp and other macroalgae species, sea slugs, bryozoans, worms, tunicates, anemones, urchins, crabs, and other crustaceans.

Kelp "forests" inhabit the lower intertidal and subtidal zones of cool western shores of continents. In the United States, kelp occurs mostly along the California coast. Kelp are dominant macroalgae that can stretch 25 m or more from bottom to surface. Kelp communities also form very productive marine ecosystems, in part because they tend to be most developed in regions of frequent nutrient upwelling from deep, nutrient-rich sources. The upwellings are caused by offshore winds that drive deeper water to the surface. These also are areas of exceptionally high phytoplankton productivity and concentrations of planktivorous fishes, many of which form important fisheries. Kelp communities are sustained by interaction between kelp feeders, such as sea urchins, and predators on sea urchins, such as sea otters. Where this interaction is disrupted, such as it was when sea otters were once overharvested for furs, the urchins and other herbivores overwhelm the kelp production.

The beach community is less diverse than rocky-shore communities because sand is much less stable, physically more uniform, and sustains less primary production. On the upper beach, a few crustacean species are adapted to feeding on the organic flotsam washed ashore. These include burrowing amphipods and crabs. The intertidal community is more diverse including small burrowing clams, annelid worms, and mole crabs, and very small invertebrates that live in the spaces among the sand grains. These interstitial organisms feed on the fine detritus that is caught in the beach sands from the wave wash. A number of shorebirds are adapted to feeding on beach invertebrates. In the subtidal zone a variety of crustaceans, echinoderms (sand dollars), molluscs, and fish species live on sandy bottoms.

Coral.   Coral reefs are tropical marine ecosystems of warm and well-illuminated seawater, which exist in association with the limestone formations of living organisms, the coral polyps of the phylum (*Coelenterata*) (Figure 9.38). They extract calcium car-

FIGURE 9.38 Coral reefs are formed from biologically secreted calcium carbonate and support some of the most productive and diverse communities on earth (courtesy of the USDI-Fish and Wildlife Service).

bonate from the seawater and secrete it layer by thin layer in a slow process that takes thousands of years to replace once the coral dies. The largest coral reef stretches over 1,000 miles along the East Coast of Australia. Not quite so extensive are the coral reefs fringing the tropical Caribbean and Hawaii. Many tropical islands are ringed by reefs. Coral reefs require a relatively narrow range of temperatures above 18° C (64° C), and water more than 100 m deep typically is too cold even where light can penetrate to that depth. The light is required by algal symbionts of the coral, which produce food through photosynthesis for themselves and the coral animals. Corals typically are not found near river mouths because the suspended matter absorbs light and the salinity falls below tolerance levels. Because of the coriolis currents, the colder water of continental west coasts is less likely to support corals than the warmer water of east coasts.

Coral reefs are among the most productive and diverse of marine ecosystems. The primary productivity derives mostly from algae living in the coral animals and with encrustation of algae on the dead coral. Little of the primary production comes from phytoplankton. A complex community of herbivores, including the coral polyps, sea urchins, clams, various fish species, and other invertebrates, is fed on by diverse carnivores including sea anemones, starfishes, crabs, carnivorous snails, and diverse fin fishes. At the top are large fish. However, as productive as coral reefs are, they are not important sources of human food resources compared to other marine environments. The diversity results in few species predominating and the coral reef structure resists use of mass harvest techniques. Coral reefs are at risk, however, in many locations because of increased turbidity, eutrophication, toxic contaminants from coastal pollutants, and overfishing (see Chapter 16). Most recently, concern has grown over global warming's possible effects on coral survival.

LITERATURE CITED

Allan, J. D. 1995. *Stream ecology.* New York: Chapman & Hall.

Alonzi, D. M. 1998. *Coastal ecosystem processes.* New York: CRC Press.

Barbour, M. G., J. H. Burk, and W. D. Pitts. 1980. *Terrestrial plant ecology.* 1st ed. Menlo Park, CA: Benjamin Cummings.

Barbour, M. G., J. H. Burk, W. D. Pitts, and M. W. Schwartz. 1999. *Terrestrial plant ecology.* 3rd ed. Menlo Park, CA: Benjamin Cummings.

Blaisdell, J. P. and R. C. Holmgren. 1984. *Managing intermountain rangelands—salt desert shrub ranges.* U.S. Dep. Agric. For. Serv. Gen. Tech. Rep. INT-163.

Branson, F. A. 1985. *Vegetation changes in western ranges.* Range Monograph 2. Society for Range Management, Denver, CO.

Buffington, L. C. and C. H. Herbel. 1965. Vegetational changes on semidesert grassland range from 1858 to 1963. *Ecological Monographs* 35:139–164.

Byrd, N. A. and C. E. Lewis. 1976. *Managing southern pine forests to produce forage for beef cattle.* U.S. Dep. Agric. For. Serv. Priv. For. Manage. Bul., Atlanta, GA.

Clary, W. P., M. B. Baker Jr., P. F. O. Connell, T. N. Johnsen Jr., and R. F. Campbell. 1974. *Effects of pinyon juniper removal on natural resource products and uses in Arizona.* U.S. Dep. Agric. For. Serv. Res. Pap. RM-128.

Cole, G. A. 1994. *Textbook of limnology.* 4th ed. Prospect Heights, IL: Waveland Press.

Colinvaux, P. A. 1993. *Ecology 2.* New York: John Wiley & Sons.

Cowardin, L. W., V. Carter, F. C. Golet, and E. T. LaRoe. 1979. *Classification of wetlands and deep water habitats of the United States.* Washington, DC: U.S. Fish and Wildlife Service.

Dasmann, R. F. 1976. *Environmental conservation.* New York: John Wiley & Sons.

Franklin, J. F. and C. T. Dyrness. 1969. *Vegetation of Oregon and Washington.* U.S. Dep. Agric. For. Serv. Res. Pap. PNW-80.

Gay, C. 1965. *Range management. How and why.* New Mexico State Univ. Coop. Ext. Circ. 376, Las Cruces, NM.

Grelen, H. E. 1978. Forest grazing in the south. *Journal of Range Management* 31:244–250.

Holechek, J. L., R. D. Pieper, and C. H. Herbel. 2001. *Range management: Principles and practices.* 3rd ed. Englewood Cliffs, NJ: Prentice-Hall.

Horne, A. J. and C. R. Goldman. 1994. *Limnology.* 2nd ed. New York: McGraw-Hill.

Hynes, H. B. N. 1970. *The ecology of running waters.* Toronto, Ontario, Canada: University of Toronto Press.

Johnsen, T. N. 1962. One-seed juniper invasion of northern Arizona grasslands. *Ecological Monographs* 32:187–207.

Jordan, William R. III. 1997. Ecological restoration and the conservation of biodiversity (pp. 371–398). In *Biodiversity II: Understanding and protecting our biological resources,* edited by Marjorie L. Reacka-Kudla, Don E. Wilson, and E. O. Wilson. Washington, DC: Joseph Henry Press.

Kuchler, A. W. 1964. *Potential natural vegetation of the conterminous United States.* Am. Geogr. Soc. Pub. 36.

Lerman, M. 1986. *Marine biology: Environment, diversity, and ecology.* Menlo Park, CA: Benjamin/Cummings.

Miller, G. T. Jr. 1990. *Resource conservation and management.* Belmont, CA: Wadsworth.

Mitsch, W. J. and J. G. Gosselink. 2000. *Wetlands.* 3rd ed. New York: Van Nostrand Reinhold.

Ricklefs, R. E. 1996. *The economy of nature: A textbook in basic ecology.* 4th ed. New York: W. H.Freeman.

Sampson, A., W. A. Chase, and D. W. Hedrick. 1951. *California grasslands and range forage grasses.* Calif. Agric. Exp. Stn. Bull. 724.

Shirley, H. L. 1973. *Forestry and its career opportunities.* 3rd ed. New York: McGraw-Hill.

Shreve, F. 1942. The desert vegetation of North America. *Botany Review* 8:195–246.

Tiner R.W., Jr. 1999. Wetland indicators: A guide to wetland identification, delineation, classification and mapping. Lewis Publishers. Boca Raton, Florida.

U.S. Department of Agriculture (USDA). 1936. *The western range.* 74th Congress, 2nd Session. Senate Document 199.

U.S. Department of Agriculture (USDA). 1972. *The nation's range resources: A forest-range environmental study.* U.S. For. Serv. Resour. Rep. 19.

U.S. Department of Agriculture (USDA). 1977. *Vegetation and environmental features of forest and range ecosystems.* Forest Serv. Agric. Handbook 475.

Vannote, R. L., G. W. Minshall, K. W. Cummins, J. R. Sedell, and C. E. Cushing. 1980. The river continuum concept. *Canadian Journal of Fisheries and Aquatic Sciences* 37:130–137.

Wetzel, R. G. 2000. *Limnology.* 3rd ed. New York: Saunders College Publishing.

Yorks, T. P., N. E. West, and K. M. Capels. 1992. Vegetation differences in desert shrublands of western Utah's pine valley between 1933 and 1989. *Journal of Range Management* 45:569–578.

The Land-Based Renewable Resources

V egetation and soil, in their various forms, determine the land-based renewable resources that support forest (Chapter 10), range (Chapter 11), farm (Chapter 12), recreation (Chapter 13), and urban uses (Chapter 14). The land-based renewable resources associated with ground and vegetation may be privately owned or managed under a public authority, which contrasts markedly with the exclusively public ownership of wildlife, fisheries, and endangered species (treated separately in the next section). Very little land is set aside for exclusive nonuse (preservation) of soils or vegetation independent of endangered species protection. Even wilderness lands are used for recreation, scientific study, generation of clean water, and a variety of other environmental services. Few land areas are dedicated entirely to one use; instead, multiple use is much more common. In addition to timber production, for example, forested land also supports range, recreation, farming (e.g., Christmas trees), low-density urban development (e.g., ground support for roads, dams, utility lines, homes, sawmills), and a variety of environmental services (e.g., clean water, climate regulation). It is similarly true that land bounded within city limits sometimes supports forest, range, farm, recreational, and even nonuse preserves. Also, pristine land converted to a new land use, such as forest to road surface, can be renewed as forest or range when demand is great enough. Several specialty areas of natural resource management are associated with these land-based resource uses including forest, range, farm, outdoor recreation, and urban land-use management. These will be discussed in this section.

CHAPTER **10**

Forests and Forestry

THE STATE OF FORESTRY

The state of the world's forests has gained greater public attention in recent years. Rain forest destruction, national forest logging, fire, and road building policies are frequent topics on the evening news. A growing number of forest species are declining from the cumulative effects of deforestation, environmental pollution, and urban encroachment. The public now has much more involvement in deciding how forests will be managed in the United States and abroad. Today's forest managers are better equipped to meet the growing demand for forest products while addressing public concerns over environmental quality. Over the past decade, forestry has been transformed as forest scientists have developed a better understanding of how ecological systems function. The central aim of forestry today is to provide a wide range of forest goods and services and to manage across broad landscapes as opposed to forest stands. It is no longer acceptable nor scientifically rational to manage forests primarily for wood products and neglect other ecosystem resources.

The key challenge is how to manage natural forests, plantations, and agroforestry systems sustainably while meeting growing demands for material goods and social services. The U.S. forest sector generates an enormous volume and range of wood products, including lumber, paper, and engineered wood products. They also provide energy, chemical derivatives, forage for livestock, wildlife habitat, various recreational opportunities, and a means of viewing nature apart from humanity's artificial landscapes and stressful lifestyles. In this chapter we will survey the forest resources, examine the ability of forests to meet our material needs, and describe how forests are managed. In addition, we will review criticisms of forest management and discuss how broader goals can be served by adjusting forest treatments. Numerous concepts will be developed that should enable the reader to more fully understand the technical aspects of forestry issues. From the onset, the reader should recognize that both forests and the science of managing them will continue to change. Forests will continue to adjust to changing environments and management will continue to adjust to society's needs and desires. Important references for this chapter include Sharpe et al. (1976), Spurr (1979), Smith (1986), Perry (1994), Kohm and Franklin (1997), Abramovitz (1998), and Abramovitz and Mattoon (1999).

World Overview

About one-half of the forests that once covered the earth are gone because of deforestation by humans (Abramovitz 1998). Unfortunately, the health and quality of many remaining forests are declining. Management of the world's forests is confronted by a variety of problems caused by increasing human populations and rising per capita material consumption. Rapid mechanization of forestry and agriculture have permitted the harvest of large forest areas as well as their conversion to farmlands, cities, dams, and other uses. Globalization of the world economies has allowed large corporations to more completely exploit forests in developing countries. In recent years huge fires in several tropical developing countries have devastated large areas and impaired human health. Additional threats to forests include pollution from automobiles and industry and the buildup of greenhouse gases.

More than timber is lost when forests disappear. Valuable nonwood products include nuts, oils, medicinals, cork, and rattan (Abramovitz and Mattoon 1999). Forests are also a primary source of biodiversity. They shelter countless organisms of indirect value to humans such as those that pollinate crops and control pests. Forest cover is critical in protecting watersheds from erosion and preventing flooding.

Although the problems confronting forests seem daunting, various new technologies provide great opportunities in forest conservation. Pressure on forests can be greatly reduced by using wood composites, improved building construction, paper recycling, improved timber processing, and wood salvage (Abramovitz and Mattoon 1999). Computerized storage of information has the potential to reduce paper consumption. Adobe and rammed earth are becoming more important alternatives to wood in home construction. Public concern about the fate of the world's forests is rising and leading to more sustainable policies in forest use (Abramovitz and Mattoon 1999).

TREE STRUCTURE AND FUNCTION

Forest resource management requires understanding some basics of tree structure and function. The differences which set trees apart from other plant life are fairly obvious. Trees are woody perennials, which continue to grow upward and outward over the course of decades, if not centuries. The major parts of a tree are shown in Figure 10.1. Lignin refers to woody material. It is the indigestible component of plant cell walls. Plants lacking lignin in their cell walls are weaker, softer, and unable to support the massive weights attained by trees as they grow upward. Trees elongate as cells divide in the growing tissues of shoot and root tips, known as meristem. This type of growth essentially produces what can be called the primary plant body. The cambial zone (Figure 10.1), found between the inner bark and outer wood, is what enables stems to continue to grow in girth over time. It is called a secondary or lateral meristem. The cambium is a very thin cell layer which produces new xylem cells toward the stem's center and new phloem cells outwardly. Xylem cells comprise tracheids in gymnosperms and tracheids and vessels in angiosperms. In simpler terms, new wood is produced inwardly and new bark outwardly. Each new layer of phloem in the inner bark crushes the previous year's phloem and makes it nonfunctional.

The stems, roots, and leaves of trees serve key functions (Figure 10.1). Roots anchor the tree while absorbing water and dissolved nutrients. The main stem and branches support the weight of leaves and flowers. Through the process of photosynthesis, leaves capture the sun's radiant energy and atmospheric carbon dioxide and convert them into carbon compounds that meet the energy demands of plant growth and tissue maintenance (Figure 10.1). Roots supply the water used in the process. Tracheids and vessels in the stems and branches provide a pipelike water and nutrient transport system between roots and leaves.

ESSENTIALS OF FORESTRY PRACTICE

Trees increase each year in height and spread of branches by adding on a new growth of twigs.

Light and heat are required by the leaves in the synthesis of nutrients obtained from the air and soil. The leaves give off moisture by transpiration.

Part of a Leaf (Vertical Cross-section)

CO_2 and O_2 CO_2 and O_2

H_2O

CROWN

TRUNK

Heartwood (inactive) gives strength.

Sapwood (xylem) carries sap from root to leaves.

Cambium (layer of cells where growth in diameter occurs) builds tissues-wood inside and bark outside.

Inner bark (phloem) carries food made in the leaves down to the branches, trunk, and roots.

Outer bark protects tree from injuries.

Cambium
Inner bark
Outer bark

Sapwood   Heartwood   Pith   Pith rays

ROOTS

Root hairs take up water containing a small quantity of minerals in solution.

Taproot

The buds, root tips, and cambium layer are the growing parts. The tree takes in oxygen over its entire surface through breathing pores on leaves, twigs, branches, trunk, and roots.

FIGURE 10.1 How a tree functions (with minor modifications, courtesy of USDA-Forest Service).

The wood of tree species differs based in part on the evolutionary relationships of the species. Two major groups are identified in tree taxonomy. Gymnosperms are plants with naked seeds, not enclosed within a fruit. Angiosperms include broad-leaved flowering trees, such as maples and elms, which are collectively called "hard-woods." Gymnosperms are a rather primitive group of trees with distant fernlike relatives that lived long before angiosperms appeared. Conifers such as pines, cedars, and junipers are cone-producing gymnosperms. Because of their primitive structure, the wood of gymnosperms is much simpler and often easier to saw than the wood of angiosperms. Such trees are also called "softwoods" by foresters. The wood of conifers is largely comprised of tracheids whereas the wood of angiosperms has many more cell types, including vessel elements. These fundamental differences have much to do with the way the wood of these groups is used. Paper, for example, requires fibers of uniform type. Pines with their massive store of uniformly shaped tracheids are well suited to paper manufacture.

For trees to survive in cold temperate climates, growth must occur when damage from tissue freezing can be avoided. Temperate trees avoid the cold by shutting down growth when daylight periods begin to shorten as winter approaches. An extremely well refined "molecular clock" enables trees to sense the changing seasons. As trees resume growth in the early spring, not all of their main parts begin growth at the same time, nor do they grow at the same rate over the frost-free season. The timing of nursery practices, outplanting, herbicide application, and treatments to enhance seed production all depend on knowledge of tree growth sequences.

FOREST PRODUCTS

Forests supply us with a variety of products besides lumber for housing (Figures 10.2 and 10.3). They are a valuable source of food (nuts, fruits, and syrup), medicines, resins, fuel wood, pulpwood (paper), and decorations (Christmas trees). Most forest lands provide valuable habitats for fish and wildlife, and many serve as critical watersheds for large cities. Recreation and livestock grazing are other important forest uses.

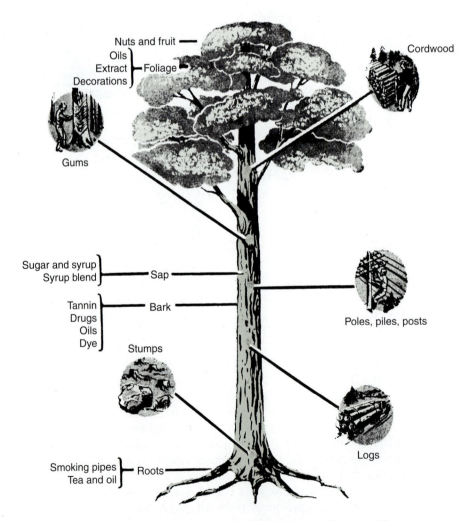

FIGURE 10.2 Some of the useful products obtained from trees (with minor modifications, courtesy of USDA-Forest Service).

PROCESSING FOREST PRODUCTS

FIGURE 10.3 Categories of various forest products (courtesy of USDA-Forest Service).

FOREST DISTRIBUTION

Forests are areas where trees are the dominant vegetation. In a general sense, they include closed canopy forests where leaves and twigs of adjacent trees touch, as well as woodlands where trees are more separated and only occasionally overlap. Forest distribution will be discussed first on a global scale (Figure 10.4). At the global scale, the distribution of forest biomes provides a means for presenting a brief overview of the world's diverse forests. For clarity, a forest biome is a major regional ecosystem with its own type of climate, vegetation, and animal life. In reality, forest biomes are not sharply separated from other biomes, but merge gradually into one another across

Tropical deciduous forest

Tropical rain forest

Midlatitude forest

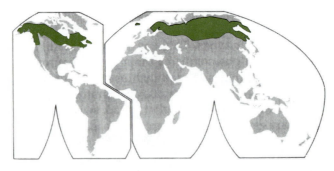

Boreal forest

FIGURE 10.4 The general distribution of the major forest types in the world (from McKnight 1990).

ecotones. We refer the reader to Chapter 9 for additional discussion of the forest and woodland biomes.

Boreal Forest (Taiga)

The boreal forest biome occurs across the high latitudes of North America and Eurasia where it extends from Scandinavia to eastern Siberia (Figure 10.4). These regions are characterized by extreme cold, soils that remain frozen beneath the surface, and a brief (50- to 100-day) frost-free season. Despite the harshness of the climate, the boreal forest is second only to the tropical moist forest in its extent (Pastor et al. 1996). Overall, it accounts for 20% of the world's forest area. Boreal forests are low in tree species diversity because of the harsh climate and their recent formation (glaciers covered the area within the past 10,000 years). Boreal forests are dominated by conifers, especially spruces, firs, and larches. Aspens (*Populus*) and birches (*Betula*) are important flowering trees. Spruces (*Picea*) and larches (*Larix*) commonly occur on poorly drained lowland soils. Pines (*Pinus*) are most common on well-drained upland soils, whereas cottonwoods (*Populus*), firs (*Abies*), willows (*Salix*), and certain species of spruce occur on the finer-textured upland soils. Boreal forests typically comprise dense stands of relatively small trees, usually less than 30 m in height. Because trees are densely spaced, the forest floor is densely shaded. Fire frequently occurs in boreal forests and strongly influences species composition, nutrient availability, and forest productivity. Forests are slow to recover from disturbance owing to the short growing season and relatively poor soils.

Temperate Deciduous Forests

At their southern boundaries, boreal forests meet temperate forests, which have greater tree species diversity and taller stature (Schulze et al. 1996). Temperate broad-leaved deciduous forests occur mostly between 30° to 50° N latitude. Major areas of these forests occur in the eastern United States, Europe, western Turkey, eastern Iran, western China, and Japan. Temperate forests are much less common in the Southern Hemisphere, but smaller areas can be found in South America, southern Africa, Australia, and New Zealand. Temperate forests are subjected to warm summers and cold winters where temperatures commonly dip to −30°C. The frost-free season is 4 to 8 months. Broad-leaved deciduous forests are concentrated in the eastern portion of North America (see Chapter 9, Figure 9.22). Maples, beeches, hickories, and oaks are among the most representative genera (Figure 10.5). Because they occur in areas long populated by humans, temperate forests have been heavily impacted by unnatural disturbance. Many tree species in temperate forests produce wood of unusually high quality.

Temperate Coniferous Forests

Temperate evergreen coniferous forests occur in a wide range of climates and include the montane forests of North America, Europe, and China (see Chapter 9). They also occur on smaller areas of the montane regions of Korea, Japan, Mexico, Nicaragua, and Guatemala. Temperate conifers tend to occur on xeric or infertile soils lacking the water and nutrients needed by deciduous species. Common genera include *Abies, Pinus, Pseudotsuga,* and *Thuja. Pinus* species occur in a wide variety of environments ranging from the hot, arid southwestern United States to cold regions of Scandinavia and Eurasia. Growth rates vary greatly depending on precipitation and soils. Forest management is intensified in regions (e.g., southeastern United States and Pacific Northwest) of high production. Even on less productive sites, biomass accumulation (wood) can be substantial over several centuries.

Temperate Mixed Forests

Mixed (deciduous plus evergreen) forests occur throughout the temperate regions where the climates are similar to those described for temperate deciduous and coniferous forests. They are most abundant in the southeastern United States, Europe, and portions of Iraq, Iran, and China. Their occurrence reflects past land-use change and local variations in soil conditions. In the southeastern United States, recently

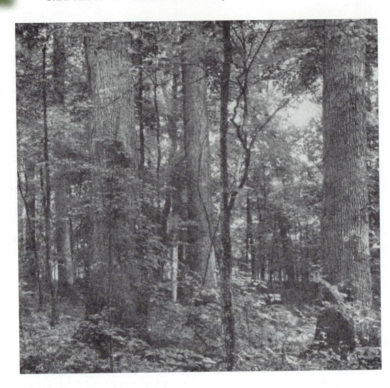

FIGURE 10.5 A deciduous forest in Indiana dominated by oaks (courtesy of USDA-Forest Service).

disturbed forests are dominated by conifers. Over time these are replaced by mixed forests, which eventually give way to broad-leaved hardwoods. In the lake states, conifers dominate the drier, less fertile soils. Broad-leaved hardwood forests occur on the more moist soils of finer texture. Mixed forests occur where soil conditions are intermediate.

Temperate Broad-Leaved Evergreen Forests

This group includes broad-leaved sclerophyll (hard-leaved) forests and broad-leaved rain forests. The broad-leaved sclerophyll forests occur in areas with a Mediterranean climate, typified by winter rain and summer drought. Broad-leaved rain forests occur in areas with humid, frost-free climates. The sclerophyll forests occur in scattered areas of the United States, around the Mediterranean, and over large areas of Asia. The greatest continuous areas still existing are the Eucalyptus forests of Australia. Temperate broad-leaved evergreen rain forests occur in Japan, Chile, New Zealand, Australia (Tasmania), and in scattered remnant patches in Asia.

Beech (*Fagus*) Nothofagus (beech) rain forests occur in stands in New Zealand and southern Chile. The sclerophyll forests in the Mediterranean area and the United States are dominated by oak (*Quercus*) species. Oak woodlands can be impressive where disturbance has been minimal, but this is rarely the case. In Australia, Eucalyptus can form closed canopy forests up to 60 m in height. Eucalyptus forests are adapted to fire. Large areas of native forests have been cleared since European settlement. Exotic softwood (*Pinus radiata*) plantations have replaced large areas of the native forests in Australia.

Tropical Evergreen Forests

Tropical evergreen forests (rain forests) comprise the largest single forest biome in the world (Figure 10.6). They are found between the Tropics of Cancer and Capricorn in areas where air temperatures are high, and they change little on a daily or seasonal basis. Rainfall is abundant (more than 1,500 mm per year) and relative humidity is uniformly high. The Amazon Basin supports the greatest area of tropical rain forest,

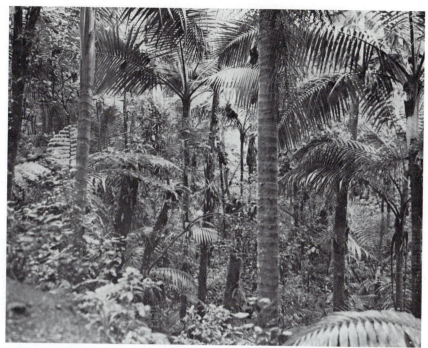

FIGURE 10.6 Tropical rain forest in Puerto Rico (courtesy of USDA-Forest Service).

but rain forests also cover the Congo Basin. Tropical forests are found in Central America where they are highly fragmented by agriculture. In Asia, they occur along the southeast coast of India, in Sri Lanka, the Malaysian Peninsula, the Indonesian archipelago, Borneo, Sarawak, and Papua New Guinea.

The stature and species composition of tropical rain forests differ according to soil type and elevation, making lowland and montane subtypes distinguishable. Rain forest canopies often have three layers. These include a layer of taller trees, which rise above the main canopy layer, and a subcanopy of smaller trees and shrubs.

Rain forest soils are typically infertile, but nutrients are cycled efficiently. Thus much of the nutrient is incorporated in biomass and is removed by intensive logging. Tropical evergreen forests are the most diverse terrestrial ecosystems on earth, with the greatest number of species per unit area. The Amazonian forests contain, for example, more than 2,500 different tree species. Rain forests are the most productive forests in the world in terms of biomass produced. Much of the production is allocated to leaf matter, which forms an energy source for highly diverse insect species.

Tropical rain forests are under constant assault by human activities (Figure 10.7). The most threatening pressures on rain forests differ across the regions (World Resources Institute 1994). In Africa, forests are cleared for agriculture to support a rapidly expanding population. In South America, shifting cultivation is a major factor, but large areas have been cleared for resettlement of immigrants, ranching, modern agriculture, industrial fuel production, logging, and hydroelectric development. Asian forests are being rapidly destroyed by population growth and abusive commercial logging operations.

Tropical Deciduous Forests

Tropical evergreen forests give way to deciduous forests where rainfall is more intermittent on a seasonal basis. Because deciduous species occur throughout the evergreen forests at least to some extent, the categorical separation has limited practical value. The factor responsible for leaf shedding is seasonal dryness. The largest areas of tropical deciduous forests are the monsoonal forests of southern and southeastern

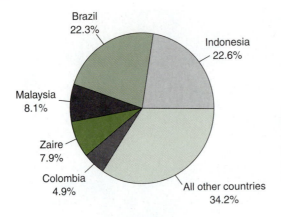

FIGURE 10.7 Percent of tropical rain forest lost (1981–1990) from the top five countries with greatest losses and all other countries combined (Food and Agriculture organization; figure from World Resources Institute 1994).

FIGURE 10.8 Tropical deciduous savanna in Sudan.

Asia. They also occupy areas bordering evergreen forests in South America and Africa (Figure 10.8).

The canopies of deciduous tropical forests are typically shorter and more open than those of tropical evergreen forests. In addition, they usually have two rather than three layers. The bottom layer is often occupied by densely growing shrubs that take advantage of the available sunlight and reduced evaporative demand. Tropical deciduous forests are being cleared to make room for grazing and agriculture and are becoming increasingly fragmented.

U.S. FOREST TYPES AND RESOURCES

Major Forest Types and Uses

The forest vegetation of the United States has been classified in numerous ways. One approach has been to map potential vegetation on the basis of climate (e.g., Küchler 1964). In contrast to Küchler's method, the Society of American Foresters' (SAF) approach identifies forest types that currently exist as opposed to those potentially existing under climax (pristine) conditions (Eyre 1980). For more intensive management purposes, forest vegetation in the western states has been classified according to habitat types as found in Daubenmire and Daubenmire (1968), Franklin and Dyrness (1973), and Pfister et al. (1977). The forest ecosystems of the United States are described in Chapter 9.

Hall (1983) combined the numerous forest cover types found in the United States to six main groups (Figure 10.9). In effect, forest trees tend to be aggregated into certain groupings or occur in mostly pure stands. SAF calls these identifiable aggregations and pure stands "forest cover types." In given areas, forest cover types can be transitory or occupy the same area for long periods. Nevertheless, several criteria must be met for a type to be recognized. Tree crowns should cover at least 25% of the area and the type must occupy a fairly large area in the aggregate, but not necessarily in a continuous manner. Many types occur infrequently and merge into others over short distances. Figure 10.10 gives the total area of each of these and other forest cover types that are available for timber harvest (i.e., unrestricted forest land).

Eastern Forests.   Eastern forests fall into six major cover types (Figure 10.9). The Northern, Central, and Southern forest types run generally in east-west bands from north to south from colder to warmer climates. The Bottomland forest follows the major river floodplains in the Southern forest and is most developed along the Mississippi River, where it divides the Central and Southern forests. Their ecology is described in Chapter 9.

All of the Northern forest provides important resources for recreational use (e.g., fishing, hunting, hiking, camping) and for watershed protection. The spruce-fir forest type produces pulpwood, construction lumber, boxes, crates, musical instruments, and Christmas trees. The white-red-jack pine forest type is an important source of pulpwood and construction lumber. The maple-beech-birch type produces high-quality wood for furniture and building interiors, flooring, crates, garden mulch, charcoal, maple syrup, pulpwood, and firewood. The aspen-birch type is a source of pulpwood, furniture parts, matches, boxes, and wooden utensils.

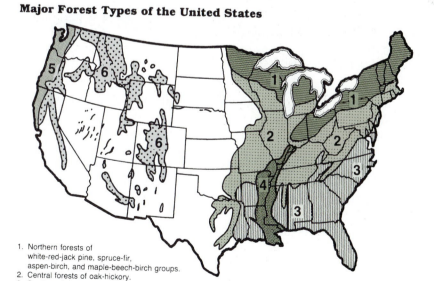

Major Forest Types of the United States

1. Northern forests of white-red-jack pine, spruce-fir, aspen-birch, and maple-beech-birch groups.
2. Central forests of oak-hickory.
3. Southern forests of: Oak-pine, loblolly-shortleaf pine, and longleaf-slash pine groups.
4. Bottomland forests of oak-gum-cypress.
5. West coast forests of: Douglas-fir, hemlock-sitka spruce, redwood, and some western hardwood groups.
6. Western interior forests of: Ponderosa pine, lodgepole pine, Douglas-fir, white pine, western larch, fir-spruce, and some western hardwood groups.

FIGURE 10.9   The six major forest cover types in the United States (courtesy of U.S. Dept. of Agriculture).

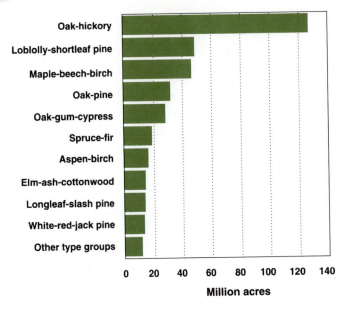

FIGURE 10.10 Forest cover-type groups on unreserved forest land in the East (from Powell et al. 1993). Cover types are determined by the dominant species in the overhead canopy (cover).

FIGURE 10.11   Longleaf pine (*Pinus palustris*) savanna in Georgia maintained by periodic burning (courtesy of USDA-Forest Service).

In the Central forest, oak-hickory is the primary forest type, which is the largest of the Eastern forest types, but much of it occurs on mountainous lands of low productivity where it is most important for watershed protection and recreation. On more productive sites, oak is used for many purposes including general construction, flooring, furniture, interior finish, millwork, barrels, ties, timbers, fence posts, pilings, pulp, and fuel. Hickory is used for lumber and fuelwood. This type also supports other tree species of superior furniture-wood quality including walnut and yellow poplar.

For reasons described in Chapter 9, the Southern forests are now dominated by the longleaf-slash pine, the loblolly-shortleaf pine, and the oak-pine types (Figure 10.11). Forest products are of primary importance in the pine-dominated types, but watershed protection is an important parallel use for the erosive soils common in these forest types. Hunting is a popular use and the more open areas are grazed by livestock. Southern pine wood is used for heavy construction such as bridges and large buildings, general lumber, pulp, plywood, utility poles, piling, railroad ties, and mine timbers. The sap is used for rosin and turpentine.

The Bottomland forest is typified by oak-gum-cypress forest type of the bottomland hardwoods and produces many of the same oak products as in the Central and

FIGURE 10.12 (a) Pacific coastal forest dominated by Douglas-fir. (b) Redwoods in the Calaveras grove in California (courtesy of USDA-Forest Service). These are among the most productive forest types, dominated by the tallest trees on earth.

Southern forests. Sweetgum is used as a plywood veneer and for furniture parts. Bald cypress is used principally for heavy building and bridge construction, exterior siding, docks, decks, and other uses requiring exceptional resistance to decay. Much of the bottomland hardwoods area has been cut for timber and cleared permanently for agriculture. Interest is rapidly increasing in restoring natural services including habitat for scarce plant and animal species and natural flood-control and water storage.

Western Forests. The western forests include diverse coniferous forest types, which can be broadly split into the highly productive West Coast forest and the less productive Western interior forests (Figure 10.9). The specific West Coast forest types include Douglas-fir and hemlock-sitka spruce ranging north from Oregon, and the redwood type limited mostly to California (Figure 10.12). While used mostly for recreation, watershed protection, and provision of habitat for scarce species, the coastal forest types also are highly valued for their wood products. Douglas-fir provides much of the general building and construction lumber and plywood used in the United States. Spruce, western hemlock and associated firs, and western cedars provide a wide variety of wood products such as lumber and plywood for general construction, beams, exterior siding, decking, interior paneling, subflooring, molding and other trim, and particleboard. Lesser quantities are used for mine timbers, planking, posts, oars, masts and spars, barrel staves, industrial crates, ladders, furniture parts, parts for pianos and guitars, garden mulch, and pulpwood. Redwood is highly valued for its weather resistance for building construction, heavy beams, bridge timbers, plywood, planks, exterior siding, doors, furniture, shakes, shingles, and posts. It is also used for interior paneling and novelties. The western hardwoods along river floodplains provide small quantities of furniture and flooring woods.

In the Western interior forests, the Douglas-fir forest type occurs at high elevations in more modest dimensions and often in close association with ponderosa pine (Figure 10.13), western white pine, lodgepole pine, larch, aspen, and western

FIGURE 10.13 Climax ponderosa pine forest in Arizona (courtesy of USDA-Forest Service). The open stand conditions reflect past natural fire, which provides grazing as well as wood product uses.

fir-spruce forest types. In these steep mountainous areas, the interior forests are extremely important for watershed protection, increasingly important for recreation and habitat protection, and also important at lower elevations for grazing. Forest products are less important than in the coastal forests, but remain significant in local economies and for certain products, especially in more accessible areas on moderate to low slopes. Douglas-fir and ponderosa pine types (and closely associated western white pine) are most important, including numerous products in addition to general construction lumber for millwork (window and door frames, shelving, molding, cabinets), subflooring, interior paneling, utility poles, posts and piling, mine timbers, railroad crossties, crates, and boxes. Western interior hardwoods are mostly shrubby and valued for game forage and cover and watershed protection. The piñon-juniper type at lowest elevations is valued most for fuelwood and pine nuts, but is more important for watershed protection, recreation (hunting mostly), and livestock grazing. People are increasingly building homes in these forest types despite the substantial risks associated with wildfire.

The piñon-juniper type consists of ecologically important woodlands scattered throughout the Southwest (Chapter 9, Figure 9.17). The trees lack impressive size, but these woodlands offer vast areas used for hiking, hunting, grazing, watershed protection, and commercially important products, such as firewood and pine nuts (see Chapter 9, Figure 9.18).

U. S. Forest Land Area

Today, 33% (737 million acres) of the total land area of the United States (Powell et al. 1993) is forested with at least 10% forest cover. North American forests have undergone considerable changes from the time early colonists began settling the land. Around 1600, about 49% of the United States was forested.

About 307 million acres of U.S. forest land have been converted to other uses since 1630 (Figure 10.14). Land clearing for agriculture has been the primary cause. Between 1850 and 1910, farmers cleared more forest than the total amount cleared in the previous 250 years. Industrialization and railroad construction also consumed

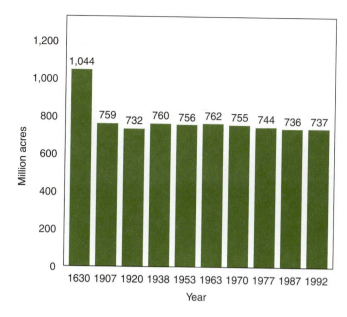

FIGURE 10.14 Trends in U.S. forest area from 1630 to 1992 (from Powell et al. 1993).

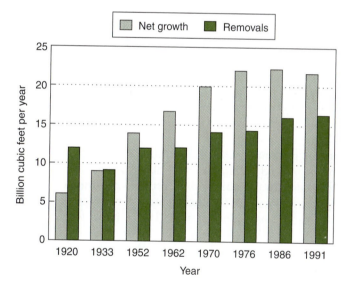

FIGURE 10.15 U.S. timber growth and removals (from Powell et al. 1993).

wood. For example, between 1870 and 1910, the huge forests of the lake states were cut to supply the material requirements of rapid industrialization. Railroads consumed nearly 25% of the wood used in the 1800s (Sedjo 1995). After 1920, as agricultural production increased per unit area, cropland area stabilized at lower total acreage, and forest production recovered slightly (Figure 10.15). Overall, the area of U.S. forests has been maintained the past 75 years (Figure 10.14). However, forest age structure has changed because secondary forests stand today where primary forest stood before. Although old stands were common in colonial America, they were limited by natural disturbances such as insect infestation, disease, and fire, as well as the actions of the native peoples (Sedjo 1995).

Timber Production

Timberland. Forests produce diverse timber-based goods (Figure 10.16). By definition, timberland is capable of producing at least 20 ft³/acre (1.4 m³/ hectare) of industrial wood per year and is not reserved from timber harvest (Powell et al. 1993). Two-thirds of the nation's forested ecosystems (198 million ha or 490 million acres)

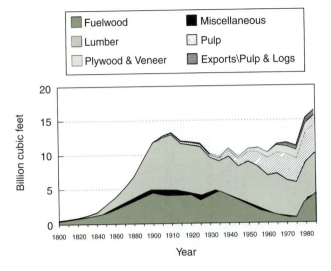

FIGURE 10.16 Domestic production of forest products, 1800–1990 (from Powell et al. 1993).

are classed as timberland. About 6% (47 million acres) of all U.S. forest land is reserved from commercial timber harvest in wilderness, parks, and other land classifications (Darr 1995). Since 1952, the area of timberland has decreased by 4% or about 19.3 million acres. This decline has resulted primarily from conversion of public timberland into wilderness or other land uses that do not permit timber harvest.

In the 1920s, timber net growth (accumulated growth after loss to mortality) was about one-half the rate of harvest (Darr 1995) (Figure 10.15). By the 1940s, improved forest net growth rates (partly because of forest protection from fire), as well as declines in harvest rates, resulted in timber net growth and harvest coming into balance (Powell et al. 1993).

Timber Productivity Across the Regions. U.S. timberlands cover a wide range of latitudes, elevations, precipitation, and soils, which greatly influence their productivity. Potential productivity measures the ability of a forest to produce wood volume. The Pacific Northwest region has some of the most productive forest land (i.e., land capable of producing 120 ft³/acre/year) (Powell et al. 1993). However, large areas in the South covered by loblolly-shortleaf pine and oak-hickory forests are also highly productive. Most of the less productive land is in the West, where forests occur at higher elevations and latitudes or in piñon-juniper woodlands, where precipitation is limiting.

Based on the amount of growing stock existing now, timber supplies should meet needs over the next 15 to 20 years. However, beyond 2020 there is considerable concern over wood shortages. By definition, growing stock includes live, sound trees suited for roundwood (timber) products. U.S. timberland contains an estimated 24.3 billion cubic meters (858 billion cubic feet) of timber, in which 92% is in growing stock. From 1952 to 1992, net volume per acre has increased 33% (Powell et al. 1993). Volume growth occurred despite harvests that make the United States the world's number-one timber producer since World War II (Sedjo 1995). The greatest gain in volume occurred in the South (104%) followed by the North (95%) and the Rocky Mountain region (27%). Recent studies indicate that the forest biomass in the northern Rocky Mountains is 30% greater now than in 1800 (Sedjo 1995). However, the Pacific Coast region has lost about 4% of its timber volume over the last 50 years.

Hardwoods and softwoods supply different types of forest products. Hardwood growing stock is concentrated in the northcentral and southeastern states whereas softwood volume is highest in the Pacific Coast states. Some 57% of the growing volume is in softwoods and 43% in hardwoods. About 66% of softwood timber is in the West and 23% in the South. At least 90% of hardwood volume is in the East.

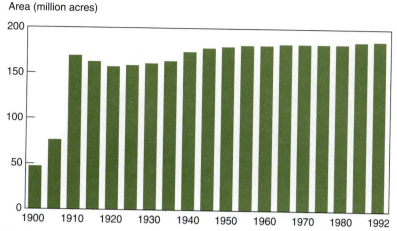

FIGURE 10.17   The total area of forest land under supervision of the USDA-Forest Service from 1900 to 1992 (courtesy of USDA-Forest Service).

Timber Removals.   Almost all of the timber harvested in the United States comes from second-growth forests (those that have regrown after an earlier clearing) or plantation forests. The undisturbed native forests of the United States have been almost completely withdrawn from timber harvest. This is because they are now part of park preservation, the wilderness system, or in other set-aside public management (Sedjo 1995). Average timber removals have increased each decade since the 1950s but volume growth has outpaced harvest from the 1950s forward (Figure 10.15). In 1991, almost 55% of all timber removals came from the forests of the South (up from 45% in 1970). Pacific Coast forests provided 23%, the North 17%, and the Rocky Mountains 5% (Powell et al. 1993). Historical timber harvests and primary uses in the United States are shown in Figure 10.16.

Forest Ownership

Some 34% of all forest land is federally owned. Most of this land is held by the U.S. Forest Service with lesser amounts under Bureau of Land Management control (Darr 1995). The Forest Service protects and manages 155 national forests, which contain 191 million acres (approximate size of Texas). About 18% of this land is designated as wilderness. The area under Forest Service management has remained somewhat constant for many decades (Figure 10.17).

The majority (58%) of U.S. timberland is privately owned. Of the 42% publicly owned, only 18% of commercial timberland is located in the National Forest System. By a wide margin, national forest land is found mainly in Pacific Coast and Rocky Mountain regions (Figure 10.18). Timber corporations own 14% of the nation's timberland area. Much of this corporation land is found in the South. Individual private ownership of forest lands is highest in the eastern United States.

FOREST MANAGEMENT

Forest Management Defined

The Society of American Foresters (SAF) defines *forestry* as "the science, the art, and practice of managing and using for human benefit the natural resources that occur on and in association with forest lands." Forestry, therefore, includes much more than just timber management. It involves watershed hydrologists, wildlife and fisheries

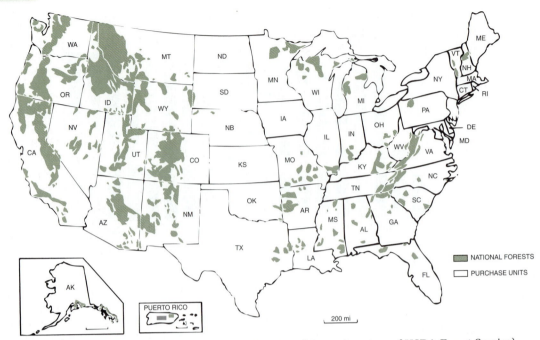

FIGURE 10.18 Location of the national forests in the United States (courtesy of USDA-Forest Service).

biologists, range scientists, and a host of other specialists, in addition to those working most closely with timber resources (silviculturalists). Recreation is a major activity on forest lands. This form of use continues to require ever increasing management on both public and private forests. In 1996, more than 800 million people visited the National Forest System to participate in recreational activities such as: camping, hiking, fishing, hunting, and wildlife viewing.

"Forest management" draws on a family of subjects and is variously interpreted to include mensuration, regulation, valuation, and preparation of working plans. It can also cover financial aspects of forestry (Sharpe et al. 1976). It is the business arm of forest science. Mensuration addresses the measurement of timbered areas, whole forests, single trees, logs, and other pieces and units of forest products. Regulation involves forest inventory and planning activities needed to determine harvest schedules. Valuation puts a monetary value on forest properties and is critical in making sound financial decisions. Although timber production remains a major objective in forest management, other objectives are becoming increasingly important to foresters and the public. In addition to traditional objectives, such as production of game, recreation, and water, other products such as nongame wildlife, biodiversity, aesthetics, and forest health are receiving increasing emphasis. As a consequence, assessment of these resources has become as much a part of forest management and public lands as forest mensuration. Forest management, therefore, is an interdisciplinary activity requiring the services of diverse specialists in addition to the traditional forestry focus on tree production.

Silviculture deals directly with managing forest growth and composition. Silviculture draws heavily on ecology, soils, dendrology, and silvics. Dendrology provides a means for classifying and recognizing trees. Silvics addresses such environmental factors as climate, slope, soil, and fire as they relate to tree growth and health. Spurr (1979) proposed three definitions of silviculture, which collectively identify the aims of silviculture: (a) the art and science of producing and tending a forest; (b) the application of silvics in treatment of a forest; and (c) the theory and practice of controlling forest establishment, composition, and growth. Overall, silviculture is at the heart of forestry. It is about decision making, and it enables people to get what they desire from forests whether it be timber or a mix of timber and other goods and services. Done correctly, silviculture avoids forest abuse and has as its principal aim ensuring that forests remain productive for future generations. Overall, the chief purpose of silviculture is to create and maintain the kind of forest that will best fulfill the objectives of the owner.

STAND MANAGEMENT

Criteria Used to Classify Forest Stands

Traditionally, silviculture has been based on the management of forest units called stands. A forest *stand* is defined as an area of forest with site condition, history, species composition, and structure of sufficient uniformity to distinguish it from adjacent areas (Figure 10.19). A stand can contain a few to several hundred acres. Stands can be grouped into forest compartments.

Stands can differ in age, species composition, density and in growth potential as determined by site quality. All of these factors determine how much time is needed for a stand to produce merchantable wood products. The segment of time between stand establishment and harvest is referred to as rotation length.

Stand Age and Size Distribution

Stand structure refers to the distribution of age and/or diameter classes and of crown classes. The distribution of age classes across the entire population of trees is important because trees of near equal age will arrive at harvest size at the same time. Age class distribution can be used to predict whether the flow of timber products from a stand will be continuous or interrupted for unacceptably long periods (Figure 10.20). The latter case might indicate some form of stand treatment to improve the harvest schedule. In even-aged stands, all live trees were established within a narrow period of time so that the difference between youngest and oldest does not exceed a specified period of time. For example, if the stand will require 100 years to reach harvest size from time of establishment, a stand can be considered even-age if all trees are within 20% (or 20 years) of one another in age. In uneven-age stands, there are by definition at least three distinct age classes (Figure 10.20). Balanced uneven-age stands contain three or more different age classes with each class occupying about the same area. Age classes are, therefore, spaced at uni-

FIGURE 10.19 A stand of jack pine in northern Michigan (courtesy of USDA-Forest Service).

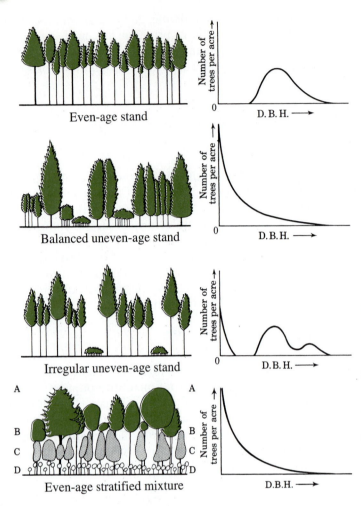

FIGURE 10.20 Typical examples of even-age and uneven-age stands. Trees are of the same species. The diameter at breast height (D.B.H.) indicates the relative size distribution of each stand condition (Smith, David M. 1986. *The Practice of Silviculture.* New York: John Wiley and Sons. Used with permission of John Wiley and Sons).

form intervals all the way from newly established reproduction to trees near harvest age. Irregular uneven-age stands do not contain all the age classes necessary to ensure that trees will arrive at harvest age at regular short intervals. Uneven-age virgin stands and culled over stands usually are irregular in age distribution.

Simple descriptions of tree sizes in stands can quickly convey the overall structure of a stand. Seedlings or seedling stands, for example, contain young trees only a few feet tall and having stem diameters less than 1 inch. A sapling stand contains trees at least 4.5 feet tall and up to 4 inches in diameter. Pole timber stands have trees 4 to 10 inches D.B.H. [diameter at breast height (4.5 feet above ground)]. Sawlog stands comprise trees suitable for conversion to lumber and are usually at least 11 inches D.B.H. Finally, an overmature stand contains trees that have declined in growth rate because of old age and loss of vigor.

Terms are also used to describe the position of individual tree canopies in the stand (Figure 10.21). This indicates which trees have the most and least resources needed for future growth:

1. Dominant—Refers to trees that extend above surrounding individuals and capture sunlight from above and around the crown.

2. Codominant—Refers to a tree that extends its crown into the canopy and receives direct sunlight from above but limited sunlight from the sides. One or more sides of a codominant tree are crowded by the crowns of dominant trees.

FIGURE 10.21   Tree canopy classification in stands. D = dominant; CD = codominant; I = intermediate; S = suppressed (adapted from Allen and Sharpe 1960).

3. Intermediate—A tree with a crown that extends into the canopy with dominant and codominant trees. These trees receive little direct sunlight from above and none from the sides. Crowns generally are small and crowded on all sides.

4. Overtopped (Suppressed)—A tree that cannot sufficiently extend its crown into the overstory and receives no direct sunlight either from above or the sides. Overtopped trees that lack shade tolerance lose vigor and die (over-topped is synonymous with suppressed).

Species Composition

Stands can be described as mixed or pure depending on species composition. In general, if 80% of a stand is composed of a single species, it can be considered a pure stand. A mixed stand would be one in which two or more species occur and one species accounts for less than 80% of the total stems in the stand.

Classification Based on Density

Density refers to the physical space that the trees take up in the stand. It is a basic measure of the growing space allowed for each tree. One method of describing density is to count the number of trees on a unit area, usually an acre. Density can then be expressed, for example, as the number of stems per acre. This expression provides a straightforward means of describing the density of seedling and sapling stands because most of the trees will be of nearly equal size. As trees get larger or where stands are uneven-aged, their diameters are less uniform. Therefore, some means must be used to account for this. Such an expression is basal area, which refers to the sum of the square feet of bole area of the individual trees on an acre (e.g., 150 ft^2/acre).

Stocking refers to the density of a stand with respect to a standard. A number of approaches have been developed for estimating stocking. Reineke (1933), for example, developed a measure of stocking that equates average tree diameter and number of trees per acre to density. His approach recognizes a concept that is fundamental to the practice of silviculture. In this concept, the total biomass of trees on a site has an upper limit and, as that limit is approached, competition for light and nutrients among trees will increase. Consequently, some trees will die as this natural thinning process occurs.

TOO MANY TREES
Yearly Growth—
None
Spacing—
5 ft. × 5 ft.
(1,700 trees
per acre)

NOT ENOUGH TREES
Yearly Growth—
One Cord per Acre
Spacing—15 ft. × 15 ft.
(170 trees per acre)

RIGHT NUMBER OF TREES
Yearly Growth—
2 Cords per Acre
Spacing—10 ft. × 10 ft.
(425 trees per acre)

FIGURE 10.22   Effect of tree spacing and competition for nutrients on annual tree growth as shown by growth rings. These three trees of the same species grew in the same general type of soil and climate and were ten years old when cut (courtesy of USDA Natural Resources Conservation Service).

Stands providing insufficient space for trees to reach merchantable size are "over-stocked" (Figure 10.22). There are plenty of individual trees in even-age pine stands, for example, but they are too small for the sawmill to process. In "understocked" stands, trees are so widely spaced that even with full growth potential realized, crown closure will not occur. With proper density management, growth will be distributed across fewer stems and mortality will not be a significant factor.

Classification Based on Site Quality

Often there is a need to express the productive capacity of a forest stand as it is determined by the ability of a given area to grow trees. The site index is the most commonly-used measure of site quality, or site productivity. It is based on the height attained by the dominant tree in a stand at some base age, usually 50 or 100 years. For trees grown over shorter rotations, the base age can be less. For example, in managing fast-growing southern pine plantations, a base age of 25 years is used.

To determine the site index of a stand, the total height and age of five to eight dominant and codominant trees are measured and compared to the average total stand height and age using base curves (Figure 10.23). Such a comparison will indicate whether a stand is high, low, or average in productive capacity since other measurements of yield follow height growth. The total height of dominant and codominant trees in a stand is the best measure of site potential because trees grow taller on good sites than on poor sites. If a base age of 100 years is used, the site index is 100 if trees on the site reach a height of 100 feet in 100 years. A site index of 80 would indicate that trees are only 80 feet tall at 100 years. Compared with the first site, this site would be less productive for growing timber. The example of a ponderosa pine site index illustrated in Figure 10.23 assumes that tree heights at different stand ages will differ in a proportional manner from one base site index to another.

We have cited several terms and criteria that can be used to describe stands. When these terms are combined, a fuller description can be obtained for communicating stand information. For example, one could say the area of interest is a fully stocked and even-age pure stand or is an uneven-age, mixed, understocked stand on a poor site.

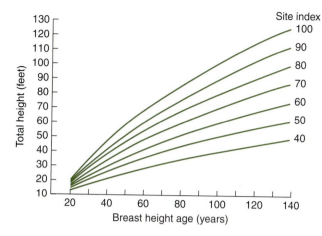

FIGURE 10.23 Site index curves for young-growth ponderosa pine in northern Arizona (from Minor 1964).

FIGURE 10.24 Schematic view of the distribution of tree crowns by tolerance class and height for four stages of succession initiated by severe disturbance: (a) Before disturbance, (b) immediately after disturbance (5 years), (c) midpoint of succession (50–75 years), (d) full recovery and return to predisturbance conditions (T = tree of tolerant species, N = tree of intermediate species, I = tree of intolerant species) (from Young 1982).

The Concept of Shade Tolerance

The concept of shade tolerance is critical to understanding silviculture and must be addressed before proceeding with stand treatments. "Shade-tolerant" species can grow in the shade of others while shade intolerant species grow poorly, if at all, under the same conditions. This fact is critically important for understanding forest succession and for successfully regenerating managed forests (Figure 10.24).

To illustrate the importance of shade tolerance, we will use the example of loblolly pine, a species of major economic importance in southern forests. Loblolly pine cannot grow in stands of larger and older trees that have a high degree of canopy closure. To establish loblolly pine, it is usually necessary to cut all of the overstory trees in a designated area to remove the influence of shade.

Additional examples of species needing large amounts of sunlight include ponderosa pine, yellow poplar, and Virginia pine. All of these species are categorized as shade intolerant. In contrast, northern hardwoods such as sugar maple and beech grow well under a relatively closed canopy and can be regenerated without concern for the shade by trees occupying the overstory (Perry 1994). Examples of western species that are shade tolerant include Engelmann spruce, redwood, and western hemlock. Between the extremes of shade tolerant and intolerant species are trees such as Douglas-fir, white oak, and sycamore.

When subjected to increasing levels of light, shade-tolerant species will reach their compensation point at lower light intensities than will intolerant species. The compensation point is met when light is sufficient to allow photosynthesis to occur at a rate equal to respiration. Above the compensation point, photosynthesis makes a net contribution to tree growth. Below this point, more energy is being consumed than is produced. Overall, this means that shade-tolerant trees can grow under light levels that would result in a net loss of energy in more light-demanding species.

Silvicultural Systems

Stand management is an intensively planned activity involving a silvicultural system. A silvicultural system is a planned program of treatments throughout the life of the stand to achieve stand structural objectives based on integrated resource management goals (see Chapter 5 for planning concepts). A silvicultural system includes integrated harvesting, regeneration, and stand-tending methods or phases. It covers all activities for the entire length of a rotation or cutting cycle. Harvesting techniques (also called regeneration methods because they are the first step in establishing the next stand) are coupled with subsequent intermediate treatments, such as thinning and weed control, to form silvicultural systems. Any reproduction, thinning, or other intermediate method applied without consideration for the other steps needed to manage a stand from establishment to harvest fails to meet the integrative requirements of silvicultural systems (Daniel et al. 1979). These are simple treatments imposed independently sometime between the establishment and the harvest phases.

Cutting and Reproduction Methods

The reproduction methods can be categorized into those that produce even- and uneven-age stands from seed and those produced by coppicing. Clear-cutting, seed-tree, and shelterwood methods are used to produce even-age stands. The selection method is used to produce uneven-age stands.

Clear-cutting. This method removes the entire stand with one cutting (Figure 10.25). Reproduction is obtained by artificial seeding and planting, by natural seeding from adjacent stands, or by regeneration from the cut trees themselves. In the strict sense, virtually all trees are cut, but at times all trees that cannot be used profitably are left and cutting may resemble a selection thinning. Species that become established naturally with full exposure are assumed to benefit from clear-cutting. Heavy partial cutting can provide the same benefit. In North America, this method has been applied extensively to Pacific Coast Douglas-fir forests. Douglas-fir is replaced by more tolerant species in the absence of wildfire. Western hemlock, western red cedar, and true firs (*Abies* spp.) are tolerant species.

Seed-Tree Method. This method calls for removing all trees except a small number of trees left to provide seed for natural restocking. Trees can be spaced to assure uniform seed distribution or grouped to provide wind protection (Figures 10.26 and 10.27).

Shelterwood Method. This method removes part of the stand to allow room for reproduction. Mature timber is removed in a series of cuttings that extend over a rela-

FIGURE 10.25 A clear-cutting of Douglas-fir where the stand of timber in the background has seeded the logged area (courtesy of American Forest Products Industries).

FIGURE 10.26 Seed-tree method (from Ministry of Forestry, British Columbia Forest Service).

tively short time (Figures 10.28 and Figure 10.29). The remaining trees shelter the new trees until they are established; then they are removed. This approach prevents the site from being dominated by species that require open exposure to sunlight and encourages even-age reproduction.

Selection Method. This method involves harvesting the oldest or largest trees in a stand at repeated intervals. The interval between cuts in the same stand is called the cutting cycle. The entire stand is never completely cut (Figures 10.30 and 10.31). There is no rotation age at which the mature crop is harvested, as is done under even-age management. Trees are cut individually or in small groups. Reproduction is an ongoing process because openings are created indefinitely. The age of maturity and the cutting cycle determine the number of age classes present in the stand. Remaining trees protect seedlings and shade-tolerant species are encouraged.

FIGURE 10.27 Application of the seed-tree method on loblolly pine forest in South Carolina (courtesy of USDA-Forest Service, Dan Todd photographer).

FIGURE 10.28 A shelterwood cutting in Northern California.

Coppice Forest Methods.   This category is addressed separately because regeneration is derived primarily from sprouts and suckers as opposed to seed (Figure 10.32). Trees may grow from dormant buds in stumps. The coppice method can be used to provide browse for herbivores such as deer.

Intermediate Treatments

An intermediate treatment is any manipulation in a stand applied between regeneration periods. Such treatments are used to ensure the desired composition, stem

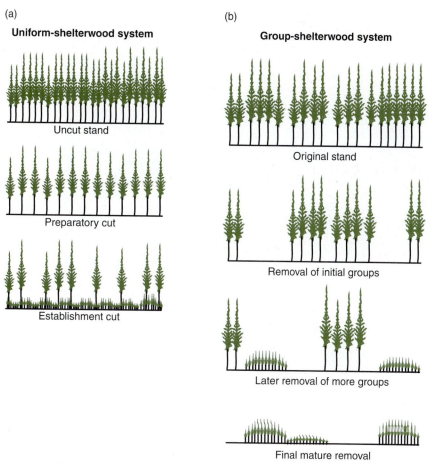

FIGURE 10.29 (a) Uniform-shelterwood and (b) group-shelterwood methods (courtesy of Ministry of Forestry, British Columbia Forest Service).

FIGURE 10.30 Single-tree-selection method (a) and group-selection method (b) (courtesy of Ministry of Forestry, British Columbia Forest Service).

FIGURE 10.31 The selection method of cutting applied to a mixed hardwood stand in Michigan (courtesy of USDA-Forest Service, Leland Prater photographer).

Coppice system

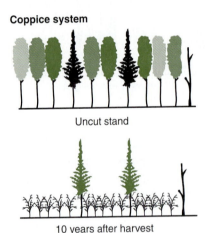

Uncut stand

10 years after harvest

FIGURE 10.32 The coppice method (from Ministry of Forestry, British Columbia Forest Service).

quality, spacing, and growth performance in a developing stand. Release cuttings cover numerous types of thinning operations done under different stand conditions. Thinning is aimed primarily at reducing stand density to increase growth of residual trees. Commercial thinning involves use of trees that would ultimately die of suppression. Thinnings are called precommercial if the trees cut are too small to be merchantable. In intensively managed stands, trees can be pruned to improve main stem form and reduce knotted wood. Fertilization can accelerate growth but is commonly used only in intensively managed stands. Prescribed burning reduces fire hazard and controls understory competition.

Emergency Cuttings or Thinnings

Salvage and sanitation cutting represent a separate category of treatments applied on an emergency basis. Salvage cutting removes trees that are dead, dying, or badly damaged, where factors other than competition are involved. Usually, the primary goal is to salvage their timber value before they decay. Sanitation thinning (or cutting) removes trees that have been attacked or are vulnerable to attack by insects or pathogens. This is done to reduce inoculum or insect populations so that the pest is less likely to spread to other trees.

FIGURE 10.33 Harvesting timber on the Lassen National Forest in California (courtesy of U.S. Dept. of Agriculture).

TIMBER HARVESTING

Timber harvesting refers to cutting, felling, and removal of wood for off-site use (Figure 10.33). Lopping is done after felling. It involves the cutting of branches and tops of large trees so that the resultant slash will lie close to the ground and decay more rapidly. A landing is any place on or adjacent to a logging site where logs are assembled for further transport. *Yarding* refers to moving logs from the point of felling to a landing. Cable yarding involves the use of a steel cable attached to a powered winch. Variations of this system include high-lead yarding and skyline yarding. A skidder is a rubber-tired machine with a cable winch or grapple used to drag logs out of the forest. *Skidding* can also involve the use of tractors, horses, or specialized logging equipment to collect felled logs at a landing. Skidding methods vary in their impact on soils and the remaining stands.

TREE PLANTING

Forests can regenerate by natural means from natural seeding or coppicing as discussed earlier. In some cases, such as heavily burned areas, seed trees may not be available. Those areas can be artificially seeded or planted to speed the recovery process. On productive forest land, planting may accelerate early growth and adequate stocking. For whatever the purpose intended, millions of acres have been either seeded or planted over recent decades (Figure 10.34). Planting is usually preceded by some type of site preparation. This may involve mechanical or herbicide treatments to reduce weed competition. On poorly drained soils, the land surface can be altered by forming beds that permit water to drain more freely from the site.

Seedlings are produced under nursery or greenhouse conditions. After a few years of growth, nursery seedlings are lifted from the soil while in the dormant condition and stored under refrigeration until planting. Their roots are bare until they are replanted. Greenhouse grown seedlings are produced in various types of containers. They require less time (months instead of years) to develop because of the optimal growing conditions provided. Their root systems are removed from the containers before planting. In contrast to bare-rooted seedlings, the roots of containerized seedlings

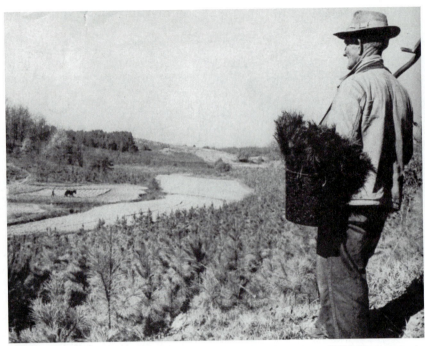

FIGURE 10.34 Pine trees planted on a hillside in Mississippi (courtesy of U.S. Department of Agriculture).

FIGURE 10.35 A mechanical tree shaker comes in handy when it is time to harvest ripe cones from a superior pine tree in a seed orchard (courtesy of U.S. Dept. of Agriculture).

remain protected by the growing media. Because they do not lose contact with the soil media, they experience less transplant shock under certain conditions. Over the years, considerable knowledge has been gained concerning which types of seedling planting stock (i.e., "stock types") will perform best under certain conditions.

Overall, successful artificial regeneration requires careful attention to the details of seed source, nursery practices, and planting practices, as well as a thorough evaluation of environmental conditions (Tappeiner et al. 1997). Seed selection zones are used to ensure that future trees are adapted to the planting site. The forest industry has made large investments in genetic improvement of planting stock. Superior tree selections made over many decades have enabled the industry to plant seedlings that grow more rapidly and produce better forest products (Figure 10.35).

FIGURE 10.36 Fighting forest fires involves a variety of equipment and skills (courtesy of USDA-Forest Service).

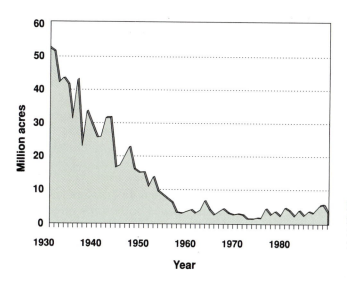

FIGURE 10.37 Trends in U.S. forest land burned by wildfire, 1930–1989 (from Powell et al. 1993).

FOREST PROTECTION

Forest Fire

Fire, insects, and disease are natural components of forests, but they cause considerable mortality and growth loss. After the great fires of the early 1900s burned massive areas and took many lives, the U.S. Forest Service and other government agencies joined forces to develop highly effective fire suppression programs (Figure 10.36). In the decades that followed, fire loss steadily declined (Figure 10.37). Fire science was also born as a forestry discipline. Better knowledge of fire behavior and fire danger ratings have greatly reduced loss of life and property from wildfire. Risks to property remain significant, however, especially where housing development has mushroomed in fire-prone forests.

FIGURE 10.38   The fir engraver beetle attacked this tree in California. The photo shows the egg galleries and larval mines 35 days after attack (courtesy of USDA-Forest Service).

Insects

The threats insects pose to forests are less obvious than fire, but mortality and growth losses can be substantial. The types of insects that threaten forests are highly diverse, but can be grouped by the type of damage they cause. Broad groups include defoliators, sapsuckers, bud, twig, seedling, and root feeders, and cone and seed insects. Bark beetles have been especially destructive over recent decades (Figure 10.38). The southern pine beetle (*Dendroctonus frontalis*) is the most damaging insect in the southeastern United States. It is responsible for killing large numbers of southern pines over wide areas. Other species of the same genus have caused extensive damage to western pines and to spruce found in Alaska.

The larvae of several moths and butterflies damage both conifers and deciduous species, but their greatest damage is to conifers. Spruce budworms infest large areas in the northeastern United States. The Douglas-fir tussock moth has caused extensive damage in the northern Rocky Mountains. The gypsy moth, prevalent in the East, primarily defoliates oaks. However, the larvae also feed on aspen, basswood, paper birch, and willow. Tent caterpillars feed on deciduous species. They live in tentlike silken webs that damage but generally do not kill infested trees. Insects that feed on cones, seed, and acorns can greatly reduce natural reproduction in some years. These are just a few examples of some better-known insect pests.

Insect control is extremely difficult because of the large areas sometimes infested, financial cost of control, and potentially adverse effects to the environment. Because insecticides are increasingly viewed as a threat to nontarget beneficial insects, other control measures must be found. The value of natural pest enemies and the promotion of better forest health are receiving much greater emphasis in modern forestry.

Disease

Numerous groups of fungi, bacteria, and viruses damage trees. Fungi are especially damaging; particularly the rusts and those fungi causing cankers, heart rot, wilts, and damping-off. Cankers mark locations where fungi have killed stem tissue. Heart rot, which devalues the wood, is especially damaging to mature stands. Wilts are caused by fungi that plug the water transport system of trees. The most notable example is Dutch elm disease. Damping-off occurs when newly emerged seedlings become infected and die. This problem is most significant in forest nurseries.

Trees can also be damaged by nematodes, which feed on tree roots or invade the vascular system causing wilt. Dwarf mistletoes (parasitic higher plants) cause extensive growth and mortality losses in conifers; especially in western pines.

Exotic Pests

Pests introduced from abroad have increasingly become a problem. They are especially damaging because native tree species have not developed natural defenses against them as they have against native pests.

Abiotic Factors

Air pollution (see Chapter 6) reduces forest health over vast areas. Particularly troublesome are the acid precipitation, heavy metals, and ozone created by automobiles and industry. Acid precipitation forms high in the clouds where emissions of sulfur dioxide and nitrogen oxides react with water, oxygen, and oxidants. Acid precipitation and other pollutants can travel far from their places of origin, causing stress to forest species with varying abilities to respond. Pollution effects on forest health are complex and difficult to separate from stress-causing factors related to drought, overstocking, and planting species in plantations outside zones where they naturally grow. Acidic deposition can reduce cold hardiness. Ozone absorbed through leaf stomata impairs cell function, reducing photosynthesis.

FOREST ECOSYSTEM MANAGEMENT

The Need for Management Change

Over the past 25 years foresters have been increasingly criticized for their forest management practices. Loud protest began in the 1970s when the public became concerned over the large clear-cuts on public lands in the western states. In the public's opinion, clear-cuts threatened the land's ability to recover, displaced wildlife, damaged fish habitat, threatened water quality, and degraded the aesthetic quality of forest landscapes (Figure 10.39). Most clear-cuts were done responsibly, but some involved little regard for slope and the fragile nature of forest soils. Across the nation concern grew over the increasing fragmentation of forests by timber cutting and urban sprawl. In the Pacific Northwest, the amount of old growth forest has declined over 50% in the last 60 years (Bolsinger and Waddell 1993). Such destruction is perceived to threaten and endanger native wildlife and fish populations (Marcot 1997). Cutting of old growth forests of the Pacific Northwest has caused one of the most fierce environmental controversies to date, epitomized by concern for survival of the spotted owl. These forests have become symbolic of the conflict between immediate material need and the need to conserve our natural resources for future generations. Forest stewardship concerns go far beyond the spotted owl controversy. They can be heard throughout the world, from tropical rain forests to the boreal forests of Siberia. In the United States, public concerns include the loss of nongame forest wildlife (Bolen and Robinson 1995); the conversion of biologically rich, mixed hardwood forests to pine plantations; the drainage of forest wetlands; and forest health deterioration resulting from fire, exotic pests, pollution, and global warming.

FIGURE 10.39 Land impaired by destructive logging practices (courtesy of USDA-Forest Service).

Ecosystem Management Defined

With the hope of regaining public trust, the U.S. Forest Service and other federal agencies have redefined the practice of forestry by putting greater emphasis on biodiversity, aesthetics, and forest health. The name given to this new approach is *ecosystem management*. Silviculture is being revised in step with policy changes and new laws passed to regulate forest management.

A major change has occurred in how forests are perceived for management purposes. Whereas silviculture previously was practiced at the stand level, the emphasis is now given to forest ecosystems and forest landscapes. A change in management scale was considered necessary to provide greater protection for natural biodiversity, watersheds, and the sustainability of natural process in general. It has provisions for monitoring levels of forest fragmentation. The adoption of geographic information systems (GIS) and computerized decision support systems (DIS) provide the tools for gathering and ciphering the vast amounts of information needed to manage at the landscape level. Thus ecosystem management is holistic and integrated management of all resources in a geographical area with large enough natural boundaries to include all of the influential factors (see Chapter 23 for a more comprehensive discussion).

Silviculture and Ecosystem Management

The Pacific Northwest will serve as a prime example for discussing how silviculture is being revised in accord with ecosystem management principles. One aim of silviculture has been to develop managed forests that more closely resemble native forests in structural diversity. Forest structure can be viewed on both horizontal and vertical planes. Horizontal structure refers to differences in plant stature that might be encountered over a given area. Where patches of trees meet shrub or grass communities across a landscape, edges (ecotones) are created. Here the horizontal structure is said to be diverse. A landscape having this structure is believed to more naturally support more wildlife species (Hunter 1990). Vertical structure is well developed where trees and shrubs of diverse heights occur together in multiple canopies. The presence of multiple canopies supports a richer assortment of forest wildlife from small mammals to predatory birds.

Several approaches have been discussed to improve the structure of managed forests. One practice being recommended is to combine thinning with extended rotations. Such proposals usually include a series of silvicultural treatments during development of the stand to ensure creation of specific structural elements.

By extending rotations 50 to 300%, trees are allowed to grow to greater maturity and to develop more structural diversity than they would under conventional (shorter) harvest schedules (Franklin et al. 1997). After some time, trees larger than those harvested today will provide an upper canopy while openings created by natural disturbance will become occupied by smaller trees and shrubs. Longer retention provides numerous additional benefits. Managed forests occupying watersheds will be disturbed less frequently, visual impairment will be reduced, and the quality of the wood harvested will be superior. In older forests, carbon is held in vegetation and soil longer from the atmosphere, thus reducing atmospheric CO_2.

Thinning had not been widely applied in the Pacific Northwest until recently because markets for small wood materials were poor and large volumes of old timber were available. Because thinning usually accelerates understory development and succession, it can improve structural diversity. It can provide even more structural improvements if shade-tolerant conifers are planted in the understory to develop multilayered stands (Tappeiner et al. 1997). Thinnings that promote the development of understory vegetation and small openings can support increased wildlife populations (Curtis 1997). At the same time, thinning can provide income during the intermediate stages of stand development.

Pruning has been advocated and used primarily to increase the amount of clear wood produced in young stands. If pruning is initiated prior to stand closure, removal of lower branches can increase the amount of light reaching the forest floor and favor development of the woody and herbaceous understory. At present, pruning is applied mostly after stand closure has occurred.

Another recommended approach is to retain aggregates of living trees that will "life boat" diverse groups of plants and animals when trees are harvested (Franklin et al. 1997). Aggregates of trees can protect forest microclimate and the forest floor, which should be left undisturbed. In this manner, beneficial fungi, mosses, lichens, and forest animals, including insects and other invertebrates, will survive and spread over time into the surrounding cut over forest area. This approach resembles the shelterwood method mentioned previously. However, the trees left behind must be sound and windfirm so they persist in the forest landscape. Dominant and codominant trees should be left uncut. They will provide critical protection and an energy source for forest fungi and other heterotrophs.

To overcome the adverse effects of forest fragmentation on wildlife, conservation biologists have generally recommended connecting "islands" of intact forests with narrow corridors of intact forest. The area surrounding the islands and the corridors may be sacrificed because of conversion to other uses. However, the retention of forest aggregates over managed landscapes, in conjunction with surrounding unharvested areas (removed from harvest), is a superior approach to maintaining biodiversity (Franklin et al. 1997). The movements of plants and animals will be less impaired and conservation will be better served. Conditions in the landscape matrix—that is, the "sea" surrounding undisturbed forest islands—are the most important factors controlling connectivity in landscapes.

The Clear-Cutting Controversy

Clear-cutting typically involves the removal of all trees from a given area in a single cutting followed by establishment of a new, even-aged stand. A clear-cut may consist of the whole stand or patches or strips of the stand (Figure 10.40). Nearly 65% of the U.S. is harvested by clear-cutting, which generally requires less cost and planning than other timber harvesting methods. Clear-cutting has other advantages that include less road building, greater timber harvest per acre, more efficient reseeding with improved stock, shorter time requirement for stand regeneration, and more efficient removal of fire, insect, and storm damaged stands (Miller 1990). If small, irregular areas are cut, clear-cutting can improve habitat for many game species such

FIGURE 10.40   Patch clear-cuts on the Kootenai National Forest in Western Montana (courtesy of USDA-Forest Service).

as deer, elk, turkey, and grouse. Clear-cutting approaches with and without consideration of wildlife are shown in Figure 10.41.

Big block clear-cutting has become an important controversy in the United States, Canada, and countries with tropical forest. It often replaces diverse, uneven-aged stands of old growth with single-species monocultures that lack resistance to disease, insects, and fire (Miller 1990). These monocultures offer poor habitat for most wildlife species. Landscape aesthetic quality is severely impaired by large block clear-cuts. Severe soil erosion, water pollution, flooding, and landslides have often resulted from large block clear-cuts on mountainous (steeply sloping) terrain (see Figure 10.39). Repeated clear-cutting can be quite damaging to the site because of soil erosion and compaction from logging equipment. Public opposition to clear-cutting has led to more emphasis on the alternative cutting methods previously discussed (seed-tree, shelterwood selection, coppice methods).

Ecosystem management emphasizes the retention of large trees, snags, and downed logs that are found naturally where disturbances such as fire or wind have occurred (Franklin et al. 1977, Tappeiner et al. 1997). These were once considered a liability because they can be hazardous to forest workers and represented a loss from forest production. They are now recognized to provide critical habitat for diverse animals, including cavity nesting birds. Consequently, federal agencies promote their continued existence in forest management plans (Thomas 1979).

Ecosystem Management and Fire

Over past decades, fire danger ratings were standardized and predictions of fire spread and intensity have become increasingly reliable. However, the natural consequences of suppressing wildfire in areas where it has played a major role in shaping forest composition and structure have recently become better understood. In some forests, particularly those in the West, fire suppression has allowed tremendous fuel loads to accumulate, thus raising the probability of catastrophic fire (Figure 10.42). Where fire once occurred at frequent intervals, stand-replacing fires became unusual.

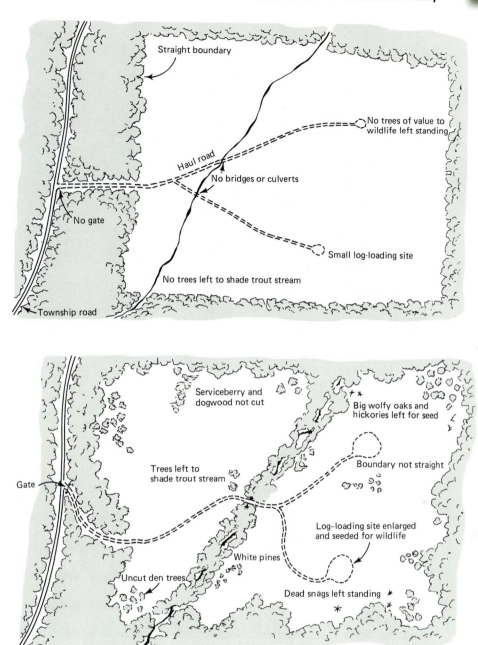

FIGURE 10.41 Clear-cut without (a) and with (b) consideration for wildlife (from Hassinger et al. 1981).

Mature trees survived as fire eliminated the understory and smaller diameter trees. In many areas the stage is presently set for all trees to be killed and for fires to be so hot that soils are damaged. Forests are also threatened by insect infestations, which continue to build in thick, overstocked stands.

In response to the adverse effects of suppression, forest and park managers increasingly view fire as nature's ally and not a force to be eliminated. An increasingly promoted strategy lets naturally ignited fires burn until they extinguish themselves. They are monitored to avoid human risks.

Prescribed fires as well as natural fires have become part of the working tools of today's fire managers, who have an ever-expanding set of land management goals

FIGURE 10.42 Catastrophic wildfires due to fuel buildup and home construction are becoming an increasing problem on forest lands in the West (courtesy of USDA-Forest Service).

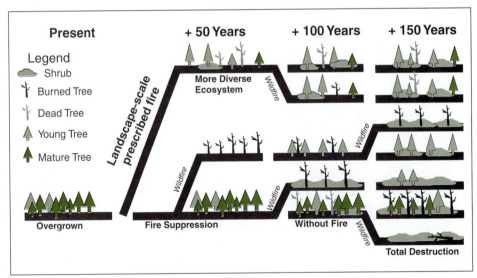

FIGURE 10.43 The risk of not conducting prescribed burns (courtesy of USDA-Forest Service).

(Figure 10.43). These goals include protection of commercial timber, managing fuels around rural developments (also called the urban interface), and reintroducing or maintaining the natural role of fire in park and wilderness ecosystems (Agee 1997). Prescribed natural fire use was severely criticized after several allowed fires in Yellowstone National Park required millions of dollars to control, and a prescribed fire escaped control and burned more than 200 residences in Los Alamos, New Mexico. However, on balance, the benefits outweigh the risks in many areas. An approach being used over much of the West is to apply thin-logging and salvage logging to avoid the site deterioration that can result from a wildfire.

Gap Dynamics

Our understanding of succession has grown from the study of forest gaps by forest ecologists. Gaps in forest canopies can differ greatly in size, ranging from a single tree fall to the clearing of large tracts of forests by hurricanes, fire, or humans. These studies increasingly point to the importance of gaps in supporting biodiversity. According to Bazzaz (1996), forests are in a state of continuous change resulting from the different patches of forest growth that develop in canopy gaps created by disturbance. Depending on gap size, patches can be trees of ages and species different from those dominating the main canopy. Branch gaps, those created when branches of existing trees are lost, might be filled simply by branches of existing trees extending into the open space. Gaps created by the loss of single trees can be filled by neighboring trees. On being released from competition, the small trees in gaps undergo accelerated growth. Large gaps and forest clearings in which much of the resident vegetation is destroyed by fire can be revegetated through seed dispersal from adjacent communities. Overall, the rate of gap creation and filling is determined by several intrinsic and extrinsic factors, and differs greatly among regions.

Forest Succession and Ecosystem Management

We will briefly discuss the relevance of forest succession to forest ecosystem management, referring the reader to Chapter 3 for a more complete discussion of succession. From the work conducted at the Hubbard Brook watershed over several decades, it appears that biomass accumulation following clear-cutting occurs in four stages: (1) reorganization, (2) aggradation, (3) transition, and (4) steady state.

The first stage, "reorganization," marks the beginning of a new stand and lasts for one to two decades. During this time, the ecosystem actually loses total biomass despite the accumulation of new living biomass in the plants. This stage is characterized by high availability of sunlight, nutrients (decaying roots and aboveground litter), and ample water (few plants are competing for it). The time necessary for completion of the reorganization stage varies greatly from one environment to another yet remains critical to ecosystem recovery (Perry 1994). When forest disturbance is severe enough to cause soil loss, the site may revert back (i.e., retrogress) to a former state requiring colonization by more primitive pioneering plants.

During the "aggradation" stage, biomass accumulates to a peak value. The aggradation stage is further characterized by increased tree mortality caused by inter- and intraspecific competition for resources (Landsberg and Glover 1997). This stage marks the point where foresters intervene to reduce competition by thinning the stand. The growth of the remaining trees is accelerated because thinning "releases" them from competition for light and nutrients. The aggrading phase can vary considerably across forest species. The long-lived conifer forests of the Pacific Northwest, for example, accumulate biomass (both living and dead) for hundreds of years, whereas this period is much briefer in forests dominated by shorter-lived trees (Perry 1994).

During the "transition" stage, biomass declines. In this stage, the importance of understory plants is renewed (Landsberg and Glover 1997) because spaces occupied by tree canopies become vacant as some trees die, forming gaps.

The "steady state" arrives when the total amount of biomass fluctuates about the mean resulting in periods when more and less than average amounts of biomass have been accumulated. This stage corresponds to the climax stage of forest succession and is characterized by large accumulations of living biomass and coarse woody debris. This stage can have considerable structural diversity (Landsberg and Glover 1997), meaning there are trees of different species and heights (some dead, some alive) and in different stages of decomposition. Consequently, it can support diverse species, including all sorts of fungi, invertebrates, cavity nesting birds, small mammals, and northern spotted owls. However, according to Botkin (1990, 1993) forests will not persist in the climax stage indefinitely as previously reported. With increasing age, the forest becomes deficient in nutrients as it continues to "leak" them. As living biomass is continually lost, the forest can revert to a shrub or grass stage.

The upshot of Botkin's description is that no forest lasts forever. Nature is not static and changeless and there is no such thing as a balance in nature resembling a pendulum that finds its point of rest and remains there.

FOREST CONSERVATION ISSUES

Deforestation and Global Warming

Two issues concerning forests are of global concern—deforestation and global warming. Deforestation is primarily a problem in countries outside the temperate zone. The conversion of forests in temperate regions to agriculture land has been reversed and forest area is now actually increasing (Figure 10.44). By contrast, the developing countries in the Tropics, where much of the earth's biodiversity exists, are still in a forest conversion mode (Sedjo 1995). Deforestation remains intense in areas such as Latin America and Africa. Major movements are underway among nongovernmental agencies to reverse forest loss. Deforestation has been the focus of numerous world conferences of one type or another over the past decade.

Environmental decline associated with intensively managed forest plantations has received considerable comment, both positive and negative. In response to criticisms, numerous strategies have been developed to make plantations more habitable to diverse wildlife groups. On balance, plantations can be intensively managed to spare natural forests from human disturbance. Forests cover about 3.4 billion hectares of the earth. To meet world wood demands, 20 to 40% of the growth from these forests must be harvested. High-yield forests, which are now 4 to 8% of Earth's forest area, might meet future world wood demands, depending on technological breakthroughs (Sedjo and Botkin 1997).

Forests could prove to be a close ally as global warming progresses (see Chapter 6). Because forests play a key role in sequestering atmospheric carbon, world forest restoration is considered a high priority by governments and environmentalists alike. The world's forests have been estimated to contain up to 80% of all aboveground terrestrial carbon and about 40% of all belowground terrestrial carbon (soil, litter, and roots) (Ciesla 1995). About 37% of this carbon is stored in low-latitude (tropical) forests, 14% in midlatitude (temperate) forests and 49% in high-latitude forests (Dixon et al. 1994).

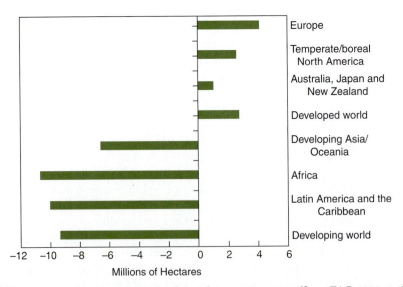

FIGURE 10.44   Losses and gains in forest land area from 1980 to 1990 (from FAO 1997 statistics).

A leading group of atmospheric scientists have predicted an increase in mean global temperature of about 1.5–4.5°C (2.7–8.1°F) in the next century in response to the growing accumulation of greenhouse gasses. Temperature changes on this scale would provide natural systems with a warmer climate than any experienced during the last 100,000 years. Some scientists say this change in the global climate could occur many times faster than it did during any previous episode. The rate of change poses a more serious threat than the magnitude of change. If global warming occurs as projected, indirect forest damage could occur over vast regions. This would involve major tree die-offs in many areas and shifts in the potential areas of growth for each tree species. Insect infestations and fire frequency will likely become more frequent. Forests already threatened by pollution, fragmentation, and poor stand condition will face yet another major threat. Sea level changes alone could pose a serious hazard to low-lying coastal forests. The mangrove forests of the tropical regions are especially vulnerable.

Wastepaper Recycling

One way to conserve forests is through greater recycling of wastepaper. Up to 50% of the world's wastepaper could be recycled, but only about 30% actually is. The United States, the world's highest consumer of paper, recycles about 30%. This compares unfavorably with most European countries, which recycle about 40%. The key to increasing recycling is requiring people to separate paper from other waste materials (Miller 1990). Otherwise, wastepaper dealers will not purchase it. Another approach is for governments to provide businesses with incentives for recycling. Currently the U.S. federal government is mandated to use recycled paper as much as possible.

Reducing Paper Use

Several approaches are available to reduce paper use. These include reduction of overpackaging, more complete use of paper by students (use of both sides), increased use of the Internet for advertising, greater purchasing of food and other goods in bulk quantities, and more use of electronic media in information transfer and storage. Product overpackaging is a major contributor to paper use and waste (Miller 1990). Oversized packages are used to make consumers think they are getting more for their money. According to Miller (1990), product packaging uses about 65% of the paper and 15% of the wood in the United States. Abramovitz and Mattoon (2000) offer a more complete discussion of how paper use can be greatly reduced by computerization and other technological developments.

CONCLUDING REMARKS

Forests play a key role in meeting our needs for material goods, but require careful management to avoid damage to all the other services they provide, including maintaining the earth's biodiversity. The forests of the United States and those of other developed countries have actually grown in area as deforestation continues at an alarming rate in countries experiencing a period of rapid economic transition. Key concerns in the United States are the condition of secondary forests and conservation of old growth forests. Forest management is in a period of rapid change because of the diverse demands of a changing society. However, its tools and concepts are becoming increasingly refined to meet a broad set of goals. Over the next decade, forests will be increasingly valued and protected in their natural state. At the same time, plantation forestry will provide an increasing share of the world's material timber needs. Global warming poses a serious threat to natural forests because some species will have too little time to adjust. Forests can, however, be used to reduce the threat.

LITERATURE CITED _____

Abramovitz, J. N. 1998. *Sustaining the world's forests. State of the world 1998.* Washington, DC: World Watch Institute.

Abramovitz, J. N. and A. T. Mattoon. 1999. *Reorienting the forest products economy. State of the world 1999.* Washington, DC: World Watch Institute.

Abramovitz, J. N. and A. T. Mattoon. 2000. *Recovering the paper landscape. State of the world 2000.* Washington, DC: World Watch Institute.

Agee, J. K. 1997. Fire management for the 21st century (pp. 191–202). In K. A. Kohm and J. F. Franklin, eds., *Creating a forestry for the 21st century: The science of ecosystem management.* Washington, DC: Island Press.

Allen, S. W. and G. W. Sharpe. 1960. *An introduction to American forestry.* 3rd ed. New York: McGraw-Hill.

Bazzaz, F. A. 1996. *Plants in changing environments.* New York: Cambridge University Press.

Bolen, E. G. and W. L. Robinson. 1995. *Wildlife ecology and management.* 3rd ed. Upper Saddle River, NJ: Prentice-Hall.

Bolsinger, C. L. and K. L. Waddell. 1993. *Area of old-growth forests in California, Oregon, Washington.* USDA Forest Service. PNW Station Resource Bulletin PNW-RB-197.

Botkin, D. B. 1990. *Discordant harmonies: A new ecology for the twenty-first century.* New York: Oxford University Press.

Botkin, D. B. 1993. *Forest dynamics: An ecological model.* New York: Oxford University Press.

Botkin, D. B. and E. A. Keller. 1995. *Environmental science: Earth as a living planet.* New York: John Wiley & Sons.

Ciesla, W. M. 1995. *Climate change, forests and forest management.* FAO Forestry Paper 126. Rome.

Connell, J. H. and R. O. Slatyer. 1977. Mechanisms of succession in natural communities and their role in community stability and organization. *American Naturalist* 111:1119–1144.

Curtis, R. O. 1997. The role of extended rotations (pp. 165–170). In K. A. Kohm and J. F. Franklin, eds., *Creating a forestry for the 21st century: The science of ecosystem management.* Washington, DC: Island Press.

Daniel, T. W., J. A. Helms, and F. S. Baker. 1979. *Principles of silviculture.* 2nd ed. New York: McGraw-Hill.

Darr, D. R. 1995. U.S. forest resources (pp. 214–215). In E. T. LaRoe, G. S. Farris, C. E. Puckett, P. D. Doran, and M. J. Mac, eds., *Our living resources: A report to the nation on the distribution, abundance, and health of U.S. plants, animals, and ecosystems.* Washington, DC: U.S. Department of the Interior, National Biological Service.

Daubenmire, R. F. and J. B. Daubenmire. 1968. *Forest vegetation of eastern Washington and northern Idaho.* Washington Agriculture Experiment Station. Tech. Bull. 60. 104 pp.

Dixon, R. K., S. Brown, R. A. Houghton, A. M. Solomon, M. C. Trexlar, and J. Wisniewski. 1994. Carbon pools and flux of global forest ecosystems. *Science* 263:185–190.

Eyre, F. H., ed. 1980. *Forest cover types of the United States and Canada.* Washington, DC: Society of American Foresters. 148 pp.

FAO. 1997. *State of the world's forests.* Rome: Food and Agriculture Organization of the United Nations. 199 pp.

Franklin, J. F., D. R. Berg, D. A. Thornburgh, and J. C. Tappeiner. 1997. Alternative silvicultural approaches to timber harvesting: Variable retention harvest systems (pp. 111–139). In K. A. Kohm and J. F. Franklin, eds., *Creating a forestry for the 21st century: The science of ecosystem management.* Washington, DC: Island Press.

Franklin, J. F and C. T. Dyrness. 1973. *Natural vegetation of Oregon and Washington.* USDA Forest Service General Technical Report PNW-8. 417 pp.

Hall, F. C. 1983. Forest lands of the continental United States (pp. 130–139). In USDA, *Using our natural resources. 1983 yearbook of agriculture.* Washington, DC: U.S. Government Printing Office. 671 pp.

Hassinger, J. C., E. Schwarz, and R. G. Wingard. 1981. *Timber sales and wildlife.* Harrisburg, PA: Pennsylvania Game Commission.

Hunter, M. L. Jr. 1990. *Wildlife, forests, and forestry.* Englewood Cliffs, NJ: Prentice-Hall.

Kohm, K. A. and J. F. Franklin, eds. 1997. *Creating a forestry for the 21st century: The science of ecosystem management.* Washington, DC: Island Press.

Küchler, A. W. 1964. *The potential natural vegetation of the conterminous United States.* American Geographical Society Special Publ. no. 36. New York: American Geographical Society.

Landsberg, J. J. and S. T. Glover. 1997. *Applications of physiological ecology to forest management.* New York: Academic Press.

Luken, J. O. 1990. *Directing ecological succession.* London: Chapman and Hall.

Marcot, B. G. 1997. Biodiversity of old forests of the west: A lesson from our elders (pp. 87–106). In K. A. Kohm and J. F. Franklin, eds., *Creating a forestry for the 21st century: The science of ecosystem management.* Washington, DC: Island Press.

McKnight, T. L. 1990. *Physical geography: A landscape appreciation.* 3rd ed. Upper Saddle River, NJ: Prentice-Hall.

Miller, G. T. 1990. *Resource conservation and management.* Belmont, CA: Wadsworth.

Ministry of Forests (B.C. Forest Service), The Province of British Columbia. 1995. Silviculture systems guidebook.

Minor, J. M. 1964. *Site-index curves for young-growth ponderosa pine in northern Arizona.* USDA, Rocky Mountain Forest and Range Experiment Station Res. Note 69. 6 pp.

Pastor, J., D. J. Mladenoff, Y. Haila, J. Bryant, and S. Payette. 1996. Biodiversity and ecosystem processes in boreal regions. In H. A. Mooney, J. H. Cushman, and E. Medina, eds., *Functional roles of biodiversity: A global perspective.* New York: John Wiley & Sons.

Perry, D. A. 1994. *Forest ecosystems.* Baltimore: Johns Hopkins University Press.

Pfister, R. D., B. L. Kovalchik, S. F. Arno, and R. C. Presby. 1977. *Forest habitat types of Montana.* USDA For. Serv. Gen. Tech. Rep. INT-34. 174 pp.

Powell, D. S., J. L. Faulkner, D. R. Darr, Z. Zhu, and D. W. MacCleery. 1993. *Forest resources of the United States, 1992.* Gen. Tech. Rep. RM-234. Fort Collins, CO: U.S. Forest Service, Rocky Mountain Forest and Range Experiment Station. 132 pp.

Reineke, L. H. 1933. Predicting a stand-density index for even-aged forests. *Journal of Agricultural Research* 46:627–638.

Schubert, G. H. and J. A. Pitcher. 1973. *A provisional seed-zone and cone-crop rating system for Arizona and New Mexico.* USDA For. Serv. Res. Pap. RM-1045. Fort Collins, CO: Rocky Mountain Forest and Range Experiment Station. 8 pp.

Schulze, E. D., F. A. Bazzaz, K. J. Nadelhoffer, T. Koike, and S. Takatsukl. 1996. Biodiversity and ecosystem function of temperate deciduous broad-leaved forests. In H. A. Mooney, J. H. Cushman, and E. Medina, eds., *Functional roles of biodiversity: A global perspective.* New York: John Wiley & Sons.

Sedjo, R. A. 1995. Forests: Conflicting signals (pp. 177–209). In R. Bailey, ed., *The true state of the planet.* New York: Free Press. 470 pp.

Sedjo, R. A. and D. B. Botkin. 1997. Using forest plantations to spare natural forests. *Environment* 39(10):14–20, 30.

Sharpe, G. W., C. W. Hendee, and S. W. Allen. 1976. *Introduction to forestry.* New York: McGraw-Hill.

Smith, D. M. 1986. *The practice of silviculture.* New York: John Wiley & Sons. 527 pp.

Spurr, S. H. 1979. Silviculture. *Scientific American* 240: 76–82, 87.

Tappeiner, J. C., D. Lavender, J. Walstad, R. O. Curtis, and D. S. DeBell. 1997. Silvicultural systems and regeneration methods: Current practices and new alternatives (pp. 151–164). In K. A. Kohm and J. F. Franklin, eds., *Creating a forestry for the 21st century: The science of ecosystem management.* Washington, DC: Island Press.

Thomas, J. W. (Tech. Ed.). 1979. *Wildlife habitats in managed forests: These Blue Mountains of Oregon and Washington.* Agriculture Handbook 553. Washington, DC: USDA Forest Service.

World Resources Institute. 1994. *World resources: 1994–95.* New York: Oxford University Press.

Young, R. A. 1982. *Introduction to forest science.* New York: John Wiley & Sons.

CHAPTER 11

Rangeland and Range Management

INTRODUCTION

The term *rangelands* refers to uncultivated land that provides forage and browse for large animals (Figure 11.1). Most rangelands in the world are characterized by low precipitation, thin soils, rugged topography, and/or cold temperatures. Generally, rangelands are managed using extensive (ecological) practices that center around controlling the kinds and numbers of grazing animals and their timing, frequency, and distribution of use. Unlike pasturelands, rangelands receive little in the way of routine agronomic inputs such as cultivation, irrigation, and fertilization. Most rangelands support native vegetation, but areas do exist in the western United States where nonnative plants such as crested wheatgrass have been established and maintained without periodic cultivation. Forage refers to standing plant material (grasses, forbs, shrubs) that is edible by livestock and wildlife. Grazing is the consumption of grasses and forbs, whereas browsing is the consumption of edible leaves and twigs from woody plants. Herbivory by grazing and browsing animals is the primary focus of rangeland management.

Historically the primary contribution of rangelands to humans has been forage for livestock. However, in countries such as the United States, where food supplies are abundant, water, wildlife, and recreation are overtaking forage as important rangeland products. Most rangelands, when properly managed, can provide all these products without being degraded. The concept of managing rangelands for a variety of uses and products is referred to as multiple use. Our discussion of range management follows Holechek (1994) and Holechek et al. (2001).

The Society for Range Management, which originated in 1948, is the primary professional organization that deals with rangelands and range management on a global basis. It publishes two journals (*Journal of Range Management* and *Rangelands*), and has published several books that are primary sources of information for range managers and scientists. These journals and books serve as primary sources of information for this chapter. Over the past 10 years the information available on range management in the United States and other countries has increased tremendously. We refer readers who wish to broaden their coverage of range management to basic textbooks by Vallentine (1989), Heady and Child (1994), and Holechek et al. (2001).

FIGURE 11.1   Rangeland in New Mexico grazed by cattle.

RANGELAND MANAGEMENT DEFINED

Range management involves the manipulation of grazing by large herbivores so both plant and animal production will be maintained or increased (Holechek 1994a). The art and science of range management developed during the early twentieth century in the western United States when it became evident that declining forage production, severe soil erosion, and heavy livestock mortality were caused by poorly controlled livestock grazing. Manipulations of the intensity, timing, and frequency of livestock grazing were the primary cornerstones of early range management. However, since the 1960s the scope has broadened to also include manipulation of fire, wildlife, human activities, and various other factors. Still, the distinguishing feature of range management is the control of livestock grazing. Presently range management is defined as the manipulation of rangeland components to obtain the optimum combination of goods and services for society on a sustained basis (Holechek et al. 2001).

Basic Range Management Concepts

Range management is based on five key concepts (Holechek et al. 2001):

1. Rangeland is a renewable source.
2. Energy from the sun can be captured by green plants, which can only be harvested by the grazing animal.
3. Rangelands supply humans with food and fiber at very low energy costs compared to those associated with cultivated lands. Ruminant animals are best adapted to use range plants. Unlike human beings, ruminants have microbes in their digestive systems that efficiently break down fiber, which is quite high in most range plants.
4. Rangeland productivity is determined by soil, topographic, and climatic characteristics.
5. A variety of "products" are harvested from rangelands including food, fiber, water, recreation, wildlife, minerals, and timber.

Generally forage production increases and more of the forage can be consumed by grazing animals without damage to the range as precipitation, temperature, and soil

FIGURE 11.2 The humid, tallgrass prairie range in central Oklahoma (top photo) can withstand a grazing intensity of around 50% use of forage. In contrast, grazing intensity on the arid Chihuahuan desert range in New Mexico (bottom photo) must be kept around 35% or less to avoid degradation.

depth increase. The flat, humid ranges in the Great Plains and Southeast are much more productive and resistant to grazing than the arid rangelands in the intermountain West. The failure to recognize that different management practices are required for humid areas compared with arid areas has been a major contribution to rangeland degradation (Figure 11.2).

TYPES OF RANGELAND

Climate, soils, topography, and human activities interact to cause differences in the types of rangeland that occur throughout the world. Basically rangelands can be categorized as grassland, desert shrubland, savanna woodland, forest, and tundra biomes (see Chapter 9). These biomes will be briefly discussed following Holechek et al. (2001).

FIGURE 11.3 Characteristics of the major range plant groups (from Gay 1965).

Grasslands

Grasslands are the most productive in terms of forage for livestock and wildlife. They occur on about 24% of the world's land surface and are dominated by plants in the family Gramineae (grasses). Grasses are characterized by a fibrous root system, hollow jointed stems, and fine, narrow leaves. Forbs, also important on grasslands, differ from grasses by having broader leaves, taproots, and solid, nonjointed stems (Figure 11.3). Shrubs and trees, a minor component of grasslands, have thick, woody above- and belowground parts. Shrubs differ from trees by branching at ground level rather than from a trunk above ground.

Climatically grasslands are favored by light, frequent rains during the summer, and dry winters. This maintains moist conditions during the growing period (summer) in the soil surface where most grass roots occur. Loamy and clay-loam soils favor grasslands because they slow moisture infiltration deep into the soil profile and keep it near the surface. Both the climate and soils of the Central Great Plains of the United States are highly favorable for grasslands (see Chapter 9).

Desert Shrublands

Desert shrublands occur under the driest climatic conditions and comprise about 30% of the world's rangeland. Generally, precipitation is erratic with infrequent storms of high intensity. Soils are often sands or sandy loams. This climate-soils combination causes the soil surface to be dry for extended periods creating unfavorable conditions for grasses. However, the coarse, deep-rooted shrubs, which have the capability to collect moisture from large portions of the soil profile, are well adapted to desert conditions.

Forests

Forests account for about 30% of the world's rangelands. Most forests are characterized by high annual precipitation because large amounts of precipitation are needed to support the greater biomass of trees compared with the biomass of grasses and shrubs. These wetter conditions cause greater leaching and lower fertility of forest soils compared with grassland soils. The clearing and burning of tropical forests to convert them into grazing land has been a major environmental concern in South American countries (see Chapter 9). There is doubt about the capability of soils in the tropical rain forests to sustain production of forage and crop plants.

Savanna Woodlands

Savanna woodlands are transition areas between grasslands and forests. They occupy about 8% of the world's rangelands and are typically characterized by scattered, low-growing trees with an understory of grasses. Many of the more humid grasslands will revert into savanna woodland without periodic fire.

Tundra

Tundra involves cold, treeless areas with low-growing grasses and shrubs in Arctic or high-elevation regions. Because the deeper portions of the soil remain frozen throughout the year (permafrost), trees do not grow well on the tundra. About 5% of the earth's surface is occupied by tundra.

HISTORICAL PERSPECTIVE

Range management as a profession had its beginning in the United States in the 1890s. In this period, biologists such as Jared Smith first defined the problems of uncontrolled livestock grazing on western rangelands. Some of the problems Smith identified included replacement of desirable forages with unpalatable plants, compaction of soil by livestock, high loss of soil during periods of torrential rain, increases in prairie dogs and jackrabbits, and reduction in grazing capacity. Smith (1899) was the first to recommend basic practices that included control of livestock numbers, range rest periods, water development, brush control, and range seeding. These have become cornerstones of modern range management.

   Published accounts of livestock grazing problems and scientific studies on rangelands generally do not exist from other parts of the world prior to 1900. Therefore many believe the science of range management was established in the United States. However, pastoral tribes in Asia and Africa have grazed livestock on rangelands for several thousand years. These herders used grazing rotation systems similar to some of the more sophisticated systems used in the United States today. Therefore, the concept of controlled livestock grazing on rangelands did not entirely originate in the United States. We refer the reader to Chapter 2 for a discussion of the history of rangeland policy in the United States.

In the twenty-first century, western rangelands have become increasingly important in providing open space, wildlife, recreation, and water for the American public. If present trends continue, livestock grazing will be deemphasized on the public rangelands in the West but will receive greater emphasis on Great Plains and eastern rangelands. During the past 15 years, many ranches in the West have subdivided into ranchettes of 40 acres or less and livestock are no longer produced. This trend will undoubtedly continue, and many more ranches will be subdivided. Public pressure has increasingly forced government agencies to emphasize recreation and wildlife on federal lands. As urbanization in the western states continues, this pressure will probably increase. The efficiency of livestock production in the West compared with the Great Plains and the East is much lower (Holechek and Hawkes 1993). The future use of rangelands in the United States will depend on what happens in other countries as well. As long as the world population grows at a rapid rate, pressure to increase livestock in developing countries also will grow. However, great opportunities exist for many of these countries to derive income from other uses of rangelands as well as livestock if appropriate government policies are implemented. This will be discussed later.

RANGELAND ECOLOGY

Grazing Effects on Range Plants

The most fundamental concept in range management centers around how excessive grazing influences the long-term status of palatable forage plants. Scientific study has established that excessive grazing of these plants causes their death and replacement by a succession of plants with lower productivity and often lower palatability; some are even poisonous (Figure 11.4).

Under moderate or light grazing levels the poisonous, unpalatable plants are at a competitive disadvantage because they invest part of their products from photosynthesis in poisonous compounds (e.g., alkaloids, oxalates, and glycosides) and appendages (e.g., spines, thorns, and stickers) that discourage defoliation rather than contribute to growth. In contrast, the palatable plants use their photosynthetic products mainly for growth in the form of roots, leaves, stems, rhizomes, stolons, and seeds. Under excessive defoliation rates, the photosynthetic capacity of the palatable plants is reduced to the point that the plants are unable to produce enough carbon compounds to maintain root systems, regeneration of leaves, respiration, and reproduction. Over time, they are replaced with unpalatable plants that have had lower rates of defoliation.

As excessive grazing continues, the palatable plants tend to be replaced by a succession of plants that increasingly are lower in palatability, lower in productivity, and often more poisonous. This process is referred to as retrogression. When defoliation is reduced to a correct rate, the palatable plants again have the competitive advantage and succession occurs back to the original or climax vegetation.

The driving environmental forces in succession are moisture (rainfall) and temperature. In wet, humid range types such as the southern pine forest in the southeastern United States or the tallgrass prairie in the eastern Great Plains, recovery after retrogression is both rapid and predictable. The climax plants will usually again

FIGURE 11.4 This fence line in New Mexico shows the effects of long-term proper grazing (left) and excessive grazing (right) on range vegetation. Productive, palatable perennial grasses dominate the range on the left, while the excessively grazed range on the right is dominated by poisonous, unpalatable forbs and shrubs.

dominate the site within 5 years if severe soil erosion has not occurred. In the drier range types such as the Chihuahuan desert, the palatable plants tend to be less resistant to grazing. Here retrogression can occur within just a few years under excessive grazing, but recovery, if it occurs, is a slow process often requiring 20 or more years. On some sites with severe soil erosion, only minor improvements have been observed after 20 or more years even under complete elimination of livestock grazing. In many areas, secondary succession to the original climax does not occur; rather a new and stable climax occupies the site.

Rangeland Condition and Trend

Range vegetation naturally shows considerable fluctuation through time because of climatic variability. However, several long-term range research studies reviewed by Holechek et al. (1999) show that sound grazing management practices maintain the desirable plants at much higher levels than poorly controlled grazing. Those plants that naturally occupy the site in the absence of severe disturbances—such as excessive grazing, fire, and cultivation—are collectively referred to as the climax. On grassland ranges, these plants are generally the most palatable and productive. However, in natural forests, desirable forage grasses and forbs are most plentiful when the trees are completely or partially removed by fire or logging.

Rangeland condition refers to the state of health of the range and is usually judged by the amount of climax vegetation that remains. J. E. Dyksterhuis in 1949 developed a system known as "the quantitative climax approach," which has been widely used to evaluate range condition. With this system, rangeland condition scores depend on the amount of remaining climax vegetation. Ranges with 76% or more of the climax are classified in excellent condition, those with 51 to 75% are in good condition, 26 to 50% are in fair condition, and 0 to 25% are in poor condition. This system has been widely applied by government agencies, such as the USDA Natural Resources Service, and works well on native grasslands.

Under the quantitative climax approach, plants are placed into categories of decreasers, increasers, or invaders depending on their response to grazing (Figure 11.5).

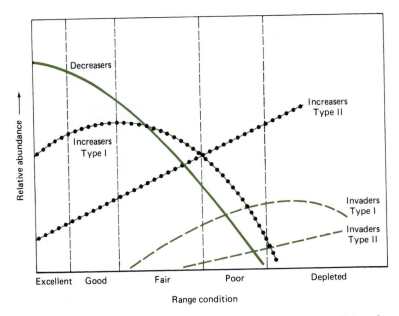

FIGURE 11.5   Changes in abundance of species groups for different range condition classes of the USDA-Natural Resources Conservation Service.

Decreasers are the highly palatable climax species that steadily decline under excessive grazing. *Increasers* are those plants that are part of the climax and initially increase as the decreasers are eliminated. Increaser I plants are distinguished from increaser II plants based on their moderate palatability and eventual decline under excessive grazing. Increaser II plants are unpalatable and in many cases poisonous; therefore, they steadily increase under excessive grazing. Invaders are those species that were unimportant members of or not part of the original climax vegetation. Invader I plants are distinguished from invader II plants by having some grazing value. Many invader plants are annuals or biannuals while decreasers and increasers are long-lived perennials.

In recent years, soil stability and suitability of the existing vegetation for the intended use have become criteria for assessing range condition, as well as the amount of climax vegetation that remains. On most ranges, limited amounts of increaser and invader plants are advantageous because they provide food and cover for certain wildlife species and often grow when decreaser plants are dormant and low in nutritional value. Although, if heavily consumed, increaser and invader plants are often toxic, moderate or small amounts in livestock diets can be nutritionally advantageous. Ranges with a mixture of different plant types generally meet the needs of livestock and wildlife better than pure grasslands or shrublands. On most grassland and desert ranges, about 60% remaining climax (good condition) gives an ideal mix of plants for livestock and wildlife and will ensure soil stability. This type of range is sustainable under sound grazing management.

Range trend refers to the rate and direction of change in range condition. Many range managers consider knowledge of trend to be the most important aspect in evaluating the effectiveness of different range management practices. Generally an upward trend is considered to be succession toward the climax or the desired plant community while a downward trend would be retrogression away from the climax or desired plant community.

In evaluating trend, climatic influences must be separated from those associated with grazing or other management practices. Exclosures (ungrazed areas) are an important tool for making these separations. Generally, grazing management is considered to be sound if vegetation improvements on the grazed area equal or exceed those in the exclosure. However, improper management would be signaled where a grazed area reveals less recovery during wet periods and more rapid retrogression in drought periods than range in an exclosure.

RANGE ANIMAL ECOLOGY

Early range research and management was concerned mostly with maintaining important range forage plants. Starting in the 1950s, increasing emphasis was placed on the nutritional needs of range livestock, their foraging behavior, and interaction with other range animals. Most recently, attention has focused on desirable rangeland wildlife species (e.g., pronghorn, mule deer, elk, white-tailed deer, and various upland gamebird species) as a result of increasing public willingness to pay for hunting and viewing these animals. Animal ecology is now as much a part of range management as plant ecology.

Comparative Digestive Systems

Knowledge of the role of different forages in meeting the nutritive needs of different range ungulates is critical for good range management. This knowledge is particularly useful for decisions regarding brush control, range seeding, grazing management, and forage allocation to different ungulates. We will discuss the role of grasses, forbs, and shrubs in meeting the nutritional requirements of range ungulates following Holechek (1984) and Holechek et al. (2001).

Forage Selection by Different Ungulates

Range ungulates can be divided into three groups based on their foraging habits. These groups include the grazers, which consume grass-dominated diets; the browsers, which consume primarily forbs and shrubs; and the intermediate feeders, which use equal amounts of grasses, forbs, and shrubs.

The Grazers.   Cattle, elk, bighorn sheep, mountain goats, musk oxen, and bison are North American ungulates classified as grazers. However, on some ranges these ungulates, with the exception of bison and musk oxen, do consume large amounts of forbs and shrubs. This occurs primarily when green grass is unavailable. These ungulates avoid shrubs high in volatile oils (e.g., junipers, rabbitbrush, and various sagebrushes) because they lack mechanisms to reduce the toxic effects of these substances.

The Browsers.   Moose, pronghorn, mule deer, domestic goats, and white-tailed deer feed primarily on forbs and shrubs throughout the year regardless of location. With the exception of domestic goats, these ungulates experience digestive upsets if forced to consume diets dominated by mature grass. This group of ungulates consumes a limited amount of grass in the spring when it is green and forbs and shrubs are unavailable. However, dry mature grass is almost completely avoided. The smaller ruminants in this group can consume large amounts of forages high in volatile oils because their small, pointed mouth parts enable them to select the portions of these plants with the lowest levels of volatile oils. In addition, the small ruminants chew their food to a much greater extent than large ruminants or monogastric animals. Apparently, fine chewing of plants high in volatile oils results in release of these substances as gasses and greatly reduces their assimilation by the animal's digestive system. If assimilated at high levels, the volatile oils found in many sagebrushes, rabbitbrushes, and junipers can be toxic to the animal.

The Intermediate Feeders.   Domestic sheep, burros, and caribou are classified as intermediate feeders. These animals have the greatest capability to adjust their feeding habits to whatever forage is available. Domestic sheep are probably better adapted to the forage resource in the intermountain West than any other ungulate because they will readily use grasses, forbs, or shrubs depending on availability. The primary problem with domestic sheep is that their short legs and relatively large body make them very susceptible to predation.

Comparative Nutritive Value of Grasses, Forbs, and Shrubs

On ranges of the western United States, the primary nutritional constraints limiting ungulate productivity are inadequate concentrations of energy, protein, phosphorus, and vitamin A in the diet. With a few localized exceptions, mineral deficiencies other than phosphorus are not a problem.

Various studies on forage nutritive quality show different forages provide different levels of critical nutrients at different times of the year. Therefore, ranges with the widest diversity of plant species provide the best nutritional conditions for domestic or wild ungulates when yearlong grazing is practiced.

The Grasses.   Grasses typically have lower crude protein, phosphorus, and lignin concentrations and higher total fiber and cellulose concentrations than do forbs and shrubs. Digestibility of grasses is generally less than forbs and shrubs. At comparable growth stages, cool-season grasses are higher in crude protein, phosphorus, and digestibility and lower in fiber than warm-season grasses. Plant fiber is digested more slowly than the cell contents. The high cellulose (digestible portion of fiber) concentration and high ratio of cellulose to lignin (indigestible portion of fiber) makes grasses best suited to large ruminants such as cattle or cecum digesters (horse) that have low nutrient requirements per unit of body weight. Leaves of grasses are nutritionally superior to stems. For this reason, short grasses are nutritionally superior to mid- and tallgrasses particularly during dormancy. Grasses are usually the component of the forage resource available in the greatest quantity unless overgrazing has been severe. Their high availability makes them important to large ruminants that have a high total forage requirement.

The Forbs.   Compared with grasses or shrubs, actively growing forbs have higher levels of crude protein, phosphorus, and digestibility, and lower levels of fiber. Leaves from deciduous shrubs are similar to forbs in nutrient content. When dormant, forbs and deciduous shrub leaves rank intermediate between grasses and evergreen shrubs in nutritive quality. Because of their low fiber levels, forbs and deciduous shrub leaves break down quickly in the rumen and permit higher intakes than grasses or evergreen shrub leaves during active growth. Forbs and deciduous shrub leaves are critical dietary components for small ruminants that require low-fiber diets, such as white-tailed deer and pronghorn.

Because many forbs are poisonous, range managers have often collectively dismissed them as undesirable. Studies with cattle, sheep, and goats show that poisonous forb problems are minimal if the range is in good condition (a high diversity of palatable plants) and grazing intensity is moderate.

The Shrubs.   Evergreen shrub leaves and buds from deciduous shrubs have lower fiber levels and higher crude protein, phosphorus, carotene (vitamin A), and digestibility levels than grasses and forbs when forage is dormant. Woody material from shrubs is highly lignified and very low in nutritive value. Therefore, grazing animals select for leaves, buds, fruits, and young twigs with low lignification. Ruminants with small mouth parts, such as goats or pronghorn, can select against woody material much better than cattle or elk. However, evergreen shrub leaves do provide an important crude protein, phosphorus, and carotene supplement to cattle and elk on many ranges when grasses are dormant. Deciduous shrubs with broad leaves, such as snowberry and ninebark, are heavily used by cattle during periods of drought. Many evergreen shrubs (e.g., oaks, sagebrushes, and junipers) have volatile oils or tannins that bind up proteins, reducing the nutritive value of the forage. Goats, deer, pronghorn, and other animals with small mouthparts use these plants most efficiently.

FIGURE 11.6 Stylized representation of the digestive anatomy and arrangement of (a) ruminant, (b) postgastric fermenter, and (c) nonruminant herbivores. 1, Esophagus; 2, Reticulum; 3, Rumen fermentation compartment; 4, Omasum; 5, Stomach (abomasum); 6, Small intestine; 7, Cecum fermentation compartment; 8, Large intestine; 9, Anus (adapted from Huston and Pinchak 1991).

Digestive Systems. Ungulates have two basic types of digestive systems that include the rumen and cecum systems (Figure 11.6). Both systems evolved to enable ungulates to digest plant fiber (plant cell walls) by microbial (bacteria and protozoa) fermentation. The fermentation processes are quite similar in both the rumen and cecum. The systems differ in that the rumen is an enlarged portion of the digestive tract that food must pass through before entering the true stomach. The cecum occurs as an enlarged portion of the large intestine that food enters after passing through the true stomach.

From the standpoint of digestive effectiveness, the rumen system has two advantages over the cecum system. First, the process of rumination (regurgitation and rechewing of forage) results in substantial reduction of particle size that provides more surface area for microbial digestion. Because food must be broken down to fine particle size to leave the rumen, retention of fiber is longer than in the cecum. This results in more complete digestion of fiber in the rumen than in the cecum because fiber digestion is a time dependent process. Second, microbes are passed from the rumen into the abomasum where they are digested and then assimilated, providing the animal with an important source of protein. Little microbial protein is assimilated by cecum digesters because microbial fermentation occurs after the food has passed through the stomach. However, research shows horses will ingest their feces when their diet is low in protein, which partially compensates for the inefficient use of microbial protein associated with cecum digestion.

The primary advantage of cecal digestion of fiber is that forage material can pass easily out of the cecum without any great reduction in particle size. Although fiber digestion in cecum digesters is less complete than in ruminants, cecum digesters compensate by consuming a much greater amount of forage. They do not have to break fiber down to a small particle size to pass it out of their system.

As might be deduced from the previous discussion, cecum digesters can subsist on lower-quality diets than ruminants. However, they must have a greater forage supply because they use the forage less efficiently. This explains why horses can survive on coarse, mature grasses better than cattle.

Large ruminants can subsist on higher fiber diets than small ruminants because they have lower nutrient requirements per unit body weight. Therefore, a large portion of the diet is typically composed of highly available forage, such as grasses for bison and cattle, or woody browse for moose. The small ruminants, such as white-tailed deer and pronghorn, must consume diets dominated by leafy material and fruits from forbs and shrubs that have high levels of crude protein, phosphorus, and digestibility, and low levels of fiber. These animals can afford to be selective for these materials because they have a low total forage demand.

Domestic sheep subsist better on grass-dominated diets than other small ruminants, such as white-tailed deer or pronghorn. This is facilitated in sheep by a relatively large rumen size in relation to their body weight.

Use of Nutritional Knowledge in Management

Until the 1980s, range management practices were often directed toward replacing forbs and shrubs with pure stands of grasses. The vast acreages of crested wheatgrass in the Great Basin and lovegrasses in the Southwest stand in support of the above statement. During active growth, pure stands of grass provide good forage for cattle or, in some cases, elk and mule deer, but poorly meet the nutritional requirements of all ungulates during most of the year. Research in the Great Basin and the Southwest show that inclusion of palatable forbs and shrubs in seeding mixtures with grasses can greatly improve livestock performance during forage dormancy and provide better habitat for small wild ungulates and other wildlife species than pure stands of grasses.

Range condition is usually based on the density and production of native, palatable perennial grass. A better criterion may be the diversity of palatable forage species. Using this criterion it may be desirable to manage for up to 20% of the yearly forage production in the form of palatable annuals. Because many annual grasses and forbs grow only in periods when perennials are dormant, a mix of perennials and annuals tends to sustain the longest period of nutritious forage.

A large number of studies involving both domestic and wild ungulates in North America are consistent in showing that forage selection changes tremendously within and between years. The nutritive quality of various forage species also fluctuates greatly within and between years. The greater the degree of forage selection a range provides domestic or wild ungulates, the more likely they will meet their nutrient needs.

Animal Suitability for Different Rangelands

Different animal species are adapted to different types of rangelands. These differences are reflected in the influence of rangeland on the animal and the effects of the animal on rangeland.

Water requirements, topography, predators, climate, parasites and disease, poisonous plants, type of forage, and economic costs and returns are the important considerations in selection of type or types of livestock for a particular range.

Cattle and horses, because of their large size, are best suited for flat, well-watered areas that are dominated by grasses. European cattle (*Bos taurus*) such as the Hereford and Angus breeds, are well adapted to cooler climates while the African cattle (*Bos indicus*), such as the Brahman breed, do well in hot, humid areas.

Sheep and goats are better adapted to hot, humid climates and rugged terrain than cattle or horses. Unlike cattle and horses, they can be watered every other day, rather than daily, and still perform well. Because of their small size and surefootedness, sheep and goats make better use of steep terrain than large animals.

In some areas, pests and diseases make it difficult to raise domestic livestock. Large areas in Africa are unsuitable for cattle because of the tsetse fly. Game ranching (raising of native ungulates for meat) shows potential as a sound alternative to domestic livestock based on evaluations by Dr. David Hopcraft in Kenya.

In some parts of the western United States, sheep ranching has been more profitable than cattle ranching. The price of lambs has usually been greater than that for calves, and wool further adds to returns from sheep. However, cattle do have the advantages of reduced susceptibility to predation and they require less labor than sheep. The cowboy persona associated with cattle has given them a special appeal.

Common-use grazing is the intentional use of rangelands by more than one type of animal. This involves some combination of animals such as cattle and sheep in the mountains of Utah or cattle and white-tailed deer in central Texas. Common-use grazing often has a number of advantages such as better distribution of animals, harvesting more of the available plant species, and reduction in risk from diversification. The disadvantages include greater labor requirements, more handling facilities, and conflicts with wildlife when certain combinations of domestic animals are grazed at excessive stocking rates.

RANGELAND MANAGEMENT

Range management centers around control of livestock grazing. The primary principles in controlling grazing are proper livestock numbers, proper livestock distribution, proper kinds of animals, and a proper system of grazing. A secondary objective in range management involves range vegetation manipulation with tools other than large animals. Practices used to achieve this objective include seeding, fertilization, and control of undesirable plants with fire, herbicides, mechanical means, or insects. Rapid increases (two- to tenfold) in forage production can occur within short time periods (1 to 5 years) using these intensive practices. Forage increases from vegetation manipulation through grazing management generally occur more gradually, but risks and costs are much lower.

Importance of Correct Stocking Rate

Selection of the correct stocking rate is the most important of all grazing management decisions from the standpoint of vegetation, livestock, wildlife, and economic return. Although this has been the most basic problem confronting ranchers and range managers since the initiation of scientific range management early in the twentieth century, specific approaches to this problem were generally unavailable until the late 1980s (see Holechek 1988). It is generally agreed that there is no substitute for experience in stocking rate decision on specific ranges. However, procedures now available will give reasonable estimates for most ranges in the United States and should work in other parts of the world.

Carrying or *grazing capacity* is a term often used when discussing stocking rate. It refers to the maximum stocking rate possible year after year without causing damage to vegetation or related resources. Although actual stocking rates may vary considerably between years because of fluctuating forage conditions, grazing capacity is generally considered to be the average number of animals that a particular range will sustain over time. In most cases, ranches are bought and sold on the basis of their grazing capacity.

Commonly, grazing capacity is expressed as the stocking rate, which is the number of animal units placed on a given area for a particular time period. An animal unit is a measure of how much forage a mature nonlactating bovine weighing about 1,000 lb will remove over a 1-year period. Various studies show ruminant animals eat about 2% of their body weight per day on a dry-matter basis. Therefore, a 1,000-lb cow requires about 7,300 lb of forage per year, 610 lb per month, and 20 lb per day. Horses and donkeys eat about 50% more than ruminants and have dry-matter intakes near 3% body weight per day. The following represent some accepted animal unit equivalents.

Animal	Weight (lb)	Animal Unit Equivalent
Cow	1,000	1.00
Steer	750	0.75
Bull	1,200	1.20
Sheep	150	0.15
Goat	100	0.10
Deer	150	0.10
Elk	700	0.70
Bison	1,800	1.80
Donkey	700	1.05
Pronghorn	120	0.12
Horse	1,200	1.80

Nearly all stocking rate studies characterize grazing intensity treatments as heavy, moderate, or light. Heavy grazing is a degree of herbage utilization that does not permit desirable forage species to maintain themselves. Moderate grazing means a degree of herbage utilization that allows the palatable species to maintain themselves

TABLE 11.1   Summary of 25 Studies on Effects of Grazing Intensity on Native Vegetation and Livestock Production in North America

	Grazing Intensity		
	Heavy	Moderate	Light
Average use of forage (%)	57	43	32
Average forage production (lb/acre)	1,175[1] (1,065)[2]	1,473[1] (1,308)[2]	1,597[1]
Forage production drought years (lb/acre)	820[1]	986[1]	1,219[1]
Range trend in ecological condition	down (92%)[3]	up (52%)[4]	up (78%)[4]
Average calf crop (%)	72[1] (77)[2]	79[1](84)[2]	82[1]
Average lamb crop (%)	78	82	87
Calf weaning wt (lb)	381[1] (422)[2]	415[1] (454)[2]	431[1]
Lamb weaning wt (lb)	57	63	—
Gain per steer (lb)	158	203	227
Steer/calf gain per day (lb)	1.83	2.15	2.30
Steer/calf gain per acre (lb)	40.0	33.8	22.4
Lamb gain per acre (lb)	26.0	20.4	13.8
Net returns per animal ($)	38.06[1] (29.00)[2]	51.57[1] (39.71)[2]	58.89[1]
Net returns per acre ($)	1.29[1] (1.72)[2]	2.61[1] (2.24)[2]	2.37[1,2]

Source: Holechek et al. 1999.
[1] Average for those studies comparing heavy, moderate, and light grazing (studies comparing only heavy and moderate grazing excluded).
[2] Average for all studies.
[3] Percentage of studies with downward trend.
[4] Percentage of studies with upward trend.

but usually does not permit them to improve their herbage-producing ability. Light grazing means a degree of herbage utilization that allows palatable species to maximize their herbage-producing ability.

The primary measure of grazing intensity used in long-term grazing studies has been percent use of palatable forage species. Although it has limitations as a measure of grazing intensity, percent use is more easily understood by ranchers and nonrange professionals than other measurements such as stubble heights, percentage of grazed plants, or minimum residues. When several years of data were collected, percent use of forage has been well related to changes in productivity of primary forage plants, livestock performance, and financial returns.

When all the North American stocking rate studies were averaged, heavy grazing averaged 57% use of primary forage species compared with 43% use for moderate and 30% use for light grazing (Table 11.1). Conventional wisdom has been that moderate stocking involves 50% use of forage. This guideline applies well in the southern pine forest, humid grasslands, and annual grasslands, but results in rangeland deterioration in the semiarid grasslands, desert, and coniferous forest rangelands. Here, research was remarkably consistent in showing that moderate grazing involved about 35 to 45% use of forage.

Conservative stocking is a term commonly used by range researchers to define a level of grazing between light and moderate, generally involving about 35% use of forage. Several researchers recommended conservative stocking over either light or moderate stocking in their summaries.

During a drought, livestock operators need to be prepared to reduce livestock numbers or provide feed supplements to prevent overgrazing. Failure to destock during and following drought periods has been one of the most serious causes of rangeland degradation in the United States and other parts of the world. For more detailed discussions on how to set stocking rates, the reader is referred to Holechek et al. (2001).

Stocking rate has more influence on vegetation and livestock productivity than does any other grazing factor (Table 11.1). When all North American studies were averaged, annual herbage production had increased by 23% when livestock use was reduced from heavy to moderate. An average increase in forage production of 36% resulted from switching from heavy to light use. When changes in forage production through time were taken into account by averaging the first and last 3 years of study for each stocking rate, heavy stocking overall resulted in a 20% decline in forage production, moderate stocking had no change, and light stocking resulted in an 8% increase. In drought years moderately stocked pastures produced 20% more forage than those heavily stocked. Forage production was 49% higher under light than heavy grazing and 24% higher under light than moderate grazing. These studies consistently showed that the greatest benefit of light or conservative stocking in terms of forage production occurred in dry years.

On some ranges, forage production was actually less under light grazing than under moderate grazing. Herbage production on most ranges can be substantially increased by switching from heavy to moderate or light grazing intensities. This is particularly true for grassland ranges.

Heavy stocking consistently caused a downward trend in ecological condition, light stocking caused an upward trend, and slight improvement occurred under moderate stocking (Table 11.1). Invariably, the decreasers (most productive and palatable forage species) showed a decline in cover under heavy stocking while they tended to increase under light stocking. The longer the time involved in the study, the more divergence there was between heavy and light stocking in terms of vegetation composition. Generally, these studies provide support for Dyksterhuis's (1949) theories of plant responses to grazing management. However, different stocking rates generally had more impact on forage production than plant composition.

Several qualitative guidelines for evaluating grazing intensity on rangelands are discussed by Holechek et al. (2001). Basically they are summarized as follows:

Heavy use. Range has a "clipped" or mowed appearance. More than half of the fair and poor forage-value plants are used. All accessible parts of the range show use, and the key areas are closely cropped. They may appear stripped if grazing is very severe. There is evidence of livestock trailing to forage.

Moderate use (proper use). About one-half of the good and fair forage-value plants are used. There is little evidence of livestock trailing. Most of the accessible range shows some use.

Light use. Only choice plants and areas are used. There is no use of poor forage plants. The range appears practically undisturbed.

On key areas average stubble heights of 30 cm to 35 cm (12 in. to 14 in.) for tallgrasses, 15 cm to 20 cm (6 in. to 8 in.) for midgrasses, and 5 cm to 8 cm (2 in. to 3 in.) for shortgrasses are recommended minimums for proper use. We refer the reader to Holechek et al. (2001) for a more detailed discussion on stubble heights and evaluation of grazing intensity.

Improving Livestock Distribution

Uneven use of rangeland by livestock has been and continues to be a major problem confronting range managers. On many ranges, improvement will occur without reduction in livestock numbers if practices to secure more uniform use are implemented. Distribution problems are most severe in arid or desert areas and in mountainous terrain. These conditions characterize most of the area west of the Rocky Mountains in the United States. Factors causing uneven use of rangelands include distance from water, rugged topography, diverse vegetation, wrong type of livestock, pests, and weather. Practices commonly used to improve livestock distribution include water development, fencing, trail building, salting, controlled burning, fertilization, and changing types or kinds of animals grazed.

Distance from Water. Poor water distribution is the chief cause of poor livestock distribution on most ranges. In arid regions of the world, water is in short supply and

FIGURE 11.7 The zone of degradation around a livestock watering point in southcentral New Mexico. Rotation of access to watering points has been effective in improving these degraded areas.

poorly distributed. Where watering points are infrequent, large sacrifice areas around watering points have often occurred. The heavy use of vegetation around watering points is well documented (Figure 11.7).

Although livestock will travel great distances to water, this is not in the best interest of the animal or the range resource. Travel increases animal energy expenditure that otherwise would go into production (weight gain, milk production) and takes away from grazing and resting time.

If animals must travel large distances between water and available forage, a series of trails will be created that gradually become larger and more numerous. These trails become water channels that cause severe erosion. The proportion of rangeland in livestock trails is a good indicator of grazing severity.

In arid areas, water-point spacing of about 3 to 4 miles apart has proven most practical when development cost and efficient use of the range are both considered. In the more humid ranges, 2- to 3-mile spacings are practical because more grazing capacity is gained relative to watering-point cost.

Failure to adjust stocking rates for distance from water has been an important cause of rangeland degradation and poor livestock performance. Various range studies show that with cattle, areas over 2 miles from water should be deleted in grazing capacity estimates. A 50% reduction in grazing capacity is required for the zone 1 to 2 miles from water, but no adjustment is necessary for the zone within a mile of water.

Salt blocks and supplemental feed in strategic places can also be useful for improving livestock distribution.

Topography. Rugged topography is the second most important cause of poor livestock distribution on rangelands. The reluctance of livestock to use steep slopes is not entirely undesirable because these areas are often fragile and valley bottoms can better withstand grazing. However, in many cases slopes serve as barriers to the use of benches and ridgetops above valley bottoms.

Livestock vary considerably in their willingness to use steep terrain. Heavy animals, such as mature cattle or horses, have difficulty in traversing steep, rocky slopes. Cattle make little use of terrain with more than 20% slope (Figure 11.8). Because of their smaller size, greater agility, and surefootedness, sheep and goats use

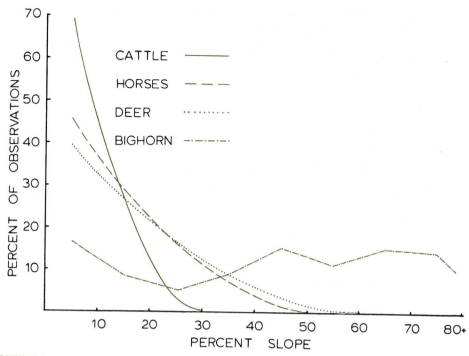

FIGURE 11.8 Relationship of slope gradient to the percent of observations of cattle, feral horses, deer, and bighorn sheep (from Ganskopp and Vavra 1987).

these areas more readily. Research shows that terrain with up to 45% slope is well used by sheep. Beyond this point, sheep use drops off rapidly. This applies to deer and goats as well.

Grazing Systems

Specialized grazing systems have been a major focus of range researchers and managers since the 1950s. During the 1950s and 1960s, deferred-rotation systems received considerable attention. Rest-rotation grazing was heavily applied on public lands in the intermountain West during the 1970s. Both deferred- and rest-rotation systems are still being used, particularly in mountainous areas. In the 1980s, short-duration (cell) grazing became the newest fad in grazing systems. Through the years, impressive claims have been made for each new system regarding increased stocking rates and livestock production. However, actual research has shown that specialized grazing systems generally give modest (10 to 30%) to no increase in grazing capacity over season-long or continuous systems (Holechek et al. 1999).

    Deferment, rest, and *rotation* are terms that receive constant use when grazing systems are discussed. *Deferment* involves the delay of grazing in a pasture until after the growth period at the time of seed maturity in the key forage species. This permits the better forage plants to gain vigor and reproduce. In *rest* systems, the range receives no use for a full year, rather than just during the growth period. This gives plants a longer period to recover from past grazing influences and provides wildlife with a pasture free from livestock use during the critical dormant period. A disadvantage of both deferment and rest is the need for increased grazing load on other pastures during the critical growth period. It is questionable whether periodic nonuse during critical periods compensates for periodic heavy use. However, deferment and rest do provide plants on sacrifice areas with some opportunity for recovery. On mountainous areas, season-long or continuous grazing generally results in degradation of areas convenient to livestock. Rotation grazing involves the movement of livestock from one pasture to another on a scheduled basis. It is the critical feature of all specialized grazing systems. The main advantage of rotation is the provision of primary grazing areas and key forage plants with periodic nonuse during the critical growing season. Systems with deferment and rest typically involve livestock rotations. Short-duration (rapid-rotation, cell, time-control) grazing is distinguished from other specialized systems by the greater frequency of nonuse and grazing rotations during the growing season.

Considerations in Grazing System Selection.

Grazing systems commonly used in the United States and other parts of the world include continuous, deferred-rotation, rest-rotation, short-duration, Merrill three-herd/four-pasture, high intensity-low frequency, best-pasture, and seasonal-suitability. Climate, topography, vegetation, kind or kinds of livestock to be grazed, wildlife needs, watershed protection, labor requirements, and developments (fence and water) are important considerations involved in grazing system selection. Specialized systems have been most useful where:

1. Terrain is rugged.
2. Wildlife is an important consideration.
3. Water distribution is poor.
4. Poor distribution of precipitation over the range occurs within years (the southwestern United States).
5. Carefully timed grazing is necessary to prevent tree damage.
6. The vegetation has low grazing resistance.

We will discuss the conditions where the various grazing systems give best results, updating the discussions of Holechek (1983) and Holechek et al. (2001).

Continuous Grazing.

Continuous grazing involves grazing a particular pasture throughout the grazing season year after year. Grazing systems other than continuous are commonly called specialized systems because scheduled moves of livestock from one pasture to another are involved.

The primary problem associated with continuous grazing is that livestock have preferred plants and areas for grazing. These plants and areas receive excessive use even under light stocking rates. The preferred areas generally occur where water, forage, and cover are in close proximity. These are often the most productive parts of the pasture.

Continuous grazing at a stocking rate that gives moderate use (40% removal of current year's growth) has given good results in the flat shortgrass prairie country of the Great Plains where watering points are usually not farther than 2 miles apart and differences in palatability between forage species is minimal. The shortgrasses, blue grama, and buffalograss are the primary forage species. They evolved with heavy grazing by bison and are quite grazing resistant. Precipitation in the shortgrass country occurs as several light rains throughout the summer months. Therefore, considerable opportunity exists for regrowth after defoliation. The flat nature of the terrain and the close proximity of watering areas minimizes the tendency of livestock to congregate and linger in the most convenient areas. Sacrifice areas can be reclaimed by temporary fencing to allow vegetation recovery. Since most of the sensitive areas occur around water, control of access to watering points can be used to provide sensitive areas with periodic nonuse.

Continuous grazing has given superior results to specialized systems in the California annual grassland type when use was moderate and practices such as salting, fencing, and water development were used to obtain proper distribution. Annual grasses need only to set seed year after year to maintain themselves, unlike perennial grasses that must store carbohydrates. Differences in palatability between most of the annual grasses are small.

Year 1	A	B
June 15–August 15	Grazed	Ungrazed
August 16–October 15	Ungrazed	Grazed

Year 2		
June 15–August 15	Ungrazed	Grazed
August 16–October 15	Grazed	Ungrazed

FIGURE 11.9 A one-herd, two-pasture deferred-rotation grazing plan used on the Starkey Range in northeastern Oregon.

In both the shortgrass and California annual grassland types, livestock performance has been better under continuous grazing than under specialized systems. This is explained by the fact that continuous grazing allows livestock to exhibit maximum forage selectivity and minimizes livestock disturbance due to gathering, trailing, and quick change in forage quality.

Deferred-Rotation Grazing.   Deferment involves delay of grazing until seed maturity of the important forage species is completed (Figure 11.9). Rotation is the movement of livestock from one pasture to another on a scheduled basis. Initial research on this system was conducted by Arthur Sampson in the Blue Mountains of northeastern Oregon in the early 1990s. The system he studied involved dividing the range into two pastures. Each pasture received deferred grazing every other year. Vegetation response under this system has been superior to continuous grazing on bunchgrass and mountain ranges in the northwestern United States and on tallgrass ranges in the eastern Great Plains. Deferred-rotation grazing provides a better opportunity for preferred plants and areas to maintain and gain vigor than continuous grazing. It works best where considerable difference exists between palatability of plants and convenience of areas for grazing. On mountain ranges, stringer meadows and riparian zones will often receive excessive use by cattle even under extra light grazing intensities while surrounding uplands will receive light or no use. The deferred-rotation system provides forage species on the lowland sacrifice areas with the opportunity to store carbohydrates and set seed every other year.

Rest-Rotation Grazing.   Rest-rotation grazing was developed by Gus Hormay of the forest service in the 1950s and 1960s. This system is unique in that one pasture receives 12 months of nonuse while the other pastures absorb the grazing load. Presently most rest-rotation schemes involve three or four pastures. Various sorts of rotation schemes are used on the grazed pastures (Figure 11.10). The problem with rest-rotation grazing is that the benefits from rest may be nullified by the extra use that occurs on the grazed pastures.

From a multiple-use standpoint, rest-rotation grazing with moderate stocking of grazed pastures has a number of advantages. Small mammals and upland game birds are provided with a good vegetative cover throughout the year on at least part of the area. In eastern Montana, waterfowl production was increased threefold when rest-rotation was compared to continuous grazing (Mundinger 1976). The rest pasture provided waterfowl with the heavy cover they need for good nesting success. Big-game animals, such as deer and elk, that generally prefer to avoid livestock are provided with an area free from disturbance and have maximum forage selectivity in the ungrazed pasture. From an aesthetic standpoint, the public prefers to see a certain amount of the range ungrazed. The major drawback with rest-rotation grazing is that livestock numbers may have to be reduced to prevent excessive use on the grazed

Pasture:

A- 6/15 To Seedripe

B- Seedripe to 10/15

C- Rested (No livestock use)

FIGURE 11.10 A rest-rotation grazing used on elk summer range at the Mount Haggin wildlife area in Montana (from Frisina 1992).

pastures. In areas with large big-game populations, the rest pasture may receive heavy use by big game.

The Merrill Three-Herd/Four-Pasture System.

In the early 1950s, Leo B. Merrill in southcentral Texas developed a grazing system involving three herds and four pastures. With this system, each pasture is grazed continuously for a year, and then given a four-month period of nonuse. The period of nonuse in each pasture has occurred during all times of the year at the end of a four-year cycle. This system has given good results where effective precipitation and plant growth can occur at any time during the year. It also works well where common use of the range by more than one grazing animal is practiced. In Texas some combination of cattle, sheep, goats, and white-tailed deer graze most ranches. Each grazed pasture is assigned a different type of livestock. Every four months the types of livestock are interchanged between pastures. White-tailed deer prefer the pasture receiving nonuse by livestock. The Merrill system has been the best studied of all the specialized grazing systems. In Texas, it is superior to continuous grazing from the standpoint of livestock, forage, and wildlife production. This system should be effective in areas where forage plants have the potential for growth throughout the year and common-use grazing is practiced. The Merrill system can be modified for areas with seasonal precipitation by dividing the ranch into four pastures and providing each pasture with growing season nonuse once every four years.

High Intensity-Low Frequency Grazing.

High intensity-low frequency (HILF) grazing differs from rapid-rotation grazing in that periods of grazing are typically longer than 2 weeks and nonuse periods are more than 60 days. This system requires at least three pastures per herd of livestock. The stocking rates must be light to moderate under this system or severe declines in livestock production and excessive defoliation will occur. This system, like short-duration grazing, works best in flat terrain with an extended growing season and forage species of similar palatability. An important objective of the HILF grazing system is to force use of old, coarse, unpalatable but grazeable forage. Because this system has negative impacts on animal nutrition and forces heavy plant defoliation, it has been largely abandoned in favor of short-duration grazing, which allows greater animal selectivity, permits lighter levels of defoliation, and prevents more of the forage from maturing prior to grazing.

Seasonal-Suitability Grazing.

Seasonal-suitability grazing involves partitioning a range into pastures based on vegetation types or condition classes. The best pasture from a nutritional standpoint is used for each season of the year. When seeded

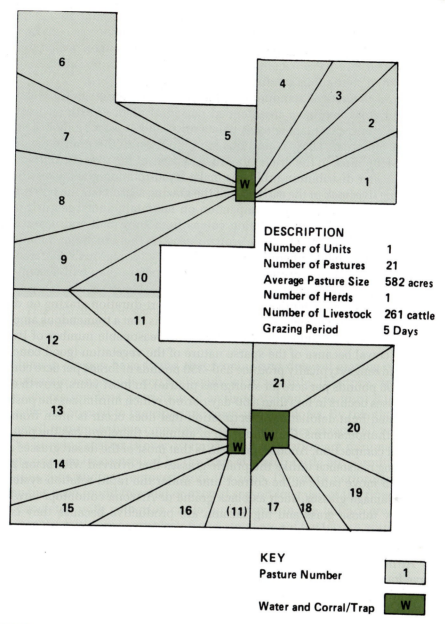

DESCRIPTION
Number of Units	1
Number of Pastures	21
Average Pasture Size	582 acres
Number of Herds	1
Number of Livestock	261 cattle
Grazing Period	5 Days

KEY
Pasture Number

Water and Corral/Trap

FIGURE 11.11 Short-duration grazing system on a small cow-calf ranch in southeastern New Mexico (from Fowler and Gray 1986).

pastures are used in conjunction with native range, this system is often called complementary grazing. In areas such as the Southwest, where local rains can cause considerable difference in forage availability between pastures, livestock are moved to pastures where green forage is available. Under these conditions the grazing scheme is called the "Best-Pasture System."

Short-Duration Grazing. Short-duration grazing was developed in Zimbabwe by Allan Savory. This system typically involves a wagon wheel arrangement of fences with water and livestock handing facilities located in the center of the grazing area (Figure 11.11). It is recommended that no fewer than eight pastures (paddocks) of equal grazing be built that radiate as spokes from the central area where the water is located. Each paddock is given a short, intensive period of grazing followed by a long period of nonuse. Ideally, the grazing period of each paddock would be 5 days or less

TABLE 11.2　Rangeland and Cattle Production Characteristics for Different Grazing Management Strategies on the Fort Stanton Range in New Mexico

	Fort Stanton Experiment Range—Cow/Calf Operation		
	Moderate Continuous	Heavy Continuous	Best-Pasture Rotation
Duration of study (years)	10	10	10
Average annual precipitation (inches)	15	114.8	15.1
Average forage production (lb/acre)	740	607	819
Total palatable forage production in 1974 drought (lb/acre)	235	103	379
Range condition	Good	Fair	Good
Acres/animal unit	67	54	54
Forage use (%)	40–45	50–55	50–55
Calf crop (%)	93	91	85
Average calf wean wt./acre (lb)	6.2	7.0	6.4
Death losses (%)	<2%	<2%	<2%
Total 10-year accumulated value ($)	388,575	358,744	382,380

Source: Adapted from Pieper et al. 1991 and Holechek et al 2001.

RANGELAND LIVESTOCK PRODUCTION

Recent Trends

In the past, "range" has been strongly associated with livestock grazing in the western United States. The traditional use of North American rangelands has been to produce meat for human consumption. However, this use—particularly in the 11 western states—is now being challenged not only by the nonranching public, but also by ranchers as they find it difficult to make a profit. Historically, annual returns on capital investment for range livestock enterprises have been low (2 to 6%) compared with returns from investments in industry (10 to 30%). In New Mexico, annual net returns on capital investment in the best of years have averaged less than 4% (Holechek et al. 2001). In most years the overall return is a little over 1%. Compared to the peak in the late 1970s, land prices in the late 1990s for ranches in the western United States have dropped 10%, operating costs have increased about 60%, and prices paid for livestock have increased about 11% (Workman and Evans 1993, Holechek et al. 2001). Based on this scenario, the economic situation for range livestock production in the western United States has been highly unfavorable. Presently, many ranchers face an economic climate that severely threatens their survival. However, other opportunities for income have emerged from enterprises such as recreation, raising nontraditional animals (such as ostriches, red deer, bison, and llamas), producing plants for landscaping, and fee hunting and fishing operations.

Importance of the West

The 11 western states support about 20% of the Nation's cattle and 50% of the Nation's sheep. Cattle numbers are highest in the Great Plains and Southeast, with the 11 western states having the highest sheep numbers. It is often unrecognized that costs associated with cattle ranching are lower in the Great Plains than in the West (Holechek and Hawkes 1993). This is because the much lower amounts of land required per animal unit reduce property taxes, fuel costs, and labor costs and permit better herd management in the Great Plains than in the West. Supplemental feed costs are less in the Great Plains because droughts are less frequent and severe. Also, active forage growth occurs for a longer period in the Great Plains. Although forage is generally low in quality in the Southeast and the natural vegetation is woodland, the high quantity of forage produced, flat terrain, long growing season, and ready availability of water make livestock production fairly efficient there.

TABLE 11.3 Herbage Production (Pounds per Acre) on Heavily and Moderately Grazed Shortgrass Prairie in Colorado During a Drought Year Compared with the 5-Year Average

Grazing Intensity	Drought Year	5-year Average	Drought Year as Percentage of Average
Excessive: 54% use of forage	312	595	52
Moderate: 37% use of forage	577	766	75
Light: 21% use of forage	609	817	75

Source: Klipple, G. E. and Costello, D. 1960. *Vegetation and cattle responses to different intensities of grazing on shortgrass ranges of the central Great Plains.* U.S. Department of Agriculture Technical Bulletin 1216.

In the most arid portions of the West where forage production is below 100 lb/acre, livestock grazing probably will be unprofitable in most years (Holechek 1992). This is because the fixed costs per animal unit (fences, roads, watering points, property taxes, and insurance) become excessive. Livestock management becomes increasingly difficult when animals are scattered over large areas. In sparsely vegetated areas the energy the animal spends seeking forage can become greater than the energy value of the forage the animal is able to consume. Stoddart and Smith (1943) stated that when more than 180 acres are required to support an animal unit for a year, grazing becomes uneconomical.

Livestock Management During Drought

Drought is a fact of life for stockmen using most rangelands. As aridity increases, severity and frequency of drought tend to increase (see Chapter 6). Climatic records show that ranchers can expect drought conditions to prevail about 3 years out of every 10 years on most western rangelands. Learning to live with drought is the major challenge confronting ranchers in most of the western United States and many parts of Africa. Advance planning is the key to drought survival (Thurow and Taylor 1999).

Inexperienced ranchers often underestimate the degree to which drought can reduce forage production. Studies from a variety of range types show drought can reduce forage production by more than 50% of the annual average (Table 11.3). The best way to deal with drought is to apply conservative stocking rates (Holechek 1999, Thurow and Taylor 1999). Under drought conditions, heavy grazing reduces forage production much more than moderate grazing. Partial confinement and drylot feeding of livestock can be advantageous under drought conditions. This allows ranchers to avoid liquidation of livestock when prices are low followed by repurchase after the drought when prices are high because of restocking by other ranchers. Another problem with liquidation is that disease-free, high-quality herds are usually difficult to obtain. Animals that have experience on a particular range perform better than those brought in from another area.

Poisonous Plant Problems

Annual livestock losses to poisonous plants average between 2 and 5% on western ranges. Heavy livestock losses to poisonous plants are generally linked with excessive grazing that reduces availability on palatable, nonpoisonous species and increases those that are poisonous. Studies from experimental ranges at a variety of locations show annual poisonous plant losses of livestock to be under 2% when stocking rates were moderate.

A few poisonous plants, such as larkspur, may be consumed by livestock when they are stocked moderately on good range. Although larkspur is quite toxic to cattle, it has little effect on sheep. Delay of grazing until larkspur dries is an effective management practice. Grazing sheep instead of cattle on range with larkspur is another approach.

Most poisonous plants are only toxic for a short period of time, usually during active growth. Many poisonous plants initiate growth earlier than nontoxic grasses,

- wildlife
- water
- aesthetics

These issues are discussed throughout the book so we will not elaborate further here. Rather, we will focus on range management opportunities.

The biggest past and present obstacles to successful range management have been the difficulty of integrating climatic, biological, financial, and political information into a management framework. New low-cost/user-friendly computers are now coming on the market that give ranchers the capability to solve this problem. Already, grazing management software programs (APSAT) have been developed for Texas rangelands that allow ranchers to make seasonal stocking rate adjustments (Kothmann and Hinnant 1992). Within just a few years, software programs will be available that help with decision making on brush control, forage allocation to livestock and wildlife, watering point spacing, and supplemental feed allocation.

The Internet (computer information highway) now provides ranchers and range managers instantaneous databases on different aspects of range management that would take several days to collect from libraries and other sources. This kind of information should make arid-land ranching a much more efficient, less risky enterprise than it has been in the past. It should change it from a natural resource to a technology-based operation.

Other uses of computers and sophisticated electronic equipment will include corral systems that monitor animal condition and sort animals for supplemental feeding, livestock ear tags that facilitate record keeping, and automated feeding systems that provide livestock with the correct amount of supplement (Holechek et al. 2001).

There are also some promising developments involving bioengineering that should positively affect rangelands in the next 20 years (Holechek et al. 2001). Instead of chemicals for noxious plant control, new herbicides will be developed from plant allelopathic toxins that do not contaminate the soil or water and do not release residues harmful to humans. Genetically engineered rumen microflora will provide range livestock with increased potential for converting range forage into useful nutrients. Genetic engineering may also improve range plants and animals, once ecological concerns are addressed. Energy used on the ranch will be largely generated by solar and wind power with the sale of the surplus to urban areas. Robotics will be used to reduce ranch manual labor requirements. Enterprise diversification into stock and bond markets, recreation, wildlife, and nontraditional animals are ways that ranchers can reduce risk and improve financial returns. Better information is needed on the long-term outcome of grazing management and range improvement practices. We believe the key to ranching survival in the twenty-first century will be the capability to effectively use biological and financial knowledge. The world trend toward free markets, less government intervention, and rapid technological advance in agriculture will make the acquisition and application of knowledge more important than ever before.

Increasing demand for outdoor recreation, water, wildlife, and open space on a shrinking land base provides a formidable challenge for future range managers. However, it provides opportunities as well. Historically, the problem of increasing demands on limited resources has created incentives for technology that greatly improved both human and natural resource conditions.

LITERATURE CITED

Clary, W. P. 1999. Stream channel and vegetation responses to late spring cattle grazing. *Journal of Range Management* 52:218–227.

Dyksterhuis, J. E. 1949. Condition and management of rangeland based on quantitative ecology. *Journal of Range Management* 2:104–115.

Elmore, W. and Kauffman. 1994. Riparian and watershed systems, degradation and restoration. In *Ecological implications of livestock herbivory in the West.* Society for Range Management, Denver, CO.

Fowler, J. M. and J. R. Gray. 1986. *Economic impacts of grazing systems during drought and nondrought years on cattle and sheep ranches.* New Mexico Agric. Exp. Stn. Bull. 725.

Frisina, M. R. 1992. Elk habitat use within a rest-rotation grazing system. *Rangelands* 14:93–96.

Ganskopp, D. and M. Vavra. 1987. Slope use by cattle, feral horses, deer, and bighorn sheep. *Northwest Science* 61:74–81.

Gay, C. 1965. *Range management: How and why.* Las Cruces, NM: New Mexico State Univ. Coop. Ext. Circ. 376.

Hanselka, C. W. and L. White. 1986. Rangeland in dry years: Drought effects on range, cattle, and management. In R. D. Brown, ed., *Livestock and wildlife management during drought.* Cesar Kleberg Wildlife Research Institute, Texas A&M University, Kingsville.

Heady, H. F. and R. D. Child. 1994. *Rangeland ecology and management.* San Francisco, CA: Westview Press.

Hess, K. 1992. *Visions upon the land.* Washington, DC: Island Press.

Hibbard, B. H. 1924. *A history of the public land policies.* New York: Macmillan.

Holechek, J. L. 1983. Considerations concerning grazing systems. *Rangelands* 5:308–311.

Holechek, J. L. 1984. Comparative contribution of grasses, forbs, and shrubs to the nutrition of range ungulates. *Rangelands* 6:245–248.

Holechek, J. L. 1988. An approach for setting the stocking rate. *Rangelands* 10:10–14.

Holechek, J. L. 1992. Financial benefits of range management practices in the Chihuahuan desert. *Rangelands* 14:229–284.

Holechek, J. L. 1993. Policy changes on federal rangelands: A perspective. *Journal of Soil and Water Conservation* 48:166–174.

Holechek, J. L. 1994a. Rangeland grazing. *Encyclopedia of Agricultural Science* 3:535–547. New York: Academic Press.

Holechek, J. L. 1994b. Financial returns from different grazing management systems in New Mexico. *Rangelands* 16:237–284.

Holechek, J. L., H. Gomez, F. Molinar, and D. Galt. 1999. Grazing studies: What we've learned. *Rangelands* 21(2):12–16.

Holechek, J. L., H. Gomez, F. Molinar, D. Galt, and R. Valdez. 2000. Short-duration grazing: The facts in 1999. *Rangelands* 22(1):18–23.

Holechek, J. L. and J. Hawkes. 1993. Desert and prairie ranching profitability. *Rangelands* 15:104–109.

Holechek, J. L., J. Hawkes, and T. Darden. 1994. Macroeconomics and cattle ranching. *Rangelands* 16:118–123.

Holechek, J. L. and K. Hess. 1994. Free market policy for public land grazing. *Rangelands* 16:63–67.

Holechek, J. L. and K. Hess. 1995. The emergency feed program. *Rangelands* 17:133–136.

Holechek, J. L. and K. Hess. 1996. *Market forces would benefit rangelands.* Forum for Appl. Res. and Public Policy 2:5–15.

Holechek, J. L., R. D. Pieper, and C. H. Herbel. 2001. *Range management principles and practices.* 4th ed. Englewood Cliffs, NJ: Prentice-Hall.

Huston, J. E. and W. E. Pinchak. 1991. Range animal nutrition. In R. K. Heitschmidt and J. W. Stuth, eds., *Grazing management.* Portland, OR: Timber Press.

Klipple, G. E. and D. F. Costello. 1960. *Vegetation and cattle responses to different intensities of grazing on shortgrass ranges of the central Great Plains.* U.S. Dept. Agric. Tech. Bull. 1216.

Knutson, R. D., J. B. Penn, and W. T. Boehm. 1990. *Agricultural and food policy.* 2nd ed. Englewood Cliffs, NJ: Prentice-Hall.

Kothmann, M. M. and R. T. Hinnant. 1992. *APSAT: An operational level grazing management model.* Proc. West. Sec. Amer. Soc. Anim. Sci. 43:354–357.

Mundinger, J. G. 1976. Waterfowl response to rest-rotation grazing. *Journal of Wildlife Management* 40:60–68.

Pieper, R. D., E. E. Parker, G. B. Donart, and J. D. Wright. 1991. *Cattle and vegetational response to four-pasture and continuous grazing systems.* New Mexico Agr. Exp. Sta. Bull. 576.

Savory, A. 1999. *Holistic management: A decision making framework.* Washington, DC: Island Press.

Shoop, M. C. and E. H. McIlvain. 1971. Why some cattlemen overgraze and some don't. *Journal of Range Management* 24:252–257.

Smith, J. G. 1899. *Grazing problems in the southwest and how to meet them.* U.S. Dep. Agric. Div. Agros. Bull 16:1–47.

Stoddart, L. A. and A. D. Smith. 1943. *Range Management.* New York: McGraw-Hill.

Thurow, T. L. and C. A. Taylor, Jr. 1999. Viewpoint: The role of drought in range management. *Journal of Range Management* 52:413–419.

U. S. Department of Agriculture (USDA). 1993. *Agricultural statistics.* Washington, DC: United States Government Printing Office.

U. S. Department of Interior (USDI). 1984. *Fifty years of public land management.* U.S. Department of Interior, Bureau of Land Management.

U. S. Department of Interior (USDI). 1993. *Public land statistics.* U.S. Department of Interior, Bureau of Land Management.

U. S. Department of Interior (USDI). 1999. *Public land statistics.* U.S. Department of Interior, Bureau of Land Management.

Vallentine, J. R. 1989. *Range development and improvements.* 3rd ed. San Diego, CA: Academic Press.

Vallentine, J. R. 1990. *Grazing management.* San Diego, CA: Academic Press.

Workman, J. P. and S. G. Evans. 1993. Utah ranches—An economic snapshot. *Rangelands* 15:253–255.

Workman, J. P., S. L. King, and J. F. Hopper. 1972. Price elasticity of demand for beef and range improvement decisions. *Journal of Range Management* 25:337–340.

Wuerthner, G. 1990. The price of wrong. *Sierra* 25:38–48.

12

Farmland and Food Production

INTRODUCTION

The capability of modern agriculture to meet the food needs of a human population that now numbers 5.6 billion and has increased by fivefold over the last 100 years is one of the greatest technological accomplishments of mankind. Since Englishman Thomas Malthus published his 1798 essay on the human population limits, a succession of like-minded observers have warned of limits to human population growth. Many have predicted that unchecked human population growth will ultimately outgrow the capacity of humans to feed themselves. So far, however, they have consistently underestimated the capability of technology to solve problems of food scarcity and natural resource depletion. To achieve that feat, however, agriculture now must use more of the earth's land, water, soil, plant, animal, and energy resources than any other human activity. It has also been a major cause of environmental degradation, including pollution and loss of biodiversity.

In developed countries technological breakthroughs involving improved crop varieties, irrigation techniques, pesticides, better tillage methods, and fertilizer usage have led to huge food surpluses, which are permitting large areas of marginal lands to be retired from farming. Other technologies are reducing environmental degradation problems previously associated with agriculture. Serious problems of food shortages and environmental degradation still plague many developing countries, and nonpoint sources of U.S. agricultural pollution remain problematic (see Chapter 7). However, we are optimistic that these challenges can be met if socioeconomic changes are made that encourage the technological developments and applications that have contributed to the food surpluses in the more developed countries. We see the present global trend toward a slowdown in the human population growth rate, rapid technology increase, and more democratic, free-market sociopolitical systems as encouraging. In developing this chapter, we have relied heavily on Miller (1990), Schiller (1994, 2000), Owen and Chiras (1995, 1998), Knutson et al. (1995, 1998), and Brown (1999, 2001).

WORLD FOOD OVERVIEW

Improvements in agriculture have increased per capita food production by roughly 25% since the mid-1970s. Real-food prices in most countries have fallen 10–30%

TABLE 12.1   Gains in World Grain Yield Per Hectare, 1950–2000

	Yield Per Hectare[1]	Annual Increase
	(tons)	(percent)
1950	1.06	
1990	2.47	2.1
2000	2.78	1.2

Source: USDA, Production, Supply, and Distribution, electronic database. Washington, DC, updated September 2000 (from Brown 2001).
[1] Yield for 1990 is three-year average.

over the past 20 years. Although great progress has been made in expanding world food supplies during the last 30 years, hunger remains a serious problem for humanity in many parts of the world. Presently about 1.1 billion (18%) of the world population (6.1 billion) are undernourished (Brown 2001). Unfortunately, there is no world plan to eliminate world hunger. At the same time world food security is threatened by rapid human population growth (80 million more people per year), shrinkage of cropland per person, climatic instability, groundwater depletion, rising energy prices, and diminishing benefits from the "green revolution." Feeding 50% more people (9 billion) expected by 2050 will be a major challenge for the world's farmers.

Most of the world's hungry live in African and Asian countries. Hunger problems are most serious in India, Bangladesh, Pakistan, Ethiopia, Nigeria, and Tanzania. Rapid human population growth is the primary factor behind hunger in these countries. As farms are subdivided among successive generations, the amount of land per person shrinks. In the previously mentioned countries, water tables are falling because of overpumping for irrigation (Brown 2000).

Between the 1950s and 1990s a steady flow of green revolution technologies rapidly raised farmland productivity. However, there is now evidence that grain yields are leveling and future increases will be more limited (Table 12.1) (Brown 2001).

Unfortunately, modern agriculture in developed countries depends on using increasing amounts of fossil fuels for cultivation, harvesting, fertilizer production, irrigation, processing, and distribution. Another major food production concern is the trend toward globalization, specialization, and urbanization. This is reducing food self-sufficiency in both developed and developing countries. Often the most productive farmlands are urbanized in developed countries. This not only shrinks the agricultural land base but necessitates massive food transportation and delivery systems to keep the expanding urban center supplied. If water, farmland, and energy resources shrink and the large urban centers continue to expand there is concern that the United States will lose its capability to be a large grain exporter.

Presently the United States plays a key role in providing developing countries with grain and preventing worldwide scarcity when climatic conditions are unfavorable (Brown 2001). A major drought in the Great Plains like the 1930s and 1950s could now cause massive food shortages in several Asian and African countries.

In the United States five factors that undermine food security are loss of farmland to urbanization, depletion of aquifers, loss of irrigation water to urbanization, heavy reliance on fossil fuels, and soil erosion (Brown 1999). There is considerable doubt that the United States over the next 15 years will have the physical and financial capability to provide free or low-cost food to developing countries on a massive scale. Future unrest over food in Asia and Africa has the potential for serious internal political turmoil and world war as more countries obtain nuclear and biological weapons.

Several technologies discussed by Brown (2001) have potential to diminish world hunger in the twenty-first century. These include double cropping, more drought-resistant grain varieties, more efficient irrigation methods, aquaculture, more efficient use of crop residues, and production of fewer fat-rich livestock products. However, most experts increasingly agree that rapid stabilization of the world's human population is the key to solving the world's hunger problem. This means encouraging two-child families in both developing and developed countries. It also means conserving farmland, reducing fossil fuel dependency, and more efficient use of water resources. We explore these issues in more detail in the remainder of this chapter and other parts of the book.

A BRIEF HISTORY OF AGRICULTURE

Over the course of world history agriculture has undergone four revolutions, each of which closely correspond to improvements in the human condition (Laetsch 1979). Our discussion of these revolutions closely follows Bolen and Robinson (1995). A more detailed discussion is provided by Chrispeels and Sadava (1977). About 10,000 years ago the domestication of wild plants initiated the first period. This happened in separate parts of the world such as China, South America, and Europe. Some of the first plants domesticated were wheat, rice, corn, oats, and potatoes, but gradually many others were added. The domestication of plants permitted the development of organized civilizations. Because people were freed from daily procurement of food, plant culture permitted specialization of labor. It changed societies from nomadic to sedentary. With permanent settlement and a reduction in the amount of time needed for acquisition of food, other important aspects of human life could develop, including art, crafts, education, religion, government, science and advanced technologies.

Exploration and discovery were the basis for the second agricultural revolution. This era dates back to the thirteenth century when Marco Polo journeyed from Europe to China on a trading mission in quest of spices. A highlight of the exploration era was Columbus' landing in America in 1492. Throughout the 1500s, a series of European explorers including Balboa, Magellan, Cabot, and Coronado traversed both the oceans and major land areas of the world accumulating previously unknown knowledge of the world's geography, wealth, and people. Land and sea trade routes were established for the exchange of agricultural products and other commodities. New World (North and South America) plants such as corn and potatoes were introduced to Europe. Sugar, coffee, tobacco, and a variety of spices became key items of trade among widely scattered countries.

The third major agricultural revolution resulted from industrialization that started in the eighteenth century with the inventions of the steam engine and the cotton gin. These inventions led to both improvements in mass transit and the substitution of machines for strenuous human and animal labor. Steamboats in the early 1800s and railroads in the mid-1800s greatly reduced the time and cost of human travel. By the 1890s, agricultural goods could be readily transported by railroad across nearly all portions of the United States. A variety of machines were widely used in farming such as reapers, combines, balers, and drills. Farm mechanization continued quite rapidly after World War I with the advent of the tractor. The replacement of labor with machines greatly boosted the supply of agricultural goods, which lowered food costs. This caused a large-scale reduction in the number of farms in the United States as average farm size increased and marginal lands were abandoned.

We are now going through a fourth revolution in agriculture based on biotechnology breakthroughs. Advances in chemistry and genetics are translating into improved crop varieties, fertilizers, herbicides, and insecticides that increase productivity far beyond those of the Industrial Revolution (Figure 12.1). Highly adapted hybrid crops and livestock are being developed for each of the specific climatic regimes, soil types, and pests that occur in different parts of the world. Agricultural technology appears to be winning the race with the growing human population by boosting food and fiber yields to previously unimagined highs.

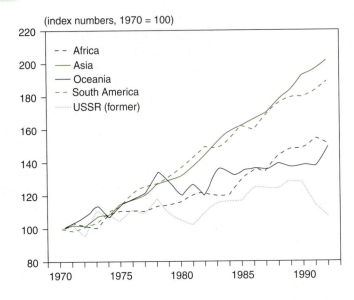

(index numbers, 1970 = 100)

Legend:
- - - Africa
—— Asia
—— Oceania
- - South America
········ USSR (former)

FIGURE 12.1 Index of food production by region, 1970–1992. Food and Agriculture Organization of the United Nations (FAO), Agrostat PC, on diskette (FAO, Rome, 1993) (from World Resources Institute 1994).

MAJOR TYPES OF AGRICULTURE

About 30 crops provide more than 80% of the world's food supply although some 80,000 plants are edible (Miller 1990). Wheat, rice, corn, and potatoes account for more than half of the world's annual food production. Livestock products such as meat, milk, eggs, and cheese are much more expensive than plant products. This is because about 90% of the energy in plants is lost when plants are consumed by livestock rather than consumed directly by people. Therefore, people in developed countries, where incomes are high, typically can afford to consume much more meat and other animal products than those in developing countries. Exceptions to this generalization occur in parts of Africa, the Middle East, and South America where herders depend primarily on livestock for their sustenance.

Basically two types of agricultural systems are used to produce crops and livestock—industrialized and subsistence agriculture (Figure 12.2). Industrialized agriculture is widely practiced in the developed countries, where specialized operators grow a single crop or one type of livestock on large areas using high inputs (fertilizer, pesticides, mechanized cultivation) (Miller 1990). In developing countries, the common form of food production is still subsistence agriculture involving small plots of land tended by individual farmers with primitive equipment pulled by draft animals. Various modifications of these two basic forms of agriculture are applied in both developing and developed countries. Using the intensive approach, large plantation farms grow coffee, sugar, bananas, and cacao in many tropical or semitropical developing countries such as Kenya, Colombia, and Brazil. On the other hand, organic farms using minimal inputs of inorganic fertilizer and pesticides are becoming common in some developed countries including the United States.

THE GREEN REVOLUTION

The basic options for increasing food production are (1) increasing yields from existing cropland and (2) bringing more land under cultivation (Miller 1990, Brown 2001). Most of the increase in world food production since 1960 has been brought about by huge crop yield increases made possible by the "green revolution."

The green revolution centers around genetic selection of highly adapted plant varieties that are responsive to large inputs of inorganic fertilizer, pesticides, and irrigation water (Chrispeels and Sadava 1977). The first green revolution plants were

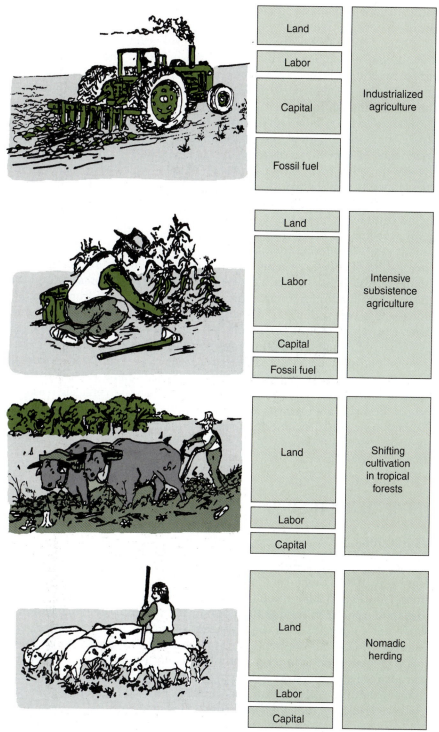

FIGURE 12.2   Relative inputs of land, labor, capital, and fossil fuel energy into major types of agricultural systems (drawing by John N. Smith, adapted from Miller 1990).

short-stemmed varieties of wheat resistant to lodging (bending) developed in the United States in the 1950s. Later in the 1970s, the same approach was used to create fast-growing dwarf varieties of rice for tropical climates in Asia. The short and stiffer stalks of the new varieties of wheat and rice enabled them to produce larger seed heads without toppling over. With appropriate water, fertilizer, and pesticide inputs, three- to fivefold increases in yields over traditional varieties were possible. Another advantage was the faster growth of green revolution plants. Multiple cropping (two or three consecutive crops within a year) became possible on the same unit of land.

Since 1960, most of the increase in world grain output has resulted from green revolution techniques (Miller 1990, Brown 2001). The primary problem with the green revolution is the requirement for large fossil fuel input to cultivate, irrigate, and manufacture pesticides and fertilizers. Agricultural use of fossil fuels has increased about six- to eightfold since 1950 (Miller 1990, World Resources Institute 1994). It now takes twice as much oil to produce a ton of grain as it did 40 years ago. Although the contribution to greenhouse warming (see Chapter 6) is relatively small, the dependence on depletable resources is not sustainable.

On the other hand, it seems probable that affordable new energy sources will be developed before the world's supply of fossil fuels is depleted (see Chapters 19 and 20). Recent developments in solar energy lend some support to the more optimistic view, but advances in fusion power have been slower than once hoped. However, policy related to future research investments cannot afford to be overly confident about easily mastering the challenges that lie ahead. Other criticisms relate to the dedication of very large areas to single crops without interspersion of other crops or natural areas. This increases the probability of pest outbreaks and other natural catastrophes. In addition to negative environmental effects from pesticides, recent evidence that pests adapt rapidly to pesticides has given rise to integrated pest management, which relies more on natural ecological processes.

Most of the potential of the green revolution has now been realized. It will be difficult for farmers to continue raising wheat, rice, and corn yields. Wheat yields in the United States have increased little since 1983 (see Table 12.1). The issue here centers around the amount of photosynthate going to the seed versus leaves and stalks. Originally, domesticated wheats put only about 20% of their photosyntates into seeds. Today's new green revolution varieties put half or more into seed formation. Theoretically, the limit that can go for seed while maintaining the rest of the plant is estimated at 60% (Brown 2001).

U.S. AGRICULTURAL SUCCESS AND PROBLEMS

A Success Story

Today less than 1% of the workforce in the United States engages in farming (Knutson et al. 1998). Since 1920, the number of farms in the United States has dropped from 7 million to 2.5 million (Figure 12.3). At the same time the farm population has dropped from 32 million to under 10 million. While the reduction in farm numbers has been a concern since the nineteenth century, it has no doubt contributed substantially to overall economic development and the rise in living standards. The release of labor from food production is one of the primary contributors to technological advance and industrialization in the United States and other developed countries. Some examples of agricultural productivity since the early 1950s summarized by Schiller (1991) include the following: annual egg production has jumped from 183 to 243 eggs per laying chicken, milk output has increased from 5,400 to 12,100 pounds per cow, wheat output has increased from 17.3 to 35.3 bushels per acre, and corn output has jumped from 39.4 to 102 bushels per acre. During the same period, farm output per hour of labor, has increased even more phenomenally by 700%. Americans eat better and at the same time spend a lower percentage of their income on food than in any other country. People in most parts of the world spend about 40% of their disposable income for food compared to about 12% in the United States (Figure 12.4).

In terms of total sales, agriculture is now the biggest industry in the United States. About 20% of the workforce, or 25 million people, work in some part of U.S. agriculture including the growing, processing, or marketing of food. Agriculture exports from the United States play a critical role in meeting food needs of several develop-

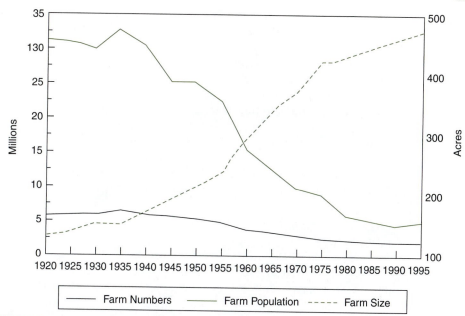

FIGURE 12.3 Farm population, farm numbers, and farm size for selected years, 1920–1995 (from Knutson et al. 1998) (*Agricultural Statistics*, Washington, DC: USDA, various issues).

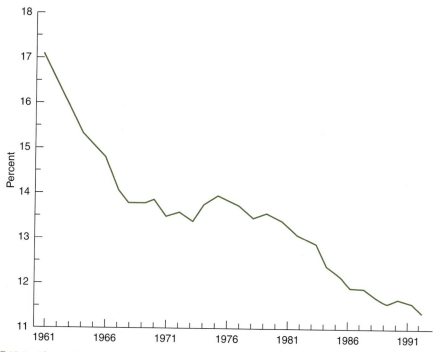

FIGURE 12.4 Share of consumer income spent on food (from Knutson et al. 1995) (Denis Dunham, *Food Cost Review* 1992, AER-672 Washington, DC: ERS/USDA, September 1993, p. 46).

ing countries particularly during periods of sociopolitical unrest such as in Somalia, in 1992–1993, and when drought conditions occur in India or China.

Although its successes are many, American agriculture confronts a number of problems that are discussed in detail by Schiller (1991), Knutson et al. (1995), and Owen et al. (1998) (Figure 12.5). These include soil erosion (see Chapter 8), increasing regulation, declining exports, urban sprawl, rising production costs, restricted water supplies, and atmospheric pollution. We will briefly discuss these issues following Owen et al. (1998).

FIGURE 12.5   This picture of a family farm in West Virginia conveys an image of prosperity. However, a variety of factors including urban sprawl, rising production costs, declining exports, and regulation now seriously threaten this type of operation (courtesy of U.S. Department of Agriculture).

Increasing Regulation

As discussed in Chapter 8, farmers must increasingly comply with federal, state, and local regulations regarding use of pesticides, use of herbicides, tillage of soil, management of wetlands, and plowing of erodible lands. Although many of these regulations are needed to protect environmental quality and human welfare, they greatly increase farming costs and can make U.S. farm products less competitive with those of other countries. Past misuse of fertilizers and pesticides has polluted groundwater supplies in some heavily farmed areas. Certain insecticides used by farmers have adversely impacted wildlife populations. Some pesticides have contaminated soils and adversely affected desirable soil bacteria that take nitrogen from the air and fix it into a form (nitrate) needed by crops.

Urban Sprawl

In the period from 1982 to 1992 the United States lost about 3 million acres (there are 382 million acres total) of its farmland to urban sprawl based on USDA Natural Resources Inventory. This rate of conversion was lower than in previous decades (1950s to 1980s) (Figure 12.6). Since 1992 the rate of conversion has accelerated to somewhere between 0.5 to 1.0 million acres per year. What makes this problem critical is the best-watered, most level, and most productive soils generally occur near cities. A tremendous increase in shopping malls, airports, factories, and highways has occurred around almost every major city during the past 40 years (see Chapter 14). This problem is made more severe by the unpredictable way in which much urban sprawl has occurred (Figure 12.7) (see Chapter 14).

Rather than developing lands around cities in tightly organized fashion, as in much of Europe, the United States has permitted random development patterns in which adjoining farmlands are broken up and cut off by an interspersion of homes, business, and public works (also see Chapter 14, Figures 14.2 and 14.15). This greatly compromises the efficiency of farms that remain because of restrictions on movement of equipment, pesticide use, fertilizer use, and raising of livestock such as hogs, chickens, and dairy cattle. This pattern is particularly severe in New England, the mid-Atlantic states, Florida, and parts of the far West.

Land Uses Occupying Over 50% of Area

LEGEND

- Developed Land
- Cropland
- Rangeland/Pasture
- Forest Land
- Mix of Two or More Land Uses
- Federal Land/Water/Non-Sampled Area

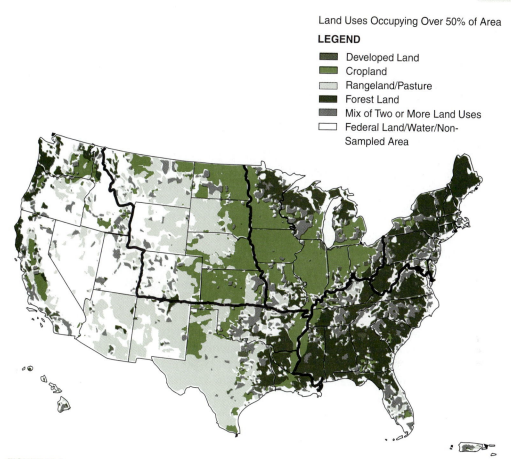

FIGURE 12.6 Dominant uses of private land in the United States in 1992 (courtesy of USDA-Natural Resources Conservation Service).

FIGURE 12.7 Urban sprawl into farmland near Phoenix, Arizona.

Letting Farmland Stay Farmland

Conversion of farmland to nonfarm uses may change land use irreversibly and alter the character of an area. It also may weaken the local agricultural economy. In heavily developed areas, loss of even a few acres of remaining farmland suggests for many the end of a way of life and separation of people from their roots.

Fifteen states, mostly in the Northeast, have enacted laws and appropriated funds to pay farmers willing to keep their land in an agricultural use. Easements stay with the land even after its sale, guaranteeing that farmland stays farmland.

Since the mid-1970s, farmland preservation laws have protected nearly 420,000 acres of farmland at a cost of almost $730 million—about $1,750 an acre. Funding for the programs has come mostly from sale of bonds and levy of sales, property, and other taxes. An additional $195 million was available early in 1996 for further purchases—$107 million in New Jersey alone.

Among the leaders in farmland protection are Maryland, which has spent about $125 million to purchase easements on 1,117,000 acres of farmland, and Pennsylvania, which has spent more than $150 million to protect almost 75,000 acres. Massachusetts and New Jersey have each spent more than $80 million to protect 35,907 acres and 27,924 acres, respectively.

The Federal Agriculture Improvement and Reform Act of 1996 established a Farmland Protection Program with a funding level of $35 million. The program will help states with farmland protection programs purchase conservation easements. Prior to the end of the 1996 federal fiscal year, $15 million were made available under the new program in 17 states through 37 individual programs. An estimated 150 to 200 farms will be signed up under these various programs.

From: A Geography of Hope. 1997. USDA-National Resources Conservation Service.

Rising Production Costs

Since the early 1970s, the costs of fuel, fertilizer, and pesticides have been rising faster than prices for most farm crops (Knutson et al. 1998). This is more a problem of overproduction of farm products rather than an excessive increase in the cost of farming inputs. Generally farm product prices have lagged behind the consumer price index since 1980. This is also true of energy, fertilizer, and pesticide prices, but to a lesser extent. The consumer price index measures the average annual changes of a standard group of needs by a typical household.

Declining Exports

Although farm exports from the United States grew rapidly in the 1970s, this trend reversed in the 1980s. Several factors including a strengthening U.S. dollar, higher interest rates, increased regulation, increased labor costs, changes in international trade laws, and higher farm input costs have tended to reduce the competitiveness of U.S. farm products with those of several other countries. Argentina, Australia, Brazil, Canada, and other countries have greatly expanded their output of several agricultural products. Their production costs are generally lower than costs in the United States. China and India, once major food importers in the 1970s, have become nearly self-sufficient because of the green revolution. For small grains such as wheat, American farmers do have a cost advantage over European countries. However, since 1974 the Europeans have generally chosen to protect their farmers with high subsidies and excessive tariffs on agricultural imports.

Atmospheric Pollution

Crop yields near major cities and highways are adversely impacted by carbon dioxide and other pollutants from vehicles using fossil fuels. The exact costs of this pollution are difficult to estimate and depend on the area and type of crop. However, losses could be as high as $5–$10 billion in years when there is little or no surplus production.

Restricted Water Supplies

In the Southwest and central Great Plains, the amount of water available for irrigation is rapidly dwindling. The greatest concern is the depletion of the Ogalalla aquifer in the plain states of Texas, Oklahoma, New Mexico, Kansas, and Nebraska (see Chapter 7). In California and Arizona, rapid urbanization and industrialization are making increased demand on limited water supplies previously used by agriculture.

Excess Capacity

In developed countries (the United States and western Europe), the biggest problem confronting farmers has been excess capacity (Knutson et al. 1995). In free-market economies, such as the United States, farm prices tend to trend downward over time because of rising productivity and low elasticity of demand. Price elasticity of demand refers to the relationship between quantity of a product demanded and price. Products with high elasticity of demand, such as automobiles and computers, show large increases in demand as prices fall. However, many agricultural products generally show inelasticity of demand in which demand is fairly constant whether prices rise or fall. This means that small increases in supply relative to demand can cause sharp drops in prices.

During the depression of the 1930s, the United States government began using price supports and cost subsidies to raise farm prices. These policies led to increased farm output requiring higher exports, government food purchases, and farmland set-aside programs to avoid a buildup of food surpluses. With the exception of the 1970s, market surpluses have characterized American agriculture since World War II. This problem has been made more severe by heavy subsidy of agriculture in western European countries (France, Spain, Germany, England, and so on) since the early 1970s. In response to the oversupply of farm products, developed countries have dumped these products on world markets at below true production costs. This has hindered agricultural progress in those developing countries that often have a competitive advantage in agricultural production of certain products such as sugar, beef, wheat, rice, and peanuts because of less regulation and lower labor costs. During the mid-1990s, the developed countries began shifting their agricultural policies towards lower price supports and the retirement of marginal farmlands through cash payments. Direct income payments to farmers decoupled from agricultural production are being considered among the ways to preserve smaller family farms that cannot compete with large corporations who practice industrialized farming.

AGRICULTURE IN DEVELOPING COUNTRIES

Subsistence agriculture is still the primary livelihood for much of the world's population. In its various forms this type of agriculture occurs on about 55% of the world's cropland. It is the primary type of agriculture in Africa, China, India, and South America. Generally subsistence agriculture involves small plots of family-owned land (mostly under 10 acres) on which a variety of crops are grown using little resource supplement.

FIGURE 12.9 Poor preparation for drought is a major cause of famine in the Sahel region of Africa. During drought, herd levels of livestock have been kept too high and more than half of the animals have perished from starvation.

underinvestment in agriculture, severe loss of topsoil from overgrazing and deforestation, and price fixing that discourages farmers from expanding output (governments keep food prices low to prevent urban unrest). More current discussions of African food problems are provided by Brown (1999, 2001).

Agriculture Prospects in Developing Countries

Famine problems have been greatest in the central portion of Africa known as the Sahel, which is on the edge of the Saharan desert. The basic problems are poor preparation for droughts, which characterize the area, and unstable socioeconomic systems, which discourage investment, savings, and infrastructure development. Because of high currency inflation rates and unsound banking systems, farmers and herders have little incentive to convert food surpluses in good years into currency that can be used to buy supplies needed in hard times. Instead they retain their crops and animals and practice a bartering system that involves exchanging a portion of their crop or a few animals for other goods when needed. During droughts, herd levels are too high and over half the animals often perish from starvation (Figure 12.9). A large amount of stored crop production is lost to weather, insects, and fungus. Unless this system is improved, the prospects for much of Africa would appear bleak at best.

Developing countries with the brightest agricultural prospects are in Latin America and Southeast Asia (Figures 12.10 and 12.11). Chile and Argentina have been particularly successful in increasing agricultural output and economic growth. Success in both countries has involved a series of free market reforms that privatized state run companies, encouraged foreign investment, stabilized the currency, removed trade barriers, reduced unnecessary regulations, lowered service on borrowed foreign capital, and permitted movement toward a more democratic government.

Latin American countries have some important advantages over African countries that are worth mentioning. These include higher rainfall, less cultural division, and a common language (Spanish). Perhaps the biggest problem in Africa is tribalism. The native peoples of Africa are fragmented into a series of tribes with different languages and customs. This has been a major barrier to economic development because of the constant quarreling among various groups. National unity

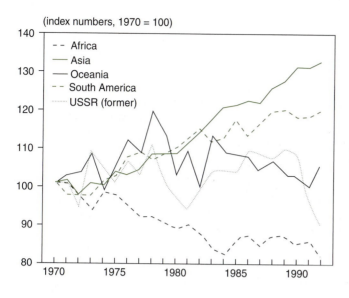

FIGURE 12.10 Index of per capita food production by region, 1970–1992 [Food and Agriculture Organization of the United States (FAO), Agrostat PC, on diskette (FAO, Rome, 1993)] (from World Resources Institute 1994).

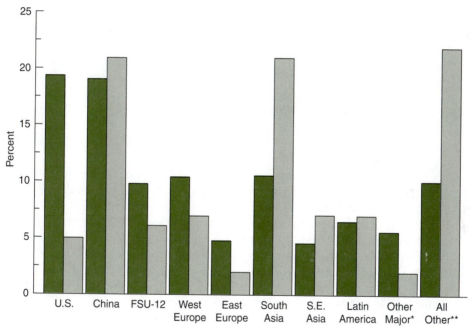

*Includes Canada, Australia, and Turkey
**Includes the Middle East and Africa

FIGURE 12.11 Share of world crop production (blue) and population (gray) by region, 1990–1992 average (from Knutson et al. 1995) [*World Agricultural Production* (Washington, DC: USDA/ERS, March 1993) and *The World Factbook: 1990* (Washington, DC: Central Intelligence Agency, 1993)].

has been a prerequisite for any country to make rapid progress in improving the human condition.

PESTICIDE CONTROVERSIES

The Pest Problem

One of the biggest controversies in modern agriculture centers around the use of pesticides (weed, rodent, and insect killers). Without question pesticides have played a crucial role in keeping world food supplies ahead of population growth. Since World

War II, pesticides have also prevented millions of deaths and untold human suffering from insect-transmitted diseases such as malaria (from mosquitoes), bubonic plague (rat fleas), typhus (fleas and lice), and sleeping sickness (tsetse fly) (Miller 1990).

An estimated 40 to 50% of the world's potential food supply is lost to pests (Miller 1990, Pedigo 1999). Without pesticides, these losses could be 30 to 50% higher. Present pesticides are an efficient, cost-effective, and seasonally safe means of controlling weeds, insects, and rodents when properly used. Other advantages include ease of shipping and handling. Pesticide companies are now developing pesticides that are safer to use and less ecologically harmful than those used over 30 years ago.

The arguments for and against pesticides have been summarized by Miller (1990) and Pedigo (1999). The main argument against pesticides is that eventually insects and weeds develop genetic resistance to them and pesticides usually kill off most of the pest's natural predators. This is particularly true for insects that have quick generation turnovers and produce large numbers of offspring. In tropical areas, insects can become resistant to a specific pesticide in less than 3 years while in temperate areas it takes about 5 years. Genetic resistance can lead to a cycle of ever-increasing costs for pesticides that are less and less effective.

Pesticides also can kill desirable wildlife species and contaminate air and water resources. Several major fish kills in the United States have been caused by misuse of pesticides. DDT, a fat-soluble, slowly degradable pesticide, interferes with reproduction in birds such as the bald eagle. This pesticide, now banned in the United States, can be increasingly concentrated in fatty tissues with upward movement through the food chain. DDT in fish is concentrated at about 10 times the level of the plankton they consume (Miller 1990). Eagles that eat these fish further concentrate the DDT. Because of this process, referred to as biological amplification, dilution is not always a solution to reducing the pesticide hazard to human health. The recent improvement of southern bald eagle status, now listed as threatened instead of endangered, is associated in part with the reduction of DDT and similar compounds in eagle habitat.

There is considerable concern that human health could be threatened by long-term, low-level exposure to pesticides. However, the Federal Food and Drug Administration requires that pesticides be intensively tested for harmful effects on humans before approved for use on farmlands. Many of the pesticides now in use can cause cancer in humans if used at extremely high levels over an extended period of time. However, a 1987 National Academy of Sciences report estimates that, at worst, about 20,000 cases of cancer a year are caused by the pesticides in food. Some of the foods most likely to contain pesticide residues include potatoes, oranges, lettuce, carrots, celery, apples, grapes, and tomatoes.

Critiques of intensive pesticide use point out there are farming alternatives that are practical and cheaper. These can be categorized into cultivation methods, plant breeding, biological pest control, allelopathic plants, insect sterilization, and insect hormones. These will be briefly discussed following Miller (1990) and Pedigo (1999).

Alternatives to Pesticides Alone

Cultivation Methods.   This approach basically involves using diversified farming instead of monotypic farming. Rather than planting one crop year after year over large areas, a variety of crops are grown in rotation systems. Hedges can be planted within and around fields to create a habitat for pest predators. Planting times can be manipulated so that major crop pests are deprived of their food source for long periods and starve to death. In addition to crop rotation, intercropping and agroforestry can be used to further increase diversity.

Biological Pest Control.   With this approach, the diseases, parasites, and natural predators of a problem pest are identified and released to obtain population reduction. Worldwide there have been about 300 successful biological pest control projects. In the northwestern United States, chrysolina beetles have been effective in controlling an aggressive weed called St. John's wort. In Australia, the prickly pear cactus has been successfully controlled with the Argentine moth bopper. A viral disease, myxomatosis, has been effective in controlling wild rabbits in Australia. In Africa, mealybugs, which cause loss of cassava crops, have been controlled by wasps. Ducks, geese, and chickens can be useful in controlling crop weeds and insects. However, they also consume certain vegetable crops. Goats have been effective in controlling some unwanted shrubs on rangelands.

Allelopathic Plants. Many plants produce substances toxic to (allelopathic) other plants that may be competitors. Through selective breeding, crop varieties of various grains can be developed that suppress weeds. Some combinations of crops planted together can be resistant to insects.

Insect Sterilization. This approach involves the use of radiation to sterilize large numbers of male insect species. They are then released and allowed to mate with wild fertile females. The screwworm, a livestock and wildlife parasite in the southwestern United States that caused several millions of dollars in damage yearly, was largely eliminated by use of sterilized males. Problems with male sterilization are high cost, prevention of reinfestation, and the difficulty of producing enough nonsterile males to ensure low reproductive success.

Insect Hormones. Chemicals extracted from the insect or made in the laboratory are strategically applied during the insect's life cycle to cause death or reproductive failure. A variation of this approach is to use insect extracts to attract the pests into traps with poison or to use attractants to help predators find pest insects. There have been some successes with hormonal growth disruptors and attractants. However, they are costly, require careful timing, and are not immediately effective.

Integrated Pest Management. Increasingly, chemical cultivation and biological methods of pest control are being combined to optimize efficiency, cost, and environmental sustainability. This approach is known as integrated pest management (IPM). IPM programs can reduce pest control costs by 50% or more, reduce pesticide use by over 50%, and sustain or increase crop yields. Among the disadvantages, IPM requires much greater knowledge than alternative methods and often is more labor-intensive. Farmers in the United States gradually shifted toward IPM in the 1990s. It is used to some degree on about 35–40% of the Nation's farmland. Other countries applying IPM include China, Brazil, and Indonesia. We refer to Pedigo (1990) for a more complete discussion of insect pest management.

SUSTAINABLE AGRICULTURE

Sustainable agriculture is popular terminology for a set of agricultural practices that are an alternative to modern monotypic, industrialized farming. Many farmers and ranchers have been part of the sustainable agriculture movement. This approach substitutes polycultures for monocultures and minimizes use of fossil fuels, inorganic fertilizer, pesticide use, and irrigation water (Miller 1990). Its philosophical roots come from a group of environmentalists and conservationists concerned that topsoil depletion and pesticide contamination under industrial farming will lead to degradation of ecosystems functions and services.

Although most features of the sustainable agriculture movement are sound under specific conditions or when used in moderation, an immediate complete switch to this type of farming would sharply lower food production and raise good prices. Breakthroughs involving genetic engineering, tillage, integrated pest management, and improved irrigation systems have greatly reduced the environmental problems associated with conventional agriculture. In the United States the need to grow crops on fragile, erodible soils has been largely eliminated by increased yields associated with the high input approach. In our view, the best industrial and subsistence practices can be integrated into sustainable agricultural practices that will best meet the food needs of society and maintain environmental integrity in the twenty-first century.

FARMLAND POLICY

The Issues

In most years, the United States government spends 15 to 25 billion dollars on farm subsidies (Figure 12.12). Since 1980, these subsidies have cost taxpayers about 350 billion dollars (Schiller 1994). With this much money, the government could have bought out every farmer in 40 states. The high cost of farm programs is of considerable concern to taxpayers. Evidence indicates many farm programs are ineffective in halting the decline in farm numbers or reducing food surpluses. A view growing among conservatives is that less government intervention in farming would be better for both farmers and food consumers.

Agriculture is one of the most competitive of all industries (Schiller 2000). This is because there are few barriers to entry. It is easier to get into farming than most other businesses although, in order to have a profitable operation, farmers need increasingly larger acreages of land, more expensive equipment, and more credit. Because of low barriers to entry, farmers can quickly expand their output during periods of high crop prices usually caused by adverse climatic conditions. For this reason profits are usually short-lived and farmers must continually invest wisely and cut costs in order to stay in business.

The huge farm production increases made possible by technology have greatly benefited consumers but created a problem for American farmers (Schiller 2000, Knutson et al. 1998). This problem exists because the demand for food is relatively inelastic. In other words, the amount of food that people can eat is limited, and well-

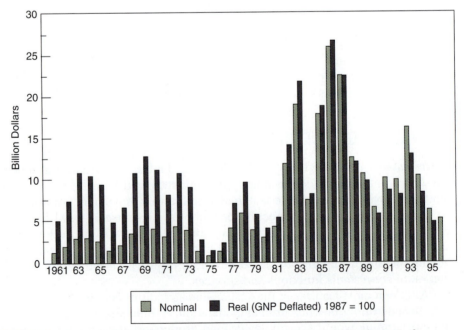

FIGURE 12.12   Nominal (blue) and real costs (gray) of government programs for agriculture, 1961–1996 (from Knutson et al. 1998) (Economic Research Service. *Agricultural Outlook*, Washington, DC: USDA, various issues).

FIGURE 12.13 Unstable corn prices. Most agricultural prices are subject to abrupt short-term changes. Notice how corn prices rose dramatically during World Wars I and II, then fell sharply. Poor harvests in the rest of the world increased demand for U.S. food in 1973–74 (from Schiller 1994) (U.S. Department of Agriculture).

fed consumers do not increase food purchase by very much when harvests are abundant. Therefore, a small increase in food supplies often leads to large drops in food prices. For example, in the periods from 1910–1914 and 1993–1995, the ratio of farm to nonfarm prices fell 50%. This ratio would have fallen even farther without foreign demand for U.S. farm products and government farm subsidies. Over the long run the ability of U.S. agriculture to produce has greatly outpaced the growth in the human population and the quantity of food demanded.

Another problem is that farmers also must confront abrupt price swings caused by changes in annual climatic conditions and world political upheaval (Schiller 1994, Figure 12.13). Favorable weather results in abundant harvests but low prices. On the other hand, droughts, freezes, insect infestations, and outbreaks of crop diseases can drastically reduce harvests but sharply push up prices. Unlike most other industries, farmers cannot quickly increase supplies when climatic adversity occurs. A year or more is required for farmers to respond by planting more land, or increasing their livestock herds when food prices go up. This lag intensifies price swings and makes it difficult for farmers to take advantage of higher prices. When corn prices are high because of unfavorable weather, farmers will plant more corn the following year. After a year of high corn prices, better weather and increased crop acreage probably will cause corn prices to drop. Individual farmers find it very difficult to avoid these booms and busts in prices.

Prior to the depression of the 1930s, the U.S. government had little involvement in agriculture. Generally the farm sector was prosperous because of expanding foreign demand, a rapidly growing human population, and a slow pace of technological advance. The 1910–1919 period was particularly favorable for American farmers because of World War I, which depressed farm output in Europe (Figure 12.14). After 1920, international trade encountered increased restriction and European demand for American farm products declined. At the same time agriculture efficiency improved, mainly from mechanization. In the depths of the depression in 1932, farm prices had fallen 75% below 1919 levels (Schiller 1994). Average farm income fell from about $2,600 to $850. Farmers pressured the U.S. Congress to take action on their problems. The main objective of the resulting law (Agricultural Adjustment Act of 1933) was to increase farm product price levels to those of the 1909–1914 period, which was considered fair or parity. The basic approach was to establish a

FIGURE 12.14   Farm prices, 1910–1940 (1910–1914 = 100). Farm prices are less stable than nonfarm prices. During the 1930s, relative farm prices fell 50%. This experience was the catalyst for government price supports and other agricultural assistance programs (from Schiller 1994).

price floor in which the government compensates farmers for the difference between the market and the parity price. The consequences of this policy have been surpluses that must be eliminated by government food purchase and giveaway, export sales, or restrictions on supply (Figure 12.15). Since the 1930s, the government has also subsidized farmers through acreage set-asides, marketing orders, import quotas, crop insurance, and cost subsidies. These will be briefly discussed.

Forms of Farm Subsidies in the United States

Acreage Set-Asides.   The surest way to increase farm prices without causing a surplus is to pay farmers to take land out of production (Schiller 1994, Knutson et al. 1995). Legislation passed in 1936 and 1938 was the basis for allotments and quotas designed to limit the supply of specific crops. Farmers have always responded to these restrictions by more intensively farming allotment acres. Quotas are often applied with allotments. They restrict the quantity of a commodity a farmer is allowed to sell. Discontent with the regimentation associated with allotments and quotas led to voluntary land retirement programs that paid producers to take land out of production. Land productivity has been used as the basis for payments. There have been three major land retirement programs since the 1930s. These include the soil bank, the payment-in-kind, and the conservation reserve programs. Our discussion of these programs follows Knutson et al. (1995, 1998).

The soil bank, first implemented in 1956, paid farmers to retire land from production for a 10-year period. By 1960, nearly 30 million acres were in the soil bank. Under this program whole farms could be retired from production. This created economic hardship for some rural communities because of a drop in commerce. The large increases in crop prices during the 1970s caused most of the soil bank lands to go back into production.

Drops in farm prices because of inflation control in the United States, a rising dollar, and increased foreign agricultural output resulted in huge grain surpluses by 1983. To deal with this problem, the USDA announced its PIK (payment-in-

(a) On the individual farmer

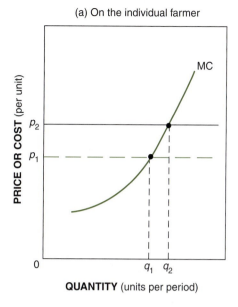

(b) On the agricultural market

FIGURE 12.15 The impact of price supports. In the absence of price supports, the price of farm products would be determined by the intersection of market supply and demand (see Chapter 4). In this case, the equilibrium price would be p_1, as shown in part b. All individual farmers would confront this price and produce up to the point where MC = p_1, as in part a. Government price supports raise the price to p_2. By offering to buy (or "loan") unlimited quantities at this price, the government shifts the demand curve facing each farmer upward. Individual farmers respond by increasing their output, resulting in a market surplus (part b) (from Schiller 1994).

kind) program, which paid farmers with grains for 80 to 95% of their farm yields in return for a 1-year retirement of their land from production. Slightly over a third (82 million acres) of the eligible cropland was removed from production in 1983 (Figure 12.16). This program was effective in temporarily lowering grain surpluses and raising farm prices, although it was expensive.

The Conservation Reserve Program, or CRP, was part of a 1985 farm bill that had dual objectives of reducing farm commodity oversupply and controlling soil erosion. It differed from the soil bank program in that only land classified as highly erodible could be retired for a 10-year period. In order to minimize adverse economic consequences to rural communities, only 25% of the land could be retired in any county. By 1995, about 35 million acres had been retired at an annual cost of $1.8 billion per year. Like the soil bank program, payments to farmers are based on land productivity. Although this program is considered costly, it has been praised for its benefits associated with soil and wildlife conservation, such as restoring wetlands. Its effectiveness in reducing oversupply of certain crops has been more controversial.

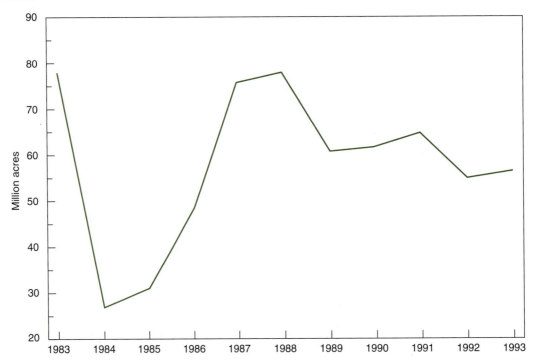

FIGURE 12.16   Quantities of farmland removed from production, 1983–1993 (from Knutson et al. 1995) [Economic Research Service, *Agricultural Resources: Cropland, Water, and Conservation*, AR-30 (Washington DC: ERS/USDA, May 1993), p. 4].

Marketing Orders.   Although the production of individual farmers is too small to raise market prices by withdrawing output, they can increase market prices through collective action (Schiller 1994). Marketing orders set minimum prices that processors must pay for agricultural commodities. Since the depression, the federal government has allowed farmers to limit the quantity of produce they market. The surpluses that result from market orders are wasted (consumed by no one) by the individual farmer. Huge amounts of oranges, lemons, nuts, raisins, and various fruits have been unnecessarily grown since the 1980s. In addition to waste, this practice adds unnecessarily to environmental degradation.

Import Quotas.   Restrictions are placed on the amount of specific farm products that can be imported from other countries (Schiller 1994). Sugar, dairy products, certain grades of beef, peanuts, and cotton can generally be produced more cheaply in other countries than in the United States. Therefore, import quotas have been imposed that protect U.S. farmers. Import taxes (tariffs or duties) on foreign agricultural products are also used to restrict their importation to the United States.

Cost Subsidies.   Cost subsidies involve direct or indirect benefits that farmers receive from the federal government, which help to reduce farm production costs (Schiller 1994). These benefits include irrigation water from federally funded reclamation projects, cash payments to cattle ranchers for feed in drought years, and partial payments for a variety of practices that include drainage, brush control, land leveling, water development, fencing, and more. These various subsidies lower production costs and encourage more output than would otherwise occur (Figure 12.17).

Crop Insurance.   Most private insurance companies generally do not offer adequate crop insurance to farmers because of excessive risk (Knutson et al. 1995). Therefore, federal crop insurance (FCI) is set up to extend insurance to all crops and give producers a choice in the level of coverage. The intent was for the federal government to subsidize

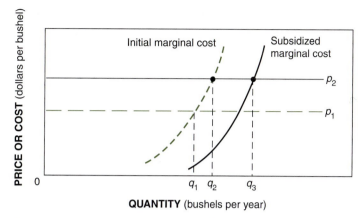

FIGURE 12.17 The impact of cost subsidies. Cost subsidies lower the marginal cost of producing at any given rate of output, thereby shifting the marginal cost curve downward. The lower marginal costs make higher rates of output more profitable and thus increase output. At price p_2, lower marginal costs increase the desired rate of output from q_2 to q_3 (from Schiller 1994).

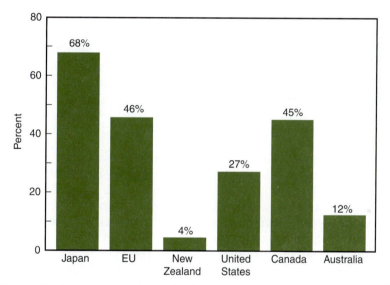

FIGURE 12.18 Levels of government intervention in agriculture, by country, 1989–1992 (from Knutson et al. 1995) [*Agricultural Policies, Markets, and Trade: Monitoring and Outlook*, 1993 (Paris, France: OECD/OCDE), p. 169].

30% of the cost of this program. The problem with FCI is that participation has been highest in high risk, nonirrigated, low-rainfall areas, and low where the opposite conditions prevail. Because this program has run at a one-half-billion-dollar annual loss during the early 1990s, there is general agreement that the program needs to be improved.

Farm Programs in Other Countries

The United States does not stand alone in protecting agriculture with federal subsidies. Most other developed countries subsidize agriculture to a greater degree than the United States (Figure 12.18). This is particularly true in France, Germany, Switzerland, Italy, Spain, Great Britain, and Japan. The reasons for farm subsidies include a national desire to maintain a healthy farm sector, maintain rural aesthetics and open space, provide food security in case of war, and maintain strong political pressure to help certain special interest groups involved in agriculture (Knutson et al. 1995). The various subsidies and agricultural protection programs are estimated to cost the average consumer in Europe 200 dollars and even higher in Japan. Some of the more costly subsidies from a consumer standpoint involve sugar in the United States, rice in Japan, vegetables in the Netherlands, and grain in Germany and France. In all these cases, other countries can produce these products at a much lower cost (have a comparative advantage), but because of government protection, inefficient producers are kept in business.

A major drawback to agricultural subsidies and protectionism in developed countries is the adverse impact they can have on developing countries (Foster 1992). Many developing African and Latin American countries have a comparative advantage in world trade in production of basic goods such as food and textiles, fossil fuels, and minerals. Their low labor costs and favorable climatic/soil conditions make them efficient producers of sugar, coffee, sorghum, peanuts, and cotton as well as various livestock products. In order to generate currency to purchase the complex goods necessary for economic development, they need to be able to sell their basic goods in world markets. However, they typically confront an international system of quotas and tariffs that reduces their opportunities to sell what they can produce at comparative advantage. On the other hand, the developed nations faced with huge surpluses of subsidized farm products habitually dump excess food on world markets at less than production costs.

Food relief is another challenging aspect of subsidized oversupply of food in developed countries. Providing food relief to developing countries experiencing famine is commonly considered humanitarian. An alternative view concludes that it does more ultimate harm than good. Massive food aid can depress local food prices and reduce food production incentives. It may also cause migration from farms to cities and create a disincentive for investment in agricultural development. Many experts in foreign development now believe food aid should only be provided on a temporary basis when some natural disaster occurs.

Trends in Current Policy

Since the mid-1990s, the United States and other developed countries have been in the process of moving toward free trade and away from agricultural subsidies and protectionism. Another proposed change in agricultural policy in the United States is the decoupling of agricultural payments to farmers from level of farm production. These changes will be briefly discussed following Knutson et al. (1995).

Trade Policy. The GATT (General Agreements on Tariffs and Trade) is the main global forum for various nations to negotiate trade agreements. It involves 92 member governments, and its main goal is to liberalize and expand trade by reducing trade barriers. As a whole, progress has been greater in liberalizing trade of industrial products than agricultural goods. This is a result of increased agricultural subsidy and protection in the developed countries since World War II. Nevertheless, there have been some breakthroughs since Uruguay hosted negotiations in 1987. These have included reductions in import barriers, reductions in direct and indirect subsidies, and minimizing the effects of unnecessary health and sanitary regulations on trade. Some of the key provisions were to reduce the 1986–1990 level of domestic subsidies by about one-third, to convert quota barriers on trade to tariffs, then to phase out the tariffs and provide increased access for products barred from specific countries such as rice in Japan.

The NAFTA (North American Free Trade Agreement), passed in 1993, is the most important recent trade agreement affecting the United States, Canada, and Mexico. Under the NAFTA, most trade barriers among the three countries will be eliminated over a 15-year period. Basically the NAFTA phases out tariff and nontariff barriers to trade, establishes rules for investment, strengthens intellectual property rights, and sets up procedures for settlement of trade disputes.

Free trade is in the interest of all countries because different countries have comparative advantages in producing different goods. Colombia is well suited for coffee production, Caribbean countries can best produce sugar, Australia and New Zealand are highly efficient producers of wool and mutton, and the United States is an efficient producer of grains and soybeans. Comparative advantages results from the ability of a country to produce and export a specific good more efficiently (lower input costs) than other countries. When free trade is maximized, different countries can specialize in producing goods where they have comparative advantage. Each country benefits because they have access to more goods at lower cost. Although the country as a whole benefits from free trade, some companies and workers in inefficient industries may be harmed in the short run because of increased competition. At the same time, free trade has often forced existing industries to become more efficient or competitive in producing goods for consumers. The U.S. government has had various trade adjustment programs to ease the hardship when economic losses occur because of trade agreements.

Disengaging. There has been increasing pressure to greatly reduce or eliminate government involvement in agriculture (Knutson et al. 1995). However, various farm groups have resisted these efforts, claiming that it would reduce their incomes and cause a reduction in farm asset values. One alternative that is receiving consideration is decoupling. It involves the separation of income payments to farmers from market prices and production. Basically under this approach all farm programs that alter a farmer's production decisions are eliminated. Decoupling to some extent transforms farm subsidies into entitlement or welfare payments. However, it largely avoids the free-market distortions that result from production subsidies. To some extent, land retirement initiatives such as the Conservation Reserve Program are a form of decoupling.

ELIMINATING GOVERNMENT INVOLVEMENT IN AGRICULTURE: THE NEW ZEALAND CASE

The case for continued government assistance to farmers centers around farming being different from other kinds of enterprises (Schiller 2000, Knutson et al. 1995). The argument is basically that unpredictable weather and other crop growing conditions make farming an unstable enterprise. Therefore, government payments to farmers are vital to protect the agricultural sector from periodic disaster.

However, many economists and congressmen now question whether farm program benefits exceed costs to taxpayers, and some think they do more harm than good. Some U.S. farmers growing unsubsidized crops—such as potatoes, fruits, and vegetables—are doing quite well. New Zealand provides some of the most convincing evidence against farm subsidies. In 1986, New Zealand ended its farm subsidy programs because of economic stagnation and huge government budget deficits (Merline 1995). Contrary to dire predictions, few farmers went bankrupt or committed suicide.

New Zealand's farmers actually prospered after the elimination of subsidies (Merline 1995). Output of some agricultural commodities fell, but production efficiency noticeably improved. There were definite environmental benefits from ending subsidies in New Zealand. Marginal land was taken out of production, and usage of fertilizer and pesticides became more judicious as their true market cost became more apparent. Jobs in the farming sector dropped, but this was merely a continuation of a trend in place long before the elimination of subsidies. Farm organizations are generally pleased with the agricultural revival in New Zealand, and there has been no movement for a return to subsidies. Farmers in New Zealand now widely believe that they are better off by being independent of the government.

According to a recent study by the USDA Economic Research Service, farmers in the United States who do not participate in government farm programs appear to be doing better than those who do. Some 63% of nonparticipating farms and 60% of participating farms in government programs had a favorable financial position. About 4% of nonparticipating farms compared with 7% of participating farms were vulnerable to insolvency (bankruptcy).

One problem with government farm programs, not always recognized as such, is that they change constantly. New farm bills are passed every 5 years and often amended. This

No matter the form or place, effective outdoor recreation restores health and motivation—it "creates anew" both physical and mental function. Even if the quality of a recreational site is unexcelled, satisfaction can be devastated by poor or uncomfortable access. Thus, comprehensive recreational management is necessarily integrated with transportation and travel management. One of the most prominent changes in the twentieth century was the degree to which outdoor recreation had become a part of most lives in the United States and other developed nations. Management of natural resources for outdoor recreation has become one of the most important emphases of public land and water management. It is also of growing interest and importance in the private sector.

THE IMPORTANCE OF OUTDOOR RECREATION

We rarely think about how important recreation has become to us during work breaks, after work, and on weekends, holidays, and vacation. Regardless of what we call it, we now take for granted the need for recreation in all forms. Studies in 1960 and 1987 showed a consistent 89 to 90% of the U.S. population participated in outdoor recreation away from the home (TFORRO 1988). Labor agreements often emphasize paid time off for rest and renewal. Newer work environments often are designed to provide short "getaways" from the grind. Many larger employers provide outdoor recreational outlets, such as parklike grounds, picnic tables, exercise areas, and pleasant walkways. Windows are important to many employees in part because they provide brief escape to the "outdoors" and seem to stimulate mental productivity. Professional meetings often are set up in attractive locations and environments designed to provide outdoor recreational opportunities as a work incentive. Cities vie for tourism dollars associated with business and purely for pleasure. Whole industries have grown up in support of recreational opportunities. Recreational tourism, including a mix of indoor and outdoor opportunities, has become one of the world's biggest businesses.

The growth of large corporate employers during the industrial and information age created conditions leading to more leisure time and a greater need for recreation. When most people worked for themselves, primarily on family farms, holidays and vacations were rare because business required continuous attention. Technology and mass production techniques greatly increased per capita production and average real income. Yet the work of family businesses was more diverse and less mind numbing than the specialized routines required of many company jobs. Work concessions were won by organized labor. It soon became apparent that productivity was especially high when workers were well rested and refreshed. This typically required shorter work days and work weeks.

By the 1940s, most employees of large companies expected a 40-hour workweek and employment benefits including a week or more of paid vacation. After World War II, family participation in outdoor recreation increased with the baby boom. Time-off benefits rose to a higher priority level in employment negotiations. As the information age emerged, many jobs became more sedentary, mentally more demanding, and often harder to leave behind "at the office." Even though the amount of reported leisure time has actually decreased in recent years, for a variety of reasons (e.g., more work commuting time), the demand for outdoor recreation has been sustained (TFORRO 1988, Cordell et al.1999). On-the-job boredom and physical inactivity continue to be the most debilitating aspects of routine work in many organizations. Outdoor recreation is one of the more effective prescriptions for many people.

ATTRIBUTES OF OUTDOOR RECREATION

Although, in the strictest sense, outdoor recreation means simply any restorative behavior done outdoors, it usually connotes a different form of activity than those commonly done inside. Benefits from outdoor recreation depend heavily on availability and quality of outdoor environment.

Outdoor recreational management centers on providing ideal outdoor environments for a wide variety of recreational activities. While provision of appropriate landscaping, site access, shelter, vegetation management, interpretive services, and facilities can greatly improve upon outdoor recreational satisfaction, the original choice of site often is key to management success. Many of the present locations accessible to either private or public recreational use were chosen decades ago and more often than not for reasons other than recreational use. Therefore, much of what recreational resource management does involves making the best of less than optimum recreational conditions.

Important management variables are natural topography, climate, ecosystem productivity and diversity, other human uses, and land-use authorities. The attributes and diversity of topography are critical for supporting diverse forms of outdoor recreation. The distribution and proportions of land and water determine the diversity of recreational activities possible at a managed site. Rugged topography is critical for certain types of recreation, such as mountain and rock climbing, skiing, back-country hiking, wilderness camping, white-water navigation, and all-terrain vehicle use. Other recreation requires somewhat less rugged terrain including most field sports, ice skating, jogging, snow sledding, cultivating gardens, bicycling, beach activity, and waterside access to docks, piers, marinas, and boat ramps. Diverse climates facilitate alpine skiing, waterskiing, ice fishing, white-water rafting, fall hunting, swimming, and sunbathing on sandy beaches. Without a diversity of natural ecosystems, many forms of outdoor recreational opportunity would be diminished or defunct, such as hunting, fishing, nature photography, and bird-watching.

RESOURCE CONFLICTS AND RESOLUTION

The history of recreational use of public resources in the United States is in part a study in resource use conflicts and their resolution. The amount of public land and water made accessible and developed for recreational use in the natural state is unexceeded anywhere else. Outdoor recreational activity has increased with vastly improved quality of life and increased diversity of recreational opportunities. However, other uses of public lands and waters have come into increasing conflict with recreation use, including timber, range, water supply, and flood management. Even different recreational groups can find themselves in conflict. At overcrowded sites, for example, those who would enjoy themselves in silent contemplation often find themselves at odds with those who prefer wild abandon in noisy group play. Recreational management has come to include making up rules, enforcing laws, and otherwise managing the conflicts that can arise in multiple-use situations.

The diversity of viewpoints relative to the appropriate limitations on the recreational use of public lands has stirred controversy. Has enough public land been set aside in the right places for present amount and distribution of recreational demand? Should private development be encouraged over public development? Should there be a cap on the total amount of land excluded from private ownership and development? How much should other authorized uses of public lands be allowed to diminish the value of recreational use and vice versa? To what extent should private concessions and other partnerships between public and private development be promoted in public policy? To what degree is recreational development compatible with emerging concerns for biodiversity protection and ecosystem management?

Recreational management often is wedged in with other natural resource management. In a multiple-resource management setting, it frequently is the last management added when budgets are fat and the first to be cut when budgets are lean. Attitudes toward recreation often exhibit less respect for recreational resources than other resources even when economics strongly suggest otherwise. Although the reasons for this are complex, some of the most basic probably extend back to a time in European and American history when having fun was viewed with great suspicion.

HISTORICAL PERSPECTIVES

Federal Management

The federal government through its various land management agencies—principally the Forest Service, Bureau of Land Management, National Park Service, and Fish and Wildlife Service—manages almost one-third of the land area of the United States (mostly in the western United States). In addition, the water resources agencies contribute greatly to water-based recreation on inland reservoirs and waterways; especially the Army Corps of Engineers, Bureau of Reclamation, Natural Resources Conservation Service, and the Tennessee Valley Authority. The acquisition of most federal lands resulted from a series of federal laws that transferred large blocks of land from the public domain to federal management. While the largest percentage of federal ownership within a state in the eastern and central states does not exceed 8% (with the exception of South Dakota with 17%), the percentage of federal ownership in the western states ranges from 35% in Washington State to 85% in Nevada. Most of this federal land acquisition occurred in the nineteenth century (see Chapter 2).

Early Attitudes

The initial philosophy of the American public toward natural resources was one which perceived superabundance and fostered conquest over nature. The wilderness was to be subdued and whatever wealth it contained was to be extracted for personal and national gain. For most people, little time was available for recreation and religious leaders often disapproved of recreational pleasure. The Spanish, French, English, and Russian interests in the New World were intent on exploiting living and nonliving resources. In the eastern United States, the fishery and logging industries were an important economic impetus. For most of the immigrant masses, however, land was cleared of timber for conversion to agriculture. The rich fertile soil, which once supported luxuriant and seemingly endless stands of forest, was the resource most sought. As the U.S. borders moved farther west, the same pattern continued—only it was the extensive grasslands that were plowed and grazed, and the original ungulate fauna replaced by domestic sheep, cattle, and horses from the Old World. The federal government acquired large tracts of public domain land mostly west of the Appalachian Mountains. While much of this land was sold at a low price or given away (see Chapter 2), a large fraction was retained under federal management. Lands east of the Rocky Mountains quickly became privately owned. By the end of the nineteenth century, most remaining public land was located in and west of the Rocky Mountains region.

The philosophical and intellectual basis in America for the preservation of nature for its enjoyment was introduced by Ralph Waldo Emerson and Henry David Thoreau before the Civil War. Nature study and tourism started to become popular. Thoreau proposed small, unspoiled parks where people could seek nature. However, it is landscape architect Frederick Law Olmsted whom we owe most for the first concepts of national and state park systems, and the basis of modern park planning. He believed humans exposed to undisturbed park settings benefited psychologically and that parks should be established for the enjoyment of all levels of society. Olmsted laid the foundation of modern park planning through his management plans for Central Park in New York City and the Yosemite State Park in California. Official establishment of natural and naturalistic city, county, state, and federal parks escalated after the U.S. Civil War.

Yosemite State Park was established by an act of Congress in 1864 on most of what is now Yosemite National Park "upon the expressed conditions that the premises

TABLE 13.1 Thousands of Days of Recreation Use Provided at Federal Recreation Areas at Intervals from 1986 through 1996

Agency	1986	1990	1994	1996
Forest Service	226,533	263,051	330,348	341,204
Corps of Engineers	144,170	189,921	204,905	212,998
National Park Service	111,014	110,205	111,197	104,286
Bureau of Land Management	23,679	43,172	46,940	72,794
Bureau of Reclamation	24,706	23,365	22,298	27,342
Fish and Wildlife Service	5,599	4,410	N/A	N/A
Tennessee Valley Authority	599	768	1,161	1,208
All Areas Reporting	536,259	634,848	710,106	758,843

Source: Cordell et al. 1999.

shall be held for public use, resort, and recreation and shall be held inalienable for all times." This state park was followed by the creation of New York State reserves at Niagara Falls and in the Adirondack and Catskill Mountains. In March 1872, President Ulysses S. Grant signed the bill that created what was to become Yellowstone National Park and which initiated the national park system. Several state and county park systems and numerous city park systems were established between 1880 and 1890. The next major growth in outdoor recreation occurred after World War II when leisure time increased and mass-produced automobiles and the expanded road system facilitated travel.

Recent Trends

The most influential modern document on outdoor recreation land-use planning was prepared by the Outdoor Recreational Resources Review Commission (ORRRC), which was established by an act of Congress. Their report, published in 1962, was a comprehensive survey of outdoor recreation in America and made outdoor recreation recommendations relative to "what policies and programs should be recommended to ensure that the needs of the present and future are adequately and efficiently met." The ORRRC noted the lack of outdoor recreation facilities near metropolitan areas and that the present recreation lands were not meeting the needs of Americans. They recommended the creation of a Bureau of Outdoor Recreation in the Department of the Interior. The Bureau was created in 1962 and was the first federal agency specifically assigned to formulate a nationwide outdoor recreation plan. Outdoor recreation greatly increased in the 1960s and 1970s. Federal expenditures for recreation increased from $75 million in 1960 to $1.4 billion in 1980. Outdoor recreation stabilized or moderately increased in the 1990s on federal lands (Table 13.1).

The next outdoor recreation report, titled *American Outdoors: The Legacy, the Challenge,* was prepared by the President's Commission on Americans Outdoors (1987). They found that Americans value the outdoors highly as a key contributor to the quality of life, providing "significant social, economic, and environmental benefits." However, because the benefits are relatively difficult to quantify in dollars, they also discovered that outdoor recreation does not compete well with public programs that can be more readily valued. Residential and commercial land development often sets the price higher than can be afforded for recreation and cut off access to areas, such as urban rivers, that can support recreation. Valuable lands near water and urban areas are increasingly unavailable, and certain wetlands, free-flowing rivers, and other recreational-support systems are becoming scarce. The Commission found that the high quality of the remaining land, water, and atmospheric resources now reserved for and needed to support outdoor recreation are at risk of degradation from user abuse and from environmental pollutants.

The Commission concluded that contending with these trends means we will have to better manage existing resources. This includes improving the quality of recreation services, the availability of resources to people of all income levels, the recreational investment environment for the private sector, the coordination of resource management and recreation programs, the information base from which to manage more effectively, the "liability crisis" that is limiting recreational opportunities, and public funding of recreational programs. Little has changed since this report was written.

RECREATIONAL MANAGEMENT

Distribution of Outdoor Recreation and Recreation Lands

The distribution of outdoor recreation with respect to the area made available for it determines to great extent the type of recreational opportunities, the intensity of use, and the management possibilities. The greatest amount of land available for recreation is private land, but the actual amount used for recreation is likely to be much less. Most private recreation occurs in urban areas, which occupy about 5% of the landscape. However, a significant amount of recreation occurs on private farmlands, mostly in the form of hunting, fishing, camping, and picnicking, and in private enterprises such as ski resorts, dude ranches, golf courses, and private campgrounds. About 14% of all outdoor recreation occurred in private surroundings according to the survey results of Cordell et al. (1999). Among the 778.4 million acres of public lands available for outdoor recreation, the federal land area is the largest (90.9%) with state (7.8%), county (0.8%), and municipal areas (0.5%) much smaller. However, 60% of the recreation occurred on local public parks and play areas, 14% on state lands, and 12% on federal lands. Most natural resource management connected to recreation occurs on federal lands, however, and most data are kept for federal land and water use.

Recreation Provided by Federal Agencies

Table 13.1 shows the days of visitation at federal sites managed by the seven federal agencies providing most outdoor recreation to the public. In 1990, the U.S. Forest Service led with 41.4% of the recreational use followed by the U.S. Army Corps of Engineers' reservoirs amounting to 29.9% of the total. The National Park Service hosted 7.4% of the visitor days while the remaining four agencies provided for 11.3%. Each of the seven federal agencies contributes uniquely to the array of federal outdoor recreation available, as summarized in the following sections.

Forest Service.   The national forests are the most important federal providers of public outdoor recreation in the United States with more than 341 million days of recreational use in 1996 (Table 13.1). Lands within this system encompass over 191 million acres and include national forest, national grasslands, and portions of the National Wilderness Preservation System, national recreation areas, wild and scenic rivers, and wildlife protected areas. The evolution of the national forest system has been one of de-emphasizing commodity values (principally timber production and livestock grazing) and emphasizing amenity values (principally recreational opportunities). This reflects the changing attitudes of the American public followed by appropriate legislative mandates (Figure 13.5).

Since 1905, the national forests have provide important opportunities for wildland and outdoor recreation, but no funding was made available to support recreation until Congress passed the Multiple Use, Sustained Yield Act of 1960. Before that recreation was viewed as incidental use, acceptable as long as it did not interfere with timber-production and watershed-protection purposes. Rapidly increasing demand for both timber and recreational opportunities led to conflicts between logging and recreation that contributed to resolution in the 1960 law. National conser-

FIGURE 13.5 Forest service lands provide vast areas for enjoying outdoor recreation (courtesy of Jack McCaw).

vation organizations were a strong force in encouraging the change in management policy. The Wilderness Act in 1964 established a national authority for wilderness protection, much of which now occurs in the Forest Service. Nearly 100 million acres of wilderness are now protected on all federal lands under the act. The National Forest Management Act of 1974 expanded multiple use to include wilderness. Other history relevant to the Forest Service is covered in Chapter 2.

National Park Service. The National Park Service manages the National Park System for its natural, historical, and recreational resources (Figure 13.6). An important objective is to use the park system to communicate natural and cultural heritage to Americans through enjoyable inspirational and recreational experiences. In 1996, it provided more than 104 million days of recreational use (Table 13.1). The National Park Service is charged with managing a variety of natural, historic, cultural, and other recreational resources in national parks, national recreation areas, national monuments, national preserves, national seashores, national lakeshores, national memorials, national historic sites, national historic parks, national battlefield parks, national rivers, and national parkways (Figure 13.7).

The 47 national parks consist of large areas that protect natural and cultural features on 47.4 million acres. The 81 national monuments protect nationally significant resources encompassing over 6.3 million acres, and more federal area was declared monument status during the last year of the Clinton administration. Although usually smaller than national parks, Death Valley National Monument in California encompasses 2,048,429 acres making it one of the top-ten largest units managed by the National Park Service. National preserves are limited to two units in the lower 49 states and 10 in Alaska. Hunting, trapping, and fishing, and the extraction of minerals and fuels are allowed in preserves under regulation. The 20 National Recreation Areas usually associated with dam impoundments in the West and encompass 3.4 million acres. The Wild and Scenic Rivers Act of 1968 authorized setting aside rivers of extraordinary recreational and other resource values in a free-flowing condition (Figure 13.8). During the same year, the National Trails Systems Act authorized establishment of a national trail system under the National Park Service leadership.

(a)

(b)

FIGURE 13.6   The National Park System in the United States includes spectacular geologic and scenic areas. Photograph (a) shows the Grand Canyon in Arizona (courtesy of the National Park Service). Photograph (b) shows White Sands National Monument in southcentral New Mexico (courtesy of Bill Walker).

Bureau of Land Management.   The Bureau of Land Management (BLM) has authority over the largest amount of public land and manages it according to the principles of multiple use and sustained yield. Most Bureau of Reclamation lands are west of the Mississippi. Recreation is one of the important uses authorized under the Federal Land Policy and Management Act. Recreational participation amounted to nearly 73 million days of recreational use in 1996 (Table 13.1). Recreational management is concentrated in 355 areas on about 10% of total holdings. There are 765 developed recreational sites.

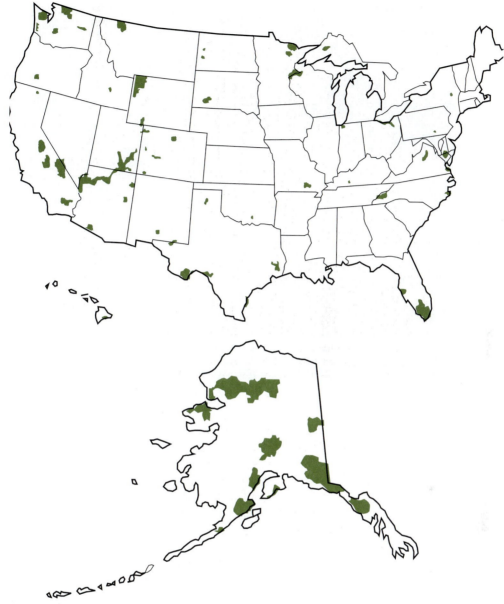

FIGURE 13.7 Primary national parks and national monuments in the United States (data from U.S. Geological Survey).

The BLM has created special management areas for areas of unique and threatened status under both congressional and internal administrative action (Zinser 1995). Some are managed as 13 national conservation areas in addition to 66 national wilderness areas. Also, there are a number of national natural landmarks. Each of these special management areas has management plans in which uses are clearly prescribed or prohibited with respect to the prominent management emphasis. The BLM has attempted to consolidate lands through trades in order to improve on special management areas.

The Fish and Wildlife Service. The Fish and Wildlife Service manages 485 wildlife refuges totaling more than 88 million acres (Zinser 1995). Under authority of the Refuge Recreation Act of 1962, a portion of many of these areas is open to wildlife-based recreation to an extent that does not interfere with the fish and wildlife protection mission of the Service and to which funding is available. An estimated

FIGURE 13.8 More than 7,300 river miles in over 720 rivers have been designated as wild, scenic, or recreational rivers (courtesy of USDA-Forest Service).

25 million people visit refuges for recreation (Zinser 1995), of which only 3% is for hunting and 16% is for fishing. In 1990 about 4.4 million recreational days were provided (Table 13.1). It is agency policy to discourage any recreational use that is not wildlife-based and is incompatible with such activities or with wildlife protection. Therefore, most visits are short and such activities as swimming, motorboating, waterskiing, sunbathing, jogging, bicycling, surfing, and horseback riding frequently are prohibited.

Federal Water Resources Agencies.   Among federal agencies, the Army Corps of Engineers is second only to the Forest Service in providing recreational opportunities, mostly at the 456 water resource development projects (mostly reservoirs) it manages for navigation, flood damage reduction, hydropower, and other purposes. In 1996 it provides over 212 million days of recreation (Table 13.1). The Corps provides more boating access and related recreational service than any other agency, and is second only to the Forest Service in recreation use of their lands and waters. Through development of the Intracoastal Canal, riverine navigation projects, and harbor maintenance the Corps also has provided for thousands of miles of recreational boating in addition to the primary commercial navigation purpose.

Other important federal providers of water-based recreation are the Bureau of Reclamation and the Tennessee Valley Authority. Most of the Bureau's 355 storage reservoirs are located where water-based recreation is highly valued—in the arid West. In 1996 it provided more than 27 million days of recreation activity (Table 13.1). The Tennessee Valley Authority has provided recreational management at its managed sites on the Tennessee River since 1969 (Zinser 1995). The last estimate of total recreational use, in 1996, indicated that more than 1 million days of recreation occurred at all sites under its management (Table 13.1).

Recreation Provided by State and Local Government

State parks hosted 666 million visits on over 10,000,000 acres of land by 1986 (TFORRO 1988). By 1996, this had expanded to over 745 million visits (Cordell et al. 1999). Although these state and local opportunities add up to a small fraction of the total public land and water area under federal authority, more than 88,000 sites existed in 1986, distributed mostly near cities. This proximity to population concentrations makes these sites much more densely visited than most federal recreation lands, thus providing about one-third of the total provision of recreational use. However, these recreational areas tend to complement rather than compete with the federal recreational lands. In contrast, with the longer-term recreation common on federal lands, there is more day use on state and local government lands.

Private Recreation Opportunities and Tourism

Much outdoor recreational management takes place on private lands. However, the opportunities and use provided on a private basis are not as well documented as for public lands, but nevertheless are very important, especially where there is little public land. Thus, private lands have tended to complement public land recreation rather than compete with it. Many concessions on public lands are operated privately under contract with public agencies including campgrounds, ski areas, marinas, hotels, restaurants, stores, and services. A need exists for updated data on recreational use of private lands to help maintain proper public recreational investments. The most complete data are more than three decades old. In 1965, more than 1 billion visits were estimated for private holdings, either free of charge or seeking profit from recreational use (TFORRO 1988). An additional 800 million visits took place under the auspices of clubs or on private lands not seeking profit for recreational use. As of 1988, about 785 million acres of private land were available for public recreation either free or for a fee (TFORRO 1988).

RECREATIONAL CHALLENGES ON PUBLIC LANDS _____

In this section we will discuss some of the major recreation challenges and approaches to multiple-use management on public lands in the United States. We will emphasize public rangelands in the western United States because of their large area and the intense controversies that have occurred over how they will be used (Wuerthner 1990, Jacobs 1991). Much of our discussion follows Holechek et al. (2001). We refer the reader to Chapter 10 for additional discussion of recreation conflicts with other uses on national forest lands and Chapters 7 and 16 for a discussion of water resources challenges.

Importance of Public Land Recreation

Recreational use of public lands in the United States has accelerated tremendously during the past 30 years. Much of the increase has resulted from the rapid human population increase in the 11 contiguous western states, combined with rising affluency and leisure time. In some areas, such as southern California, central Colorado, western Arizona, central Texas, and central Oregon, recreation has become a far more important use of public rangeland than livestock grazing. If present trends of urbanization continue in the 11 western states, it appears likely the recreational value will exceed livestock grazing value on most rangelands within the next 15 to 20 years.

TABLE 13.2   Types of Rangeland Recreational Uses

Activities with Minor Impacts on Rangeland	Activities with Major Impacts on Rangeland
Hiking	Hunting
Camping	Horseback riding
Fishing	Trail biking
Skiing	Motorcycling
Boating	Jeeping
Rock hounding	Dune-buggy riding
Tubing	Home building
Canoeing	
Mountain climbing	
Relic hunting	
Bird-watching	
Picnicking	

Source: Holechek et al. 2001.

FIGURE 13.9   Thousands of recreational homesites are sold every year in the western United States, severely impacting aesthetic, wildlife, and recreational values in many areas.

Recreational use of public lands includes a variety of activities (Table 13.2). The economic benefits from many of these activities, such as camping and hiking, can be estimated (see Chapter 4), but the negative impacts of recreation can be as difficult to quantify as the negative impacts of grazing. Some forms of recreation, such as off-road vehicle travel and trail biking, can be as destructive as overgrazing if uncontrolled.

Subdividing Private Grazing Lands

The sale of recreational homesites has had the most severe impact of all the wildland recreational uses in the western United States. Literally thousands of recreational homesites are being sold every year in each of the western states (Figure 13.9). Although the total area lost each year is relatively small [about 1 million acres (404,694 ha)] compared to the total land base, this trend is having a major impact on the local range livestock industry in many areas.

Large-scale conversion of agricultural and rangeland to housing is occurring in many mountain valleys of the Intermountain West. These lands are critical in terms of providing forage for livestock during the winter when snow makes grazing infeasible at the higher elevations. Without winter-base property, grazing of surrounding summer range, typically in federal ownership, becomes impractical. The people who build homes in the mountain valleys generally come from the urban areas and use these second homes primarily for vacation purposes. They use the surrounding summer ranges for recreational pursuits. Under these conditions, recreation replaces livestock production as the primary use on the remaining federally owned rangeland.

The subdivision of mountain valleys can severely limit big-game populations; particularly mule deer and elk. These animals are directly affected by the loss of important winter ranges. Indirectly, they are subjected to greater stress on remaining wildlands as the result of increased human activity that goes with subdivisions, such as snowmobiling, woodcutting, and hiking.

To maintain aesthetic values, clean air, wildlife populations, and other amenities that make mountain valleys so desirable, restrictions on private land use may be necessary. Otherwise, the very qualities that make these areas so attractive may be destroyed.

Practices that provide economic incentives to landowners to retain their land in agriculture show some promise. In Oregon, for example, state legislation has reduced property taxes for land kept in agriculture. This coupled with zoning laws that restrict subdivision in farming and ranching areas have kept a good balance between development and agriculture. Similar approaches to those used in Oregon will probably be applied to other western states as they become more urbanized. However, some people question whether government should benefit comparatively few landowners and recreationists at greater cost to the taxpaying public. Such decisions require careful analysis of the benefits and costs (see Chapter 4).

Agriculture on the Urban Interface

Large zones of rapid urbanization occurred around many cities in the 1990s (see Chapter 14). The problems of farming and ranching on the urban-rangeland interface are considered by Huntsinger and Hopkinson (1996). Many people moving to rural areas do not understand rural customs and activities just as many rural people do not understand urban sensibilities and behavior. Farmers and ranchers on the urban-rangeland interface often complain about marauding dogs, vandalism, trespass, carelessness with gates and fences, increased liability costs, and introduction of exotic plants (Hart 1991, Forero et al. 1992). In turn, suburban and resort residents often complain about the use of prescribed burning, predator control, and weed control. Stray livestock can cause property damage and vehicle accidents. Commuter traffic may be blocked by slow ranch vehicles. The new neighbors also may complain about livestock odors and the threat of pollution from pesticide or fertilizer applications.

For these and other reasons, better planned development is critical to ranchers as well as society at large. Huntsinger and Hopkinson (1996) believe environmentalists should be more tolerant in their rangeland appearance expectations while ranchers should rethink their concept of property ownership. They note that debates over the ecological impacts of grazing, wildlife management imperatives, and riparian zone restoration become moot when grass is replaced by concrete.

Huntsinger and Hopkinson (1996) point out that sustaining rangeland ecosystems in the future will be as much a social-economic process as an ecological one. They provide examples from Marin County, California, of how successful planning and alliance building between ranchers and environmentalists prevented development of a large farming-ranching area. Near San Francisco a combination of zoning, conservation easements, tax relief for farmers and ranchers, community leadership, and recognition of the heritage value of rural lifestyles all played a part in the success. They noted that a similar pattern was emerging in other parts of the West. They believe carefully planned ranching can play a vital role in conserving biodiversity in many threatened range landscapes. This planning process should include ecological, social, and economic considerations. The land market is driving the conflict over scarce, desirable land. Where the market does not adequately reflect all costs, the government needs to assure otherwise through careful benefit-cost analysis (see Chapter 4). Although environmentalists and ranchers differ in many beliefs, Huntsinger and Hopkinson (1996) point out that both groups generally want to preserve open space and share a love of nature. When environmentalists were given more exposure to ranchers and their way of life, many came to see ranching as compatible with their goals.

Scenic Beauty and Range Management

The impact of multiple-use management on scenic beauty in the United States has been an important concern in recent years because of the public's increased sensitivity to the environment and its great role in the decision-making processes on public lands. Range management practices have been especially controversial. Using photographs of selected public land use conditions and management practices, Sanderson et al. (1986) investigated the attitudes of dispersed recreationists in the Malhuer National Forest of eastern Oregon. From the results, they concluded that recreationists responded most favorably to environmental scenes where range management practices were least visible, such as fences, and livestock were grazed well within the pasture grazing capacity. Less favored was full livestock use of forage by using range management methods that evenly dispersed livestock. Least favored was intensive management of environment and livestock to maximize forage production and sustainable forage use consistent with multiple-use objectives.

Different recreational user groups had different general perspectives. The average angler, for example, rated aesthetic quality lower than the average camper and was more likely to accept fences than other recreationists because they often are used to exclude livestock from streams. Hunters, on the other hand, reacted negatively to road closures and other restrictions to hunting access. Campers reacted more favorably to livestock in open pastures than in forests, apparently because they considered forests to be more suitable for camping. Most campers did not want livestock near campsites.

Sanderson et al. (1986) concluded that recreationists do perceive differences in visual quality, although differently depending on the type of recreation and the experience of the recreationist. The more experienced were more likely to accept range management practices. Even so, the public in general is unaware of the requirements for managing forest and range environments for increased forage. The majority (84%) were unfamiliar with the concept of multiple use or statutes governing its use, and 91% had no idea what "more intensive management of the range resource" meant, thinking it meant grazing more cattle rather than improving the forage.

While the public generally tolerates range management activities, it generally has little understanding of what they involve. In confirmation, Mitchell et al. (1996) found about the same fraction of visitors (33–34%) to a national forest in Colorado were affected positively as were affected negatively by livestock grazing. Those using dispersed campsites where livestock roamed were more critical than those in developed campsites where livestock were excluded. Mitchell et al. (1996) concluded that there was little objection to livestock if they were kept out of developed campgrounds, riparian areas used for fishing, and other areas typically used for dispersed camping.

Public Opinion and Management of Federal Rangelands

Public opinions on grazing issues have become an important concern of federal rangeland managers. A survey conducted by Brunson and Steel (1994) showed U.S. residents place little confidence in livestock growers, miners, and the energy-extraction industry. They tend to believe more protection should be given to wildlife and fish resources. In another survey, Brunson and Steel (1996) found only weak opinion differences on range issues between the eastern and western United States. However, attitudes differed between urbanized areas and rural regions, which could eventually result in political power loss for rural land protection constituencies in the western United States.

This survey was consistent with the results of Sanderson et al. (1986) and Brunson and Steel (1994), which all indicate the need to better inform the public on range management issues. At the same time many new and divergent interpretations of multiple-use management on public lands are being advanced that must be considered by modern range managers. Many range scientists have come to believe that systems are needed that provide more sustainable, environmentally sensitive range-livestock production as alternatives to traditional systems (Box 1995, Heitschmidt et al. 1996, Vavra 1996). Meeting this need will be an important challenge for range scientists and managers in the twenty-first century. Many ranching operations are marginally successful and any additional management cost may threaten their continuance, unless they take innovative steps.

Recreation and Ranching

Advantages to Ranchers. The urbanization of the western United States might be advantageous to the ranching industry. Returns from livestock on western ranches as a percentage of capital investment have been low (they have averaged about 1 to 3% during the past 25 years). However, the increased human population in the West provides economic opportunity for ranchers who are willing to diversify their enterprise. See Chapter 11 and 15 for a discussion on the potential of fee hunting. Other forms of recreation, such as packing trips, horseback riding, sight-seeing trips, dude ranching, and fishing, are proving to be highly lucrative in certain areas. Many ranches have built ponds, stocked them with fish, and charged either daily fishing fees or so much per fish. They often also rent cabins and provide meals to vacationing tourists. In some cases, these enterprises bring in far more net income than the sale of livestock. It appears that recreational enterprises will increasingly displace livestock as the main source of income from ranches in many parts of the western United States.

Disadvantages to Ranchers. Ranchers using public rangelands generally view recreation negatively, often with good reason. Increased human disturbance can reduce animal performance. Increased incidences of vandalism, fire, and losses of livestock from theft, traffic collision, and shooting usually accompany heavier recreational use. In some cases, camping occurs around livestock watering points and livestock not used to human activity may be reluctant to water when this occurs.

Managing Recreation Costs on Public Lands

Many segments of the public now believe that federal agencies should charge fees for recreational use of public grazing lands. These fees could then be used to mitigate some of the negative effects of recreation on livestock production. In reality, many recreational uses—particularly those forms involving unregulated off-road vehicle travel—can affect the range more severely than livestock grazing or logging (Figure 13.10). To prevent further destruction, regulation of recreation on public lands will be increasingly necessary. It seems only reasonable that recreationists using public land be charged a fee for their activities just as ranchers are charged for grazing. However, many recreationists believe this grazing fee is too low and management is ineffective for protecting recreational interest. Ideally, fees should be in proportion to recreation management costs.

FIGURE 13.10   Damage to public rangelands in southern California from uncontrolled travel by four-wheel-drive vehicles.

Basic principles used in managing livestock grazing can also be used in managing recreational use (Heady and Vaux 1969). Proper stocking rate can be modified to proper numbers of people or vehicles for a particular recreational activity. Proper distribution and timing of use is just as important with most recreational activities as it is with grazing. This is particularly true for off-road vehicle travel.

Off-road recreational vehicle use has had measurable negative impact on forest and range soils and vegetation. Payne et al. (1983) used a typical four-wheel-drive vehicle to determine that up to eight trips during the growing season in the same tracks did little long-term damage to vegetation and soils when the soils were dry. Damage was much more severe on wet soils, and, under any conditions, damage was much more likely to persist over eight trips in the same tracks. Management that encouraged not driving in the same tracks was recommended for dry growing-season conditions. However, where forage was reserved for winter grazing they recommended driving in the same track during winter to minimize breaking down the dried grass. They also recommended preventing all off-road driving during wet periods. Eckert et al. (1979) found that four-wheel-drive trucks had more impact than motorcycles on reducing runoff infiltration rate into the soil and increasing soil erosion. Such studies have resulted in much more regulation of off-road vehicle use on public lands. Even greater restriction is expected if recent trends in public use continue.

Conflict Resolution in Multiple-Use Decisions

In recent years, public land management agencies have become heavily burdened with resolving controversies among various interest groups regarding natural resource use and management. Capability in conflict resolution has become of critical importance in managing public lands. Coordinated Resource Management Planning (CRMP) has been widely used in the western United States to deal with multiple-use conflicts. CRMP brings together public and private interests to resolve multiple-use and land management conflicts through reasoned-scientific analysis, bargaining, and compromise (Anderson 1991).

The participants in CRMP have a common goal in correcting past land management mistakes and making informed decisions about management alternatives based on trade-off analyses, including who gains and who loses. It has been effective in resolving many conflicts on public rangelands (Anderson 1991, McClure 1992, Grizzle 1992). However, quantitative decisions based on economic trade-offs are usually complex because many natural resource products on public lands are free and their values vary among interest groups (Anderson 1981). However, other than biodiversity values, most natural resource values can be quantified either through direct market means or indirect, non-market valuation (see Chapter 4). CRMP guidelines that have developed from 40 years of experience are available from Cleary and Phillippi (1993).

Another approach called "dispute resolution" is discussed by Torell (1994). It involves bringing together parties in disagreement to participate in joint decision-making processes that seek win/win solutions and avoid litigation. An objective third party is commonly recruited to assist in resolving conflicts. A study of environmental disputes found that 78% of the cases where alternative dispute resolution techniques were used resulted in settlement (Bingham 1986).

OUTDOOR RECREATIONAL MANAGEMENT

Interfacing People With Resources

The goal of outdoor recreational management is to provide optimum conditions for people to obtain outdoor recreational opportunities. Outdoor recreational managers usually are responsible for the management of a site or a group of sites linked to a specific geographic area, which may be privately or publicly owned. Site managers must understand principles of landscape and habitat management pertinent to recreation and other uses of the area. They must also be informed of user needs with respect to resource qualities, access, facilities, interferences, costs, and availability of compatible and synergistic activities. Before all else, recreational management requires careful planning. The main difference between managers and biologists in natural resource management for recreation is the understanding of recreational management principles. Fish and wildlife "management" is especially likely to underestimate the importance of recreational management principles and overemphasize the importance of biology.

Recreational Planning

Recreational management begins with development of a management plan with goals, inventory information, analysis and forecasts of use, management objectives, a prescription for implementation based on the most cost-effective alternative, and an evaluation of management effectiveness (see Chapter 5 for general planning principles). Most recreational management plans combine geographic and event plans. They usually emphasize identification of resource use areas, access to those areas, placement of supportive facilities, various landscape modifications for improving resource utility, compatibility with other uses, timing of events, and specific locations of organized events. Recreational management planning starts with choosing the location of the site to be developed and managed. If, as is increasingly the case, the plan is for a site that already has been identified and developed for recreational use, the conditions left by previous site managers become important constraints for future planning possibilities.

Management plans are guided by the mission and goals of the management organization or group of organizations involved in a coordinated or integrated planning process. Information for planning usually is gained through one of several approaches. A common approach is to use existing studies of resource condition and recreational activities at the site. This is an inexpensive and efficient approach as long as the conditions at the site are similar and sites have not greatly changed. Reliance on generic studies, or studies of sites elsewhere, often risks being at variance with conditions at the site. Management targets may be misidentified such as the numbers and locations of campsites, docks, trails, roads, boat ramps, toilets, and food concessions. Old surveys may be out of date because both the landscape and human preferences changed more rapidly than expected.

Recreational sites rarely exist outside some larger context of recreational opportunities. A site developed 20 years ago in a rural setting may now be surrounded by suburban growth. While previously prized as a primitive wildland recreational resource, the site now may have resource attributes more compatible with homes and traffic in close proximity. The site may have been the only location with water-based recreation and camping and now is surrounded by new reservoirs and campgrounds under the same or different management authorities. It is important to know if the targeted site is a high priority destination of recreational users, only average, or gets used only when preferred sites in the vicinity are filled. For commercial recreational development, what the competition is going to do will affect the income of all competing parties.

Management objectives are identified based on resource supply and demand relationships and the interactions of recreational use with other uses. First, a general vision of the desired future condition of the site is developed, which often requires integration with other land management goals and objectives. An effective approach is to develop several alternative plans for converting the vision to reality and then selecting the most cost-effective plan. Objectives are specified as clearly as possible. They often include specifications for landscape shaping, planting, flooding, draining, other habitat modifications, and construction details needed for contractors to bid on and carry out the necessary work.

Recreational Economics

A chronic dilemma for natural resource managers is judging the extent to which they should budget and invest in recreational improvements relative to other resource development. For example, if a public reservoir serves multiple uses, including recreational fishing, boating, and swimming, how much should recreation dictate the water level maintained in the reservoir over hydropower generation and water supply? What are the trade-offs? It is virtually impossible to compare numbers of fish caught to kilowatts of energy produced or gallons of water delivered without transforming those outputs into some comparable measure of value, such as money (see Chapter 4).

While it is relatively easy to estimate the market value of a kilowatt of energy or a gallon of water delivered, recreational value may be more problematic depending on whether the site is private or public. If the site is privately owned, access is controlled, a fee is charged for recreation, and the profit made from recreation at any particular water level can be readily compared to the profits obtained from hydropower and water supply. Then the optimum water level for profit maximization from all three services can be estimated through use of economic planning tools.

If the site is public, the approach taken to estimate benefits from recreation depends on the site attributes and public policy. Where access can be controlled and a user-pay policy exists, management emphases can be decided on the basis of the total revenue derived from water, electricity, and recreational uses. Localized areas around lakes and reservoirs often are amenable and especially attractive as recreation sites. Where access cannot be controlled, there is no practical way to establish a recreation market. The benefit derived from recreation must be estimated indirectly from surveys of user willingness to pay. The approach preferred by many economists is the travel-cost method (Loomis 1993). In this approach, the willingness of users to pay a "price" for recreating at the site is based on the travel costs incurred. The effect water level or other site management has in increasing or decreasing the willingness to use the site for recreation can then be compared to effects on other services provided at the site.

Models have been developed for estimating recreational use based on site qualities under different management strategies (e.g., Ward et al. 1997). An important variable in estimating the willingness to recreate at a site is the availability of substitute sites with the same kind of recreational services. Substitute sites may be visited instead when they have similar quality but cost less to visit or when they offer better recreation at similar travel cost. Such models allow assessment of complex interactions resulting from management in a system of sites spread throughout a region. They can facilitate coordination of different public authorities for the most efficient provision of recreation. Cole and Ward (1996), for example, demonstrate this capacity for public sportfishing.

Integrative Management Planning

The primary elements in the recreational management system are the natural resources, the visitors, and the management plan (Jubenville 1978). High-quality recreational experiences that satisfy visitors require careful integration of these components (see Chapter 5 for more detailed planning concepts).

Integrative management planning for outdoor recreational use takes into consideration the plans of other agencies and private enterprise so as to avoid over- or underdevelopment of opportunities relative to anticipated demand. Integrative management planning typically is regional in perspective. The average distances traveled to participate in outdoor recreation often exceed two hours. With high speed highways in the vicinity of a population center, the region may easily include 30,000 square miles of alternative weekend recreational opportunities. Within that region, a wide variety of federal, state, county, municipal, and private authorities are likely to provide recreational opportunities. Although outdoor recreation is an increasingly important aspect of quality of life for Americans and people in other nations, past planning has not emphasized integrated approaches.

Outdoor recreation is highly interactive with transportation access to the recreational resources and with transportation planning. Transportation choices may include plane, train, and boat. However, once within a region, the automobile is by far the preferred means of travel to outdoor recreational locations. Sight-seeing by car is one of the most popular outdoor recreational activities (TFORRO 1988). Sight-seeing commonly includes specific target locations as travel stops or an array of possible stops any of which might be suitable as whim and physical need dictate. Regional metropolitan or state planning often revolves around transportation needs. Outdoor recreation is an important consideration. In many U.S. locations, the main travel use of rural roads is for recreational purposes and the main source of outside income is based on recreational demand.

Recreational Resource Managers

The recreational resource manager must have broad understanding of the needs of the visitor, the site's natural resources, and the services needed to make the social and resource environments satisfying (Jubenville 1978). Therefore, the primary components of recreational management are visitor, natural resources, and service management. Visitor management involves controlling distribution of use, safety, and access to information and education. It requires understanding of visitor perceptions, needs hierarchy, and style of participation. Natural resource management is the protection and enhancement of the landscape, soils, waters, vegetation, geological formations, and other natural resources for recreational purposes. Service management takes many forms, but emphasizes provision of access to resources and basic facilities needed to meet basic visitor needs.

Visitor Management

People seek different things from recreation, but all need food, shelter, and safety appropriate to the recreational experience. Most desire being part of a recreational partnership or group. Some recreators spend a lot on equipment and supplies usually in response to some perceived recreational status or competency level they seek to attain. Others desire a very personal sense of recreational satisfaction often independent of group dynamics.

Differences in behavioral style are important. For instance, a participant may be motivated to participate in hunting for the meat, a trophy, or mastering the technology. Camping attracts a wide variety of recreational types ranging from wilderness camping to transient vehicular camping. Perceptions of experience can vary. Therefore, feedback between users and managers is important in assessing the perception of the experience. Was it too crowded or the experiences not intense enough or too sporadic? Was the information service sufficient? Were there unexpected, threatening hazards? Were opportunities diverse enough to satisfy the whole group?

A primary strategy for managing visitors is to control their distribution. Controlling access and use of sites is a critical management responsibility because crowding affects user satisfaction and landscape condition. Use may be redirected or dispersed in a number of ways. Regional and project-area planning is the most effective means for controlling visitor access and distribution. Recreational use zoning is an effective way to sort out compatible and incompatible uses. This can be facilitated through information services and types of access and facilities ranging from highly developed roads, food services, and utilities—often on small public holdings or on the edges of larger public lands—to primitive and wilderness conditions more typically located in a large site's interior (Figure 13.11).

Improper original location of access and facilities will make any subsequent visitor management more difficult. Intensively used sites may require restriction or even site closure. While limited use requires expensive supervision, site closure is the last resort. Regional and on-site information provision is the most important avenue to visitor management. Where economically justified, visitors should be provided information in advance about any change in services and/or access to recreational locations.

An effective means of rationing use is through access-fee charges, a means often advocated by resource economists. Some people object to this, believing in principle that public lands belong equally to all. However, the largest cost for outdoor recreation typically is associated with travel costs and equipment. Fees are a minor consideration for most.

FIGURE 13.11 A well-designed and highly accessible state park in Oregon, which accommodates both day trips and overnight campers (courtesy of U.S. Government).

Natural Resource Management

Intensity of Use.   On large federal and state public lands, much recreation management must be integrated with other specialty areas including forestry, range, wildlife, and fisheries management. A number of management strategies are applied depending on purposes and circumstances. Recreational use may be dispersed or concentrated on developed sites to protect the natural resources from degradation and erosion. Intensively used sites may be culturally treated through fertilization and irrigation to increase their natural resistance to human use. Area use may be rationed by establishing capacity limits on continuously used sites and by closing and rotating site use. Once sites are degraded, they may be naturally restored or enhanced through cultivation techniques. Certain intensively used areas may be hard-surfaced in aesthetically compatible ways. Unintentional destruction by visitors sometimes is a problem, which often is corrected through information and education, or through simple prevention of access, such as vehicle barriers. Law enforcement has become important at most intensively used outdoor recreational areas.

Vegetation.   Vegetation management of turf, shrubs, and trees is an important aspect of recreational resource management in moderate- to high-use areas. Turf may be improved by thinning shade trees, irrigation, fertilization, and other treatments. For special use areas, such as golf courses and various field sports, more intensive and specialized turf management practices are applied. Trees and shrubs often need to be monitored for disease and trimmed or removed to prevent spread. Vegetation also needs to be managed for fire control through encouraging certain species, thinning, and creation of firebreaks. The natural history of fire in the region needs to be incorporated into the management. Monitoring is done with a variety of scientifically developed techniques using photography, fixed point surveys, plots, line transects, and other approaches. Typical attributes measured include the percent of vegetative cover and a ranking of plant vigor. Recreational managers typically rely on a diverse array of professional support services such as the Cooperative Extension Service associated with land grant universities, state departments of conservation, and state soil testing labs.

Access.   For areas managed to highlight natural beauty, access to visual resources is an important part of recreational service. Placement of road, trail, and waterway access with vista views that do not degrade the value of the resource is an important planning activity at national parks and other public lands. This planning often requires classifying landscapes in terms of dominant features and their arrangements and with respect to the qualities users seek in visual resources. On the ground, vegetation often needs to be managed not only to allow viewing, but to frame it with complementary foreground, much as art is presented in a museum.

Hazards.   Outdoor experiences include hazards, some of which are necessary for a fully satisfying recreational experience. Among natural hazards are cliffs, landslides, avalanches, electrical storms, high winds, desiccation, falling trees and branches, dangerous wildlife, disease, freezing conditions, deep water, currents, tidal changes, flooding, and natural obstructions in water, trails, and roads. The most common anthropogenic hazards include old mine shafts and mine high walls, abandoned buildings, wells, poor road design, and the behavior of other visitors. While many hazards are accepted and anticipated, others are not tolerated. A hazard-free recreational experience is viewed by many participants as sterile. However, hazards caused by human engineering and mismanagement are much less likely to be accepted than hazards associated with nature. Acceptance of hazards in one situation does not necessarily carry over to other situations depending on the age, experience, and attitudes of participants.

  Hazard management objectives, which first need to be defined, depend on the recreational services demanded of the site. Hazards are identified next and evaluated for their degree of threat, and mitigation strategies are adopted appropriate for the conditions. A basic decision is whether to manipulate the use of the site or the site's resources. In some cases, facilities may require relocation to reduce hazard encounters. Provision of information in appropriate locations and form is a basic management tactic. The recreation manager has to consider liability as well as recreational quality in seeking the optimum level of hazard management, especially in private provision of recreational opportunity.

Information Service Management

Information services inform recreation participants of opportunity availability, special events, use restrictions, other sources of information, and hazards. Regional services are aimed at resource users before they reach the site. The information provided helps them to choose among site alternatives. The most comprehensive services, such as national and regional catalogs of recreational campgounds, often include both private and public recreational opportunities with detailed summary descriptions of all facilities and services provided, as well as important restrictions. Other forms of regional information may be provided through mass media (typically expensive), public hearings, and group contacts. The Internet offers opportunities for providing much more detail on public use areas.

Visitor centers are the most common means for providing information at large recreational areas, such as national parks. Decentralized information stations often are used at local campgrounds, trailheads, boat ramps, and other staging points for recreational activity. These may be little more than an information shed or box that protects maps, pamphlets, and notices from the weather. The most explicit information is provided by large and unavoidable signs. In some locations, radios and other gadgetry have been used, but remain rare because of expense and frequent failure. The design of information packages is a critical aspect of communication. Avoidance of information overload is important and loose paper often becomes litter.

Interpretive service is a special category of information service designed to elicit more from the recreational resources. Interpretive services are most commonly employed where natural and cultural history are primary objectives of the recreational use. At many large sites, reference publications often are provided for the more serious student. Interpretive services usually inform visitors about what they may otherwise overlook at the site; often inform about management goals, such as preservation of natural, cultural, and historical features; and sometimes inform about the organization and its mission.

FUTURE DEMANDS FOR OUTDOOR RECREATION

Several factors impact the future demands for outdoor recreation in the United States. Demographic changes in the American population have implications for managing outdoor recreation, in particular the increase in minority ethnic groups and increases in the older age groups. Racial minorities will significantly increase population percentage and total numbers. During the period between 1990 and 2025, minorities (principally those of Spanish and African origin) will increase from 24 to 35% of the population and will continue increasing well into the next century. The greatest growth will occur in the Spanish-descendent population. The Hispanic population increased 44% from 1980 to 1990. About 22 million Hispanics now live in the United States, with 13 million in the southwestern United States. The Hispanic population in the United States was estimated to be 31 million in 2000 and is projected to be 81 million in 2050, while the non-Hispanic white population will constitute 53% of the U.S. population.

Relative to the age of the U.S. population, the median age increased from 23 in 1900 to 30 in 1980, and is expected to reach 41 by the year 2025. The percentage of individuals 65 and older will increase from 12.6% in 1990 to 21% in 2025, with a

FIGURE 13.12   There is growing demand by joggers, hikers, and pleasure walkers for safe and aesthetically pleasing natural areas within or adjacent to urban areas (courtesy of USDA-Forest Service).

marked increase after 2010. Also, urban residents continue to increase with only moderate rates of growth in rural states such as Montana, Idaho, and Wyoming. The increasing urbanization of the American population will require additional recreation facilities near urban areas, and oriented more toward an aging society.

Based on recent trends reported by Cordell et al. (1999) the recreational activities with the greatest increase in near-future demand will be bird-watching, hiking, backpacking, downhill skiing, and primitive-area camping (Figure 13.12). By 2050, the most popular recreational activities will be sight-seeing, walking for pleasure, pleasure driving, pool swimming, picnicking, day hiking, family gatherings, bicycle riding, photography, stream/lake/ocean swimming, wildlife observation, visiting historic sites, and camping.

Relative to social, economic, and environmental implications of demand-supply comparisons, the following are projected by Cordell et al. (1990):

1. The social characteristics of selected multicounty communities across the United States can be compared with the available recreation opportunities to yield information on social imbalances. In general, Americans who are elderly, less educated, part of a racial minority, economically disadvantaged, disabled, or living in cities have fewer opportunities to participate in resource-based recreation than do others.

2. The uneven distribution of opportunities can have adverse social effects including reduced family stability, more crime and juvenile delinquency, less opportunity for social bonding, more social conflict, and slower ethnic and cultural assimilation.

3. Increased economic opportunities for the private sector are projected for several categories of recreation. These include investments in developed recreation areas and the provision of associated goods, services, and information. Increased government revenue generated by user fees is expected to be offset by higher management costs for dispersed recreation.

4. Impacts on natural systems from most outdoor recreation and wilderness uses are minimal compared to more consumptive uses such as lumbering or

mining. Recreational impacts such as soil compaction and erosion are generally local in nature and the greatest damage occurs during the initial use of an area.

5. Outdoor recreation and wilderness use can benefit natural systems through improved aesthetic quality, greater environmental awareness, and preservation of natural systems. For example, demand for water opportunities has generated pressure in governments and industry to improve water quality in rivers, especially near urban areas.

The obstacles hindering attainment of opportunities in outdoor recreation include the following:

1. A major problem is the imbalance between recreation and wilderness land distribution (mostly in the West) and the population distribution (mostly in the East).

2. Private landowners are often hesitant to provide access to their land for public use without economic incentives or protection of the uses for which they own the land.

3. Insufficient funding, information, cooperation, and coordination among agencies contribute to problems in reducing the recreation-wilderness supply-demand gap (Cordell et al. 1990).

These challenges are some of the same ones confronting recreation development throughout its history. Particularly important in the future will be the inclusion of private landowners in providing additional opportunities for recreation participants. Recreation is a basic component of integrated natural resources management. The inclusion of recreational components in integrated ecosystem planning involving state, federal, and private entities are essential to the sustainable use of natural resources and the provision of recreational facilities for future generations.

LITERATURE CITED

Anderson, E. W. 1991. Innovations in coordinated resource management planning. *Journal of Soil and Water Conservation* 46:411–414.

Anderson, T. L. 1981. *Multiple use management: Pie slicing vs. pie enlarging.* Workshop on political and legal aspects of range management. Jackson Hole, WY: Natural Resources Council.

Bingham, G. 1986. *Resolving environmental disputes, a decade of experience.* Washington, DC: The Conservation Foundation, p. 43.

Box, T. W. 1995. A viewpoint: Range managers and the tragedy of the commons. *Rangelands* 17:83–85.

Brunson, M. W. and B. S. Steel. 1994. National public attitudes towards federal rangeland managements. *Rangelands* 16:77–81.

Brunson, M. W. and B. S. Steel. 1996. Sources of variation in attitudes and beliefs about federal rangeland management. *Journal of Range Management* 49:69–75.

Chavez, D. J. (Tech. Coord.). 1994. *Proceedings of the second symposium on social aspects and recreation research.* General Technical Report. PSW-GTR-156. USDA Forest Service, Albany, CA: Pacific Southwest Research Station. 186 pp.

Cleary, C. R. and D. Phillippi. 1993. *Coordinated resource management guidelines.* Denver, CO: Society for Range Management.

Cole, R. A. and F. A. Ward. 1996. Sustaining reservoir fisheries and angler benefits through comprehensive management analysis. In L. E. Miranda and D. R. Devries, eds., *Multidimensional approaches to reservoir fisheries management.* Bethesda, MD: American Fisheries Society.

Cordell, H. K., J. C. Bergstrom, L. A. Hartman, and D. B. K. English. 1990. *An analysis of the outdoor recreation and wilderness situation in the United States: 1989–2040.* General Technical Report RM-189. Fort Collins, CO: Rocky Mountain Forest and Range Experiment Station, USDA Forest Service.

Cordell, H. K., C. J. Betz, J. M. Bowker, D. B. K. English, S. H. Mou, J. C. Bergstrom, R. J. Teasley, M. A. Tarrant, and J. Loomis, 1999. *Outdoor recreation in American: A national assessment of demand and supply trends.* Sagamore Publishing. www.sagamorepub.com.

Dwyer, J. F. 1994. *Customer diversity and the future demand for outdoor recreation.* General Technical Report RM-252. Fort Collins, CO: Rocky Mountain Forest and Range Experiment Station.

Eckert, R. W. Jr., M. K. Wood, W. H. Blackburn, and F. F. Peterson. 1979. Impact of off-road vehicles on infiltration and sediment production of two desert soils. *Journal of Range Management* 32:394–398.

Forero, L., L. Hutsinger, and W. J. Clawson. 1992. Land use change in three San Francisco Bay Area counties: Implications for ranching at the urban fringe. *Journal of Soil and Water Conservation* 47:475–480.

Grizzle, G. 1992. Coordinated resource management planning in New Mexico. *Rangelands* 14:272–274.

Hart, J. 1991. *Farming on the edge.* Berkeley: University of California Press.

Heady, H. F. and H. J. Vaux. 1969. Must history repeat? *Journal of Range Management* 22:209–210.

Heitschmidt, R. K., R. E. Short, and E. E. Grings. 1996. Ecosystems, sustainability, and animal agriculture. *Journal of Animal Science* 74:1395–1405.

Holechek, J. L., R. D. Pieper, and C. H. Herbel. 2001. *Range management principles and practices.* 4th ed. Upper Saddle River, NJ: Prentice-Hall.

Huntsinger, L. and R. Hopkinson. 1996. Viewpoint: Sustaining rangeland landscape: A social and ecological process. *Journal of Range Management* 46:167–173.

Jacobs, L. 1991. *Waste of the West: Public lands ranching.* Lynn Jacobs, P.O. Box 5874, Tucson, AZ.

Jubenville, Alan. 1978. *Outdoor recreation management.* Philadelphia, PA: W. B. Saunders.

Loomis, J. 1993. *Integrated public lands management: Principles and applications to national forests, parks, wildlife refuges, and BLM lands.* New York: Columbia University Press.

McClure, N. R. 1992. Ranchers and resources reaping benefits of CRM. *Rangelands* 14:249–250.

Mitchell, J. E., G. N. Wallace, and M. D. Wells. 1996. Visitor perceptions about cattle grazing on national forest land. *Journal of Range Management* 49:81–86.

Outdoor Recreation Resources Review Commission. 1962. *Outdoor recreation for America.* Washington, DC: U.S. Government Printing Office.

Payne, G. F., J.W. Foster, and W. C. Leininger. 1983. Vehicle impacts of northern Great Plains range vegetation. *Journal of Range Management* 36:327–331.

President's Commission on Americans Outdoors. 1986. *Report and recommendations to the President of the United States.* Washington, DC: U.S. Government Printing Office.

President's Commission on Americans Outdoors. *Americans outdoors: The legacy, the challenge.* 1987. Washington, DC: Island Press.

Sanderson, H. R., R. A. Meganck, and K. C. Gibbs. 1986. Effects of overland flow on water relations, erosion, and soil water percolation on a Mojave Desert landscape. *Soil Science Society of America Journal* 53:1567–1572.

Task Force on Outdoor Recreation Resources and Opportunities. 1988. *Outdoor recreation in a nation of communities: Action plan for Americans outdoors.* Superintendent of Documents. Washington, DC: U.S. Government Printing Office.

Torell, D. J. 1994. Viewpoint: Alternative dispute resolution in public land management. *Journal of Range Management* 47:70–74.

U.S. Department of Agriculture (USDA). 1980. *Report of the Forest Service: Fiscal year 1985.* Washington, DC: United States Department of Agriculture, Forest Service.

U.S. Department of Agriculture (USDA). 1992. *Report of the Forest Service: Fiscal year 1992.* Washington, DC: United States Department of Agriculture, Forest Service.

U.S. Water Resources Council (USWRC). 1978. *The nation's water resources, 1975–2000.* Second National Water Assessment. Vol. I. Summary. Superintendent of Documents 052-045-00051-7. Washington, DC: U.S. Government Printing Office.

Vavra, M. 1996. Sustainability of animal production systems: An ecological perspective. *Journal of Animal Science* 74:1418–1423.

Ward, F. A., R. A. Cole, R. A. Deitner, and K. Green-Hammond. 1997. Limiting environmental program contradictions: A demand systems application to fishery management. *American Journal of Agricultural Economics* 79:803–813.

Wuerthner, G. 1990. The price is wrong. *Sierra* 25:38–48.

Zinser, C. I. 1995. *Outdoor recreation.* New York: John Wiley and Sons.

CHAPTER 14

Urban Land-Use Management

THE URBAN ECOSYSTEM

Resources and Services Expectations

Chances are good that you are now located in an urban building or vehicle engineered according to human design. Even though only 5–6% of U.S. landscape has been urbanized, a large majority of people live in towns and cities with more than 2,500 people. Virtually all live in an engineered environment of buildings, vehicles, and supporting infrastructure that forms a very extensive interface with the more natural landscape. We inhabitants of the urban landscape take for granted the provision of diverse services provided from natural resources and human ingenuity. They do not come cheap, however. Urban land resources are the most highly valued of natural resources.

We expect safe water out of our taps, clean air upon opening a window, heat when the thermostat is turned up, light to read by, music or a movie at the flick of a switch, and routine garbage removal. We want ready and affordable access to food, clothing, furnishings, building materials, and other goods provided from natural resources usually produced somewhere well beyond the city limits. Provision of these basic goods and services is normally taken for granted and happens as a consequence of well-designed urban land use and associated natural resource management.

Chances also are good that you very recently participated in an outdoor activity of some kind—perhaps no more than a commute to work, a shopping trip, a basketball game in a backyard, or an outing at a nearby park or nature reserve. Your choice of activity would have been decided by your activity needs and preferences, the time you had available, proximity to suitable sites, available modes of travel, expense, and quality of the environment in terms of health, safety, recreation, and aesthetics. Most of us take for granted the design of urban land use and how much it affects the quality of our daily lives. The ease of getting to desired destinations outside the home and workplace contributes importantly to our quality of life. Chronic worry and stress on congested and unsafe streets and walkways is a common urban complaint.

Fries (1997) suggests that Americans have a long history of ambivalence or even hostility toward cities, as evidenced by Thomas Jefferson's (an architect among

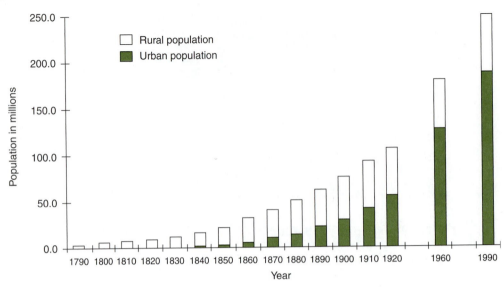

FIGURE 14.1 Urban growth in the United States from 1790–1990 (from Judd and Swanstrom 1994 and with permission of Cullingworth 1997).

other things) likening of cities to sores on the human body. Despite their doubts, Americans have steadily moved from rural to urban landscapes (Figure 14.1). How well these habitats serve human function depends on how effectively individuals and private and public institutions interact in planning land resource use in the urban environment (Fowler 1992, Kaiser et al. 1995). The provision of basic goods and services safely, in a nurturing environment, depends on city and natural resource managers working closely with private enterprise. We refer the reader to Kaiser et al. (1995), Garvin (1996), and Cullingworth (1997) for additional reading about urban planning.

THE URBAN ECOSYSTEM

Material and Energy Flux

Cities are ecosystems. Like other ecosystems, cities have structures and functions that depend on management of energy and material resource inputs and waste outputs (Bridgeman et al. 1995, Hough 1995). Cities are the ultimate destination of most extracted natural resources. The structures of cities are developed from inputs of materials including mineral, wood, oil (e.g., plastics, asphalt), metal, natural life-forms, and other physical resources. Cities function from reliable and continuous inputs of food, energy, electricity, fuel and solar energy, water, air, and land resources. Efficient functioning requires dampening the extremes associated with flood, drought, snow, storms, fire, and other disturbances. Resource conversion to useful products also results in waste products, which are concentrated where resource use is most intense—in the cities. Wastes are eliminated in gaseous, liquid, and solid forms into air, water, and earth. Improper elimination threatens physical and mental health and renewal of needed natural resources. Waste elimination solutions are found in more efficient resource use, waste assimilation, and waste dispersion.

The quality of ecosystem services supporting urban life depends on how well ecosystem structure and functions are managed. The urban landscape operates fundamentally like any other ecosystem, but is at the high end of the spectrum of human influence and need. That influence and need extends deeply into rural environments and sources of urban natural resources. The problems of natural resource supply and demand concentrate in urban landscapes more than any other type of ecosystem.

FIGURE 14.2 Urban areas increasingly extend into farm, forest, range, and wild recreational lands (courtesy of U.S. government).

Much of what natural resource managers do in urban landscapes is to regulate use, prevent abuse, and contribute to design of urban form and function. Managers of urban natural resources often complement the activities of environmental engineers who develop artificial approaches to augment natural ecological services, such as waste treatment facilities.

Extent of Urban Impacts

A high flight over almost anywhere in the United States outside Alaska reveals an interconnected network of villages, towns, cities, and megalopolises linked by highways, railroads, navigational waterways, and utility right-of-ways. Urban infrastructure extends like roots deep into the land's resources. The miles of roads on U.S. forest lands alone greatly exceed that of the entire interstate highway system. These extensions of travel and utility corridors from the urban ecosystem enable the extraction of agricultural, mineral, forest, and other natural resources and outdoor recreational use. The interspersion of urban resource demand and rural resource supplies contributes significantly to transportation needs and associated environmental impacts.

Although land classified as urban area is only about 5–6% of the land surface, the extensions of urban development into rural areas via roads, pipelines, waterways, atmosphere, utility right-of-ways, transmission lines, and railroads extend urban demands and impacts far beyond the city limits (Figure 14.2). (NRC 1992, Hough 1995, Landsberg 1981). The natural resource effects of urban development are especially intense, not only as sources of resource demand and destinations for extracted resources, but also because of their widespread impacts on air, water, and land resources (Hough 1995) (Figure 14.3). In Chicago, for example, 45% of the land surface is impervious concrete and other hard surfaces (NRC 1992). This impervious surface

FIGURE 14.3   High fractions of impervious material are an aspect of densely populated urban areas and cause increased amount and intensity of contaminated storm runoff (courtesy of U.S. government).

alone dramatically changes the dynamics of water movement and energy flow through the urban ecosystem, altering the quality of land, water, and air, often far from the urban setting. The urban "metabolism" generates much of the carbon dioxide and other gasses now believed to influence the radiation and thermal balance of the biosphere (Houghton et al. 1996).

Integrating Natural Ecological Services into Urban Design

Environmentally sensitive urban design is the most fundamental approach to managing natural resources in cities to sustain a high quality of urban life (Fowler 1992, Roseland 1998, Beatley and Manning 1997). To do that effectively, the ecological services provided by healthy ecosystems need to be identified, protected, and restored wherever it is cost-effective. Much of past urban development proceeded in ignorance of ecosystem services, sometimes requiring expensive engineering alternatives. Where ecological services are insufficient, environmental engineering can be adapted to natural process to increase efficiency of those ecological functions most in demand. Many of the engineering procedures used for environmental improvement attempt to improve upon natural processes. Sewage treatment plants, for example, collapse the natural process of decomposition into a smaller space where it is enhanced through artificial aeration and sedimentation. However, spatial efficiency comes at a high cost in energy requirements, which contribute to other waste problems. Wherever more natural ecosystem services can be used, they generate a cost-savings.

Natural ecosystem services are most cost-effectively provided when land, water, wetland, and atmospheric systems remain intact. The ecological functions supporting services of urban ecosystems often are overwhelmed by over use and abuse of the landscape, which is common in poorly managed urban landscapes. Careful land use planning is the key to improved provision of needed urban services. Clean water, for example, is provided by natural filtration and sequestering of contaminants and prevention of erosion by vegetated watersheds and wetlands. Those areas that form recharge zones for groundwater aquifers serving cities often are particularly sensitive to disturbance. Coastal wetlands, river floodplains, and barrier beaches can serve to protect urban ports and shores from storm and flood damage (Hudson 1996). Coastal wetlands also generate organic detritus, which supports coastal fisheries. By infil-

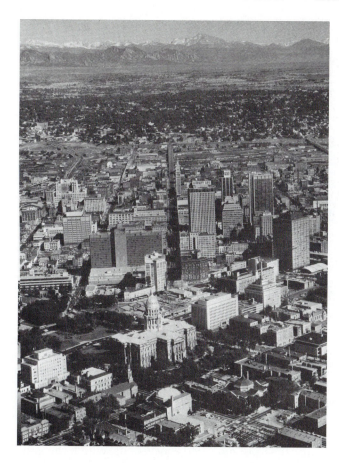

FIGURE 14.4 Urban form greatly influences the quality of life of most Americans (courtesy of U.S. government).

trating water, uncompacted soils free of impermeable concrete or other hard surface reduce damaging storm run off (Hough 1995, Perry and Vanderklein 1996). Natural vegetation absorbs noise and carbon dioxide (an important greenhouse gas), regenerates oxygen, dampens climate extremes, and forms habitat for diverse wildlife-based recreation and other outdoor recreation. Where wild communities remain intact and are managed carefully, diverse wildlife often provides natural controls of noxious pest species that flourish in habitats with low biological diversity. However, rapid and poorly planned urban sprawl, with insufficient consideration of all costs, continues in most parts of the United States.

Urban Form and Function

Urban form largely determines urban function and quality of life for human habitats (Figure 14.4). The best urban form optimally links private residences, workplaces, stores, and other service structures through a public infrastructure of roads, walkways, railways, waterways, sewerage, electric and fuel utilities, recreational areas, and greenways. It functions to sustain safe, healthy, and positively stimulating environments. As urban form materializes, it both provides and constrains choices for the next generation. Much of urban form carries over functions that may have been more appropriate for past culture. The history of urban development is informative because it reveals how cities can become prisoner to their past, unable to adapt, and finally dysfunctional.

The rise of modern urban structure began thousands of years ago in basic urban form that persists to this day (Mumford 1961, Morris 1994). Urbanization accelerated during the industrial revolution as people moved from sparsely populated agricultural land to concentrate at new employment centers. The cities that resulted were

FIGURE 14.5 A strip of woods along a river, long ago dedicated as a park, provides both open space relief and flood water storage in a densely settled area (courtesy of U.S. government).

compact because most people had to walk to work. Also, land speculation was an early profitable investment in the United States. More money was made from subdividing into closely spaced small units. Archeology reveals that even the earliest cities tended to rely on geometric design with an emphasis on straight lines and sharp angles that form an open grid for maximum access in and out of neighborhoods (Ward-Perkins 1974, Morris 1994).

Human communities occupy the most modified ecosystems, which can grade into adjacent rural and wilderness ecosystems in ways compatible with, or in sharp disruption of, ecosystem structure and function. Well-designed and maintained parks, gardens, and yards soften the harder edges of urban existence (Hough 1995, Nasar 1998). Natural parks can be designed into the urban landscape to thread through and extend to the urban fringes (Figure 14.5). When designed well, these natural spaces can provide recreation as well as contribute to other ecological services such as cleaner air, cleaner water, and less soil erosion. However, excessive urban sprawl with its intense impact on natural process has led to widespread degradation of natural functions. The main causes of dysfunction are environmental contamination, environmental fragmentation, and widespread conversion of watersheds to inert and impermeable structure (asphalt and concrete).

The high contact area with landscapes makes travel and transport corridors (roads, waterways, railroads, airports) a major source of abiotic and biotic contaminants for land, air, and water. Roads are especially problematic because their impact is so extensive. Travel corridors are major routes for contamination by exotic plants, petroleum products, ozone, noxious aerosols, dust, sediment, diverse spilled chemicals, and litter (Hough 1995). In addition, they are one of the main ways that wildlife habitats are split into fragments, threatening some species and contributing to wildlife-related driver safety problems. Poor transportation planning results in greater environmental degradation and human stress.

Integrating Urban Form with Natural Functions

Much of the history of urban development in the twentieth century has been in search of urban form more compatible with natural function. Cities have adapted mostly by extending outward from the stark artifice of city centers into rural outskirts through suburbanization. A main motivator has been a desire to achieve a

FIGURE 14.6 This photograph demonstrates both the loss of space and the high amount of infrastructure needed to serve urban sprawl (courtesy of U.S. government).

more parklike environment where the best of nature is captured and the worst is eliminated or controlled. The results have not always been satisfying because of competing demands for space and costly demands for infrastructure to serve urban sprawl (Figure 14.6).

As the United States became more wealthy, more attention was placed on the aesthetics of urban form including more parks, parkways, and cultural centers. The conservation movement was gathering influential backing along with the belief that efficient utility of resources could be compatible with natural beauty and culture, even in cities. This vision emerged in curving boulevards that replaced the street grid as a more aesthetically pleasing way to link especially auspicious locations, such as public parks, museums, and auditoriums (Garvin 1996).

Three architects were particularly influential in developing this vision of urban form: Frederick Law Olmsted, Frank Lloyd Wright, and Charles-Edouard Jeanneret, who used the professional name Le Corbusier (Fowler 1992). Olmsted, a landscape architect, initiated the urban metamorphosis as early as 1857 when he completed construction of Central Park in New York City. Central Park was impressive enough to establish both Olmsted and city parks as fixtures in numerous U.S. urban plans. By the turn of the century he had designed many other parks, parkways, and several park-like planned communities, which were the forerunner of present-day suburban developments (Figure 14.7). Wright emphasized a need for individual space in residences and natural integration with nature, even for more modest homes. He increasingly influenced many suburban developers after the 1920s. The French architect, Le Corbusier, emphasized revitalization of the downtown areas with tall high-rise residences and commercial buildings surrounded by open parklike environments. He influenced urban renewal programs, some of which proved

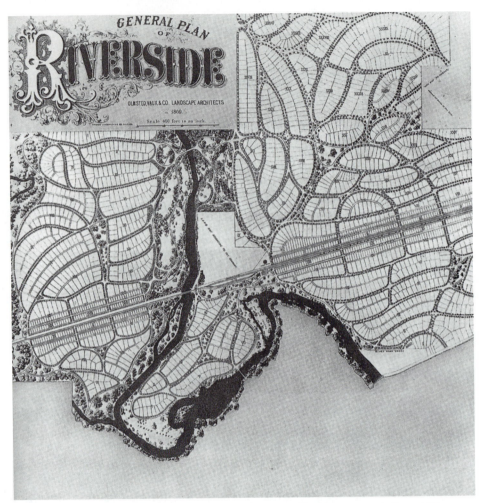

FIGURE 14.7 Frederick Law Olmsted, an early landscape architect, was very influential in establishing contemporary concepts of parks and parkways. Many modern suburban developments have adopted Olmsted's concepts and propagated them across metropolitan landscapes. The plan shown above was proposed by Olmsted and partners in 1869 (courtesy of the National Park Service).

to be inappropriate for use, and also influenced the now widespread development of naturally landscaped commercial centers in suburbia.

Following the lead of Olmsted, Wright, and Le Corbusier, parks, parkways, and park-like yards became a central theme for planning a more natural urban world that takes advantage of natural ecosystem services. Parks and parkways placed along flood-prone stream drainages, for example, can serve as corridors for highways, recreation use, and floodways when needed. Even very narrow parkways with little capacity for recreation and wildlife use can absorb highway sound, reduce side glare and distraction, protect against erosion, act as water recharge and filtration areas, and possibly provide some aesthetic comfort for travelers. Where compatible, some parkways also provide paths for bicyclists, joggers, and walkers. However, high-speed traffic in narrow parkways often forces all other parkway uses into secondary and less satisfying roles. Numerous parkways, built in a more leisurely time, are no longer well adapted to satisfying multiple use because the demands of the automobile culture now take precedence.

The largest U.S. urban parks promote natural landscapes of wild vegetation or naturalistic landscapes of undulating topography, lawns, gardens, and thinned out woods based on nature. Water is typically a central element in most large parks. Like parkways, parks often are managed for many purposes. Water, for example, is an important natural element that often is integrated into parks for flood and pollution control purposes as well as recreation and aesthetics. In recent years, cities like Portland, Oregon, Austin, Texas, and the greater New York region have developed plans that protect natural land from suburban development (Beatley and Manning 1997). A mix of strategies are being applied that include purchase and reservation of natural areas for watershed and low-density uses, such as lands protected in the New Jersey pine barrens for groundwater recharge and other low-intensity uses. New Jersey is considering zoning plans for protecting remaining rural lands for traditional uses.

History shows that without careful land-use planning in urban landscapes, costly adjustments often are needed to sustain necessary functions. The costs have to great extent been borne by the taxpayer in the form of new and rehabilitated public infrastructure. Despite the obvious need, urban planners often are considered as necessary evils in the United States and elsewhere (Fowler 1992), where regional planning often is viewed as an infringement on private property rights. However, it is a necessary one given the tendency for the marketplace to overlook long-term community needs. As a consequence, the role of natural resources management in urban planning often is contentious and political.

DEVELOPMENT OF URBAN INFRASTRUCTURE

A Partnership of Public and Private Interests

Private Independence. Development of urban infrastructure is costly. The most highly valued land and water resource use is for private development of residences, businesses, and the public urban infrastructure that serves them. In contrast with wild lands and waters, which are predominantly owned by the public, ownership of urban and agricultural landscapes is mostly private. Because of its high social value, demand for urban development usually establishes the prevailing land prices around existing centers of urban development. Natural ecological services typically require large land areas, such as for producing food and clean water, absorbing floodwater, and providing wildlife and fish habitats. Much of the nation's richest farmland already has become urbanized, and much more is likely to be if the pattern of recent urban growth continues unabated (see Chapter 12).

Most Americans have placed a high value on economic, religious, and political independence. Rights to private property and freedom of movement are important ideals. The promise of plentiful land and the opportunity to live on it and develop it were important reasons for immigrants to come to the United States. The typical American dream was, and continues to be, independent land ownership centered around home and family, freedom for personal beliefs and expression, opportunity for wealth accumulation, and mobility. By the mid-twentieth century, the American dream for the average citizen was a privately owned, appliance-filled family home with a big yard, a car in the garage, and enough personal security, wealth, and leisure time to enjoy them. The Federal Housing Act of 1949 extended this dream nearly to the status of a right by declaring its intent to provide a decent home and environment for every American (Garvin 1996).

Public Interdependence. American private well-being also depends on publicly owned community infrastructure—most notably parks, roads, and water resource systems. Starting with the earliest attempts, urban design in the United States has been an exercise in providing an appropriate balance between private development and public service needs (Judd 1988). The town square often was, and still is, common-use space in the geographical center of town life. Town centers frequently became the first city parks, which were surrounded by buildings dedicated to key businesses, town government, education, and religion (Figure 14.8). As people gained more opportunity for relaxation and recreation, public parks became more important resources.

FIGURE 14.8   Public buildings, historic structures, and a landscaped plaza are united to form a community focal point in Alexandria, VA (courtesy of U.S. government).

Early Americans accepted travel and even came to seek it for pleasure once invention produced a comfortable means. The concept of parkways emerged, blending public transportation corridors with public recreation and creating an aesthetic sense of connection and community (Garvin 1996). American taste for pleasurable travel extended well beyond the city limits to enjoy some of the most diverse and awe-inspiring terrain in the world. Infatuation with the "great outdoors" grew with the improvement of the automobile. Similarly, Americans supported public development of navigation, water storage, and flood control in their waterways. Later, they would accept with little hesitation public water treatment systems for protection of public health and the environment.

Public and Private Coordination.   The value of private real estate usually depends on the extent to which publicly funded and managed infrastructure is provided in coordination with private development. Public lands typically sustain the engineered infrastructure required for contemporary urban functions—the public buildings, streets, highways, parking lots, airports, railroads, waterways, dams, walkways, bridges, tunnels, utilities, water supply, sewerage facilities, and public recreational facilities. The value of private real estate has had much to do with how the public infrastructure was developed. However, provision of infrastructure also has had major impact on real estate value and on ultimate public welfare. Too frequently, special interest wealth influenced infrastructure for private gain without much concern for the public interest.

Poor planning coordination in the past has contributed to many of the urban resource management problems of the present. Many of these problems will require decades to solve even under the best of circumstances. Much past development of private land and public service has progressed incrementally with minimum concern for cumulative regional impact. Past and present urban expansion is the consequence of planning that integrated public subsidy with private investment. Tension between national and local interests in the United States continues to make regional planning a strenuous process (Fowler 1992, Kaiser et al. 1995), but it remains the most promising way to sustain and improve upon quality of urban life.

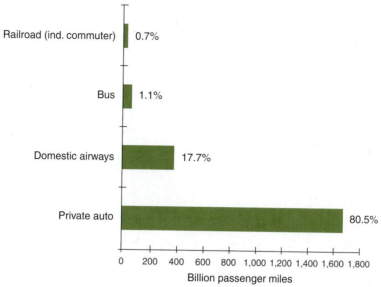

FIGURE 14.9 In 1992, the automobile exceeded all other forms of passenger transport in the United States (with permission from Cullingworth 1997).

Transportation

The development of modern urban infrastructure has been dominated by public policy in support of transportation. Many of the most impressive inventions and developments by Americans have responded to American predilection for travel. The means we have chosen for travel have shaped the modern urban landscape at least as much as the buildings in which Americans live, work, learn, shop, govern, pray, and play (Figure 14.9). More than any other form of transportation, the automobile has come to dictate the design of most U.S. cities (Jakle 1990, Langdon 1994, Roseland 1998). Excessive urban spread with its intense impact on natural process has led to widespread degradation of natural function, much because of the need for automobile transport. The main causes of urban dysfunction are environmental contamination, landscape fragmentation, and soil conversion into water-impermeable surfaces.

Where urban environments were more consolidated, as in Europe, there was less space available for the demands of the automobile. But for many U.S. cities, congested, noisy, and exhaust-polluted streets became the norm (Figure 14.10). Many European countries were already densely settled before cars became commonplace. Their larger cities are older than those in the United States. Historical appreciation and high cost were great enough to prevent extensive remodeling around the automobile. As a consequence, Europeans have relied more on a network of mass transport within and between cities. Some of the larger cities of the United States continued to build on their investment in railroad mass transportation. The first subway systems were constructed late in the nineteenth century in Europe and the United States. Later, insufficient resource wealth also dictated an emphasis on mass transportation systems in various other countries.

The automobile was invented at a time in U.S. history when a frontier mentality and unfettered entrepreneurial spirit still motivated independent thought and action. Coordinated planning and design processes received little consideration. Americans had long been a mobile society that solved local resource depletion problems simply by moving to unexploited resources—usually westward and off the beaten path. The U.S. populace was still more rural and small village than big city. Many of the first buyers of automobiles were farmers seeking more frequent connection to cities. Farmer's market and fresh "truck crops" quickly became staples of city life.

FIGURE 14.10 Congestion in downtown Pittsburgh in 1937. Congestion was a major problem facing cities at that time (courtesy of Carnegie Library in Pittsburgh).

These were a predecessor of the modern supermarket that now supplies diverse foods, often from sources thousands of miles away.

By World War I, a sizable fraction of U.S. society could afford a car, but prices remained out of reach of the majority. That was changed by Henry Ford and other automakers through the development of mass-production techniques (Jakle 1990). Cars, trucks, tractors, and many other manufactured items became much more affordable. Private enterprise also developed more questionable strategies for construction of infrastructure based around the internal combustion engine. General Motors purchased electric trolley companies and then removed the trolley tracks and replaced them with internal combustion buses (Jakle 1990). Clearing the streets of trolleys made automobile travel much easier and encouraged more auto commuting farther from the city center (Figure 14.11).

As more people bought cars, industry successfully lobbied for public investment in highway transportation designed for internal-combustion traffic. The electric trolleys were gradually displaced. Because of uncoordinated local control, highway development created a patchwork of roads of variable quality. The federal role in influencing highway system development was justified based on the need for improved interstate commerce. The Federal Highway Act of 1916 was the first major involvement of federal government in highway expansion since early nineteenth-century investments in a national highway westward. By offering matching funds for highway construction, this act coerced states to create state highway departments with authority to oversee and replace local road construction. Car ownership boomed during the economically expansive period following World War I and the lure of suburban development increased.

Utilities

The wealth that powered suburban development also influenced local and state governments to invest in the infrastructure along the public highways. Our modern con-

FIGURE 14.11   Chicago 1909. Electric street cars, although fuel efficient, impeded automobile traffic. General Motors bought up street car companies and converted them to buses with the intent of favoring the internal combustion engine (courtesy of Chicago Historical Society).

cept of utility networks grew out of that convenience. The government subsidized private enterprises by providing access through the public lands associated with roads. Electric and telephone lines usually were strung up along the highways at much less cost to private utilities than would have occurred without public highways. As the need for traffic lights and suburban electrification increased, electric-power services increasingly became an indispensable part of the urban infrastructure. Some private water supply utilities also began to provide treated water along highway right-of-ways, followed by natural gas lines mostly after World War II.

Street flooding was another problem that had to be addressed. The right-of-ways beneath highways also were convenient for draining storm water runoff. At first, water supply and water-borne waste removal were locally provided through wells, latrines, and septic waste fields. Some private water supply services developed where urban development occurred in locally deficient areas. But local worry about dependable water supply and water-borne disease grew into widespread public health concerns. As a consequence, local governments gradually assumed most water supply and waste removal services. Sewerage pipes were added to the infrastructure buried beneath the streets. Garbage removal initially depended on individual disposal responsibility and hired services. Garbage dumps began to degrade the landscape as people dumped virtually anywhere to get rid of solid wastes. As public health and aesthetic concerns grew, public management of solid wastes became the rule.

Housing Subsidies

Before the depression of the 1930s, virtually all home finance was privately managed. The depression slowed urban expansion less than might have been the case without government intervention and plentiful oil resources. In response to the depression's impact on housing construction, the Federal Housing Administration Act of 1934 increased mortgage insurance and made mortgages much easier to obtain (Cullingworth 1997). Before the act, a mortgage usually covered only one-third to one-half of the

FIGURE 14.12 In no other city are superhighways more a part of the landscape than in Los Angeles where, as in many urban areas, they relieved congestion at the expense of air quality (the haze in the background is smog) and fragmentation of neighborhoods (courtesy of the Environmental Protection Agency).

house cost and had to be refinanced every five to 10 years. The act enabled house purchases with mortgages up to 30 years with only a 5% down payment by the buyer. No down payment was required from World War II veterans. Housing starts in the suburbs increased rapidly in the midst of the depression years and accelerated after the war. Cheap fuel sustained the use of cars throughout this period. Burgeoning oil, trucking, and intercity busing industries joined the automotive industry to continue to influence government investment in the highway system. In addition to other public works, such as dams, highway building was also accelerated during the depression years using public funds. It encouraged employment while building an improved infrastructure for future expansion of private enterprise.

The Superhighways

During the depression, the government became much more active in building public works and often focused on project objectives with single-minded intensity. Science and technology were widely accepted as the way to a more satisfying future. Much private land was claimed and purchased for eminent domain needs pertaining to public transportation. Robert Moses, in New York City, was one of the more effective public transportation managers of the time (Caro 1974). Like most other regional planners focused on one land use, he advocated development of roads for transportation benefits without in-depth consideration of their ultimate effect on other aspects of lifestyle (curiously, Moses never drove himself) (Figure 14.12).

The new superhighways were designed and developed through close association of automotive engineers and executives with highly focused planning objectives. For many decades, the typical superhighway systems were designed as if to discourage mass transportation and pedestrian traffic. New highways frequently sliced through neighborhoods creating isolated urban backwaters and worsening decay (Muller 1990) . They concentrated urban noise and air pollution. In addition to providing the means for suburban expansion and more rapid traffic flow, they encouraged urban conditions that stimulated even greater urban flight to the suburbs (Figure 14.12).

FIGURE 14.13 The development of high-occupancy vehicle (HOV) lanes to encourage less traffic congestion has had disappointing results. Urban demands make car pooling a time-consuming and demanding task that is not a very attractive alternative to slower traffic in low-occupancy lanes (courtesy of the Environmental Protection Agency).

During the Cold War following World War II, federal support of highways was justified for the rapid evacuation of nuclear targets and to facilitate movement of troops and retaliatory missiles. This initiative received enthusiastic support from the auto, oil, trucking, and busing industries. The Interstate Highway Act of 1956 subsidized further urban decentralization along interstate routes. This act was followed by the Instate Highway Act, which funded and coordinated minimum access "beltways" around city centers. The beltways were needed to relieve the interstate highways from the escalation of traffic that was occurring on them. Most metropolitan areas of that time favored growth, and beltways provided additional access to existing suburbs and a stimulus to develop more suburbs.

In a typical expression of public works and private land development, the beltways reshaped land development almost overnight. Business complexes sprang up around the beltway forming nuclei for new "edge" cities (Garvin 1996). Much of the public supported these massive engineering projects—especially those developed to relieve the traffic-choked highways first built in the 1920s. Many of those earlier highways had become magnets for commercial development and cross-traffic leading to subdivision housing. Thus a reinforcing feedback between public highway development and private suburban development began and continues to dominate city planning and function. From 1950 to 1992, the number of cars in the United States quadrupled. Now about 80% of all passenger traffic is in cars. Most travel to work is by car, and most commuters travel alone. To reduce commuter traffic, large amounts of money have been invested in special travel lanes for high-occupancy vehicles in the larger metropolitan areas of the United States with the intent of reducing traffic (Figure 14.13). The effect overall has been disappointing, however. Commuting contributes only about 20% to total automotive travel, with similar contributions each from pleasure driving, shopping, and personal needs (Cullingworth 1997).

POPULATION REDISTRIBUTION AND URBAN CHANGE

Growth and Decline of the Urban Center

The industrial revolution of the eighteenth and nineteenth centuries accelerated the urban growth concentrated around city-center industry and business. This continued until after World War II into the present. As cities continued to grow, they diversified into identifiable neighborhoods and districts, which began to replace the city center as discrete places of residence and sources of specialty services (e.g., financial, theater, warehouse districts). However, most central cities continued to thrive until after World War II despite the nearby neighborhood decay. Starting in the early twentieth century, skyscraper construction and rail transportation rejuvenated city centers by multiplying useful work space in the city center. It facilitated efficient access and networking among growing businesses. The city center continued to be a hub where activities converged in the "controlled chaos" of a "people center" dependant mostly on walking (Goldfield 1977).

U.S. city centers attracted people from many other parts of the world as well as rural America. The influx of mostly low-income people overwhelmed housing and urban infrastructure (Cullingworth 1997). The consequences of overcrowding and underemployment caused city residents who could afford alternative housing to move to suburbs, taking needed tax revenue with them. Federal legislation in support of more affordable mortgages encouraged urban flight to the suburbs. Mortgage restrictions discouraged purchase of older houses, especially in urban neighborhoods with risky futures. They diverted prosperous people, who might have otherwise stayed, to new housing in the suburbs. As urban decay advanced, disease, fire, and crime progressively crippled older parts of the city center (Judd 1988).

Residential neighborhoods varied in health according to income, with the poorest neighborhoods typically remaining on the industrial and warehouse fringes around the city center. While the public needs of many neighborhoods were well served, the poorest people were least represented and ended up in the least tolerable ghettos. As they became more wealthy, a succession of people moved from poorer neighborhoods to newer neighborhoods nearer the city edge or beyond. The poorest remained trapped in decline and grew in number, especially as the tenant farmer system of the South gradually collapsed during the first half of the twentieth century. During that time, many unskilled people moved off the land to the cities in search of scarce employment (Lemann 1991).

The Role of the Automobile

The automobile was a "catalyst" for decline of many city centers, starting before the depression and rapidly accelerating after World War II (Goldfield 1977). Those cities most able to sustain a healthy city center, such as New York, had developed efficient mass transportation systems early in the century. Cities more dependent on the automobile suffered more severe and persistent declines. The private car required too much space to maintain areas dedicated to walking and mass transport. Car exhaust also overwhelmed assimilation by the urban atmosphere. Tall buildings trapped fuel exhausts, especially where natural atmospheric inversions caused stagnant air (Goldfield 1977). Air quality worsened even in cities with little industry like Washington, D.C., and Los Angeles.

It became more expensive for large businesses to attract people to downtown locations. They began to leave for the new satellite cities growing rapidly along major intersections with superhighway beltways. With the establishment of satellite cities along the beltways (Figure 14.14), a new form of centralized retailing came into its own—the shopping mall (Cullingworth 1997, Garvin 1996). Satellite cities attracted local business centers and light industry, especially in aeronautics, computers, com-

FIGURE 14.14 Satellite, or "edge," cities grew up rapidly after the interstate highway system was developed. Tysons Corner, VA, pictured here in 1989 on the beltway around Washington, D.C., was little more than a country store in 1935 (courtesy of Fairfax County Public Library Photographic Archive).

munications, and other "high-tech" employment of the information age. These industries attracted employees from the older industrial centers, which were in decline as a consequence of cheaper labor associated with international competition. Defense spending supported the cost of moving workers from old industrial city centers to new beltway locations. Many older heavy industries closed, leaving blighted "brownfields" of rusting abandoned buildings and debris with few prospects for desirable land development.

Urban Sprawl and Downtown Renewal

Suburban Sprawl. Urban decentralization was especially problematic for cities. It also created problems in the suburbs, which grew twice as fast as the central cities during the 1980s (Cullingworth 1997). Many of these metropolitan areas have coalesced into a sprawling "megapolis" where the city landscape sprawls over hundreds of square miles (e.g., Phoenix, Arizona; Los Angeles, California; Atlanta, Georgia; and from New York City to Washington, D.C.). Goldfield (1977) called the lure of suburbia the "Golden Fleece of the pastoral ideal." Suburbanites paid in part for their urban independence through utter dependency on cars and associated noisy, congested, aesthetically distasteful thoroughfares with numerous costly side effects.

Roads and traffic became a major source of air pollution, suburban and urban fragmentation, and competition for needed mass transportation and urban redevelopment resources. Kaiser et al. (1995) summarized this as follows: "Low density sprawl with its leapfrog development pattern is seen to be not only ugly and wasteful of land but also fiscally inefficient, environmentally harmful, socially isolating, and exclusionary." Numerous books have been written on the social and environmental problems associated with urban decentralization (e.g., Langdon 1994, Kunzler 1996). Sprawl is typically regional, extending beyond the limits of urban jurisdiction, and even state jurisdiction (Figure 14.15). This makes city governments unsuited managers of sprawl. Some state governments have tried to address the problem with mixed success (Kaiser et al. 1995).

Among the more pernicious problems in suburban environments is the sense of isolation that often befalls people in the midst of thousands of people. Because suburban arteries are often high-speed thoroughfares with crossing lights spaced far apart, the roads are often threatening barriers to pedestrians, forcing reliance on vehicles (Kunzler 1996). Children need to be chauffeured or bused most places. Many suburban areas have limited outlets, forcing people to drive the same routes repetitiously and farther than necessary past monotonously similar buildings (Langdon 1994). Much driving is done alone because people are so spread out. Retail services have become centralized in large shopping malls or multiservice stores, which can only be reached comfortably by vehicle.

The protection of public open space and the provision of improved infrastructure are becoming increasingly problematic as land prices increase and the public resists

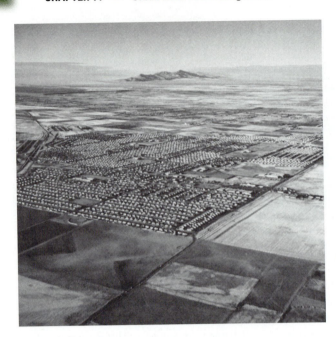

FIGURE 14.15 Urban sprawl displaces irrigated farmlands in Utah (courtesy of U.S. government).

further conversion of private land to public control. Local governments have won constitutional approval for zoning use of private lands (Cullingworth 1997), which provides a valuable land-use planning tool. Zoning has often promoted urban sprawl and exclusionary policies resulting in large minimum-size lots and wide separation of homes from services (Cullingworth 1997). Public investment in infrastructure is becoming more costly per person as the distance from service centers increases and the density of people served decreases. Much of the existing highway infrastructure is aging to a point requiring major repair and rehabilitation—a costly process in both taxes and driver time. As concern for private property rights has grown, it is becoming increasingly difficult for private lands to be condemned and purchased for rehabilitation of public infrastructure. The wealthier people seek solutions by moving farther out or, increasingly, back into renewed city centers. The rush to suburbia has shown signs of slowing in recent years, but the economic forces behind further development remain intense. The wealthiest, at least, are likely to continue to move to the fringes, repeating a pattern that started well over a century ago.

Urban Renewal.   Urban renewal and rehabilitation are measures taken to counteract urban decay. The basic strategy of urban renewal has been to eliminate urban blight through destroying an old structure and rebuilding on the cleared land. The strategy of rehabilitation is preserving the best of the old urban structure and renovating it for new use.

Urban renewal started with the Housing Act of 1949, which sought to provide affordable decent housing and environment for every American. Early urban renewal translated into whole-scale physical destruction rather than restoration of what was left of decaying neighborhoods. In some locations, renewal got no farther than clearing the land for private redevelopment, which was reluctant to invest in many locations. This was because of a perceived lack of interest by middle-class homeowners who, for the most part, wanted a suburban location (Cullingworth 1997). In other locations, government subsidized the construction of large high-rise structures surrounded by open, park-like areas in the manner of LeCorbusier. Without a sense of neighborhood and community function, and with high unemployment rates, some of these buildings became high-rise ghettos permeated and surrounded by crime and devastation. Even though many renewal projects were quite successful, the disappointing results of some highly publicized renewal projects have cooled government support.

Cities have learned that successful redevelopment requires a complete integration of social function with the structural development. It is important to create a mix of residences and community services that provide the physical support for neighborhood rebirth. Safe parks, playgrounds, and other open spaces also are important components for success. Judicious investment in new mass transportation routes has helped gain middle class interest in complete neighborhood renovation. Problems remain, however, in placing people of lower income into the process of private restoration.

Urban rehabilitation of city centers began to take shape in the 1950s and 1960s in cities such as Minneapolis. The focus was on downtown revitalization and mass transportation links from neighborhoods to downtown areas (Goldfield 1977). In many cases, they restricted auto travel or eliminated it entirely from downtown pedestrian malls, while providing necessary parking on the periphery. Other successes include more effective bus or rail mass transportation, such as in Washington, D.C., and attractive downtown centers, such as the river walk at San Antonio, the old-town Bourbon Street area in New Orleans, the wharfs in San Francisco, and the riverfronts in New York and Philadelphia. In the last two decades, the construction of public aquariums has been an especially effective way to rehabilitate once decaying urban waterfront areas, such as at Baltimore, Boston, Camden, Chattanooga, Monterey, New Orleans, and Tampa.

URBAN LAND-USE PLANNING

Historic Land-Use Development

Land-use planning in the United States has been and continues to be mostly a local process (Cullingworth 1997). If city persistence is a clue, the most effective urban planning has always taken advantage of the natural resources and trade made accessible through transportation. The ghost towns of the United States are typically at a dead end or along a little used highway. Most large cities are strategically situated between major resource supply and demand centers. Chicago, for instance, grew rapidly as it received grain and livestock by rail, which were processed and distributed using the Great Lakes and the Mississippi River system. East Coast cities shipped products overseas. Much city planning from the beginning took place to assure someone or some group obtained greater wealth based on trade, land speculation, or land development.

The extent to which urban planning has become a formal and comprehensive process varies tremendously across the nation, depending much on local history and attitude toward government. As far back as the early colonial governments, urban planning has been dominated by a wariness of strong central government (see Chapter 2). The major evidence of extensive government influence in urban planning before the twentieth century shows up in government capitals. Washington, D.C., as the seat of the federal government, came closest in the United States to being completely planned in advance in response to the recommendation of President Washington. The French planner, L'Enfant, envisioned a basic grid superimposed on radiating boulevards and traffic circles linking houses of government and monuments (Fries 1997). The design created many odd-shaped plots of land and intersections, which may have contributed to slow city development as transportation changed. However, while at first viewed as an embarrassment, the U.S. capital has since become a world attraction because of its monuments, museums, and unique urban design (Zelinsky 1990).

Other early federal influence on urban planning stemmed from the way the public domain lands were organized for transfer to private ownership. The practicality of organizing land ownership and development on a grid was institutionalized by the U.S. federal government for the national survey soon after the nation formed. The national grid formed square-mile townships composed of 640 acres each. These township boundaries defined the basic urban pattern for many Midwestern cities.

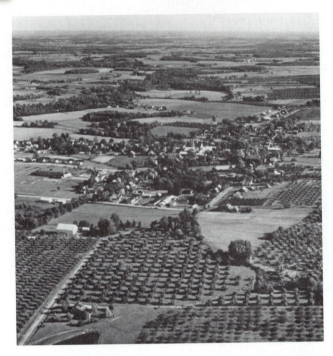

FIGURE 14.16   This landscape pattern of villages surrounded by farmland in the northeastern United States was typical of many communities until the 1960s when rapid urban expansion began to occur (courtesy of U.S. government).

Local planning was accomplished more through reaction to undesirable trends than through proactive prevention of such trends. Local urban regulations started immediately following settlement to assure dependable commerce and safe conditions. Regulations were developed, for example, to prevent devastating fire, mired streets, depleted wood, depleted wildlife and fishery resources, and to assure that local food needs would be met. As time passed, new cities learned lessons from older cities regarding fire, flood, mud, disease, and other resource misuse. In addition, local governments came to depend more on state and federal government for disaster relief, especially after the Great Depression of the 1930s. Such aid relieved local governments of the need for more careful land-use planning for cities in the paths of flood, fire, earthquake, hurricane, tornado, and other natural events.

Most existing urban lands in the United States developed without a comprehensive plan. They expanded incrementally on private lands at the margin of the existing city. In contrast with Europe, where the largest cities became centers of civilization and bases for government and public policy, religion, and culture, most large U.S. cities became seats of private enterprise and entrepreneurial spirit (Mumford 1961) (Figure 14.16). As urban growth began to be influenced more strongly by the automobile in the early 1900s, the first urban regional planning agencies were formed in Los Angeles and Chicago. Urban regional planners grew influential enough to encourage comprehensive planning initiatives at the federal level during the Great Depression. They emphasized developing sustainable and self-contained cities that linked comfortably with the natural environment. This was, however, eliminated before any substantial progress was made. The first zoning laws also originated during the early 1900s and, although contested, have withstood the test of constitutionality in providing for the general welfare. Although zoning is a means for organizing district conformity, and not planning per se, it has been used in numerous ways to meet planning objectives (Cullingworth 1997).

Gaps materialized routinely between the development of private lands and public infrastructure before laws were adopted to facilitate integration. Even so, gaps still remain common as evidenced by incomplete superhighways, typically placed in areas of dense population with diverse stakeholder concerns, such as in Memphis,

Boston, Baltimore, Washington, D.C., and central New Jersey. Private developments of wood frame houses continue to be built immediately adjacent to fire disclimax ecosystems on public lands. Although local governments have learned from past mistakes, the learning process has been a slowly evolving one. Problems show up especially at times of "natural catastrophe" during flood, storm, earthquake, and fire. Despite all that has been learned, the mistakes of the past continue to catch up with the future and often are repeated.

CONTEMPORARY URBAN LAND-USE PLANNING

Planning Professionals

Contemporary land-use planning is a highly developed discipline with a large professional membership. Much has been learned and published about urban planning. According to Kaiser et al. (1995), effective land-use planning considers all interests while attempting "to integrate land use, transportation, public service, environmental protection, hazard mitigation, public finance, historic preservation, and other related functions into a comprehensive plan that accounts for the connections among these areas." Much of the experience of the regional planning process has been gained through urban land-use management. Kaiser et al. (1995) likened urban land-use planning to a "big-stakes game of serious multiparty competition over an area's future land use pattern." Planning helps create the rules of the game from existing law and agreement of stakeholders in the planning process. Although regional urban planning has been a highly contentious process with mixed success, comprehensive regional planning is gaining influence. The growth of environmental, transportation, and social restrictions at all levels of government has encouraged comprehensive planning (Cullingworth 1997). The alternatives have proven too costly in the long run.

One of the basic objectives of urban land-use planning is that benefits will accrue to some degree to each of the many stakeholders. In providing for those interests, it is the intent of planners to generate a livable environment. This is not an easy task, given the fragmentation of planning authorities, marked differences in attitudes and goals among different planning authorities, and rapid urban growth in many areas. Contemporary urban-use planning now deals routinely with tensions among environmental, market, and social-use activists. Increasingly, the intent is to "strike a balance" among the interests in managing land-use changes through sustainable development. However, there remains great regional variation in commitment to land-use plans that will sustain benefits far into the future.

Nongovernment Stakeholders

In countries such as the United States, where much of the land is privately owned, private interest is the primary motivator of changes in land-use patterns (Kaiser et al. 1995). Among those interests are stakeholders in three main categories: market-value interests, neighborhood resident interests, and environmental interests.

Market-value interests are motivated by profit and are concerned about land-use decisions that will affect their investments. These interests range from the local homeowner concerned about resale value to the large land development organizations concerned about profit for themselves and their shareholders. Because land development itself typically creates a need for adaptive change, development unavoidably influences subsequent development. Neighborhood stakeholders are concerned about sustaining or improving the social values of their neighborhood that make it a satisfying place to live. Most environmental stakeholders are concerned about the protection or restoration of those ecological services that support a healthy, interesting, and aesthetically pleasing environment. The aim of some environmentalists is to stop and even roll back further urban growth, but most environmental groups target more specific concerns.

The challenge of urban planning is to plan land use to benefit all stakeholder interests as much as possible within the conflicting demands placed on land use. Although many interests have been included in most recent urban land-use planning, one of three interest areas usually dominate. Thus the development of certain communities is obviously dominated by real estate markets while other development is dominated by urban environmental concerns and by neighborhood protection, restoration, and rehabilitation.

Government Role in Planning

The Regional Planning Dilemma.   Local government is the primary arbiter for resolving conflict among competing stakeholder interests. However, local governments have only partial authority over regional planning problems. The effectiveness of local urban planning very much depends on the actions of federal and state governments. The diversity of approaches among local governments often subverts any comprehensive planning process that might leap ahead of private land speculation. In small towns and cities, urban sprawl is of little concern. However, the sprawl surrounding larger cities often extends over different municipal, county, and state governments requiring cumbersome intergovernmental planning commissions.

Regional planning has been most successful where federal and state agencies provide assistance or regulation, as for river-basin water supply, navigation, federal highway, and environmental needs. Respecting states rights, the federal government in the United States has done relatively little to regulate urban sprawl, leaving it to local interests to "muddle through." However, the lack of government coordination has aggravated regional land-use problems. For example, federal housing policy inadvertently promoted urban decentralization and associated problems in part to protect the mortgage lenders' interests and to ensure tax collection. The federal investment in superhighway transportation in support of national commerce and security purposes also facilitated local shortsighted planning that contributed to undesirable attributes of urban sprawl. More recently, the federal government has attempted a more positive effect, but usually indirectly through environmental, social, housing, and transportation regulation. The general principle here is to establish national standards for protecting the public interest and allow local flexibility in meeting those standards.

Federal Government.   Federal government effects urban change through regulation, providing grants in aid, withholding grants, and through development of national infrastructure. Federal government has been influential by setting and enforcing national environmental standards, providing transportation funding and regulation, and by supporting community development and public housing. In recent years its influence has moved more toward setting standards that must be met and away from direct funding of corrective actions. This has been transferred to the responsibility of states and local government (Kaiser et al. 1995). The Coastal Zone Management Act of 1972 is one example of how federal legislation influences regional planning when natural resources are being severely threatened by reckless development. The Clean Air Act (1970) and Clean Water Act (1972 and later amendments) are outstanding examples of influence through regulation and funding for municipal treatment systems. Federal government continues to use grants in aid to encourage local decisions that are in the national interest. The Intermodal Surface Transportation Efficiency Act of 1991, for example, encourages links between land-use plans and transportation plans, environmental protection, energy conservation, and maintenance of transportation infrastructure.

State Government. The states are important players whose roles in urban land-use planning has increased in recent years (Kaiser et al. 1995). The states split out into three categories. The first group of states has enough interest in the problems associated with urban sprawl to enact legislation that controls urban growth. This group includes some of the more crowded and fastest-growing states as well as states with strong ecological values such as Hawaii, Florida, Oregon, Washington, New Jersey, Vermont, California, Rhode Island, Maryland, and Georgia (Cullingworth 1997). A second group is debating problems, considering about what to do with them, and developing legislation. This group includes Colorado, Massachusetts, New York, North Carolina, Virginia, West Virginia, and Pennsylvania (Kaiser et al. 1995). Most of these states have legislated some form of comprehensive planning law with goals that municipalities must recognize in their own planning process. However, the effectiveness of these laws varies greatly among states and many are undergoing amendment as the forces of private and public interests vie for satisfactory balance. Most other states either perceive no problems with urban growth, are dominated by philosophies that discourage interference with market-driven development, or are just beginning to feel the consequences of unplanned growth.

Guidance for Integrative Planning. Environmental concerns have encouraged a regional integration approach to protecting natural resources while providing for sustainable development. Early guidelines for environment integration into land use were offered by McHarg (1969), who developed a coherent approach to land suitability analysis in land-use planning. His approach examined basic land and water attributes as foundations for environmental development and protection. The carrying capacity of the urban habitat has been analyzed in an approach to estimating limits to growth (Clark 1981). Planning for environmental concerns has resulted in numerous land preservation or regulated-use decisions to establish and maintain nature preserves, watersheds, groundwater aquifers, flood plains, and buffer areas around pleasing vistas, parks, and other protected lands. Recently the comprehensive concept of sustainable development has emerged (The President's Council on Sustainable Development 1996, Roseland 1998). One goal is to identify the level and intensity of land use that can be sustained without environmental damage while providing for present and future economic and social needs. In this view, the costs of protecting and restoring environmental service associated with ecosystem stability and diversity are most likely to be successfully borne by a healthy economy and high quality of life.

MANAGING THE URBAN ECOSYSTEM

Integrating Urban Ecosystem Services

The most challenging part of natural resource management may exist in the urban landscape. Natural resource professionals are involved in the management of public space that provides the supporting infrastructure for the urban landscape—the highways, pathways, waterways, public railways, greenways (natural vegetation corridors), and recreational areas. Contemporary natural resource concerns often concentrate on the interfaces between urban and wild landscapes where environmental stress is especially acute. This is the arena of regional planners, landscape architects, highway and waterways engineers, environmental engineers and regulators, public health professionals, and urban forest, wetland, fisheries, wildlife, parks, and recreation managers. They usually work in interdisciplinary planning teams to coordinate their areas of specialization into effective sustained use of natural resources.

FIGURE 14.17 A public park that is integrated into a primary business district in Oakland, California (courtesy of U.S. government).

Land

Aesthetics.   Much of the livability in cities is based on the appeal of urban land development, or, as Nasar (1998) refers to it, the city evaluative image. Although the functional performance of the city as a place to live is greatly influenced by the arrangement of structural landmarks, paths, districts, edges, and nodes, the feelings and meanings linked to those structures also are important. The convenience of linkages of courthouses, libraries, schools, museums, monuments, and religion centers by paths (roads and walkways) to residential and business districts is of basic importance. Also important is the degree of transition from one area to another (edge) and the appropriate provision of activity centers (nodes), such as shopping malls, community centers, and entertainment and recreational outlets.

Nasar (1998) found that the "likability" of urban settings depends on a number of criteria including naturalness, upkeep, openness, historic significance, orderliness, and complexity. Most people respond positively to an urban environment that incorporates open vistas, natural areas, and artificial structures in a way complex enough to develop interest, but not so complex as to confuse and threaten. They also like neat and orderly neighborhoods with a sense of continuity among structures including identifiable belonging to a historic period. Aesthetic evaluative techniques need further improvement, but show promise for providing needed input information to planning process. Desirable urban land-use patterns need to include aesthetic appeal. This means integrating natural ecosystems, much through vegetation management, into urban landscapes (Figure 14.17).

Transportation.   Some of the most heavily used public resources exist in transportation right-of-ways. Congestion alone is estimated to cost $40 billion annually in lost benefits (Cullingworth 1997). Transport and utility corridors linking urban ecosystems form extensive ecological edges and barriers. They may be developed and managed to promote the protection of resources, such as habitat for endangered species, watershed, and wetlands, or they may be haphazardly developed and mismanaged to social detriment. Transportation right-of-ways typically create sharp changes in vegetation and landform, often fragmenting wildlife habitat. They also form conduits for toxic spills, dissolved contaminants, and suspensions of eroded particulates into streams, lakes, wetlands, and coastal zones.

Vehicular collision with wildlife escalates where human travel corridors intersect with wildlife travel corridors, posing threats to both wildlife and people. Design that provides safe alternatives for both drivers and animals is an important aspect of contemporary right-of-way management. Development of natural corridors, or "green belts," in urban areas for recreation and other ecological services intensifies the need for careful management of undesirable encounters between nature and humanity. Storms, floods, and other natural stresses frequently exert the greatest damage along travel corridors. The management of natural resource degradation along transport, utility, and natural corridors starts with comprehensive planning based on understanding of the many interfaces between the urban ecosystem and more natural ecosystems.

Natural resource professionals associated with transportation often are involved in meeting resource protection requirements. Through auspices of the National Environmental Protection Act (1969), any projects funded with federal money need to determine the environmental impact and seek the least environmentally degrading choice within similar cost ranges. The Endangered Species Act of 1973 and similar state laws require planning around listed federal and state endangered species. The Clean Water Act of 1972 and later amendments established provisions for protecting national waters and wetlands. It requires biologists, hydrologists, and soil scientists familiar with ecosystem processes to assure regulations are met on all highway development. All unavoidable wetland damage must be compensated through creation or restoration of similar wetlands. Some states have even more stringent laws.

The Intermodal Surface Transportation Efficiency Act of 1991 requires the states to prepare their own transportation plans and integrate environmental considerations into the plan. The plans must address issues such as congestion management, land-use effects of transportation, consistency among transportation and land-use plans, and environmental, energy, and social effects of transportation development. One important section of the act restricts federal funds for highway development where the Clean Air Act is violated. Methods for reducing automobile traffic have to be proposed. Through that act, highway funds are available for mitigating impacts on wetlands and other wildlife habitat, preservation of historic sites, air quality improvement, highway beautification, construction of bicycle and pedestrian facilities, and alternative forms of ground transportation. To get the money, a 25-year transportation plan must be developed.

Urban Outdoor Recreation and Natural Aesthetics.

Urban outdoor recreation includes many of the natural resources services usually associated with park development and management. A short list of recreational opportunities includes development and management of facilities for public swimming, boating and marinas, bicycling, hiking, dog walking, fishing, camping, golf, tennis courts, baseball, football, basketball, lawn bowling, skating, running, archery, horseback riding, motorized off-road vehicles, sight-seeing, nature study, picnicking, and lounging. Many larger park systems attempt to provide most of these opportunities in addition to maintaining heavily used traffic arteries.

The primary challenge for recreational managers is how to develop limited space for diverse use in a safe environment. Many outdoor activities are inherently dangerous to nonparticipants who find themselves in the path of a vehicle, golf ball, boat, arrow, skater, bicycler, or nervous dog. Recreational planners work closely with landscape architects and construction engineers to place tracks, paths, roads, courts, rinks, links, diamonds, lawns, restrooms, benches, waterways, service buildings, vegetation screens, earth barriers, bridges, various service concessions, and other structures in ways that minimize collision of uses while maximizing user satisfaction as a whole. Facilities need to be developed with enough challenge to satisfy skilled users while remaining safe enough for novices. Regulation and direction are unavoidable necessities. Signs need to be obvious yet unobtrusive, informative but concise, and universally understandable for multilingual users.

A variety of specialists are involved with recreational management including urban foresters, horticulturists, fisheries and wildlife managers, environmental scientists, educators, sports trainers, law enforcement specialists, architects, and engineers. They typically provide services for a network of parks and related recreational facilities. Recreational managers often are involved in urban planning. Because air and water quality are especially critical in park settings, they often are trained in environmental assessment.

Urban Wildlife Management.
In barren city centers and industrial areas before the turn of the last century, virtually all vegetation was trampled, cleared away, burned for fuel, or otherwise used. Wildlife were reduced to a few highly tolerant undesirable species. Most of these species evolved with people in Eurasia, where they ate their grain and castoffs and nested in or on their buildings. Prominent among them were rock doves (pigeons), house sparrows, starlings, house mice, and Norway rats. They were often joined by feral domestic cats and dogs, which adapted to the frequently abundant rodent and bird populations.

These few species of urban wildlife flourished especially before cars and trucks replaced horses, whose feces contained digested grain that formed the basis of a food chain. Waste grain and other foods also were common at granaries, wharfs, railroad yards, stables, markets, and city dumps. As medical understanding of disease transmission improved, especially for such worrisome diseases as bubonic plague, urban wildlife became more of a concern. Cities began to spend public money in education and control programs. Even after World War II, urban wildlife management frequently connoted public health management.

As open land and natural vegetation has become more integrated into the urban environment, some native forms of wildlife have adapted to the urban scene. Certain wildlife species have always been well adapted to the farm-village habitats of humanity wherever some semblance of natural habitat persisted. However, before hunting was regulated, many species were less common near human habitation than now. Rapid suburbanization following World War II resulted in increased encounters between people and wildlife, especially where large areas of natural vegetation remained among homes and developments. In addition to the Eurasian imports, suburban backyards frequently are visited by numerous native birds including robins, house finches, mockingbirds, chickadees, titmice, cardinals, crows, grackles, mourning doves, hummingbirds, and more.

Where wild parklands enter city centers, especially along waterways, many other species may be encountered. These commonly include rabbits, squirrels, skunks, weasels, opossums, raccoons, fox, coyotes, deer, muskrats, various bats and small rodents; geese, seagulls, ducks, herons, and other birds; turtles, snakes, lizards, frogs, and salamanders; and a great diversity of insects and other invertebrates. Where developments have included boat slips and ponds in otherwise suitable habitat, alligators, water snakes, and aggressive turtles are increasingly encountered. Some of the most desirable homes back up to parks and other wild space. In such areas, even wilderness wildlife such as bears, mountain lions, and moose may be encountered.

Urban wildlife have become very abundant in some locations—a boon to many and a nuisance to others. A number of factors have contributed to their increase. Because "green space" is desirable, especially when it threads through cities and connects to wilder lands at city edges, a greater diversity of urban wildlife has become more common. Large areas of "slum" housing were cleared and never redeveloped in some locations. Abandoned industrial sites have colonized with natural vegetation following the closure of many factories. Lethal applications of pesticides are less commonly used now and hunting is outlawed in most urban and suburban locations. Most cities require tight control of dogs.

In addition to the interest they add, urban wildlife cause accidents, damage yards and homes, transmit disease, inflict bodily harm, and frighten some people. Large animals, such as deer and alligators, cause traffic accidents that are sometimes lethal. Beaver have become common once again after near total extirpation in the eastern United States. Skunks, raccoons, and opossums are frequent visitors to Fido's food bowl on the back porch, where they dump garbage cans and sometimes harm small pets. Even larger pets are sometimes killed and consumed by coyotes, mountain lions, and alligators on suburban margins. Burrowing muskrats break through pond liners and drain them. Moles, muskrats, and mice burrow into lawns. Rodents, deer, and rabbits eat gardens. Children try to catch animals of all kinds, sometimes getting bit in the process. Rabies, although a relatively rare threat, is a continuing concern. Canada geese defecate all over park and yard lawns. Birds are sucked into jet aircraft and may occasionally cause a crash. Poisonous snakes are encountered in new developments with a lot of wildland. Even harmless snakes remain a phobia for many people.

The urban wildlife manager uses many tools to develop compatibility of wildlife and human needs. Among the most effective is appropriate design of urban space, which uses knowledge of wildlife and habitats to confine wildlife to desirable areas. Education and, where necessary, regulations are important means for assuring that people behave responsibly with respect to waste, pet, and yard management affecting wildlife pest problems. As a last resort, wildlife management requires direct control by removal of offending wildlife using safe and ethical methods.

Land Wastes.   One of the more vexing problems of urban areas is solid waste disposal (Hough 1995). Solid wastes have increased at a rate greater than the growth in consumer purchasing power. Much of the solid waste problem derives from a lifestyle that places a premium on time savings. Because most people in households now work outside the home there has been a disproportionate increase in preprepared and packaged foods. Elimination of disposable diapers alone is estimated to cost $4 billion yearly, nearly equal to their entire market value (Cullingworth 1997). In addition to the metal, plastic, paper, cardboard, wood, garden, and food wastes generated by the average household, industrial and infrastructural wastes require environmentally compatible disposal. Debris from demolished buildings, roads, and utilities needs to be recycled or eliminated. Fly ash from coal-fired power plants, slag wastes from various industrial processes, and sewage treatment sludge all need disposal. The primary ways to dispose of materials include land dumping, ocean dumping, incineration, and recycled use.

Environmental concerns have greatly altered the way solid wastes are managed. The burning dumps that commonly blackened urban fringes through the 1960s are no longer tolerated under modern environmental laws protecting land, water, and air. Most dumping in marine environments is prohibited except where materials might contribute to development of desired structure, such as artificial reefs. Some landfills used to bury chemical wastes now require expensive treatment or isolation under superfund legislation. Landfills need to be managed to assure groundwater and surface water quality is not degraded. Restored land use must be compatible with the conditions in the landfill (Freedman 1995, Cullingworth 1997). Space suitable for landfills is rapidly disappearing in many urban areas. The NIMBY (Not In My Back Yard) phenomenon is common. Some eastern municipalities ship garbage to western rural locations where space is sold for landfill use. Even rural disposal has become controversial in many locations.

Incineration is safe for uncontaminated organic matter. One of the major problems with incineration of sewage sludge is contamination by toxic metals such as mercury, which can be dispersed through the atmosphere (Freedman 1995). Use of sewage sludge for agricultural soil enrichment is controversial because microbial copper, zinc, cadmium, and nickel contaminants could exist (Freedman 1995). Yard and park organic matter contributes much to solid waste and often is intensified by overuse of fertilizers (Hough 1995).

Recycling has greatest utility where there are associated savings. Recycling is especially profitable for metals, glass, and to a lesser extent for paper products. It also makes sense to recycle cuttings and leaf litter from yards and parks. The old automobile junkyards that once dotted the landscape have for the most part disappeared as the scrap has become valuable for metal recycling. Aluminum cans and whole glass containers are much less commonly encountered in litter along highways because of their recycling value. However, for the most part, the bulkiest nonorganic matter remains problematic. In some urban areas that have instituted recycling programs, material is accumulating in storage areas faster than demand can reuse it. However, major advances have been made in recycling and more can be expected in the future.

Hazardous waste is the most difficult form of land disposal problem to manage. Discovery of contaminants leaking into basements from an old waste pit at Love Canal in Niagara Falls, New York, precipitated wider investigation of such conditions around the country. There was, it turned out, no easy way to determine how many such sites existed. Congress passed what has been popularly known as the Superfund Act in 1980 to deal with this problem. It authorized the EPA to develop a priority list starting with 400 sites. Now there is careful documentation of hazardous waste production, use, and disposal designed to avoid future problems. One of the earliest results of the Superfund Act was closure of many landfills too costly to comply with federal law. Instead wastes were treated with the aid of funding under a 1984 amendment. One of the critical provisions of this law is that those responsible for improper hazardous waste would also pay for its cleanup. Since cleanup costs can reach billions of dollars, this has led to contentious and costly legal battles. One of the loopholes in the present law is that household wastes are excluded from the hazardous waste category. This makes hazardous materials typical of most households a problem for waste management.

Water

Water Supply.   Water supply is as important to city ecosystem function as it is to natural ecosystems. The influence of urbanization on resources far beyond city limits is no where clearer than in managing water supplies for navigation and consumption. Water is used for many domestic, industrial, and outdoor purposes consuming hundreds of gallons per capita per day. Water is needed for cooking, cleaning, drinking, industrial processes, food preparation, irrigation of lawns and gardens, fire fighting, cooling, waste removal, transportation, recreation, and sustaining natural aquatic ecosystems. Clean local sources of water are valuable because water purification, or importing it from afar, are expensive (Perry and Vanderklein 1996). Thus water management has been a dominant theme of urban resource management nearly as long as people have inhabited cities.

Water management remains inadequate in many locations of the world. Over half of the world's population may be affected by water-borne diseases resulting from poor water supply management (Perry and Vanderklein 1996). Even though the worst problems are in underdeveloped countries, water supply continues to rank high among urban resource management problems in industrial nations. Water uses for transportation, waste dilution, and consumption become particularly intense in city areas. Therefore, the acquisition of clean water usually requires that cities go to remote sources.

Most sources of water contamination are local, but in regions of sprawling suburbia and other intense land use, contamination may originate in remote parts of watersheds. The cumulative effects of rural and urban water contamination have increased costs of treatment and acquisition of clean water. Among the more difficult problems caused by navigation is contamination from oil and derivatives, most of which occur from normal traffic. Occasionally disastrous large spills occur, such as the *Exxon Valdez* spill in Alaska coastal waters, which had widespread economic and ecological impacts. Millions of gallons of fuel are transported annually on railroads and roads to urban destinations and threaten spills into lakes, rivers, and estuaries.

Water dirtied by inorganic and organic sediment needs to be treated by settling and filtration through beds of fine sand or other media (see Chapter 7). Disinfection is done by chemical oxidation, usually with chlorine or ozone. Culturally eutrophied water often has a fishy or musty taste. This is difficult to remove through physical and chemical engineering, and prevention in the first place is often more cost-effective. Water that is contaminated with metals, nitrates, arsenic, and syn-

FIGURE 14.18 Created wetlands in urban areas are increasingly being used to provide recreation, wildlife habitat, and water treatment facilities, often with several integrated services (courtesy of U.S. Department of Agriculture).

thetic compounds requires additional expensive treatment through chemical precipitation and specific filtration and absorption media. Increasingly, contamination from organic compounds, such as gasoline and oil, is being treated with bacteria that decompose the contaminants to harmless elements.

Urban water management is greatly influenced by federal and state laws. The federal Clean Water Act of 1972 and later amendments have been especially important for establishing the ultimate objectives of water quality management and regulation. The act seeks "restoration and maintenance of the physical, chemical, and biological integrity of the nation's waters." This act has had positive influence, mostly through regulation and financing massive development of waste treatment plants. Still, many plants are inadequately funded for optimum operation (Cullingworth 1997, Adler 1993). The act also established regulation of wetland filling, which minimizes the overall negative impact on wetlands. Some cities have integrated wetland maintenance with water treated by managing the natural capacity of wetlands to trap nutrients and other wastes (Figure 14.18). Because many urban areas are situated near water and wetlands, the Clean Water Act will have an important impact on future land use and urban development.

Groundwater. Because it has been naturally filtered and only rarely contaminated by natural sources of pollutants, unimpaired groundwater is especially valuable wherever it is close to the surface and in large aquifers. However, many groundwater aquifers have been contaminated, especially near the ground surface (Perry and Vanderklein 1996). Drilling and pumping from deeper groundwaters are more expensive. Demands on local groundwater and surface water have exceeded supply in many large metropolitan regions, requiring conservation practices, recycling, and importation of clean water from outside city limits. Overpumping at rates faster than new water can percolate into the aquifer results in lowered water tables, higher pumping costs, and eventual use of the entire available aquifer. Overpumping near ocean coasts results in salt water intrusion, which is costly to desalinate. It can cause land subsidence leading to building and infrastructural damage. Other sources of salinity contamination of groundwater include improper use of deicing salts and agricultural irrigation in dry environments. Overpumping also can dry up wetlands and springs that serve diverse needs, including support of threatened and endangered species. Unlike federal regulation of surface waters, groundwater regulation is left up to the states.

Chemical contamination of groundwaters in urban areas can come from many sources (see Chapter 7). Common ones are from septic tanks, industrial waste disposal lagoons, urban landfills, mines, buried storage tanks for chemical wastes, fuels from gas stations and home heating, and chemical spills of diverse kinds. Hydrocarbon compounds are the largest type of pollutant because they are so widespread in our internal-combustion culture. Nitrate fertilizer seepage from local lawns or nearby agricultural use also is a source of groundwater pollution. Sewage, gas, and other pipes collapse and leak, especially in older urban areas. Bacterial contamination from broken pipes, old septic systems, past landfills, and graveyards is a major problem in some cities. Abandoned and poorly maintained wells can be a route of contaminated flow into groundwaters. Aquifer protection is an important concern for many urban areas. They usually approach these problems through setting standards for quality, regulating land management practices, and forcing polluters to pay cleanup costs. Especially sensitive aquifers include those that lie very close to the surface, are in karst formations of fissured limestone, and areas of volcanic rock.

Surface Water.   Another cheap source of clean water is from groundwater that has emerged through springs into streams and lakes. Many urban areas long ago contaminated their local surface waters. Land purchase and water transport to the city was cheaper than intensive clean up. Because much of New York City is on impermeable rock with little groundwater, it has purchased many acres of undisturbed watersheds and water supply reservoirs. Urban water-supply watersheds and reservoirs usually are protected from any degrading use and managed for compatible uses, such as wildlife refuges.

But many urban areas have little recourse other than to use waters that have been affected by local land use. Thus waste treatment required in the Clean Water Act contributes to water treatment savings in downstream communities. Threats to continuous supply of urban water during droughts have increased as regional urban development has intensified. Another reason for investing in groundwater for urban use is its dependability during drought. For those many areas of the country that must augment municipal water supplies, the protection of water yield from watersheds is critical.

In arid lands of the western United States, cities compete for water with agricultural users. Water law in most western states treats water as a private good associated with irrigable land (see Chapter 7). Cities have bought control of watersheds or the purchase of water rights mostly from agricultural users. Los Angeles has been particularly dependent on water produced elsewhere. It began buying and diverting water from the Owens River valley some 200 miles away early in the twentieth century. Resulting problems with dust pollution in what used to be Owens Lake may ultimately result in a return of some water to control dust pollution. This follows a California Supreme Court decision with respect to Los Angeles' right to acquire water rights and divert water that would damage the Mono Lake ecosystem, a unique wildlife habitat. In so doing, the court said all water rights remained subject to review and revision to assure the public trust is protected (Getches 1993).

Waste Assimilation, Transport, and Contamination.   Untreated wastes used to be dumped directly into streams, lakes, and ocean coastal waters. Systems of drains were developed in many cities simply to collect sewage for dumping directly to natural waters without treatment. With growing medical recognition of waste hazards in the last century, urban sewage systems began to appear in larger cities. In more rural areas, individual septic systems were used to treat wastewater. However, as rural population density intensified, leakage and contamination from individual systems became a widespread problem, requiring extension of centralized treatment systems into the urban sprawl. By the 1920s and 1930s, there was growing scientific awareness of the extent of water contamination from organic wastes. Many miles of urban and industrial waters were devoid of intolerant organisms including virtually all fish. The first treatment plants were designed to remove organic waste and to chlorinate for control of pathogens, but much of the organic matter remained in the wastewater. Because organic waste treatment focused on oxygen restoration, the Biological Oxygen Demand (usually referred to as BOD) was used as the gauge for treatment need and success.

After World War II industrial growth and increasing use of chemicals for various purposes, including fertilizer applications, contributed to rapidly worsening water quality. Conditions in some urban river settings were so bad from oil pollutants that they caught fire—the most infamous example occurring in downtown Cleveland, Ohio. Inorganic nutrients remained a problem in treated sewage and in urban runoff, contributing to cultural eutrophication (nutrient enrichment). This caused excessive water "blooms" of phytoplankton, many of which were foul tasting, somewhat toxic, and further depleted oxygen. Following the Clean Water Act of 1972, the United States embarked on a mission of complete cleanup, but was most successful with point sources of such pollutants. The EPA administered a construction program for sewage treatment plants with greater capacity to remove BOD and inorganic nutrients, such as phosphorus and nitrogen. In some locations additional treatment is needed to protect pristine environments such as Lake Tahoe in Nevada. Wetlands have been shown to be effective in waste water "polishing" (Mitsch and Gosselink 2000). Although there has been much improvement in waste removal from natural waters, contamination still remains and many of its sources are urban.

Urban Runoff. Surface runoff from rain and snow in over half the urban areas is drained away from the sewage treatment systems to natural drainages (Perry and Vanderklein 1996). Runoff from streets, parking lots, roofs, and lawns is contaminated with a variety of nutrient and toxic materials. Lawn pesticides and fertilizers, waste oil and gasoline, road salts, vehicle emission particles, various solvent cleaning agents, inorganic suspended solids, heavy metals, and various pathogens all are flushed into storm drainage systems. Combined storm water and sewage systems exist for about 20% of the U.S. population compared to more than half in Europe. These systems are rarely built with the capacity to contain excessive storm runoff. The treatment plants often spill untreated sewage. Whenever it rains over 0.75 inches in New York City, for example, 500 million gallons of untreated sewage is flushed into the receiving waters (Perry and Vanderklein 1996).

The most common approach to managing urban runoff is detainment in settling basins and to encourage infiltration into the soil. These need to be large enough to hold water for a day or more to have significant settling effect (Perry and Vanderklein 1996). Percolation basins also risk groundwater contamination if not carefully placed. A strategy now gaining increased attention relies more on the natural filtration service of the urban watershed by reducing the extent of continuous impervious surface using porous concrete, grass swales in place of concrete gutters, and more natural material spaced intermittently with concrete surfaces.

Waste Heat. Thermal pollution occurs when heated water is released to the environment in a way that reduces the valued functioning of aquatic ecosystems. The largest source of waste heat is from cooling systems of steam-generated electric power plants. Cooling water effluent is either mixed rapidly into large water bodies or is enclosed in cooling ponds to avoid exposure of natural ecosystems. Cooling water ponds often require barriers of various kinds to prevent fish impingement and death. Attempts have been made to use heated water for aquaculture and other low-grade uses of heat, but with marginal success. Cooling ponds are more readily included in some forms of recreational use, such as fishing and waterfowl management.

Flood Damage Control and Navigation and Reclamation. The U.S. Army Corps of Engineers (the Corps) is the lead agency authorized to promote river, harbor, and intracoastal several degrees above that of surrounding natural forest. In contrast, a suburban area in a desert biome may be cooler than the natural ecosystem. Nonvegetated urban areas typically have lower humidity than vegetated areas (Landsberg 1991, Bridgeman et al. 1995). Integration of vegetation and water may provide humidity and air movement closer to optimum levels for human activity than surrounding drier or wetter ecosystems. By supplying water to desert urban landscapes, the resulting trees, gardens, lawns, and surface water cool and humidify the urban landscape. Because of the space dedicated to concrete, asphalt, and other building materials in a wetter urban landscape, the city is warmer but possibly less humid than surrounding forest ecosystems, depending on airflow through the city.

Auxiliary Forms of Energy. Solar energy provides light for an average of half a day—an inconvenience for an animal that requires about eight hours of sleep and cannot see well in the dark. Also, the cavelike structures that humans typically inhabit often are dark even during daylight. Until the invention of electric light by Edison about a century ago, auxiliary light was provided by an open flame. Flames enclosed in furnaces and stoves have been effective means for heating buildings and water, generating steam, and cooking foods. They continue in such service fueled mostly by natural gas, coal, oil, and other fossil fuels, which must be imported from remote sources by the city. Industrial and home use of oil, wood, and coal for heating caused massive emission of incompletely burned particulates until conversion to natural gas mostly during the last few decades.

Open flames are inefficient sources of light and risk fire much more than electric light, which was adapted to city illumination soon after invention. Electricity also was an effective power supply for household appliances. Electric fans were among the more important inventions for encouraging greater evaporative cooling. Two of the most important appliances are electric refrigerators and air-conditioning. They did not become widely affordable until after World War II. They accelerated growth of the southern cities in the United States as urban growth slowed in the northern cities. Before World War II, the icebox was a primary means for storing food. Electricity is now so widely used to power appliances, it is difficult to imagine household life without it. The urban demand for electricity has increased exponentially since its invention.

Atmospheric Contamination. Boorstin (1973) recounts an exuberant promotion of the purity of Los Angeles' air in 1874. How things have changed! Before improvements in the 1960s and 1970s, air quality in many urban areas often was unhealthy and sometimes deadly (Figure 14.20). It was not uncommon in some of the industrial cities of the northeastern United States for street lights to remain on most of the day because of the darkness caused by aerial particulates (Freedman 1995). The widespread burning of coal especially contributed to particulate pollution as well as to gaseous emissions. Coal heated homes and workplaces, provided fuel for electric power generation, and was the main energy source for heavy industry and the railroads. Particulate conditions have much improved in the United States as employment shifted away from heavy industry to light industry; cleaner atomic energy, fuel oils, and gas replaced coal; and regulation of atmospheric emissions became more technically feasible and effective. The particulate atmospheric problems of a half century ago in the United States still persist in developing countries. Heavy industry is now concentrated in these countries and they continue to rely on dirty fuels with little regulation of emission.

The demand for cleaner sources of power for electric generation and heating, such as wind and solar energy, was thought to be incentive for altering the scope of cities, returning them to walking size (Goldfield 1977). However, recent history indicates no such trend. The emphasis instead has been on centralizing electricity production in large power plants and cleaning up their emissions. Other conservation measures include extending natural gas supply plumbing to households and workplace, and emphasizing energy conservation through building codes and economic incentives. Power plants, factories, and automobiles are the major sources of air quality problems.

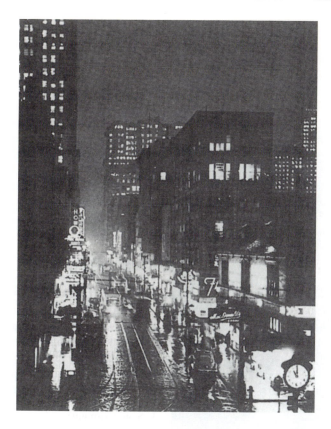

FIGURE 14.20 Pittsburgh at 11:00 A.M. in 1945 darkened by industrial air pollution. Clean air legislation following World War II eliminated this extreme form of particulate air pollution (courtesy of Allshevy Conference in Community Development).

The major gases produced are carbon monoxide (CO), carbon dioxide (CO_2), sulfur dioxide (SO_2), hydrogen sulfide (H_2S), ammonia (NH_3), nitric oxide (NO), nitrous dioxide (NO_2), nitrous oxide (N_2O), ozone (O_3), and various hydrocarbons. Carbon monoxide ranks highest among U.S. toxic emissions. It is a highly toxic by-product of fossil-fuel combustion, which is oxidized rapidly to carbon dioxide and rarely becomes life threatening in the open environment. Carbon dioxide is generated from combustion of organic matter in any form and from respiration of plants. It is also absorbed by plants and fixed in biomass. Thus, widespread removal of plant biomass contributes to the reservoir of carbon dioxide in the atmosphere. Global emissions of carbon dioxide from fuel combustion have increased by a factor of more than 40 since 1860 (Freedman 1995). The reduction in forest biomass and combustion of carbon dioxide have about equally contributed to the observed increases (see Chapter 6).

The accumulation of carbon dioxide is believed to be causing the greenhouse effect (heat increased in the atmosphere), resulting in rising global temperatures (Freedman 1995, Chapter 6). Other gasses, mostly from urban sources, may also contribute to the greenhouse effect. There remain many unanswered questions about the link between greenhouse gasses and temperature change on the earth's surface. Depending on the extent to which the earth actually warms from future green house gas emission, sea levels, agricultural production, natural fire occurrence, biodiversity, and many other earth attributes could be affected.

Sulfurous gasses contribute to acid rain and are toxic enough to slow plant growth. The sulfur gasses are emitted from fossil fuel combustion. Anthracite coal has a particularly high concentration and western softer coals have lower contents (see Chapter 19). Oil products have somewhat lower concentrations. Power plants and manufacturing are particularly important sources of sulfur gasses although some sulfur is removed from smokestack effluents to prevent formation of acid rain. Concentrations of sulfurous gasses average about 500 times greater in urban areas than in remote locations (Freedman 1995). Because of high emission stacks and changing winds, the effects of emission gasses are typically dispersed over relatively large areas.

Bridgeman, H., A. Warner, and J. Dodson. 1995. *Urban biophysical environments.* New York: Oxford University Press.

Caro, R. A. 1974. *The power broker: Robert Moses and the fall of New York.* New York: Vintage.

Clark, J. 1981. The search for natural limits to growth. In Judith de Neufville, ed., *The land use policy debate in the United States.* New York: Plenum Press.

Committee on Science, Engineering, and Public Policy. 1992. *Policy implications of greenhouse warming.* Washington, DC: National Academy Press.

Cullingworth, B. 1997. *Planning in the U.S.A.: Policies, issues and processes.* New York: Routledge.

Daily, G. C., ed. 1997. *Nature's services: Societal dependence on natural ecosystems.* Washington, DC: Island Press.

Dramstad, W. E., J. D. Olson, and R. T. T. Forman. 1996. *Landscape ecology principles in landscape architecture and land-use planning.* Washington, DC: Island Press.

Fowler, E. P. 1992. *Building cities that work.* Montreal, Quebec, Canada: McGill-Queens University Press.

Freedman, B. 1995. *Environmental ecology: The ecological effects of pollution, disturbance, and other stresses.* 2nd ed. New York: Academic Press.

Fries, S. D. 1997. *The urban idea in colonial America.* Philadelphia, PA: Temple University Press.

Garvin, A. 1996. *The American city: What works, what doesn't.* New York: McGraw-Hill.

Getches, D. H. 1993. Water resources: A wider world (pp. 124–161). In Lawrence J. MacDonald and Sarah F. Bates, eds., *Natural resources policy and law: Trends and directions.* Washington, DC: Island Press.

Goldfield, D. R. 1977. Recovering the american city: Minneapolis as a case study (pp. 456–497). In J. Cairnes, Jr., K. L. Dickson, and E. E. Herricks, eds., *Recovery and restoration of damaged ecosystems.* Charlottesville, VA: University Press of Virginia.

Hough, M. 1995. *Cities and natural process.* New York: Routledge.

Houghton, J. T., L. G. Meira Filho, B. A. Callander, N. Harris, A. Kattenberg, and K. Maskell, eds. 1996. *Climate change 1995: The science of climate change.* New York: Cambridge University Press.

Hudson, B. J. 1996. *Cities on the shore: The urban littoral frontier.* New York: Pinter.

Jakle, J. A. 1990. Landscapes redesigned for the automobile (pp. 293–310). In M. P. Conzen, ed., *The making of the American landscape.* New York: Routledge.

Judd, D. R. 1988. *The politics of American cities: Private power and public policy.* 3rd ed. Glenview, IL: Scott Foresman.

Judd, D. R. and T. Swanstrom. 1994. *City politics: Private power and public policy.* New York: HarperCollins.

Kaiser, E. J., D. R. Godshalk, and F. Stuart Chapin Jr. 1995. *Urban land use planning.* 4th ed. Urbana, IL: University of Illinois Press.

Kunzler, J. H. 1996. *Home from nowhere: Remaking our everyday world for the twenty-first century.* New York: Simon and Schuster.

Landsberg, H. 1981. *The urban climate.* Volume 28, International Geophysics Series. New York: Academic Press.

Langdon, P. 1994. *A better place to live: Reshaping the American suburb.* Amherst, MA: University of Massachusetts Press.

Lemann, N. 1991. *The promised land: The great black migration and how it changed America.* New York: Alfred A. Knopf, Inc.

McHarg, I. 1969. *Design with nature.* Garden City, NJ: Doubleday.

Mitsch, W. J. and J. G. Gosselink. 2000. *Wetlands.* New York: John Wiley & Sons.

Morris, A. E. J. 1994. *History of the urban form: Before the industrial revolution.* New York: Longman Scientific and Technical and John Wiley and Sons.

Muller, E. K. 1990. The Americanization of the city (pp. 269–292). In M. P. Conzen, ed., *The making of the American landscape.* New York: Routledge.

Mumford, L. 1961. *The city in history: Its origins, its transformations, and its prospects.* New York: Harcourt Brace (Penguin edition).

National Research Council. 1992. *Restoration of aquatic ecosystems.* Washington, DC: National Academy Press.

Nasar, J. L. 1998. *The evaluative image of the city.* Thousand Oaks, CA: Sage Publications.

Perry, J. and E. Vanderklein. 1996. *Water quality: Management of a natural resource.* Cambridge, MA: Blackwell Science.

Roseland, M. 1998. *Toward sustainable communities: Resources for citizens and their governments.* 2nd ed. Stoney Creek, CT: New Society Publishers.

Smith, Z. A. 1995. *The environmental policy paradox.* 2nd ed. Englewood Cliffs, NJ: Prentice-Hall.

Steiner, F. 1991. *The living landscape: An ecological approach to landscape planning.* New York: McGraw-Hill.

The President's Council On Sustainable Development. 1996. *Sustainable America: A new consensus for prosperity, opportunity, and a healthy environment for the future.* Washington, DC: U.S. Government Printing Office, Superintendent of Documents.

Ward-Perkins, J. B. 1974. *Cities of ancient Greece and Italy: Planning in classical antiquity.* New York: George Braziller.

Zelinsky, W. 1990. The imprint of central authority (pp. 311–334). In M. P. Conzen, ed., *The making of the American landscape.* New York: Routledge.

FIGURE 15.1 Restoration of wolves in areas where they were extirpated has been contentious because of the differing values attached to wolves by sectors of the American public (courtesy of Lynn Starnes).

population protection to protection of the individual animal. They oppose the use of animals in research, the fur industry, recreational hunting, or other activities that might result in animal pain, suffering, and death.

Although most Americans no longer depend on wild animals for their livelihood, wildlife continues to be very much a part of their cultural, political, judicial, and economic life (Kellert 1996). The common controversies caused by wildlife derive from an incomplete understanding of wildlife management needs by people who are both positively and negatively affected by wildlife. This incomplete understanding derives from both economic and ecological ignorance and sometimes results in policy inconsistency. Wildlife are controversial in part because human relationships with wildlife have changed rapidly and unevenly across society. Conflicts can even arise between groups holding positive but different wildlife values such as hunters and animal rights groups, as well as between groups in conflict over the benefit and the damage associated with wildlife. Recent suburban sprawl into wildlife habitat has established a new awareness, respect, and fear for the damage wildlife can cause (see Chapter 14). Because wild animals are a publicly owned resource in the United States, the resolution of conflicts over them typically involves government employees charged with their management (Grosse 1997).

Wildlife Authority

Most of the laws now pertaining to hunting and other wildlife-related regulations are set by the respective states, which have jurisdiction over resident wildlife. State wildlife agencies are usually directed by a commission of one to several members, usually appointed by the governor. The commissioners oversee the administrative organization of the state wildlife agencies, which are headed by directors typically hired by the commission. The states set seasons, limits, and license fees for harvesting game birds, mammals (including furbearers), and fish.

Federal agencies have regulatory powers over migratory birds, managing national refuges, coordinating endangered species programs, administering federal aid to states, and negotiating international wildlife agreements. The lead federal wildlife agency is the U.S. Fish and Wildlife Service within the Department of the Interior. One of its important responsibilities is managing the 452 federal refuges located in 49 states with a total area of 92 million acres. Refuges had 30 million visitors in 1996 (Wildlife Management Institute 1996). The National Marine Fisheries Service within the Department of Commerce is the major agency responsible for conducting research on and managing marine vertebrates (fishes, sea turtles, whales, and seals, among others) and marine invertebrates. It is the lead agency in managing offshore wildlife resource development. Most of the wildlife on other federal public lands are the responsibility of the Bureau of Land Management (Department of the Interior), the U.S.

Forest Service (Department of Agriculture), and the Department of Defense. However, state regulations regarding harvest and other take also apply to federal public lands (Grosse 1997, Bean and Rowland 1993).

WHAT IS WILDLIFE?

The Professional Concept of Wildlife

Wildlife in the broadest sense includes all wild animals. This definition excludes domestic animals—although feral animals (domestic animals that survive without the direct assistance of humans), such as horses and burros that roam freely in public lands, are a management concern of wildlife biologists. Some biologists extend the concept of wildlife to plants; however, the usual usage restricts the term wildlife to terrestrial vertebrates (most mammals, birds, and reptiles) and semiaquatic vertebrates (amphibians, waterfowl, walruses, otters, muskrats, beavers, and seals, among others), whereas fisheries refer to aquatic animals harvested for food, sport, and other uses such as fish, crustaceans (shrimp, crayfish, lobsters), mollusks (squid, clams, oysters), marine turtles, and cetaceans (whales, dolphins, and porpoises). Wildlife typically refers to those vertebrate species that depend on atmospheric sources of oxygen in at least the adult life stage. In practice, wildlife management has been reserved for terrestrial ecosystems while fisheries management addresses aquatic ecosystems. The two areas of management focus overlap at the water's edge where many wildlife species use aquatic resources.

Early government concepts of wildlife were closely associated with commercial and sport hunting values. Some state agencies continue to be called "game and fish" departments, reflecting the continuing influence of sport hunting in the wildlife concept. However, the Endangered Species Act, passed in 1973, greatly changed how wildlife is defined in professional practice and in the public vernacular. This law defined the term fish or wildlife as any member of the animal kingdom, with certain exceptions. Insects classed as pests with "overwhelming and overriding" risks to humanity were exempted from the act. States have also had to expand their protection authority to a broad inclusion of many species. Some states have assumed management authority over all wild vertebrates and an increasing number of invertebrates and wild plants. However, domestic animals and privately owned animals usually are exempt, with certain exceptions. Privately owned elk, other deer, and other native animals, normally free-ranging, continue to come under a wildlife authority in many state agencies. Owners must have a permit to keep them captive and sell them for meat and other uses. Exotic animals that have become naturalized in the wild and are considered game species, such as the ring-necked pheasant, are typically placed under a wildlife (or game) authority. The Wild Horses and Burros Act of 1971 extended federal protection to "wild free-roaming horses and burros" because a large portion of the public viewed them as part of our national heritage despite the damage they might cause to native wildlife habitats (Grosse 1997). Thus the legal definition of wildlife has changed dramatically over the past century and is likely to change further in the future.

The Public Concept of Wildlife

The sense that a species must have a net positive value to be "classified" as wildlife remains part of the generally accepted public concept of wildlife. It explains why many people do not consider pests and vermin to be wildlife. This differentiation, based on the negative and positive values assigned by people, is the source of many wildlife controversies. Coyotes, for example, are viewed by many as a valued part of our natural and cultural heritage while others see them only as livestock killers. While many people exclude most insects, most people will include butterflies and other aesthetically pleasing invertebrates among wildlife. These are among the invertebrates most valued by hobbyist collectors and fit the positive aesthetic value typically associated with wildlife (Kellert 1996).

The wildlife concept has evolved from one almost entirely associated with the positive values of wildlife as game to a much more inclusive one. The public concept of wildlife is continuing to evolve and the boundaries determining inclusion and exclusion continue to change. For most people, wildlife conjures positive feelings associated with a "natural" setting, even including naturally planted backyards. Professional attention has strongly leaned toward natural settings as well, reflecting where the public has directed funding for management. Species once thought to be useless or vermin gained wildlife status as they became more widely recognized for their ecological and social values. At some point in the future, all wild organisms may eventually be valued for the roles they play in ecosystem processes. Then they will be managed collectively to sustain their ecological services as well as their commercial and recreational values. It is doubtful though that all wildlife species will be equally valued. Conflicts will continue to arise over their management because of the different values assigned to wildlife by different sectors of the public.

MANAGEMENT PHILOSOPHY: WILDLIFE CONSERVATION

The prevailing management philosophy is wildlife conservation, which advocates the sustainable use of native wild animals and their habitats for present and future generations of human societies. Whereas individual populations under private control may not be managed sustainably, such as some populations managed for meat or hide production on commercial farms (e.g., alligator farms, venison farms), government-managed wildlife populations typically are managed for sustainable use. In public management, even when native species are viewed as pests, total eradication is rarely contemplated. Rather the species is managed to reduce damage while sustaining population viability. On the other hand, exotic species (animals introduced in areas where they do not naturally occur) may be considered for eradication, especially where they have altered the natural ecosystem in undesirable ways.

Wildlife conservation is a social process that focuses on maintaining and improving the productivity of wildlife habitat. Wildlife conservation involves planning, research, population and habitat assessment and manipulation, education, public relations, administration (logistics), establishing and enforcing wildlife law (e.g., wildlife harvest regulations), and other forms of management. Wildlife management involves the application of scientifically based knowledge to sustain production of wild animals for recreational, commercial, and protection purposes (e.g., endangered species). Wildlife management often involves manipulating populations or their habitats to achieve human objectives, such as providing hunting, fishing, viewing, or wildlife-product opportunities, protecting the options for use by future generations, or reducing animal damage. Wildlife management may be private or public, targeted for the select few, or broadly directed toward a large sector of the public. It applies population and habitat research results to the management of wildlife, habitats, and human use.

Wildlife professionals in the earliest years of management were concerned principally with game birds and mammals—those wild vertebrates that provided hunting recreation and required a state or federal license or permit to harvest. Wildlife professionals now manage a much greater variety of wildlife including nongame, endangered, exotic, and pest species (Figure 15.2). The enlargement of concern arose as urban development spread into more natural landscapes, more harmful human actions impacted wildlife and their habitats, and people sought less consumptive use (e.g., hunting and trapping) and more nonconsumptive use (e.g., feeding, photography) of wildlife.

The measurement of management effectiveness in satisfying conflicting demands depends on estimation of costs and benefits associated with wildlife. Effective man-

FIGURE 15.2 Threatened and endangered nongame animals, such as the bald eagle shown here, now receive protection and management consideration in the United States (courtesy of Jack McCaw).

agement of wildlife resources improves human health and nutrition, economic prosperity, leisure enjoyment, and future options for wildlife-based goods and services. Management that provides the greatest total benefit for the costs depends on determining who and how much each public segment will benefit from and pay for wildlife protection, enhancement, and damage control. The inability to convert all wildlife values into easily compared monetary terms underlies many of the controversies in wildlife management.

Values of Wildlife

Wildlife have aesthetic, social, commercial, educational, ecological, and recreational values. During most of the history of human existence, wildlife and plants provided all of the necessities for human survival. It has only been within the last 10,000 years that humans began to depend on agriculture, industrial development, and fossil-fuels as sources of food, fiber, fuel, and medicine. Previously, fishing, hunting, and plant gathering were the only human activities involving natural resource use. Indeed, there are still human populations, even in North America, whose primary source of meat protein is from wildlife populations.

Many of the controversial aspects of wildlife conservation are rooted in a dynamic history of wildlife-societal relationships and the diversity of values that have emerged out of that history. In the United States, wildlife is now less valued for its commercial consumptive uses—that is, for meat, hide, fur, oils, down, and other goods—than for its recreational consumptive and nonconsumptive uses. Trapping has greatly declined since World War II as fewer people think it justified for what they perceive to be luxury goods. Recreational hunting has become less popular as more people question the ethics of killing for pleasure and fewer young people find the opportunity to participate in this American heritage (Dunlap 1988).

While some wildlife benefits are easy to assess, others are more difficult, and some are not yet possible to estimate and to compare in monetary terms. Commercial benefits are easily estimated through the market place. Recreational and aesthetic benefits can be estimated through indirect economic methods that estimate public willingness

FIGURE 15.4 In the late 1800s many wildlife species, such as the egret in the photograph, now protected by law, were subjected to unregulated market hunting that nearly resulted in their extinction (courtesy of Jack McCaw).

FIGURE 15.5 Large mammals were not the only wildlife extirpated in the Great Plains. Prairie dogs once occurred in the millions in grasslands but were eradicated because they were viewed as incompatible with livestock husbandry (courtesy of New Mexico Department of Game and Fish).

bill presented to President Grant for protection of bison from market hunters was pocket vetoed with the intent of starving plains Indians into submission. Settlers converted much of the prairie to cropland and livestock range, degrading habitat suitability for many prairie species (Figure 15.5). Only the establishment of Yellowstone National Park in 1872 saved the bison from extinction in the United States (Trefethen 1975, Wagner 1978).

Personal consumption of wildlife contributed significantly to the decline of some species. However, unregulated market hunting was blamed for the most devastating damage in the last half of the 1800s. Early environmentalists such as Henry David

Thoreau, and later John Muir, called attention to the accelerating changes in the landscape and wildlife populations. However, wealthy sportsmen were most instrumental in outlawing market hunting and establishing the system of state fish and wildlife agencies that now enforce most wildlife laws throughout the nation (Reiger 1975). The first state agencies for game protection were established before the Civil War. Most states and territories had "game departments" by the early twentieth century. The game laws particularly restricted market hunting, which steadily declined in the United States. The last significant market hunting, for alligator hides and curios, decimated alligator populations. The alligator recovered rapidly once it gained protection in various states and under the Endangered Species Act. This was also true for early game species after the market carnage was stopped at the turn of the century. Until 1900, the federal government played little role in wildlife management, other than to create a few national parks where wildlife protection was authorized, but not always strongly enforced. The first federal law to significantly strengthen enforcement of state game protection laws was the Lacey Act of 1900. The Lacey Act made it a federal offense to transport illegally killed game across state lines and greatly contributed to the demise of market hunting (Grosse 1997).

During most of the twentieth century the commercial development of wildlife goods has been limited to game-farm and ranch production where animal husbandry was overseen by state agencies. The commercial production of wildlife for meat and hides has continued. Important commercial species include elk, other deer, pheasant, quail, mink, fox, nutria, and alligators. Fur farming has declined since fur demand for fashionable clothing diminished. The trapping of wild furbearers also has declined markedly and may be entirely outlawed in individual states as antitrapping groups gain influence from an increasingly sympathetic public. However, trapping remains an important source of income for certain rural communities in Canada and Alaska. Hunting remains an important source of protein in subsistence cultures. The commercial production of wildlife goods from native species has become quite specialized and is a now a small part of the estimated economic value of wildlife. However, the personal use of wildlife for food and fiber remains a significant part of the recreational hunting experience (Bolen and Robinson 1999).

Milestone Federal Wildlife Legislation

1900. Lacey Act. Prohibited the transportation of illegally taken wildlife, fishes, plants, and other organisms across state borders and prohibited the importation of certain exotic species.

1913. Migratory Bird Act. Federal government assumed regulatory powers over migratory birds.

1918. Migratory Bird Treaty Act. Provided for coordination between United States and Canada in managing migratory birds and later amended to include other nations.

1931. Predatory Mammal Control Program. Authorized the Department of Agriculture to study and control predatory mammals causing damage to crops and livestock.

1934. Migratory Bird Hunting Stamp. Required that waterfowl hunters purchase a duck stamp and monies generated be spent on wetland conservation programs.

1934. Fish and Wildlife Coordination Act. Authorized the Department of the Interior to assure the welfare of fish and wildlife in water development programs initiated or licensed by federal agencies.

1935. Creation of the Cooperative Wildlife Research Units. The units conducted research and established graduate programs in wildlife science at state universities.

1937. Federal Aid in Wildlife Restoration Act. Imposed a federal excise tax on sporting arms and ammunition with proceeds to be distributed to states for wildlife-related projects.

FIGURE 15.9 Exotic species have been introduced both accidentally and purposefully. Large mammal species, such as oryx introduced from Africa for sport-hunting purposes, have the potential to cause damage to native species and habitats. This oryx has been fitted with a radio collar (note antennae) to monitor its movements.

After World War II, the demand for hunting jumped and with it a demand for exotic species (Figure 15.9). Numerous exotic deer, sheep, and antelope species, successfully introduced on fenced-in private lands in Texas, have escaped. The nutria, a South American wetland rodent, was introduced for its fur in Louisiana, where it escaped from captivity. It now is a widespread wetland nuisance in warmer regions of the southeastern United States. Other isolated populations of exotic species occur elsewhere in the United States. Government agencies for the most part now disparage exotic introductions, considering them more of a risk than a benefit.

Management Perspectives in the Twentieth Century

Wildlife management in the first half of the twentieth century emphasized the establishment and enforcement of game laws and the destruction of game predators (Bavin 1978). Wildlife legislation was largely a reaction to changing perceptions of the American public toward wild animals. The most influential wildlife law following the Lacey Act was the Migratory Bird Treaty Act of 1918, which initiated protection of migratory birds, outlawed market hunting, and established regulations on season lengths and bag limits. It also facilitated the coordination of hunting seasons and regulations for harvest of migratory birds among the states and among countries, starting with Canada. The treaty was later expanded to include Mexico, Japan, and the former Soviet Union. Except for migratory birds, however, wildlife management remained under state control (Grosse 1997).

Another management emphasis was the reintroduction of wildlife to areas where they had been locally extirpated, or into an entirely new range. Elk, turkey, bighorn, pronghorn antelope, bison, whitetail deer, black bears, beaver, and many other native species were transferred from refugial populations. Many introductions were successful, hastening the recovery of game species. After World War II, the restoration of game species had reached a point where habitat was the major limiting factor. In some cases, game animals were causing damage to private property and threatening human health.

The personnel of the early state game departments were typically few and often were expected to cover huge areas to assure that harvest regulations were enforced.

They also were charged with determining the population status of diverse species. To contend, they enlisted the aid of volunteer wardens from among sportsmen's groups and other government agencies. Many of the rangers in the United States Forest Service became game wardens on the lands they managed. Among them was a forester named Aldo Leopold, who would lead in the development of the new discipline of wildlife science and write the first wildlife management book, *Game Management* (Leopold 1933).

Using his forestry background, Leopold emphasized the importance of habitat management as the basis of wildlife production. He recognized that once populations returned to the "carrying capacity" of the land, only changes in habitat could alter the number of wildlife produced. Realizing that most wildlife species require a mix of different vegetation and habitat attributes, he was the first to point out the importance of the arrangements and relative amounts of different plant communities in determining wildlife production.

Leopold's (1949) book, *A Sand County Almanac,* published a year after his death, is a classic that established a land ethic based on a respect for natural process. Leopold emphasized that wildlife was a renewable resource that could be replenished through sound knowledge of its relationships to its habitat. He believed humans had an ethical responsibility to respect the integrity, stability, and beauty of natural communities. His holistic approach integrated management of wild animals, agricultural practices, and other natural resources on public and private lands. *A Sand County Almanac* is recommended reading for an understanding of human relationships and ethical responsibilities with respect to the natural world. Leopold established one of the first university programs in wildlife science at the University of Wisconsin.

During the 1920s and 1930s, a small group of wildlife professionals began to coordinate their influence on legislation and eventually formed the Wildlife Society in 1937. Through the Fish and Wildlife Coordination Act of 1934, federal law established that wildlife resources be managed cooperatively by federal agencies. The Federal Aid in Wildlife Restoration Act of 1937 levied an excise tax on sporting arms and ammunition to raise money to support wildlife research, land acquisition, and habitat management (Trefethen 1975).

Ding Darling was politically very influential. In 1934, he became the director of the government bureau that was the predecessor of the Fish and Wildlife Service. He had much to do with originating duck stamps as a source of funds for purchasing wetland habitat for waterfowl. In addition, he was the main political force behind creation of the Cooperative Wildlife Research Units at state universities. These became the main source of professional wildlife researchers and managers during the period following World War II when interest in hunting and other wildlife-based recreation escalated.

During the 1960s, human interest and awareness in the welfare of the environment and wildlife increased. Rachel Carson's book, *Silent Spring,* warned of the serious damage to humans and wildlife caused by the indiscriminate use of persistent pesticides such as DDT. Numerous conservation groups led by the National Audubon Society, Wilderness Society, Sierra Club. The Nature Conservancy, and the National Wildlife Federation joined with other sectors of the American public to lobby the federal government for improved environmental legislation. The Wilderness Act of 1964 directed the Forest Service, Fish and Wildlife Service, and the National Park Service to take action to preserve wilderness areas. The Endangered Species Act of 1973 and subsequent amendments directed federal agencies to protect and restore endangered species and provided expanded funding toward this goal. Research on predator, prey, and wildlife habitat relationships greatly expanded the database for managing wildlife populations and habitats (Bolen and Robinson 1999, Bean and Rowland 1993).

After years of fluctuating increases following the drought years in the 1930s, duck populations declined more than 30% during the decade between 1970 and 1980. They dropped from 44 million birds in 1972 to 28 million in 1980. This was due largely to declining waterfowl habitat. In 1986, the United States and Canada initiated a joint waterfowl conservation plan named the North American Waterfowl Management Plan

FIGURE 15.10   Plantings around prairie potholes preserved under the North American Waterfowl Management Plan have contributed to recovery of waterfowl and several other bird species (courtesy of the Natural Resources Conservation Service).

FIGURE 15.11   Ducks and geese have increased significantly since 1990. There are 30 species of ducks, geese, and swans in the United States (courtesy of New Mexico Department of Game and Fish).

(Figure 15.10). The goals were to protect and create 6 million acres of wetland habitat and increase waterfowl numbers to 62 million breeding ducks with a fall flight of 100 million birds (Figure 15.11). The fall flight in 1986 was only 73 million birds. By 1997, duck populations had significantly increased and only two species remained below the goals of the plan. The fall flights in 1997 and 1998 numbered over 100 million birds. Canada goose populations have increased to the point that they are considered a nuisance in some areas (Ankney 1996). Those mammal and game-bird species that have significantly increased do well in habitats with early to midsuccessional stages, usually associated with human-induced changes.

The North American Waterfowl Management Plan: Waterfowl for the Future

Waterfowl are a precious natural resource, important to hunters, bird-watchers, and naturalists. But numbers of some of our most popular duck species are at an all-time low. Millions of acres of wetlands have already been destroyed by agriculture, urban development, and industry, and more are threatened each year. *Waterfowl need help.*

What's being done? The United States and Canada have joined forces to reverse the decline in certain populations of ducks and geese. In 1986, United States and Canadian officials signed a far-reaching document: the **North American Waterfowl Management Plan.** Today, that historic agreement has become an innovative international partnership in wildlife conservation. The plan has inspired cooperation between federal, provincial, and state governments as well as private conservation agencies in the two countries.

Work has already begun. The plan has established headquarters in both the U.S. and Canada. At the grassroots level, partners have formed **"joint ventures,"** composed of representatives from public and private organizations. *In the U.S., six priority joint ventures are underway:*

- Atlantic Coast
- Gulf Coast
- Lower Mississippi Valley
- Central Valley
- Lower Great Lakes-St. Lawrence Basin
- Prairie Pothole

In Canada, there are four joint ventures:

- Arctic Geese
- Prairie Habitat
- Black Duck
- Eastern Habitat

Partners in the joint ventures are carrying out specific tasks—**developing economic incentives to change land-use practices, striking agreements with private landowners, improving water management, and sponsoring research studies.** Land acquisition is only one facet of the joint venture initiatives.

The plan, with its 15-year horizon to 2000, **establishes specific objectives to restore duck populations to the levels of the 1970s.** It aims for breeding populations of 62 million that should produce a fall flight of 100 million birds. Attaining these objectives would mean that 2.2 million hunters could harvest about 20 million ducks annually. **The plan also lists population objectives for geese and swans.**

Habitat management objectives are equally ambitious. **The plan targets critical waterfowl breeding, staging, and wintering areas in both countries.** In the **Prairie Pothole Region,** cooperators intend to protect and improve 1.1 million acres of habitat. In the **Lower Mississippi Valley and Gulf Coast,** the objectives total 686,000 acres. **California's Central Valley** has an objective of 80,000 acres. Along the **Atlantic Coast,** 50,000 acres will be protected and improved. In the **Lower Great Lakes-St. Lawrence Basin,** more than 10,000 acres have been identified. In Canada, the objective is to protect and improve 3.6 million acres in the prairie Provinces, 60,000 acres in the Lower Great Lakes-St. Lawrence Basin, and 10,000 acres in the Atlantic Coast.

Although the plan focuses on waterfowl, other species will benefit too. Waterfowl habitat offers shelter for a variety of other waterbirds, shorebirds, songbirds, small mammals, and resident game species. Wetland communities also produce invertebrates and plants that are important foods for fish and wildlife. These ecosystems act as pollution filters, floodwater stores, and erosion controls.

The price tag on this unprecedented habitat protection and enhancement program is an estimated $1.5 billion. Federal and state governments will need help securing this sum and are looking to the private sector for support. The plan calls for $1 billion to be spent in Canada, 75% of which is scheduled to come from U.S. sources.

So far, the North American Waterfowl Management Plan has involved Canada and the U.S., but both countries are pursuing a working agreement with Mexico.

Source: The North American Waterfowl Management Plan, U.S. Fish and Wildlife Service.

The result is not always optimal for the sustainability of wildlife or for the benefit of all of the public. Those commissions that accept public input through scientifically based survey procedures are more likely to accurately assess public demand. Despite a trend toward more use of scientific evidence, a political process that responds to pressure from special interests continues to be a factor in state and federal management. The most effective managers are those who manage based on clearly set principles, are respectful of the rights of others, can clearly and objectively articulate the relevant scientific information, and are tactful and patient with those who are not easily persuaded.

Categorizing Wildlife for Management

Wildlife are categorized by important management attributes. Big game (large mammals such as hoofed mammals and bears) and small game (squirrel, rabbits, red fox, mink, raccoon, and others), upland game birds (turkey, quail, pheasants, and others inhabiting terrestrial areas), waterfowl (ducks, geese, swans), shorebirds (small wading birds such as plovers, snipes, and avocets), raptors (hawks and owls), songbirds, small mammals, predators, and furbearers are important wildlife categories. Furbearers are those wildlife with pelts of commercial value that are not usually hunted for sport purposes. Like hunted species, they are harvested. However, their use is consumptive. Consumptive use and nonconsumptive use differ in approach and intensity of management needed. Nongame species are those noted for purposes other than hunting or commercial uses, such as songbirds, avian predators, most reptiles, most rodents, most amphibians, and certain predatory mammals. Some nongame animals may be killed without limit, such as most rodents and coyotes (in many states). Such nongame species have been viewed in the past mostly as pests with little positive value. As the ecological value of all species becomes more scientifically demonstrable, regulations are likely to tighten even on pest species.

Habitat Management

The wildlife ecologist collects habitat data and predicts what will happen under a given set of circumstances to both the habitat and the dependant populations. Wildlife ecologists can theorize regarding the long-term consequences of management actions or lack of actions based on data gathered in a scientific manner. In managing the habitat of wildlife populations, wildlife managers confront a host of challenges because of specific habitat requirements for each species. For example, species inhabiting a forest may favor certain variables such as a specific tree species composition, age of trees, tree density and diameter, and presence or absence of a vegetation understory. The carrying capacity of habitat for each species differs because of their unique demands. Management of habitats for a number of wildlife species usually means some condition less than optimum for each species needs to be identified (Bookhout 1994, Patton 1997).

Habitat preservation, creation, and restoration are the key elements in sustaining wildlife populations. A key goal is population sustainability at some level determined by human demand for wildlife and by ecological limits. A major objective of habitat management is to identify and mitigate those factors in the habitat that limit wildlife populations. Providing these necessities at the proper time and the proper amounts is the key to successful wildlife management. To survive and reproduce, terrestrial wildlife populations must be provided with food, water, and shelter, such as escape terrain. Habitat management is an indirect approach but the benefits are greater and longer lasting than any other practice.

Habitat Supports Food Sources.
The habitat supports a biotic community including food organisms. Food must be available in sufficient quantity and suitable quality to provide all the needs of each species. Young may require different and more nutritious foods than adults. Young nesting birds require a diet high in insects because they provide the high protein content necessary for rapid growth. In temperate re-

FIGURE 15.15 Healthy wildlife habitats provide food, water, space, and cover for a variety of species. Note the diversity of plants and how these desert mule deer blend in with their background (courtesy of New Mexico Department of Game and Fish).

gions with freezing temperatures, animals may require a high caloric diet in winter to offset loss of body heat. During fall most large mammals build up fat reserves that can comprise up to 20% of their total body weight. They rely on fat reserves during days of inclement weather when food is difficult to obtain.

Water. Most terrestrial species require free water obtainable from standing bodies of water or dew. Some may require water only during certain times of the year, such as during the dry season. Some can obtain their water requirements from vegetation with high water content, such as cacti. Small quantities of widely available water better meet the needs of wildlife than a few areas with large quantities of water. Water becomes more of a management tool where it is scarce in arid and semiarid habitats. Careful provision of water where it is limiting may add substantially to the habitat carrying capacity.

Habitat As Cover. Wildlife use different types of cover for shelter to escape predation; feed in security; rest; protect themselves from cold, heat, wind, and precipitation; reproduce; and nest and breed. Cover is provided by the physical structure of earth and vegetation (Figure 15.15). Especially important attributes of wildlife cover are its dimensions, opacity, effect on air circulation, and resistence to destruction by natural forces.

Rock, soil, and snow provide important cover for burrowing, dens, and retreats. As the structure of the burrowing medium varies from optimum, the amount of energy and time that must be devoted to burrowing diverts from other life processes. Many species sustain burrows year round for many life functions while others maintain them part of the year, usually for winter protection, hibernation, and breeding. Availability of snow of the right depth and quality may be critical for survival of overwintering birds and mammals because it provides insulation from severe cold. For desert wildlife, burrows provide insulation from extreme heat and cold as well as a more humid retreat for water conservation. Some species rely on other species to provide burrows. Prairie dog burrows, for example, may be occupied by burrowing owls,

FIGURE 15.18   Wild turkeys in central Texas taking advantage of edge effect. They use the more open areas for feeding and the more heavily wooded areas for escape cover.

TABLE 15.1   Examples of North American Mammals and Birds Associated with Climax and Seral Communities

Climax Communities		Seral Communities	
Mammals	Birds	Mammals	Birds
Caribou	Prairie chicken	White-tailed deer	Bobwhite quail
Mountain goat	Spotted owl	Mule deer	Mourning dove
Bighorn sheep	Spruce grouse	Pronghorn antelope	Ruffed grouse
Bison	Ptarmigan	Elk	Scaled quail
Musk ox	Mearns quail	Prairie dog	Red-tailed hawk
Grizzly bear	Meadowlark	Jackrabbit	Lark bunting
Polar bear	Aplomado falcon	Coyote	Loggerhead shrike
Wolf	Cassins sparrow	Fox	Brownheaded cowbird

Wildlife Associations with Successional Communities.   Community succession is a basic concept in wildlife habitat management (see Chapter 2). Succession refers to the sequence of communities that replace one another on a given area. If, for example, a forest area is cleared of vegetation by fire or other natural or human-induced change, a series of vegetation types occur. This begins with pioneer stages (low seral) of initial vegetation growth (small forbs and grasses), which are replaced by more mature communities (midseral and high-seral stages composed of shrubs and trees) until the original dominant vegetation community or climax vegetation reestablishes itself; in this case, a forest. Wildlife can be classified according to the successional stage they prefer. Wildlife biologists use this knowledge to modify habitats to accommodate and increase the numbers of specific species being managed.

Wilderness or climax wildlife are those species that prefer undisturbed habitats (Table 15.1). Species in this category occur in wilderness areas, national parks and national forests and include the grizzly bear, polar bear, mountain goat, caribou, musk ox, wolf, mountain lion, lynx, and wolverine. These species are most harmed by human land use. Therefore, suitable habitat usually occurs far removed from human impacts. The grizzly bear, caribou, and mountain sheep are in this category be-

FIGURE 15.19 White-tailed deer are woodland species that respond well to human-induced habitat changes. In this case logging has increased edge effect and food supplies for both deer and moose (from Hunter 1990).

cause they fail to thrive in close contact with humans. In the case of grizzly bears, it is because they prey on livestock and can be dangerous to humans. Bighorn sheep are susceptible to diseases transmitted by livestock. Mountain goats and polar bears require such rough or extreme habitats that they are automatically wilderness species. Mountain lions are more likely than other wilderness species to move into areas where suburbs abut wilderness. Here they can occasionally threaten human life.

Forest and range (midstage in vegetation succession) wildlife consist of species inhabiting lands compatible with forestry and livestock interests (Figure 15.19). These respond well to human-induced changes in habitats. Many common game species are in this category, such as white-tailed deer, mule deer, elk, moose, turkey, ruffed grouse, scaled quail, black bear, pronghorn antelope, several species of tree squirrels, and alligators (Figure 15.19). White-tailed deer, turkey, raccoons, and some grouse do well in landscapes interspersed with forest patches and cultivated areas. White-tailed deer, squirrels, and raccoons are among the forest species most likely to invade suburban habitats of cities. Traveling by waterways into backyard boat slips and parks, alligators have become a nuisance and even life threatening in some southern suburban locations. Mule deer, pronghorn, and elk tolerate moderate grazing by cattle. Most of these species do best when the stages of vegetational succession are below climax (Wagner 1978).

Farm wildlife are adapted to the earliest stages of vegetation succession and exhibit a high tolerance for habitat modifications, including urban habitats. They typically have small home ranges. These species present the least habitat problems because they do well in agriculturally developed areas. They do require interspersed areas of natural vegetation free of cultivation for escape cover, loafing, and raising young. Examples of these species are bobwhite quail, mourning doves, ring-necked pheasant, cottontail rabbit, fox squirrel, opossum, and red fox (Burger 1978).

FIGURE 15.21 National wildlife refuges in the United States (courtesy of USDA-Fish and Wildlife Service).

FIGURE 15.22 Federal rangelands under control of the Bureau of Land Management and U.S. Forest Service are now managed to provide habitat for wildlife as well as livestock. The proper juxtaposition and interspersion of habitat components are critical for meeting the needs of diverse wildlife populations, such as these mule deer (courtesy of New Mexico Department of Game and Fish).

by shooting, trapping, snaring, or other means. Whether or not an animal is killed, the population effect is the same when it is removed from the population—it contributes to the population loss rate (see Chapter 3 for details on population dynamics). Wildlife is a renewable natural resource, meaning that mortality can be replenished by natality (birthrate) if the population is not overexploited and if required habitats are provided. Sustaining a wildlife population at a specified number requires balancing the birthrate (natality) and the death rate (mortality). Such management can result in increased densities.

Wildlife managers must objectively determine whether harvest through hunting or other means decreases population abundance below a targeted sustainable number. Based on many studies, wildlife agencies have determined that removing "a surplus" of animals from some populations is not detrimental to their long-term sustainability. Studies of quail have shown that population numbers fluctuate regardless of population size. A certain portion of the population will die each year from disease, active predation, and other causes and hence hunting a population in the fall removes those animals that would normally die of other causes during the winter. Of course, regardless of the cause of death in a natural setting, the dead animal usually is consumed by other animals.

When humans harvest wildlife, they necessarily alter the flow of energy through the ecosystem and decrease the productivity of other species. Responsible wildlife management assures that this effect does not cause decline in diversity and extinctions. To avoid this, a larger ecosystem view of effected populations is required rather than the traditional focus on the viability of the harvested species.

Present agricultural practices and urbanization have greatly modified the natural functioning of wild populations (Flather et al. 1994). The elimination of key predators, such as wolves, has been particularly important. Under these conditions, some species of animals such as white-tailed deer overpopulate areas, leading to depletion of their food resources and eventual decrease in the quality and abundance of the population. Deer in a borderline nutritional state associated with overpopulation have enough food for subsistence only and are highly vulnerable to die-offs caused by disease or other stress. Disturbance of ecosystem controls, such as predation, often result in more erratic population fluctuations with longer periods of high and low population abundance than what occurred before the disturbance. It also often results in habitat degradation by overly abundant prey species.

In urban areas, overpopulation of wildlife leads to high incidence of animal-caused damage (see also Chapter 14). The increase in elk in some urban areas of Alberta, Canada, and moose in New England has led to human injuries. These problem elk and moose have had to be removed. In the eastern United States, large populations of black bears and white-tailed deer in some areas have caused concern because of damage to property and human well-being. A large source of human injury is wildlife collision with automobiles. The increased incidence of Lyme disease, a virus spread to humans by tick bites, has been aided by the high populations of deer and other wildlife, which harbor the ticks. The threat of rabies has increased with urban sprawl. This is because the abundance of garbage increases the incidence with which pets and humans encounter potentially rabid animals, such as raccoons and skunks.

Under such circumstances, managing wild populations without hunting or some other form of lethal removal is difficult and often prohibitively expensive. However, alternatives such as animal trapping and transfer frequently are used where hunting is not acceptable. Animal transfers are expensive and often only shift problems because resident wildlife already occupy the available habitat. So far, methods used to induce infertility to avoid outright killing have proven impractical (too costly). While an increasingly vocal minority are repelled by any human killing of wildlife, the ecological and economic realities lead the majority to accept the most cost-effective safe and humane control available—and that is usually some form of killing.

The elk herd in the Grand Teton region of Wyoming provides a rural example of human-induced changes that have resulted in overpopulation of ungulates (hoofed mammals such as deer). The original wintering range of elk is now occupied by cattle ranches and agricultural development. The elk spend summers in national parks, but in winter they seek food in private lands, which causes conflicts with the landowners. A hunting program has been established to remove excess animals. Bison often wander outside the boundaries of Yellowstone National Park. They also come into conflict with private landowners. Because this population of bison has been shown to be possible transmitters of brucellosis to domestic livestock, they have been shot when they leave the park (Bolen and Robinson 1999). Elephant populations in Africa have caused similar conflicts because they destroy agricultural crops.

There is increasing evidence that remote sources of habitat change can have major effects on wildlife abundance. For example, the increasing probability of continued climatic warming has many implications for wildlife management. A good example of how remote effects can combine to cause population changes is provided by a recent study of snow geese (Figure 15.23). Because of changes in habitat and harvest management, snow goose populations have reached densities that are degrading their northern tundra nesting and rearing habitat in Canada. Most of the population increase has come as a consequence of much improved winter habitat in the south-central United States where there is a history of increasingly mild winters and conversion of habitat area to agricultural production. Grains, which are consumed by the geese, have improved their reproductive condition and increased their egg clutch size once they reach the nesting grounds. The changes in winter temperature may be a consequence of global warming caused by atmospheric management policies. A combination of decreased interest in hunting and conservative harvest limits set by the states has prevented increased mortality from balancing the increased natality. Researchers predict that the population will eventually crash when the food resources for goslings are totally overwhelmed. The dimension of management problems encountered with snow geese shows how important a total ecosystem perspective is in wildlife management.

Harvest strategies have shifted from maximum sustained to optimum sustained yield. Initially, the objective in harvest management was to maximize sustained

FIGURE 15.23 Snow goose populations in North America have increased to levels that are now endangering nesting habitat. Global warming may be a factor in their increase (courtesy of Jack McCaw).

yield over a long period of time without decreasing future harvest. This concept was based primarily on the production of goods, such as from market hunting and commercial fisheries (see Chapter 16 for further development). Maximum sustained yield was a difficult management objective to attain because potential yields often vary annually, depending on ecological conditions. Only when managers have complete data on population structure and size, which is often not the case, can a reasonably precise estimate of maximum yield be made. Misjudgments frequently result in overharvest and prolonged reduction of allowable population yield. A philosophy of optimum sustained yield replaces maximum sustained production of specific goods, such as meat. This allows balancing all goods and services of benefit to society over the long run and allows for error in estimating sustainable yield.

Stocking and Reintroduction.   The reintroduction of game species had been for the most part a highly successful aspect of management in the early years of state agency programs. After World War II, reintroduction had become mostly a thing of the past as most of the extirpated populations were restored wherever suitable habitat remained. During those early years, another common activity was the continuous stocking of game birds in areas that underwent high natural or hunting mortality (Allen 1984). However, scientific study soon revealed that stocking usually was too costly for the human benefit produced. Except for some intensively hunted private lands, routine stocking was discontinued.

Following passage of the Endangered Species Act, reintroduction into suitable habitat became one of the approaches for recovery of endangered species populations. However, recent attempts to reintroduce wolves in Yellowstone National Park, eastern Arizona, and in isolated locations in the southeastern United States have been quite controversial. Discussion of grizzly bear reintroduction also generates controversy. Although habitat protection and restoration are emphasized in endangered species management, reintroduction will remain an important part of recovery programs where suitable habitat exists.

COMMERCIALIZATION AND WILDLIFE MANAGEMENT

Wildlife commercialization (private sector economic gain from exploitation of wildlife resources) is a controversial issue. Some contend that commercialization, in the form of paid hunting and the sale of meat and wildlife by-products, threatens the public ownership of wildlife and our whole system of wildlife management. Others contend that only the wealthy nations, such as the United States, can afford the luxury of limited wildlife use mostly for recreation. The commercial value of wildlife is a reality in many parts of the world, including the United States. In some cases commercial development is the only impetus to conserve wildlife resources.

Although some countries raise significant income from ecotourism and other nonconsumptive use of wildlife, such as nature viewing in national parks, these activities alone do not guarantee the sustainability of wildlife. In many parts of the world, wildlife must be able to generate economic gains for the general populace. Welfare conservation, the dependence on foreign aid or domestic government funding to support conservation projects, is only a short-term solution. In Africa, large areas are being used to produce wildlife for harvest as a means for economically justifying land use where wildlife is the principal product.

Where wildlife remains nonmarketable in economically underdeveloped regions of the world, rural populations and landowners will not participate in conservation programs. Wildlife habitats will continue to be degraded in favor of agricultural development and wildlife will continue to be viewed as an impediment to economic gain. Wildlife use can be particularly appropriate where other forms of land use can be detrimental, such as in marginal lands easily eroded and not favorable to agricultural development or livestock grazing. Wildlife commercialization can enhance biodiversity by maintaining natural habitats outside protected areas while enhancing rural economic development.

Commercialization of wildlife in the United States is legally practiced in a variety of enterprises. In leasing programs, hunters purchase exclusive rights to hunt on a private property for an agreed upon time. Leases are available for big game, waterfowl, game birds, such as pheasants and quail, and sportfishing (Figure 15.24). In some cases, income from hunting enterprises exceed that of livestock production. Property owners can set quotas for the number of animals collected, but must abide by state and federal laws that restrict the methods of hunting, bag limits, and seasons, with some exceptions.

In Texas, where many exotic species are available for hunting on private land, the state has no laws governing exotic harvest and the landowner is given full discretion. In California, a special program known as Ranch for Wildlife permits landowners to develop a wildlife management plan detailing the wildlife and fish enhancement programs. If approved by the California Fish and Game Department, the landowner is licensed to operate a private wildlife program, which allows the landowner to charge for hunting and fishing under special game seasons and limits approved by the department. In some cases, depending on game population numbers, seasons and game bags may be more liberal than statewide regulations. This and other programs stimulate private participation in wildlife conservation and permits the landowner the flexibility needed to manage wildlife and fishery resources for increased income. It also increases hunter participation of hunting license sales and sustainable management of wildlife and fishery resources on private lands. In Texas ranches with quality trophy buck populations, seasonal leases per acre range from $5 to $7. A 4- to 5-day trophy deer hunt can be priced from $2,000 to $5,000. Landowners in Sonora, Mexico, charge $50,000 for a desert wild sheep hunting permit (Figure 15.25). In desert areas such as Sonora, where cattle production is highly risky because of the uncertainty of forage production, big game hunting enterprises are an important source of supplemental income. In some cases, landowners have abandoned cattle production and are concentrating on wildlife ranching enterprises.

FIGURE 15.24 Hunting leases on private land are becoming more important as a source of income to landowners. Modern agriculture and hunting are generally compatible. In the photograph grain stubble has been left to provide cover for pheasants and good shooting for hunters in a private club (courtesy of U.S. Dept. of Agriculture).

FIGURE 15.25 Wildlife can be a lucrative source of income to private landowners. Bighorn sheep hunts have been auctioned for several hundred thousand dollars. Some landowners in arid areas of Sonora, Mexico, have abandoned livestock production and devoted their efforts into more profitable wildlife enterprises (courtesy of the Bighorn Institute).

Game ranching (production of wildlife on open grazing lands) and game farming (intensive husbandry of wildlife for commercial purposes), often using exotic species, are commercially viable enterprises in several countries. In the United States, fallow deer have been farmed in at least two states. Game farms can produce large numbers of deer in a small area. One Texas farm has 3,500 fallow deer on about 400 acres. A pivot irrigation system provides for the growth of fresh forage, and diets are supplemented with high protein and energy foods. In New Zealand, European red deer on game farms exceed 1 million animals. The sales of red deer antlers and meat now exceed $300 million.

Texas has imported more species of exotic ungulates (hoofed mammals) than any other state or country. Some of these exotics and their hybrids, such as ibex (Asian wild goat) and domestic goat crosses, are used in game ranching operations. In 1963, a state survey revealed 13 exotic species totaling 13,000 head. By 1988, there were 67 species totaling 164,257 animals on 486 ranches. Six species had established wild free-roaming populations: nilgai antelope, blackbuck antelope, and axis deer from India, sika deer from Japan, fallow deer from Europe, and aoudad or Barbary sheep from northern Africa. In some ranches, exotics became more numerous and were considered to have greater priority than native species because of the high prices some hunters are willing to pay for hunting trophy animals. Game ranchers have come to realize that native species such as trophy white-tailed and mule deer command higher prices than common exotics. Nonetheless, exotics will remain a viable income option for Texas ranchers (Mungall and Sheffield 1994).

Marketing can be a year-round option because exotics in most counties are exempt from state regulations. They can be incorporated in multispecies grazing systems with livestock enterprises. Negative aspects of exotic game ranching include transfer of diseases to native wildlife and livestock and competition for forage resources with native species. Some states, viewing the potential dangers of exotics, have prohibited game farming and the use of exotics in game ranching.

CHALLENGES AND TRENDS IN WILDLIFE MANAGEMENT

Wildlife managers in the United States have been especially successful in establishing a system of laws and enforcement that have sustained game populations and prevented extinction of several species. Public land management has greatly improved over the past century, integrating wildlife with other forms of land use. Wildlife management has been less successful in protecting and restoring habitat for nongame wildlife, especially on private lands, and protecting wildlife from habitat contaminants of various kinds (Flather and Hoekstra 1989). As a consequence, numerous wildlife species are threatened or endangered.

Many wildlife species require large tracts of land to maintain viable populations. Mountain lions, for example, require hundreds of square miles to acquire the necessary resources for survival and reproduction. Where land is private, public agencies have not been as effective as certain nongovernment organizations, such as the Nature Conservancy, in maintaining habitat. Migratory birds can travel thousands of miles during migration to and from their nesting grounds. This requires the coordinated management of federal and state wildlife agencies and those of other countries, and private landowners (Agee and Johnson 1988).

However, even coordination among federal agencies is difficult and unqualified success stories appear rare. One of the better examples of apparent success is the North American Waterfowl Management Plan, which has significantly advanced coordination of wetland habitat development across nations along migratory pathways. All too often, federal agencies do not coordinate their management plans, resulting in different goals, which sometimes are counterconstructive to wildlife management programs and wasteful economically (Wright 1992). A recent movement toward what is called ecosystem management is an attempt to better integrate agency activities (see Chapter 23).

The ever-increasing competition for natural resources between the needs of wildlife-based benefits and other benefits is a constant concern. Water diversion and groundwater pumping for irrigated agriculture and human and industrial use will mean less water for wildlife. Wildlife managers will also have to deal with the toxic trace elements and pesticides from agricultural and industrial contamination (Langner and Flather 1994). They continue to pollute ecosystems even after the Federal Water Pollution Control Act of 1972. But the water pollution threats are worse in other parts of the world. Pollution of inland seas and coastal estuaries, combined with overfishing, is diminishing the world's fish catch. The reduction in marine fish is causing increased concern for marine mammals and birds also dependent on fish.

Millions of acres of undeveloped lands are annually converted to urban, transportation, and specialized commercial uses (see Chapter 14). Industrial expansion, urban sprawl, road construction, water resource development, and other factors have often eroded the quality and quantity of habitat to the detriment of wildlife species (Flather et al. 1994). Only the collective cooperation of resource managers, landowners, concerned conservation organizations, and general public can reverse the continued decline of wildlife habitats. Progress in recent decades has shifted from an emphasis on protection of environmental resources, including wildlife, to restoring degraded ecosystems of significant environmental value.

Most progress has been made on public lands. However, of the 2.26 billion acres of land in the United States, 60% is privately owned. The percentage is much higher in the eastern United States than in the West. Private land ownership and control still remains a paramount pursuit of most Americans. Obviously, private lands must be included if programs to benefit wildlife and other renewable natural resources are to succeed. Proper policy incentives for wise land use and wildlife conservation must be provided to landowners or they will be unwilling to make necessary monetary investments. It is difficult to convince farmers that they should not drain a wetland, clear a shrubland, or plow a grassland if it does not translate into more cash returns and less financial disparity in their budgets. Similarly, it is difficult to convince real estate developers that provisions for wildlife should be included in regional plans if they do not translate into profit. Changes in knowledge about the costs of uninformed and unregulated private land development start with education of each citizen.

Natural resources managers, farmers, ranchers, agriculturists, and land developers must understand the needs and concerns of each other, integrate planning for more effective management—and compromise when necessary for the larger good (Marcot et al. 1994). This requires an investment in objective research and diversification of wildlife conservation and management concepts. Diversified wildlife management can be accomplished through flexibility and cooperation.

Two major types of management are practiced in North America: protective management and multiple-use management. Protective management refers to land that is locked into parks and preserves and is dedicated to a single use or highly compatible multiple uses (Figure 15.26). Multiple-use management is commonly practiced by most agencies involved with wildlife resource management. In this system, the benefits and values of various resources are balanced against one another, including timber, livestock grazing, mining, energy extraction, and wildlife resources. In the past, multiple-use management often resulted in independent resource management proposals and resolution of conflicts that became evident as the implications of different proposals become clear. In such management, most of the planning process is conducted independently and then aligned at relatively late planning stages. As Chapter 23 discusses, recent analyses of the management process reveal a need for more integration of wildlife programs with those for timber, rangeland, water, fishery, recreation, wilderness, and mineral development through an integrated planning process. Educational and technical assistance programs that aid private landowners to improve habitats should be encouraged in order to promote the enhancement and sustainability of wildlife resources.

FIGURE 15.26 The availability of accessible wild landscapes with abundant and diverse wildlife populations is a national heritage that will require the cooperation of the public and private sectors to maintain its aesthetic and economic values for future generations.

Although the majority of the public is "pro" wildlife, they are woefully uneducated about the values of wildlife resources, the dynamics of natural populations and ecosystems, and the costs of management that sustain wildlife. They typically relate more to the welfare of individual animals than to the sustainability of populations in ecosystems. There is growing sentiment against any killing of wildlife and a general shift toward emphasis on nonconsumptive wildlife recreation. This has contributed to a gradual decline in funds derived from hunting licenses over the past two decades. Because that was a primary source of funds for management and research, including numerous nongame issues, supportive research and management funding are not keeping pace with need. Ecological and social education needs to be made more effective in primary and secondary schools if the public is to become more knowledgeable about funding needs for research, management, and land acquisition. Even though the state and federal agencies responsible for wildlife and fisheries management are staffed by knowledgeable, dedicated personnel, it will require the cooperative efforts of all citizens to ensure that the animals and their habitats are sustained. The professional resource managers of the future also will need to be better communicators and educators of the public as well as technically competent in managing habitat and populations.

LITERATURE CITED

Adams, L. W. 1994. *Urban wildlife habitats: A landscape perspective.* Minneapolis, MN: University of Minnesota Press.

Agee, J. K. and D. R. Johnson, eds. 1988. *Ecosystem management for parks and wilderness.* Seattle, WA: University of Washington Press.

Allen, D. 1984. *Our wildlife legacy.* Rev. ed. New York: Funk and Wagnalls.

Anderson, S. H. 1991. *Managing our wildlife resources.* 3rd ed. Upper Saddle River, NJ: Prentice-Hall.

Ankney, C. D. 1996. An embarrassment of riches: Too many geese. *Journal of Wildlife Management* 60:217–223.

Bavin, C. R. 1978. Wildlife law enforcement (pp. 350–364). In H. P. Brokaw. ed., *Wildlife and America.* Washington, DC: U.S. Government Printing Office.

Bean, M. J. and M. J. Rowland. 1993. *The evolution of wildlife law.* 3rd ed. Westport, CT: Praeger.

Bolen, E. G. and W. L. Robinson. 1999. *Wildlife ecology and management.* 4th ed. Upper Saddle River, NJ: Prentice-Hall.

Bookhout, A., ed. 1994. *Research and management techniques for wildlife and habitats.* Bethesda, MD: The Wildlife Society.

Burger, G. V. 1978. Agriculture and wildlife (pp. 89–107). In H. P. Brokaw, ed., *Wildlife and America.* Washington, DC: U.S. Government Printing Office.

Council for Wildlife Conservation and Education. 1992. *The hunter in conservation.* Washington, DC: Wildlife Management Institute.

Cronon, W. 1983. *Changes in the land: Indians, colonists and the ecology of New England.* New York: Hill and Wang.

Dunlap, T. R. 1988. *Saving America's wildlife.* Princeton, NJ: Princeton University Press. 222 pp.

Dunn, C. P., F. Stearns, G. R. Guntenspergen, and D. M. Sharpe. 1993. Ecological benefits of the Conservation Reserve Program. *Conservation Biology* 7:132–139.

Fazio, J. R. and D. L. Gilbert. 1981. *Public relations and communications for natural resource managers.* Dubuque, IA: Kendall/Hunt.

Flather, C. H. and T. W. Hoekstra. 1989. *An analysis of the wildlife and fish situation in the United States: 1989–2040.* Ft. Collins, CO: USDA Forest Service General Technical Report RM-239.

Flather, C. H., L. A. Joyce, and C. A. Bloomgarden. 1994. *Species endangerment patterns in the United States.* Ft. Collins, CO: USDA Forest Service General Technical Report RM-241.

Grosse, N. J. 1997. *The protection and management of our natural resources wildlife and habitat.* 2nd ed. Dobbs Ferry, NY: Oceana Publications.

Hornaday, W. T. 1913. *Our vanishing wildlife, its extermination and preservation.* New York: New York Zoological Society.

Hughes, J. D. 1996. *North American Indian ecology.* El Paso, TX: Texas Western Press.

Hunter, M. L., Jr. 1990. Wildlife, forests, and forestry. Englewood Cliffs, NJ: Prentice-Hall.

Johnson, F. R. 1987. Wildlife benefits and economic values (pp. 219–228). In H. Kallman, chief ed., *Restoring America's wildlife, 1937–1987.* Washington, DC: USDA, Fish and Wildlife Service.

Kellert, S. R. 1996. *The value of life.* Washington, DC: Shearwater Books and Covelo, CA: Island Press.

Kimball, T. L. and R. E. Johnson. 1978. The richness of American wildlife (pp. 3–17). In H. P. Brokaw, ed., *Wildlife and America.* Washington, DC: U.S. Government Printing Office.

Langner, L. L. and C. H. Flather. 1994. *Biological diversity: Status and trends in the United States.* Fort Collins, CO: USDA Rocky Mountain Forest and Range Experimental Station, General Technical Report RM-244.

Leopold, A. 1933. *Game management.* New York,: Charles Scribner's Sons.

Leopold, A. 1949. *A Sand County almanac.* New York: Oxford University Press.

Marcot, B. G. M., M. J. Wisdom, H. W. Li, and G. C. Castillo. 1994. *Managing for featured, threatened, endangered, and sensitive species and unique habitats for ecosystem sustainability.* Portland, OR: USDA, Forest Service Pacific Northwest Research Station General Technical Report PNW-GTR-329.

Martin, T. E. 1992. Landscape considerations for viable populations and biological diversity. *Transactions of the North American Wildlife and Natural Resources Conference* 57:283–291.

Matthiessen, P. 1987. *Wildlife in America.* Rev. ed. New York: Viking-Penguin.

Morrison, M. L., B. C. Marcot, and R. William Mannan. 1998. *Wildlife-habitat relationships.* 2nd ed. Madison, WI: University of Wisconsin Press.

Mungall, E. C. and W. J. Sheffield. 1994. *Exotics on the range, the Texas example.* College Station, TX: Texas A&M University Press.

Patton, D. R. 1997. *Wildlife habitat relationships in forested ecosystems.* Rev. ed. Portland, OR: Timber Press.

Payne, N. F. and F. C. Bryant. 1994. *Techniques for wildlife habitat management of uplands.* New York: McGraw-Hill.

Reese, K. P. and J. T. Ratti. 1988. Edge effect: A concept under scrutiny. *Transactions of the North American Wildlife and Natural Resources Conference* 53:127–136.

Reiger, J. F. 1975. *American sportsmen and the origins of conservation.* New York: Winchester Press.

Robbins, C. S. 1991. Managing suburban forest fragments for birds (pp. 253–264). In D. J. Decker et al., eds., *Challenges in the conservation of biological resources: A practitioner's guide.* Boulder, CO: Westview Press.

Scalet, C. G., L. D. Flake, and D. W. Willis. 1996. *Introduction to wildlife and fisheries.* New York: W. H. Freeman.

Trefethen, J. B. 1975. *An American crusade for wildlife.* New York: Winchester Press.

U.S.D.I., Fish and Wildlife Service and U.S. Department of Commerce, Bureau of Census. 1997. *1996 national survey of fishing, hunting, and wildlife-associated recreation.* Washington, DC: U.S. Government Printing Office.

Wagner, F. H. 1978. Livestock grazing and the livestock industry (pp. 121–145). In H. P. Brokaw, ed., *Wildlife and America.* Washington, DC: U.S. Government Printing Office.

Wildlife Management Institute. 1996. *Outdoor News Bulletin.* 50:3.

Wright, R. G. 1992. *Wildlife research and management in the national parks.* Urbana, IL: University of Illinois Press.

16

Fishery Conservation and Management

THE FISHERY RESOURCE

Fisheries exist wherever aquatic animals are actually or potentially harvested or captured by humans. Most harvest of finfishes (true fishes) is for food or other goods. However, some sport fisheries are nonconsumptive and many fisheries are for animals other than finfish. A catch-and-return sportfishery is one example of a nonconsumptive fishery. The least consumptive fisheries "capture" fish on camera or in viewing. Fishery capture more usually relies on various nets, traps, hooks, lines, spears, harpoons, arrows, snares, poisons, anesthetics, clubs, and even bare hands. Some types of fishing have much in common with hunting except they occur in aquatic habitats. Fisheries based on the use of harpoons and arrows are examples.

Modern fishery science and management in the United States originated from fish culture over a century ago. Fish culture, now more commonly known as aquaculture, involves intensive management of waters to produce fish for commercial markets and for public management of sport fisheries. From aquacultural roots, the American Fisheries Society has grown to be the principal professional organization in the United States, representing diverse aquatic interests and concerns. It is involved with all areas of consumptive and nonconsumptive use of fishery resources including preservation of fishery stocks for future resource renewal (Figure 16.1).

Since antiquity aquatic resources have held great meaning for diverse cultures. Fishing implements are common among the artifacts of prehistoric cultures. Early Egyptian, Greek, Assyrian, Chinese, and Roman art and writing depict fishing by net, spear, rod, and line (Radcliffe 1921, Waterman 1975; Figure 16.2). The world's aquatic resources are now valued for food and other goods, for recreational experience, and for their wondrous biodiversity.

The roots of the word "fishery" reach back to when most aquatic animals were casually classified as "fish." Excepting birds, fisheries include resource populations from all vertebrate classes and many phyla of invertebrates. Vertebrate fisheries include the true fishes, certain turtles and frogs, porpoises, and whales. Among invertebrate animals, the crustaceans (crabs, lobsters, shrimp, crayfish) and molluscs

FIGURE 16.1 The Coho salmon is an important recreational and commercial fish species that needs to immigrate upstream as adults to spawn. Although some natural barriers can be bypassed, high dams have contributed to declines by impeding passage (courtesy of USDI-Fish and Wildlife Service).

FIGURE 16.2 Depiction of Egyptians fishing about 4,000 years ago by various means, including rod and line (courtesy of USDI-Fish and Wildlife Service).

(clams, oysters, scallops, mussels, squid, octopus, snails) are among the most important fisheries in the United States and elsewhere. These two phyla contain most of the shellfish species. Sponges, jellyfishes, corals, segmented worms, and sea urchins make up the remainder of marine invertebrate fisheries. Freshwater invertebrate fisheries include crayfish, freshwater shrimp, and freshwater mussels.

While the finfishes dominate aquatic diversity (finfish species outnumber the total of all other vertebrate species), weight harvested, and total economic yield, the most commercially valued category is invertebrate—the shrimps. Marine turtles, certain marine mammals, and certain salmoniform fishes are high-profile threatened and endangered species among the many aquatic species at risk. Important types of freshwater fish, marine fish, and shellfish are shown in Figures 16.3 and 16.4.

The first human interest in fisheries probably was for subsistence food and clothing. Subsistence fishing evolved into commercial food and recreational trade, which support employment income estimated at nearly $30 billion annually in the United States alone. Recently, with increased recognition of widespread degradation of aquatic resources, the concept of protecting resource options for future potential use has become a predominant concern in fishery management. This chap-

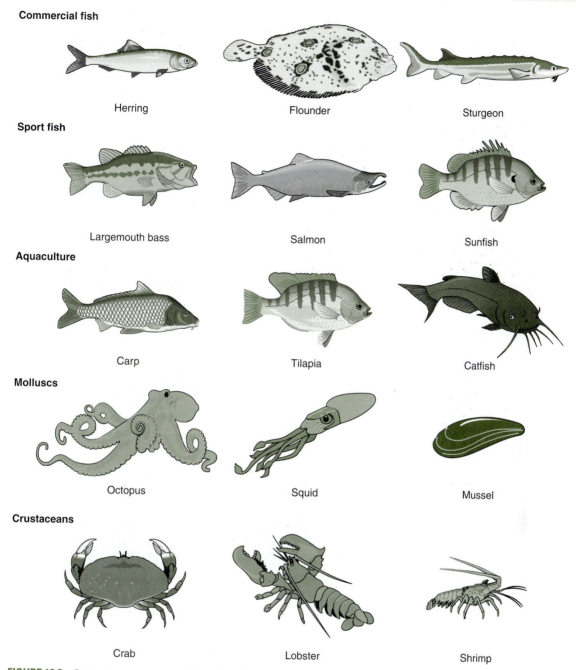

FIGURE 16.3 Some important examples of vertebrates and invertebrates that contribute to fisheries, including species emphasized in aquaculture.

ter follows this lineage of management concern starting with fishing for food and other goods, then fishing for recreation and, finally, biodiversity concerns. The chapter ends with a summary of different occupational activities in fishery resource management.

FISHING FOR FOOD AND OTHER GOODS

Cultural Importance

Aquatic resources have provided important sources of protein, clothing, medicines, fuels, jewelry, animal feeds, and other goods for many cultures of the world (Miller and Johnson 1989). Subsistence fishing remains an important theme in sustaining the

Marine reptiles

Sea snake   Marine iguana   Leather back turtle

Marine mammals

Sperm whale   Manatee   Walrus

FIGURE 16.4   Some important marine mammals and reptiles, a few species of which have composed fisheries (e.g., sperm whales and sea turtles). Note: manatees are marginal marine species that inhabit both estuarial and totally fresh waters.

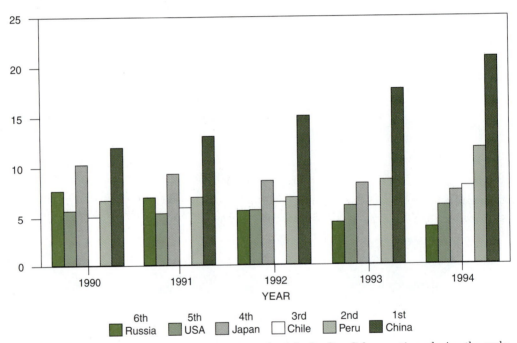

FIGURE 16.5   The world's commercial catch for each of the leading fishery nations during the early 1990s (from National Marine Fisheries Service 1995).

cultural integrity of remaining nonindustrial societies (Miller and Johnson 1989). For many other societies, fishing has become an important basis for commerce and monetary income. Commercial fishing was a well-established enterprise by the time of the Roman Empire (Radcliffe 1921).

Fisheries provide more protein for people than terrestrial animals (Miller and Johnson 1989). Until recently, use of unexploited fisheries kept pace with growth of the human population. However, the limits of the world's aquatic food resources are now being reached, even in the vastness of the open sea (McGinn 1997, Johnson 1989). The world's commercial catch by the leading fishery nations is shown in Figure 16.5.

Early Fishery Allocation

The contemporary means by which food fisheries are allocated is based on a tradition of resource use that started over thousands of years ago in Europe and Asia. Monarchs held claim to territory for themselves and their subjects, reserving the most desirable resources for royalty. In medieval Europe all land initially belonged to "the crown" for use at its discretion. Private lands came from grants of crown land to independent ownership. Most of these lands were granted to aristocracy who then controlled access and fishery use of inclusive waters. Larger navigable waters were more freely used for transportation and commerce including fishing. Commerce corridors became common-use areas for all citizens early in the evolution of nations from kingdoms and estates.

Early food fisheries remained close to land but eventual fishery depletion encouraged expansion into deeper waters where gear was limited mostly to long handlines holding hundreds of baited hooks. Crews from Western Europe fished off the North American shore soon after Columbus landed (Waterman 1975, Kurlansky 1997). Conflicts over fishing rights grew as oceanic fishing expanded, contributing to the growth of national navies, larger ships, more far-reaching exploration, and new territorial claims. International tension has continued to accompany oceanic fishing throughout its long history.

North American Fisheries

Native American fisheries were well established when Europeans first colonized the Americas (Figure 16.6). Much of the early exploration financed by England was motivated by natural resources and the fishery resource of the northeastern Atlantic coast was found to be exceptionally bountiful. The Pilgrims came to New England in part to fish for food and commerce (Kurlansky 1997). By the eighteenth century, commercial fisheries supported numerous coastal economies in colonial North America. Most North American commercial fishing remained close to the East Coast until the nineteenth century, when it expanded rapidly along all coasts, inland, and over the open seas. Whale and finfish (mostly cod and haddock) became mainstays for employment in New England and maritime coastal communities. Lobster fishing later became established there in the nineteenth century. The steam engine gradually replaced sails during the nineteenth and early twentieth centuries, extending the daily reach from home ports. Fisheries thrived along the lower Atlantic and Gulf Coasts based on shrimp, crab, clam, oyster, shad, sea basses, and other finfishes.

After the early decimation of fur seals and marine otters on the West Coast by Russian crews, salmon, halibut, herring, and oyster fisheries contributed significantly to local economies of newly established Pacific Coast communities. Fishing industries became established on the Sacramento and Columbia rivers and other West Coast locations soon after canning technology was developed in the middle 1800s. Later in the nineteenth century, salmon were efficiently caught in wheel nets, which cycled and dumped migrating salmon into attached barges. Halibut and salmon depletion resulted in formation of the International Halibut Commission in 1923 (Bell 1970) and the Pacific Salmon Commission in 1937 (Larkin 1970).

The Great Lakes and larger interior rivers also supported sizable settlements based on freshwater fisheries started in the early nineteenth century. Lake sturgeon populations plummeted rapidly followed by whitefish, chubs, and lake trout late in the nineteenth century. Although inland fisheries were mostly based on finfishes, significant shellfisheries developed for crayfish, freshwater mussel shell, and pearls in midwestern and southern rivers (Pennak 1978).

FIGURE 16.6 A sixteenth-century artist's depiction of native North Americans fishing with spears and a fish trap (upper left) (courtesy of USDI-Fish and Wildlife Service).

Technological Revolution

From the late nineteenth to the early twentieth century was a period of exceptional technical innovation that transformed marine fisheries. Internal combustion engines, hydraulic systems, electric generators, refrigeration systems, telephones, radios, and plastics were invented. These advances in technology initiated profound change in twentieth-century lifestyle, condition of the aquatic environment, and fishing efficiency. Steam power and electrical refrigeration were especially important because they allowed marine fisheries to extend far from shore. Harvest grew many fold, greatly benefiting numerous cultures in many Nations. However, some fish populations began to decline from the combined effects of habitat deterioration and overfishing.

Fishery habitat problems increased rapidly as the United States. industrialized in the late 1800s. Problems were caused by water project engineering, sediment from watershed abuse, industrial contaminants, and, after World War II, applications of artificially developed fertilizers and pesticides. Although environmental conditions have improved because of environmental laws, major habitat problems still remain. In addition to impeding fish migration, dams, canals, dikes, and other water control structures changed water quality, altered nutrient movements, and modified river flow characteristics. Environmental regulation of the newly emerging oil, plastics, and other chemical industries was weak at best. They released toxic materials directly into public waters. Soil erosion and sedimentation over fish habitat accelerated as agricultural technology expanded.

The once impressive inland fisheries of the Great Lakes suffered from damaging invasions of lampreys and alewives caused by placement of locks and canals in the Great Lakes (Smith 1970). Salmon abundances continued to decrease as dams accumulated on West Coast spawning rivers. Numerous spring and riverine habitats were threatened or eliminated by groundwater pumping and irrigation diversion. Many lake, riverine, and estuarine fisheries declined from sedimentation, chemical pollution, and eutrophication (see Chapter 3).

After World War II, marine fisheries rapidly expanded because of technical advances including sonar, improved marine communications and navigation equipment, and synthetic materials for larger and more efficient gear (Figure 16.7). By the 1970s, commercial fishers from diverse nations were using huge factory boats and fishing gear. Fish were harvested, processed, frozen, and stored for return to port and sale. The fraction of fish traded in the international market increased from under 10% before World War II to 20% in 1950 and over 40% at present (Miller and Johnson 1989).

Technological advances affected size of catch much more than selectivity of catch. Large masses of nonmarketable species, called by-catch, were killed and dumped overboard (Wise 1984, National Research Council 1999). As the most desired commercial species were depleted, new species were selectively harvested using trawls, seines, gill nets, pound nets, and various other means. An effect of efficient harvest of marketable fish and mortality of by-catch was a reduction in food resources for fish-feeding fishes, mammals, and birds (Northridge 1984). Porpoises, sea turtles, and other "nontargeted" species were increasingly caught and killed.

Reaching Food Fishery Limits

Many advances were made in understanding ecological process during the 1950s and 1960s. One of the most basic discoveries was that aquatic communities retained only 5 to 15% of biomass energy originally fixed in photosynthesis with each transfer of energy from one feeding level to the next (see Chapter 3). Based on estimates of plankton production, ecological efficiency, and the species judged suitable for harvest, Rhyther (1959) estimated that sustainable yield of food fisheries worldwide was about 100 million metric tons. Present trends (Figure 16.8) suggest that world yield is approaching that limit. The fisheries have remained stable for the past decade, varying around 90 million tons (FAO 2000). The per capita harvest rate also has already leveled and may decline in the future (McGinn 1997) (Figure 16.9). Other than China, the fish catch per person has been in decline since the 1980s.

Soon after World War II, fishery professionals realized that many specific marine finfisheries had been or were becoming overexploited even as the total fishery yield continued to increase (Miller and Johnson 1989). Sardine populations "crashed" and the viability of several whale species was questioned (Ahlstrom and Radovich 1970). This exploitive trend was an outcome of what Hardin (1968) called the "tragedy of the commons" (see Chapter 2). This occurs when renewable resources are used in an area held in common by many people without clear responsibility for the future of the resources. If one person does not exploit the resource, another probably will.

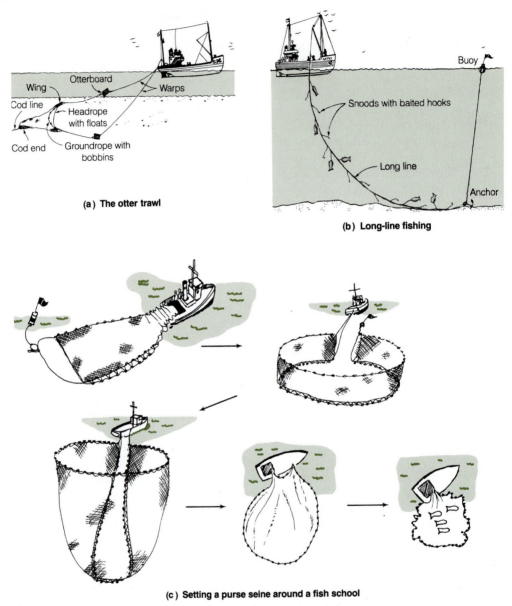

(a) The otter trawl

(b) Long-line fishing

(c) Setting a purse seine around a fish school

FIGURE 16.7   Common methods of commercial fishing (from Lerman 1986). The otter trawl (a) is towed just above bottom to capture shrimp and bottom finfish. Long-line fishing (b) requires the baiting of many hooks tied along lines extending thousands of feet. Purse seining (c) involves setting a long, deep net around and under a school of fish, then drawing the catch in.

There is an inclination to exploit rather than to regulate use for resource sustainability. The need for more harvest control was clear. Starting in the 1970s, fishing nations declared sovereignty over larger fishing territories, usually extending them 200 miles offshore. Each nation established and enforced its own regulations for fish harvest method and amount. For the United States, this was authorized by the Magnuson Fishery Conservation and Management Act of 1976, which required regional fishery plans that complied with national standards. The 200-mile limit forced more competition for limited resources within the U.S. territory, increasing tensions among commercial, recreational, and biodiversity interests.

Following WWII, the yield from wild inland freshwater fisheries of the United States and elsewhere decreased and aquacultural production increased. The Great Lakes fisheries reached an economic low in the 1950s (Smith 1970). Introductions of

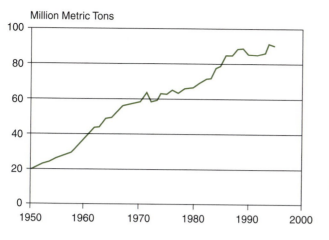

FIGURE 16.8 World fish catch, 1950–1995 (from McGinn 1997).

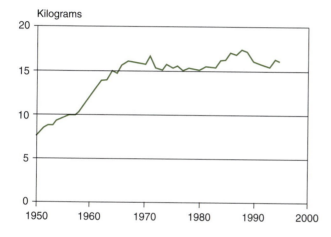

FIGURE 16.9 World fish catch per person, 1950–1995 (from McGinn 1997).

West Coast salmon into the Great Lakes established an attractive recreational fishery. However, commercial food fisheries remain a tenuous enterprise to this day. Technological improvements favoring oceanic exploitation and alternatives to fishery products also contributed to inland commercial fisheries' decline. Development of large electric freezers permitted open ocean fisheries to better compete with small inland fisheries. Freshwater mussel fisheries declined rapidly once superior synthetic materials were invented to make buttons and related items (Pennak 1978). The world's commercial fisheries are now about 10% freshwater and 90% marine.

By the early 1990s the commercial catch in the Great Lakes had become a very small part of the total. The commercial harvest in Alaska exceeded all other U.S. harvests in weight (Figure 16.10). The high harvest in the Gulf includes diverse species dominated by shrimp and menhaden, a herring used for animal feed. The Chesapeake Bay is especially recognized for its shellfish and New England for cod and lobster.

World harvest continued to climb through most of the 1970s and 1980s, reaching a value of $53.4 billion ($4.6 billion for 4.5% of the harvest in the United States) by 1999. However, the rate of increase in worldwide harvest began to decrease after 1970 and now appears to be leveling (McGinn 1997). Signs of decline now appear in many stocks of the world's marine fisheries (FAO 2000) and in U.S. waters (National Research Council). The National Marine Fishery Service (2000b) recently determined that 46% of the known offshore stocks in U.S. coastal waters were overexploited.

Despite the 200-mile limit, the United States and Canada have been forced to restrict or eliminate harvest for many of their most profitable wild commercial fisheries until they recover. The Northeast fisheries were particularly depleted by the mid-1990s

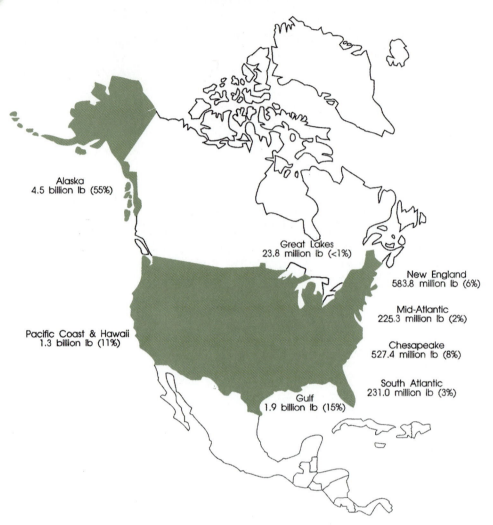

FIGURE 16.10 U.S. commercial fishery landings by region in 1999 (from National Marine Fishery Service 2000a).

(mostly Cod, haddock, and yellowtail Flounder), but show early signs of recovery. Harvest restriction has resulted in local economic hardship and higher prices for prime seafood. The means of controlling commercial operator entry into the fishery is being reevaluated as are the government subsidies provided to commercial fishers by both governments.

The impacts of repeated trawling over huge areas and fishing mile-long gill nets is now appreciated after years of denial. Appropriate management should allow greater total sustained yield than was possible under previous fishery conditions. By-catch remains an inefficient waste of important resources, which may be reduced by improved gear. Unconventional species, such as krill, may eventually contribute more to future fisheries if economical harvest techniques and demand develop (Wise 1984). Because krill and other abundant unconventional species also provide food for many other species, the ecological implications of such harvest are being considered. The use of smaller forage species for fish meal and oil already comprises one-third of the world's fisheries.

The results of a long-term study along the Alaskan coast added evidence to the growing probability that commercial fishing is contributing to complex changes in ocean systems (Estes et al. 1998), leading to a decline of marine species. Sea otter populations recently declined sharply over much of the Alaskan coast after a century of recovery from near extinction. The decline is likely due to predation by orcas (killer whales) following a population collapse of the orca's preferred foods—seals and sea lions. The seal and sea lion collapse is linked to the loss of their fish prey, which has been associated with commercial fishing and climate change. Because sea otters consume sea urchins, which eat kelp, the kelp-based communities are in decline.

Recently, Jackson et al. (2001) reviewd paleoecological, archaeological, historical, and scientific information to analyze the "top-down" ecological effects of overfishing on coastal ecosystems. The effects worked from upper trophic levels down to lower trophic levels in the ecosystem. They linked documented large declines in aquatic vertebrate and vertebrate biomass with dramatic changes in kelp forests, coral reefs, and estuaries. The results of these effects generally confirmed the findings of Estes et al. (1998) for kelp forests. Two other important implications for fisheries pertain to estuaries and coral reefs. In estuaries the great reduction of phytoplankton-feeding oysters by fishing is implicated in the widespread abundance of phytoplankton, which cause oxygen depletion, toxic conditions, and other symptoms of eutrophication. At coral reefs the recent massive die-offs of coral are linked to the buildup of algae on coral surfaces and coral predators. Overharvesting of fish that feed on algae and on coral predators appears to be the cause of these problems.

The Aquacultural Potential

There has been much discussion about the potential of aquaculture to replace wild fishery resources. Fish farming has been an important practice in Asian cultures thousands of years. Aquaculture was described in Chinese writing over 2,000 years ago (Miller and Johnson 1989, Bowen 1970). It is most important in Asia, where in 1994, 87% of the world's aquacultural yield occurred. Aquaculture outside Asia was initiated in the early nineteenth century. Food markets for catfish and trout developed rapidly in the United States following WWII (Bowen 1970).

Aquaculture is now the fastest growing sector of commercial yield, providing nearly 26% of the total world fish consumption (FAO 2000; Figure 16.11). China leads the world in increased production. Aquaculture of food fishes has increased in importance as wild fisheries harvest has leveled. Outside Asia, aquaculture concentrates on product quality more than quantity. In the United States, trout culture began for the market soon after the Civil War (Scott 1875). Warm-water culture was less successful except for catfish culture, which thrived in the southern United States after WWII when the interstate transportation system was completed and truck refrigeration was perfected. More recently, both West Coast and Atlantic salmon have contributed to U.S. aquacultural production. Crayfish, tilapia, freshwater shrimp, and oysters are cultured at lower levels of production.

Elsewhere in the world, carp and other large minnow species are widely cultured in Asia and Eastern Europe. Some tropical species show promise for growing product acceptance in Africa and Asia. Worldwide, 58.7% of all aquacultures is in

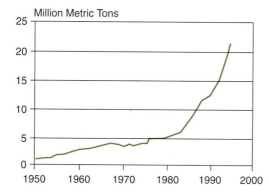

FIGURE 16.11 Aquacultural production for the world (from McGinn 1997).

freshwater, 6.3% is in brackish water, and 35.0% is in marine environments. Mariculture (marine aquaculture) has grown rapidly in recent decades, especially for salmon, oysters, shrimp, and other high-quality products. Atlantic salmon are cultured in Scandinavia, the United States, and Canada. Asian mariculture has been an important source of regional protein for centuries. Shrimp are now the major species exported from Asia. A special area of aquaculture is the tropical fish pet trade. Present world retail trade is $3 billion (FAO 2000), but only a fraction are cultured fish.

As technology improves, aquaculture should contribute substantially more to human protein consumption, especially where labor and environmental costs are low. However, aquaculture is not a panacea for overexploitation of wild fishery resources of high quality (Wise 1984, Goldberg et al. 2000). The products must be of high enough quality to warrant prices required for profit. Aquaculture in the United States requires high facilities investment, high operations costs, and high cleanup costs to protect the environment. Only a small fraction of aquatic species are likely to prove suitable for aquaculture in the foreseeable future. A growing concern is that netpen-cultured fish will escape and negatively affect the genetic makeup of wild stocks or transmit disease. The state of Alaska has prohibited all netpen culture for these and other reasons.

RECREATIONAL FISHING

Cultural Significance

Most of the value of recreational fishing is in the experience rather than in food value. Fishing appears to have been a common form of fun and relaxation for many cultures. Rod, reel, and line were first depicted in a thirteenth-century Chinese painting (McClane 1974). The first extensive account of recreational fishing has been attributed to Dame Juliana Berners in an essay written in 1496 and entitled "A Treatyse Of Fysshinge Wyth An Angle." Recreational fishing was common among the European affluent well before Isaac Walton wrote the first edition of his famous treatise on sportfishing in 1653 (Walton 1676). His book was to become the philosophical basis for sportfishing among the educated, as signaled by its title: *The Complete Angler or the Contemplative Man's Recreation.*

Although food fishing was common in colonial North America, sportfishing was neither common nor frequently advocated. An exception was Captain John Smith, who grew rich from commercial fishing for cod (Kurlansky 1997) and admitted to fishing for fun (Waterman 1975). Leisure time was a rare luxury for the working-class society. The Puritans disparaged pleasurable use of time. Presaging twentieth-century ethical concerns, some religious leaders claimed that pain caused as a consequence of sport was sacrilege.

Attitudes toward relaxation and pleasure gradually changed as human welfare improved. By the 1840s, many Americans admitted they enjoyed the sport as well as the harvest for food. Unlike Europe, even the "working class" owned land adjacent to fishable water or knew someone who did. In keeping with European tradition, large waters were open to common use including fishing. Many people fished both for food and for a break in their daily routines (Figure 16.12).

Growth of Sportfishing in the United States

After the U.S. Civil War, tourism based on sportfishing grew at the remaining wilderness areas. The rapidly growing railroad network facilitated sportfishing. Tourist "destinations" grew up around wilderness resorts in the "north woods" of Maine, New York, Michigan, eastern Canada, the Rocky Mountains, and in Florida. The first National Parks were set aside from remaining wilderness lands. Unspoiled stream,

FIGURE 16.12 Eel spearing at Setauket, Long Island, is shown in this painting by William S. Mount, 1845 (courtesy of U.S. Department of Interior).

pond, estuary, and coastal fishing remained close to large cities like New York even though habitat-destructive pollution was becoming more common (Scott 1875). By the early 1900s, much of the western United States had been declared public land with open access to anglers and other recreationists.

Technology had much to do with the expansion of sportfishing in the United States. It freed up leisure time for the working classes. Following the invention of the internal combustion engine, all forms of sportfishing continued to grow in popularity as automobiles, roads, and boat motors improved and became more affordable. The first molded plastic fishing lures and rod parts were developed. The costs of fishing equipment dropped as manufacturing methods improved.

After WWII, families looked forward to a promising and enjoyable future. They demonstrated their optimism by expanding their family sizes and by recreating more outdoors. Sportfishing was the ideal outdoor family pastime for the babyboom generation, especially when combined with boating and other outdoor recreation. Sportfishing gear further improved and anglers became better educated by the new invention, television. Development of the interstate highway system provided easy access to many more fishing sites. Fishing intensity increased rapidly at many locations (Figure 16.13).

The combination of growing participation, more sophisticated participants, and more effective gear increased angler efficiency and the need for improved sportfishery management. Starting in the 19th century, anglers formed clubs for comradery and policy influence. Effective coalitions were formed in the Izaak Walton League, Trout Unlimited, the Sportfishing Institute, and other organizations, often with support of the sportfishing industry. In early response to growing demand, Congress passed the Dingell-Johnson Act in 1950, which provided federal aid for sportfishery management. It was administered by the U.S. Fish and Wildlife Service and authorized sales taxes on fishing equipment and supplies. Revenues were distributed to the states according to a formula based on licenses sold (60%) and geographic area (40%).

Government also invested heavily in water-based public works programs, which generated a 40% increase in the surface area of fishable water excluding Alaska and the Great Lakes (Jenkins 1970). The public water projects were constructed for numerous primary uses such as flood control, irrigation, navigation, and electricity generation. Public policy often required management for multiple use including recreation.

FIGURE 16.13   Anglers fishing at an especially good location in the Northwest illustrate the increasing intensity of fishing that occurred after World War II at many locations (courtesy of U.S. Fish and Wildlife Service).

The large reservoirs, canals, and other waterways were designed and operated by the Bureau of Reclamation, Army Corps of Engineers, Tennessee Valley Authority, and various public utilities. The Soil Conservation Service (now the Natural Resources Conservation Service) aided construction of nearly one million small reservoirs, on private lands. Through these diverse water management agencies and the U.S. Fish and Wildlife Service, the federal government formed intricate management relationships with state management agencies and private groups and individuals.

The postwar expansion of reservoir habitat, highway access, and management revenues helped fishery managers keep up with burgeoning sportfishing interest and increasing fishing efficiency. After reaching a low in the 1960s, water quality began to improve following comprehensive legislation in the early 1970s under the lead of the U.S. Environmental Protection Agency. Previously degraded habitat once again became suitable for sportfishing. Provision of amenities like campgrounds, boat ramps, and sanitary facilities encouraged even more sportfishing interest. At the same time fish habitat expansion peaked, leveled, and then began its present slow decline. To keep up with recreational demand, the Dingell-Johnson Act was amended in 1984 (Wallop-Breaux Amendment) to authorize expanded tax revenue collection with special provisions for improved boating access, services, and aquatic education.

The Limits to Sportfishing

With important help from massive reservoir construction, recreational fisheries management in the United States has sustained generally good sportfishing. The number of anglers in the United States has increased over the past 40 years (Figure 16.14). In 1991, anglers in the United States alone spent $25 billion on sportfishing, rivaling the world trade in commercial fishing. However, there are indications that limits to quality sportfishing are being reached even as demand for more sportfishing grows.

Most suitable reservoir sites have been developed. Numerous existing water projects are viewed as more of a liability than an asset because of undesirable environmental effects on biodiversity. Some of the most used sportfishery management strategies are now questioned because of undesirable effects on biodiversity (Williams et al. 1989). Many reservoirs are filling with sediment and more water is being diverted to uses incompatible with sportfishing. In coastal waters, competition with food fisheries could become a limiting factor. Deteriorated water quality re-

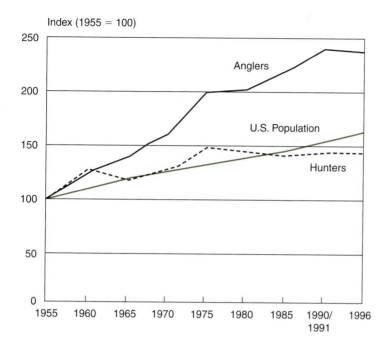

Index (1955 = 100)

FIGURE 16.14 Relative increase in sportfishing participation compared to hunting and the total U.S. population from 1955 to 1996 (courtesy of U.S. Department of Interior).

mains a limiting factor despite water quality improvements in the past several decades. The net result is the beginning of a decline in total habitat surface area and increased competition with other resource uses. Although remaining habitat can be upgraded to higher quality with improved environmental regulation, it will need more intense and smarter management to keep pace with demands from competing interests. One of the most important competing interests is protection of biodiversity.

FISHERIES BIODIVERSITY ISSUES

A Growing Concern

Long before the emergence of widespread environmental awareness in the 1960s, individuals and agencies became concerned about loss of biodiversity associated with aquatic resources. Fur seals and sea otters were decimated during the eighteenth and nineteenth centuries. Precipitous declines and extinction of unique finfish species were especially evident in the upper Great Lakes. Here a diverse assemblage of salmonid fishes known as chubs were negatively impacted by overfishing and invasion by the marine lamprey through locks and canals (Smith 1970). Desert fishes were growing scarce because of habitat alteration and water diversion (Minckley and Douglas 1991).

Although relatively few marine fish and invertebrates are at risk, marine mammals and turtles have been exceptionally vulnerable to overharvest. Several species of whales have been declared endangered. Until devices were required to allow sea turtle escapement, many turtles were trapped and drowned in fishing trawls (Figure 16.15). Accumulated evidence indicates that huge drift nets set in the oceans are a hazard to many large species, including endangered whales (Federal Register 1997), and drift nets have been prohibited in many locations.

A high fraction of native freshwater fish species became endangered or extinct over the past century. Williams et al. (1989) listed 254 living fish species and subspecies of inland waters within the United States as threatened or endangered. Nehlsen et al. (1991) identified risk to 214 salmonid stocks on the Pacific Coast. In Canada and the United States, 48% of all crayfish (Taylor et al. 1996) and 72% of mussel species (Williams et al. 1993) are at risk of extinction or already extinct.

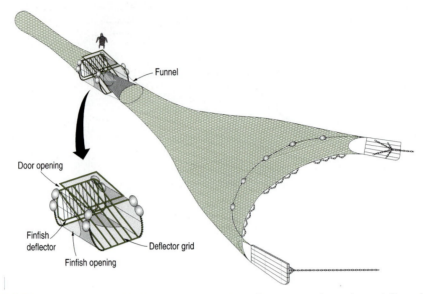

FIGURE 16.15 The turtle excluder device (TED) has been demonstrated to substantially reduce accidental catch and killing of endangered marine turtles (courtesy of USDI-Fish and Wildlife Service).

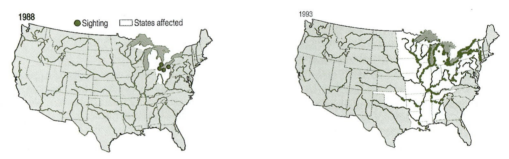

FIGURE 16.16 The zebra mussel is an exotic species introduced accidentally from Eurasia. Once introduced, it spread rapidly and added to the risk of endangerment of natural species (courtesy of USDI-Fish and Wildlife Service).

Causes include introductions of exotic species, physical habitat changes, chemical pollution, and river system modification for flood control, irrigation, and navigation purposes (Figure 16.16).

Growing concern about genetic loss through extinction resulted in passage of the Endangered Species Act of 1973. The act initiated revolutionary change in the practice of aquatic resource management. The primary intent of the Endangered Species Act is the conservation of species for their diverse human-based values (see Chapter 17). This anthropocentric basis for the law is consistent with the U.S. Constitution and a century-long tradition of utilitarian conservation based on future human needs. In practice, the protection of "species" has included unique populations within species, called stocks by fisheries professionals. Especially important at this time are threats to the diverse salmonid stocks of the U.S. West Coast. The U.S. Fish and Wildlife Service and the National Marine Fishery Service are lead organizations in protection of aquatic species.

Biodiversity Values

The construction and popularity of numerous public aquariums during the past decade reflects great public interest in the biodiversity of the world's waters. However, aesthetic, scientific, and educational value make up only a part of the value held in the genetic diversity of aquatic life. Diversity also has ecological value to the extent that it contributes to maintaining important ecosystem functions needed for human welfare. There are increasing concerns that various fish-eating mammals and birds are threatened by the commercial harvest of their foods or are contaminated with materials in their food that may reduce their reproduction rates.

Perhaps most importantly, protection of global biodiversity protects the options of future resource users (Bishop 1987, Loomis and White 1996). Once a species is extinct, there is no use option left. Unlike the self-centered focus of conservation in the early twentieth century, the Endangered Species Act enlarged the conservation perspective to *potential* resource use. The broad definition of use here includes all consumptive and nonconsumptive use, including the costs to preserve species simply to know they continue to exist.

Society has shown increasing inclination to pay for protection of what has been called "intrinsic worth" held in living organisms. The "existence value" (Bishop 1987) of this concept of intrinsic worth, independent of human value, is an important motivator for species protection (Deacon and Deacon 1991, Holmes 1991). Some argue that intrinsic value is priceless and therefore should not be considered among competing interests in determining human benefits. However, the biocentric ethic behind this approach to valuation has yet to become well established in human societies. Economic methods to estimate values associated with endangered species and biodiversity have been difficult to develop, but are becoming more available and reliable (Loomis and White 1996).

FISHERY SCIENCE AND MANAGEMENT

Early Management Emphases

Knowledge of fishery management had begun to accrue by the time medieval European landowners closed their waters and lands to public hunting and fishing. Like hunting, the early sportfishing in inland waters was reserved for "the crown" and anyone receiving land grants including water. Fishing prohibition enforced by wardens was designed to discourage subsistence fishing by hungry peasants. The perceived need for regulating access to small waters indicates the ease with which inland streams and ponds could be depleted of their aquatic resources.

Diverse Native American fisheries existed before Europeans colonized the New World. Not all Native Americans fished, but the traditions of certain cultures were intricately linked with the animals they caught. They promoted resource sustainability through personal self-regulating belief systems, usually based on spiritual respect for the harvested animals. Europeans brought new fishing methods and an exploitive culture somewhat less respectful of the animals as "brother beings." They relied on a social system of written laws, which were established in response to problems, usually as a last resort.

Local food fishery depletions caused protective regulation among East Coast settlements as early as the seventeenth century, soon after European colonization. It was based in a legal tradition established in Europe centuries before, but was decided more democratically through town government instead of by the lord of the manor. Only local and spotty management remained in effect, however, until after the U.S. Civil War.

Linnaeus's development of an organized classification of life forms spurred a growing scientific interest in natural history and species inventory. A prolonged period of natural resource exploration and inventory followed and grew as travel improved with better boats, roads, and the railroad. Much of the earliest investigation was financed privately. Government financing grew slowly. A number of initial surveys were conducted during exploration of new transportation routes, starting with the Lewis and Clark Expedition in 1804–06, and continuing throughout the nineteenth century. Other

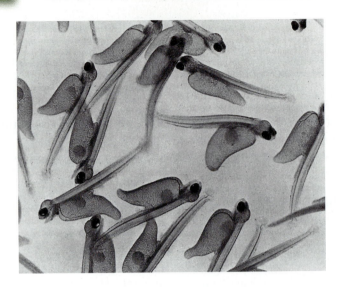

FIGURE 16.17 Fish stocking of fry (shown here), small "fingerlings," or larger sizes has been an important management strategy for over a century (courtesy of USDI-Fish and Wildlife Service).

studies were conducted through colleges and museums. The Philadelphia Academy of Science was an early leader (McHugh 1970) followed by Harvard College, which initiated marine surveys sponsored by the U.S. Coastal Survey (McClane 1974). Many of the scientists associated with museums and educational institutions contributed to the development of modern fishery science and management late in the nineteenth century.

Modern Fishery Management

Concerns.   By the time of the U.S. Civil War, fish stocks were locally depleted near many settled areas and popular tourist resorts. The Commission of Fish and Fisheries, later to become the U.S. Bureau of Fisheries, was created in 1871 by Congress to address the issue of declining fisheries (Thompson 1970). The commission initiated inventories of fishery resources, concentrating on marine resources, which continued over the next several decades. Except for the Great Lakes, inland inventories were left more to the discretion of individual states, educational institutions, and private interests.

Much of the loss of sportfisheries was blamed on indiscriminate harvest methods, such as poisons and gill nets, which were used for market fishing. Between 1857 and 1871, ten states formed fish commissions to regulate fishing and to reduce impacts of dams (Thompson 1970). By 1875, New York had outlawed use of poisons and required fish sluices to be built in dams where migratory fishes occurred (Scott 1875). For certain species, seasons were closed to sportfishing and gear was limited to hook and line angling. Other states and Canadian provinces enacted similar laws. By the turn of the century, most states and provinces had authorized commissions and supporting agencies to establish and enforce game and fish regulations.

Stocking and Introduction.   Following the lead of European advances, stocking of fish fry and small "fingerlings" was advocated as an effective way to recover and sustain fisheries (Scott 1875), and is retained today for certain fisheries (Figure 16.17). Hatcheries initiated large-scale production in the United States just after the Civil War. Fish were cultured to stock depleted habitats and to serve as a base for new species introductions. By 1870, about 200 individuals were operating private hatcheries in the United States. They met to form the American Fish Culturist's Association, the precursor of what was to become the American Fisheries Society in 1884 (Thompson 1970).

The fish culturists were effective activists. Scott (1875) suggested: "Every farmer should have a trout preserve" for the hatching of eggs, stocking, and sale to others. In addition to trout, some 28 other species were successfully spawned, hatched, and stocked with fish fry into various waters where stocks of Atlantic salmon, American

FIGURE 16.18 Many fish species were introduced to waters they did not previously inhabit. After 1973 the Endangered Species Act slowed the process down because of negative effects on rare native species. However, accidental introduction remains a big problem, such as by release of live bait (courtesy of USDI-Fish and Wildlife Service).

shad, and striped bass had declined (Bowen 1970). These anadromous species, which migrated from the sea to spawn, were severely depleted in northern coastal streams where industrial mills built numerous dams. The fish culturists spurred federal legislation to assess and improve fishery conditions.

In 1873, the Commission on Fish and Fisheries was authorized by the U.S. Congress to support aquaculture, fry stocking, and introduction of desirable species. More than 90% of the commission's activity was devoted to aquaculture, an emphasis that continued for the next six decades at a federal facility near Woods Hole, Massachusetts.

Local anglers, fish culturists, and entrepreneurs sought to "improve" native fisheries by widely introducing European species to U.S. waters. The Federal Fish Commissioner promoted and facilitated widespread introduction of carp and other exotic species, especially along the railroads, which burgeoned after the Civil War. In addition to the European brown trout, many native sport species were widely introduced to new locations. They included brook trout, rainbow trout, largemouth bass, northern pike, muskellunge, Atlantic salmon, and channel catfish. The eggs of striped bass and American shad, both Atlantic Coast natives, were successfully introduced into California coastal waters soon after the transcontinental railroad was completed in 1869 (Figure 16.18).

Habitat Quality Issues.   By the end of the nineteenth century, water pollution was intense in settled watersheds. Causes included poor livestock management, poor logging practices, untreated sewage, slaughterhouses, and flour, saw, pulp, textile, metal, and industrial mills. Mill dams blocked fish movement in many streams.

The American Fish Culturist's Association lobbied for legislation to curb pollution and provide fish ladders around dams on coastal rivers, but emphasized stocking and regulation of indiscriminate harvest as the primary needs (Scott 1875). The first comprehensive statewide and federal pollution laws were enacted about the turn of the century, outlawing certain types of waste dumping into public waters. Investigation of fish kills from pollutants is essential to preventing further pollution and kills (Figure 16.19).

Public Land Management.   Expansion of public sportfisheries became a popular management goal for newly created land management agencies. The first national parks and state reserves were set aside before 1900, and the national forests followed soon after. Stocking to sustain recreational fishing was widely promoted even in national parks. The negative effects of stocking on native aquatic communities were unknown at the time. The conservation of genetic information never received consideration because

FIGURE 16.19   Sleuthing the source of fish kills like the one shown here is an important aspect of fishery management associated with pollution prevention (courtesy of USDI-Fish and Wildlife Service).

the genetic basis of evolution was yet to be elaborated. Wide availability of managed public lands and waters, particularly in the West, helped encourage continued growth of sportfishing as human population, means of travel, and leisure time increased.

Scientific Management.   Most of the main research themes of fishery biology were initiated before World War I (McHugh 1970). The earliest research concentrated on classifying the different species of fish, which started in the United States soon after Linnaeus developed a satisfactory classification system. Research also grew out of aquacultural needs for understanding the early life stages of fish, starting early in the nineteenth century in Europe. Nutritional, health, and predation studies were conducted in association with fish culture studies. The hypothesis that most fish mortality occurs during the first few weeks of life was first proposed in 1914. It focused attention on the suitability of habitat for the earliest life stages.

In addition to exploratory studies on distribution and life history, the first successful studies of fish movement by marking fish and recapturing them was conducted by the Commission of Fish and Fisheries in 1873. Age and growth studies, the foundation of modern fish population dynamics, were initiated about the turn of the century. Studies of fish population dynamics started with Petersen's (1903) concept of a sustained population at equilibrium between fish death rate (including fishing mortality) and fish birthrate. The earliest physiology and behavior studies were reported in 1913 because of concern for water pollution effects. About the same time, the first research into differences among fish stocks within a species was based on fish morphology.

In 1918, the American Fisheries Society began a campaign to develop fisheries science curricula in universities, much like those previously developed for agricultural colleges (Carlander 1970). By then, a school of fisheries had been established at the University of Washington and courses had been started at Cornell University. Before then, most training had been done by apprenticeship under early fish culturists with little formal education. Fishery science grew slowly at first and was almost entirely biological in orientation. Based on membership in the American Fisheries Society, the number of fishery professionals remained steady from 1920 to the end of World War II (Benson 1970).

After WWII, the number of professionals increased rapidly through the 1960s (Benson 1970) and has slowed during recent years. Numerous college programs were developed (Carlander 1970). The growth of fishery science was closely associated

FIGURE 16.20 Fishery management education was at first highly focused on fishery biology, but increasingly a human dimension has been added. Many managers now must educate the public about their professional duties (courtesy of the Fish and Wildlife Service).

with development of federal aid for sportfisheries management, growth of food-fish aquaculture, sea grant, and environmental and endangered species legislation. These new sources of funding became available mostly between 1950 and 1984, the period of most rapid professional growth.

Fishery education became highly focused on biological processes to the extent that the title "fishery biologist" was nearly synonymous with fishery scientist (Figure 16.20). The social side of fisheries received little educational attention and remains underdeveloped to this day. This educational orientation had much to do with the development of principles guiding fishery management.

Management for Biodiversity. Until the Endangered Species Act, the attention of fishery management agencies was directed mostly at consumptive use, in keeping with a long tradition established in the last century. The federal act was quickly followed by companion state laws that authorized federal fund expenditures, state enforcement, and listing of additional state-protected species. Numerous fish species were included on the original lists of threatened and endangered species. In this way, the authorites of state and federal fishery management agencies were substantially expanded and diversified, resulting in a greater diversity of human interests served by each of them.

Because consumptive resource management activities sometimes conflicted with the intents of the Endangered Species Act, the new authority caused significant changes in agency goals and operations. Introductions of exotic organisms, continuous stocking of nonnative stocks, increased reservoir surface area, and various pollutants contributed to native stock declines. Some exotic species are superior competitors and predators primarily responsible for reduced populations of certain native species (Williams et al. 1989). In addition to introduced sportfish, bait-bucket transfer introduced highly competitive minnow species. Certain crayfish species were introduced to control aquatic plants, which interfered with sportfish harvest (Figure 16.21). They also competed with native crayfish (Williams et al. 1993). Through inbreeding, stocking programs have homogenized the native stock diversity of a number of species. The great increase in sportfishery habitat, associated with reservoirs and water diversion, also reduced and fragmented pristine habitat for many native species (Minckley and Douglas 1991).

Key Management Principles

Sustaining Harvest Value. The underlying rationale for fishery science and management is the assurance of fishery resource sustainability for human benefit. This concept was distilled with greater understanding of the relationship between fishery harvest and fishery renewal rates. The management principle of fishing for maximum sustained yield evolved out of quantitative understanding of fishery population dynamics in the 1930s (McHugh 1970). It postulated that the maximum yield of

FIGURE 16.21 Crayfish are important invertebrate members of freshwater ecosystems. Management of crayfish is needed to assure rare species are sustained and common ones are not allowed to invade areas where they can do damage to other crayfish species or to desirable natural aquatic vegetation (courtesy of USDI-Fish and Wildlife Service).

a stock could be sustained indefinitely at some stock number less than the unfished stock number. For a number of stocks, the maximum sustained yield appeared to exist when fishing reduced the original stock number by half (Figure 12.5). Assuming no interaction among stocks, the maximum yield sustained by all fisheries would result from the sum of the individual stocks. The concept of maximum sustained yield proved to be an inadequate guiding principle, however.

Over time it became clear that fishing changed stock quality by reducing average age and size of fish in the remaining population. Because management based on maximum sustained yield did not consider changes in yield quality, it was inconsistent with the economic value of the fisheries. In the early 1940s, a discourse began between biologists and economists in a debate about the main objective of fishery management (McHugh 1970). Was it to be a biological target of maximum sustained yield of fish biomass or was it to be an economic target of maximum sustained human welfare based on optimum sustained yield?

Management based on the principle of optimum sustained yield considers social factors in addition to total yield such as the quality of catch and, for recreational fishing, the fishing experience. It holds that the maximum value sustained from a fishery occurs at some optimum yield, which typically is less than the maximum sustained yield (Figures 16.22 and 16.23). An intellectual shift toward the concept of optimum sustained yield occurred following a 1928 symposium (Roedel 1975). However, management of both food and recreational fisheries has had difficulty moving away from the biological concept of maximum sustained yield. Partly responsible is the past biological training in the fishery profession, which has emphasized resource quantities over economic qualities. Another reason is the greater complexity involved in determining optimum sustained yield.

Maximum sustained yield as expressed in Roedel (1975) emphasizes food and recreational fishery values. However, the concerns legislated in the Endangered Species Act moved the fishery profession toward a refined concept of optimum sustained yield. This shift is based less on quantified economic theory and more on the professional imperative to ensure species viability is sustained through time.

Habitat Management.   Since the early recognition of chemical and physical impacts on fisheries, habitat improvement has been a cornerstone principle for fishery management. All other management is contingent on habitat suitability. Good references on management of fish habitat, including the influential watershed, are provided in Kohler and Hubert (1993), Nielsen and Johnson (1983), Stroud (1992), and Williams

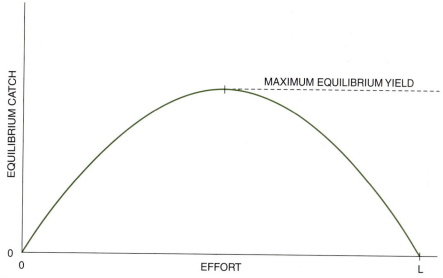

FIGURE 16.22 The theoretical average catch at equilibrium with fishing effort. The maximum equilibrium yield is indicated halfway between no fishing and fishing to catch extinction. The maximum equilibrium yield is the maximum sustained yield (from Roedel 1975).

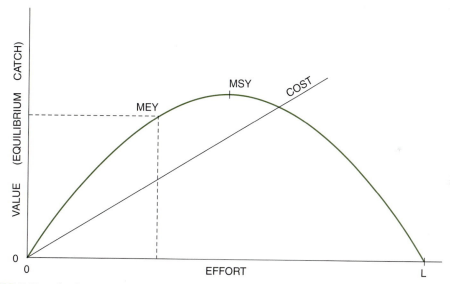

FIGURE 16.23 The theoretical maximum economic yield, which occurs at an optimum sustained yield less than the maximum sustained yield. The difference is caused by a number of factors such as the average size of fish caught (from Roedel 1975).

et al. (1997). Millions of habitat modifications have been made using public funds, including construction of thousands of small lakes and ponds under programs administered by the U.S. Natural Resources Conservation Service, the U.S. Fish and Wildlife Service, and state agencies. However, the largest habitat effects are associated with water development projects justified for other primary purposes such as water supply, flood control, navigation, hydropower, and environmental protection, and also for secondary uses such as sportfishing. In addition to the environmental restoration authority first assigned to the U.S. Army Corps of Engineers in 1986, the Sustainable Fisheries Act of 1996 and the Estuary Restoration Act of 2000 authorized the National Marine Fishery Service and cooperating agencies to restore degraded estuarine and marine habitat.

In recent decades, a more holistic approach to habitat management has been advocated by government and nongovernment organizations. During the 1950s and

1960s, degraded water quality was the main limitation to better fishing in existing waters. While some techniques existed for on-site treatment, they were expensive and of limited effect. Much more comprehensive management at the source was needed; therefore, sportfishing interests combined with other environmental interests to ensure passage of the Clean Water Act of 1972. It has been the single most important force in assuring suitable water quality for fish populations. Before the act, contamination with toxic materials and periodic or sustained oxygen depletion were much more common. However, nutrient-caused eutrophication from nonpoint sources, such as excess fertilizers from crops and lawns, continues to be a problem (see Chapter 7), causing dense plant growths and low oxygen concentrations. Expensive treatments can be used locally, such as artificial aeration and chemical treatment to precipitate nutrients. In large reservoirs, drains can sometimes be engineered to allow greater control of downstream water quality. Riparian vegetation can also be managed to intercept nutrient-rich runoff, reduce bank erosion, and control shade and temperature of stream habitats. Fencing to control livestock access often is required.

The construction of small and large reservoirs greatly augmented surface area suitable for sportfishing, but it negatively impacted many stream and river species. Numerous endangered species are now associated with impasses to fish and other organism movements resulting in much corrective engineering in the form of fish ladders and other bypasses around turbines. Other management techniques center on controlling sediment fill rate and massive growths of aquatic plants in lakes, usually resulting from accelerated watershed erosion and nutrient runoff. The solutions range from engineering of sediment-control structures to restoring natural watershed processes.

A wide variety of artificial structures for sediment retention, bank stabilization, flow-diversion, fish spawning, and fish retreats have been designed and installed to improve both degraded and suboptimal natural habitat. Water-level and discharge fluctuations have been managed to limited degree, usually in and downstream from reservoirs. Lake sediments may be mechanically dredged and plant growth controlled with mechanical harvesters, herbicides, or introduced herbivores (e.g., grass carp, crayfish). These expensive methods have had mixed success, however, and where problems originate from watershed changes, restoration of more natural watershed conditions may prove more cost effective. In recent years, a more integrated watershed or other ecosystem approach has been increasingly advocated.

Ecosystem Management.   The great complexity of managing for diverse aquatic resource needs, including compliance with the Endangered Species Act, has led the responsible federal agencies and many state agencies toward a more holistic approach to management. This new approach, often referred to as ecosystem management, emphasizes the close association of ecological and social processes. It seeks to resolve conflicts among various social sectors with stakes in aquatic management decisions. Most agencies concerned with aquatic resource management have used a watershed approach. This is because watersheds are among the most common means by which aquatic ecosystems are bounded. Within the watershed boundaries, habitat and aquatic communities are viewed more holistically as the basic unit of management. Ecosystem management integrates habitat and population management, and emphasizes adaptive management via assessment of results and refinement when needed.

Watershed abuse has lead to increased nutrient, sediment, and pesticide concentrations. The need for more comprehensive understanding is widely recognized as social demands on resources trend more toward collision with resource sustainability goals. Industrial development, including atmospheric contamination of rain and snow, has altered the basic chemistry of watersheds and contributed to toxicity and food web contamination. Because they are near the end of the chain of these effects, estuaries often are drastically affected by contaminants. Many ecosystems also suffer from widespread purposeful and accidental introduction of highly competitive exotic species. These exotics have contributed to the decline of numerous native species, which evolved separately from the exotics and were not adapted to live with them. Human-caused environmental impacts tend to favor fewer species with wide environmental tolerances.

The stakeholders affected by the management of watershed ecosystems include many diverse groups whose actions might be regulated because they affect water quantity and quality. These stakeholders include people associated with timber harvest, crop culture, urban development and maintenance, range grazing, mining, transportation networks, utilities operations, and virtually all other human activities, both private and public. Seeking maximum sustained benefit out of optimum provision for each of these diverse and often conflicting interests is the central challenge of fisheries management. In this context, the following section describes some of the common activities of future fishery professionals.

THE FISHERY PROFESSIONAL

Changing Expectations

Aquatic resource professionals consistently have focused their efforts on sustaining aquatic resources. The range of their focus has greatly widened as the complexity of resource management has grown. If formally trained at all, most of the first fishery professionals were trained as fish culturists and ichthyologists. Only a handful of U.S. universities offered fisheries programs before WWII (Carlander 1970). Now the number offering several specialized courses in fisheries exceeds 100.

The concerns of aquatic resource professionals have diversified and expanded beyond the narrow focus on harvestable stocks. However, "fishery" remains firmly entrenched in professional titles and optimum sustainable yield (or potential for yield) remains a central management concept. As the complexity of management for diverse interests has grown, fishery professionals increasingly find themselves in the company of professionals from other disciplines, such as oceanography, limnology, hydrology, forestry, range science, soil science, wildlife science, economics, and political science. The profession now relies on interdisciplinary team approaches to solve many pervasive, critical, and complex problems. In the United States, the lead fisheries professional organization remains the American Fisheries Society. It functions to enhance communications among fishery professionals and policy makers through technical publication, conferences, lobbies and other services. It supports numerous sections representing geographical regions and subdisciplines. The society is in the forefront of changing expectations for fishery resource management.

The activities of many contemporary fishery scientists are only remotely connected to fish catch and harvest. Many are concerned about sustaining aquatic populations and ecosystems for other reasons. Scuba diving, beach combing, and aquarium keeping, for example, are pastimes that need not be consumptive for enjoyment of aquatic resources. As more is learned about the role of different species in sustaining aquatic functions, species are increasingly managed for their ecological value. A prevalent contemporary employment focus is on protection of threatened and endangered species. One of the more rapidly growing aspects of the fishery profession in recent years has been the "human dimension," including sociology, economics, ethics, law, and conflict resolution.

Fishery professionals typically are engaged in one or more of several core activities. These include (1) bioassessment and management; (2) habitat assessment and management; (3) resource demand assessment and management; (4) resource user satisfaction assessment; (5) management effectiveness assessment; (6) management planning, coordination, and administration; (7) original research; and (8) education. Some aquatic resource professionals are specialists; particularly those engaged in original research. Others are generalists, participating in numerous activities. Fishery professionals also may become involved in various support activities such as development of new habitat and law enforcement. However, these support services often do not require a fishery education as long as fishery professionals guide operations.

Bioassessment and Management

Bioassessment.   The status of all kinds of aquatic species are assessed through surveys of distribution, abundance, age and size structure, birthrate, growth rate, death rate, nutritional status, and health (Nielsen and Johnson 1983, Schreck and Moyle 1990, Kohler and Hubert 1993). Analytical skills are prerequisite, including at least an elementary understanding of mathematics and statistics. Some fishery professionals become biometricians, but most work with statisticians to accomplish work objectives. Most professionals in this area have earned an M.S. or Ph.D. degree.

The main objective of most bioassessments is to identify changes in resource status as a consequence of management, habitat modification, change in harvest, or suspicion of species decline and endangerment. Scientific surveys have adapted methods used in commercial harvest as well as new methods such as electrofishing. In the past, commercial fish and sportfishes have been the main objects of such surveys. Now diverse vertebrate, invertebrate, plant, and microbial populations may be surveyed because they are ecological indicators of ecosystem productivity, sustainability, and condition.

Management.   Past management of aquatic populations has relied on introductions of new species, reduction of less desirable species, and routine stocking of desirable species. Stocking of popular sportfish species remains the backbone of biomanagement, although its effect on genetic diversity of native stocks is an increasing concern (Winter and Hughes 1997). Aquatic species are now rarely introduced to new habitats because of their potential negative impacts on native fish. Reduction of abundance of certain competitor and predator species to favor desirable species may have merit in some circumstances when the cost-benefit ratio is favorable. Such management may be pursued to protect endangered species as well as to increase recreational or commercial fish species.

Habitat Assessment and Management

Assessment.   Habitat assessment involves measurement of habitat extent and suitability for various species. It relies on the sciences of limnology, oceanography, and wetland science (Wetzel 1983, Cole 1994, Horne and Goldman 1994, Mitsch and Gosselink 2000). Different aquatic communities are broadly associated with habitat features based on temperature, salinity, pH, flow velocity, plant nutrients, bottom condition, water clarity, the dependability of oxygen concentration, and the concentrations of toxic materials. Although environmental quality has generally improved since the low point in the 1960s, it remains problematic in numerous waters.

Habitat assessment also examines the amounts and arrangements of habitats needed for the main life stages from eggs to reproducing adults. This is especially critical for management decisions. Habitat assessment often requires interactions with hydrologists, chemists, engineers, and others familiar with the relationships among climate, watersheds, and aquatic habitats. The dimensions of habitat assessment have increased as understanding of ecosystem process has improved.

Management.   Management of aquatic habitat may be as simple as adding bottom structure, such as artificial reefs or spawning gravel for protective cover or reproduction (Kohler and Hubert 1993) (Figure 16.24). It also may involve engineering more complex structures including dams and fish ladders over dams, or structures that stop the passage of unwanted species. On the other hand, more old engineered structures are being modified or removed to restore natural connections among habitats. Many degraded habitats can be cost-effectively aerated to increase oxygen con-

1. Wing deflector
2. Bank cover
3. Half-log
4. Riprap
5. Cattle crossing
6. Brush removal
7. Brush shelter

FIGURE 16.24 Fish habitat improvements (from Owen et al. 1998).

centration, fertilized to increase productivity, or dredged to remove sediment or aquatic plants.

In its most complex form, habitat management integrates the management of whole watersheds. Sustaining an optimum balance of surface water and groundwater contributions to aquatic habitats, controlling erosion of sediment and nutrients, sustaining optimum habitat shapes and sizes, and controlling pollution are all parts of the management process (see Chapter 7). Habitat management requires close coordination and cooperation among people responsible for management of various parts of watershed ecosystems. Because habitat degradation is the major cause of aquatic resource decline, habitat management often is the key to recovery.

Resource Demand Assessment and Management

Assessment. Aquatic resource demand typically is evaluated by monitoring resource use. It includes assessing the amount of fishing (the fishing effort) and fish death caused by harvest and by-catch. Hooking mortality of returned sportfish is an important consideration. In on-site surveys, the condition of both harvested and returned fish is assessed including sizes, ages, and general health (Sigler and Sigler 1990, Schreck and Moyle 1990, Kohler and Hubert 1993). Differences in the characteristics of caught fish from those left in the habitat are examined for characteristics and trends that could indicate reduced sustainability or other important change.

In sportfisheries, the resource demand sometimes is estimated indirectly through angler reports of what they harvested. This method provides an inexpensive and comprehensive means for assessing fishing activity. But because of less control over accuracy, it usually is used in combination with on-site assessments by professionals (Pollock et al. 1994).

Management. Resource demand is managed mostly through setting regulations and through public education. Regulation setting requires knowledge of fish population behavior, population dynamics, and renewal rates as well as knowledge about the acceptability of the regulation to the user public and the difficulty entailed in enforcement (Kohler and Hubert 1993). Regulations may close habitats for specified times, restrict the numbers and sizes harvested, or restrict the fishing methods used.

Resource users often are educated about the rationale behind regulations, which sometimes encourages self-monitoring in place of regulations. Education may be used to improve fishing efficiency while reducing unnecessary mortality. Many management programs report where the sportfishing is currently most rewarding and provide helpful hints on how anglers can be more successful at underused sites. Education also may be used to develop demand for underused species and reduce demand for overused species.

Resource User Satisfaction

Knowledge of resource user satisfaction helps to assess management effectiveness. Satisfaction is an expression of the value obtained from resource use. A high satisfaction index implies highly rewarding use. Various levels of sophistication are used to assess resource user satisfaction, but most depend on well-designed mail, telephone, or on-site surveys.

Satisfaction surveys become especially useful when satisfaction estimates can be compared to the management costs. That kind of information allows managers to redistribute their management revenues toward the most satisfying use. In that way they maximize benefits to the resource user relative to management costs.

One of the recurrent questions associated with benefit-cost analyses using satisfaction indices is the comparability of satisfaction indices among various activities. Is the basic value of a day of trout fishing on a remote wilderness lake the same as a day of angling for sunfish at the local pond? Answering this or similar questions requires an economic estimate of angler willingness to pay for their fishing. Therefore, fishery science increasingly has an economic orientation.

Management Administration

Much of what modern aquatic resource professionals do involves coordination with other professionals and with the public (Kohler and Hubert 1993). This aspect of the profession demands good communication and interpersonal skills as well as a thorough knowledge of aquatic resource ecology and sociology. As more species become scarce and their demand increases, conflicts over resource use and allocation will intensify.

Conflict resolution is becoming a critical part of public service because of growing and diverse demands for limited resource supplies. Often at the heart of conflict is inadequate knowledge regarding how various stakeholders will be affected by implementation of natural resource decisions. Economic analysis, although imperfect, can help quantify the various degrees of benefit gain or loss by various stakeholder groups.

Much of what a fishery professional does involves teamwork and coordination with people of diverse educational backgrounds. Communication skills are critical. In addition to clear, complete, and concise technical writing skills, fishery professionals need strong and public speaking and social skills. They need to be able to work effectively with people having diverse values. They also need to be careful planners, considering alternative pathways to desired ends. A critical aspect of future aquatic resource management will involve the development of attainable goals for future ecosystem condition and the planning of strategies to reach these goals.

Original Research

Original research is conducted to develop new understanding of ecological, social, and managerial processes. It is the means by which assessment techniques are improved. Original research in fisheries is conducted mostly in research universities and in government laboratories. Most original research is led by a "principal investigator" who has earned a Ph.D. in a highly specialized area. The principal investigator often supervises others who actually carry out the specialized activities involved in research. These research technicians frequently have earned an M.S. degree, but some find positions with a B.S. degree. They often have relatively little to do with research design or communication of results, but are experts on specific methods. Research in universities usually is conducted with graduate assistants because education of future researchers is a primary mission.

Many of the easiest research problems have been well addressed, if not entirely solved. Therefore, research is becoming more interdisciplinary and team-oriented, as the dimensions of primary problems become better known. To be successful, team researchers need to develop communication and interpersonal skills in addition to the specialized skills required in their specialties. This usually means learning more about the other disciplines.

Education

Most fishery professionals have some opportunity to educate other professionals and the lay public. However, relatively few teach in universities or other formal educational environments. The most effective educators are themselves broadly enough educated to understand the diverse needs and interests of students.

In the fisheries professions, university education is closely tied to original research. With careful planning, research and teaching can merge to produce greater strengths in both areas. The best educators serve as models for their graduates. They are enthusiastic, optimistic, and adaptive while remaining realistic about professional challenges. One of the greatest challenges for fishery educators is to develop and use their specialized skills in a more integrated and interpersonal learning environment. They play a key role in teaching students, by example, how to work effectively in groups of diverse people.

FUTURE FISHERY ISSUES

A quick review of the predominant issues in fishery resource management shows recurrent themes in a variety of aquatic ecosystems settings. They include the following issues and concerns.

■ The sustainability of aquatic resources is threatened by growing human demands on aquatic ecosystems and by inadequate past management. Fishery professionals need to address diverse challenges to maximize human benefit while sustaining resource renewal.

■ The negative effects of environmental change in fishery ecosystems have been cumulative and interactive. Predictive understanding and more effective management requires a more holistic ecosystems approach.

■ Understanding of the ecosystem process, including cultural influences, remains too qualitative to accurately predict the beneficial consequences of costly "corrective" management proposals. Adaptive management, which incorporates research into the management process, is a suitable approach for analyzing management effectiveness and improving benefits estimation.

■ The negative effects of overfishing have continued to worsen in many locations despite warning and regulation. This has occurred mostly because (1) resource demand exceeds optimum sustained yield, (2) short-term emphasis on employment outweighed long-term sustainability, and (3) by-catch mortality was poorly controlled.

■ There is a need for accelerated assessment of biodiversity, especially in regions of rapid environmental change, and improved protection of resource options before they become threatened with extinction.

■ All relevant private and public benefits and costs associated with fishery management have not been estimated objectively and thoroughly enough for sound decision making. Greater investment is needed in benefits estimation and cost accounting in all areas of fishery management, including biodiversity.

■ A professional belief that recovery of ecosystem "integrity" is the most appropriate means for obtaining optimum sustained benefit has gained considerable "scientific" credence. However, little empirical information is available to determine appropriate recovery targets. This requires a greater investigation of the relationship of biodiversity to ecosystem structure and functions.

■ There needs to be improved education of fishery professionals in the United States, emphasizing interdisciplinary integration in a systems approach to management. Agencies need to be encouraged to work in an integrated way toward the ideal of improved general public welfare.

These fishery issues need to be addressed by a combination of research, technical, policy, education, and public relations strategies. Students who elect to enter the fisheries profession can look forward to personal involvement in at least some of these issues over the next several decades.

LITERATURE CITED

Ahlstrom, E. H. and J. Radovich. 1970. Management of the Pacific sardine (pp. 183–193). In N. G. Benson, ed., *A century of fisheries in North America.* Washington, DC: American Fisheries Society.

Bell, F. H. 1970. Management of Pacific halibut (pp. 209–221). In N. G. Benson, ed., *A century of fisheries in North America.* Washington, DC: American Fisheries Society.

Benson, N. G. 1970. The American Fisheries Society, 1920–1970 (pp. 13–24). In N. G. Benson, ed., *A century of fisheries in North America.* Washington, DC: American Fisheries Society.

Bishop, R. C. 1987. Economic values defined (pp. 24–33). In Daniel J. Decker and Gary R. Goff, eds., *Valuing wildlife: Economic and social perspectives.* Boulder, CO: Westview Press.

Bowen, J. T. 1970. A history of fish culture as related to the development of fishery programs (pp. 71–93). In N. G. Benson, ed., *A century of fisheries in North America.* Washington, DC: American Fisheries Society.

Cairns, M. S. and R. T. Lackey. 1992. Biodiversity and management of natural resources: The issues. *Fisheries* 17(3):6–10.

Carlander, K. D. 1970. Fishery education and training (pp. 57–70). In N. G. Benson, ed., *A century of fisheries in North America.* Washington, DC: American Fisheries Society.

Cole, G. 1994. *Textbook of limnology.* 4th ed. Prospect Heights, IL: Waveland Press.

Deacon, C. and J. E. Deacon. 1991. Ethics, federal legislation, and litigation in the battle against extinction (pp. 109–121). In W. L. Minckley and J. E. Deacon, eds., *Battle against extinction: Native fish management in the American West.* Tucson, AZ: The University of Arizona Press.

Estes, J. A., M. T. Tinker, T. M. Williams, and D. F. Doak. 1998. Killer whale predation on sea otters linking oceanic and nearshore ecosystems. *Science* 282:473–475.

Food and Agriculture Organization (FAO). 2000. Production yearbook, United Nations FAO Statistics Series No. 52, Rome, Italy.

Federal Register. 1997. *Endangered fish or wildlife: Special prohibitions.* North Atlantic Right Whale Protection. Proposed rule; temporary closure of fishery. National Marine Fisheries Service. Vol. 62, No. 212, 59335.

Goldberg, R. J., M. S. Elliott, and R. L. Naylor. 2001. *Marine aguaculture in the United States: Environmental impacts and policy options.* Arlington, VA: Pen Oceans Commission.

Hardin, G. 1968. Tragedy of the commons. *Science* 162:1243–1248.

Holmes, R. III. 1991. Fishes in the desert: Paradox and responsibility (pp. 93–108). In W. L. Minckley and J. E. Deacon, eds., *Battle against extinction: Native fish management in the American West.* Tucson, AZ: The University of Arizona Press.

Horne, A. J. and C. R. Goldman. 1994. *Limnology.* 2nd ed. New York: McGraw-Hill.

Jackson, J.B.C. et al. 2001. Historical overfishing and the recent collapse of coastal ecosystems. *Science* 293: 629–638.

Jenkins, R. 1970. Reservoir management (pp. 173–182). In N. G. Benson, ed., *A century of fisheries in North America.* Washington, DC: American Fisheries Society.

Johnson, F. G. 1989. Fisheries: Harvesting life from water (pp. 2–9). In F. G. Johnson and R. G. Stickney, eds., *Fisheries: Harvesting life from water.* Dubuque, IA: Kendall/Hunt.

Kohler, C. C. and W. A. Hubert, eds. 1993. *Inland fisheries management in North America.* Bethesda, MD: American Fisheries Society.

Kurlansky, M. 1997. *Cod: A biography of the fish that changed the world.* New York: Penguin Putnam.

Larkin, P. A. 1970. Management of Pacific salmon of North America. (pp. 223–236). In N. G. Benson, ed., *A century of fisheries in North America.* Washington, DC: American Fisheries Society.

Lerman, M. 1986. *Marine biology.* Menlo Park, CA: Benjamin/Cummings.

Loomis, J. B. and D. S. White. 1996. Economic values of increasingly rare and endangered fish. *Fisheries* 21(11):6–11.

McClane, A. J., ed. 1974. *McClane's new standard fishing encyclopedia.* New York: Holt, Rinehart and Winston.

McGinn, A. 1997. Global fish catch remains steady (pp. 32–33). In Linda Stark, ed., *Vital signs: The environmental trends that are shaping our future.* New York: W.W. Norton & Company.

McHugh, J. L. 1970. Trends in fishery research. In N. G. Benson, ed., *A century of fisheries in North America.* Washington, DC: American Fisheries Society.

Miller, G. T. 1990. *Resource conservation and management.* Belmont, CA: Wadsworth.

Miller, M. L. and F. G. Johnson. 1989. Fish and people (pp. 10–23). In F. G. Johnson and R. R. Stickney, eds., *Fisheries: Harvesting life from water.* Dubuque, IA: Kendall/Hunt.

Minckley, W. L. and M. E. Douglas. 1991. Discovery and extinction of western fishes: A blink of the eye in geologic time (pp. 7–18). In W. L. Minckley and J. E. Deacon, eds., *Battle against extinction: Native fish management in the American West.* Tucson, AZ: The University of Arizona Press.

Mitsch, W. J. And J. G. Gosselink. 2000. *Wetlands.* 3rd ed. New York: Van Nostrand Reinhold.

National Marine Fisheries Service, National Oceanic and Atmospheric Administration. U.S. Department of Commerce. 2000a. *Fisheries of the United States, 1995.* Silver Spring, Maryland.

National Marine Fisheries Service, National Oceanic and Atmospheric Administration. U.S. Department of Commerce. 2000b. *Our living oceans: Report on the status of U.S. living marine resources, 1995.* Silver Spring, Maryland.

National Research Council. 1999. Washington DC: Sustaining marine fisheries. National Academy Press.

Nehlsen, W., J. E. Williams, and J. E. Lichatowich. 1991. Pacific salmon at the crossroads: Stocks at risk from California, Oregon, Idaho, and Washington. *Fisheries* 16:4–21.

Nielsen, L. A. and D. L. Johnson, eds. 1983. *Fisheries techniques.* Bethesda, MD: American Fisheries Society.

Northridge, S. P. 1984. *World review of interactions between marine mammals and fisheries.* FAO Technical Paper 251. Rome, Italy: Food and Agriculture Organization of the United Nations.

Owen, O. S., D. D. Chiras, and J. P. Reganold. 1998. *Natural resource conservation.* 7th ed. Upper Saddle River, NJ: Prentice-Hall.

Pennak, R. W. 1978. *Freshwater invertebrates of the United States.* 2nd ed. New York: John Wiley & Sons.

Petersen, C. G. J. 1903. What is overfishing? *Journal of Marine Biology Association.* (U.K.) 6:587–594.

Pollock, K. H., C. M. Jones, and T. L. Brown. 1994. *Angler survey methods and their applications in fisheries management.* Bethesda, MD: American Fisheries Society Special Publication 25.

Radcliffe, W. 1921. *Fishing from the earliest times.* Unchanged reprint of the edition. Chicago, IL: Ares Publishers.

Rhyther, J. H. 1959. Potential productivity of the sea. *Science* 130:602–608.

Roedel, P. M., ed. 1975. *Optimum sustainable yield as a concept in fisheries management.* Proceedings of a Symposium. Special Publication No. 9. Washington, DC: American Fisheries Society.

Schreck, C. B. and P. B. Moyle, eds. 1990. *Methods for fish biology.* Bethesda, MD: American Fisheries Society.

Scott, G. C. 1875. *Fishing in American waters.* New York: The American News Company.

Sigler, W. F. and J. W. Sigler 1990. *Recreational fisheries management, theory, and application.* Reno, NV: University of Nevada Press.

Smith, S. H. 1970. Trends in fishery management of the Great Lakes. In N. G. Benson, ed., *A century of fisheries in North America.* Bethesda, MD: American Fisheries Society.

Stroud, R. H., ed. 1992. *Fisheries Management and watershed development.* American Fisheries Society Symposium 13. Washington, D.C.

Taylor, C. A., M. L. Warren Jr., J. F. Fitzpatrick Jr., H. H. Hobbs III, R. F. Jezerinac, W. L. Pflieger, and H. W. Robison. 1996. Conservation status of crayfishes of the United States and Canada. *Fisheries* 21:25–38.

Thompson, P. E. 1970. The first fifty years—the exciting ones (pp. 1–11). In N. G. Benson, ed., *A century of fisheries in North America.* Bethesda, MD: American Fisheries Society.

U.S. Fish and Wildlife Service and U.S. Bureau of the Census. 1993. *National survey of fishing, hunting, and wildlife-associated recreation.* Washington, DC: U.S. Government Printing Office.

Walton, I. 1676. *The complete angler or the contemplative man's recreation.* Reprinted in 1985 with modernized spelling. London: Harrap Limited.

Waterman, C. F. 1975. *Fishing in America.* New York: Holt, Rinehart and Winston.

Wetzel, R. G. 1983. *Limnology.* 2nd ed. New York: Saunders College Publishing.

Williams, J. 1962. *Oceanography.* Boston, MA: Little, Brown and Company.

Williams, J. E., J. E. Johnson, D. A. Hendrickson, S. Contreras Balderas, J. D. Williams, M. Navarro-Mendoza, D. E. McAllister, and J. E. Deacon. 1989. Fishes of North America, endangered, threatened and of special concern. *Fisheries* 14:2–20.

Williams, J. D., M. L. Warren Jr., K. S. Cummings, J. L. Harris, and R. J. Neves. 1993. Conservation status of freshwater mussels of the United States and Canada. *Fisheries* 18:6–22.

Williams, J. E., C. A. Wood, and M. P. Dombeck, eds. 1997. *Watershed restoration: Principles and practices.* Bethesda, MD: American Fisheries Society.

Winter, B. D. and R. M. Hughes. 1997. Biodiversity position statement—corrected version. *Fisheries* 22(3):16–23.

Wise, J. P. 1984. The future of food from the sea (pp. 113–127). In Julian L. Simon and Herman Kahn, eds., *The resourceful earth.* New York: Basil Blackwell.

Biodiversity and Endangered Species Management

THE DILEMMA

The myriad of life forms inhabiting the earth have been a constant source of wonder and wealth to human societies. The diversity of life is the result of the evolutionary process that began some 2.5 billion years ago when the first forms of life appeared on the earth. *Homo sapiens* is a late arrival—our direct human ancestors appeared only a few million years ago in Africa and from there spread throughout the world. Recent evidence indicates that humans reached North America more than 13,000 and possibly as much as 30,000 years ago from Asia via the Bering Land Bridge, which connected Siberia and Alaska (Hopkins et al. 1982). Humans evolved the skills and technology necessary to fully exploit the earth's natural resources, often with catastrophic consequences for life on earth. By the end of the twentieth century, the diversity of the earth's natural bounty had been threatened to an alarming degree by overexploitation, uncontrolled technology, and continued human population growth and resource demand. This degradation has forced human societies to evaluate their impacts and to develop corrective resource management policies.

Following years of neglect, biodiversity protection and restoration is now recognized in law and convention as necessary for the long-term survival and economic viability of human societies. Humanity is developing an ethic for more responsible interaction with all living organisms, which places a higher priority on sustainability of renewable resources than on immediate economic gain (Kellert 1996). Maintaining biodiversity depends heavily on the few who disproportionately control most of the earth's resources.

However, economic wealth in and of itself should not be blamed for poor conservation ethics. Wealth enables education, in-depth reflection, and influential action. In addition to self-interest, the drive for wealth is the engine of investment in scientific understanding, universal education, and the democratic process. Without some degree of wealth, ignorance, despair, and self-serving urgency tend to dominate human action. However, the unconstrained rush for wealth is the main enticement are extinct, endangered, threatened, or declining. In the Colorado River system alone, 34% of native fishes are extinct, endangered, or threatened.

Biodiversity Services and Value

Genes, species, populations, communities, and ecosystems form the bases of a network of interactive, interdependent, and easily disrupted associations. When maintained in a healthy condition, they can provide sustainable natural resources and economic viability for human populations. Wild organisms interacting in ecosystems provide many services vital to the well-being of humans.

Detoxification. The breakdown of toxic materials is one essential service. About 90% of the 12 billion tons of organic wastes, of which 130 million tons are human excreta and 2 billion tons are manure produced by livestock, is decomposed by the interactive complex of soil and water invertebrates. These include earthworms, protozoans, nematodes, fungi, and bacteria. Microorganisms aid in disintegration and neutralization of more than half a million tons of pesticides, 4 million tons of soaps and detergents, 18 million tons of ammonia, 12 million tons of alkalies, and 4 million tons of acids produced annually by U.S. agriculture, industry, and homes. Of the more than one-half million tons of pesticides applied to crops in the United States, less than 1% reach the target pests. Their disintegration by soil organisms is vital to reducing pollution. New varieties of microorganisms are being developed to aid in facilitating breakdown of the 7 million barrels of oil spilled into the environment annually (Pimental 1981, Daily 1997b).

Nitrification, Pollination, and Pest Control. Wild organisms perform many other valued services. An estimated 15 billion tons of nitrogen, vital for plant growth and production of high protein foods worth an estimated $3.5 billion, is biologically fixed by soil microbes. Cross-pollination by insects, for which no substitutes are available, is essential in the production of 90 U.S. crops worth nearly $4 billion. The value of nine additional crops, valued at $4.5 billion, is enhanced by insect pollination. Thousands of wild plants also depend on insects and other animals for pollination (Figure 17.2). Naturally occurring and introduced predators, such as insects and parasites, are essential in controlling pest insect populations of forestry and agriculture crops worth billions of dollars (Pimental 1981, Daily 1997b).

Materials Supply. The integrity of natural ecosystems and their associated biodiversity are critical in the storage and cycling of global material resources. Intact communities sustain healthy root systems that resist soil erosion and nutrient loss. The biomass of intact forests stores a large fraction of the biosphere's carbon, which is now accumulating instead in the atmosphere, where it appears to be causing climatic warming.

Water Services. Intact watershed ecosystems store water and release it at moderate rates, thereby reducing flood impacts on ecosystems and human communities downstream. Upland watersheds combine with the unimpeded functioning of floodplain ecosystems to absorb and transform flood impacts into useful forest and wildlife productivity. Further, they protect human property and health. In arid parts of the world, riparian ecosystems provide a disproportional high fraction of the natural production and biodiversity.

Sustaining Ecosystem Functions. Certain species have outsized keystone impacts on ecosystem function and their loss from overharvest or other selective process has great impact on the continuity of ecosystem function. No where does this appear more evident than in the heavily fished estuaries and coastal waters of the United States. The massive reduction of many coastal marine species by overfishing has been implicated as the indirect cause of many dramatic changes in marine coastal

FIGURE 17.2 Unknown to most people many animal species are important in maintaining natural ecosystems. Bats, such as this species from the American Tropics, are the second largest order of mammals. Some species are important pollinators and disseminators of seeds.

environments leading to increased die-offs of kelp forests, coral reefs, and sea grasses, and the eutrophication of estuaries (Jackson et al. 2001). The loss of sea turtles, sea cows (now extinct), and manatees may be in part responsible for massive die-offs of sea grasses. Also, the impact of fishing on whales, seals, and crocodiles, as well as predatory fishes, has contributed to estuarine eutrophication by altering food chains and reducing organisms that control algae.

Potential Resource Value. Genetic diversity is a natural library of information waiting to be "read" for the scientific insights that are held within (Wilson 1988). The genetic diversity contained in wild organisms is the basis for future food, medicine, and other resource development with an inestimable contribution to the future world economy. Sustaining genetic diversity preserves options that would otherwise be denied.

The genes of many wild forms related to domestic crop species possess resistance to parasites and pests. Their genes can be used to impart resistance to cultivated forms. New crops with greater tolerance to heat, drought, and salinity can also be developed from wild forms and aid in greater crop production and greater human welfare. The benefits of genetic diversity were understood by pre-Columbian southwestern cultures. They planted several varieties of maize to assure successful crops under a variety of weather conditions.

Wild animals provide a wealth of genetic diversity for creating new species of domestic animals. They can be crossed with domestic animals to produce new, more adaptable and productive animals from wild forms (Figure 17.3). All of the major domestic terrestrial animals produced in North America originated in the Old World, except for the turkey. However, other native species could be developed to advantage for use in North America, such as the bison. Because bison coevolved with North American native plants, they are better adapted than domestic sheep and cattle for making efficient use of native rangeland plants. Bison also are better adapted to the climatic regimes where they evolved.

The world depends on less than 10 animal species (sheep, cattle, chickens, and pigs, among others) for most of the world's terrestrial animal food production. Less than 20 crops account for 90% of the world's plant food production. None of these

FIGURE 17.3 Eland, shown here, have been recently domesticated from wild stock. Many wild animal species have the potential for becoming sources of new domesticated animals.

species are native to North America north of Mexico. Thousands of wild plants and animals could potentially be used to produce new domesticated forms better adapted to local conditions. They are more productive, and require less fertilization, pesticide application, and tillage. Because of the high reliance on a few domesticated species for food, human populations are highly vulnerable to crop devastations. Preventing the further loss of species is a global priority because wild or semiwild forms of potential economic importance could become extinct, resulting in great future economic losses (McNeely et al. 1990).

Commodities.   Medicines are another category of beneficial goods derived from wild organisms. Poisonous marine vertebrates and invertebrates provide antimicrobial, antiviral, cardioactive, and neurophysiological substances. Many plants, fungi, and other life forms also contribute massively to our store of medications. These substances are essential in providing better health services and they generate billions of dollars for the pharmaceutical industry. Many animals and plants have medicinal properties that have yet to be developed or are unknown. Wild animals also provide oils, ornamental by-products, fuel, feathers, and pets worth billions of dollars in international markets (Daily 1997a, 1997b).

ENDANGERED SPECIES

What Are Endangered Species?

Endangered species are at risk of extinction and require protective management to avoid extinction. A species is considered endangered if its death rate consistently exceeds its birthrate; if it is incapable of adapting to natural or human-induced environmental changes; if much of its habitat is threatened by destruction or serious disturbance; or if its survival is threatened by the unwanted introduction of other species through predation, competition, disease, or environmental pollution. A threatened species is scarce throughout its range and will probably not become extinct as long as conditions are managed to remain stable and favorable.

There were 902 (349 animals and 553 plants) species listed as endangered and 233 (117 animals and 116 plants) listed as threatened by the United States Fish and

Wildlife Service (USFWS) as of April 1998. Only eight have been removed from the list since 1973 and 18 species have been upgraded to threatened status.

The main source of information on endangered species is the International Union for Conservation of Nature and Natural Resources (IUCN), a nongovernmental conservation organization headquartered in Gland, Switzerland. The *IUCN Red List of Threatened Animals,* a publication that is periodically updated, presents information on the current status of globally threatened vertebrates and invertebrates, exclusive of microorganisms. The 1996 edition lists 5,205 species that are threatened with extinction. The listed species are those experiencing significant population declines, those restricted in distribution, those with very low population numbers, or those exhibiting any combination of these characteristics. A large percentage of the vertebrates are threatened with extinction, including 25% of mammals, 11% of birds, 20% of reptiles, 25% of amphibians, and 34% of fish. Countries with the largest number of endangered mammals are Indonesia (128), China (75), and India (75), all of which also support extremely high human populations. Countries with the largest number of known endangered birds are Indonesia (104), Brazil (103), and China (90). Countries with the largest number of endangered fishes are the United States (123), Mexico (86), and Indonesia (60).

In the United States, the federal agency charged with maintaining a list of endangered species is the Fish and Wildlife Service. Of the species listed as threatened or endangered, 10% were considered to have increasing populations, 38% were considered declining, 31% were estimated to have stable populations, 19% had unknown status, and 2% were believed to be extinct. The greatest percentage of species with declining populations or unknown status were invertebrates (81%), fish (51%), and amphibians and reptiles (79%). This indicates that smaller species and aquatic species may be at a greater risk of extinction.

Forest ecosystems were associated with the greatest number of listed species (312), followed by rangeland (271), water (244 dominated by fish and mollusks), and barren land (beaches, dry salt flats, exposed rock, areas disturbed by mining, and other habitats) (176 species). Wetlands supported fewer (155) threatened and endangered species than other habitats. However, wetlands—which comprise only 5% of the contiguous United States—support a disproportionally high number of listed species (30% of listed animal species and about 15% of listed plants) (Flather et al. 1994). Of all the habitats, riverine habitats hold the highest concentrations of endangered species in the United States.

Of the factors contributing to species endangerment in the United States, habitat losses and fragmentation associated with intensification of land and water use are the most important. Second and third in importance are interspecific (between species) interactions, especially those associated with introduced species, and human overuse associated with the harvest, collection, or commercial trade of species. Agricultural development is a prominent factor for endangerment among most species, particularly mammals and birds. The primary factor impacting fishes and birds is human-introduced exotic species. Commercial exploitation and collecting are particularly detrimental to snails, reptiles, crustaceans, amphibians, and plants (Wilcove and Bean 1994).

Causes of Extinction

The five main factors contributing to species extinction are habitat modification, natural causes, unregulated hunting, introduced predators, and nonpredatory exotics. Natural causes are part of the evolutionary process. Animals become overspecialized and are unable to adapt to environmental changes or cannot compete with other evolving organisms. Natural catastrophic events—such as sudden climatic changes, earthquakes, or volcanic eruptions—can also be involved in extinctions. From a management standpoint, human-caused extinction is the greater concern and these causes are discussed in more detail on the pages that follow.

(a)

(b)

FIGURE 17.4 Suburban sprawl (a) and dams (b) have been two primary factors associated with habitat fragmentation and species endangerment in the United States and other parts of the world (courtesy of U.S. government).

Habitat Modification.   Habitat modification by humans is the primary threat to biodiversity (Figure 17.4). Such modification occurs through logging, draining of wetlands, agricultural and industrial development, reservoir construction, urban sprawl, pollution, and other changes (Flather et al. 1994). Of the recent extinctions involving habitat loss, humans are probably responsible for over 90%.

The most significant factors noted in the *Red List* (IUCN 1997) are habitat reduction, fragmentation, and degradation. Natural systems are being replaced by human-dominated landscapes resulting from unrestrained economic development in conjunction with human population increases. Once widespread wildlife populations are being fragmented into small, isolated subpopulations (Figure 17.5). For those animals listed as threatened or endangered, habitat loss or degradation have been identified as one of the main reasons for listing of 75% of mammals, 44% of birds, 68% of reptiles, 58% of amphibians, 55% of fishes, 52% of crustaceans, 47% of insects, 95% of clams and oysters, and 50% of snails. Unplanned and unrestricted exploitation of many of the larger terrestrial vertebrates and marine fishes in the *Red List* has resulted in severe population declines.

Increasing human development impacts around rivers, lakes, and oceans are continuing to change the nature of aquatic areas and their aquatic forms. Extensive

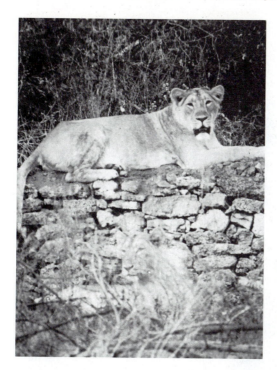

FIGURE 17.5 Asiatic lions are now restricted to a lone population fragment in the Gir Forest of western India. Lions originally occurred from southern Africa to India. Populations have been extirpated from the region between India and central Africa.

stream channelization, construction of dams and reservoirs, and conversion of wetlands have been principal components in altering aquatic habitats (see also Chapter 7). Forty fish forms, including 27 species and 3 genera, became extinct in North America, and of these, 9 became extinct after 1964 (Williams and Miller 1990). The American Fisheries Society listed 364 North American fish species and subspecies as either threatened (114 species), endangered (103 species), or of special concern (147 species) (Cairns and Lackey 1992).

Unregulated Hunting and Fishing. Excessive and unregulated hunting includes hunting for commercial purposes such as in the cases of the great auk and passenger pigeon, and those targeted for control because they transmit diseases or are pests such as the Carolina parakeet. Demand for specific animal parts for folk medicines and ornamentation, mostly in Asia, has caused the excessive killing of rhinoceros, tigers, and other animals (Figure 17.6). Harvest exploitation has been specifically identified as one of the main reasons for the threatened status of listed species among 13% of mammals, 7% of birds, 31% of reptiles, and 68% of marine fishes.

Introduced Predators. Introduced predators can be highly destructive because native faunas have not evolved defenses to cope with the new predator especially in small, isolated areas such as islands. Some introduced predators, such as the Indian mongoose in the Caribbean Islands and Hawaii, were released to control pests, such as rats and snakes, but instead concentrated on the more vulnerable native prey species. Introduction of Nile perch into African rift valley lakes has led to the disappearance of large fractions of the native fish.

Even the seemingly harmless domestic cat is a wildlife conservation dilemma. An estimated 60 million pet cats lived in urban and rural regions of the United States alone in 1990, and another 40 million were semiwild or free ranging. More than a billion small mammals and hundreds of millions of birds are killed by rural cats alone each year (Bolen and Robinson 1999). This estimate does not include those wild animals, such as songbirds, killed each year by urban and suburban cats.

FIGURE 17.6 In many countries, species are exploited without regard for their continued existence. This snow leopard skin was being offered for sale in a Chinese market even though protected within the country.

Nonpredatory Exotics.   Exotic organisms can be detrimental to native species by displacing native animals, altering ecosystems, introducing foreign diseases and parasites, hybridizing, and diverting funds from native wildlife programs for control and eradication. Nonpredatory exotics include introduced carriers of diseases and parasites that can infect native species, which have no or low resistance (Figure 17.7). Even the Native American human population was vulnerable to diseases introduced by Europeans, such as smallpox, which resulted in the deaths of many thousands of people.

The introduction of exotics also has been a serious threat in aquatic systems. Hybridization of exotic forms, such as cutthroat and rainbow trout, with native forms, such as Gila trout, have genetically altered native species. Competition and diseases associated with exotics, plus the alteration of ecological relationships between species, threaten entire aquatic ecosystems.

Exceptional Vulnerability of Island Species.   Island forms are particularly vulnerable to exotic predators because they evolved in isolation free from predators and human interference. Feral cats have been particularly destructive when introduced on islands previously devoid of land predators. One of the best examples of dramatic island change following human colonization is Hawaii, where unique plant and animal life has been greatly diminished.

In the Hawaiian Islands as many as 98 bird species became extinct between A.D. 400 when Polynesians first colonized the islands and the first European contacts in 1778. Another 17 species became extinct in Hawaii after European settlement. These extinctions were a result of the clearing of forests for agricultural development, alteration of native vegetation, exotic plant introduction, hunting, predation by introduced species (rodents, exotic birds, domestic cats, and the Indian mongoose), and introduced avian diseases. Half of Hawaii's endemic (original distribution restricted to the islands) birds have become extinct or been greatly reduced since 1880 (Royte and Jones 1995). Similar extinctions have occurred on many other landmasses in-

FIGURE 17.7 Exotic species can pose significant problems. The rock dove or feral pigeon shown in this photo has become abundant throughout the United States. It carries diseases that can be harmful to humans and native birds.

cluding New Zealand, Australia, and Madagascar where, after colonization by European cultures, domestic species such as pigs, sheep, goats, and rabbits degraded the native vegetation (Reid and Miller 1989).

ENDANGERED SPECIES POLICY AND MANAGEMENT

Legislation and Treaties

Endangered Species Act. The strongest and most direct legislation pertaining to endangered species in the United States is the Endangered Species Act of 1973. It gave broad powers to the U.S. federal government in regulating actions affecting endangered animals and plants. The Endangered Species Act made it unlawful to be in possession of endangered species anywhere in the United States, provided a formal structure for listing and management of endangered species, required all federal agencies to use their existing authorities to conserve listed species, and provided a means for citizens to bring suit against any federal agency for failure to meet its obligations as specified in the legislation (Grosse 1997).

The Endangered Species Protection Act of 1966 authorized the Secretary of the Interior to develop a list of species facing extinction. It also sponsored research and authorized acquisition of terrestrial habitats to protect endangered species. The Endangered Species Conservation Act of 1969 added protection for federal lands and activities. The Endangered Species Act of 1973 superceded the two previous laws and, with subsequent amendments, has become the federal law directing the identification of threatened and endangered species. It regulates the taking and capture of all listed animals and plants, including foreign species, as well as their interstate and foreign commerce. It provides for designation of critical habitats, land acquisition, and financial assistance to states and foreign governments. In addition to public lands, it extended federal authority over private-land use practices that could potentially affect endangered species. Projects that are a threat to an endangered species or its critical habitat must be altered to remove the threat or canceled. The act directs the government to become involved in international efforts to protect species by conducting law enforcement investigations and prohibiting the importation of endangered and threatened species. The U.S. Fish and Wildlife Service is responsible for land and freshwater species and the National Marine Fisheries Service in the Department of Commerce is responsible for marine species (Grosse 1997).

The USFWS may nominate a species for listing, but an individual or a private organization may also petition to initiate the listing process. The act provides protection for species threatened or endangered by one or more of the following factors: (1) present or threatened destruction, modification, or curtailment of the species' habitat or range; (2) overuse for commercial, sporting, scientific, or educational purposes; (3) disease or predation; (4) inadequacy of existing regulatory mechanisms for preventing decline or degradation of habitat; and (5) other natural or man-made factors affecting its continued existence. Threatened species are those likely to become endangered within the foreseeable future.

The term endangered species as defined in the law includes any species, subspecies, or population that is in danger of extinction throughout all or a significant portion of its range. This allows for listing of unique forms in portions of a species' distribution. For example, it was possible to specifically designate the subspecies of mountain sheep in southern California as endangered. This subspecies is not endangered in other parts of its distribution and this mountain sheep species occurs over a wide area of the United States. Individual states may also designate endangered and threatened populations under authority of state laws (George et al. 1998).

The Secretary of the Interior and Secretary of Commerce have the power to initiate actions that promote recovery of species. In order to restore a species to a nonendangered status, a recovery plan must be initiated for each species listed. Recovery plans are usually developed by wildlife biologists and resource specialists with expertise on the species being considered, or with other relevant expertise, and who form recovery teams. Recovery teams make recommendations on the steps necessary to remove a species from the endangered list. Removal from the list means the species is no longer in need of protection and therefore considered recovered. Conclusive evidence of extinction also leads to removal from the list.

Recovery plans have been developed for many species. They may include captive breeding programs, such as for the California condor, peregrine falcon, whooping crane, Apache trout, masked bobwhite, black-footed ferret, and Mexican wolf. Once propagated, species are released in the wild, although not always successfully. Harvest prohibition has been most successful for species that had been previously overharvested, such as sea otters, fur seals, beavers, snowy egrets, and American alligators. Translocation (moving animals from one area to another) programs were important in reestablishing populations once locally exterminated by hunting, including white-tailed deer, turkeys, elk, pronghorn antelope, bison, and bighorn sheep. However, most of those transplants occurred before genetic variation was understood. Unique populations were in some cases lost, such as Merriam's elk in the southwestern United States. Maintenance or development of proper habitat and acquisition of essential habitats are other management strategies for rehabilitating endangered species. The creation of the Arkansas National Wildlife Refuge in 1937 on the Texas coast was of vital importance in saving the whooping crane from extinction. Among the animals the Fish and Wildlife Service has recently downlisted or plans to delist in the near future because they have recovered to stable, viable population sizes are the bald eagle, American peregrine falcon, brown pelican, island night lizard, and Ash Meadows amargosa pupfish.

The Endangered Species Act of 1973 precluded consideration of economic impacts. However, in 1978 it was amended to include the possibility of appeal to an Endangered Species Committee of high-level federal administrators, who might decide that the costs of protection were too high for the perceived benefits. Appeal to this committee, sometimes referred to irreverently as the "God Squad," has often been debated but rarely made.

Habitat Conservation Plans. An amendment to the Endangered Species Act in 1982 allows a Habitat Conservation Plan to be developed by a landowner whose land use has been identified as critical habitat of a threatened or endangered species (Beatley 1994). In exchange, the U.S. Fish and Wildlife Service issues an incidental take permit. The plan must be approved by the Fish and Wildlife Service and is developed under the oversight of a committee of informed stakeholders. This amendment was passed to contend with the increasing conflicts that developed among environmental groups, developers, and landowners over endangered species issues. The intent of Habitat Conservation Plans is to allow land use while assuring protection of the species despite some incidental loss from the population. The plan must meet certain standards. It must spell out the steps for minimizing incidental take, the impact that would result from incidental take, alternative plans and why they were not adopted, and the source of funding for carrying out the plan. For a plan to be approved, the applicant must assure minimum effect of incidental kill, funding to carry out the plan, and convincing evidence that the incidental take will not have a negative impact on species viability.

Beatley (1994) reviewed the progress of habitat conservation planning with respect to urban development in three of the fastest growing states: California, Texas, and Florida. He provided a sample of species that have been addressed through Habitat Conservation Plans, including fringe-toed lizard (California), American crocodile (Florida), desert tortoise (California), several kangaroo rats (California), Key Largo woodrat (Florida), Buena Vista lake shrew (California), San Joaquin kit fox (California), several butterflies (California and Florida), several songbirds (California and Texas), and several plants (California and Texas). Strategies typically include reservation of land for habitat protection, habitat restoration, and management providing special needs (e.g., removal of exotic species and competitor control).

Cites. The most influential international approach to endangered species conservation is the Convention on International Trade in Endangered Species (CITES), which was signed in the United States in 1973. International trade can be an important factor in the decline and extinction of some species if trade promotes overhunting for furs or hides, for food products, for exhibition, for sport, for scientific experimentation, or for other purposes. The convention addressed the protection of species from the standpoint of how trade impacted the status of a particular species in the wild in its native country.

Overexploited, imported, or exported species are eligible for protection under the convention (Grosse 1997). The participating parties provided for appendices listing endangered and potentially endangered species. Stringent controls were directed at regulating trade activities of endangered species. All trade of such species, their parts and derivatives, including manufactured products, requires permits from both the importing and exporting countries. As of 1997, 134 countries had signed the treaty. CITES aided in curtailing the rampant trade in a host of endangered species including cheetahs, other spotted cats, tigers, elephants, rhinos, reptiles, parrots, and many other birds. Nonetheless, trade in endangered species continues at an estimated $5 billion to $10 billion per year, and requires the persistent concerted efforts of the international community to limit its destructiveness.

Management Needed to Maintain Biodiversity

Halting further degradation and fragmentation of existing natural habitats and ecosystems is essential for the conservation of biotic resources. Protection of intact natural areas and restoration of degraded habitat are key components of protecting biological diversity from further erosion. A fundamental cause of habitat degradation is human population growth beyond the capacity of landscapes to support biodiversity. Limiting human population growth is essential for sustaining biodiversity. Development of more efficient uses of natural resources without degrading biodiversity values is also a fundamental strategy. In the United States, Dahl (2000) estimated that wetlands have declined 53% (221 million acres in the 1780s to 104 million acres today). The surface area of wetlands in the 48 contiguous states has been reduced from over 10% to 5%. The conversion of forests and grasslands to agricultural uses has also greatly diminished the native fauna and flora. Only 1% of the original 10 million km^2 of tallgrass prairie in the eastern Great Plains remains in natural vegetation. Certain forest types have substantially decreased, particularly bottomland hardwoods of the lower Mississippi Delta and riparian forests in the semiarid to arid West. These forest types supported communities rich in biodiversity and produced timber products in high demand (Flather and Hoekstra 1989).

Because federal lands provide habitats for a large proportion of native wildlife, federal agencies must be major participants in conserving biodiversity. Of the 660 listed species for which data are available, 24% occurred on U.S. Forest Service lands, 17% occurred on lands managed by the Bureau of Land Management, and 26% occurred on Department of Defense lands. Noss and Cooperrider (1994) provide guidelines for managing forests, rangelands, and aquatic ecosystems. Reserve networks and monitoring programs are important features of a national biodiversity conservation strategy.

Most federally owned lands occur in the West. Lack of federally owned land limits the government's role in conserving diversity in the eastern United States. Even in the West, many elements of diversity occur outside of federally protected lands. Species that require very large tracts of undeveloped land occur principally on federal lands. However, these protected areas—primarily national parks and national forests—may not be sufficient in size and interconnection to ensure the long-term viability of large mammals such as grizzly bears, gray wolves, Florida cougars, and wolverines. In areas where little public land exists, habitat fragmentation is a major problem even for small species.

Except for habitat conservation planning, the Endangered Species Act discourages private landowners from creating, restoring, or enhancing habitats for endangered species. Private landowners fear greater regulatory actions if endangered species are found on their properties. Incentives for landowner participation in managing endangered species could include the option of partially deferring estate taxes and allowing participants to claim tax deductions for costs associated with habitat management practices. A conservation strategy that entices more participation of private landowners is essential and vital in curtailing the loss of biodiversity (Wilcove et al. 1994). Anderson and Leal (1991) describe alternatives that would reward landowners for protection of threatened and endangered species habitat.

ENDANGERED ECOSYSTEMS

One of the declared purposes of the Endangered Species Act is to "provide a means whereby the ecosystems upon which endangered species and threatened species depend may be conserved." The primary strategy used to attain this end has been identification of critical habitat and its protection through regulation and habitat conservation planning. The focus is on critical habitat determination based predominantly on the patterns of past and existing distribution of the threatened species and its associated habitat. Ecoregional identification of sensitive, threatened, and endangered species (Stein et al. 2000, Abell et al. 1998) shows the distribution of ecosystems of exceptional concern. One recent effective approach is the use of Geographical Information Systems (GIS) to map land uses, individual population distributions, and habitat as indicated by remote sensory signatures of vegetation and geological features. Using an overlay approach, those areas of greatest unique biodiversity and threatening existing or potential land uses can be identified for priority attention.

However, overattention to a pattern approach has been criticized as insufficient without careful attention to the preservation and restoration of processes that sustain diversity (Smith et al. 1993, Risser 1995). The Nature Conservancy has developed a planning handbook to guide the process.

The prevalent forces determining the interactive dynamics among species within ecosystems often originate from outside a community including storms, fire, atmospheric changes, flooding, drought, off-site changes in habitat of migratory species, and colonization (invasion) by new or already established species. The evolutionary role of these forcing factors can be clearly seen in the invasion of island ecosystems by new species. The slow rate of natural invasion has resulted in the unique biodiversity of many island ecosystems. However, the accelerated rate of invasion caused by humans has greatly reduced that unique biodiversity. Invasion in itself is not the problem; it is the extent that the rate and intensity varies from historic conditions. The periodicity and intensity of forcing factors from outside ecosystems have been profoundly important in determining the unique character of ecosystems. Ecosystems are most endangered where the rates and intensities of those forces have been greatly modified either by accident or by human design.

Ecosystem processes are adapted to these influential forces. The unique character of ecosystems, and of many of the unique species within them, often depends on the extent to which the forces continue to operate as they have during past evolutionary history. Noss and Cooperrider (1994) have pointed out the importance of identifying endangered ecosystems as a precursor to the prevention of individual species endangerment. Certain wetland and prairie ecosystems are among the most deserving candidates of endangered ecosystem status. In prairie ecosystems the deprivation of fire as a natural forcing factor has been critical in determining biodiversity and the ability of an ecosystem to recover from drought. In wetlands and rivers, the periodicity and intensity of flooding is often a critical forcing factor for sustaining ecosystem processes. Of course, both wildfire and unrestrained floods are threatening to human property and health. They require control where those benefits have higher priority than protection or restoration of natural ecosystems.

Future recovery of existing endangered species is likely to emphasize the importance of protecting and restoring fundamental ecosystem processes, including restoring those forces that were most influential in sustaining the unique attributes of ecosystems. This will be challenging for natural resource managers because past political decisions about land ownership and control have placed little emphasis on maintaining natural ecosystems and their processes. The floodplain communities of large rivers, for example, are most often privately owned and influenced by artificial river control and hydropower structures. Similarly, only small remnants of tall grass prairie remain in isolated islands of protected land in otherwise developed or developing landscapes.

Past natural resource management tactics, such as flood and fire prevention, usually will have a higher priority to society than sustaining pristine ecosystem processes. However, there often are alternatives that can reduce threats to biodiversity while sustaining the most crucial artificial forcing factors. Research on biodiversity maintenance is increasingly turning its attention to these large landscape processes. Conflicts between the public interest and individual interest in property protections and development will continue to be a fundamentally divisive issue.

LITERATURE CITED

Abell, R., D.M. Olson, E. Dinersteih, P. T. Hueley, W. Cichbzum, S. Walters, W. Wettengel, T. Allnut, and C. J. Loucks. 1998. *A conservation assessment of the freshwater regions of North America.* Washington DC: World Wildlife Fund.

Anderson, T. L. and D. L. Leal. 1991. *Free market environmentalism.* Boulder, CO: Westview Press.

Beatley, T. 1994. *Habitat conservation planning: Endangered species and urban growth.* Austin, TX: University of Texas Press.

Bolen, E.G. and W. L. Robinson. 1999. *Wildlife Ecology and Management*, 4th ed. Prentice Hall, Upper Saddle River, NJ.

Cairns, M. A. and R. T. Lackey. 1992. Biodiversity and management of natural resources: The issues. *Fisheries* (Bethesda) 17:6–10.

Cox, G. W. 1997. *Conservation biology.* 2nd ed. Dubuque, IA: Wm. C. Brown. 362 pp.

Dahl, T. C. 2000. *Status and trends of wetlands in the Coterminous United States, 1986–1997.* U.S. Dept. of Interior, Fish and Wildlife Service. Washington, DC.

Daily, G. C., ed. 1997a. *Nature's services: Societal dependence on natural ecosystems.* Covelo, CA: Island Press. 392 pp.

Daily, G. C. 1997b. Introduction: What are ecosystem services? (pp. 1–10). In G. C. Daily, ed., *Nature's services.* Covelo, CA: Island Press.

DeLong, D. C. Jr. 1996. Defining biodiversity. *Wildlife Society Bulletin* 24:738–749.

Flather, C. H. and T. W. Hoekstra. 1989. *An analysis of the wildlife and fish situation in the United States: 1989–2040.* USDA Forest Service General Technical Report RM-178. Fort Collins, CO: Rocky Mountain Forest and Range Experiment Station. 146 pp.

Flather, C. H., L. A. Joyce, and C. A. Bloomgarden. 1994. *Species endangerment patterns in the United States.* USDA Forest Service General Technical Report RM-178. Fort Collins, CO: Rocky Mountain Forest and Range Experimental Station. 24 pp.

George, S., W. J. Snape, and M. Senatore. 1998. *State endangered species acts: Past, present, and future.* Washington, DC: Defenders of Wildlife and Albuquerque, NM: Center for Wildlife Law. 133 pp.

Grosse, W. J. 1997. *The protection and management of our natural resources, wildlife, and habitat.* Dobbs Ferry, NY: Oceana Publications.

Hopkins et al., eds. 1982. *Paleoecology of Beringia.* New York: Academic Press.

IUCN. 1997. 1996 IUCN red list of threatened animals. Gland, Switzerland: IUCN. 368 pp.

Jackson, J. B. C. et al. 2001. Historical overfishing and the recent collapse of coastal ecosystems. *Science* 293: 629–638.

Kellert, S. R. 1996. *The value of life: Biological diversity and human society.* Washington, DC: Island Press. 263 pp.

Langner, L. L. and C. H. Flather. 1994. *Biological diversity in the United States.* USDA Forest Service General Technical Report RM-244. Fort Collins, CO: Rocky Mountain Forest and Range Experimental Station. 24 pp.

Marcot, B. G. M., M. J. Wisdom, H. W. Li, and G. C. Castillo. 1994. *Managing for featured, threatened, endangered, and sensitive species and unique habitats for ecosystem sustainability.* USDA Forest Service General Technical Report PNW-GTR-329. Portland, OR: Pacific Northwest Research Station.

McNeely, J. A., K. R. Miller, W. V. Reed, R. A. Mittermeier, and T. B. Wemer. 1990. *Conserving the world's biological diversity.* Gland, Switzerland: IUCN. 193 pp.

Meffe, G. K. and C. R. Carroll. 1997. *Principles of conservation biology.* 2nd ed. Sunderland, MA: Sinauer Associates.

Morrell, V. 1999. The variety of life. *National Geographic* 195:6–31.

Noss, R. F. and A. Y. Cooperrider. 1994. *Saving nature's legacy: Protecting and restoring biodiversity.* Washington, DC: Island Press.

Pimental, D. 1981. *Biological diversity and environmental quality* (pp. 44–47). In Proceedings of the U.S. Strategy Conference on Biological Diversity. Department of State Publication 9262, Washington, DC.

Primack, R. B. 1998. *Conservation biology.* 2nd ed. Sunderland, MA: Sinauer Associates.

Reaka-Kudla, M. L., D. E. Wilson, and E. O. Wilson, eds. 1997. *Biodiversity II.* Washington, DC: Joseph Henry Press.

Reid, W. V. and K. R. Miller. 1989. *Keeping options alive, the scientific basis for conserving biodiversity.* New York: World Resources Institute. 128 pp.

Risser, P. G. 1995. Biodiversity and ecosystem function. *Conservation Biology* 9:742–746.

Royte, E. and C. Jones. 1995. Hawaii's vanishing species. *National Geographic* 188:3–37.

Smith, T. B., M. W. Bruford, and R. K. Wayne. 1993. The preservation of process: The missing element of conservation programs. *Biodiversity Letters* 1:164–167.

Stein, D.A., L.S. Kutner, and J. S. Adams, eds. 2000. *Precious: Status of biodiversity in the United States.* New York: Oxford University Press.

West, N. E., ed. 1995. Biodiversity on rangelands. Logan, UT: College of Natural Resources, Utah State University. 114 pp.

Wilcove, D. S. and M. J. Bean, eds. 1994. *The big kill: Declining biodiversity in America's lakes and rivers.* Washington, DC: Environmental Defense Fund. 275 pp.

Wilcove, D. S., M. J. Bean, R. Bonnie, and M. McMillan. 1994. *Toward a more effective Endangered Species Act for private land.* Washington, DC: Environmental Defense Fund. 20 pp.

Williams, J. E. and R. R. Miller. 1990. Conservation status of the North American fish fauna in fresh water. *Journal of Fish Biology* 37:79–85.

Wilson, E. O., ed. 1988. *Biodiversity.* Washington, DC: National Academy Press. 521 pp.

Wilson, E. O. 1992. *The diversity of life.* Cambridge, MA: Belknap Press of Harvard University Press. 424 pp.

CHAPTER 18

Mineral Resources

INTRODUCTION

The purpose of this chapter is to provide an overview of Earth's crust resources that supply our material needs. To comprehend the nature of our dependence on these resources we will discuss their origin, classification, properties, abundance, and use. Our focus will be nonfuel minerals (including metal minerals and nonmetallic resources mined from the earth). Coal, oil, oil shale, sand tar, and uranium will be discussed in Chapter 19 on nonrenewable energy resources. The availability of these resources is often determined by cyclic processes working over time well beyond our immediate comprehension. Our use of these resources is of critical importance in determining the sustainability of global and local economies. In this chapter, we provide an overview of mineral availability, mining, and environmental problems and refer the reader to Kesler (1994) for more comprehensive coverage of these subjects. Gardner and Sampat (1999) offer a detailed discussion of recycling and materials disposal.

GEOLOGICAL FOUNDATIONS

The Realm of Minerals and Rocks

Minerals and rocks provide us with a host of material goods, many of which escape our attention. To understand their origin, we must provide a framework of definitions and classification suited to our needs. By definition, a rock is a naturally occurring solid containing one or more minerals. A mineral is a naturally occurring inorganic solid with specific physical properties and a consistent internal structure. In composition, minerals can range from those containing a single element, such as native copper, to those comprising several elements. A mineral can be nonmetallic such as salt, gypsum, or sand, or metallic such as aluminum or copper. Strangely, water in its frozen form satisfies this definition owing to the crystalline structure of ice, whereas liquid water does not. Also, coal is routinely listed in tables of economic minerals, but in fact is not a mineral. It more correctly is an organic sedimentary rock.

For identification purposes, geologists rely on several mineral properties. These include color, crystal form, specific hardness and density, luster, and cleavage. Crystal form reflects atomic structure of the mineral and refers to the assemblage of flat faces that make up the surface. Hardness indicates resistance to scratching or abrasion and determines which materials can be used to efficiently cut hard metals or

581

drill into rock. Cleavage refers to the manner in which minerals split along planes determined by their crystal structure. For example, cubic minerals such as halite often display three mutually orthogonal cleavage directions. Mica has a perfect basal cleavage in one direction and splits into thin sheets. Feldspars commonly show two strong cleavages.

To understand the chemical and physical properties of minerals, one must examine the ways the atoms they contain bond to one another. Bonding refers to the way electrons are shared or released by atoms. It can occur in three basic ways, which include ionic, covalent, and Van der Waal's bonding. Ionic bonding is a linkage formed by transferring or shifting electrons from one atom to another. Ionically bonded minerals are moderately hard, have high melting points, and dissolve easily in water (for example, halite or NaCl). Metal atoms are joined by shells of electrons overlapping them. This results in a solid swarm of electrons that are free to move about and which enable metals to conduct electricity and transmit heat efficiently. Because of the dense arrangement of their electrons, metals can be twisted or beaten into thin sheets without breaking, and can be drawn into thin wire. Covalent bonds are formed by sharing electrons and create the strongest types of chemical bonds. Diamonds are extremely hard owing to the covalent bonds in their structure. Covalently bonded minerals do not readily dissolve in water, have high boiling points, and do not conduct electricity. The weakest of the chemical bonds is Van der Waal's bonding maintained by the weak residual charges between sheets of atoms. Minerals with this type of bonding, such as the micas, can be split into sheets using only a fingernail. The individual sheets are held by covalent bonds and are stronger. Like mica, most minerals exhibit more than one kind of bonding.

Concerning classification, all minerals can be grouped into either silicates or nonsilicates. Because the two most abundant elements in the earth's crust are oxygen and silicon, silicates (silicon-oxygen compounds) form the most abundant group of minerals. Most minerals in Earth's crust are silicates. Silicate minerals are further classified according to the structural arrangement of the silicate anions. The feldspar silicate group is most abundant and represents a family of closely related minerals having sodium, potassium, or calcium in addition to silicon and oxygen. Ferromagnesian silicates are rich in magnesium or iron. Soil scientists are especially concerned with the mineralogical properties of clay minerals found in the mica group. In soils, clay minerals can greatly affect moisture and nutrient retention. Talc, found in any drugstore, is a member of the same group.

Nonsilicates include minerals that can be grouped by common chemical characteristics. These include oxides, sulfides, sulfates, carbonates, halides, and native elements. Native copper, graphite, and diamond are examples of native elements. Gypsum is calcium sulfate, hematite is iron oxide, and dolomite is one of the carbonate forms. Lead is obtained from galena, a sulfide mineral. Other examples of nonsilicates and silicates, and their sometimes complex mineralogical structures and elemental compositions, are discussed more fully in geology texts (see Hoeffer 1995).

The formation of economic minerals can be understood by examining the fundamental structure of Earth, including how crust is created and destroyed by plate tectonic activities. This information is necessary to understand how rocks and minerals are created and recycled over eons of time. Our understanding of Earth's dynamic geological processes has grown considerably in recent decades. Earth usually is depicted as consisting of four concentric layers differing in composition and density (Figure 18.1). These layers include Earth's inner core, outer core, mantle, and crust. Within Earth's inner core, the pressures are so great that the iron and nickel contents are solid despite high temperature. In the outer core, temperature and pressure are so balanced that matter is molten and exists as a liquid. Because their contents are believed to be similar, it is added pressure that separates the inner and outer core layers.

Geologists divide the mantle and crust into three distinct regions. The mesosphere ("intermediate or middle sphere") exists from the core-mantle boundary to a depth 350 km beneath Earth's surface. In this region, rock is so highly compressed it exists as a solid even though the temperature is high. Within the upper mantle, from

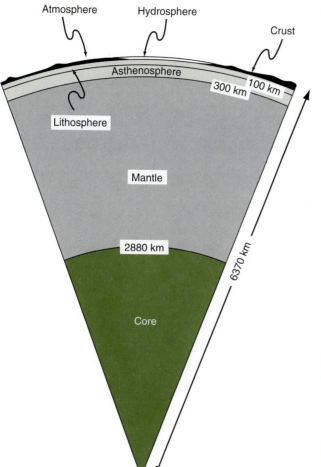

FIGURE 18.1 Distribution of the lithosphere, hydrosphere, and atmosphere (except for that part in the hydrosphere, the biosphere is too small to be shown at this scale) and major divisions of the solid earth (from Kesler 1994).

a depth of 350 km to 100 km, is a region where the balance between temperature and pressure is such that rocks have relatively little strength. In this region, the asthenosphere ("weak sphere"), igneous rock is easily deformed. Above the asthenosphere is the region composed of upper mantle and crust called the lithosphere ("rock sphere"). Here rocks are stronger and more rigid. It is rock strength, not rock composition, that results in regional distinctions. The crust, or rigid outermost layer of Earth, consists of continental crust and oceanic crust. The crust varies in thickness from about 56 km (35 miles) under the continents to 5 km (3 miles) under the ocean basins. Earth's crust consists mainly of alumino-silicates. The granitic materials of the continental crust are less dense than the basaltic materials of oceanic crust.

Plate Tectonics

The theory of plate tectonics has become widely accepted and is said to have revolutionized our thinking about some of the most fundamental issues in geology. According to this theory, the lithosphere is broken into plates which can slide over the asthenosphere. This movement is possible because the asthenosphere is soft and flows easily. The mechanism by which Earth's plates move involves the formation of convection cells, which develop as hot mantle material. This rises from the earth's interior, moves laterally, then cools, and begins to sink (Figure 18.2). The energy driving this movement comes from hot convection currents in the mantle. Sources of the earth's interior heat are residual heat resulting from the earth's cooling from a molten state, heat produced by the decay of radioactive isotopes, and heat produced by crystallization of the core's liquid portion.

Plate Tectonics and Mineral Deposit Environments

FIGURE 18.2   Plate tectonic environments showing their relation to ore-forming magmatic and hydrothermal processes (from Kesler 1994).

Plate Boundaries

Actions occurring at the margins of plates explain how new lithosphere is produced and how rocks are recycled. Along divergent plate boundaries, or spreading centers, plates are moving apart. Spreading centers can involve oceanic to oceanic plate margins and continental to continental plate margins. Along these margins, basaltic lava eruptions and earthquake activity are common and new lithosphere forms. The mid-Atlantic ridge marks the location of ocean-to-ocean plate margins where the lithosphere is being pushed upward by rising hot mantle. Along such ridges, convection currents bring up hot rock from deep in the mantle and the asthenosphere becomes hot enough to begin melting. The magma that forms beneath the ocean ridge rises upward to the top of the lithosphere where it cools and hardens to form ocean crust. This new oceanic crust moves away at right angles from both sides of the ridge (horizontally), carrying the continents with it. The rift valleys of Africa mark the locations of divergent boundaries where fragments of continental crust are moving apart.

Convergent plate boundaries are areas where two plates move toward each other, forcing either development of a subduction zone or a collision zone. In a subduction zone, one plate capped with less dense continental crust will override the denser oceanic plate, which sinks along the contact margin. The entire oceanic plate becomes increasingly hotter as it descends deeper into the earth. As the plates grind against each other, temperatures increase even more. Where the subducting plate comes into contact with the asthenosphere, melting occurs and magma is created. As the less dense magma makes its way upward, it either erupts on the earth's surface and solidifies as extrusive igneous rock, or solidifies within the earth to become intrusive igneous rock. New metamorphic rock is created where rock near the subduction zone does not melt because of the pressure present. In collision zones, continental crust is not recycled into the mantle because this crust is lighter and less dense than the contents of the mantle beneath. In short, as continental plates move toward each other, their crust is too buoyant to be dragged downward on top of the sinking lithosphere. Instead, collision zones push up massive mountain ranges, such as the Himalayas. Transform plate boundaries are areas where two plates are sliding past each other, resulting in earthquake activity but no volcanic activity. California's San Andreas fault is an example of this type of boundary.

The Rock Cycle

Plate tectonics offers an explanation of how rocks are created and destroyed as they are subjected to extreme pressures and heat (Judson et al. 1987, McKnight 1990,

Igneous

Extrusive

Intrusive

Sedimentary

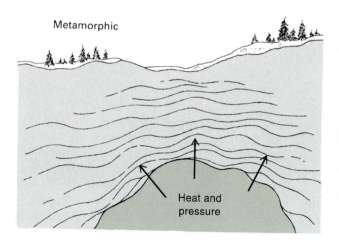

Metamorphic

Heat and pressure

FIGURE 18.3 The three basic types of rocks include igneous rocks formed from cooling magma, sedimentary rocks formed from consolidation of deposited particles, and metamorphic rocks produced when heat and pressure act on preexisting rocks (from McKnight 1990).

Kessler 1994). A more complete explanation of rock genesis is to be found in the rock cycle, which calls upon tectonic activities and aboveground factors including the actions of water, ice, wind, and gravity. The rock cycle (Figure 18.3) is a simplified description of the events leading to the formation, alteration, destruction, and reformation of the three major rock groups: igneous, sedimentary, and metamorphic. In the beginning, all rocks on Earth were igneous. Igneous rocks formed as the magma of early Earth cooled and solidified. Today there are many kinds of igneous rocks and their characteristics vary. However, they can be placed in one of two basic groups based on the conditions under which they become solidified. Extrusive rocks arrive on the earth's surface while still in a molten state. Lava, for example, forms new extrusive igneous rocks as it is spewed from volcanoes. Intrusive igneous rocks solidify as they cool slowly beneath the earth while being insulated by nonmagmatic material. Igneous rock such as granite and basalt are created when magma wells up from the earth's mantle and solidifies as it cools above or beneath the surface.

As mountains comprised of igneous rock are weathered by the actions of wind, rain, ice, and acidic solutions created by vegetation, the particles are transported to lower elevations. Sediments transported by rivers and streams to the bottom of ocean floors can accumulate in thick layers over time. The lower deposits are exposed to great pressure due to the sheer weight of the matter above and this causes their particles to adhere and interlock. As silica, calcium carbonate, and iron oxide precipitate from the water into the pore spaces, these cementing agents and the action of pressure consolidate the sediments into sedimentary rock. As rock forms, minerals actually recrystallize as forms that will accommodate a more confined space. The conversion of sediment into solid rock by cementation, compaction, or crystallization is referred to as lithification.

Sedimentary rocks are classified on the basis of mode of formation: mechanical, chemical, or organic. Mechanically accumulated sedimentary rocks are composed of fragments of preexisting rocks in the form of boulders, gravel, sand, silt, and clay. Shale (composed of silt and clay) and sandstone are examples. Sedimentary rocks formed by the deposition or precipitation of soluble materials are said to be chemically accumulated. Calcium carbonate is a common component of such rocks and limestone is a widespread example. Organically accumulated sedimentary rocks are formed from the remains of dead plants or animals. Limestone produced from the skeletal remains of coral and other lime secreting sea animal serves as an example.

Metamorphic rocks were originally something else. Over eons of time, deposits of sedimentary and igneous rocks became increasingly buried until intense heat and pressures transformed them into metamorphic rocks. Metamorphism, the transformation of rocks either in texture or mineral composition, can also be influenced by chemically active solutions. Some rocks change in a predictable fashion. Limestone usually becomes marble, sandstone becomes quartzite, and shale often becomes slate. The rock cycle begins again as metamorphic rocks either melt or are uplifted to the earth's surface. Evidence of the rock cycle can be found in various places around the world. For example, new igneous rock material is found at divergent plate boundaries (midocean ridges) and at convergent plate boundaries (volcanic island arcs).

Mineral Deposits

The hydrothermal activity occurring at tectonic plate margins provides some of the richest deposits of economically important metallic ores. Hydrothermal vents occur on the ocean floor along ridges where plates separate. As hot, mineral-laden water rises through fractured rocks and spews out of hydrothermal vents, it deposits metal sulfides around the vents as the water cools. Deposits that are associated with divergent plate boundaries, such as midocean ridges and rift zones, include chemically precipitated copper, lead, and zinc sulfide deposits found within ocean floor sediments.

Where tectonic plates come together, igneous rocks saturated with seawater undergo partial melting with the descent of oceanic lithosphere at a subduction zone. The combination of heat, pressure, and resulting partial melting facilitates the mobilization of metals, which are concentrated and ascend as components of the magma. The metal-rich fluids are eventually released (or escape) from the magma, and the metals are deposited in a host rock. The world's great disseminated copper deposits are found in plutons that have formed above subducting plates. All bodies of intrusive igneous rock, regardless of shape or size, are called plutons. The rocks formed are of intermediate composition, formed by the partial melting of oceanic basalt that was enriched in copper.

Mineral deposits also form where plate movements allow magma to penetrate the earth's surface and form volcanos. As the molten igneous rock inside the volcano cools, minerals with high densities crystallize first and sink to the bottom. Less dense minerals crystallize later and are found near the top of the rock deposit.

Ore deposits are often found along the contact between igneous rocks and the rocks they intrude. This area is characterized by contact metamorphism, which is caused by interactions between heat, pressure, and chemically active fluids of the cooling magma in contact with the surrounding country rock (preexisting solid rock). Contact metamorphic deposits form lead, copper, zinc, and silver deposits when hydrothermal fluids from a pluton, an igneous intrusive mass, rich in these metals replaces the country rock (usually limestone). Hydrothermal fluids carrying metallic ions can also enter fractures or cavities in solid rock, then cool and precipitate these metals as vein deposits. Such deposits can contain gold, silver, lead, zinc, or tin.

In some cases an entire igneous rock mass contains disseminated crystals that may be economically recovered. Perhaps the best-known example of this is the occurrence of diamond crystals found in a coarse-grained igneous rock called kimberlite. It typically is found as a pipe-shaped body of rock that decreases in diameter with depth. Almost the entire kimberlite pipe is the ore deposit, and the diamond crystals are disseminated throughout the rock.

Igneous and metamorphic processes are responsible for producing much of the stone used in the construction industry. Granite, basalt, marble (metamorphosed limestone), slate (metamorphosed shale), and quartzite (metamorphosed sandstone), along with other rocks, are quarried to produce crushed rock and dimension stone in the United States.

Sedimentary processes also concentrate minerals in deposits. The processes that weather and erode igneous rocks, for example, can deposit them as sediments while removing unwanted materials. Weathering can concentrate some materials to the point that they can be extracted at a profit. Intensive weathering of residual soils (laterite) derived from aluminum-rich igneous rocks can concentrate relatively insoluble hydrated oxides of aluminum and iron. The more soluble elements such as silica, calcium, and sodium are selectively removed by soil and biological processes. Weathering and erosion also deposit materials in ocean and streambeds at lower elevations. Sand and gravel, for example, are sedimentary deposits of fine-grained rock. Mineral deposits called "placers" are formed when flowing water separates heavier mineral particles (such as gold) from sediment and drops them on streambeds with little water flow and turbulence. Placer deposits of both gold and diamonds are concentrated by wave action and other coastal processes. Beach sands and near-shore deposits are mined in Africa and other places to acquire these precious minerals.

Rivers and streams usually carry large quantities of material derived from the weathering of rocks. When mineral-laden water pours into shallow marine basins or lakes that become isolated by tectonic activity (uplift), massive amounts of "evaporite" minerals can remain after the water evaporates. In other cases, large inland lakes with no outlets essentially dry up. Examples of minerals found as "evaporite deposits" include potassium and sodium salt, gypsum, sodium, and calcium carbonate, sulfate, borate, nitrate, and limited iodine and strontium compounds. Salt water (brines) derived from wells, thermal springs, inland salt lakes, and the ocean can yield iodine, calcium chloride, and magnesium.

Mineral Distribution and Abundance

The U.S. Geological Survey has adopted specific terminology to provide a uniform description of supplies of actual and potential mineral resources (Figure 18.4). It divides the estimated total resources of a given mineral into identified and undiscovered categories. Identified resources are mineral deposits that have a known location, quality, and quantity. Undiscovered resources are those believed to exist on the basis of geological information, but quality and quantity are unknown.

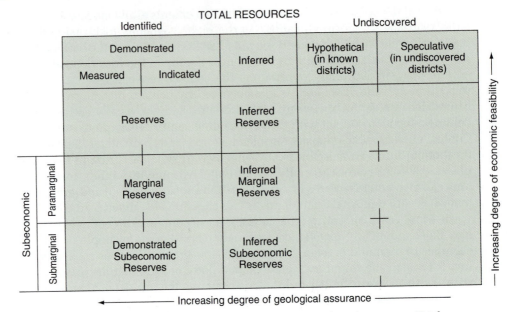

FIGURE 18.4 Mineral resources can be classified as reserves and total resources. Total resources include all the potentially mineable discovered and undiscovered deposits in the world (from U.S. Geological Survey).

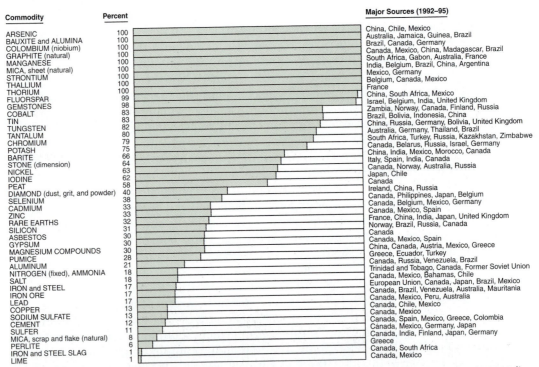

FIGURE 18.5 U.S. net import reliance on selected minerals as a percentage of total consumption in 1996 (in descending order of importance). The shaded area represents net imports (from U.S. Geological Survey).

The amount of a mineral reserve does not remain fixed because economic conditions, exploration effort, and technology change over time. For example, when energy is cheap, marginal ore deposits can become economical and the reserve base expands. The base can shrink when the reverse is true. The system also allows nonprofitable identified resources to be added to the reserve base when ore market values increase.

Because mineral deposits are not evenly deposited in the earth's crust, some nations have rather large supplies of most economic minerals while others must obtain some or most from foreign markets. Figure 18.5 shows the position of the United

States in the world mineral market by identifying minerals obtained mostly or exclusively by importing. Some minerals are imported into the United States because they can be obtained more cheaply elsewhere, while others are simply absent.

Strategic and Critical Minerals

Minerals that can interrupt industrial production when supplies are inadequate are called critical minerals. For the purpose of national defense, some minerals are defined as strategic minerals. Metals needed directly or indirectly to produce arms can be classified as strategic. Examples include some scarce metals needed in large quantity to produce high-strength metal or electronic components.

MINING AND MINERAL EXTRACTION

What is an Ore?

An ore is any naturally occurring material from which a mineral or minerals can be extracted profitably. A mineral deposit or part of a mineral deposit, consisting of ore, is an ore body.

Ore bodies range in size from a few tons to more than a billion tons. Even to the nongeologist, the word ore has a somewhat straightforward meaning. It indicates that something of value is concentrated above average crustal amounts. Metals generally considered scarce include copper, lead, zinc, gold, silver, platinum, uranium, mercury, and molybdenum. In the abundant class we have iron, aluminum, chromium, manganese, titanium, and magnesium. Nonmetallic resources generally are not called ores, but are classified as industrial rocks and minerals.

Surface Mining

Open Pits.   In general, deposits within a hundred meters of the surface are extracted from open-pit mines and those at greater depth come from underground mines. Open pits account for 90% of the ore mined in the United States but only about 50% of the ore mined in the rest of the world. Iron, copper, and lead ores are commonly obtained from open pits (Figure 18.6).

FIGURE 18.6   An open-pit copper mine near Silver City, New Mexico.

Deposits with intimate mixtures of ore and worthless rock are usually mined by bulk extraction methods whereas rich zones of ore are often mined by more selective methods. Open-pit mining is almost always less selective than underground methods. Because the cost and difficulty of mining increase with depth, extremely deep ore deposits are not mineable at a profit by any method.

Open-pit mines typically remove overburden (the worthless rock that overlies ore) and extract the ore. Rock often must be broken by blasting before it can be removed. The ratio between the volumes of ore and the overburden that must be handled is an important factor in determining the feasibility of mining an ore. For the sake of creating a safe workplace, pits are much wider at the top than at the bottom. One of the largest open-pit mines in the world, the Bingham Mine just outside Salt Lake City, Utah, removes more than 225,000 metric ton of rock each day, only 20% of which is ore.

Strip mining is a special open-pit method that is used on shallow, flat-lying ore bodies such as coal. Strip mining is actually one of the most environmentally acceptable forms of mining because it fills the pit and restores the land surface to its original form. Coal extraction will be discussed more fully in Chapter 19.

Quarries.   Sand and gravel are removed from thousands of small pits in many parts of the country. Building rocks such as limestone, granite, and marble are taken from larger pits called quarries.

Dredging.   Dredging is a special type of open-pit mining that is used where the inflow of water is too high to be pumped out economically, and the ore is sufficiently unconsolidated to be dug without blasting. Dredging employs chain buckets and draglines to scrape up surface deposits covered with water. For example, it removes sand from streambeds and gravel from placer deposits in streams. Hydraulic mining is a cross between dredging and conventional mining in which a jet of water is used to disaggregate the rock and wash it into a processing facility.

Subsurface Mining

Some mineral deposits lie so deep that surface mining becomes impractical. These deep beds or veins are removed by subsurface mining. For some metallic ores, miners dig a deep vertical shaft, blast tunnels and rooms, and transport the ore to the surface. Certain soluble minerals such as salt are removed from underground deposits by solution subsurface mining. Compared to surface mining, proportionally more underground mining is done in developing countries where labor costs are lower.

Processing

Extracted minerals usually need to be processed before they can be used, such as copper (Figure 18.7). Rocks taken from quarries may simply require crushing, grinding, and sizing. Metal ore requires considerably more work to remove impurities. In some cases the purified mineral is converted to a different chemical form by smelting or other chemical processes. Usually these processes free a metallic element from the oxygen, sulfur, or other elements with which it is combined in an ore. For example, aluminum is found in ore form as aluminum oxide (Al_2O_3). After the ore is purified and melted, electrical current is passed through the molten oxide to convert it to aluminum metal and oxygen gas. Copper and zinc often are found combined with sulfur, which is removed in smelters by heat or chemical reactions. The physical separation of a mineral from its ore is referred to as beneficiation.

Mining
Open-pit mine
150 tons of ore

Overburden
Over 350 tons

Milling
Crushing
Grinding
Concentrating
3 tons of concentrates

Tailings pond
Approx. 145 tons

Smelting
Roasting
Smelting Furnace
Converting
1 ton of blister copper

1.8 tons of SO_2 gas
or
2.7 tons of H_2SO_4

Slag
1.8 tons

Refining
Refining furnace
Electrolytic refining
Gold
Silver

1 ton of refined copper

Fabricating facilities

FIGURE 18.7   The principal stages of the copper production process (courtesy of U.S. government).

Integrated mills are the traditional method for making steel from raw iron ore. There are three main components to an integrated mill. Coke ovens bake coal to produce coke. A blast furnace makes iron from iron ore, coke, and limestone. A basic oxygen furnace then converts the iron into steel. Minimills (small- to medium-sized steel mills) have lower capital investment and operating cost than integrated mills because they use electric arc furnaces. However, they do not have the facilities necessary to process raw iron ore like integrated mills. Instead they use scrap steel as their primary raw material. While these units have traditionally produced lower grades of steel, they are now competing with integrated units for flat-rolled steel and specialty grades such as stainless steel.

IMPORTANT METALLIC MINERALS

Iron

Numerous metals are known and used, but the most important ones are iron, copper, aluminum, lead, zinc, gold, and silver. Of these, iron must be considered the single most important metal of the modern world.

In fact, about 95% of all metal consumed in the world is iron. More than 70 countries make steel basically from iron. World iron ore and steel production is worth $350 to $400 billion annually, slightly less than world oil production (Kesler 1994). Steel is widely used in constructing buildings, highways, industrial plants, and other structures. Iron and steel products still make up about 75% of the weight of an automobile. However, this figure will continue to fall as lighter materials are substituted to improve fuel consumption.

Ferro-alloy metals are mixed with iron to increase the strength, hardness, and corrosion resistance of steel (Table 18.1). Alloy steel accounts for about 15% of world iron production. Ferro-alloy metals include chromium, manganese, nickel, silicon, cobalt, molybdenum, vanadium, tungsten, columbium, and tellurium. They permit steel to be used in a wide variety of applications, thus expanding its market and displacing use of aluminum and other metals (Schottman 1985). The total value of primary world ferro-alloy metals in these intermediate forms was slightly more than $30 billion in the early 1990s, almost 10% that of iron and steel (Kesler 1994). The United States is especially vulnerable to supply shortages of alloy metals. For example, the United States has virtually no manganese reserves and is currently importing all of the manganese it uses. Iron is the fourth most abundant mineral (5%) in the earth's crust and can be formed by sedimentary, hydrothermal, and igneous processes. The chief ores of iron are hematite (Fe_2O_3) and magnetite (Fe_3O_4) (Hoeffer 1995). More than 50 countries produce iron ore (Kuck 1988). Although small iron deposits occur in almost all countries of the world, large deposits have a more restricted distribution.

Aluminum

Aluminum is the second most abundant metallic element in the earth's crust after silicon. However, it is a comparatively new industrial metal that has been produced in commercial quantities for just over 100 years (Plunkert 1993). Aluminum is lightweight, resistant to atmospheric corrosion, a good conductor of electricity, and strong when alloyed with other metals. This makes aluminum metal ideal for the manufacture of cars, trucks, aircraft, trains, and ships. It is also used to construct bridges and to produce aluminum siding, mobile homes, wiring, beverage cans, and foil wrap. In

TABLE 18.1   Alloy Elements Used in Steel

Element	Function in Steel
Aluminum	Deoxidation and grain-size control
Chromium	Hardenability, high-temperature strength, corrosion resistance
Cobalt	High-temperature hardness (HTH)
Columbium	As-rolled strength
Copper	Corrosion resistance and precipitation hardening
Lead	Machinability
Manganese	Deoxidation, sulfur control, hardenability
Molybdenum	Hardenability, HTH, temper brittleness control
Nickel	Hardenability and low-temperature toughness
Rare earths	Inclusion control, ductility, toughness
Silicon	Deoxidation, electric properties
Sulfur	Machinability
Tungsten	HTH and hardenability
Vanadium	Grain-size control, hardenability, HTH

Sources: Kesler 1994 and U.S. Bureau of Mines, *Mineral Facts and Problems.*

the United States, the container and packaging industry is the largest consumer of aluminum metal, accounting for about 35% of apparent consumption. The main ore from which aluminum is obtained is bauxite (Al_2O_3 ZnH_2O). Australia is a leader in bauxite exports. Aluminum has a sedimentary origin. It is the end product of intense chemical weathering of aluminum-bearing minerals (primarily feldspars).

Copper

Owing to its conducting properties, about 70% of the copper produced is used in the electrical industry (Hoeffer 1995). Most copper ores are sulfides. The principal ones include native copper (Cu), chalcopyrite ($CuFeS_2$), and chalcocite (Cu_2S). Copper can form as an early crystallization mineral of a mafic magma or may be hydrothermal in origin (Hoeffer 1995).

Lead

Refined lead is a soft, heavy metal which was one of the first metals used by man. It has a low melting point, which makes it among the easiest metals to cast, is the most corrosion-resistant common metal, and has unusual electrical properties. About 70% of lead produced finds its way to the transportation industry where it is used to make batteries, fuel tanks, solder, seals, and bearings. Batteries are a main end use. Galena (PbS) is the chief ore of lead. Lead is of hydrothermal origin and is commonly associated with copper and zinc deposits.

Zinc

Zinc is the fourth most widely used metal after iron, aluminum, and copper. About three-fourths is used in metal form and one-fourth in compound form. More than 90% of the metal is used for galvanizing steel and for alloys (Jolly 1993b). The remainder is used to produce dust, oxide, and various chemicals. Most metal products find widespread use in the automotive, construction, electrical, and machinery sectors of the economy. The chief ore of zinc is the mineral sphalerite (ZnS). Zinc is found in association with lead and most lead mines also extract zinc.

Gold and Silver

Gold and silver are among the precious metals. Throughout world history they have been used as the most basic forms of money. Most of the newly refined gold that is fabricated today goes into the manufacture of jewelry. However, because of its superior electrical conductivity and resistance to corrosion, gold is an essential industrial metal. It performs critical functions in computers, communications equipment, spacecraft, and jet aircraft engines. Its use continues in dentistry, but its use as a form of money is now restricted to collectable coins. Gold is usually found unconsolidated with other minerals in the form of nuggets or grains. Hydrothermal veins and placer deposits are main collecting sites. Silver is less expensive and is used in coins, jewelry, tableware, and many other products. Until the 1960s it was used in U.S. coins, but rising prices encouraged speculation and the melting of coins into other products. Silver is found in the native state and in sulfide ores. Most silver is obtained as a by-product of lead and copper mining.

Nonmetallic Resources

For our purposes, nonmetallic resources include industrial rocks and minerals. Notable examples include:

Sand and Gravel. These are used in the preparation of concrete for highway and building construction. Pure quartz sand is used to make window glass.

Building and Crushed Stone. Limestone and granite are typically removed from quarries in blocks to construct buildings. Crushed stone (mostly limestone) is used in forming roadbeds during highway construction.

Rock Salt. Rock salt is a coarse, crystalline halite that forms as an evaporite. It can be obtained from underground mines or extracted from seawater. Rock salt finds its way to the home as table salt.

Gypsum. Gypsum is widely and heavily used to make wallboard and plaster. It is formed as a sedimentary evaporite mineral.

Clay. Clay is used to produce numerous products including ceramics. It results from chemical weathering of rock-forming minerals.

Agricultural Fertilizers. Phosphate, nitrate, and potassium are used to create crop fertilizers. Phosphate is produced from deposits composed of the remains of certain marine organisms. Nitrate and potassium can form as evaporites.

RECYCLING

One of the most attractive means of mineral conservation is recycling (reuse) of materials from discarded industrial products. It is doubly attractive, because it can mean partial recycling of the energy used in producing the materials in the first place. Aluminum is the most striking example. As indicated previously, the energy cost of extracting the metal from bauxite is very high. When aluminum is recycled, 90 to 95% of the energy is saved. Recycling is nothing new. It has been an important industrial activity for decades, but both the energy crunch and environmental considerations have brought efforts to improve and broaden it.

Whether mineral materials can be recycled depends on their patterns of use, which may be either dissipative or conservative. The burning of coal, oil, and natural gas for transport, heating, or power generation, for example, is dissipative. The principal products, carbon dioxide and water, are useless as sources of further energy. Phosphate, potash, and nitrogen compounds spread on fields as fertilizer are largely dissipated and cannot be recycled. Very little of the nonmetallic minerals used can be recycled, although recycling of glass is becoming more important. Concrete and asphalt, for example, are being recycled in road making. The costs of recycling nonmetals are generally far beyond their value in reuse.

Recycling Metals

Opportunities to recycle metals stand in sharp contrast to those presented by nonmetal minerals. Recycled metals become increasingly important as resources are depleted in the United States. Because metal recycling is cost-effective, progress continues to be made in recycling technology. There is also a growing public awareness concerning energy and material conservation to sustain the planet. Overall, the metal recycling industry has grown to the extent that by the year 2010, almost half of the world's metal will be recycled. We will provide a brief discussion of this industry focusing on some individual metals that are used primarily in metallic form.

Seventy-three of the 90 naturally occurring elements that are found in the earth's crust are metals or metalloids. The quantities of metals used annually in the United States range from about 100 million metric tons (Mmt) as in the case of steel, to a few kilograms, as occurs with osmium. In 1990, secondary (recycled) metals produced in the United States were worth $37 billion, only $2 billion less than the value of primary, newly mined metals.

TABLE 18.2   Recycling of Metals in the United States in 1990

Metal	Secondary (Metal Content Metric Tons)	Percent of Total Secondary Metal	Value of Secondary Metal (Million Dollars)	Percent of Value of Total Secondary Metal
Iron (inc. steel)	55,500,000	91	25,000	68
Aluminum	2,400,000	4	3,900	11
Copper	1,300,000	2	3,600	10
Lead	920,000	2	930	2
Zinc	340,000	1	560	2
Manganese	60,000	—	3	—
Magnesium	54,000	—	170	—
Nickel	25,000	—	220	—
Antimony	20,400	—	35	—
Titanium	15,000	—	140	—
Tin	7,800	—	65	—
Molybdenum	3,000	—	17	—
Tungsten	2,200	—	9	—
Silver	1,700	—	260	
Cobalt	1,600	—	30	—
Cadmium	700		5	
Selenium	100		1	—
Vanadium	100	—	2	—
Chromium	90	—	580	
Mercury	90	—	<1	—
Platinum Group	71	—	660	2
Tantalum	50	—	18	—
Gold	49	—	600	2
Totals (rounded)	60,600,000	100	36,800	100

Source: Bureau of Mines, U.S. Department of Interior (1993).

The form in which metals are recovered varies widely. The precious metals, for example, usually are separated from other alloy metals and refined as individual metals. The ferro alloys, however, largely remain in steel as it is reprocessed. Twenty-two metals for which secondary recovery is important, in terms of quantity and/or value, are shown in Table 18.2. The list includes the six platinum metals represented as a group. In both quantity and value, iron and its myriad steel alloys are more important than all other metals combined. Five metals—iron, aluminum, copper, lead, and zinc—account for well over 99% of the quantity and just over 92% of the value of all secondary metal produced.

The amounts of secondary metal shown in Table 18.2 represent metal obtained from both new and old scrap. The relative contribution of the two types of scrap ranges from nearly all new scrap (e.g., titanium) to nearly all old scrap (e.g., the platinum group).

The contribution of secondary metal to total metal consumed ranges from greater than 70% (lead) down to about 5% (vanadium). Many factors determine the extent of secondary recovery of individual metals, including unit value, the size of the pool-in-use, the cost of collection and transport of scrap, the cost of metallurgical processing, the cost of metal disposal if it is not recycled, and the cost to the environment of nonrecovery.

The platinum metals are also intensively recycled, mainly because of their extremely high unit value. Although the amounts of secondary platinum and other precious metals are scant compared to steel, they account for 4% of the value of all secondary metals.

Iron and Steel

Over 50% of the iron and steel used each year in the United States is obtained from scrap. In the 1960s, 35 to 40% of production came from scrap (Houck 1993). About

80% is recycled at steel mills and foundries in the U.S. In addition, steel mills and foundries recycle scrap generated within their own plants.

About 20% of the iron and steel scrap recovered is exported for recycling in other countries. This makes the U.S. the largest exporter of scrap iron and steel.

Iron and steel recycling is a major industry with more than 5,000 establishments throughout the country purchasing steel from individuals and businesses and preparing it for reprocessing mills and foundries.

The demolition of steel buildings and industrial plants creates significant amounts of recyclable scrap. Large numbers of old railroad cars are also recycled each year. Junked automobiles are the largest source of obsolete or postconsumer scrap. Except in the largest cities, virtually all automobiles that are junked pass through an automobile dismantling operation before being sold to a shredder operator. The dismantler removes any salable parts from the car, including reusable parts and readily removable nonferrous metal scrap parts. Examples of salable nonferrous metal scrap are the battery, radiator, heavy copper parts such as the alternator, and the catalytic converter, which contains recyclable precious metals. Municipal solid waste contains as much as 10% steel, consisting primarily of used steel or bimetallic (steel body-aluminum lid) beverage cans and steel food cans. The steel is mostly tin plate. Many municipalities separate the steel and bale it for recycling.

Iron and steel scrap is recycled by remelting and casting into semifinished steel forms or castings. Much more scrap is melted in electric furnaces than in any other type of furnace. In 1990, well over one-half the scrap recycled in the United States was melted in electric furnaces. When steel is recycled, care is taken to change alloy steels containing recoverable elements into new steel of same or similar alloy composition. Stainless steel contains at least 10.5% and usually 18% or more chromium. Most stainless steel also contains from 6 to 11% nickel and some contains from 1 to 3% molybdenum. More than 70% of the nickel recycled each year comes from stainless steel scrap and alloy steel scrap (Houck 1993).

Aluminum

About 40% of the domestic supply of aluminum metal is metal recovered from both purchased new and old aluminum scrap (Plunkert 1993). Recycling aluminum saves 95% of the energy needed to make aluminum from bauxite ore. In 1991, 2.3 million metric tons of metal valued at an estimated $3.0 billion was recovered from both new and old aluminum scrap.

Since opening the nation's first consumer-based recycling center in 1968, Reynolds Aluminum Company has recycled more than 7.3 billion pounds of aluminum. According to Reynolds, that is enough aluminum to fill 4 million garbage trucks had they not recycled it. They also report that two out of every three aluminum cans are recycled annually. This is higher than the recycling figures for any other beverage container (such as plastic, glass, or steel) and almost double the recycling rate for newspapers. Recycled aluminum cans are almost always made into new cans. It takes 4 pounds of bauxite to make 1 pound of aluminum, and all of it must be imported from other countries. Since 1968, recycling at Reynolds' centers has eliminated the need to mine 30 billion pounds of bauxite.

Copper

In 1991 copper from old and new unalloyed and alloyed scrap was valued at $2.2 billion and accounted for 44% of domestic consumption. Most old scrap must be reprocessed to form pure copper. The rate of old scrap recovery is limited somewhat by copper's use in products that remain in service for a long time, such as home wiring, automobiles, and electrical plants (Jolly 1993a).

Lead

Lead recovered from old scrap accounts for about 67% of the domestic demand in the United States (Edelstein 1993). The chief source of old scrap in industrialized countries is lead-acid storage batteries. Scrap lead is also obtained from obsolete manufactured products. Smelters are a source for new lead scrap. A notable feature of the lead industry is the significant reduction in dissipative uses of lead in gasoline, pigments, ammunition, and chemicals. In 1972, dissipative uses consumed 30% of the total domestic lead demand, but only 11% in 1991. This reduction largely resulted from regulated phaseouts of lead in the interest of environmental quality.

ENVIRONMENTAL CONCERNS WITH MINING ACTIVITIES

Mining activities have been the focus of public and media attention for some time. The pressures to curb mining abuses are steadily rising. Criticisms have addressed nearly all aspects of the industry, from exploration activities to the energy needed to create market products. Each of these will be addressed.

Geological Exploration

In their quest for new deposits, geologists explore areas where there is evidence for the occurrence of specific mineral forming processes. Mineral exploration usually has limited impacts on the environment as access roads, survey lines, and deposits are sampled. However, in sensitive areas, such as wetlands or animal breeding areas, there is greater concern.

Mining Extraction

The Problem of Disturbance.
Mining rearranges the landscape and this does not go unnoticed by the public (Figure 18.8). Adverse effects, however, go beyond the cosmetic features of landscapes because mining destroys habitat and, temporarily at least, causes ecosystem function to cease. Soils once permeable to precipitation become water repellant in some cases and soil microbes and fauna important to ecosystem functioning are lost. Disturbance extends beyond the boundary of the mine proper as roads are constructed to carry massive loads of ore. Dust from the entire operation can often be seen for miles. In coastal areas, dredging can damage aquatic and riparian ecosystems. It can effectively change the drainage patterns adversely impacting bottomland forest species through vegetation changes. On the positive side, mining operations involve less land area than one might suspect. Mines occupy about 3,700 km^2 in the United States, or 0.26% of the land area (Barney 1980).

Corporate proposals to begin mining on federally owned land near Yellowstone National Park brought public complaints serious enough to involve the President of the United States. Although the mining concerns felt the area was fair game because it had been previously mined, public outrage was sufficient to stop the project. Mining conflicts involving federal lands arise as corporations make claims to their right protected under the General Mining Law of 1872. The original intent of the law was to promote mineral exploration and development in the western United States, to offer an opportunity to obtain a clear title to mines already being worked, and to help settle land disputes. The law granted free access to individuals and corporations to prospect for minerals on public lands, and allowed them, on discovery, to stake a claim on the deposit. Recent public objections to mining operations near Yellowstone National Park characterized the 1872 law as an outmoded and damaging force, serving corporate interests at the expense of ecosystem preservation.

FIGURE 18.8   Surface-mined land left with no protective cover prior to reclamation (courtesy of U.S. government).

Toxicity and Contamination Problems.   The waste material remaining after ore has been pulverized and concentrated for processing is referred to as tailings. Mismanagement of this material decades ago has created difficult problems today. Many mines were abandoned before the existence of waste regulations. These old mines themselves are safety hazards. However, the real concern centers on mine wastes, which contain metals and other potentially toxic chemical compounds that are being dispersed into the surrounding surface, air, and groundwaters. Problems associated with the mining and processing of sulfide ore deposits have been especially troublesome. As the sulfides oxidize and mix with water, sulfuric acid is produced. This further accelerates the weathering process, including the release of numerous toxic metals such as copper, lead, and cadmium. Cleanup of these wastes could cost as much as $70 billion. Some sites will require complete isolation of wastes in new landfills underlain by impermeable clay barriers.

Mining wastes are regulated in the United States by the Environmental Protection Agency (EPA) and related state agencies under provisions of the Clean Water Act, Clean Air Act, Comprehensive Environmental Response, Compensation and Liability Act, and Toxic Substance Control Act (see Chapter 2). Regulations now generally require that tailings ponds be lined with plastic or impermeable clay to contain the toxins (Kesler 1994). Abandoned tailings piles are dried out, sealed with impermeable clay to keep them dry, covered with soil, and seeded with plants. Impoundment dams are built to hold the leached water in a pond. The potential for groundwater contamination sometimes requires the installation of monitoring wells. The U.S. Geological Survey, although not a regulatory agency, routinely works with communities to investigate existing or potential problems from mine-related contamination.

Mine Reclamation. For some time it has been necessary to obtain approval of reclamation plans and environmental impact statements before mining begins. Bonds are routinely secured to assure successful reclamation. As a minimum, the land surface must be restored to an acceptable contour to prevent acid mine drainage caused by the weathering of rock and unmined ore (Johnson and Paone 1982, Kesler 1994, Carlson and Swisher 1987). Bonds can also be used to cover the costs of correcting environmental damage such as stream pollution, should it occur. In some cases, sites have been converted to recreational areas or nature parks containing wetlands. Other examples are less impressive and there is uncertainty regarding whether arid land areas can be returned to original vegetation within decades after mining operations cease.

Funding for reclamation of old, abandoned sites has been restricted. Whereas considerable federal funding has gone into the cleanup of chemical wastes, reclamation of mined areas has received less government funding.

Mining the Ocean. Environmental problems have developed as countries have either initiated mining ocean floors or made plans to do so. Objections center on ownership of such deposits and on enforcement of measures to avoid pollution effects. The mining of manganese nodules serves as an example of an economically important ocean mining activity (Baturin 1987, Cronan 1992). Nodules the size of a golf ball or softball consist of thin, concentric layers of manganese and iron oxide minerals. In addition to manganese and iron, the nodules contain minor but economically important amounts of nickel, cobalt, copper, and other metals. Most nodules are found in areas of the deep-sea floor in water depths of 5 to 7 kilometers (Kesler 1994). Nodules can also occur at the bottoms of some large lakes.

Mineral Processing

Smelters. Smelting is of great environmental concern because it produces gasses and dust. Depending on specific operations, arsenic, mercury, zinc, and other toxics can be released. Smelting releases sulfur dioxide, a corrosive gas, that combines with atmospheric moisture and oxygen to form sulfuric acid. The effects of acid depositions can be rather specific, resulting in the death of plants or aquatic animals, or rather diffuse and difficult to measure in short time frames. Diffuse effects might include soil acidification that adversely affects soil microbes, which contribute to plant nutrient supplies.

Efforts to reduce smelter emissions have centered on controlling the SO_2 content of flue gas. Consequently, smelter emissions have decreased steadily since reaching peak values in the 1970s. Modern smelters can recover 90 to 99% of their sulfur emissions, depending on date of construction. Smelter particulate emissions peaked around 1950 and have decreased steadily since.

Human Health and Injury. A persistent concern is human injury and death caused by mining accidents or the chronic exposure to mine dust in one form or another. Mining regulations now address these hazards and companies take elaborate steps to prevent employee injury. Nevertheless, fatalities do occur when rock, equipment, and operators fail. Mine fires pose a serious threat in some industries, with fatalities usually being caused by carbon monoxide poisoning. In addition, mine dust can cause a host of lung problems. Silicosis is a lung disease caused by insufficient dust control in silica-rich mines. In the energy sector, coal miners suffer black lung disease and uranium mine workers have high rates of lung cancer. Due to improved safety measures, fatalities and injury issues have been in steady retreat in the United States for years. Miners working in developing countries have been less fortunate.

Future Mineral Availability

The availability of minerals to future generations is an important concern in natural resource management. Their future availability is difficult to judge owing to the manifold factors that can affect their production. A simplistic approach is to divide the quantity reported to be available by the annual rate of consumption and project when the supply will run out. However, this approach disregards the role of economics, technology, and environmental issues in shaping current availability.

Overall, one can say that material well-being in the United States derived from nonfuel minerals has improved over this century. This has occurred despite the fact that we have become increasingly dependent upon other nations. Some experts assume that minerals will remain accessible and that we will continue to prosper. Others believe that several rapidly developing nations, such as China, will need more mineral resources and that we will have less at our disposal. Some believe our society needs to self-impose limitations on our consumption in the interest of ecological sustainability, regardless of external factors (Gardner and Sampat 1999).

Optimists look to the past and present as a means of projecting our materials future. They point to the dire predictions concerning mineral shortages that have been sidestepped by advances in technology and expanding free markets. In fact, numerous metals are becoming less expensive to the consumer. Also, it is becoming increasingly possible to engineer substitutes for materials in short supply and to react to increasing mineral prices by substituting cheaper materials. Plastics have been widely substituted for several metals in the last 25 years. Substitutions and technological innovations continue to downsize or "dematerialize" the mass of minerals used in products. For example, lightweight optical fibers with 30 to 40 times the carrying capacity of conventional materials are replacing copper in many segments of the telecommunications infrastructure. Similarly, cars, computers, containers, and other products are becoming lighter and often smaller as performance and durability improve. In the steel industry, innovations in powder metallurgy, thin casting, directional solidification, and drop and cold forging have allowed savings up to 50% of material inputs over the past 20 years. Since the early nineteenth century, the ratio of weight to power in industrial boilers has decreased almost 100 times. Overall, the basic issue concerning mineral resources is not how soon they will become exhausted, but whether progress will continue to be made in free trade and technology.

LITERATURE CITED

Barney, G. O. 1980. *The global 2000 report to the President of the United States.* New York: Pergamon Press. 360 pp.

Baturin, G. N. 1987. *The geochemistry of manganese and manganese nodules in the ocean.* Hingham, MA: Kluwer Academic Publishers. 356 pp.

Bureau of Mines, United States Department of Interior. 1993. *Recycled minerals in the United States.* Special Publication. 76 pp.

Carlson, C. L. and J. H. Swisher. 1987. *Innovative approaches to mined land reclamation.* Carbondale, IL: Southern Illinois University Press. 752 pp.

Cronan, D. S. 1992. *Marine minerals in the exclusive economic zones.* London: Chapman and Hall. 292 pp.

Edelstein, D. L. 1993. Lead (pp. 33–35). In *Recycled minerals in the United States.* Bureau of Mines, United States Department of Interior. Special Publication.

Gardner, G. and P. Sampat. 1999. Forging a sustainable materials economy. *State of the World 2000.* Washington, DC: World Watch Institute.

Hoeffer, R. L. 1995. *Physical geology.* Springhouse, PA: Springhouse Corporation. 151 pp.

Houck, G. W. 1993. Iron and steel (pp. 27–31). In *Recycled minerals in the United States.* Bureau of Mines, United States Department of Interior. Special Publication.

Johnson, W. and J. Paone. 1982. *Land utilization and reclamation in the mining industry, 1930–1980.* U.S. Bureau of Mines Information Circular, 8862. 22 pp.

Jolly, J. H. 1993a. Copper (pp. 17–22). In *Recycled minerals in the United States.* Bureau of Mines, United States Department of Interior. Special Publication.

Jolly, J. H. 1993b. Zinc (pp. 69–72). In *Recycled minerals in the United States.* Bureau of Mines, United States Department of Interior. Special Publication.

Judson, S., M. E. Kauffman, and L. D. Leet. 1987. *Physical geology.* 7th ed. Englewood Cliffs, NJ: Prentice-Hall. 484 pp.

Kesler, S. E. 1994. *Mineral resources, economics and the environment.* New York: Macmillan. 391 pp.

Kuck, P. H. 1988. Iron ore (pp. 412–423). In *Mineral yearbook.* U.S. Bureau of Mines, United States Department of Interior.

McAlester, A. L. and E. A. Hay. 1975. *Physical geology: Principles and perspectives.* Englewood Cliffs, NJ: Prentice-Hall. 439 pp.

McKnight, T. L. 1990. *Physical geology: A landscape appreciation.* 3rd ed. Englewood Cliffs, NJ: Prentice-Hall. 610 pp.

Plunkert, P. A. 1993. Aluminum (pp. 1–4). In *Recycled minerals in the United States.* Bureau of Mines, United States Department of Interior. Special Publication.

Schottman, F. J. 1985. Iron and steel (pp. 405–424). In *Mineral facts and problems.* U.S. Bureau of Mines, United States Department of Interior.

CHAPTER 19

Nonrenewable Energy Resources

INTRODUCTION

Energy is the capacity to do work. With unlimited energy the human mind becomes the limit to what can be built, manufactured, or chemically created. With abundant, low-cost energy, for example, desalination plants could transform enough seawater to freshwater to meet the needs of people living in arid lands throughout the globe. Of course, energy is costly and this limits economic development. One could say that we cannot well understand global politics and economics without understanding how nations differ in their energy resources. Similarly, energy is linked directly to major changes in the ways civilizations advance, from primitive people carrying fire with them as they pursue food to nations going to war to protect their energy supplies.

For the present, however, our energy concerns are more than academic and are directed more to the future than the past. This is so because growing dependence on fossil fuels appears to be altering Earth's climate (Karl et al. 1996, Meyers and Schipper 1992; Chapter 6). Our use of energy and the future of the planet have become closely intertwined and solutions are needed to some of the most complex problems faced by humankind. As these problems are discussed and acted upon by local as well as national governments, our lives may be affected. We may be asked to pay more for energy, directly or indirectly, or be rewarded for making energy substitutions. We may be called upon to make informed judgments and to adjust individual priorities. Whatever comes our way, we will be better prepared if we understand how energy is provided and consumed. The aim of this chapter and the next is to provide that basic understanding.

WORLD OVERVIEW

Proven world reserves of oil, gas, and coal are estimated to be 43, 61, and 228 years, respectively, at current consumption rates based on a recent Cato Institute study. Probable reserves (includes those undiscovered) of oil, gas, and coal are forecast to be 114, 200, and 1,884 years at present usage. Unconventional fossil fuel sources such as oil tars and shale oil might extend the day of reckoning decades or centuries into the future if new extraction and refinement technologies can be developed.

However, extraction is becoming a bigger problem with proven oil reserves. Many experts are concerned that rising world demand coupled with depletion of readily extractable oil reserves will cause world scarcity by 2018.

During the period from 1987 to 1999 world fossil fuel (gas, oil, coal) use increased 17%. In the developing countries fossil fuel use increased 44% compared to a 3% increase in the developed countries. In the United States, fossil fuel use increased about 6%. The Asia-Pacific region (India, China, South Korea) has experienced the most rapid recent growth in fossil fuel use. This increase has been driven by exploding use of motor vehicles. The United States now imports 58% of its oil due to diminishing reserves. Gasoline prices have risen rapidly worldwide since 1999 as a result of supply reductions by oil-exporting countries, lack of refinery capacity, and increased demand. Although reserves of fossil fuels are finite and rapidly declining, dependency on these nonrenewable energy resources is greater than ever before and growing. At the same time there is growing concern over global warming caused by fossil fuel use (see Chapter 6).

Various strategies to deal with pressing world energy problems are discussed in detail by Flavin and Dunn (1999), Dunn and Flavin (2000), Dunn (2001), and Sheehan (2001). Based on these reviews, fuel cells, wind power, solar power, hydropower, geothermal power, and biomass energy collectively have great potential to replace fossil fuels as the world's primary energy source (see Chapter 20). However, changes in transportation systems, urban design, and population stabilization will also be necessary to avoid a major energy crisis in the twenty-first century. Conversion to renewable energy sources and away from fossil fuel use will be essential in slowing climatic change (Dunn 2001). In this and the following chapter we will discuss the various energy sources and their potential to meet future needs.

ENERGY USE

Historical Perspective

Primitive humans used fire for warmth, protection, harvesting game, and eventually for meal preparation. The carbon records created by human fire pits across the land are used to date ancient landscapes. As nomadic people settled, fuelwood provided a reliable energy source. As time progressed, wood energy was put to other uses. First there was pottery making and with the dawning of the Bronze Age, primitive metallurgy. In the Mediterranean area, the dependence on fuelwood became so great that vast areas were deforested and became severely eroded. Deforestation led to local fuel crises and forced some of the earliest attempts to recycle metal resources. To reduce the energy needed to produce weapons and primitive tools, bronze was recycled to avoid working directly with ore. In 300 B.C., so much wood was being used for metal smelting that the Roman Senate limited mining (Simon 1996). Numerous examples of Roman architecture also show the need to conserve energy. Glass windows were used to capture solar energy and bathhouses were positioned so that they were passively heated by the sun.

Wood remained the principal source of energy well beyond the Middle Ages. Wood shortages continued to occur, depending on local needs. In England, the shortage of wood used for charcoal in the casting of iron became so acute that, in 1599, Parliament passed a law against the cutting of trees for coke used in iron making. The pressure on wood forced an innovation in the smelting process. When it was discovered that blowing machines could be used to eliminate metal impurities caused by smelting with coal, the stage was set for the upcoming Industrial Revolution (Simon 1996). The invention of the steam engine in the late eighteenth century lessened dependence on animals, wind, and water but did not reduce the use of wood energy.

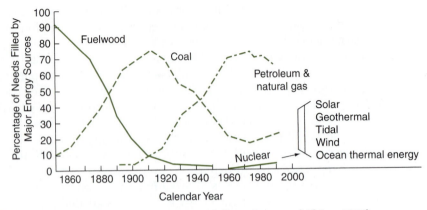

FIGURE 19.1 U.S. energy consumption since 1850 (from Owen and Chiras 1995).

Coal mining began in the United States around 1860 and, in 1880, wood gave way to coal as the dominant source for energy in the United States (Figure 19.1). To this day, coal remains a key source of electrical power.

Oil made its appearance on the energy scene in 1859 when Edwin Drake drilled the first producing oil well in Titusville, Pennsylvania. More than 10 years later (1870), Standard Oil entered the refining business to produce kerosene for lighting purposes. The future of oil seemed threatened when Thomas Edison invented the lightbulb in 1882. It was just a brief time (1896), however, before Daimler and Benz invented the automobile in Germany and we have been hooked on oil ever since. In the early 1940s, petroleum and natural gas gained the dominant role in energy consumption.

The first self-sustaining chain reaction and consequent controlled release of nuclear energy was achieved in 1942. Within two years, the first atomic bomb was detonated and the Nuclear Age began. In 1954, the Atomic Energy Act was amended to permit the use of nuclear energy for producing electricity. The nuclear industry has yet to play a dominant role in U.S. energy consumption, but maintains a significant and controversial presence in electrical power generation.

Energy Use Today

Energy use can be expressed in numerous ways. The basic unit in the metric system is the joule, which is the force of 1 newton applied over a distance of 1 meter. In expressing the massive amounts of energy consumed by nations, the unit commonly discussed is the exajoule, or a billion billion joules. For further simplification an exajoule is roughly equal to one quadrillion, or 1,015 British Thermal Units (BTUs). We, therefore, can reference energy production or consumption in numbers of quads of energy. In 1995, crude oil and natural gas plant liquids (NGPL) produced the greatest share of the world's energy supply, 39.6% (Figure 19.2). Coal produced 25.3% and natural gas 21.5%. Hydroelectric power and nuclear power were of much less importance (7.2 and 6.5%, respectively).

In the United States, coal accounts for 31% of the energy produced followed by natural gas (26.9%), crude oil (18.9%), and others (EIA 1997). With few exceptions, our dependence on petroleum has risen steadily over the past 45 years (Figure 19.3a). Because the growth in nuclear power and other energy sources has not been sufficient to offset our growing energy demands, the United States consumes more energy than it produces (Figure 19.3b). This has been the case since the 1950s, but each year for the past decade marked an even greater imbalance.

In 1996, petroleum (38%), natural gas (24.1%), and coal (22.4%) contributed 79% of the energy consumed in the United States (Figure 19.3c). Energy derived from nuclear power (7.7%) and all forms of renewable power, including hydropower, provided less than 8% (EIA 1997).

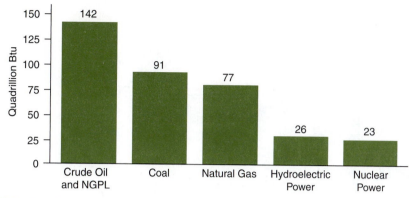

FIGURE 19.2 World primary energy production by source (from EIA 1997).

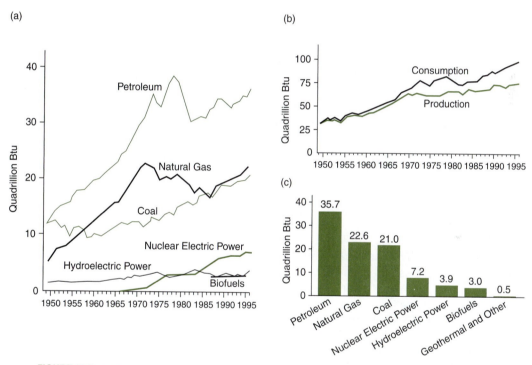

FIGURE 19.3 U.S. energy consumption (a) by source from 1949 to 1996, (b) with energy production from 1949 to 1996, and (c) by source in 1996 (from EIA 1997).

Electrical Energy

As individuals we depend on two forms of energy to meet our daily needs. Gasoline fuels our automobiles and electricity meets a host of everyday needs. The electricity reaching us by way of the electrical grid may have been produced by several means, but the basic principle is the same for all. Regardless of the primary energy source, the generation of electricity requires rotation within a generator. Steam-electric generating units burn fossil fuels such as coal, natural gas, and petroleum. The steam turns a turbine that produces electricity by means of an electrical generator. Natural gas and petroleum are also burned in gas turbine generators where the hot gases produced from combustion are used to turn the turbine, which in turn spins the generator to produce electricity. Petroleum is also burned in generating units with internal-combustion engines. As combustion occurs, the engine turns the generator shaft and mechanical

Fossil fuel power plant

Nuclear power plant
boiling-water reactor (BWR)

FIGURE 19.4 Electric power plants fueled by coal and nuclear energy work on the same principle. Heat boils water to generate steam pressure, which drives a turbine that generates electricity. To maintain a cycle in the closed system, the steam is condensed by cooling and returned to the boiler (from Owen and Chiras 1995).

energy is thereby transformed into electrical energy. Nuclear power plants seem complex but, like fossil fuels, produce steam that drives turbines (Figure 19.4). Turbines are also driven by hydropower as water falls from one level to the next. Wind generators similarly involve the rotation of blades or propellers.

As electricity flows to where it is needed, more than one primary energy source can feed the commercial grid. We measure the output of electrical energy in kilowatts and megawatts (capacity) or kilowatt-hours and megawatt-hours (demand).

In 1994, fossil fuel accounted for 62% of the world's production of electricity, followed by hydropower (19.22%) and nuclear energy (17.4%). Renewable energy sources other than hydropower, such as wind and photovoltaics, provided less than 1%. Compared to the United States, Western Europe derives a greater share of its total energy from nuclear power (31%). In Central and South America, hydropower generates 76% of the electricity used.

In the United States, utilities generated 88% of the electricity consumed in 1996. Coal was the fuel used to generate the largest share (56%) of electricity at electric utilities (Figure 19.5). Nonutility power producers used more natural gas than coal and, because they operate on a much smaller scale, are able to use more renewable energy sources. Nuclear energy generated 22% of the electricity produced at utilities.

Energy in the Home, Business, and Industry

The electrical energy consumed by American homes and business places has grown steadily over the last 45 years (Figure 19.6), owing to population growth and the growing numbers of appliances Americans operate. Homes and businesses use other

FIGURE 19.5 Electric power generation by energy sources serving utilities and nonutility power producers in 1996 (from EIA 1997).

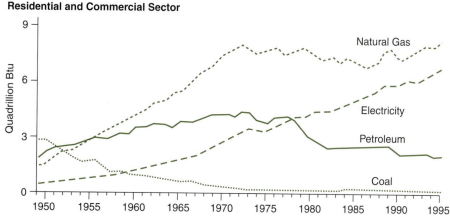

FIGURE 19.6 Energy consumption by residential and commercial use in the U.S., 1949 to 1995 (from EIA 1995).

forms of energy and when data are compared in thermal units, more natural gas energy is consumed than is electrical energy. On a per household basis, home energy consumption was actually less in 1993 than it was in 1978. This came about as homeowners, builders, and appliance manufacturers made progress in improving energy efficiency in the face of rising energy costs. In 1993, home space heating consumed the greatest amount of energy (53.2%), followed by appliances (24%), water heating (18.3%), and air-conditioning (4.5%) (EIA 1995). On average, each individual consumes three times as much energy now as in 1850. There are now 11-fold more people in the United States than in 1850.

For most of the 1980s, homes and businesses used as much energy as the industrial sector (Figure 19.7). From 1950 to 1980 the industrial sector consumed substantially more. In the face of rising energy prices, industry also took energy conservation measures. Another key point is that our nation is becoming less energy intensive, as measured by the energy needed for each unit of gross domestic product. The greatest reduction in energy intensity followed the period of high oil prices of the 1970s. Since 1985, energy intensity has changed little. The industrial sector's use of electricity continues to grow, but coal consumption is considerably less in 1995 than in 1975. The use of coal in industry has actually been in steady decline since the 1950s. From 1975 to 1995, petroleum and natural gas were consumed at roughly comparable rates by industry. The petroleum and coal products industry use the greatest energy when compared to other industry types. Primary metal and paper production each consume less than one-half the energy used by petroleum or chemical product industries (EIA 1995).

By End-Use Sector

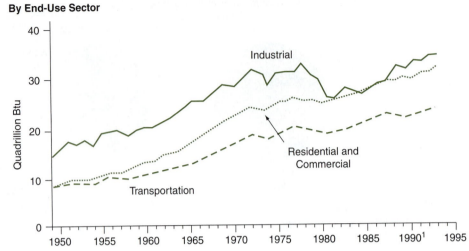

FIGURE 19.7 Energy consumption by end-use sector, 1949 to 1995 (from EIA 1997).

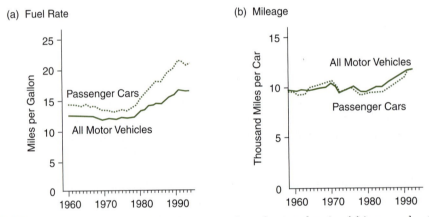

FIGURE 19.8 U.S. motor vehicle statistics from 1960 through 1994 showing (a) improved auto fuel efficiency achieved from 1980 to 1990 and (b) increasing use of automobiles according to miles traveled.

Energy Used for Transportation

The number of automobiles, trucks, buses, and motorcycles registered in the United States is about 210 million. One-third this many vehicles were on our roads in 1960. Our consumption of motor oil has likewise increased. Now we need about 10 million barrels per day to keep moving. The volumetric equivalent of one barrel of oil is 42 U.S. gallons. To reduce greenhouse gas emissions and our dependency on foreign oil, the automobile industry has followed government dictates and built more fuel-efficient vehicles (Figure 19.8a). However, Americans have chosen to invest their energy savings in driving more miles (Figure 19.8b). The popularity of less-fuel-efficient sport utility vehicles also has contributed to higher consumption. Therefore, our national daily consumption has not been reduced.

ENERGY PRODUCTION

The United States draws from diverse energy resources to meet its present energy needs. The finite nature of some energy sources has lead many experts to believe that adjustments will be needed to meet future energy demands. There is also the criticism

TABLE 19.1   A Comparison of Coal Carbon and Energy Rank by Type

Coal Type	Carbon Content (%)	Energy Released on Combustion (10^6 Btu per Ton)	Approximate Age (10^6 Years)
Anthracite	85 to 90	22 to 28	60
Bituminous	60 to 85	19 to 30	100
Subbituminous	55 to 60	16 to 24	300
Lignite	50 to 55	9 to 17	350

that our heavy use of fossil fuels is affecting the planet adversely. Before we can grapple with these limitations and concerns, we need to understand how energy is produced. This question will be addressed as we discuss energy resources individually.

Coal

Coal Formation.   Coal is an organic rock as opposed to most other rocks such as clays and sandstone, which are inorganic. It contains mostly carbon (C), but also hydrogen (H), oxygen (O), sulfur (S), and nitrogen (N), as well as some inorganic constituents (minerals) and water (H_2O). Coal was formed some tens to hundreds of million years ago as plant material accumulated in swampy environments. Because oxygen is limited in swamps, plant materials do not undergo complete decomposition to carbon dioxide and peat is formed instead. Peat is a yellowish to black, usually fibrous material in which plant fragments are still recognizable. The peat thus formed is subsequently changed into brown coal. As the coalification process continues, cellulose, the dominant material of plants, is chemically converted to CO_2, H_2O, methane (CH_4), and carbon. The first three of these products are volatile and are expelled whereas the carbon remains. Therefore, as coalification proceeds, carbon content increases relative to the other substances just identified.

Coal types are ranked according to the relative amounts of carbon and energy they contain. From the lowest to highest rank more time is required for formation (Table 19.1). The basic process of fuel formation is the same for all the ranks. Owing to their higher energy content, bituminous (soft coal) and anthracite (hard coal) are the stages of coal most often mined, processed, and used as fuel. However, even dried peat is used as a fuel in some countries and lignite can be gasified to produce a clean burning natural gas or liquefied to produce liquid petroleum fractions. The costs associated with gasification and liquefaction processes prohibit their use at current petroleum prices.

Coal contains varying quantities of sulfur, which is released to the atmosphere when coal-fired power plants produce electricity. In terms of sulfur content by weight, coal is generally classified as low when it contains 1% or less sulfur. Medium sulfur coal contains 1 to 3% sulfur and high sulfur coal contains more than 3%, as measured on a percent by weight basis. Unfortunately, the low-rank coals typically contain much less sulfur than the bituminous coals. Because bituminous coals have relatively high heating values, most electric power plants prefer to use them. Ash is another undesirable component of coal because it reduces energy content and can be a source of environmental pollution. The ash finding its way into coal heaps can be the result of inadvertent contamination in the mining process or local geology. Moisture content can also vary from one coal source to another. Moisture reduces energy content. Coal rank, sulfur category, and ash and moisture content collectively determine the monetary value of coal.

The Distribution and Abundance of Coal.   Most of the world's coal exists in the Northern Hemisphere. About 85% of the world's coal resources are located in the United States, the area represented by the former Soviet Union, and China. The United States has far more coal than gas and petroleum. Thus, if supplies of gas and petroleum become scarce, coal could be turned to as our major fossil fuel energy source.

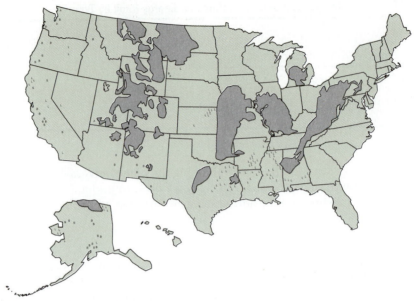

FIGURE 19.9   Coal fields of the United States (from EIA 1996).

For the present, coal is the only major U.S. fuel resource that provides a positive trade balance. In 1995, the value of coal exports was $3.3 billion.

A recent estimate of demonstrated reserve base (DRB) of coal resources is 496 billion short tons (EIA 1996) (Figure 19.9). However, almost half of the base is either inaccessible or is likely to be lost in the mining process.

Nearly half (49%) of the DRB is found in the West, with the vast majority in Montana and Wyoming. Eastern Wyoming is best positioned to meet our needs for low sulfur coal. Appalachia contains about 22% of the DRB and the nation's interior region contains about 29% of the DRB.

It is difficult to estimate how long our domestic coal will last. Based on the 1994 production rate, which involves some exports, and the DRB information previously discussed it should last around 250 years. However, as was discussed in Chapter 18, technology, economics, and environmental concerns are changing at such a rapid pace that estimates remain crude at best.

Coal Mining and Production.   The amount of coal extracted from U.S. mines has increased steadily since the 1960s and in 1994 exceeded 1 billion short tons (Figure 19.10). By a wide margin, bituminous coal has been the coal type most heavily mined. Before the early 1970s, the major fraction of coal was extracted by underground mining. Since then, surface mining has provided a greater and steadily increasing share (Figure 19.10). At present, surface mining accounts for 60% of the coal removed nationwide. In the western states, surface mining accounts for 90% of coal production.

Underground mining, widely used in most eastern coal fields, generally employs one of two methods (Figure 19.11). The "room and pillar method" involves cutting "rooms" into the coal seam leaving large pillars of unmined coal to help support the ceiling (Radovic and Schobert 1992). Because the pillars are left standing, about 40% of the coal seam is not extracted. The longwall method represents an attempt to overcome this limitation. It employs mechanized mining equipment to carve the coal from the seam so that the ceiling created can collapse as the equipment retreats. Although this method recovers almost all of the coal, physical limitations sometimes will not allow use of all needed equipment.

Surface mining has extraction efficiencies of 80 to 90% (Hall et al. 1992). Compared with underground methods, it uses considerably less labor to extract a given

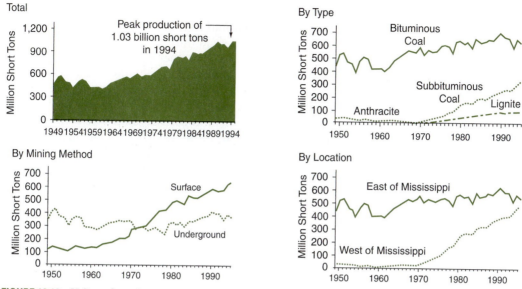

FIGURE 19.10 U.S. coal production, 1949 to 1995, showing production over time (top left), by mining method (lower left), by type of coal (top right), and by location (lower right) (from EIA 1996).

Coal Mining Methods

FIGURE 19.11 Coal mining methods include (a) strip mining, (b) contour mining, (c) room and pillar mining, (d) and longwall mining (from Kesler 1994).

amount of coal. The growth of the western coal industry and its reliance on surface mining explains why both mining efficiency and coal production have increased nationally. While eastern coal production increased only slightly since 1970, western coal production has climbed rapidly and, in 1995, nearly equaled eastern production. Almost all of the lignite and subbituminous coal are obtained from strip mines.

In the eastern United States, contour surface mining is used to extract near-surface coal occurring in hilly country. A wedge of the hill is actually cut away by removing the overburden starting at the outcrop and then proceeding along the contour of the bed in the hillside. Bulldozers and steam shovels are used to remove the overburden covering the coal seams. In the West, flatter terrain permits area strip mining. In many respects it is similar to contour mining, but, as strip mining proceeds, the site is excavated one strip at a time and the overburden from one strip is used to fill the one previously cut.

Although strip mining is highly efficient, its adverse impacts on the environment do not go unchallenged. Strip mining temporarily, at least, disturbs surface and groundwater flow patterns. It causes an abrupt increase in the presence of soil chemical elements in the surface landscape. Without close supervision, some toxic elements can migrate off-site through seepage, runoff, or sediment transport, depending on their mobility and form.

Fertile top soil and overburden must be removed to expose the coal. If the top soil is not properly handled, replaced, and successfully reseeded after mining, large barren areas remain. The water needed to assure successful revegetation success at some western locations is another source of controversy. However, in the driest areas, the more important issue may be whether once productive shrublands will remain reclaimed after operations are abandoned. Mines located near plentiful water sometimes return land to a higher value as viewed by local communities (Figure 19.12) (also see Chapter 2, Figure 2.21).

In 1977, the U.S. Congress passed the Surface Mining Control and Reclamation Act, which requires mining companies to reclaim their land. Under this act land

FIGURE 19.12 Strip-mined rangeland near Colstrip, Montana, which has been revegetated with a mixture of perennial grasses, forbs, and shrubs. Productivity of this land for livestock equals or exceeds that prior to mining.

mined must be returned to its original contour and revegetated. It also requires control of on-site erosion and chemical contamination of nearby lakes and streams. This was a major step toward reducing the environmental impacts of surface mining. However, the funds provided by a federal tax on coal may not be sufficient to fully restore the disturbed lands (Owen and Chiras 1995).

Issues surrounding coal miner health and safety also cause public concern. Although fatal accidents have steadily decreased, owing to mechanization and improved safety awareness, some risk will always remain. More than 100 miners are killed in mine-related incidents every year in the United States One serious threat is the possibility of finely powdered coal dust being ignited to produce a mine explosion. The chronic risk from dust inhalation has been substantially reduced in underground mines, but the danger of black lung disease still remains a serious threat to miner health.

Coal Transportation. Most coal is transported by railway but waterways, highways, or slurry pipelines are also used. Shipments abroad involve ships or barges. Coal transportation is expensive and, depending on rail distance, can double the mine price.

Slurry pipelines have been offered as an alternative method for some situations (Radovic and Schobert 1992). This method involves mixing pulverized coal with roughly an equal weight of water and pumping the resulting slurry through pipelines from the mine to the point of use. One such pipeline is currently in use in the United States. Slurry pipelines are subject to two criticisms. For one, the use of water in water-scarce areas offends the public. For another, with current combustion technology, the coal must be separated from the water before it can be used. More important is the problem of treating the impure water before it can be discharged to the environment (Radovic and Schobert 1992). Finally, the transportation industry serving the coal industry would suffer. Coal is one of the railroad industry's primary clients, accounting for almost 20% of freight revenues (Hall et al. 1992).

Coal Utilization. Eight percent of the domestic coal produced in 1993 was exported. Less than 1% of the coal used was imported. In the United States, coal is used in three ways: combustion, carbonization, and conversion (gasification or liquefaction).

Coal combustion is by far the largest and the most important use of coal. More than 85% of the coal consumed domestically in 1995 went to electric utilities. About 8% went to fuel industrial uses (primarily manufacturing), and nearly 4% was used to produce metallurgical coke (EIA 1996). Coal is combined with oxygen to release its energy. The oldest and still most commonly used form of coal combustion is the pulverized coal-fired boiler furnace, used by most utilities in the generation of electricity.

Coal combustion presents the problem of dealing with the ash and the emissions of CO_2, sulfur, and nitrogen oxides created in the burning process. Ash consists of silica, iron, aluminum, and other noncombustible matter that are contained in coal. Ash increases the weight of coal, adds to the cost of handling, and can affect its burning characteristics (Radovic and Schobert 1992). Carbon dioxide is one of the four major greenhouse gases that threaten climate stability (see Chapter 6). In 1993, coal consumption contributed 35% of the CO_2 released in the United States (EIA 1996). CO_2 accounts for 50 to 60% of the greenhouse warming effect. The sulfur released from electric power plants is subsequently converted to SO_2. Both SO_2 and nitrogen oxide cause acid rain and the latter also contributes as much as 5% to the greenhouse effect.

In the United States, the sulfur content of coals varies from 0.2 to 10% and, unlike carbon and energy content, is not directly related to coal type. Most of the low-sulfur coal in the United States is found west of the Mississippi River. However, more western coal has to be burned to produce the same energy. Removal of the sulfur is not an easy task. The sulfur attached to coal can be removed by washing and other coal preparation methods. Currently, 40% of the bituminous coal used for power generation is cleaned in some manner. All bituminous coal is crushed to provide uniform size, to

FIGURE 19.13   Fluidized bed combustion. Crushed coal mixed with limestone blown into a furnace. The limestone reacts with sulfur dioxide gasses, thus removing most of them from the smoke stack. Air turbulence within the furnace promotes thorough and therefore efficient combustion (from Owen and Chiras 1995).

increase surface area for combustion, and to remove impurities (Hall et al. 1992). The organic sulfur contained within the coal itself requires post combustion removal, usually by employing scrubbers or liquid filters that put exhaust fumes through a spray of water containing lime. The sulfur reacting with the lime is converted to calcium sulfate, which can be removed as a solid. A newer innovation for sulfur removal involves injecting crushed limestone into a boiler containing powdered coal and combusting the coal at a lower temperature. This reduces the release of both sulfur and nitrogen oxides, another serious pollutant contributing to acid deposition. Fluidized bed combustion (Figure 19.13) can remove 90% of the sulfur, 50% of the nitrogen oxides, and 20% of the CO_2. This method involves blowing pulverized coal into a bed of inert ash or limestone that is fluidized (held in suspension) by the uniform injection of hot air through the bottom of the steam generator. When fluidization occurs, the bed of material expands and exhibits the properties of a liquid. As the coal is burned the sulfur it releases reacts with the limestone and falls to the bottom of the generator as calcium sulfite. Sulfur removal then becomes a straight forward process.

The Clean Air Act of 1990 addressed acid deposition by mandating reductions in both sulfur dioxide and nitrogen oxide levels. It mandated total sulfur emissions in 2000 reduced 10 million tons below 1980 levels. Before 1995, a utility releasing less sulfur than allowed could sell the difference in its allowance to another utility. Allowance trading became a major market activity that continues today, but with the restriction that no new allowances are being given. This puts a permanent cap on the total amount of sulfur released from the coal power industry (see also Chapter 4).

The coal carbonization process essentially launched the industrial revolution of the nineteenth century. When coke derived from bituminous coals replaced wood charcoal, the quality of iron and steel products improved greatly. The close proximity of productive bituminous coal mines to the city of Pittsburgh enabled it to become the steel capital of the world. The carbonization process involves heating coal to about 1,000°C in the absence of O_2. If O_2 is present, combustion transforms the coal to ash. Without O_2, only the volatile contents are removed and the remaining product is the fixed carbon, which is called coke (Radovic and Schobert 1992). Coking coals are created from a select group of bituminous coals. Coke (C) is then used in the blast furnace to convert iron ore (FeO) to iron (Fe) as shown by a simple chemical equation:

$$FeO + C = Fe + CO$$

Iron is then converted to steel. Sales of coking coals have declined in the United States as the nation's steel industry has downsized. However, this decline in coal consumption has been more than offset by the rising needs for coal in electric power generation (Kesler 1994). About 60% of the coal exported by the United States is high-quality metallurgical grade (Hall et al. 1992).

Coal can be converted to a cleaner and more convenient fuel through either gasification or liquefaction processes. Coal gasification converts coal into synthetic natural gas. Gasification involves the burning of coal at very high temperatures in the presence of hydrogen to produce a gas. This process can yield gasses with varying degrees of heating values, but few can compete with natural gas (Hall et al. 1992). In the direct liquefaction process, pulverized coal is slurried then mixed with hydrogen in the presence of a catalyst under moderate temperature (455°C) and pressure. The vapor and liquid phases produced are then cooled to separate the products and refined to remove any by-products in the liquid fuel. Germany liquified coal during World War II to meet its petroleum needs when supplies were interrupted. Another process, indirect liquefaction, first converts coal into a gaseous state using a coal gasifier, then recombines the gaseous molecules into liquid products. Indirect liquefaction has several possible routes into the marketplace. Ultimately, stand-alone plants may be built to produce synthetic liquids as substitutes for crude oil. However, within only a few years, indirect liquefaction may become an integral part of advanced electricity-generating power plants. Inherent to the process is the opportunity to store energy from the power plant during off-peak periods when demand for electricity is low. Many of tomorrow's power plants will likely use coal gasifiers to generate combustible gas for high-efficiency gas turbines. When the gas turbines are not using all of the gas from the gasifiers, it could be sent to an indirect liquefaction unit. There it would be converted to methanol and either stored for use later as a turbine fuel or sold commercially.

Overall, gasification and liquefaction technology are commercially available today in several forms, and technology is improving. When the price of oil increases coal conversion becomes more attractive. However, when one reviews the environmental impacts of creating synfuels, there are several potential problems. Some argue that the widespread use of synfuels would further accelerate surface mining and that production requires huge amounts of water. Whatever course is taken, coal represents a tremendous energy source because every ton of coal is the equivalent of more than five barrels of crude oil.

Oil and Gas

Oil and gas were formed from the fats and other lipids accumulated as debris of plants, bacteria, phytoplankton, and zooplankton were buried and preserved, along with mud and silt, in ancient marine sediments (Kesler 1994). Microorganisms began converting lipid-rich organic matter to methane as these materials were deposited. However, the richest deposits of oil and natural gas were formed in extensive coastal marine environments with too little oxygen in bottom sediments to support thorough decomposition.

Over long periods of time, kerogen was formed by the leaching action of groundwaters. The organic matter escaping decomposition was more deeply buried beneath sediments. It was converted to kerogen by leaching from groundwaters. As land subsided from the separation of Earth's plates (discussed in Chapter 18), sediment-bearing rock strata were exposed to greater temperatures. At temperatures from 50° to 60°C, oil formed. At 100°C most of the oil contained in the kerogen was released and further reactions created natural gas.

As plates collided, immense forces built mountain chains, crumpled rocks into folds, and thrust rock strata over each other to form complex structures. The sites where rock strata formed arches or domes (i.e., anticlines) became oil-rich pockets. Large, economically recoverable amounts of oil can be trapped in pools or remain in the pores of reservoir rock, porous sandstone, or limestone. Oil not contained in this manner can move freely to the surface where it is further transformed and lost as a recoverable fossil energy resource. Areas such as the La Brea Pits in California mark sites where oil continues to ooze to the surface today. Oil is found in large quantities in the Persian Gulf area because salt deposited by an evaporating sea effectively caps the upward movement of oil. In addition, the region's aridity prevents water from reaching and destroying the resource.

Oil Field Exploration.   Oil accumulations, in one form or another, have been found in every continent except Antarctica. Oil does not accumulate as an enormous underground lake. Rather, it has backed up in the pores of reservoir rock, or has accumulated in the space beneath an impermeable structure or stratum. Very often, oil is found in seemingly solid rock, which on close inspection contains minute spaces or pores. Exploration for oil and gas is done mostly by seismic methods. Shock waves from an explosion or vibrating source are sent into the ground from a land-based unit or ship. Further, more conclusive evaluation of potentially productive sites must be done by drilling. Drilling is an extremely expensive process especially costly when no oil and gas are found. Sometimes only gas is found but oil will always be associated with at least some gas.

Several factors determine the value of a crude oil deposit. The location and depth of the oil reservoir are important, as is the hardness of the intervening rock layers that must be drilled. A deposit capable of producing large quantities of gasoline is especially valued. Much of the oil contained in such deposits consists of relatively short carbon chains (i.e., five to nine carbon atoms) and has very low sulfur content. Finally, the current market price of oil is always a key consideration.

Exploration is decided by the prevailing economic conditions created by supply and demand. Exploration boomed in the early 1980s in response to higher oil prices created in part by the oil shortages of the 1970s. Shortages arose as OPEC (Organization of Petroleum Exporting Countries) tried to force higher world oil prices. The percent of successful versus failed exploration wells has actually increased since 1970. Exploration involved more wells as opposed to drilling individual wells to greater depth. Efforts to find oil were successful, despite the fact that our oil resources have been steadily depleted. However, the gushers of yesteryear are no more.

Oil and Gas Production.   Efforts to pump oil are directed by the ways it is trapped beneath the earth. The U.S. production of crude oil peaked in 1970 (Figure 19.14a) and has been in uninterrupted decline since 1985. The addition of the Alaskan oil field forestalled the decline in total production, but even its production has fallen since the mid-1980s. The overall decrease in total production is due to the falling productivity of individual wells (Figure 19.14b) rather than a decline in the total number of producing wells. Each day the average U.S. well produces about 12 barrels of oil. In Saudi Arabia, each well averages 10,000 barrels per day! Years ago, U.S. oil wells relied to some extent on water and little gas pressure to force oil to the surface. Now, efforts to keep wells productive frequently must involve sophisticated oil recovery techniques, such as steam injection, to decrease viscosity and increase flow. Secondary recovery methods extract an additional 10 to 20% of the available oil from a well (Radovic and Schobert 1992).

(a)

By Geographic Location

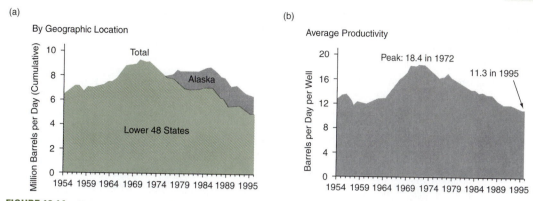

FIGURE 19.14 U.S. crude oil production, 1954–1995 showing (a) production including Alaska's contribution relative to the lower states, and (b) the peak and declining production per well of U.S. oil wells (from EIA 1995).

FIGURE 19.15 Offshore oil production requires expensive platforms.

Today, most U.S. oil production is concentrated in only five states. Roughly 70% of the oil produced in the United States in 1989 (about 3 billion barrels) came from Alaska, Texas, California, Louisiana, and Oklahoma. The oil fields of northwestern Pennsylvania, where it all started in 1859, have long been depleted. Currently, major U.S. oil drilling projects are located offshore (Gulf of Mexico, California) and overseas.

Offshore production today provides more than a million barrels of oil each day and has done so since the late 1960s. Production requires the use of expensive platforms (Figure 19.15). Some reflect ingenious engineering and, in scale, may rival the heights of tall city buildings. Some oil occurs in pools underneath the seafloor (in the Point Arguello field in California, for example). Extracting that oil from the earth requires undersea drilling.

Oil and Gas Processing. Crude oil is not a chemical compound but rather a mixture of chemical compounds. Some are as simple as methane (CH_4); some are as complex as $C_{85}H_{60}$. Most of the compounds in petroleum contain from five to about 20 carbon atoms (Radovic and Schobert 1992). Crude oils vary, however, depending on the geological processes that led to their formation. Some are referred to as old-deep oils,

U.S. ENERGY SOURCES AND END USES, 1995

Primary energy sources (percent of total)

End use consumption (percent of total)

Work achieved

Waste

Total 100% (93.4 quadrillion BTUs per year)

Oil (refined petroleum products)   37.5

Natural gas   20.5

Coal   23.4

Nuclear power   7.7

Water power   7.1

Wood and other 3.7

Transportation 26.5

Industrial 38.1

Residential Commercial (space and hot water heating, lighting, motors) 35.4

Electric power generation consumes 36.3 percent of primary energy

Energy conversion

FIGURE 19.16 Refining of crude oil. Major components are removed at various levels, depending on their boiling points, in a giant distillation column (from Nebel and Wright 1996).

meaning they have been acted upon over long time periods and were subjected to high temperatures beneath the earth because of burial depth. By contrast one could speak of young-shallow oils that are highly viscous, high-density materials with a high sulfur content. Consequently, they require more processing to substitute for treatments not provided by nature. Nevertheless, compared with coal the elemental composition of crude oils from different parts of the world shows less variation.

As indicated by the simple chemical formulas just given, crude oils are mostly mixtures of hydrocarbons. In general they contain about 83 to 87% carbon and 11 to 16% hydrogen. They also contain up to 7% oxygen and nitrogen and 4% sulfur and therefore contain dissolved gasses as well as solids and liquids.

Crude petroleum must be processed, or refined, to untap the energy locked up in its molecular structures. The initial phase of refining involves fractional distillation to separate useful products (Figure 19.16). Much of the crude oil present is vaporized by heating. The vapor is passed to a distillation tower where it cools, with lighter molecules, such as gasoline condensing high in the tower and heavier molecules, such as fuel oil, condensing at lower levels. The remaining unvaporized heavy oil is passed to a vacuum distillation tower, where it too is vaporized and condensed into fractions. In the second step, cracking, heavy molecules from distillation are heated under pressure and broken down into smaller molecules that can be used in gasoline and other light hydrocarbon products. The third step, reforming, is a similar process in which the actual molecular structure of each product is changed to make it more acceptable to today's markets (Kesler 1994).

Individual refineries can differ as to the types of oil they process and the products they create, but they all usually employ these three basic steps (Radovic and Schobert 1992). Products recovered from these processes include natural gas, gasoline, diesel fuel, jet fuel, light oils (low viscosity), heavy oils (high viscosity), asphalt, fuel oil, gasoline, kerosene, lubricating oils, naphthas, paraffins, petroleum coke, petroleum jelly, and wax, as well as feedstock for petrochemical manufacture. The most important refinery product, unleaded gasoline, contains less than 0.05 grams of lead and 0.005 grams of phosphorus per gallon. In addition to transportation fuels, oil is used for home heating and for generating electricity. The fraction of petroleum used for polymeric materials is comparatively small but greatly impacts our lives. Plastics such as polyethylene, polyvinylchloride (PVC), formica, and polystyrene are all derived from petroleum. Numerous synthetic textiles (e.g., dacron, orlon, nylon, and polyesters), synthetic rubber products and some pharmaceuticals, cosmetics, and detergents are derived from petroleum.

Natural gas was produced by decomposing plant and animal remains that were buried in the earth for millions of years by sedimentary deposits. It is therefore a fossil fuel like coal and oil and is often found associated with them. As the gas leaves the well and is transported, it cools and the heavier hydrocarbons liquefy. The gas and crude oil separate as this mixture enters a field separator and is subjected to less pressure. Natural gas can then be drawn off from the top of the separator. Both the condensate and gas are useful products. Natural gas removed from field separators still may contain hydrocarbons heavier than methane and may be further processed to recover natural gas liquids (NGLs). Products recovered from the NGLs include ethane, propane, butanes, and additional natural gas (Leffler 1979). Processing also removes nonhydrocarbon gasses, which can be sufficiently abundant to constitute resources in their own right. The removal of essentially all hydrogen sulfide (H_2S) is of particular concern because it is toxic and reacts with moisture in pipelines to create highly corrosive sulfuric acid.

Natural gas currently supplies about 24% of U.S. energy needs. It is used primarily for heating buildings, home cooking, industrial processes, and generating electricity. It presents several advantages over other fossil fuels. It is easier to transport than coal because it can be moved throughout a piped network extending across international boundaries. More importantly, it is a cleaner burning fuel because it contains less carbon relative to hydrogen, is nearly ash free, and has very little sulfur and nitrogen.

Because of the declining availability of domestic oil, the industry has been evaluating alternative supplies. Oil shales and tar sands are among the candidate substitutes, but their use involves major compromises. For beginners, both are examples of oil resources that were not fully developed over their geologic past. Oil shale is a sedimentary rock that was formed millions of years ago from the mud at the bottoms of lakes. It contains kerogen, the same solid organic material contained in rocks that produced oil when subjected to higher temperature and pressure. Tar sands contain a thick, oily residue called bitumen. The kerogen and bitumen contained in these resources can be extracted and refined like petroleum. The industrial processes required, however, are energy intensive. Another drawback is that existing deposits will not greatly extend our energy supplies. Tar sands deposits are not extensive and only about one billion barrels could be recovered from U.S. deposits. With current technologies, 80 to 300 billion barrels of oil could be extracted from oil shale thus providing the U.S. with no more than 30 years of oil. Finally, the use of oil shale and tar sands could be one more step toward environmental degradation. For example, the most valuable oil shale deposits in the United States occur in Colorado, Wyoming, and Utah where sufficient water for processing is in short supply (Gleick 1994).

The Future of Oil and Natural Gas.
Some would characterize the twentieth century as a period of rapid economic development in which the energy needed for industrial expansion was based on oil resources. Tremendous economic development, especially in

the United States, Europe, and Japan, was facilitated by low oil prices. However, it is obvious that our relationship with oil will face forced revisions sometime in the not too distant future. One necessary adjustment will be to find alternate sources of energy as the U.S. becomes increasingly dependent on foreign oil supplies. Granted, the United States currently is not facing an oil crisis and this condition will prevail as long as it is able to purchase foreign oil at modest prices. In 1995 the United States actually purchased more oil from Venezuela than from Saudi Arabia, whose contribution was matched by Canada. Imports from Mexico, Nigeria, and the United Kingdom were also significant.

A key factor is how long developing countries will be able to supply U.S. needs. Asian countries such as China are becoming oil dependent themselves and may soon begin competing with the United States for world supplies. More specifically, it is anticipated that world oil consumption will rise to more than 74 million barrels within the next few years and should reach a range of 89 to 99 million barrels a day by 2015 (EIA 1995). Also, the number of automobiles around the world has doubled since 1973 and is approaching 600 million. By some estimates the world oil supply will last less than 45 years. An optimist would say that developing countries have not intensively explored their natural resources and undiscovered oil could greatly extend oil supplies. A pessimist might say that the world demand for oil will outpace population growth, thus shortening the life of oil in the global economy to less than 50 years.

In 1994, the United States obtained 45% of the oil it consumed from foreign sources. By 2005, net imports will account for about 57% (EIA 1995). Once again, the vast stores of oil found in the Middle East will likely be critical in meeting U.S. energy needs in the near future. By 2015, barring unforseen developments, OPEC will export about 52 million barrels each day to the United States. The United States has given itself some protection against the oil supply disruptions by creating an oil stockpile, the Strategic Petroleum Reserve. However, it can postpone economic turmoil for only a few months.

The outlook for natural gas is better than for oil, both in the United States and abroad. New studies suggest that the total U.S. reserves could be seven times greater than previously estimated and could last about 60 years. Estimates of world natural gas vary, but proven and undiscovered reserves will last about 104 years at the current rate of consumption. Oil shale or tar sand might be substituted for natural gas in some places. However, their distribution is too fragmented for them to become dominant world fossil fuel sources without global cooperation in investment and production.

Fuel Conversion.   As discussed, the earth's fossil fuel energy can be separated into vapors (natural gas), liquids (crude oil), and solids (coal, oil shale, and tar sand). The United States, the former Soviet Union, and China together possess more than 80% of the ultimately recoverable resources. The most important of these fossil fuels are natural gas, petroleum, and coal. However, as domestic and world oil supplies of these resources fade, transitions to other energy resources will be essential. No one knows exactly when these transitions will begin because of the complex of economical, political, and technological factors potentially involved. However, one can look to existing resource uses to identify the most probable possibilities.

To minimize the greenhouse effect, our most immediate need is to move away from fuels that release high amounts of carbon per unit of energy expended. This statement applies to both transportation and electric power generation. Natural gas (methane) is a likely substitute candidate because of its high hydrogen to carbon content. However, methane itself is a stronger greenhouse gas than CO_2. Small leaks in extensive production and delivery systems could be major contributors to atmospheric pollution.

Overall, the United States has more coal than oil or natural gas and more natural gas than oil. Coal conversion to gas or liquids, as previously discussed, therefore remains a strong possibility. Because conversion either adds hydrogen or removes carbon, depending on the process used, it favorably alters the hydrogen to carbon ratio and creates a cleaner fuel. To date, industrial development along these lines in the

United States has been discouraging. The Great Plains Coal Gasification Plant in Beulah, North Dakota, was completed in the mid-1980s. However, construction and operating costs were higher than anticipated and its product cannot compete financially in today's energy market place. South Africa provides a more encouraging example (Kesler 1994). There, political embargoes on oil imports stemming from apartheid forced the conversion of its abundant coal resource to liquid fuel. Three plants owned by SASOL employ 32,000 people and produce 130 different products including gasoline, diesel fuel, jet fuel, and motor oil. Its plants are supplied by the largest underground coal mining system in the world.

In the United States, propane-like methane (natural gas) is also a very clean burning fuel and existing automobiles can be retrofitted with the equipment needed to use it. No modifications are needed to burn ethanol, which can be added to conventional gasoline to create a cleaner fuel ("gasohol"). Ethanol and other fuel alcohols can be obtained from sources other than fossil fuels (see Chapter 20).

Environmental Problems Associated with Exploration and Production.
Oil exploration and production can create several environmental problems. Numerous exploration wells are necessary to estimate the extent of an oil deposit. The initial phases of oil field exploration development therefore can involve considerable road construction and land disturbance. Native wildlife and wilderness can be seriously affected, especially if work is done recklessly. Currently, regulations and company practices are done more carefully, but site disturbance simply cannot be avoided. Improvements in drilling technology have actually reduced the impact of drilling on the environment and large quantities of gas are no longer flared in the night sky. By today's standards this would be wasteful and, in fact, many oil fields have lost most of their associated gas.

The greatest threat posed by oil and gas extraction is the escape of underground fluids and land subsidence. The escape of fluid from offshore drilling operations is actually a rare event. Only one well of the 30,000 drilled between 1964 and 1991 has blown out (Kesler 1994). This period includes the time period in which offshore exploration peaked in the mid-1980s (EIA 1997).

On land, the most troublesome fluid that comes from oil and gas wells is brine—i.e., fluid with concentrated salts. Some brines contain enough sodium, potassium, bromine, or related elements to be of commercial interest. However, most simply create a disposal problem and are injected back into the sedimentary strata from which they came. The early practice of collecting the brine in unlined ponds called oil pits is no longer permitted.

The removal of shallow oil deposits can cause the overlying land surface to subside. Modern reservoir engineering techniques designed to remove oil more efficiently have minimized this problem, but low-lying areas are vulnerable to even small subsidence effects (Kesler 1994).

Oil transportation involves domestic pipelines but also a growing dependence on international shipping. About 200,000 miles of oil pipelines exist in the United States. The 800-mile Trans-Alaskan pipeline running from Alaska's "north slope" at Prudhoe Bay to the ship port at Valdez is best known to the public. Its construction caused serious protest in the pipeline's planning stages, owing to the delicate ecosystems potentially affected by its construction. Ocean-crossing oil supertankers provide an economical means of oil transport. One tanker can carry as much as 50 million barrels. This is enough to meet U.S. oil needs for three days. However, the rupture of a tanker hull can create an ecological catastrophe costing billions to clean up (Figure 19.17). The much publicized *Exxon Valdez* spill in 1989 created haunting images of birds, marine mammals, and fish being lost and rescued. Much of the damage caused by spills goes without notice. The larval and egg stages of fish are especially sensitive to the toxic effects of oil, and a surface slick may have its greatest impact on immature fish. Populations of most commercially important species of fish are extremely variable and a spill at just the wrong time can greatly affect their numbers.

FIGURE 19.17   A common murre covered with oil (courtesy of U.S. Fish and Wildlife Service; photo by Jill Parker).

The extraction of oil or its transportation in environmentally sensitive areas continues to receive justified public attention. As exploration presses into more remote areas to locate oil supplies, we will increasingly be forced to find a balance between the risk of foreign dependence on energy and potential damage to sensitive ecosystems. Two cases illustrate the dilemmas at hand. One involves oil company desires to obtain oil from Alaska's arctic slope and islands which are located near North America's largest oil resource, the Prudhoe Bay oil field. Protection of ecosystems against the intrusive nature of oil field development continues to block oil field development. This is most evident in ongoing attempts to open the Arctic National Wildlife Refuge to drilling. Along the coastline of California a protest recently arose over development of the Point Arguello field, near Santa Barbara. Similarly, the state of Florida has successfully blocked plans to open offshore Gulf of Mexico waters to oil development. Such cases justify considerable debate and thorough scientific study.

The facts are not always accurately stated by either overzealous environmentalists or the oil industry. For example, a major leak in an undersea pipe could create severe ecological damage. However, as discussed earlier, such events are rare and the public is not well aware of natural oil emissions from the marine floor.

Transportation Fuel Issues.   The United States is forced to move ahead with oil development largely because of its excessive dependence on automobiles. The public wants to be free of automobile pollution but does not want to sacrifice the loss of personal freedom the automobile provides. The long lines at gas stations during the very brief oil crisis we experienced in the 1970s demonstrate the extent of our dependence on gasoline. The only people spared this brief period of forced trip planning and high gas prices were those who relied exclusively on mass transportation. It follows that part of the transportation problem can and must be improved by providing more means of public transit. In areas where public transit would be a poor financial investment, solutions must be found elsewhere. From 1974 to 1985, the fuel economy of U.S. cars almost doubled, which represents real but insufficient progress. Fuel modifications and emission retarding devices were also developed. Recent progress in automotive design gives hope that more advances are on the horizon.

Cars are becoming increasingly lighter as the auto industry learns to substitute lightweight, composite materials for more conventional steel. The ability to mold these materials in aerodynamically efficient car bodies is in a state of rapid advancement. Finally, low resistance tires, more aerodynamic designs, reductions in vehicle weight, and advances in electric battery technology are providing realistic

FIGURE 19.18 A nuclear power plant. Note the conspicuous cooling tower, which is necessary to cool and recondense the steam from the turbines before returning it to the boiler (courtesy of U.S. Dept. of Interior).

opportunities to convert autos to electric power (Schiffer 1994, Sperling 1995). It is likely that a completely new kind of vehicle will begin to appear in automobile showrooms within the next few years. This new class of automobiles will operate more cleanly, and will be quieter and more responsive than current models (Flavin and Lenssen 1994). Both electric and electric-hybrid autos are being designed. However, efforts to redesign the automobile and to spur public transportation will be wasted if people are unwilling to revise their individual driving habits and a recent trend toward preferences for heavier vehicles that are excluded from fuel efficiency requirements. Other nations will increasingly encounter the blessings and burdens associated with the automobile. Most projections show rapid increases in the number of automobiles driven in Asia. This will increase demand for fossil fuel and the greenhouse effect simultaneously.

Nuclear Energy

The Nuclear Alternative. As fossil fuel resources decline, we must look elsewhere for energy resources. Nuclear power is a possible alternative (Figure 19.18). Nuclear power seems capable of eliminating or greatly reducing two of our most pressing problems: fading energy resources and global warming, as related to the combustion of fossil fuels. However, one must understand the basics of the resources and industry involved before drawing conclusions on its viability as an energy solution.

Nuclear power stations and fossil-fueled power stations of similar capacity have many features in common. Both require heat to produce the steam needed to drive turbines and generators. In a nuclear power station, however, the nuclear fission replaces the burning of coal or oil. Nuclear fission reactions involve the basic particles located in the nuclei of heavy elements (i.e., elements having large numbers of protons and neutrons in their nuclei). More specifically, nuclear reactions usually involve ^{235}uranium, which is referred to as an isotope. Atoms of the same element with different atomic weights are called isotopes. Elements are defined by the number of protons and, therefore, all uranium atoms have 92 protons. The superscript number appearing before an element's name gives the atomic mass (the sum of protons and neutrons) of a given isotope. Because a single proton and a single neutron both have an atomic weight of about one, the number also indicates the total number of these particles present in the isotope.

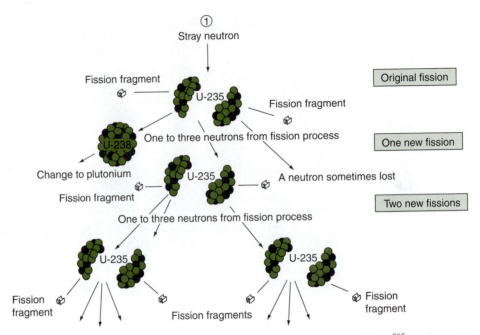

FIGURE 19.19 Nuclear fission involving a chain reaction. A neutron fly off by a ^{235}uranium atom strikes another which undergoes fission, thus releasing fragments or daughter nuclei and one to three neutrons. The process continues as these neutrons cause other ^{235}uranium nuclei to split (from Owen and Chiras 1995).

In a nuclear reactor, a free neutron collides with ^{235}uranium causing the isotope to split and release neutrons (Figure 19.19). Energy is created as the reaction proceeds because some matter is lost in the fission process. Because more neutrons are released from a fission event than are needed to induce the event, more and more uranium nuclei split in what is described as a "chain reaction."

To provide useful power, the rate of the chain reaction must be controlled. This can be achieved in nuclear reactors by using a "moderator" such as ordinary or "light water," which absorbs the free neutrons emitted as opposed to fissioning. Therefore, it can be said that the chain reaction is controlled by diluting the fissionable uranium atoms with water or other types of nonfissionable atoms. All of these fission fragments created by the fission process have high kinetic energy, which is converted after collisions with other nuclei into the random motion of thermal energy.

In nature uranium exists as ^{235}uranium or ^{238}uranium. The isotope that serves the U.S. nuclear industry, ^{235}uranium, is much rarer in nature than the alternate isotope. The ^{238}uranium accounts for 99.3% of the uranium naturally present in Earth's crust and ^{235}uranium only 0.7%. The ^{238}uranium is not considered a primary nuclear fuel because it nearly always reacts with a free neutron by absorbing it rather than by fissioning.

In U.S. nuclear reactors, uranium fuel is housed in an array of reactor "fuel assemblies." Before the initial start-up of a nuclear power reactor, the core must be loaded with fresh nuclear fuel. The fuel is basically a reservoir from which energy is extracted as long as a chain reaction can be sustained. The reactor itself is housed by a containment building that is sturdy enough to withstand a limited explosion. The aim is to confine any steam or gas explosions that might otherwise release radioactive water to the environment (Mounfield 1991). Assemblies comprise fuel rods created by encasing ^{235}uranium pellets in zirconium alloy. The heat produced by nuclear fission in the reactor core is carried away by the water which, as stated, serves also as a moderator.

The process varies depending on the type of reactor. In nuclear plants that involve a "boiling-water reactor," water is converted to steam, which directly drives a steam turbine connected to an electric generator. The steam is then condensed to water and pumped back to the reactor to begin another cycle. In plants employing a "pressurized-water reactor," a device called a "steam generator" prevents radioactive water from directly contacting the turbine. Heat from water that has passed through the reactor core is transferred to water confined in a secondary loop. As heat is added, the steam produced in the secondary loop directly drives a turbine. In the reactor core, cooling water is kept at a very high pressure and is heated to some 600°C.

The boiling-water reactor is potentially more troublesome than the pressurized water reactor because the water serving as coolant and moderator can become radioactive before reaching the turbine. Contamination is possible because slight leaks can occur in the thin cladding of the fuel rods, and from radioactivity induced just outside the cladding. Because the steam that drives the turbine is potentially radioactive, great care must be exerted to avoid steam leaks in the turbine itself. The pressurized water reactor eliminates this problem altogether by confining the water circulating through the reactor core to a separate loop. However, the inclusion of a second loop reduces operational efficiency below that provided by the boiling-water reactor. Other reactor designs have also been developed and used abroad. These include the pressurized heavy-water reactor, the gas-cooled reactor, and the advanced gas-cooled reactor.

Because the coolant water in light-water reactors absorbs free neutrons, the amount of the fissile ^{235}uranium present must be concentrated above its natural level of 0.7%. Were this step omitted, it would be impossible to sustain a chain reaction. The concentration of ^{235}uranium steadily decreases as a reactor operates and energy is produced. At the same time the fertile $_{238}$uranium nuclei present are constantly being converted into fissile ^{239}Pu nuclei, some of which will, in turn, undergo fission and produce energy. While these reactions are taking place, the concentration of neutron-absorbing fission products (also called "poisons") increases within the nuclear fuel assemblies. After years of operation, the concentrations of fissile nuclei and poisons in fuel rods become altered to the extent that the chain reaction will not proceed. In other words, free neutrons are being absorbed or lost faster than they are created by fission events. A reactor reaching this condition must be shut down and refueled.

As one might guess, intervals between refueling are related to the amount of fissile ^{235}uranium loaded into the fuel assemblies at the beginning of the reactor's fuel cycle. Between refueling events, there is a period over which a reactor could operate at full output before a fission chain reaction would cease to be sustained. Its energy reservoir at any given time can therefore be said to contain so many full-power days. The plant's "capacity factor" is defined as the ratio of its actual level of operation to the maximum, full-power level of operation for which it is designed. The rate of fission at any given time is regulated by inserting or removing control rods, which absorb neutrons. Control rods can be inserted or withdrawn from around the fuel, thereby slowing or increasing the rate of fission. A variety of systems are in place to automatically stop the reaction and to provide emergency cooling.

We have just described the fueling factors associated with the converter type reactors presently used in the United States. However, there is another category, the breeder reactor, which is restricted from use in the U.S. because of security reasons. Breeder reactors get their name from their ability to breed, or create, fuel as they operate. Breeder reactors can be fueled with the more common uranium isotope, ^{238}uranium, and plutonium. The plutonium undergoes fission, much like the ^{235}uranium in a light-water reactor, producing neutrons that can be absorbed by ^{238}uranium to create ^{239}uranium, which eventually becomes plutonium. This reactor requires fast neutrons for absorption by ^{238}uranium, so there is no moderator. The coolant for a breeder reactor is a liquid metal, currently sodium.

The Nuclear Fuel Cycle.

The complete nuclear fuel cycle for a typical light-water reactor goes well beyond the boundaries of the nuclear reactor. The cycle is routinely described as having a "front end," which includes the processes required to prepare

Front End

Back End

Fuel Fabrication

Reactor

Interim Storage

Enrichment

Plutonium

Conversion

Uranium

Spent Fuel Reprocessing

Milling

Mining

Final Disposition

→ Fuel cycle as it currently operates in the United States.
⋯→ Fuel cycle as it currently operates with spent reprocessing in some foreign
   countries and/or final waste storage.

FIGURE 19.20   The nuclear fuel cycle (from EIA 1995).

nuclear fuel for the reactor. The "back end" of the cycle includes the processes required to properly manage the spent nuclear fuel. Because the unused uranium and plutonium in the spent fuel can be reprocessed to recover useful nuclear fuel, we are speaking of a cycle as opposed to a chain of events (see Figure 19.20). Commercial recycling in the United States has limited applications, but this is not so in all countries.

The front end of the cycle involves several phases starting with exploration to identify resources and reserves. Uranium ore must be discovered by drilling and other exploratory techniques. Ore deposits believed to exist through technical assessment, but are as yet undiscovered and, therefore, unquantified, are called resources. A body of uranium ore is classified as a reserve once the quantity is known and extraction costs are acceptable by current practice.

The next phase is mining. The methods used to mine uranium ore are similar or identical to those used to extract ores of other metals. However, uranium can also be obtained by solution mining and as a by-product of phosphate mining. In the United States, the content of ores typically ranges from 0.05 to 0.3% uranium oxide (^{308}U). In other countries, ore concentrations of ^{308}U can be less, but in general are of a higher grade. Countries with high-grade deposits include Australia, Canada, South Africa, Niger, Namibia, Brazil, France, and Gabon. China and Russia may also have substantial deposits of uranium.

Milling follows mining. This process involves crushing and grinding the uranium-bearing ore followed by chemical extraction to obtain uranium oxide. This product is also called uranium concentrate or "yellow cake" (^{308}U), which is 85% pure uranium in an isotopic mixture of 0.7% fissionable ^{235}U and 99.3% nonfissionable ^{238}U.

Milling generally is done near the mining operation. Depending on circumstances, the yellow cake can be marketed like other energy resources.

The next step in the cycle is the chemical conversion of ^{308}U to uranium hexafluoride (^{6}UF), which is transformed from a solid to a gas if it is elevated above room temperature.

Enrichment involves concentrating ^{235}uranium so that enough fuel can be loaded to sustain a nuclear chain reaction in a light-water reactor. One method by which the necessary material can be concentrated, or enriched, involves gaseous diffusion. This consists of passing a "feed stream" of uranium hexaflouride (^{6}UF) gas through a long series of diffusion barriers that pass ^{235}uranium at a faster rate than the heavier ^{238}uranium atoms. This differential treatment progressively increases the percentage of ^{235}uranium in the "product stream." The "waste stream" contains the depleted uranium (that is, uranium having a ^{235}uranium concentration below the natural concentration of 0.7%). The enrichment process requires considerable energy. Alternative means for enrichment include gas centrifuge separation, as done in Europe by commercial operations and a laser separation technique being developed.

The final phase in the front end is fabrication. After the enriched ^{6}UF is changed to an oxide, it is pressed into pellets of ceramic uranium dioxide. The pellets are then encased in tubes of zirconium alloy or stainless steel, which will resist corrosion. After the tubes, (also called rods or elements) are mounted into assemblies the processed fuel product is ready for loading into a reactor.

The back end of the cycle involves fewer steps starting with interim storage. After spent fuel is removed from a reactor, it is stored either at the reactor site or at a facility specifically designed to manage highly radioactive waste. At the reactor site, spent fuel rods are usually stored in water, which serves two purposes. It provides continual cooling for rods still giving off heat created by residual radiation, and shields workers from the radiation otherwise emitted.

Reprocessing follows interim storage. Spent fuel contains significant amounts of fissile (^{235}U, ^{239}Pu), fertile (^{238}U), and other radioactive materials. A "fertile" is capable of producing a fissile isotope. All of these radioactive materials are potentially useful to the nuclear industry and can be chemically recovered from the spent fuel and recycled if the nation in question permits it. At present, spent fuel obtained from nuclear power plants in Europe and Japan is being reprocessed to recover nuclear fuel. One of the issues pertaining to recycling is, of course, national security against atomic terrorism.

The last step is waste disposal. The safe disposal and storage of spent fuel and other high-level nuclear wastes is a highly controversial subject. Nevertheless, these hazardous wastes must be isolated from the environment until the radioactivity they contain decays to a safe level. Under the Nuclear Waste Policy Act of 1982, as amended, the Department of Energy has responsibility for the development of the waste disposal system for spent nuclear fuel and high-level radioactive waste. Current plans call for the ultimate disposal of the wastes in solid form in licensed, deep, stable geologic structures. More will be said about this later.

The Current Status of Nuclear Power.

Nuclear reactors essentially generate electricity by capturing the thermal energy released from their reactor core. Today, nuclear power accounts for over one-fifth of international electricity generation. Since 1972, their contribution to U.S. power generation has grown steadily (Figure 19.21).

At the end of 1995, 437 commercial nuclear units were operating in 31 countries throughout the world. The United States leads the world in power generation with its nearest competitor being France (Figure 19.22). In 1995, the 109 nuclear power plants operating in the United States met 20% of the nation's demand for electricity. By comparison, France used nuclear reactors to produce 76% of its electricity. In the United States, nuclear power plants are more numerous in the eastern half of the nation (Figure 19.23).

Nuclear and Total Net Generation of Electricity, 1957–1995

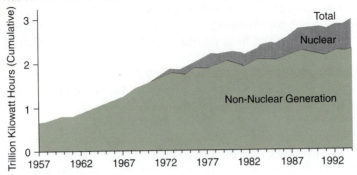

FIGURE 19.21   The relative contribution of nuclear power to the generation of electricity from 1957 to 1992 (from EIA 1995).

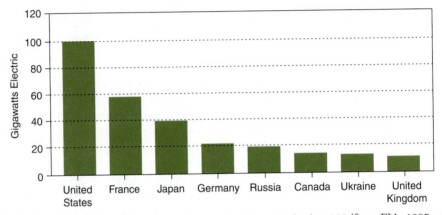

FIGURE 19.22   Nations with the largest nuclear generating capacity in 1995 (from EIA, 1995 with the original source being: International Atomic Energy Agency, "Nuclear Power Reactors in the World," Vienna, Austria, April 1996).

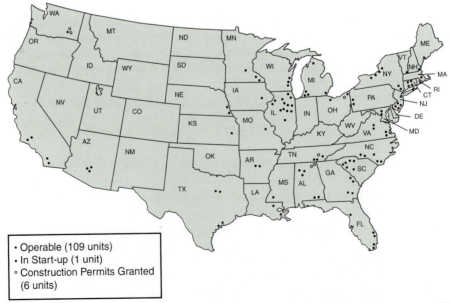

- Operable (109 units)
- In Start-up (1 unit)
- Construction Permits Granted (6 units)

FIGURE 19.23   Nuclear power plants operating in the United States as of December 31, 1995 (from EIA 1995).

The Future of Nuclear Power in the United States.
When the feasibility of the nuclear fission reaction was confirmed in 1939, it was immediately realized that nuclear energy had the potential to supply tremendous amounts of power. In the 1950s, President Eisenhower believed that nuclear energy would someday be so plentiful that its consumption would not be fully measured. Unfortunately, the nuclear industry suffered some setbacks along the way and now its future is altogether uncertain. The U.S. nuclear industry probably will, at best, maintain its contribution to electric power generation in future decades. Several factors have challenged the nuclear industry and collectively weakened its appeal. These factors can be grouped into issues pertaining to economics, public safety, and environmental quality.

Economic Issues.
Economic concerns stem from the marginal ability of the nuclear industry to compete with other power industries. The capital-intensive nature of nuclear-power plant construction requires billions in investment dollars. Too often construction costs substantially exceeded initial projections. This problem became worse, not better, over time. The factors responsible for escalating cost include construction delays, midcourse design changes, siting and licensing controversies, and increased labor costs (Holdren 1992). Construction costs in the United States increased after 1979, the year of the accident at the Three Mile Island plant in Pennsylvania. Prior to that time, the 63 completed plants required an average construction time of 6.3 years each. From 1979 through 1989, however, the average construction time rose to 11 years for the 47 plants that were built. Whereas coal-fired power plants and other utilities also experienced increasing construction costs, they were less severe, making nuclear power comparatively less attractive to investors.

Operating costs have also been a problem. Rather than providing mostly uninterrupted power, plants were too frequently shut down for repairs. One measure of operational efficiency is load factor, which is the ratio between the electricity that a plant actually produces and the amount that it was designed to produce. The average load factor in the United States through 1987 was less than 60%, indicating that many of the plants experienced frequent shutdowns. In contrast, Canada's load factor in 1987 was 78.2%. At that time, Belgium led the world with an average lifetime load factor of 81%. Recent trends in the United States toward deregulation and privatization of electricity supply systems have increased pressure on nuclear plant operators to be economically competitive with other generating technologies. For this reason and others, load factor and plant capacity have steadily improved since 1987. Now the industry operates near the 75% load factor level (Figure 19.24).

After a period of substandard performance, some plants have been closed, or decommissioned. The elimination of such plants improves average load factor, but involves, again, tremendous costs to contain radioactive materials. When plants were

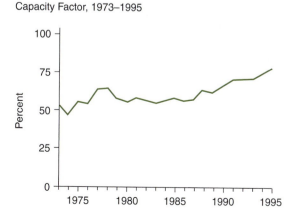

Capacity Factor, 1973–1995

FIGURE 19.24 Nuclear power plant capacity factor from 1973 through 1995 (from EIA 1995).

constructed, it was anticipated that they would eventually return investments after 40 years of service. Plants prematurely decommissioned have failed to reach the mature years of profitability.

NONRENEWABLE ENERGY AND THE ENVIRONMENT

Nuclear Plant Accidents

At present, the public perception of nuclear safety limits potential for growth in the nuclear industry. In 1970, only about 25% of the public opposed building new nuclear plants. From 1975 to 1979 less than 35% of the public opposed the industry. However, after the accident at Three Mile Island in 1979, public support faded even more, and by 1985, 67% of the public opposed nuclear industry growth (Holdren 1992). The accident at Chernobyl, Ukraine, in 1986 further reinforced public distrust.

In the accident at Three Mile Island, 27 tons of fuel melted in the base of the pressurized water containment vessel. Although no significant leaks to the environment were reported, the public remains skeptical about the amount of radioactive material actually reaching the atmosphere (Figure 19.25). Because the adverse effects of radiation exposure appear years after it occurs, surrounding communities remained concerned. The total cost of the cleanup has not been established, but exceeded $1 billion six years after the accident (Hall et al. 1992).

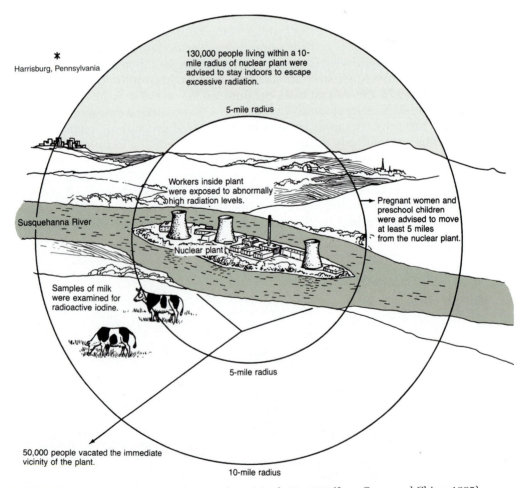

FIGURE 19.25   Three Mile Island nuclear accident, March 28, 1979 (from Owen and Chiras 1995).

The effects of the Chernobyl accident were more obvious. Twenty people were killed immediately at the site and millions of people were exposed to measurable quantities of radiation. The failed reactor was a light-water graphite reactor with a graphite core and no shielding structure (Edwards 1994). By U.S. and western standards, the design itself was substandard and clearly would be unacceptable in the United States. A fundamental flaw was the absence of a heavily reinforced containment building to limit radiation exposure if an accident should occur. The same design was used in the construction of many Soviet reactors. Ironically, a safety training exercise triggered the tragedy. The reactor was operating outside its own stated safety regulations when the accident occurred. In the accident, the water coolant pump failed and as the core temperature rose to some 3,000°C, the uranium fuel melted. Several explosions removed the top of the building over the reactor and the graphite ignited. The radioactive particles released to the atmosphere were carried all over northern Europe.

When the United States and former Soviet Union mended their relations, there was a major cutback in atomic weaponry. In the interest of world safety, the United States has extended its nuclear policy to one of improving nuclear safety in Russia and Ukraine. In the initial stages of U.S. assistance, it became clear that Soviet nuclear power plants lacked adequate operational and maintenance procedures, quality assurance programs, and reliable construction designs. A report by the World Bank and International Energy Agency identified 25 high-risk reactors in Russia, Ukraine, Armenia, Bulgaria, Slovakia, and Lithuania. These need to be upgraded or replaced at costs estimated to be as high as $24 billion. U.S. interests go beyond the immediate need of protecting the world against radiation exposure. In Ukraine, for example, nuclear energy was consumed regionally while petroleum energy was exported to provide profitable international trade. In the face of a collapsing economy Ukraine and other presently independent states may return to cold war politics and threaten global nuclear destruction.

The Chernobyl accident was the worst disaster in the history of nuclear power (Figure 19.26). The world community is still being challenged by its long-term economic, environmental, biological, and psychological impacts. However, by some assessments reactors are considered safe because accidents are rare and emissions of radioactive material are very low under most conditions. In fact, there have only

FIGURE 19.26 Aerial view of the Soviet nuclear reactor building in Chernobyl after the accident (from Owen and Chiras 1995).

been about 10 accidents where reactors have released radioactive particles into the atmosphere.

It will be hard to repair public distrust in nuclear power plants and to overcome the financial difficulties of the industry. However, this situation may be altered because of factors inside and outside the industry. Inside the industry, plans are to renovate reactor designs to improve safety and reduce operating costs. Newer designs call for smaller plants with passive ("brain dead") controls, which take over in the event of cooling failure. Improvements in computer controls allow reductions in otherwise extensive wiring systems. Moreover, designs allow 50% fewer welds in the structure that encloses the fuel and its cooling water. Structures therefore will be less prone to failure. By simplifying designs and scaling down plant size, the industry is making an overt attempt to reduce capital investment and to reduce construction delays. According to the Electric Power Research Institute (EPRI), a group funded by electric utilities in the United States and some foreign countries, reactors should be designed to last for 50 years, to have a load factor of 87%, and to be built in less than four and one-half years. The industry hopes to satisfy these criteria as it seeks to restore investor appeal.

Outside the industry, environmental problems created by burning fossil fuels are reaching the point where some measure of risk may, sooner or later, be forced on the public to avoid economic turmoil. The diminishing supplies of U.S. fossil fuel resources will also force major adjustments and will become a jointly considered factor in setting national policy. Abroad, a developing world will similarly struggle to fund energy sources.

About half of the CO_2 released to the atmosphere over the past 150 years was produced by burning fossil fuels, particularly coal. Without nuclear power, the world would have to rely almost entirely on fossil fuels, especially coal, to meet electricity demands. Emissions of carbon dioxide from burning fossil fuels are calculated worldwide to be about 20,000 million tons per year, of which about 45% comes from coal and 40% from oil. Coal-fired electricity generation produces nearly twice as much CO_2 as natural gas per unit of power, but nuclear power contributes no CO_2. If all of the world's nuclear power were replaced by coal-fired power, CO_2 emissions would increase by one-third.

Because the United States has more extensive supplies of coal than oil or natural gas (EIA 1996), coal provides the logical alternative to nuclear energy. A 1,000 megawatt coal-fired electrical generation plant operating at 75% capacity has a typical fuel requirement of about 2.3 million tons of coal a year. A nuclear power reactor of the same capacity (after its initial fuel loading of uranium) has an annual requirement of about 25 tons of fuel. Compared to U.S. nuclear power, coal-fueled power generation in the United States causes seven times more fatalities per unit of electricity generated. Another factor often omitted in energy discussions is that coal burning often releases more radioactivity than the equivalent nuclear power. Finally, the types of uranium and plutonium used for bombs and electric power generation differ greatly. Bomb-grade uranium is highly enriched (>90% ^{235}U, instead of about 3%) and bomb-grade plutonium must be fairly pure (>90%). The real risk created by building a plant is therefore one of environmental contamination as occurred in Ukraine.

According to the United Nations projections, more than 90% of world population growth in the foreseeable future will be in the less-developed countries. Furthermore, 90% of that growth will be concentrated in areas where nuclear energy would be practical in engineering terms. This explains why China is adding 10 reactors to its three already in use. China's power generation requirement is expected to almost double from 1994 to 2010. Much of the energy needed will be obtained from nuclear power plants. Like many developing countries, China's energy demands are growing even faster than the population (Levine et al. 1992). Energy shortages are aggravated further by the existence of outmoded coal-burning technology, which inefficiently uses resources and contributes avoidable pollution. Although nuclear energy is now marginally competitive in the United States, this is not so in many places abroad. Some countries lack fossil fuels and the cost of coal transportation rules against its use.

Terrorism

The threat of nuclear terrorism cannot be ignored by the world's nuclear community. This threat has been greatly exacerbated by the September 11, 2001 attack on the World Trade Center in New York City. According to some experts, security standards are too lax (Berkhout and Feiveson 1993, Stansfield 1997). The potential for problems in this area is another mark against the growth of the nuclear industry, here and abroad.

All nuclear power plants use or generate uranium and plutonium that can be employed to manufacture nuclear weapons. In this context, nuclear proponents point to the fact that nuclear power plants have not been a source of nuclear weapons development. In Europe, however, the number of arrests for nuclear trafficking has increased (Williams and Woessner 1996). In particular, there is the concern that organized crime in Russia will gain access to nuclear materials that by some descriptions posed less risk to the world when positioned on missile warheads (Bukharin 1996). The covert nature of smuggling and terrorism may rule out a precise assessment of the threat posed by criminal use of nuclear materials. To reduce access to bomb-capable materials, American Presidents have steered the domestic industry away from breeder reactors and have called on the global community to do the same. Nevertheless, enriched uranium and plutonium are part of the fuel reprocessing industry in Europe, at least on a limited scale. The United States is assisting Russia and Ukraine with security controls to eliminate theft. An outright ban on the production of weapon-grade plutonium is under mutual study.

Environmental Degradation

Contamination of the environment by nuclear waste remains a hot topic today and is an added disincentive to nuclear industry growth (Holdren 1992). Some view the threats posed by waste to be greater than those posed by nuclear power plants. They cite the fact that nuclear plants do not pose the threat of a nuclear explosion because the fuels are insufficiently concentrated and plant contamination, as occurred in Ukraine, is unusually rare. To some at least, the risk of contamination caused by a transportation accident or waste burial represents a substantial threat to them and their ecosystems.

Such fears are not groundless. There are examples of human death and health impairment associated with management of waste materials. These center on the improper disposal of milling wastes that caused avoidable suffering among the Native Americans living near uranium spoils in the Southwest. They center also on the families exposed to radon gas when spoil materials were used for residential landfills. In Colorado, people actually lived above radioactive material put there to level their homesite. Hopefully, the nuclear industry is now beyond this type of recklessness, but the public rightfully refuses to put safety much beyond their immediate control.

The problems just cited were created by mismanagement in the front end of the nuclear energy cycle and fall under the category of mining and milling. A substantial number of safeguards now regulate spoil materials. Today, the back end of the nuclear energy cycle is the focus of attention.

Nuclear plants generate both low-level and high-level radioactive waste. Low-level waste is material with expected radioactivity decay to safe levels in less than 50 years. It can be buried in shallow landfills. Among the materials included in low-level wastes are isotopes of nickel and cobalt from corroding pipes and small amounts of radioactive contaminants discarded with worker-protective clothing and tools. High-level waste obtained from spent commercial fuel requires centuries to decay to safe levels of radioactivity. It must be stored in sealed containers because of the intensity of its ionizing radiation. The volume of high-level waste produced outside the power industry is 20 times greater but contains less radioactivity.

Spent fuel rods are viewed as especially hazardous and contaminations caused from their improper disposal raise serious concerns. Such releases would be more threatening than accidents involving extraction and processing because most of the radionuclides produced during these processes occur naturally. Nuclides become more radioactive after undergoing fission in plant operation. On the positive side, the costs of dealing with this high-level waste are built into electricity tariffs. For instance, in the United States, consumers pay 0.1 cents per kilowatt-hour, which utilities put into a special fund. So far more than $11 billion has been collected.

The public, nuclear industry, and federal and state governments have invested large sums of time and money as the waste issue has gone before Congress, state legislatures, and courts. For the moment, Yucca Mountain in Nevada is scheduled to become a permanent depository site for high-level radioactive waste. Opponents of the site claim that the mountain may not be safe for the thousands of years that the high-level waste must be isolated. Further, the site may not meet the regulations of the Nuclear Regulatory Commission and the Environmental Protection Agency.

The issue of high-level wastes seems all but solved. Spent fuel is currently stored at nuclear plants in pools of water.

The Low-Level Waste Policy Act of 1985 stated that all states either develop their own site or form a compact with other states to select a disposal site of mutual satisfaction. However, this has often been delayed and deadlines have been repeatedly extended.

The Waste Isolation Pilot Plant (or WIPP) located in southeastern New Mexico is designed to dispose of transuranic radioactive waste left from the research and production of nuclear weapons. Transuranic elements have an atomic number higher than that of uranium (atomic number 92) and all such elements are produced artificially and are radioactive. The WIPP site was also hotly debated for years before it became a reality. Nevertheless, project facilities include disposal rooms excavated in an ancient, stable salt formation 2,150 feet underground. The facility began receiving waste in 1998.

THE IMMEDIATE FUTURE

As one can see, nuclear energy has been surrounded by controversy and disappointments arising not only from public fear but also its inability to compete in the energy sector for private financing. Some would say it has been forced to limp along and suffer failure because government support was lifted before the fledgling industry took root. Public confidence must be restored and this will require continued improvements in plant capacity and safety. A just appraisal also calls on the public to examine the natural sources of radiation present and radiation risks created outside the energy sector from medical exposures, the military, and chemical industry. Even smoke detectors designed to improve home fire safety rely on small quantities of radioactivity.

Fusion energy will vastly eliminate the threat of contamination and will one day draw upon a vast resource found in the sea. However, this technology is decades away because of the challenge of containing reactions that melt metal. Some form of magnetic containment is most promising. In the interim, we must find substitutes for fading sources of fossil fuels, which may be altering the world's climate. Based on present plans, the immediate future for nuclear energy is decommissioning existing plants. We will also have to replace the power lost from nuclear operating licenses that began expiring in the year 2000. Natural gas provides a short-term solution while helping with the problem of global warming. In the future, other substitutes must be found and renewable energy sources are among the possibilities (see Chapter 20).

LITERATURE CITED

Berkhout, F. and H. Feiveson. 1993. Securing nuclear materials in a changing world. *Annual Review Energy Environment* 18:631–665.

Bukharin, O. 1996. Security of fissle materials in Russia. *Annual Review Energy Environment* 21:467–496.

Dunn, S. 2001. Decarbonizing the energy economy. *State of the World 2001.* Washington, DC: World Watch Institute.

Dunn, S. and C. Flavin. 2000. Sizing up micropower. *State of the World 2000.* Washington, DC: World Watch Institute.

Edwards, M. 1994. Chernobyl: Living with the monster. *National Geographic* 186(2):100–115.

Energy Information Administration. 1995. *Annual Energy Review 1995.* DOE/EIA–0384(95).

Energy Information Administration. 1996. *U.S. coal reserves: A review and update.* U.S. Department of Energy. DOE/EIA–0529(95), p. 83.

Energy Information Administration. 1997. *Annual Energy Review 1996.* DOE/EIA–0364(96).

Flavin, C. and S. Dunn. 1999. Reinventing the energy system. *State of the World 1999.* Washington, DC: World Watch Institute.

Flavin, C. and N. Lenssen. 1994. Reshaping the power industry. In L. Starke, ed., *State of the World 1994.* New York: W. W. Norton.

Gleick, P. H. 1994. Water and energy. *Annual Review Energy Environment* 19:267–299.

Hall, C. A. S., C. J. Cleveland, and R. Kaufmann. 1992. Energy and resource quality. In *The economy and economic process.* University Press of Colorado. 577 pp.

Holdren, J. P. 1992. Radioactive-waste management in the United States: Evolving policy prospects and dilemmas. *Annual Review Energy Environment* 17:235–259.

Karl, T. R., R. W. Knight, D. R. Easterling, and R. G. Quayle. 1996. Indices of climate change for the United States. *Bulletin of the American Meteorological Society* 77(2):279–292.

Kesler, S. E. 1994. *Mineral resources, economics, and the environment.* New York: MacMillan. 391 pp.

Leffler, W. L. 1979. *Petroleum refining for the non-technical person.* Tulsa, OK: Petroleum Publishing Co. 159 pp.

Levine, M. D., F. Liu, and J. F. Stinton. 1992. China's energy system. *Annual Review Energy Environment* 17:405–435.

Meyers, S. and L. Schipper. 1992. World energy use in the 1970s and 1980s: Exploring the challenges. *Annual Review Energy Environment* 17:463–505.

Mounfield, P. R. 1991. *World nuclear power.* New York: Routledge.

Nebel, B. J. and R. T. Wright. 1996. *Environmental science.* Upper Saddle River, NJ: Prentice-Hall. 698 pp.

Owen, O. S. and D. D. Chiras. 1995. *Natural resource conservation.* Upper Saddle River, NJ: Prentice-Hall. 586 pp.

Radovic, L. R. and H. H. Schobert. 1992. *Energy and fuels in society.* New York: McGraw-Hill. (College Custom Publishing Series). 422 pp.

Schiffer, M. B. 1994. *Taking charge: The electric automobile in America.* Washington, DC: Smithsonian Institution Press.

Sheehan, M. 2001. Making better transportation choices. *State of the World 2001.* Washington, DC: World Watch Institute.

Simon, J. 1996. *The ultimate resource.* 2nd ed. Princeton, NJ: Princeton University Press.

Sperling, D. 1995. *Future drive: Electric vehicles and sustainable transportation.* Washington, DC: Island Press.

Stansfield, T. S. 1997. *Caging the nuclear genie: An American challenge for global security.* Boulder, CO: Westview Press.

Williams, P. and P. N. Woessner. 1996. The real threat of nuclear smuggling. *Scientific American* 274(1):40–44.

CHAPTER **20**

Renewable Energy: The Sustainable Path to a Secure Energy Future

INTRODUCTION

The United States has two basic energy problems. It is becoming less able to meet its needs for oil as its domestic reserves decline. At the same time, the world is bracing itself for climatic change due to carbon dioxide increase from fossil fuel combustion. The United States contributes a disproportionate amount of greenhouse gasses to the atmosphere. The one thing that seems certain is that this will force change in the emphasis the United States gives to energy issues. In addition, it is highly unlikely that there will be a quick fix because the issues are multinational and often pit technologically advanced countries against those on the cusp of rapid economic development. This was obvious as nations aired their views on greenhouse gas emissions during the United Nations conference (Kyoto Conference) on global warming held in Kyoto, Japan, in December 1997. Before this historic meeting, the U.S. position was one of calling for voluntary emissions reduction programs and monitoring. The United States considered a reduction of emissions amounting to about 25% by 2012. Since then, the United States has recanted to reexamine its position, much to the dismay of other nations because the U.S. contribution to world CO_2 emission is so large (25%). By taking no action, it will dump 25% more carbon dioxide into the atmosphere by 2012 (Figure 20.1). Furthermore, if the United States continues along its current path it will become more dependent on foreign oil and, by some estimates, will invite economic calamity. This argument is not without merit because two-thirds of the world's oil is projected to come from the Middle East by the year 2020.

A number of changes will be needed for the United States to reduce impacts on climate change and set the stage for energy sufficiency (Thurlow 1990). It seems likely that one response will be to move more aggressively into renewable energy. Today, renewable energy provides only about 8% of the energy consumed in the United States (Figure 20.2). Advocates of renewable energy argue that this figure is far too low and that renewable energy has great potential for solving the energy dilemma. We will address this issue in some detail in this chapter. Flavin and Dunn (1999), Dunn and Flavin (2000), and Dunn (2001), offer additional discussions.

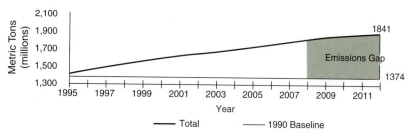

FIGURE 20.1 Graph showing U.S. carbon emissions with and without a federal commitment to return to 1990 levels as discussed during the 1997 Kyoto Conference (from EIA 1997a).

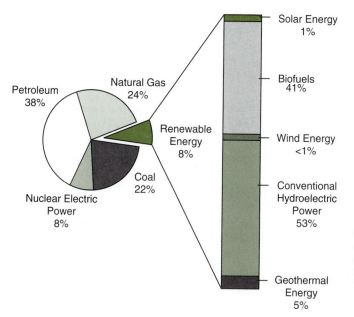

FIGURE 20.2 The contribution of renewable energy to total U.S. energy needs in 1996 (from EIA 1997b).

Renewable energy has many advantages. These became more obvious when U.S. energy prices soared in the 1970s and the American public became better informed about environmental issues. One effective response to the energy crisis was simply to conserve home and transportation energy. However, conservation alone will not meet our future needs because the world's population and economies continue to grow.

The single most attractive feature of renewable energy is that it is supplied mostly by the sun and is therefore limitless. The sun drives the wind and part of the hydrologic cycle, which provides hydropower. Also, solar power can be captured as heat, which can be used directly for power generation or converted with no intermediate steps to electricity. The sun also drives photosynthesis, which can be used to create renewable liquid fuels for transportation. Beyond the sun, only gravitational energy remains as a renewable energy source. It is provided through hydropower (falling water) and by the moon's gravitational effect in tidal energy.

Despite the impressive potential of renewable energy sources, some energy analysts regard them as impractical and exotic. History suggests otherwise. In the early nineteenth century, the most common source of energy in the United States was firewood. In the same era, stream flow powered the milling of grain and lumber, and wind was used to pump water and propel ship cargo and passengers. Also, greater attention was given to the features of home construction that affected light and temperature comfort. As fossil fuels spurred on modernization, toilsome and time-consuming practices were mostly, but not entirely, abandoned and forgotten. Unfortunately, renewable energy resources were forgotten with the exception of stream power and domestic wood fuel. Although we no longer see stream-powered grist mills, hydroelectric power was harnessed in the early 1900s and remains an important energy source. Also, some 5 million households burn wood as their primary source of heat (Brower 1992), while the pulp and paper industry meets over half of its energy needs by burning process wastes.

RENEWABLE ENERGY SOURCES

Solar Energy

Uses of solar energy range from simple to complex, depending on modes of energy capture, transmission, and utilization. Passive solar heating is on the simple end of the scale and involves designing a building's structure to capture and store solar energy (Bevington and Rosenfeld 1990). Although passive solar designs can greatly reduce energy use in new buildings, little or no extra cost is required. Significant energy reduction in homes and office buildings is no trivial issue. About one-third to one-half of all energy produced in industrialized countries goes toward heating and cooling buildings (Brower 1992). In Boulder, Colorado, all new residential construction must include some passive solar heating.

Energy savings are greatly multiplied when solar designs are combined with efficiency measures such as insulation and low-emissivity glass. Passive solar design is site specific but generally involves orienting a building toward the south to capture the winter sun. South-facing window areas are used for winter heating and natural ventilation. In general, 75% of the heating requirements for a building occur at night. Therefore, designs frequently employ wall or floor materials that can store heat in their mass. Landscaping can be an integral part of the design and can afford seasonal shading and solar exposure as needed to further improve living comfort (Figure 20.3). Overhead shading can reduce a typical home's heating and cooling costs by 15 to 25%. The main constraints on passive solar design are economic and technical in nature. However, the small additional cost of passive design is returned over the life of the home. Materials for passive design are being improved at a steady pace.

Active Versus Passive Heating.   Solar collectors are discrete units that collect, store, and distribute solar energy for water heating, space heating, and space cooling (Figure 20.4). Most use pumps or fans to circulate hot water or air through ductwork or plumbing (Figure 20.5). However, some rely entirely on natural convective currents created by temperature gradients. Systems without motorized components are

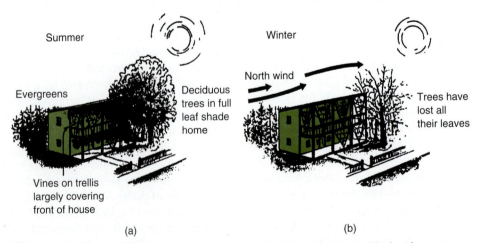

FIGURE 20.3 Landscaping can improve home energy efficiency. In summer (a), deciduous trees or vines can shade a home to reduce interior heat loads. In winter (b), the absence of leaves allows sunlight to provide additional warmth (from Nebel and Wright 1996).

FIGURE 20.4 A flat-plate solar collector. Sunlight is converted to heat as it is absorbed by a black surface. A clear glass or plastic window covering the surface traps heat once sunlight energy enters. Air or water is heated as it passes over and through tubes embedded in the black surface (from Nebel and Wright 1996).

FIGURE 20.5 Solar panels provide the primary source of heat for this modular home (courtesy of U.S. Dept. of Energy).

properly called passive systems. Active systems have been used to provide energy to numerous types of buildings, including those serving farms, universities, and hospitals. Often the purpose is to supplement rather than replace more conventional methods of controlling building temperature.

Solar systems are well suited to meeting the hot water demands of home and commercial buildings. Solar water heating units often consist of a series of plastic or metallic pipes that are mounted on building roofs (Figure 20.5). The hot water they produce can substantially reduce energy costs. Some systems can be retrofitted to existing buildings and are becoming increasingly attractive to homeowners as deregulation weakens competition from other energy sources. Solar water heaters are frequently seen in developing countries. The majority of homes in Israel use solar water heaters.

Solar-Thermal Concentrating Systems.

Solar energy can be used to heat and vaporize water or another working fluid to produce electricity. Several approaches have been used for concentrating the solar energy needed. The power tower approach uses an array of sun-tracking mirror assemblies that concentrate sunlight onto a tower-mounted central receiver in which a circulating fluid is heated and used to produce energy (Dracker and De Laquil 1996). The Department of Energy's interest in central receiver systems, such as the power tower approach, has remained high over the years. Experimental systems have been erected in the United States and numerous foreign countries.

The solar trough has received both research and commercial attention (Figure 20.6). The curved and highly reflective inner surface of the trough concentrates sunlight onto a line receiver, a vacuum-sealed metal or glass pipe containing fluid. The pipe runs the

FIGURE 20.6 A solar trough. Note how the curved reflector concentrates sunlight on the pipe positioned at the trough's center. The oil heated within the pipe can be used to boil water and generate steam for driving a conventional electrical generator (from Nebel and Wright 1996).

length of the trough and the fluid contained is heated to temperatures as high as 400°C (750°F) (Brower 1992). A commercial electric generating plant installed in the Mojave Desert of California grouped large numbers of troughs together to generate heat.

In another approach, parabolic mirrors (heliostats) are used to concentrate solar energy for the purpose of steam generation. The steam is used to generate electricity. Parabolic dishes can also be used for manufacturing metals, glass, cement, and other materials because of the intense heat that can be generated. Such temperatures cannot be obtained from parabolic trough systems. Dishes are also more efficient at generating electricity. Electrical applications employ numerous dishes. The fluid heat gathered by the centrally positioned receiver of each dish is transferred to a central location for steam generation.

Dish/engine systems add an electrical generator to the center of the dish. This technology uses a stretched-membrane parabolic dish to focus solar energy onto a receiver filled with sodium. As intense heat vaporizes the metal, the vapor condenses on the heater head of the attached Sterling-cycle engine. The transferred heat causes the helium gas inside the engine to expand. As the gas expands and contracts in a cyclic manner, the engine's piston is driven to and fro. The piston in turn propels an alternator, which generates electricity. The use of the engine overcomes a serious limitation of solar power—the need to supply power when sunlight is minimal to nonexistent. An uninterrupted power supply is possible because the engine can be powered by conventional fuels when solar energy is absent.

The most novel feature of the Sterling engine is its reliance on external combustion. The engines commonly used involve internal combustion and require fossil fuels in some form. The Sterling engine is not a new idea. Robert Sterling was given a patent for the idea in 1816. Dish/engine systems are especially attractive for remote applications and for village communities in Third World countries because of the amounts of electricity they generate. In addition, the engines require little maintenance because they are self-lubricating and, unlike portable generators, run quietly. Sandia Laboratory in Albuquerque, New Mexico, is but one of several institutions attempting to further the dish/engine concept. Because the net solar-electric efficiency

FIGURE 20.7 Panels used to generate solar electricity (courtesy of U.S. Department of Energy).

of dish/engine designs is 29% or better, it offers one of the best approaches among solar thermal technologies. Several companies in the United States and Europe are developing dish/engine systems for both remote applications and grid-connected, distributed-power applications (Dracker and De Laquil 1996).

Technologies involving parabolic troughs, central receivers, dishes, or dish/engine systems have their greatest appeal in desert areas with intense sun. Because deserts are generally sparsely populated, such technology will have limited value on the national scale. Power could be transmitted to large cities beyond deserts. However, much of the power generated would be lost to internal resistance as it is transmitted over long distances. Regionally, it can substantially decrease reliance on fossil fuels in the southwestern United States, parts of Australia, the Middle East, and Africa.

Photovoltaics.   The basic unit in a photovoltaic system is a solid-state device called the solar cell (or photovoltaic cell) (Weinberg and Williams 1990). The unique feature of the solar cell is its ability to convert sunlight directly to electricity without any intermediate steps involving heat transfer. Photovoltaic cells are made of semiconductors and light-sensitive metals, such as selenium. As electrons in the cell become exited by sunlight, an electrical current is generated. Commercially useful quantities of electricity can be generated by coupling panels of solar cells in vast arrays (Figure 20.7). Solar cells operate best in full sun but can also function on cloudy days. Under the best conditions, photovoltaic cells can convert about 17% of the sun's energy to electricity. Photovoltaics are used across the globe to provide electricity to remote areas. Solar electricity can compete with conventional power in remote areas because the latter would have to be transmitted over large distances at higher cost.

In developing countries, photovoltaics are being strongly encouraged and financially sponsored by the United Nations and individual donor countries. The U.N. estimates that more than two million village communities lack power for basic necessities such as pumping well water. Lighting photovoltaics can bring village communities into national communication networks, thus giving village people a more active voice in democratic governments. Education can also be disseminated to remote areas. In industrialized countries, photovoltaics power remote telephones, navigation beacons, electric fences, field data recording devices, and a host of other types of equipment. These applications are cost-effective in the absence of an opportunity to cheaply tap an electric grid. The use of photovoltaic cells in the NASA space program over past decades has been a source for technology development and publicity.

The Future of Solar Energy.   Solar energy must clear some obstacles before making deep penetration into the commercial energy market. One major hurdle is the present inability of solar energy to compete economically with fossil fuels. Photovoltaics are especially affected by high manufacturing costs. As long as solar manufacturing operates at its present volume, the cost of individual units will remain high and so will power plant construction. Under existing technology, considerable work must be done by hand. The preparation of crystalline silicon cells, for example, is a tedious process, which involves growing single crystals and cutting them into wafers. Methods involving materials more suited to mass manufacturing are being developed, but progress has been slow under the present market conditions. As long as other fuel sources remain more economical, photovoltaics will be confined to the fringe of the energy sector in developed countries. There are cases, however, where the addition of solar power to an existing electrical grid can eliminate the need for a new and larger conventional power plant. For example, Pacific Gas and Electric in northern California has found that the higher cost of photovoltaic systems is justified today in areas where they effectively reduce peak loads on transmission and distribution networks that must otherwise be upgraded at even greater cost (Brower 1992). Also attractive to grid managers is the ability of solar units to provide consistent power, to perform in a predictable manner, and to be relatively free from module failure.

Existing tax codes also detract from broad market acceptance of solar technologies. Businesses can deduct fuel energy costs but not money invested in capital-intensive solar energy. Immediately following the energy crunch of the 1970s, numerous states provided tax exemptions for solar applications, but many of these were phased out over time. What appears to be missing in tax formulations are the environmental benefits that accrue to the public at large as renewable energy replaces fossil fuels. Solar energy proponents are quick to point out the subsidy paid to the petroleum industry in defending foreign oil fields.

One additional obstacle to application of solar technology is energy storage. This problem exists because of the intermittent nature of sunlight. Photovoltaic systems only generate power and have no storage capacity. Dish/engine systems can keep operating if provided combustible fuel. Here storage is less critical. A more encompassing view toward energy storage issues might be to consider any energy contribution from nonpolluting means a practical benefit. This seems especially true for systems capable of supplementing peak loads.

The future of solar energy in the United States rests mostly in the hands of the federal government. Private enterprise has sufficient capital to fund research on basic components, but power plant demonstration projects are mostly beyond its financial reach. Some financial support from the public at large will, therefore, be needed to promote more immediate progress. Government support is slowly returning to levels comparable to the period immediately following the oil crunch, but overall funding remains low relative to the overall task. Without serious setbacks, it appears that solar energy will continue slow penetration into the energy market. By the year 2030, it will provide a substantial contribution to the total energy consumed.

Wind Energy

The Practicality of Wind Energy.   Atmospheric circulation involves massive exchanges of energy provided by the sun. On a global scale, winds are caused by the temperature

FIGURE 20.8 Cattle drinking at a stock tank near Grants, New Mexico, which is filled with water by wind power.

differences between the earth's latitudes and deflection caused by the rotation of the earth (see Chapter 6). Dry air in the vicinity of 30 degrees north and 30 degrees south sinks and flows toward the equator where it replaces rising hot air. This pattern of air movement is called Hadley circulation. At midlatitudes, between 30 and 70 degrees latitude north and south, air flows toward the poles and is deflected westward, creating a wavelike pattern known as the Rosby circulation (Grubb and Meyer 1993). Many smaller circulation patterns are superimposed on the global systems as a result of regional variations in temperature and pressure. Within regions, mountains and valleys can intensify winds as they are channeled across the earth's topography. Large bodies of water can influence wind direction as well as intensity. Scientists use very sophisticated models to predict winds. For our purposes, we need only to know that wind energy can be captured by wind turbines, which convert motion energy to electricity.

Wind provided a practical source of energy long before the petroleum era. The remains of ancient sailing vessels recovered from the depths of existing lakes or marine shorelines by archaeologists attest to the importance of wind energy to much earlier civilizations. Some date back to 5000 B.C. Archaeologists have also provided evidence of the existence in China of simple windmills dating back at least 2,000 years. On American soil, millions of windmills were erected during the late nineteenth century as the West was settled (Figure 20.8). They provided a simple but critical means of pumping water on farms and ranches. By the turn of the century, wind-driven systems generated electricity in small rural communities. By the 1930s, power grids extended into most of these communities and wind generators were abandoned. As rural electrification programs developed, only the most isolated windmills remained for pumping water.

Until the 1970s, there was little interest in reviving wind power. After the 1973 OPEC oil scare, wind, like other alternative energy sources, experienced a rebirth of attention. California became the focal point of wind energy development from 1981 to 1986. More than 15,000 utility-scale wind turbines were installed with a total peak capacity of 1,300 MW. Over the same period, small wind turbines were installed throughout the United States, mainly in rural homes.

In response to the somewhat abrupt entry of government wind industry incentives, developers went to work quickly to bring wind turbines to the market. In the absence of advanced testing and experience, poorly designed machines were sold and installed. The reputation of the wind power industry suffered as machines broke down and failed to deliver the power predicted. Engineers soon found that sites had to be chosen with more selective criteria and that variable wind created unforseen stresses

FIGURE 20.9   Wind turbines on a "wind farm" in the Mojave Desert near Palm Springs, California. California leads other states in electricity generation from wind.

on equipment. In some cases, units actually flew apart, thus endangering onlookers. In addition to technical problems, there were dishonest salesmen who sought only the generous tax credits offered by a hopeful and trusting public. Only in recent years has the image of the industry been somewhat restored.

The late 1980s presented more challenges to the wind industry as tax credits were eliminated and fossil fuel prices fell. Consequently many companies went out of business. The early years of the auto industry were similarly marked by the gradual elimination of pioneering manufacturers.

The 1990s registered a major reversal in the wind industry's image. This occurred as wind generators became increasingly efficient and durable, and as capital and operating costs were reduced to the extent that wind could compete with other energy sources. By 1996, the cost of wind-powered electricity had dropped to only 5 cents per kilowatt-hour. Projections suggest that a further reduction to about 2 cents by 2010 might be attainable. The United States presently leads the world in installed wind capacity. In 1994, California had about 16,000 wind turbines and led all other states in electricity generation by a vast margin (Figure 20.9). Its wind power industry was built when natural gas prices and tax incentives were high. In fact, California contains less than 1% of the total U.S. wind energy potential and at least 16 states have a wind resource base that is equal or greater. Overall, wind is currently among the fastest growing renewable energy sources and this growth is occurring in numerous states.

Wind Resources.   The amount of wind energy theoretically available for use in the United States is enormous. Of course, only a small fraction of this resource can be exploited because of physical constraints on available land and the efficiency of energy extraction. However, the present amount of electricity generated from wind in the United States could be easily doubled.

Many regions of the country offer at least some usable wind resources. The Great Plains states have abundant wind resources, followed by other parts of the Midwest, the West, and the Northeast. Compared to other regions, the southern United States has less potential for wind energy development. The windiest areas in the United States tend to occur along the East and West Coasts, along the Rocky and Appalachian Mountains, and in a comparatively expansive belt embracing the Great Plains and 12 states. Kansas and Texas, followed by North Dakota, have the greatest potential power output for wind generating capability. Montana and Wyoming are also rich in wind resources. Overall, at least 37 states have defined potential for utility-scale wind energy development.

Wind power classes are used to express the wind energy available to specific areas. These range from class 1 (the least amount of energy) to class 7 (the greatest amount of energy). The classification process takes into account average wind speed and its variation over time, and the average density of the air. Wind speed is especially critical because wind energy production is proportional to the cube of the wind speed. This means that a 10% increase in wind speed will yield about a 30% increase in the output of a wind turbine. Because air density decreases with altitude, wind speeds at high altitude sites must exceed those at lower sites to be competitive for development (Rosenberg 1993). In general, areas identified as class 4 and above are regarded as potentially economical for power production with existing technology. Good wind areas cover about 6% of the contiguous U.S. land area. They have the potential to supply more than one and a half times the current electricity consumption of the United States.

Factors that determine the feasibility of actually harvesting wind power go beyond wind factors. The availability of land for development and the distance to existing transmission lines are especially important. Unfortunately, some of the most promising concentrations of wind energy are found in areas having low population densities and are far removed from major transmission lines.

Wind Technology.   Most modern wind turbines resemble traditional farm windmills while others follow a vertical-axis design causing them to resemble an overscaled eggbeater. The latter design was developed by the French inventor Darieus, and windmills of this type bear his name. Wind turbines are deceptively simple machines consisting of blades, rotor, transmission, electrical generator, and control system all mounted on a tower. The tower is needed to prevent topographical irregularities, trees, and buildings from impeding wind speed.

The present generation of wind turbines is characterized by advanced aerodynamic blade shapes, use of highly advanced materials in blade construction, and sophisticated control equipment (Sorensen 1995). Most turbines are made with three aerodynamic glass- or carbon-fiber blades 20–30 meters long.

The blades are mounted on a driveshaft, which usually incorporates a gearbox transmission to provide a constant rotor speed. The choice of three blades is a compromise between high-power output and the stability and lifetime of the blades (Sorensen 1995).

Wind turbines are available in a variety of sizes that can be matched to power ratings for commercial and private demands. The largest models can have propellers that span more than the length of a football field. Such models can provide power to more than 1,000 homes. The largest wind turbine built in the United States today (Zond Z-40, 550 kW), stands over 140 feet tall, uses three blades each over 65 feet long, and in windy locations can produce as much energy each year as is consumed by more than 200 homes.

As engineers and scientists gained experience in wind power generation, research attention was redirected to intermediate sizes, which could be more easily financed and maintained in clusters called "wind farms." Intermediate-sized wind turbines capable of producing 100 to 400 kilowatts of electricity serve the needs of the bulk electricity market. Wind farms can employ thousands of machines on towers 100–165 feet tall (Figure 20.9). Clustering takes advantage of unusually good wind sites, eases construction, and gives utilities an opportunity to more efficiently manage the operation. Electricity from wind farms is fed into the local utility grid and is distributed to customers just as it is with conventional power plants. In densely populated countries in western Europe, wind farms have been built offshore to overcome public objections to their presence in populated areas.

Small turbines generate 1 to 50 kW of electricity. Units having 1 to 10 kW capacity are ideal for generating electricity in rural or underdeveloped areas without an electricity grid. In developing countries, they compete favorably with photovoltaic systems, diesel generators, and grid extension. In the United States, small wind turbines can be attractive to rural homeowners and farmers who want to reduce their electricity bills. A small home-sized wind machine has rotors between 8 and 25 feet in diameter and usually stands about 30 feet tall. It can supply the power needs of an all-electric home or small business. Small wind turbines require less maintenance than larger and more complex machines. In most small wind turbines the rotor drives the generator directly and gears subjected to wear are avoided, thus reducing maintenance.

In the early years of the wind power industry, turbine life span was considerably shorter than predicted and operation and maintenance costs were unacceptably high. By adopting more rugged designs and more advanced materials, systems became highly reliable. New designs improved the efficiency of energy capture while reducing stress on wind turbine components, including drive-train hardware. Vibration fatigue was greatly reduced and the addition of aerodynamic braking protected turbines, which tended to "run away" during high winds. As machine durability and performance improved, the cost of wind power declined greatly. Improvements in the manufacturing process played a major role in improving power productivity. As time progressed, companies became more specialized. Some began concentrating on blades or controls while others became industry leaders in tower and gear mechanisms. Only the more mundane parts of the process, such as foundation and site work, are left to local contractors (Sorensen 1995).

Future of the Wind Power Industry.

Wind energy brings numerous benefits to the energy scene. Most obvious perhaps is renewability of the resource and the relative absence of nonpolluting features. Regarding domestic fuel availability, wind power adds diversity to the fuel mix and therefore serves as a hedge against supply interruptions. The environmental concerns appear trivial compared to the problems created by fossil fuels.

Two factors that have constrained wind industry growth are cost and the intermittent nature of wind energy supply. Even though the cost of wind power has decreased dramatically in the past 10 years, further reductions will be needed to offset financing difficulties. Wind technology requires a higher initial investment than fossil-fueled generators. About 80% of the cost is the machinery, with the balance being the site preparation and installation. In states where the public has insisted on including external costs, wind power plants have gained a solid footing.

Like solar energy, wind power flows are interrupted and vary in intensity. The value of electricity generated by a wind farm is therefore reduced if winds occur outside periods of peak load. Many experts believe wind power will continue to play an incidental role in the nation's electricity supply until reliable and low-cost energy storage becomes widely available. Some argue, however, that answers to our complex energy problems will not be found in a single energy source of any type and that wind energy can reduce our immediate dependence on fossil fuels. An additional constraint on wind energy growth is the distance of remote energy resources from power grids. Nevertheless, numerous states are able to harvest regionally concentrated wind resources and are in the process of constructing and planning plants.

The overall expectation is that technical and economic issues will be solved in at least some regions and that the contribution of wind power to our total energy demand will continue to grow. The Department of Energy forecasts a 600% increase in wind energy use in the nation in the next 15 years. By the middle of the twenty-first century, wind could be producing 10% of U.S. electricity or as much as hydroelectric dams produce today. Some fear that growth may expand to the extent that local communities will begin protesting wind farms as they would coal-fired power plants. Public objections have centered on the loss of aesthetic appeal in natural landscapes, the noise created by wind turbine blades, and the threat posed by turbines to birds. Some have even raised the question of the influence wind farms might impose on global atmospheric circulation, but experts can reliably counter this argument.

On an international scale, countries such as the United Kingdom, Canada, Denmark, and Germany are also pursuing numerous wind projects. Also, the small wind turbine market is showing impressive growth in sales to developing countries.

FIGURE 20.10 The Fort Randall Dam was built by the Army Corps of Engineers for power production and other uses (courtesy of U.S. Army Corps of Engineers).

Hydropower

Present Status. In 1900, 4% of our delivered energy supply (in thermal units) was in the form of hydroelectric power. Eighty years later it is still about 4%, although much larger in absolute terms (Hall et al. 1992). Today, hydropower provides more than 50% (50.6% in 1996) of the nation's renewable energy (EIA 1997c). It accounts for slightly more than 10% of the electricity generated in the United States.

Hydroelectric plants convert the potential energy of stored (or impounded) water into electric energy (Figure 20.10). Numerous hydraulic turbines generate electricity as they are spun by water discharged from a dammed river. In some cases the water used for power generation is returned from a lower to higher reservoir to make energy available during periods of peak electrical demand (Moreira and Poole 1993). In effect, energy of lower value is invested to produce energy of higher value. Pumped storage plants, therefore, sometimes intentionally produce less but more valuable energy.

The Near Future of U.S. Hydropower Plants. Hydropower generation offers several advantages over fossil fuels, and hydropower dams often provide positive flood control, navigation, and recreation benefits (see Chapter 7). Hydropower does not produce air and water contaminants, including greenhouse gasses. However, it is unlikely that new hydropower plant production will pave the way to growth in the industry. This view is supported by the fact that most highly productive dam sites have already been built. Another constraint is the intensification of public objections to dam construction over the years. One objection is that dam construction reroutes and often alters the water quality of natural flows, and that valuable natural ecosystems are lost in the process. Species can also be lost or threatened. Another objection stems from conflicts over the water resource itself. As water is impounded, some is lost in the form of reservoir surface evaporation and lands once fed by natural flows are cut off from their critical water source. The evaporation of water is directly related to the surface area of the body of water and varies with temperature, wind conditions, and humidity. Evaporative losses are, in fact, a form of consumptive water use, which further aggravates water conflicts among agricultural users, industrial users, commercial users, and environmental advocates (Gleick 1994). Hydropower also depends on continuity of water storage. The drought experienced by the northwestern United States in 2000–2001, reduced generating capacity and contributed to high electric costs in California.

Today, public resistance poses a challenge to the relicensing of existing dam sites as well as new dam construction. All considered, growth in hydropower production will most likely occur in the form of modifying existing plants to improve operational efficiency. Many dams built in the past were not equipped to their maximum potential because of insufficient local demand for power. Also, in the past, economic disincentives existed for long distance electric power transmission. Many old facilities were retired during times of less expensive energy. Other dams were built solely for flood control, irrigation, and recreation purposes. With the increased dollar value of electricity and new automatic equipment, it is now economically feasible to equip such sites with power generation equipment and to restore or rebuild old power equipment at retired sites (Hall et al. 1992).

The International Scene.   Numerous developed nations face similar constraints in new dam construction. In Canada, for example, the adverse impacts of hydroelectric development, including displacement of population from flood plains and disruption of fisheries, have led to the suspension of a number of major projects in recent times. However, public resistance increased after Canada had extensively developed its hydroelectric power resources.

In contrast, undeveloped countries are actively developing hydropower and two-thirds of all energy growth is projected to occur in newly industrializing economies. In Central and South America, hydropower is the dominant source of electricity generation. In 1995, more than 80% of the region's net electricity consumption was attributed to hydropower. In Latin America, 85% of growth in electric capacity from 1990 to 2000 was from hydropower. More dams are being built in Brazil where hydropower already provides 95% of its total energy. Argentina plans to add several small- to medium-scale hydroelectric projects in the near future. Closer to home, Mexico's hydroelectric capacity could double by the year 2010.

The greatest growth is now occurring in Asia and other regions that have seen little dam construction to date. Currently, more than 150 hydroelectric projects are under construction in China. The most prominent is the $30 billion Three Gorges Dam project. Construction began in 1994 and is expected to be completed in 2009. However, the project has been severely criticized in China and abroad and financing has been difficult. Nevertheless, work continues at that site and at several others being constructed on a much smaller scale. India has also initiated dam building at an aggressive pace, but again not without public protest (e.g., the Narmadu Dam project). The environmental issues and high costs involved in developing India's hydropower resource make it unlikely that substantial development will occur there. Both China and India depend more on coal for electricity generation than other developing Asian nations. Nepal cites environmental problems as the primary reason for its desire to build a hydroelectric station on the Arun River. The Nepalese government argues that forests, their main energy source, are being depleted and that its extensive hydroelectric potential can be employed to protect its environment while furthering economic development. Numerous other Asian countries are building dams. Hydroelectric projects are planned for Borneo (Malaysia), Vietnam, Cambodia, and Thailand. In Africa, hydropower advocates promote the abundant hydroelectric potential existing in sub-Saharan Africa.

Objections to dam construction in developing countries often center on resettlement issues. Too often the people forced to abandon their homes and land are given insufficient voice in how resettlement is to occur. Consequently, the adverse social and economic effects tend to be underestimated (Gutman 1994).

Biomass

Biomass represents a broad category of energy sources and has been the predominant energy source for all but a fraction of recorded history. Biomass includes all forms of nonfossil plant and animal matter that can provide energy. Entire dried plants, plant parts, plant oils, animal fat, animal dung, and other forms of biomass

FIGURE 20.11 Animal dung is an important energy source in some developing countries (courtesy of Agency for International Development).

can serve as energy resources (Figure 20.11). The energy obtained from biomass can be used for numerous purposes including heat production, electricity generation, and transportation fuel.

In 1996, biomass supplied 43% of the renewable energy consumed in the United States (EIA 1997c). Biomass energy use steadily declined in the United States until the 1960s, but since then it has expanded to equal about 4% of the energy demand. On a global scale, biomass ranks fourth as an energy resource, providing about 15% of the world's energy needs (Larson 1993). Biomass energy is especially important to the developing world. For example, cooking with wood accounts for about 60% of all energy consumed in the African Sahel (Aubrecht 1995).

Biomass energy resources can be divided into two basic categories: (1) forest- or agriculture-derived plant materials and (2) animal or plant by-products. These will be discussed.

Energy from Forests. When home heating bills began to soar in the 1970s, U.S. households near forested areas began using wood energy more aggressively. Wood-burning stoves regained popularity and were modified to improve home heating efficiency. At the same time, the wood products industry began using more and more of its own waste products to meet both its heating and electrical needs. Today, the paper and pulp industry satisfies more than half of its energy needs using mill wastes. Growth in this type of activity is largely responsible for the gains in biomass energy consumption in the U.S. in recent decades.

Energy can be obtained from forests and plantations in numerous ways (Graham et al. 1993). Logging residues composed of tree tops, larger branches, and dead trees represent one of many forest energy resources. Only about 60% of the production of wood in the United States is harvested. The rest, which eventually falls to the ground and decomposes, represents a potential energy resource. While some of that should remain in the forest to sustain mineral cycles and habitat, some could be used for fuel. Timber-producing forests and plantations are often thinned to improve the growth and quality of remaining trees. The thinnings have limited lumber value but can supply energy. As harvested timber goes through saw mills some material is lost as edgings or uneven slab material. This and the sawdust produced contain useful energy. As lumber reaches milling plants, more of the original material is lost but can be recovered for energy. About one-third to one-half of all hardwood lumber consumed in the United States is used to construct pallets for shipping. Worn-out pallets and wood in demolition and construction waste can provide a concentrated energy source. In some cases, trees are grown in intensively managed plantations for the purpose of producing energy. Species suited to short rotation energy production are fast growing and include—but are not limited to—poplars, willows, and eucalyptus.

Energy from Agriculture.   Corn and other grain crops can supply plant sugars, which can yield alcohol as they ferment. Distillation is used to further concentrate ethanol, which can be used as a biofuel. Euphorbia and sunflowers are among plants that produce oils that can be used in a diesel engine.

Crop- and Food-Processing Wastes.   Energy can be obtained from semiprocessed agricultural and food-processing wastes. Bagasse, a crushed plant waste product created in large quantities in sugarcane processing regions, can be an important energy source. By burning bagasse residues, the sugar industry captures energy for plant processing and avoids open-air burning, which fouls the atmosphere. Bagasse supplies about 9% of Hawaii's electricity. Waste can also be used to produce biofuel as is done when cheese wastes are converted to alcohol. The main advantage of these approaches is that the energy resource is already concentrated and energy processing substantially reduces industry pollution. With ample raw material, energy processing costs are mostly limited to construction of an energy plant. Theoretically, corn and other grains could supply around 3 billion gallons of ethanol annually.

Municipal and Industrial Waste.   About half the material contained in municipal and industrial wastes is combustible. The heat content of this material is only about a third that of high quality coal, but this waste is abundant. Because wastes are often more expensive to dispose than to burn, energy extraction can be economically rewarding. Also, any practice which reduces the bulk of waste material otherwise sent to landfills potentially benefits the environment. Almost 140 waste-to-energy facilities were in operation in 1990. Most of these were mass-burn facilities—which consume raw waste—but some used what is called refuse-derived fuel (RDF). This is essentially chopped-up or pelletized waste with most of the metal, glass, and other inorganic matter removed. In contrast to unsorted waste, RDF can be burned in conventional boilers and contains fewer toxic materials that can enter the air. RDF can also be burned with wood chips, coal, or both, in cofired facilities. The pulp and paper industry is the leading user of wastes for energy in the United States.

Capturing Biomass Energy.   Biomass-based power systems can be applied to a diverse set of energy needs. Biomass can be used for generating power for individual households or village communities, for electricity or steam generation in agricultural applications, or for utility-scale electricity generation. Some biomass systems are rather simple while others are more complex. New types of systems are being developed for industry and transportation as proven methods continue to be improved.

Electricity from Biomass.   Power generation is perhaps the most readily accessible route to the large-scale expansion of biomass for energy. The technologies for biomass conversion to electricity are direct combustion, gasification, and pyrolysis. Using forced air for more thorough oxidation, direct combustion generates hot flue gasses, which are used to produce steam in the heat exchange sections of boilers. Gasification and pyrolysis have been developed to produce different types of fuels from carbon-rich biomass, which can improve power generation. Gasification was discussed earlier as a means of processing coal energy. The biomass gasification process is similar to coal gasification, described in Chapter 19, but instead involves the conversion of the carbon contained in wood to Syngas. The mixture of carbon monoxide and hydrogen contained in Syngas can be used for either space and industrial process heat or for powering gas turbines, which generate electricity. Biomass gasification/generating systems could be less expensive and more efficient than conventional wood-fired, steam-electric power plants. Pyrolysis is defined as the thermal destruction of organic materials in the absence of oxygen. In pyrolysis processes, indirect heating is also used to convert biomass to a mixture of gasses and organic vapors. Finally, a number of major utilities are evaluating the co-firing of biomass in existing coal-fired power stations because of the environmental benefits that may accrue.

The biomass used to fuel these processes can come from a number of sources, including herbaceous as well as woody plants grown specifically for energy production. For efficiency, the biomass needs to be somewhat uniform to reduce handling, storage, and processing costs. "Energy plantations" of fast-growing tree species have received considerable attention in this regard.

Biofuels. Ethanol alcohol can be produced by biologically catalyzed reactions. In much the same way that sugars are fermented into beverage ethanol by various organisms, such as yeast and bacteria, sugars can be extracted from sugar crops, such as sugarcane, and fermented into alcohol. For crops such as corn, starch is first broken down to simple glucose sugars by acids or enzymes known as amylases (Wyman et al. 1993).

Brazil became the world leader in ethanol production when it began using sugar from its sugar industry to offset rising trade deficits associated with imported oil (Goldemberg and Monaco 1993). Ethanol consumption equaled gasoline consumption on a volume basis in 1990. The majority of new cars sold in Brazil today are designed to use ethanol (Larson 1993).

Today, two forms of ethanol are produced from biomass: anhydrous (100% ethanol) and hydrous (containing about 5% water). Anhydrous ethanol can be mixed with gasoline to create gasohol, a product sold at filling stations. Gasohol can contain up to a maximum ethanol content of 20%. The addition of ethanol actually boosts fuel octane rating. Hydrous ethanol must be used in engines designed specifically for ethanol. The amount of anhydrous ethanol currently produced in the United States corresponds to about 8% of total gasoline used in the United States (Larson 1993). Grain surpluses, principally corn, are the main source of feedstock for the ethanol industry.

Cellulosic ethanol is one of the most promising technological options available to reduce transportation-sector greenhouse-gas emissions (Lynd 1996). Without government intervention, it is unlikely that ethanol would have evolved as a transportation fuel in either the United States or Brazil. A tax exemption, grain production subsidies, and government-sponsored research sheltered ethanol development in the United States. In Brazil an abundant supply of sugarcane and high import oil costs drove the government to find a domestic fuel substitute that remains comparatively strong.

Biogas (roughly half methane and half CO_2) is produced by the biological process of anaerobic digestion of wet organic matter. Almost any biomass except lignin (a component of wood) can be converted to biogas. Anaerobic digesters are made out of concrete, steel, brick, or plastic. In developing countries biogas generators, fed with dung and other waste, provide village energy. In industrialized countries animal wastes, crop residues, carbon-laden industrial processing by-products, and landfill material have been widely used to provide power and eliminate waste. China is served by millions of household and community digesters (Levine et al. 1992). An estimated 5,000 digesters are installed in industrialized countries (Larson 1993). These are typically located at large livestock processing facilities (stockyards) and municipal sewage treatment plants. Generally, anaerobic digestion is economical for on-site power production wherever sufficient wastes are available and there is a ready demand for power (Brower 1992).

The Future of Biomass Energy. The advantages and disadvantages of biomass energy are as diverse as the energy applications themselves. Many of the approaches discussed improve or protect the environment by displacing the use of polluting fossil fuels. By extending our energy menu biomass fuels provide hope for finding renewable fuel substitutes. On an industrial scale, the use of plantations and agriculture crops offer several benefits. Plantations can be grown on lands that are marginally productive for agriculture and biofuel grain crops can provide much needed economic benefit to rural communities. Biofuel and plantation energy advocates point to the opportunity to produce energy on some of the lands presently excluded from production by government programs. Today, roughly 33 million hectares of cropland are idle in the United States. The U.S. Department of Agriculture projects that some 52 million hectares will be idled by 2030. Replacing annual row crop production with longer-rotation woody crop production might help revitalize worn-out agricultural soils. Regarding atmospheric CO_2, energy plantations would not result in a CO_2 net gain because this greenhouse gas would be incorporated into new plant growth as it is combusted.

Critics approach the subject of industrial-scale biomass production with less optimism. They cite the potentially polluting effects of applying more fertilizers and herbicides to both grain and tree crops, and problems associated with the expansion of crop monocultures. They also cite the narrow profit margins associated with energy crops when all energy costs are correctly tallied (Larson 1993). Some also question the use of grains for fuel when millions go hungry around the world.

The biofuel industry would actually be better served if plant material other than grain crops could be used. Biofuel would probably be cheaper and would rest on a more stable economic platform because grain prices fluctuate. However, plant parts containing relatively more complex forms of carbons simply do not ferment satisfactorily. Biotechnology may one day provide the ideal mix of microbes for producing ethanol from lignocellulosic biomass, but this presently is not the case. The practice of converting waste to energy has been criticized because of pollution and site location issues. The use of biogas digester technology is, perhaps, least objectionable because it provides the most basic energy needs to impoverished millions. One could say that even dung should be put to a higher purpose. However, it does not lose its crop nutritive value when used for biogas generation and can be withdrawn for fertilizing subsistence crops.

Overall, the future of biomass energy is tied to comparative cost advantages that are likely to occur as petroleum energy costs climb and as environmental issues are resolved on a case-by-case basis. In both developing and industrialized countries, biomass is used much less efficiently than is technically possible and economically feasible. Most likely, gasification practices will continue to replace less efficient burning practices. Further technology improvements will continue to reduce pollution problems which, at present, are much less serious than those associated with fossil fuels. In comparison, wood and other biomass fuels contain less sulfur. Technology will also overcome many critical biofuel conversion limitations. In the long run, the availability of land will be a primary constraint. Reasonable estimates assume that crop residues and energy plantations by the year 2050 can supply 30% of the world's energy needs (Larson 1993).

Geothermal Energy

Geothermal Status.   Beneath the earth's surface, temperature increases 1°F for every 75 feet increase in depth. In certain areas, however, temperatures rise much more quickly owing to subsurface irregularities that have brought molten material closer to the crust. Such areas contain enormous amounts of heat which can be used as a reliable energy source. The source of the heat—or geothermal energy—is molten rock in the earth's interior and the radioactive decay of material contained beneath earth's crust. America's first inhabitants, Paleo-Indians, were first to use some forms of geothermal energy. They warmed themselves in hot springs, which occur in areas where groundwater heated by the earth's interior flows to the surface in pools of bubbling water. They, as well as early European explorers, surely took note of the spectacular geyser displays, or jets of water and steam, which like hot springs, occur in areas where the earth's subsurface heat is especially close to the surface. Such areas are called hydrothermal convection zones.

Heated groundwater trapped by impervious rock layers awaited detection by those first able to drill deep enough to release the steam and water superheated by the molten rock beneath. Areas containing trapped geothermal energy are called geo-pressurized zones. It was eventually determined that energy was also potentially acquired from hot rock zones where magma heats overlying rock.

Most geothermal energy today comes from hydrothermal convection zones because the near surface energy source can be developed without elaborate engineering. At several well known locations throughout the world, steam or hot water from such zones is a primary energy source for an entire community. In Reykjavik, the capital city of Iceland, energy captured from a hydrothermal convection zone is used to heat homes, buildings, public baths, and greenhouses.

Power Production.   The steam or hot water provided by geothermal energy can be used to generate electrical power as was first demonstrated in Larderello, Italy, in 1904. Geothermal electricity is now produced in numerous countries throughout the world, including New Zealand, Japan, Mexico, and the Philippines as well as the countries already mentioned. It is no coincidence that these countries have regions of overlying areas known for their volcanic activity. Geothermal energy originates in the earth's interior, where the hottest fluids and rocks at accessible depths are associated with recent volcanic activity. One of the world's largest geothermal projects, "the Geysers," is located in northern California on the slope of an extinct volcano.

Geothermal resources occur as hydrothermal fluids, hot dry rock, geopressured brines, magma, and ambient ground heat. To date, only hydrothermal fluids have been developed commercially for power generation.

Hydrothermal Fluids.   Three technologies are used to convert hydrothermal fluids to electricity. The method applied in each case is determined by the physical state of the heated water available. If the fluid is wholly or mostly steam, conventional steam turbines are used to generate power (Figure 20.12). When the fluid is above 200°C—and is primarily water—flash steam technology is usually employed. This method involves spraying the fluid into a tank held at a much lower pressure than the fluid. This causes some of the fluid to vaporize, or flash, to steam. Once converted to steam, the fluid can be used to power turbines and generate electricity. For water less than 200°C, a secondary or working fluid must be used to convert the energy contained in the water. The hot geothermal fluid vaporizes the working fluid which then drives a turbine and generator.

Steam resources are the easiest to harness for power generation, but are rare. Only one steam field, the Geysers, has been commercially developed in the United States. It follows that hot water plants—using high- or moderate-temperature geothermal fluids—are more common and are the major source of geothermal power in both the United States and the world. At present, hot-water plants operate in California, Hawaii, Nevada, and Utah.

Low- to moderate-temperature (20°C to 150°C) geothermal resources are widespread in the United States and are used to provide direct heat for homes and industry. By definition, direct use of geothermal resources does not involve power plants or heat pumps. The heat provided is used directly for warming buildings and greenhouse, for industrial processes and a host of other purposes. Direct-use projects generally rely on resource temperatures between 38°C to 149°C. District heating supplies multiple users with hot water from a utility plant or well field. Heat pumps can also be used to acquire energy from low to moderate temperature geothermal resources. Ground-source heat pumps take advantage of the relatively constant temperature of the earth's interior as a heat source in winter and a heat sink in summer.

Geopressured Brines and Magma.   Geopressured brines are hot, pressurized, methane-rich waters found in layers of sandstone 10,000 to 20,000 feet beneath the

FIGURE 20.12   A geothermal power plant (from Owen and Chiras 1995).

surface (Carless 1993). Geopressured systems may one day provide a source for power generation. However, the great depth of these sources and pressures involved require engineering technology that is not cost effective at present. About 20% of the geothermal energy found in the U.S. exists in the form of geopressured reservoirs.

Hot Dry Rock.   Some areas sit above subsurface geologic formations that are abnormally hot but contain little or no water. To unlock this energy source, water must be pumped into hot rock zones then pumped to the surface for heating or electrical power generation. Hot, dry rock resources may supply a significant part of U.S. electric power needs for centuries once technology is developed to make the engineering practices needed commercially acceptable.

The Pros and Cons of Geothermal Energy.   Compared to conventional energy sources, geothermal energy offers several advantages. Geothermal electricity is clean, reliable, and—depending on location—cost-effective. Geothermal power is cleaner than coal-fired and nuclear power plants. The economics of geothermal power can be attractive because less capital investment is needed. Compared to coal-fired plants, geothermal plants can be 40% less expensive. The capital needed for a nuclear power plant can be three times greater. In addition, geothermal power plants are more reliable than conventional power plants. The new steam plants at the Geysers are operable more than 99% of the time. Finally, the many forms of geothermal energy represent a resource that is both abundant and secure.

Criticisms of geothermal technology center on the chemical contaminants it releases into the environment. Geothermal steam and water frequently contain salts and other contaminants, such as sulfur compounds, which are leached from bedrock minerals. The gaseous pollutants sometimes occurring in the steam—such as hydrogen sulfide, carbon dioxide, ammonia, and methane—require pollution control devices. The corrosive properties of geothermal fluids create both engineering and environmental problems that raise concerns and project expense.

Another limiting feature is the incapacity of steam to retain heat over long transportation distances. This explains why district heating projects are not more geographically extensive. The location of geothermal resources is another issue. Geothermal resources are scattered throughout the more sparsely populated West. Finally, geothermal energy may not always be sustainable. The Geysers in California and a project in New Zealand have both suffered serious losses of power production energy over time. Such cases have negative impacts on the industry's investment image.

The Future of Geothermal Energy.

In the United States, geothermal electric power generation totals about 2,200 MW or about the same as four large nuclear power plants. Overall, the current production of geothermal energy from all uses puts this resource third among renewables. Its behind hydropower and biomass but ahead of solar and wind.

Compared to conventional energy, the potential for geothermal energy has been barely tapped and geological exploration continues to locate new resources. Although its development will be confined somewhat within geographic boundaries, geothermal energy represents a significant source for power generation and heat. Also, the cost of generating geothermal energy is anticipated to continue a decreasing trend as technology improves.

Additional Renewable Energy Sources

Tidal Power.

Massive amounts of energy are potentially available from the sea. As we watch large waves pounding a coastline we witness, for example, the action of tremendous forces. The energy potentially available from tidal changes is less striking but, along with wave energy, attracts the attention of energy experts. The technology used to harness tidal energy is rather simple and resembles engineering approaches used to generate electricity from major rivers. A feature that is critical to both is the ability to capture energy from massive amounts of falling water. It follows that both approaches require dams to be constructed. To obtain energy from tidal surges, a dam is built across the mouth of a bay and turbines are mounted in the structure. The in-rushing tide generates power as it passes through turbines. As the tide shifts, the blades can be reversed to capture the energy of the return flow.

The world's first tidal-electric installation was built in 1966 in the La Rance estuary in France. A similar plant was constructed in Russia. The incentives for building such plants are obvious. Tidal power production does not require massive capital, produces no pollutants, and is sustainable. However, the sites capable of producing tides high enough to justify power development are scarce. Only about 24 good sites exist worldwide. To be useful for power production, shoreline topographic features must channel the tide to superior heights. The only location in North America with impressive tides is the Bay of Funday. Here shoreline features channel a tide that approaches 16 meters, the largest in the world. Simply too few sites have such massive tides. Throughout most of the world, the small difference existing between low and high tide (less than two feet) is less than needed for driving turbines. Also, there are environmental and practical concerns. Depending on location, tidal dams can interfere with fish migrations, impede boat traffic, trap sediments, and alter the critical balances of fresh- and saltwater in estuaries. Overall, tidal power is not likely to become a significant contributor to our energy future.

Ocean Thermal Energy Conversion.

The ocean stores vast amounts of energy received from the sun. Ocean thermal-energy conversion (OTEC) technology is directed toward recovery of this energy for power production. The basic aim of OTEC is to exploit the steep temperature gradient found in tropical waters. Project engineering involves floating platforms that draw water from the warmest and coldest depths the location can offer. Pipes extending to a depth of 900 meters (3,000 feet) can be used to exploit cold water. The warm water pumped from immediately beneath the surface platform is used to convert ammonia into a gas, which drives the blades of an electric turbine. The cold water pipe supplies the cooling needed to recondense the gas so that the energy exchange cycle is maintained. Ammonia, the working fluid, is needed because of its low boiling temperature, which is much lower than water. The electric power produced from an OTEC plant can be transferred to shore for various end uses. Another approach, the open or Claude cycle, uses warm seawater to create steam. The warm seawater is sufficient to create steam only because the system employs a vacuum vessel as the evaporator. Because the pressure maintained in the vessel is only 0.03 times atmospheric pressure, the boiling temperature of water is lowered and low pressure steam is produced.

Although OTEC offers the possibility of sustainable energy production, there are serious limitations. Operational efficiency is a key issue because much of the energy obtained must be reinvested in pumping large amounts of water through the system. For example, an experimental plant constructed off the shore of Hawaii used 80% of the energy generated to pump water to the platform. Another limitation is the relative scarcity of locations suitable for positioning OTEC plants. Suitable locations for U.S. development include the Gulf Coast and waters near the shores of Hawaii, Guam, and Puerto Rico.

Again, there are environmental concerns. As cold waters are heated at the OTEC platform, the dissolved carbon dioxide they once contained is released to the atmosphere. Localized effects potentially include disturbance to fisheries and disruptions in the marine food chain stemming from the redistribution of nutrients found in deep waters. The prospects for OTEC therefore seem limited in the broad scope of sustainable energy sources. Extreme care will be needed to avoid ecological disturbances where engineering criteria can be satisfied.

Hydrogen.   Hydrogen is believed by many energy experts to be one of the main renewable energy fuels of the future. Its renewability derives from the fact that it can be produced from energy provided by solar, biomass, and other sustainable energy resources. At present, hydrogen is produced primarily by steam reforming of natural gas. Electrolysis is used to obtain extremely pure hydrogen for various pharmaceutical and industrial applications. This process employs electricity to dissociate, or split, water into its hydrogen and oxygen components. Both of these methods are too costly for wide energy applications.

As a renewable fuel, hydrogen may one day be produced on a commercial scale by renewable technologies that can be grouped into three types of processes: photobiological, photoelectrochemical, and thermochemical. Photobiological and photoelectrochemical processes generally would use sunlight to split water into hydrogen and oxygen. Thermochemical processes, including gasification and pyrolysis systems, would use heat to produce hydrogen from sources such as biomass and solid waste.

Some obstacles must be overcome before photobiological technologies can become commercially viable. Most photobiological systems use the natural activity of bacteria and chlorophyll-containing green algae to split water and produce hydrogen (Sandia Laboratories 1995). Current research efforts center on developing strains of microbes, which more efficiently channel the energy obtained from the sun to the reaction responsible for splitting water to produce hydrogen. Photoelectrochemical (PEC) methods employ photovoltaic technology to acquire the energy needed and their energy conversion efficiencies are being improved. Thermal processes such as fast pyrolysis are being tested as a means of using renewable biomass to produce hydrogen. When biomass is heated to moderately high temperature and is vaporized, the vapors can be catalytically reformed to produce hydrogen.

On combustion, hydrogen does not produce particulates, carbon dioxide, and sulfur oxides. Instead, combustion yields energy, water, and a small amount of nitrogen dioxide which is produced when the heat produced causes oxygen and nitrogen in the air to combine. Hydrogen has numerous potential applications, including electrical power generation, space heating, and transportation. Furthermore, the versatility of hydrogen will enable sustainable energy technology to advance on the domestic energy scene more rapidly than would be the case if electricity were the sole means for carrying renewable energy from source to end use.

The ability of hydrogen to serve as a transportation fuel has been demonstrated for many years in the space industry. Liquid hydrogen and liquid oxygen are combined as propulsion fuel for the space shuttle and other rockets. Onboard fuel cells using hydrogen and oxygen also provide most of the shuttle's electric power. Most of the hydrogen used to fuel the nation's space shuttle program is made from natural gas by the steam reforming process. The technology involved is not directly applicable to automobiles. Nevertheless, hydrogen will power pollution-free vehicles of the future. All of the technologies necessary to achieve this goal have been proven in concept. Work is underway to improve the practicality of energy storage and to reduce the costs of the technology for renewably generated hydrogen. In the interim, hydrogen will be increasingly used on a limited scale as an automotive fuel and to demonstrate technological if not economic feasibility. Advanced internal combustion engines will power ultraclean hydrogen transportation vehicles until efficient fuel cells become commercially available. The use of liquid hydrogen to fuel commercial jetliners has already been demonstrated. For cities with heavy jet traffic, liquid hydrogen will contribute substantially to cleaner air.

Fuel Cells. Fuel cells are devices that convert hydrogen gas directly into low-voltage direct-current electricity. Fuel cells show great promise as replacements for internal combustion engines. The fuel cell is an electrochemical device that splits hydrogen into ions that either run along an electrode or combine with oxygen, producing electricity with water (Dunn and Flavin 2000). It was invented by the British physicist William R. Grove in 1839. Fuel cells have been deployed extensively in the U.S. space program, but until recently cost has prevented widespread commercial use.

Advantages of fuel cells over combustion generators are discussed by Dunn and Flavin (2000). Fuel cells are silent and their only by-product is water if used with hydrogen fuel. They can also be built small enough to power electronic devices such as laptop computers or cell phones. If natural gas is used as a hydrogen source, fuel cells produce some pollutants (nitrogen oxides and carbon dioxide), but at lower quantities than internal combustion engines. Because fuel cells have no moving parts they require little maintenance and are more reliable than conventional power plants.

The world's leading automakers are accelerating the commercial use of fuel cells: Toyota, Ford, Volkswagen and Daimler-Chrysler have proclaimed fuel cells as likely replacements for the internal combustion engine. Four automakers plan to have fuel-cell cars in their showrooms by 2004. Various other commercial and residential uses are planned for fuel cells. Major companies, including General Electric, are actively pursuing the fuel-cell power market. Dunn and Flavin (2000) provide a more complete discussion of fuel cells and other types of micropower.

CONCLUDING REMARKS

As discussed, there are numerous options for producing future energy in a sustainable manner (Figure 20.13). This should allay fears that our lives will be greatly affected as the fossil energy age draws to a close over coming decades. The primary issue before us is how to phase in renewable technologies. Some argue that fossil fuel should be heavily taxed to discourage greenhouse gas emissions. To support their arguments, they cite the environmental costs of fossil fuels which are not included in the price paid for them. Renewable energy advocates have also objected to the way fossil and nuclear fuels concentrate capital and political clout. They prefer a less centralized approach that is more user friendly. The best arguments for a decentralized energy strategy were expressed two decades ago by Lovins (1977), who proposed a "soft path" to energy sufficiency. Many of Lovins's arguments remain valid today. His arguments against nuclear energy appear even stronger now because major accidents have occurred since his remarks were published. Another source of complaint among soft path advocates is the rather low level of financial support provided by our government. In the wake of the 1970s oil

Energy Information Administration. 1997d. *Early release of the annual energy outlook.* (AEO98).

Flavin, C. and S. Dunn. 1999. Reinventing the energy system. *State of the World 1999.* Washinton, DC: World Watch Institute.

Gleick, P. H. 1994. Water and energy. *Annual Review Energy Environment* 19:267–299.

Goldemberg, J. and L. C. Monaco. 1993. The Brazilian fuel-alcohol program, (pp. 841–863). In T. B. Johansson, H. Kelly, A. K. N. Reddy, R. H. Williams, and L. Burnham, eds., *Renewable energy: Sources for fuels and electricity.* Cavelo, CA: Island Press.

Graham, R. L., L. L. Wright, and A. F. Turhollow. 1993. Short rotation woody crops: Opportunities to mitigate atmospheric carbon dioxide buildup. *Climate Change* 22:223–238.

Grubb, M. J. and N. I. Meyer. 1993. Wind energy: Resources, systems, and regional strategies, (pp. 157–212). In T. B. Johansson, H. Kelly, A. K. N. Reddy, R. H. Williams, and L. Burnham, eds., *Renewable energy: Sources for fuels and electricity.* Cavelo, CA: Island Press.

Gutman, P. S. 1994. Involuntary resettlement in hydropower projects. *Annual Review Energy Environment* 19:189–210.

Hall, C. A. S., C. J. Cleveland, and R. Kaufmann. 1992. *Energy and resource quality. The economy and economic process.* Niwot, CO: University Press of Colorado.

Johansson, T. B., H. Kelly, A. K. N. Reddy, and R. H. Williams. 1993. Renewable fuels and electricity for a growing world economy: Defining and achieving the potential. In T. B. Johansson, H. Kelly, A. K. N. Reddy, R. H. Williams, and L. Burnham, eds., *Renewable energy: Sources for fuels and electricity.* Cavelo, CA: Island Press.

Larson. E. 1993. Technology for electricity and fuel production from biomass. *Annual Review Energy Environment* 18:567–630.

Levine, M. D., F. Liu, and J. F. Stinton. 1992. China's energy system. *Annual Review Energy Environment* 17:405–435.

Lovins, A. B. 1977. *Soft energy paths: Towards a durable peace.* Cambridge, MA: Ballinger Publishing.

Lynd, L. R. 1996. Overview and evaluation of fuel ethanol from cellulosic biomass: Technology, economics, the environment and policy. *Annual Review Energy Environment* 21:403–465.

Moreira, J. R. and A. D. Poole. 1993. Hydropower and its constraints, (pp. 73–119). In T. B. Johansson, H. Kelly, A. K. N. Reddy, R. H. Williams, and L. Burnham, eds., *Renewable energy: Sources for fuels and electricity.* Cavelo, CA: Island Press.

Nebel, B. J. and R. T. Wright. 1996. *Environmental science.* 5th ed. Upper Saddle River, NJ: Prentice-Hall.

Nebel, B. J. and R. T. Wright. 1998. *Environmental science.* 6th ed. Upper Saddle River, NJ: Prentice-Hall.

Owen, O. S. and D. D. Chiras. 1995. *Natural resource conservation.* Upper Saddle River, NJ: Prentice-Hall.

Rosenberg, P. 1993. *The alternative energy handbook.* Lilburn, GA: The Fairmont Press.

Sandia National Laboratories. 1995. *Advanced hydrogen technologies.* U.S. Department of Energy. DOE/GO–10095–065.

Sorensen, B. 1995. History of, and recent progress in, wind energy utilization. *Annual Review Energy Environment* 20:387–424.

Thurlow, G. 1990. Implications for policy formulation (pp. 73–76). In G. Thurlow, ed., *Technological responses to the greenhouse effect: Wyatt committee report No. 23.* New York: Elsevier Applied Science.

Weinberg, C. J. and R. H. Williams. 1990. Energy from the sun. *Scientific American* 263:147–151.

Wyman, C. E., R. L. Bain, N. D. Hinman, and D. J. Stevens. 1993. Ethanol and methanol from cellulosic biomass (pp. 865–923). In T. B. Johansson, H. Kelly, A. K. N. Reddy, R. H. Williams, and L. Burnham, eds., *Renewable energy: Sources for fuels and electricity.* Cavelo, CA: Island Press.

6

Integration of Natural Resources Management

This book has provided a comprehensive introduction to the principles of natural resources management so far as applied primarily in the geographical and political context of the United States. The history of natural resource development and management, summarized here, reveals how profoundly natural resource development has shaped human culture and improved material well-being. However, increasing resource demands and scientific awareness also have made it clear that humanity is having a tremendous impact on natural resource sustainability worldwide. Individual chapters have shown that management systems often do not stop at national borders. While ecosystems and their management details vary from those in the United States, the scientific process and management principles generally apply worldwide. In this section we address some basic social, economic, and ecological issues now facing the globe. We also summarize trends pointing toward greater integration of economies, ecological understanding, and management approaches to a more sustainable natural resource management worldwide. Chapter 21 is an introduction to some of the problems associated with past international resource management, with emphasis on food and fiber production and related trade and distribution. Chapter 22 discusses world economic systems emphasizing the general advantages of market economies over command economies in improving human living conditions. As all of the world seeks the material welfare enjoyed by the present minority in developed nations, the demands on atmospheric and biospheric resources will increase, particularly if the sustainability of resource development does not improve. U.S. national policies applying to international resource management are especially significant because of the disproportional influence they have on international affairs and global resources. Chapter 23 summarizes recent policy pertaining to the goals of sustainable development and growing emphasis on integrated resource management as the means to achieve that goal. Chapter 23 also discusses the importance of technology in the attainment of sustainable development and improved living conditions.

CHAPTER **21**

Natural Resources and International Development

INTRODUCTION

Great differences exist between the per capita incomes of developed and developing countries (Schiller 1994). About 3 billion people, or over half of the world population, live in countries where the average income is under $500 per year. In contrast, the average income in the United States was about $26,000 per year in 1996. Unfortunately while living standards continue to improve in the developed countries, they are becoming worse in many of the least developed countries; particularly those in Africa (Figure 21.1). At the same time, certain countries with intermediate levels of development in the Pacific Rim (Korea, Singapore, Taiwan) and portions of Latin America (Chile, Argentina) are experiencing rapid improvement in their living standards.

We will discuss why living standards and economic progress vary widely among different countries. The role of natural resource availability and management in economic development will be considered along with policies that are most likely to promote sustainable development. The main themes of this chapter have been developed mostly from Brown and Wolfe (1985), Miller, (1990), Foster (1992), and Schiller (1994, 2000).

PROBLEMS WITH THIRD WORLD DEVELOPMENT

Most affluent countries have experienced poverty sometime in their past. From the fall of the Roman Empire around A.D. 450 until the mid-1700s, improvements in the human condition were almost nonexistent. Then the Industrial Revolution, starting in Europe (particularly Great Britain), triggered rapid economic growth. The key feature of the Industrial Revolution was the substitution of machines for human labor. The invention of the steam engine in the late 1700s was the most crucial breakthrough in this economic transformation that permitted labor to become more specialized and more efficient. At the same time, it reduced the need for large portions of the human population to work in agriculture (see Chapter 2).

same time, this is seldom possible. Experience has shown that the most successful developing countries concentrated their resources in industries where they have the greatest comparative advantage in natural resources, labor, and management (Schiller 1994). In other words, growth in targeted industries should not be greatly limited by shortages of labor, land, technology, management, or some other input. Ideally the targeted industries should be linked to other industries the country could develop. For example, development of the timber resources is compatible with furniture manufacture. Growth in the cattle production industry should accompany development of a meat processing and packing industry.

Another choice that confronts developing countries is whether to emphasize production for domestic consumption or for export. Generally the most successful countries have heavily depended on exports for economic growth (Schiller 1994). This has the advantages of generating foreign currency that can be used to purchase complex goods and technology from other countries and it permits tapping into markets that are much larger than those existing internally. Japan, South Korea, and Taiwan are a few examples of countries that have exported themselves to prosperity since World War II.

The disadvantages of relying on exports include barriers to market access in developed countries, difficulty in producing goods of comparable quality to those in developed countries, and the erratic nature of world demand for exportable products (Schiller 1994). Trade restriction problems have been greatest with textiles, shoes, steel, clothes, sugar, peanuts, and other basic commodities where developed countries such as the United States have established industries that lack a comparative advantage. Here domestic producers apply political pressure for trade barriers to keep out foreign competition.

Quotas in the United States have reduced export earnings of Argentina and Australia from beef production and of the Philippines and several Caribbean countries from sugar production. However, in certain cases a developing country may be given a favored nation status that allows them more access to U.S. markets.

Coffee is an agricultural product that cannot be easily grown in the United States and many other developed countries (Schiller 1994). Therefore, it is well suited as an export crop by several African and Latin American countries. Since Brazil is a major coffee producer, its crop can have great impact on other countries. When frost or other factors reduce Brazil's coffee crop, as in 1994, coffee prices go up and other coffee-producing countries benefit. However, when growing conditions are ideal in Brazil, coffee supplies expand, reducing export earnings in countries such as Haiti. Too much dependence on one export leads to sequences of boom and bust that can cause erratic long-term growth and political instability.

Cartels have been formed to stabilize export earnings (Schiller 1994). They basically attempt to limit exports in years of excess supply. The best-known cartel involves the oil-producing countries (OPEC) dominated by Saudi Arabia. Cartels have had only limited success because those countries with faltering economies are strongly tempted to underprice and exceed their voluntary quotas during periods of oversupply.

Rather than take the risks of international trade, many developing countries have concentrated on domestic markets. With the exception of agricultural products, this approach is usually flawed because only a few goods can be produced on a competitive basis with developed countries (Schiller 1994). Domestic demand seldom justifies heavy capital expenditures for production of goods that will be higher priced and of inferior quality to imports.

NATURAL RESOURCES VERSUS ENTREPRENEURSHIP

Without question, countries that are well endowed with natural resources—such as Canada, Australia, New Zealand, Brazil, Argentina, and South Africa—have great advantages in economic development over countries where natural resources are scarce relative to resource demands. However, the successes of Japan, South Korea, and, most recently, China indicate that human entrepreneurial resources are more important than physical resources in economic development.

Hernando De Soto, in his best-selling book *The Other Path*, has been one of the most convincing advocates of unlocking the power of markets and people (Schiller 1994). After years of financial success in Europe, he returned to his native Peru and observed that it had a thriving underground economy in spite of repressive government policies. Although Peru had some trade in drugs, most of it centered around essential consumer goods. Government regulation of prices, trade, and other business activities had caused most businesses to go underground. De Soto advocates government adoption of policies that encourage entrepreneurial activities through privatization, reductions in regulation, elimination of price controls, legal protection of property, and infrastructure development. De Soto's book has become the model for market-oriented reforms in Peru, Argentina, Mexico, Russia, and India.

The basic point from De Soto's work is that the greatest natural resource is human ambition, ingenuity, and creativity (Schiller 1994). All things being equal, those countries possessing the highest levels of physical natural resources have the advantage in international development. However countries that adopt democratic, market-oriented development policies can overcome many natural resource limitations. Japan and South Korea are examples of countries that have become world economic powers with limited natural resources.

INTERNATIONAL POLICY

The 1990s were characterized by a gradual shift from centrally planned to market-oriented economies by most of the world's developing countries. Those countries that made the most progress with this shift generally experienced the highest levels of per capita economic growth. At the same time many of those countries reduced rates of environmental degradation and natural resource depletion because production efficiency was higher under market-oriented systems (Chandler 1987). While markets alone cannot assure environmental and natural resource sustainability, they do impose some self-administered checks on natural resource waste. This is because the resource user has strong incentives for efficiency.

Although in theory centrally planned economies can make resource conservation a high priority, in practice this did not happen (Chandler 1987). Prior to its disintegration in 1991, the former Soviet Union had one of the poorest natural resource conservation records of any country in the world. Their energy consumption rate per unit of gross national product (GNP) was about 50% higher while their sulfur dioxide emissions per unit of GNP were more than twice those in the United States. Soil erosion rates on Soviet farmlands far exceeded those in the United States and in Western European countries. Prior to market-oriented reforms in the 1970s, China experienced severe forest and cropland degradation and natural resource depletion. Although these problems continue in China, they are now less severe.

Because capital is in short supply in developing countries, loans are important to their development. The World Bank is the primary lending agency to developing countries. It also provides technical assistance to improve the success of the projects it finances. Multinational banks and individual countries also provide similar financial and technical assistance. Developing countries prefer loans with few strings attached. However recent history shows the most successful loans have been those predicated on monetary, market, and environmental reforms.

In the past, developing countries often borrowed beyond their capability to repay their debts (Schiller 1994). Because these debts must be paid back in hard currencies such as U.S. dollars export-related projects, such as factories, have taken precedence over projects providing domestic infrastructure such as schools, roads, and sewage systems.

TABLE 21.2   Foreign Aid in Relation to GDP, 1990

Country	Official Development Assistance	
	Millions of U.S. Dollars	As a Percentage of GDP
United States	$11,394	0.21
Japan	9,069	0.31
France	9,380	0.79
Germany	6,320	0.42
United Kingdom	2,638	0.27
Canada	2,470	0.44
Netherlands	2,592	0.94
Sweden	2,012	0.90
Australia	955	0.34
Norway	1,205	1.17
Denmark	1,171	0.93
Switzerland	750	0.31
New Zealand	95	0.23
Total foreign aid	$56,632	Average 0.36

Source: World Bank and Schiller 1994; total includes other developed countries.

In the early 1980s, a worldwide recession reduced the capability of Mexico, Brazil, Peru, and other large borrowers to service their debt (Schiller 1994). In response to this crisis, the developed creditor nations adopted a resolution to reduce their exports to developing countries, to increase their imports from developing countries, to place greater emphasis on investment in developing countries, and to increase the level of outright gifts in foreign aid to developing countries.

Naturally, developing countries would like more foreign aid in the form of gifts rather than loans. Foreign aid data in Table 21.2 show that about 57 billion dollars are given to developing countries each year. Because of the low amounts given and the strings that are usually attached, developing countries are now asking for "trade" not aid (Schiller 1994). In other words, they wish to export more of their goods so they can finance internal investment themselves.

The objective of the North American Free Trade Agreement among the United States, Mexico, and Canada is a step toward the trade-over-aid approach. This agreement not only opens U.S. markets to Mexico, but it also has important environmental provisions relating to air quality, water quality, and land use.

Latin American countries—particularly Chile, Argentina, Mexico, and Peru—have demonstrated that market reforms can turn around stagnant economic growth, inflation, and capital flight (Stein 1994). Economic changes in the 1990s involving cutting tax rates, business privatization, lowering of trade barriers, stabilization of currency, and easing restrictions on foreign investment turned these nations away from the abyss and toward prosperity. However, those changes involved human and environmental costs that must also be recognized (Roodman 2001).

THIRD WORLD DEBT CRISIS

The world debt crisis centers around overborrowing by developing countries and overlending by international financial institutions such as the World Bank and the International Monetary Fund. Many poor developing countries are now spending more on foreign debt payments than on basic social services for their citizens. As an example, Zambia devoted 40% of its national budget to foreign debt payments but only 7% to basic social services (Roodman 2001). The debt crisis centers around the basic problem that many of the loans went into unsound development projects. Roads, dams, agricultural projects, schools, and factories funded by debt often failed to stimulate and improve the economy enough to make repayment possible. In some cases the borrowed money was squandered and/or embezzled by corrupt and/or incompetent dictatorial governments. A brief history of third world debt problems is provided by Roodman (2001).

Recently, the debt crisis has been linked with environmental degradation in several countries. Although the issue is complicated, heavy debt levels encourage economic policies that contribute to rain forest clearing, farming of erosion-prone hillsides, industrial pollution, and unsound mining practices (Roodman 2001). The need for exports to service foreign debt has resulted in lax environmental regulation and poor enforcement of existing laws in many developing countries.

Various reforms and changes are now being proposed and tested by international lenders with regard to third world debt. Debt-for-nature programs involve exchanging debt for control of nature reserves, environmental cleanup, and rain forest protection. Environmental groups have made these kinds of deals with Bolivia, Costa Rica, the Philippines, Poland, and several other countries. Some of the other reforms include lending selectively to countries that show commitment to reducing poverty and environmental protection, granting partial loan forgiveness for environmental and economic reforms, donor governments basing funds provided to lending agencies on the basis of their past loan performances, and providing more grants.

We refer the reader to Roodman (2001) for a more complete discussion on these reforms.

LITERATURE CITED

Brown, L. R. and E. C. Wolf. 1985. *Reversing Africa's decline.* World Watch Papers 65. Washington, DC: World Watch Institute.

Chandler, W. U. 1987. Designing sustainable economies. In *State of the World.* New York: World Watch Institute.

Foster, P. 1992. *The world food problem.* Boulder, CO: Lynne Rienner Publishing.

Holechek, J. L., R. D. Pieper, and C. Herbel. 1998. *Range management principles and practices.* 3rd ed. Upper Saddle River, NJ: Prentice-Hall.

Investor's Business Daily. 1995. Economics and culture. Section B1. September 22, 1995.

Investor's Business Daily. 1996. Guide to the markets. New York: John Wiley & Sons.

Jacobsen, J. 1983. *Promoting population stabilization incentives for small families.* World Watch Paper 54. New York: World Watch Institute.

Miller, G. T. 1990. *Resource conservation and management.* Belmont, CA: Wadsworth.

Roodman, D. M. 2001. Ending the debt crisis. *State of the World 2001.* Washington, DC: World Watch Institute.

Schiller, B. 1994. *The economy today.* 6th ed. New York: McGraw-Hill.

Schiller, B. R. 2000. *The economy today.* 8th ed. New York: McGraw-Hill.

Stein, R. S. 1994. Why Latin America is thriving. *Investor's Business Daily,* January 4, 1994.

CHAPTER 22

Economics and Economic Systems

INTRODUCTION

Because natural resources and other material goods and services are available in limited supplies, but human wants are unlimited, trade-offs or choices must be made regarding what combination of goods and services will be produced by any society. The science of how to allocate scarce resources among competing uses is called **economics** (Schiller 1994). The science of economics is relatively young, originating when Adam Smith wrote *The Wealth of Nations* in 1776, which described the benefits of market-driven economies.

In this chapter, we will provide a basic overview of economics and economic systems. We explore the issue of natural resources scarcity versus technological scarcity in constraining improvements in human welfare. Finally economic approaches and systems that will meet the needs of the rapidly changing world of the twenty-first century will be explored. While this chapter extends beyond the narrow field of natural resource economics, it illustrates how economic systems enfold natural resource management systems and determine the efficiency of natural resource use and abuse. Our discussion has drawn heavily from Davidson and Rees-Mogg (1987, 1993), Schiller (1994, 2000), and *Investor's Business Daily* (1996).

ECONOMIC TERMINOLOGY

To further develop our discussion of basic economics, it is necessary to define some of the commonly used terminology. The definitions of economic terms in the glossary come from *The Economy Today* (Schiller 2000), which we consider to be an excellent basic textbook on both macroeconomics and microeconomics. **Macroeconomics** focuses on the economy as a whole while **microeconomics** deals with the behavior of individual components of the economy. While Chapter 4 addressed benefit-cost microeconomics, this chapter addresses macroeconomics.

BASIC ECONOMIC PRINCIPLES

The two most basic principles of economic science are that resources are scarce and that economic decisions involve opportunity costs (Schiller 1994). **Opportunity**

costs are goods and services foregone when the decision is made to produce more of one item and less of another. The allocation of scarce resources to obtain four basic social goals is the primary purpose of economic science. These four social goals aimed at maximizing human welfare include (Schiller 1994):

1. Economic growth—This involves expanding the supply of goods and services available to the human population.
2. Equitable distribution of income—This involves developing social-political systems that provide everyone with the opportunity to obtain the material goods and services necessary for a quality lifestyle.
3. Full employment—This involves the application of social-political systems that permit everyone capable of working to have a job.
4. Price stability—This involves managing the economy so that periods of rapidly rising or falling prices for goods and services are minimized or avoided.

No socioeconomic system in history has been able to achieve all of the above goals for an extended period of time (20 years or more). However, the United States experienced a period of prosperity from 1953 to 1973 that probably represents the historical pinnacle in terms of near simultaneous achievement of all four economic goals.

Generally, market-oriented systems with minimal government intervention maximize efficiency in terms of providing society with the optimal combination goods and services it desires and can afford (see Chapter 4). Both economic growth and price stability tend to be best achieved with market-oriented systems. However equitable distribution of income and full employment are to some extent compromised under market economies. Without some degree of regulation by government severe environmental damage is probable under any economic system. This is because environmental protection is a transaction externality (also see Chapter 4).

An **externality** is a cost of an economic activity borne by a third party not part of the original transaction. During the "smokestack era" of the late 1800s to the mid-1900s, pollution of rivers and the air by factories in the northeastern United States was rampant. Government regulation in the 1970s brought this problem under control. Soil erosion from farming is an externality that is born more by future than present generations. Here again government intervention has been necessary to control the problem. Chapter 4 describes the problem in more detail and offers solutions based on market incentives.

Market Theories

Adam Smith in 1776 formulated theories that the basic economic questions of what, how, and for whom goods and services are to be produced is best decided by the "invisible hand" (Figure 22.1). This invisible hand was a characterization of how markets operate based on indirect communication between producers and consumers (*Investor's Business Daily* 1996). Under the "invisible hand," increased demand by society for more of something causes prices to rise at the marketplace or shortages will develop. Higher prices send a signal to producers to increase output. Profit potential provides an incentive for producers to respond to this signal. Moreover, failure to sell all the output of some other good generally causes inventories to build up and prices to fall signaling producers to reduce production (see Chapter 4 for additional discussion). Through these "market mechanisms," members of society get the optimum combination of goods and services relative to their incomes. Individuals who have the ability and desire to pay for a good will be able to get it in a market economy. Producers will use the most efficient methods of production to maximize profits. Competition from other producers will tend to keep prices and supply in balance with demand and lead to improvements in the production process.

FIGURE 22.1   Adam Smith, an English economist, formulated basic theories on free markets in 1776. His work titled *An Inquiry Into the Nature and Cause of the Wealth of Nations* is the foundation for the philosophy and workings of market economies (from The Bettmann Archive, IBD 1996).

Adam Smith believed that price signals and consumer demand at the marketplace are far more effective than government in allocating resources. Under market-oriented (capitalistic) economies as envisioned by Smith the factors of production (land, equipment, labor, management) are controlled by individuals rather than the state. Adam Smith's ideas provided the basis for the capitalistic system of government adopted in the United States after independence from Great Britain was won in 1783. Figure 22.2 shows the increase in inventions and improvement in living standards associated with the development of market economies in Western Europe and the United States in the 1700s.

The Marx Alternative to Markets

An economic alternative to market oriented economics was theorized by Karl Marx when he published *Das Kapital* in 1867 (Schiller 1994). Marx developed the rationale for centrally planned or **command economies**—which are the polar opposite of market economies. Under the system theorized by Marx, the factors of production (land, labor, equipment, management) would be controlled by the government and the government would make all production and distribution decisions for society. This type of system, in which the government owns all nonlabor means of production and decides how resources will be allocated, has been referred to as either **communism** or **socialism**.

Communism has many elements of socialism (Schiller 1994). It differs in that theoretically everyone is supposed to share in production and consumption according to their individual abilities and needs. The basic concept of communism is a stateless, classless economy with no private property. Under socialism the state owns all the nonlabor means of production and exercises control over resource allocation. However, resources are only partially distributed according to each person's effort (Schiller 1994). In reality, communism is merely a concept that has never been achieved by any country, including the former Soviet Union and modern China.

Karl Marx thought government should take strong steps to correct the inequities he observed in England and other European countries in the mid-1800s. A few rich people enjoyed extravagant lifestyles while large numbers of people had inadequate housing and suffered periodically from starvation under the market-oriented

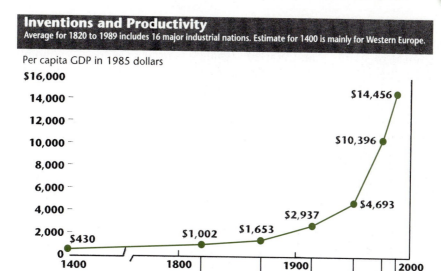

Inventions and Productivity
Average for 1820 to 1989 includes 16 major industrial nations. Estimate for 1400 is mainly for Western Europe.

Per capita GDP in 1985 dollars

Source: Data from *Dynamic Forces in Capitalist Development: A Long-Run Comparative View,* Angus Maddison, Oxford University Press, Walton Street, Oxford, 1991.

Year	Invention	Inventor
1440	Printing with movable type	Gutenberg
1642–1671	Calculating machines	Pascal, Leibniz
1764	Spinning Jenny	Hargreaves
1785	Power loom	Cartwright
1825	Steam locomotive	Stephenson
1837	Telegraph	Morse
1856	Bessemer steel process	Bessemer
1865	Antiseptics in surgery	Lister
1876	Telephone	Bell
1877	Internal combustion engine	Otto
1879	Electric light	Edison, Swan
1903	Airplane	Wright
1909	Bakelite (plastic)	Baekeland
1925	Television	Zworykin, Farnsworth
1948	Transistor	Shockley, Brattain, Bardeen
1952	Commercial computer	
1957	Laser	Gould
1958	Integrated circuit	
1965	Minicomputer	
1970	Microprocessor	
1973	Personal computer	
1980s	Genetic engineering	
1990s	World Wide Web	

FIGURE 22.2   The rise of per capita GDP and increase in inventions as market economies began to develop in the 1700s in western Europe and the United States (from IBD 1996).

economies that existed at the time. Marx theorized that central planners could produce and distribute a more desirable combination of goods than that achieved under markets. Methods in production could take into account environmental, employment, and other social needs without being determined by cost alone.

Mixed Economies

Various compromises between the extremes of market and command economies have been applied in a wide variety of countries over the last 100 years. So far no economic system has fulfilled all economic needs. However, without question some systems have been more effective than others.

Today no nation relies completely on market or central planning mechanisms in production and allocation decisions (Schiller 1994). Instead, a combination of government regulations and market signals are used to achieve economic and social goals. This integrated type of system is referred to as a mixed economy. The United States is considered to be the most market-oriented economy in the world. Great Britain, Japan, Singapore, Canada, and Australia are considered to be market oriented, but their taxation and regulation levels are higher than in the United States. Russia and China have been attempting to privatize many centrally planned operations such as farming, mining, and marketing, but central planning still strongly dominates in the manufacturing sectors. In Latin America and Europe the state controls many key enterprises, but private markets flourish in agriculture and in the distribution of goods. European economies tend to be the most evenly balanced or mixed.

The degree of taxation is the primary determinant of whether a country has a market, command, or mixed economy. In the United States, people on average pay about 36% of their gross income to government in taxes. In China, the average citizen pays over 70% of his income to the government in taxes. In mixed economies such as Germany, France, Spain, and Sweden, the average citizen pays 40 to 65% of their income in taxes. The government then redistributes the bulk of this income back to the population according to need for healthcare, food, housing, and retirement.

It is now well recognized that both market and centrally planned economies can fail without proper direction from government. The 1930s depression in the United States—when unemployment hit 25% and economic output fell by 10%—is the most impressive example of market failure in the twentieth century (Schiller 1994). The collapse of communism in the former Soviet Union in 1989 provides a primary example of central-planning failure. The challenge for society is to select the appropriate balance of market incentives and government taxation and regulation. It is important to understand economic interventions available to the government and how they work. Historically, the ability of a society to improve its well being has depended far more on its political system and economic management than on its level of natural resources.

The Most Successful Economy in the World

U.S. Economic Status.   In many ways the United States can be called the most successful economy in the history of the world. This generally applies in development of technology, improved living conditions for the poor as well as the rich, improved human health, and improved opportunities for social mobility. The success of the United States economy has been attributed to a political-economic system that emphasizes democracy, fair play, competition, opportunity, individualism, and capitalism. We will now focus on the United States in our discussion of economic systems.

People in the United States today enjoy a living standard that would have been almost unimaginable 100 years ago (IBD 1996; Figure 22.3). People on average live about 30 years longer, eat better food, and experience much less physical hardship. We have developed more material comforts and conveniences—such as pocket televisions, cellular phones, fax machines, laptop computers, and the Internet—in the last 100 years than in all of previously recorded history. Still, cures for many diseases (AIDS, cancer, Alzheimer's disease, and more) allude us and violent crime and drug use are serious problems throughout the country. Loss of open space, atmospheric and water pollution, and species extinction are new challenges that were of minor concern in the year 1900.

Economic Indicators.   Ten basic indicators are commonly used to access the economic health of the economy in the United States at any point in time. These include the following:

1. Gross domestic product
2. Unemployment rate

How Sweet It Is

	1970	1990
Average size of a new home	1,500	2,080
New homes with central a/c	34%	76%
People using computers	<100,000	75.9 mil.
Households with color TV	33.9%	96.1%
Households with cable TV	4 mil.	55 mil.
Households with VCRs	0	67 mil.
Households with two or more vehicles	29.3%	54%
Median household net worth (real)	$24,217	$48,887
Housing units lacking complete plumbing	6.9%	1.1%
Homes lacking a telephone	13%	5.2%
Households owning a microwave oven	<1%	78.8%
Heart transplant procedures	<10	2,125*
Average work week	37.1 hrs.	34.5 hrs.
Average daily time working in the home	3.9 hrs.	3.5 hrs.
Work time to buy gas for 100-mile trip	49 min.	31 min.*
Annual paid vacation and holidays	15.5 days	22.5 days
Number of people retired from work	13.3 mil.	25.3 mil.
Women in the workforce	31.5%	56.6%
Recreational boats owned	8.8 mil.	16 mil.
Manufacturers' shipments of RVs	30,300	226,500
Adult softball teams	29,000	188,000
Recreational golfers	11.2 mil.	27.8 mil.
Attendance at symphonies and orchestras	12.7 mil.	43.6 mil.
Americans finishing high school	51.9%	77.9%
Americans finishing four years of college	13.5%	24.4%
Employee benefits as a share of payroll	13.5%	24.4%
Life expectancy at birth (years)	70.8	75.4
Death rate by natural causes (per 100,000)	714.3	520.2

*Figures are for 1991

Source: Federal Reserve Bank of Dallas

FIGURE 22.3 A comparison of material wealth in the United State in 1970 and 1990 illustrates gains in the living standard of an average U.S. citizen (from IBD 1996).

3. Inflation rate

4. Interest rates

5. Consumer debt level

6. Consumer confidence

7. Stock market

8. Housing starts

9. Retail sales

10. Durable good orders

By any measure the United States is the world's largest economy. In 1996 it produced nearly $7 trillion of output, which accounts for nearly one-fourth of the world's entire output (IBD 1996). A recent historical overview of the United States economy is provided in Table 22.1. Gross domestic product (GDP) is used as the primary measure of economic growth in the United States and other parts of the world. It is an indicator of how much the average person would get if all output were divided evenly among the population. Historically, the United States economy as measured by GDP has grown at a rate of 3% per year. Because economic growth has exceeded population growth, GDP per capita is three times today what it was in 1900 (Figure 22.4). During the 1990s, the difference between GDP and population growth was 1.5%. Assuming it persists, per capita incomes will double in roughly 40–50 years. In 1996, the per capita GDP in the United States was about $26,000, which is more than five times the world average ($4,800).

TABLE 22.1   Overview of the United States Economy Since 1950

	Real GDP Growth	Unemployment	Consumer Price Index	Prime Interest Rate	Real Interest Rate	Discount Interest Rate	Return on S&P 500 Stock Index
				%			
1950	8.7	5.3	5.9	2.07	−3.8	1.59	22
1951	8.8	3.3	6.0	2.56	−3.4	1.75	16
1952	4.3	3.0	0.8	3.00	2.2	1.75	12
1953	3.7	2.9	0.7	3.17	2.5	1.99	−7
1954	−0.7	5.5	−0.7	3.05	2.4	1.60	45
1955	5.6	4.4	0.4	3.16	2.8	1.89	26
1956	2.0	4.1	3.0	3.77	0.8	2.77	3
1957	1.8	4.3	2.9	4.20	1.3	3.12	−14
1958	−0.5	6.8	1.8	3.83	2.0	2.15	38
1959	5.5	5.5	1.7	4.48	2.8	3.36	8
1960	2.2	5.5	1.4	4.82	3.4	3.53	−3
1961	2.1	6.7	0.7	4.50	3.8	3.00	23
1962	6.0	5.5	1.3	4.50	3.2	3.00	12
1963	4.3	5.7	1.6	4.50	2.9	3.23	19
1964	5.8	5.2	1.0	4.50	3.5	3.55	13
1965	6.4	4.5	1.9	4.54	2.6	4.04	9
1966	6.4	3.8	3.5	5.63	2.1	4.50	−13
1967	2.6	3.8	3.0	5.61	2.6	4.19	20
1968	4.7	3.6	4.7	6.30	1.6	5.16	8
1969	3.0	3.5	6.2	7.96	1.8	5.87	−11
1970	0	4.9	5.6	7.91	2.3	5.95	0
1971	3.3	5.9	3.3	5.72	2.4	4.88	11
1972	5.4	5.6	3.4	5.25	1.8	4.50	16
1973	5.7	4.9	8.7	8.03	−0.7	6.44	−17
1974	−0.4	5.6	12.3	10.81	−1.5	7.83	−30
1975	−0.6	8.5	6.9	7.86	1.0	6.25	32
1976	5.6	7.7	4.9	6.84	1.9	5.50	19
1977	4.9	7.1	6.7	6.83	0.1	5.46	−12
1978	5.0	6.1	9.0	9.06	0.1	7.46	1
1979	2.9	5.8	13.3	12.67	−0.6	10.28	12
1980	−0.3	7.1	12.5	15.27	2.8	11.77	26
1981	2.5	7.6	8.9	18.87	10.0	13.42	−10
1982	−2.1	9.7	3.8	14.86	11.1	11.02	15
1983	4.0	9.6	3.8	10.79	7.0	8.50	17
1984	7.0	7.5	3.9	12.04	8.1	8.80	1
1985	3.6	7.2	3.8	9.93	6.1	7.69	26
1986	3.1	7.0	1.1	8.83	7.7	6.33	15
1987	2.9	6.2	4.4	8.21	3.8	5.66	2
1988	3.8	5.5	4.6	9.32	4.7	6.20	12
1989	3.4	5.3	4.6	10.87	6.3	6.93	27
1990	1.2	5.5	5.4	10.01	4.6	6.98	−7
1991	−0.9	6.7	4.1	8.00	3.9	5.45	26
1992	2.7	7.4	2.9	6.00	3.4	3.25	4
1993	2.3	6.3	2.7	6.00	3.3	3.00	7
1994	3.5	5.8	2.7	7.20	4.5	3.88	−2
1995	2.3	5.6	2.5	8.83	6.3	5.00	34
1996	3.4	5.3	3.3	8.27	5.0	5.13	20
1997	3.9	4.6	1.7	8.44	6.7	5.00	31
1998	3.9	4.5	1.6	8.35	6.8	4.92	27
1999	4.3	4.2	2.6	8.12	5.5	4.75	20
2000	3.6	4.0	3.3	9.00	5.7	5.50	−10

Source: Adapted from Schiller 2000.

PROBLEMS WITH MARKET ECONOMIES: THE BUSINESS CYCLE

While the United States is considered to have developed the most successful economy in the history of the world, there have been several periods of severe hardship. Many of today's senior citizens still have vivid memories of the 1930s depression

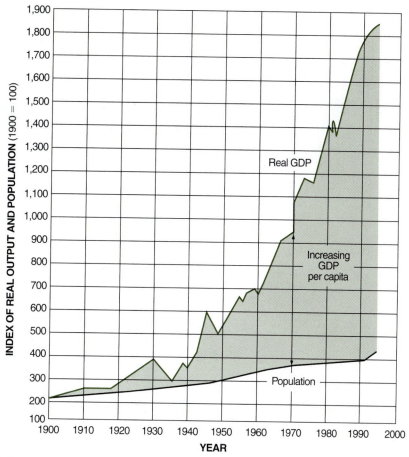

FIGURE 22.4 Output of goods and services and population growth since 1900 in the United States (U.S. Dept. of Labor, adapted from Schiller 1994, 2000).

when unemployment rates were over 20% and there was massive financial upheaval (bankruptcies) in the farming, banking, and industrial sectors (Schiller 1994). The Great Depression stands out for its longevity. Generally it is considered to have started with the stock market crash in October of 1929 and ended when the United States became involved in World War II in December 1941.

During the 1920s, the United States seemed to have achieved utopia through its market-oriented system and minimal regulation by government. The economy rapidly expanded for eight years in a row (1921–1929). A variety of new products and technologies such as radios, cars, movies, and various appliances were rapidly improving the human condition and jobs were plentiful. However, the stock market crash and depression in the 1930s caused a major rethinking of economics and government's role in managing the economy.

Stages of the Business Cycle

The rise and fall of gross domestic product that periodically occurs in market economies such as in the 1920s and 1930s is called the **business cycle** (Figure 22.5). The four basic parts of business cycles include recession, depression, recovery, and boom. When gross domestic product and employment are declining, the economy is said to be in recession. When this condition becomes severe, it is called a depression.

Each of the six basic stages of most business cycles favors different classes of assets (commodities, stocks, bonds, cash, real estate) (Stoken 1984, Pring 1992). During Stage 1 at the bottom of a slump, business becomes leaner and more productive by

FIGURE 22.5   Business cycles in U.S. history from 1930 to 1995 (from Schiller 1994).

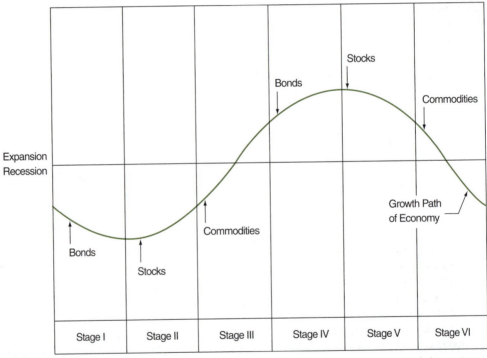

FIGURE 22.6   The relationship between the stages of the idealized business cycle and the various asset classes (from Pring 1992).

eliminating unprofitable operations and reducing labors costs (Figure 22.6). In this period, consumer demand is low because of concerns over debt, high unemployment, and high interest rates. Stocks, real estate, and commodities are depressed but high-quality bond prices are up.

Austerity leads to Stage 2, when capital accumulation and lack of credit demand pushes interest rates lower. This causes a mild increase in economic activity. Bankers are cautious because they have just gone through a period of bankruptcies and fore-closures. Therefore, only those businessmen with the highest credit ratings have access to capital. Financial assets, primarily stocks and bonds, do well in Stage 2, but prices of real estate and commodities remain depressed.

Stage 3 marks the beginning of the recovery in commodity and real asset prices. This is due to reduction in inventories and depletion of consumer goods. Bond prices tend to be flat, but stock prices increase because of improvements in corporate earnings. Real estate and commodity prices start to increase in this stage.

Stage 4 brings a high level of confidence about the future of the economy. Consumer and business spending causes interest rates to rise rapidly, which depresses bond prices. Growing inflation and easy credit cause real estate and commodity prices to shoot upwards.

Stage 5 brings the peak in the business cycle. Here optimism about the future has led to recklessness. Credit is too easy to obtain, which causes high inflation and encourages poor business decision making. Commodity prices and real estate peak in this period because of both real and speculative demand. Real demand results from workers experiencing increased wages and access to easy credit. They are in a position to upgrade their standard of living. Speculative demand results from inflation pressures that causes investors to shift into real assets as a hedge against devaluation of the currency. Money flows out of the stock market into short-term money market funds that provide high yields. Long-term bonds are in disfavor due to fears of increasing inflation.

Stage 6 is characterized by a crash in commodity and real estate prices and a general economic downturn due to an oversupply of goods financed by excessive debt. The prosperity of Stages 4 and 5 causes recklessness and overoptimism in bankers, producers, and consumers. The only way the boom can be sustained is with excessively loose credit. If the Federal Reserve maintains the discount rate (cost of money to banks) below the inflation rate, the money supply increases at a more rapid rate than the expansion of the economy. This occurred during the 1970s (Table 22.1). Historically, overly loose credit has always caused devaluation of a nation's currency and collapse of its bond market (Davidson and Rees-Mogg 1993). Debtors are always favored over creditors when the government takes the inflationary approach by making the real cost of money negative (prime interest rate minus consumer price index). Investment turns to speculation in real estate, gold, precious metals, chinese ceramics, and other scarce commodities as a hedge against currency devaluation, rather than into creation of real wealth through product development and improved production efficiency.

To contain inflation the Federal Reserve can raise the discount rate well above the inflation rate (typically measured by the consumer price index). This forces bankers to withhold credit, which in turn slows product demand. Commodity and real estate prices fall in response to tighter credit and an oversupply of goods. Falling prices are accentuated by bankruptcies of heavily indebted businesses and consumers that now meet their financial obligations with lower collateral (falling real estate) and less income (lower wages, lower employment levels).

Depressions in the United States

Severe or great depressions involving natural resources have occurred in the United States in the 1780s, 1830s, 1870s, and 1930s. These will be discussed following Batra (1988) and Schultz (1972).

1780s Depression.

A severe economic downturn occurred after the American revolutionary army—commanded by George Washington—defeated the British in Yorktown in October 1781. The situation intensified with the signing of the peace treaty in Paris in 1783 that formally ended the Revolutionary War. This depression was caused mainly by a huge deficit in the balance of trade (Batra 1988). After the Revolutionary War, an influx of British manufactured goods into the United States was caused by heavy demand for reconstruction of the newly free nation. Prices of American goods (primarily food and fiber resources) were depressed because of protectionist policies by Britain and France and expanded domestic supplies in America due to the ending of the war. This problem gradually subsided, starting in 1788, because Americans were given improved access to European markets created by wars (e.g., the French Revolution) that reduced Europe's farm production and increased its demand for American exports. The depression of the 1780s nearly destroyed the newly formed government in the United States. The leadership of President George Washington was critical in holding the country together.

1830s Depression.   During the 1830s, cotton production boomed in the southeastern United States following the invention of new textile manufacturing technology in England. High cotton prices led to massive land speculation. In the later stages of this boom in 1836, large amounts of land were bought with the sole idea of selling it at a higher price. The banks, which were all private and unregulated at the time, fueled the speculation by printing currency notes that were not backed by silver or gold. To control the speculation, President Andrew Jackson put into effect policies requiring all loans in currency notes (paper money) to be backed by hard money (gold and silver). At the same time, the cotton supply from the United States caught up with cotton demand by England. The combination of credit shrinkage and cotton demand cutbacks caused massive insolvency among the land speculators, cotton producers, and several large banks. European financing stopped, causing massive bankruptcy throughout the South. Eventually, financial chaos spread to the entire country. The Mexican War of 1845 finally ended this depression.

1870s Depression.   The worst depression of the 1800s began in 1873 and lasted until 1879. It was brought about by unrestrained speculation in railroad construction and a winding down of post–Civil War reconstruction. The root cause was an overheated demand for new resource development and transport to markets. It spurred overexpansion of the money supply and credit by the federal government during and after the Civil War. By 1873, total credit was six times greater than total cash. In September 1873, it became apparent that earnings from all of the bad investments, particularly in the railroads, could not support the debt levels. Stocks on the New York Exchange began a precipitous crash. In 1874, massive bankruptcy resulted in the railroads and connected industries such as steel rolling mills, machine shops, and foundries. Unemployment was widespread. Hunger and homelessness were beyond the relief capabilities of the charities and churches. Ultimately the farm sector was impacted because of falling food and clothing prices resulting from shrinking demand.

1930s Depression.   The depression of the 1930s was the most severe in the history of the United States in terms of length (1929–1941) and depth (18–25% unemployment). The causes of the 1930s depression are more complex than those in the early history of the country. During the 1920s, capitalism in the United States was more restrained than at the turn of the century. President Theodore Roosevelt had put into effect powerful antitrust legislation to break up the monopolies in the railroad, oil, and steel industries. This resulted in more competition and caused wealth to be less concentrated in a few hands. Banking legislation was passed to set up the Federal Reserve Banking System in 1913, which supposedly would restrain speculation. Under this system, the government restricts the amount of money loaned by private banks. They must keep a portion of their deposits in reserve with a central bank controlled by the federal government (Federal Reserve Bank). The nation's money supply and access to credit can supposedly be controlled by the Federal Reserve Bank through changes in the interest rates the central bank charges member banks and through changes in reserve requirements for the member banks (Schiller 1994).

Nevertheless, the root cause of the 1930s depression was excessive credit extended in response to agressive business and consumer demand, which resulted in commodity overproduction and massive bankruptcy during the 1920s. During the 1920s there was a flood of new products such as automobiles and home appliances. At the same time the housing industry boomed. Purchases of homes, automobiles, and appliances were based on easy access to credit. This was particularly true during the latter part of the expansion after 1926. Innovations in debt finance such as installment loan purchases and new home mortgages were used to extend the expansion phase of the business cycle beyond its rational limits. Surpluses of raw resources and finished goods—such as oil, tin, rubber, copper, steel, wheat, corn, leather, houses, and autos—began to pile up as consumer demands for new commodities were satisfied. Employers had little choice but to lay off workers, who also were consumers in debt. Basically by 1929, the limits to credit expansion had been reached. Consumers had gone deeply into debt to buy material goods while producers had gone into debt to expand their supply. Without employer and employee income to pay off the debts, divesture and bankruptcy were unavoidable.

The depression progressed rapidly because of the way the stock market and trade were regulated by government. During the last half of the 1920s, a huge boom was underway in financial assets, primarily stocks. This was initially based on the growing profits for companies in the new industries, such as General Motors and Radio Corporation. However, as the expansion continued, the stock market became increasingly speculative. In other words, investors bought stocks based more on the idea that previous price increases would continue rather than on the basis of future earning prospects. At the same time, deterioration of investment conditions in Europe, following World War I, helped concentrate speculation in American financial assets. The problem of stock speculation was furthered by the banks and brokerage houses that required investors only maintain a 10% equity position in stocks they purchased. This practice is referred to as buying on margin.

By 1929, the economic expansion that started in 1922 on sound fundamentals had turned into a speculative stock market binge. The reckoning that caused investors to confront reality came from two events. The first occurred when the New York Federal Reserve Bank raised its discount rate from 5 to 6% to slow down stock market speculation. This caused all interest rates to move upward and thereby tightened access to credit. The second event was the announcement of the Smoot-Hawley Tariff Act.

The day the stock market collapsed—Black Thursday, October 24, 1929—was the day President Herbert Hoover announced he would sign the Smoot-Hawley Tariff Act. This act would increase effective tariff duties on imports by nearly 50%. Naturally, other countries retaliated by imposing barriers to imports from the United States. In other words, savvy investors understood Smoot-Hawley was bad for business domestically and abroad. It severely reduced foreign buying of U.S. goods; particularly farm products. Some economists believe passage of the Smoot-Hawley Act was the primary reason the worldwide depression of the 1930s became so severe (Davidson and Rees-Mogg 1987).

The stock market crash in October of 1929 started a process of credit shrinkage that became self-perpetuating. In 1932, the stock market had lost 95% of the 1929 peak value. As equities fell in value, banks had to convert securities into cash to meet customer withdrawals. This reduced their lending capability. Demand was reduced because of lower credit availability from the financial institutions and because common stocks had been the primary assets of many consumers. The fractional reserve banking established in 1913 turned the leverage of the banking system inside out. Bank failures caused massive withdrawals by consumers. Under deflationary conditions it was rational for consumers to postpone purchases. Because deflation raised real interest rates sharply any investment in productive capacity was a losing proposition. The best returns came from buying government bonds. In the 1930–32 period, the real returns from buying government bonds was 6 to 13%, which is quite high by historical standards.

Natural resource-based industries such as farming and energy (oil) were hit the hardest by the deflation of the 1930s. Part of the problem was caused by severe overexpansion in commodity production capacity prior to the depression and part was due to foreign retaliation in response to the Smoot-Hawley Act, which many farmers had so strongly supported.

Keynesian Economic Approach

The failure of market systems in the 1930s led to a new school of economic thinking based on the views of Sir John Maynard Keynes's, a British economist (Schiller 1994). From Keynes's point of view, the depression of the 1930s was caused by a shrinkage of demand. He believed the best way to counteract this problem was for government to reduce taxes and spend large sums of money on public works and "safety net" programs to help the needy. It was thought this approach would stimulate the economy and reduce social unrest. Once business activity resumed and the economy was growing at a healthy pace, the government could keep the economy from "overheating" by reducing its spending and/or raising taxes.

The ideas of John Maynard Keynes were applied by President Roosevelt in the mid-1930s when a variety of federal government public works, safety net, and conservation programs were put into effect. At the same time, Social Security was initiated to provide retirement benefits to the elderly. These programs accelerated the United States' movement away from an unregulated market economy to a more mixed economy.

Although it is generally agreed that Keynesian programs implemented by President Roosevelt in the mid-1930s helped alleviate the hardship of the depression, it was the United States entry into World War II that really ended it. World War II set up the United States for a long period of prosperity that lasted into the early 1970s. During World War II the production capacity of most of the world was destroyed or severely impaired, but the United States was completely untouched. By the end of the war, the United States had the best factories and technology in the world and there was huge pent-up demand both within and outside of the country. Consumer debt had been nearly eliminated by the inflation and forced savings that resulted from the wartime economy. New efficient technologies had largely replaced those that were old and obsolete. Although wars cause great human suffering and hardship, they also can bring economic renewal and revival.

From 1950 until 1973 only two mild recessions occurred, in 1954 and 1958 (Table 22.1). During this period, the United States experienced strong economic growth, low unemployment, and price stability. The Keynesian economic policies being used by the government appeared to have finally tamed the business cycle. However, the Korean War (1950–1953) and the Vietnam War (1965–1972) both provided strong stimulation when the U.S. economy might have otherwise faltered.

Stagflation in the 1970s. In the early 1970s, a new combination of factors came together that severely dampened the previous prosperity. Rather than depression, the outcome of these factors was a condition called "stagflation." It involved rising prices, rising unemployment, and a low level of economic growth (Schiller 1994). During the 1970s, demand boomed for basic natural resources such as oil, gold, silver, and land. Farm product prices, particularly for grain, made large increases. This was a decade when scarcity, pollution, and human overpopulation were considered major threats to the future of mankind (Ehrlich 1968, Meadows et al. 1972, Brown 1983). Much of contemporary thought on natural resource management came out of the 1970s. We will make the case that the economic stagflation of the 1970s was caused more by economic failure than by natural resource exhaustion. Davidson and Rees-Mogg (1987), Schiller (1994), and IBD (1996) are primary sources of information pertaining to this issue.

The basic cause of the stagflation of the 1970s centered around the difficulties that the U.S. government encountered in the 1960s' funding of the Vietnam War and the "War on Poverty" under Lyndon Johnson's presidency (1963–1968). Rather than raise taxes to pay for the Vietnam War and the "great society" social programs, the government under Presidents Johnson and Nixon basically chose to fund them through printing the money. At the same time, a variety of environmental laws were passed that raised the production costs for many basic industries. During this same period, Japan and West Germany emerged as strong competitors with the United States in production of automobiles, electronics, and appliances. By the 1960s, the Japanese and Germans had replaced factories destroyed during World War II with those using the most modern innovations. In contrast, much of the industrial capacity of the United States depended on factories that had minor upgrading since the 1940s. Therefore, the United States increasingly was forced to give up world market share of several manufacturing industries to Japan and West Germany and became more dependent on agricultural exports.

The expansion of the money supply relative to GDP created pressure for U.S. currency (dollars) to be devalued against other currencies, particularly the German mark and Japanese yen. Based on an agreement reached in 1944 (Bretton Woods Agreement), the International Monetary Fund (IMF) was established. The IMF fixed all exchange rates in terms of the dollar and the dollar in terms of gold. The dollar became the world's principal reserve currency and the only currency that could be directly redeemed in gold. This allowed the U.S. government to increase the supply of dollars to cover up to a point its excessive spending during the 1960s. That point was reached in 1971 when the United States could no longer meet the demands for redemption of dollars with gold. In August of 1971, President Richard Nixon took the dollar off the gold standard, which set the stage for floating exchange rates (devaluation of the dollar against other currencies). This set the stage for the series of events that caused the "stagflation" of the 1970s and early 1980s.

The devaluation of the dollar caused U.S. exports—primarily farm products—to become more competitive on world markets while at the same time its primary import—oil—became much more expensive. In order to protect themselves against the declining purchasing power of the dollar, the mideastern oil producing nations formed a cartel (OPEC) and demanded higher oil prices. At the same time climatic adversity in China, India, and Russia in the early 1970s depressed world grain production and greatly elevated U.S. grain exports.

The combination of increased world competition, dollar devaluation, oil supply shocks, world climatic adversity, loose monetary policy by the Federal Reserve, and increased regulation of U.S. industry during the 1970s all explain the high levels of inflation, high unemployment, and reduced level of economic growth compared with the 1940s, 1950s, and 1960s.

In contrast to a depression, under inflationary conditions natural resource-based industries such as farming, mining, energy extraction, and forest products flourish. Gold went from $35 an ounce in 1971 to over $800 an ounce at the high in early 1980. Before the dollar's devaluation in 1971, crude oil fluctuated around $2 to $3 per barrel. By the early 1980s, it shot up to $35 per barrel. Wheat, corn, soybeans, cotton, and cattle are some of the farm commodities that showed large price runups during the inflation of the 1970s.

One factor often overlooked as a cause of the natural resource supply shocks of the 1970s was world communism (Davidson and Rees-Mogg 1987). This was the decade when Marxist centrally planned governments were at their peak on a global basis. In China and the former Soviet Union, both agricultural and oil production were severely depressed because of repressive, inept governments that discouraged human initiative and relied on antiquated production technology and equipment. Without question, the extreme inefficiency in use of natural resources and human labor by the communistic countries in the 1970s contributed to the natural resource supply and demand imbalances of the era.

The Economy of the 1980s and 1990s

In 1980, Ronald Reagan—a Republican—was elected president of the United States. He had many very different ideas on how to run the American economy compared with either his Democratic (Jimmy Carter) or Republican (Richard Nixon, Gerald Ford) predecessors of the 1970s. The Reagan strategy centered around restraining the money supply to control inflation, deregulation, lower taxes, encourage immigration, and expand military spending. This mix of programs has been referred to as "supply side" economics. The most drastic element of his program was using the Federal Reserve Bank to restrain inflation by elevating the discount rate (Table 22.1). Throughout the 1980s, this policy of high real interest rates kept the dollar strong and contained inflation. It caused money to flow into financial assets (stocks and bonds) and out of natural resource assets such as gold, farmland, and oil, and other commodities that are inflation hedges.

The strong dollar caused U.S. farm products to be expensive relative to those of other countries in world markets such as Canada, Argentina, and Australia. During the 1980s and 1990s world climatic conditions were quite favorable for food production. In this same period new technology and market-oriented reforms in China, India, Argentina, and several other countries resulted in a great boost in world agricultural production. Deregulation of the oil industry in the United States in conjunction with improved efficiency in energy use caused precipitous drops in oil prices. Between 1980 and 1983, oil dropped from $35 to $9 a barrel. Overproduction was so severe in the farming sector (see Chapter 7) that farmers were given surplus grain if they agreed not to plant part of their land in 1983. In the Food Security Act of 1985, a major provision was to pay farmers to retire their more erodible lands under the Conservation Reserve Program. Although this program had important conservation benefits, its main objective was to reduce the oversupply of wheat, corn, and other farm products. This program has been continued under the 1990 and 1995 farm bills (see Chapters 8 and 12).

Most of the economic policies of the Reagan administration were retained by President George Bush (1989–1992) and President Bill Clinton (1993–2000). So far these policies (monetary and regulatory restraint, free trade, loose immigration laws) have been effective in prolonging economic growth, minimizing unemployment, and keeping prices stable. A major criticism of both the Reagan and Bush administrations was the high-level deficit spending. The major departures of the Clinton administration from policies of Reagan and Bush were fiscal policy restraint (reduced federal budget deficits) and some modest increases in taxes on business and capital gains. However, capital gains taxes were lowered by Congress in 1997. All three presidents have been effective in creating conditions that encourage technological development and application.

Currently, the biggest economic concerns are the high debt levels of American consumers, local governments, and the federal government, and a demographic shift toward a higher proportion of older people in the population. Many economists have found great similarities between the 1990s and the 1920s (Batra 1988, Davidson and Rees-Mogg 1987). In both periods, taxes were relatively low on the rich, technological growth was rapid, unskilled workers had little improvement in their wages, agricultural prices were depressed, there was a large runup in the stock market, and consumers went heavily into debt to buy houses and new products.

It is our point of view that macroeconomic policies applied in the United States during the 1980s and 1990s have indeed moderated the excesses associated with business cycles of the past. We believe that the global trend toward accelerated technological advance, market economies, and free trade—at least in the near term—will prevent another severe depression such as in the 1930s. However, we also believe recessions and short-term depressions will always be a part of market economies. Further we feel they serve a useful function that is often overlooked. This will be discussed.

CREATIVE DESTRUCTION AND HUMAN PROGRESS

Any serious study of the origin of the U.S. Constitution shows its authors believed that cycles of growth and decay were necessary to maintain a nation's vitality and moral fabric (IBD 1996). A basic premise was that periods of hard times tend to make people strong and responsible while waste and corruption are inevitable after a long

FIGURE 22.7 Joseph Schumpeter, an early twentieth-century U.S. economist, saw capitalism as a dynamic system beset by periodic crises he termed "waves of creative destruction" (from IBD, The Bettmann Archive, 1996).

run of prosperity. Fundamental parts of early American ideology emphasized freedom to fail, as well as to succeed, and strong belief in the ability of the individual rather than the state.

From its beginning anyone who developed new products in the United States could easily market them without interference from the government (IBD 1996). In contrast, innovation under the totalitarian governments that had existed in Europe since the Dark Ages often resulted in punishment because it was perceived to cause political instability. As an example, in 1579 the German inventor of a new technique for weaving ribbons was ordered strangled because it threatened the economic security of the ruling class.

New technology, when applied, causes old technology to be discarded (IBD 1996). Under democratic market economies this results in wealth constantly being transferred among individuals based on their contribution rather than their heredity. It means at anytime that part of the population contributing the most desired products will experience disproportional prosperity to those whose products have become obsolete. In the United States, the buggy whip and carriage industries were allowed to disappear at the turn of the century when automobiles replaced the horse as the primary means of transportation. The electric light made life easier for people but it destroyed the oil lamp industry.

In the 1920s, the economist Joseph Schumpeter, used the term "creative destruction" to characterize the wave of product cycles and innovation that occur under democratic market economies (Figure 22.7). Schumpeter argued that innovations tended to appear in clusters facilitated by entrepreneurs (Davidson and Rees-Mogg 1987). The new businesses increase the demand for the means of production and labor. This impacts the old businesses by causing their costs to rise and shifting demand away from their products. As consumer preference shifts toward the new products, the older businesses are forced into consolidation and/or insolvency. The profits from the new businesses encourage competition that leads to improvement in the production process. Soon the lure of easy profits causes an overexpansion of the new businesses. When consumers' need for the new products are satisfied and business has overbuilt production capacity, depression sets in to level the excess and encourage the innovations that lead to the next recovery.

From Schumpeter's perspective, human progress depended heavily on permitting business cycles to work with minimal interference from government. He understood that established business had a vested interest in keeping new products and services off the market. Hence the entrepreneur was an essential factor in overcoming the various roadblocks thrown up by these vested interests. To Schumpeter, stabilized market systems as envisioned by today's economists were a contradiction. The great progress that results from waves of "creative destruction" is only possible under a chaotic, uncontrolled economic system that allows the new to replace the old.

PROBLEMS WITH CENTRALLY PLANNED ECONOMIES

We believe our economic discussion would be incomplete without some analysis of why communism failed in the former Soviet Union. This was true to a lesser extent in China under Mao Tse Tung. After Mao's death in 1976, a gradual process of reform was initiated that has made China's transition to a mixed economy much smoother than in Russia.

Centrally planned economies, such as the former Soviet Union, promised equitable distribution of income and full employment. However, the actual reality of centrally planned economies was that the higher-level bureaucrats enjoyed lavish lifestyles while the common citizen lived in severe poverty when compared to the United States and most other market-oriented countries (Schiller 1994). While nearly everyone in the former Soviet Union held a job, they often held unneeded jobs and/or were highly dissatisfied with their occupation. Although the former Soviet Union could keep prices for basic goods stable and at low levels relative to incomes, most citizens shopped in near empty stores and stood in line for hours to buy bread or shoes, or other necessities. Low-quality goods, scarcity of basic goods such as soap, sugar, and hamburger, and poor systems for distribution of goods have been serious problems for all centrally planned economies (Schiller 1994). Since the 1970s, the poorest people in the United States have had far better health care, food, transportation, and housing than the middle class in the centrally planned economies of the world.

The basic problems with centrally planned economies center around human nature and the difficulties that occur when the government tries to control all aspects of the production/distribution process (Schiller 1994). The hard reality is that what works in theory often does not work in practice.

The idea of everyone contributing what he can based on his abilities but taking only what he needs is at the core of the communist viewpoint (Schiller 1994). However, the problem centers around human motivation. Much more training and mental discipline are required to attain the skills of a rocket scientist or biochemist than to be a janitor or a farm field worker. Without the incentives of higher pay and greater prestige, few people want to go through the rigorous four to six years of training required to be an engineer, doctor, or lawyer. Another communist failure centered around lack of tangible reward for good work habits and willingness to take risk.

Satter (1989) provides an excellent analysis of agricultural failure in the former Soviet Union. Under the Czar in 1900 before the communist revolution, Russia had became the largest grain exporter in the world. By 1989, after 62 years of central planning, Russia was the largest grain importer in the world. It had more land devoted to cereal production than any other country and roughly 25% of its workforce was in agriculture compared with 3% in the United States. Yet Russia's grain output per acre was little more than half the production of the United States.

In the former communistic Soviet Union, all that mattered was the production plan. Regional officials told the chairman of each collective farm what to plant and how much. Every aspect of farming was to be directed rigidly like a military operation without the individual farmer making a single decision. Although the Soviet government sank nearly $1 trillion into agriculture between 1971 and 1985, there was practically no increase in production.

Under the Soviet system, farmers were often forced to plant inappropriate crops and comply with rigid deadlines for sowing and harvesting. When the plan said it was time to plow, attempts were made to plow even when the ground was so wet that the tractors became stuck in mud. When the plan said time to harvest, crops were cut—ready or not.

Satter (1989) reported that in every place he visited in Russia, farmers were doing the absolute minimum. They fulfilled the plan for sowing, but ignored it for fertilization because that was harder to monitor. Instead of spreading manure, they burned it so no one would know it was not used. Chemical fertilizer was commonly dumped in the streams.

In characterizing the farmworker, Satter (1989) described a pattern of perpetual idleness and alcoholism. Perhaps most interesting was the work pattern of tractor drivers paid by the acre to plow fields. Near the roads they would cut furrows 9 inches deep, but as soon as they were far enough into the field to avoid detection by the farm chairman they would lift the plow, race the engine, and cut 2-inch furrows with harmful consequences to the crop. Although nearly everyone seemed to make some attempt at working in the morning, by afternoon it was common to observe groups of people standing around smoking and talking.

Satter (1989) found that crops that did get harvested in Russia ran a gauntlet of barriers before reaching consumers. These included a lack of trucks to haul crops to the cities, primitive roads that were mostly unpaved, and a lack of suitable storage facilities. Because of poor storage facilities, nearly a third of the harvested grain, fruit, and vegetable production was lost to spoilage.

Prior to its disintegration in 1989, the former Soviet Union made a token attempt to deal with its worsening food situation by allowing farmers to work small tracts on their own. The problem with this scheme was that it operated within the existing collective-farm system that controlled access to equipment, feed, fertilizer, and seed. Satter (1989) describes how one small farmer was able to prosper under the new system of partial privatization but at the same time suffered because of the disapproval of his neighbors. He had to contend with verbal abuse, intimidation, and even physical assaults from a society that had been brainwashed for years to equate personal ambition with greed and prosperity with profiteering.

Environmental Problems in the Former Soviet Union

Environmental pollution has been a major criticism of market-oriented economies. However, major reductions in air and water pollution levels have been achieved in the United States over the past 25 years as a result of government regulations, improved production technologies, and tax incentives (see Chapter 4). In contrast, pollution problems in most communistic countries—particularly Russia and China—increased in the same period (Chandler 1987, French 1991). One part of the problem centers around failure of the centrally planned economies to replace old inefficient production processes and equipment with new and improved ones. Another problem is the lack of accountability of any organization or individual for environmental damage.

By the 1980s throughout Russia and China, it was nearly impossible to find a major river that was unpolluted (French 1991). Water use in steel mills was 50 to 150% higher than in the United States. Sulfur dioxide and nitrogen oxide emissions approached lethal levels in some areas. Toxic emissions of lead and fluoride were permitted in quantities hazardous to human health.

One outcome of environmental decline in Russia has been a shortened average human life span compared with market-oriented economies such as the United

TABLE 22.2   Human Life Expectancy at Birth for Selected Countries

	Life Expectancy At Birth	
Noncommunist Countries	1970–1975	1990–1995
United States	71.3	75.9
Great Britain	72.0	76.2
France	72.4	76.9
Spain	72.9	77.6
Canada	73.1	77.4
Mexico	62.9	70.3
Brazil	59.8	66.2
Japan	73.3	78.7
Sweden	74.7	77.9
Communist Countries		
Russia	X	70.0
China	63.2	70.9
Cuba	70.9	75.7
North Korea	61.5	70.8
Vietnam	50.3	63.8

Source: United States Government Population Division.

States, Great Britain, or Canada (Table 22.2). Reductions in infant mortality in the former Soviet Union have been much lower than those in the United States, Canada, Japan, and the Western European countries over the past 25 years (World Resources Institute 1994).

The Collapse of Communism

Because of the various problems previously discussed, the economic collapse of communism was certain by the late 1980s. Just prior to the disintegration of the Soviet Union in 1989, the average Soviet citizen had no telephone, no car, few appliances, and crowded housing (Schiller 1994). Basic necessities—such as soap, sugar, salt, and meat—were rationed and often unavailable. Even the Soviet government categorized 40% of its human population as poor. This forced the realization that communism had been a failure and resulted in the new commonwealth of independent states replacing the former Soviet Union.

Today all the various countries in eastern Europe that comprised the former Soviet Union are restructuring and applying various degrees of market reforms (Schiller 1994). The transition from communistic rule to mixed or market economies has not been smooth. Political, judicial, and financial changes are accompanying the economic changes. Converting homes, industries, and land into private ownership has been difficult. The lack of management skills needed to convert to and run profitable private enterprises is a serious basic problem. A major impediment in Russia has been the lack of an institutional structure to facilitate market transactions and private property rights (Schiller 1994). The clumsy conversion of inefficient state-owned enterprises into productive, profitable operations has caused massive unemployment and factory shutdowns. Implementation of a fair and workable tax system has been another daunting challenge. This in part contributed to the near bankruptcy of the government in 1998. Failure to privatize many large, unprofitable state-owned enterprises continues to drain government resources.

In spite of these setbacks in Russia, market reforms are gaining ground. Russia has retained its new democratic form of government, and the 1996 election showed the majority want democratic market reforms to continue and do not want to return to the communistic system. If the process of "creative destruction" can go on for another 5 or 10 years without political and social upheaval, many economists believe

Russia and most of its former states will begin to experience economic growth and prosperity. Because of its vast oil, mineral, forest, and farmland resources, whatever happens in Russia will be critical to the economic welfare of the rest of the world in the twenty-first century.

PROBLEMS WITH MIXED ECONOMIES

Sweden, Austria, Denmark, the Netherlands, France, and Germany are examples of the mixed economies in Europe that provide a stronger safety net (unemployment, health care, retirement, education, benefits) than the United States. In order to pay for these benefits, the average worker is taxed at a rate between 45% and 65%, depending on country, compared with 36% in the United States (Stein 1993). The United Kingdom (Great Britain), which is the most market oriented of the European economies, has an average tax rate of 39%.

The primary problems with the mixed economies in Europe have been a low level of economic growth, low growth in jobs, high levels of unemployment, and more rapidly rising government debt levels than in the United States (Stein 1993). GDP growth has averaged about 1.5% for European mixed economies over the past 10 years compared with 2.6% for the United States. Unemployment across the mixed economies of Europe has averaged 7–12% compared with 6% in the United States. The United Kingdom, with the lowest tax rate of the European economies, has had the best economic growth rate (about 2.0%) but the 10-year unemployment rate is near 8%.

The generous safety nets and higher taxation levels in the European mixed economies have reduced the incentives to work. Unemployed workers in western Europe can collect 40 to 65% of their salary in government compensation (Stein 1993).

Perhaps most alarming is the trend toward increasing levels of unemployment in the European mixed economies. Unemployment data from the European Union show an upward drift from 7–8% in the 1980s to 10–12% in the 1990s (Stein 1993). Apparently the generous unemployment benefits reduce incentives to find work once workers are laid off. Long-term unemployment causes them to gradually lose their skills and work ethic. This makes them less employable. Studies indicate employers prefer to hire workers with the least amount of time unemployed.

The conservative viewpoint is that too much economic security will destroy American character (Davidson and Rees-Mogg 1993, Oliver 1994a, b). Individualism has been a key element in the economic success of the U.S. Basically, individuals take responsibilities for their actions with little dependency on the state. They expect to fully enjoy the rewards from their thrift and enterprise, but at the same time they accept personal accountability for reckless risk taking and hard luck.

In contrast the egalitarian, or liberal, viewpoint centers around the idea that the state should be responsible for the welfare of its citizens and should equalize to some extent good fortunes of some with the misfortunes of others (Oliver 1994a, b). In other words, the state is expected to protect people from the consequences of their own decisions.

Another concern with egalitarian or socialistic type of economic system is that taxes needed to pay for public security reduce the pool of capital (savings) available for technological development and enterprise expansion (risk taking) in private enterprises (Davidson and Rees-Mogg 1993). The high debt levels and low economic growth rates in the mixed European economies to some extent support this concern.

Defenders of the present system in the United States argue that provision of opportunity to the disadvantaged is important in maintaining social stability and vitality in any society. Human productivity is likely to be higher if they can take limited risks knowing there is some degree of government support if they fail.

The degree to which success should be taxed and misfortune should be subsidized remains a controversial question. Current thinking by some leading market economists in America, such as Milton Friedman and Gary Becker, is that both taxes and the safety net should be reduced. People should have the option of determining their own degree of economic security. Those wanting more security would live a more austere life, have higher savings, and more catastrophic insurance. Under this approach taxation levels reduced to somewhere between 15 and 25% of average income would provide essential government services such as law enforcement, defense, education, and infrastructure with some help to the poor in terms of food and shelter. Charities and churches would assume a bigger role in helping the needy as was the case during the 1800s and early 1900s. Under this approach, the safety net would be similar to that during the early 1960s before President Lyndon Johnson's "war on poverty."

Presently, Congress appears to be moving in this direction with welfare reform. However, basic dedication to changes in government social security and health care programs rules out tax reductions much below the present 36% level in the foreseeable future.

IMPORTANCE OF INTERNATIONAL TRADE AND COMPETITION

International Trade

We consider removal of trade barriers to be one of the most critical factors in improving global human welfare. This is because world natural resources are unevenly distributed. Countries such as the United States and Canada that are well suited for wheat, cattle, and timber production have little capacity to produce coffee, bananas, cocoa, or rubber. Russia and other countries in the former Soviet Union have vast reserves of minerals, timber, and fossil fuels, but lack the technologies to extract and transport these resources to factories for conversion into usable products. Densely populated countries—such as Japan, Germany, and Great Britain—have low levels of mineral, timber, and energy resources, but have high levels of technological and capital resources. Through free-trade capital, management, and labor resources can move to where they will be most efficiently deployed. Different countries gain access to products and services they would not have if each country tried to be self-sufficient. At the same time, the competition that results from free trade creates strong incentives for industries to use the best production processes and locate where production costs will be lowest. Coffee, aluminum, chrome, tin, tea, and rubber are just a few of the natural resource–based commodities that we do not have in the United States but are a critical part of our welfare (Schiller 1994).

Another important concept in free trade centers around the benefits of specialization (Schiller 1994). It is much more efficient for a country or an individual to produce a few things and let the marketplace supply the rest. A key concept of Adam Smith's classic work *The Wealth of Nations* is that the organization of labor to produce different things in mass quantities is the key to rapid human progress. Specialization allows different companies and countries to produce different kinds of computers, automobiles, telephones, televisions, and more.

However, the question that comes up is why should a country import things that it also exports. The answer centers around product differentiation and efficiency. Consumers have access to more options and variety if they can buy either Japanese or U.S. cars. This also applies to television sets, computers, cameras, food, and other commodities. It is also true that each country is capable of producing all of these goods, but the quantity that can be most efficiently produced will vary considerably between countries.

Schiller (1994) provides an example using bread and wine, which are both produced in the United States and France. He demonstrates that France—which is best at producing wine—and the United States—which is best at producing bread—both benefit by engaging in trade. Although each country will continue to produce both bread and wine, under international trade their outputs of these products will be adjusted so each country benefits. Schiller (1994) makes the point that, by engaging in trade of goods both countries can produce, consumers in each country benefit by getting more of both products at lower cost.

Comparative advantage refers to the ability of a country to produce specific goods at a lower cost than another country (Schiller 1994). The United States has the most land highly suitable for wheat farming, but France has the most land well suited for growing grapes. By growing crops that are most suited to land type and engaging in trade, each country does better than if each used poorly suited lands to grow certain crops because they do not engage in trade (e.g., land well suited for grapes is not necessarily a good place to grow wheat). French wine and bread do not taste exactly the same as American wine and bread. Consumers in both countries undoubtedly enjoy the variety that comes from international trade.

International Competition

By engaging in international trade, countries force businesses in their countries to become more efficient and responsive to consumer needs. The automobile industries in the United States and Japan provide an example of how consumers can benefit from international trade and competition. Until the late 1960s, three U.S. companies (General Motors, Ford, and Chrysler) built most of the cars sold in North America. However, during the 1970s, the type of car demanded by U.S. citizens changed from large, heavy, and fuel inefficient to the opposite. This was due to increased gas prices, a flattening of consumer income growth, and the desire for cars that were more environmentally friendly. The small, compact Japanese cars met this requirement. During the 1970s, the American automobile companies lost more than a third of their market share to the Japanese. However, during the 1980s and 1990s, the American companies made major improvements in their cars partly in response to Japanese competition. The outcome of international trade in the automobile industry is that American consumers have access to cheaper, better made, more fuel-efficient cars that come in a wider variety of sizes and styles and last longer than they did 30 years ago.

International Trade Policies. Without question, free trade results in increased consumption and lower prices. However, individual companies and corporations that have inferior products apply constant political pressure for trade restrictions in the forms of quotas, tariffs, embargos, and other barriers. As we have previously pointed out, the Smoot-Hawley Tariff Act of 1930 contributed much to the Great Depression. The Smoot-Hawley Act raised tariffs to an average 60% and eliminated most imports (Schiller 1994). This act was designed to raise domestic employment and demand for goods. However, it transferred the unemployment problem to other countries and made them less able to purchase U.S. goods. This hardship caused other countries to retaliate with trade restrictions of their own.

U.S. imports from other countries fell 29% while exports fell 33% in the 1930 to 1931 period (Schiller 1994). Worldwide trade declined about 30%, causing massive unemployment and economic retrenchment. Since 1934, the United States and most other countries have pursued policies geared toward reducing trade barriers.

Since World War II, broad trade agreements that apply across industries and countries have been the objective of world trade negotiations (Schiller 1994). The General Agreement on Tariffs and Trade (GATT) was signed by 23 of the world's largest trading countries in 1947. Its objective is to pursue free trade and market access for all of its members (now 107 nations). Initially GATT focused on manufactured goods, but more recently agricultural products, music, computer software, movies, and copyrighted books have been included. Although many trade barriers still exist, GATT has resulted in definite progress toward free trade.

The North American Free Trade Agreement (NAFTA), signed in 1992, involves trade agreements among the United States, Canada, and Mexico. Its goal is the elimination of all trade barriers among these countries within 15 years.

environmental stability, and genetic information. Because of incomplete under-standing of human impacts on natural ecosystem functions, future improvement in human welfare is being increasingly jeopardized by environmental degradation. The world's natural heritage teeters at the edge of rapid and irretrievable loss.

Because of unprecedented human population numbers coupled with awesome technology, humankind is impacting the world on a scale far surpassing all previous centuries combined. Nowhere is this more clear at a local level than when rural land suitable for forestry, range, outdoor recreation, or wilderness biodiversity is con-verted to urban use. In the global view, climate change, biodiversity loss, and re-source sustainability have become the major environmental concerns (Lubchenco et al. 1991).

While technology advances have done much to diffuse concerns about nonre-newable resources, the greatest concerns are now more broadly based in sustaining renewable resources for the services they provide. Trends in technological advance promise a brighter future for human welfare (e.g., Naisbitt and Aburdene 1990), but the future of humanity will also depend on understanding of ecological, economic, and political systems. The natural resource managers of the twenty-first century must learn from past error and build on past success to meet future challenges.

The world is now a much smaller place than it was just a century ago. What once took months and even years to accomplish in communication and travel now takes nanoseconds to hours. No significant national economy can stand completely inde-pendent. Economic depression in Southeast Asia, turmoil in Russia, instability in Brazil, and consolidation in European currencies make headline news in the United States because of their potential implications for the world economy.

Through collective global output, carbon dioxide concentrations have increased over the last century to the highest levels in 160,000 years, based on studies of gla-cial ice cores. The extent to which the United States may be pressured through in-ternational politics to cut back emissions of greenhouse gasses is closely related to rain forest management in South America and industrialization in Asia and other parts of the world. The more forest, peat bogs and other natural carbon sinks are con-verted to carbon dioxide, the more fossil fuel use will need to be slowed, if stabi-lization of greenhouse emissions becomes a dedicated international policy objective.

The rate of biodiversity loss is unprecedented in the history of humankind, espe-cially in underdeveloped nations. Poorer countries often can be coerced through trade sanctions by richer countries to protect endangered species or to reduce global carbon emissions. However, the morality of such policies is questionable. When ecosystems are protected through international law, it is often unclear who benefits, who bares the costs, and how to proceed fairly. Sound decision making will depend increasingly on the quality of information provided by natural resource ecologists, economists, policy specialists, planners, and managers working with each other and with other disciplines.

In this chapter, we describe new strategies involving a more integrated ecosys-tems approach to natural resource management, referred to as sustainable develop-ment. Then we explore the role of technology in natural resource management. In closing, we briefly consider some influential trends for the future.

SUSTAINABLE DEVELOPMENT

Defining Sustainable Development

Sustainable development involves integrating the management of ecological sus-tainability and economic development to obtain continuous improvement in human welfare on a worldwide basis. This global concept of sustainable development has rapidly gained the attention of natural resource policy analysts. It is likely to be one of the important concepts guiding resource management in the twenty-first century.

Literature referring to "sustainability" has proliferated during the past decade with many variations in its meaning. The concept continues to evolve as knowledge of possibilities and limit grow. Thought-provoking discussions on sustainable development are presented by Muschett (1997) and Daly (1997).

The President's Council on Sustainable Development (1996) published a report with goals and recommended actions oriented toward U.S. national development. The report encapsulated much of contemporary thinking about sustainable development. The council included a cross section of U.S. leaders representing diverse governmental, business, environmental, civil rights, labor, and cultural interests. Their report identified many of the important natural resource management issues in the decades to come. The council accepted the definition of sustainable development used by the World Commission on Environment and Development (1987): ". . . to meet the needs of the present without compromising the ability of future generations to meet their own needs" (page iv, President's Council on Sustainable Development 1996).

National Sustainable Development Goals

The council set the desired future condition of the United States in its vision statement: "A sustainable United States will have a growing economy that provides equitable opportunities for satisfying livelihoods and a safe, healthy, high quality of life for current and future generations. Our nation will protect its environment, its natural resource base, and the functions and viability of natural systems on which all life depends" (page iv, President's Council on Sustainable Development 1996). Toward accomplishing that vision, the council established 10 goals among which several pertain directly to natural resources management.

Leading the list of goals is achieving a healthy environment for all and an economy of sustained growth that provides a high quality of life for all. Natural resources are to be used, conserved, protected, and restored with assurance of long-term benefits. In addition, the United States needs to "move toward stabilization of U.S. population" and lead implementation of global sustainable development policies in the international arena. The global intent of sustainable development is to maintain a rate of resource development into beneficial products that will ultimately provide a high quality of life worldwide. With specific attention to ecosystem management, the President's Council on Sustainable Development (1996, page 119) recommends that the United States "enhance, restore, and sustain the health, productivity, and biodiversity of terrestrial and aquatic ecosystems through cooperative efforts to use the best ecological, social and economic information to manage natural resources."

Underlying the concept of a sustainable economy, as espoused in the council's report, is the assumption that resources are plentiful enough for continued growth of personal wealth in the United States and in other nations of the world. Because the world's national economies are now so closely linked, the concept of sustainable development is unworkable in anything but a global perspective. It is doubtful that any nation, no matter how economically strong, can sustain development indefinitely in a world that otherwise fails to do so. Although an ultimate aim of international sustainable development is improved distribution of the world's natural resource–based wealth, the rate at which that can happen is unclear—as are all of the trade-offs that might be required. Natural resources must be developed and used much more efficiently everywhere if the developing nations are to economically catch up.

Worldwide economic sustainability to a large extent depends on the assumption that the science and technology for maintaining resource supply will continue to advance faster than the worldwide demand for resources. The dimensions and uncertainty of this achievement has lead some to conclude that sooner or later a sustainable world economy will require a lower level of material demand (e.g., Daly 1997). Natural resources analysts will better clarify what is possible and what is not in the coming decades as advances in science and technology close the gap in uncertain knowledge. Sustainability will require greater integration of economists and ecologists in conceptualizing the systems that maintain and improve human welfare.

NATIONAL STRATEGIES FOR SUSTAINABLE DEVELOPMENT

The strategies presented by the President's Council on Sustainable Development (1996) emphasize cooperative partnerships among governments at all levels. Government should promote collaborative multistakeholder approaches to natural resources management. Formal education should "emphasize systems thinking and interdisciplinary approaches . . . and pursue experimental, hands-on learning at all levels . . ." (page 74). The council emphasizes integration of environmental, economic, and policy needs through public involvement and objective science. It also encourages more use of market incentives for attaining environmental goals and encouraging greater flexibility for private business in demonstrating accountability.

The council promotes the concept of extended product responsibility for environmental effects, which is a shared responsibility among manufacturers, suppliers, users, and disposers of the products. It promotes industrial ecology, which studies how to integrate solutions of production inefficiency and waste generation. In the view of the council, government should facilitate a system of collaborative regional planning "accounts" that transcends jurisdictional boundaries and estimates the benefits and costs of alternative land uses. Government also should work with business to improve the design of sustainable human communities and associated landscapes with emphasis on providing alternatives to urban sprawl, new highway construction, and equitableness of public works delivered to communities of different economic status. The council's report emphasizes the importance of personal stewardship in use of the nation's natural resources. They promote voluntary collaborative approaches to protection, restoration, and monitoring of natural resources and resolution of conflicts.

In managing these natural resources, the council recommended that agencies at all levels of government should facilitate local use of ecosystem approaches by eliminating administrative obstacles and by providing information and technical assistance. Agencies should also encourage development of shared goals and responsibilities for restoring damaged ecosystems and by developing collaborative partnerships and compatible databases. Within the ecosystem context, government should encourage sustainable development of agriculture, forests, fisheries, and biodiversity and more efficient use of depletable resources through recycling and other strategies.

Sustainable Development and Conservation

The concept of sustainable development is an outgrowth of the early twentieth-century concept of resource conservation in the United States and a large body of subsequent conservation law. An assortment of laws pertaining to public land management in the 1960s and 1970s were precursors to the larger global view that is now emerging. While the conservation philosophies of the past focused on public lands management and agricultural practices, sustainable development emphasizes more responsible use of all privately held resources. Sustainable development extends conservation principles from the national to the global arena. It places greater emphasis on a systems approach and on public policy that works through market incentives.

Although the concepts of sustainable development have roots extending back a century or more, ecological and economic sciences are now more advanced. However, many controversial points in scientific understanding and human idealism remain to be resolved. One particularly troublesome difficulty is the inability to compare ecosystem services that can be priced in economic terms with those that cannot, such as biodiversity. Another problem is determining how the costs of managing for sustainability can be fairly distributed among all public interests. A particular problem in the United States is the private land "takings" issue associated with lost economic value of land when its use becomes restricted through law.

Information Deficiencies and Sustainable Development

The rate of sustainable development that can be achieved is contingent on resource status. A complete inventory of resource supply and use rate is yet to be accomplished, even in the developed world. This deficiency of information is especially great for environmental resources; particularly biodiversity. However, some progress has been made and the results are unsettling. Costanza et al. (1997), in an attempt to draw attention to the value of natural services, developed an "order-of-magnitude" estimate of ecosystem service values that rivals the total economic activity of the world (the Gross World Product). Regardless of unavoidable inaccuracies, they made the point that present measures of economic activity do not fully account for the loss of existing environmental services that occur in natural resource development. Much more ecological and economic research needs completion before an accurate accounting is possible and the risks of unsound decisions can be assessed with assurance.

Among the most debated environmental costs are those associated with the loss of global biodiversity (Wilson 1988, Noss and Cooperider 1994). Two-thirds of the world's genetic information could be lost sometime in the twenty-first century based on estimated extinction rates. With that loss goes the raw material of many potential new products, such as medications, fibers, foods, lubricants, and natural agents of pest control. The potential for genetic engineering depends on the natural occurrence of desirable genes. An international conference in Rio de Janeiro demonstrated the difficulties inherent in global biodiversity protection. Underdeveloped nations, mostly in the Tropics, hold claim to the greatest reservoirs of biodiversity. They want some form of compensation for beneficial discoveries based on unique chemical, physical, or biological attributes of native species in their countries. Nations with low diversity and more potential to develop new products from nonnative sources, have been less enthusiastic about sharing the benefits of genetic discoveries, including the United States.

Biodiversity is important in reducing environmental disturbances, such as flooding, fires, and storms. Although the services provided by diverse communities are well documented—such as stabilization of eroding soils and storage of nutrients—their economic values are difficult to fully quantify. We have barely developed a preliminary list of the ecosystem services provided (e.g., Daily 1997). We know little about the cost-effectiveness of natural services compared with engineered alternatives. Much has been said, for example, about the importance of floodplain and wetlands in maintaining water supply and trapping nutrients and contaminants. However, there is as yet no complete assessment of how natural services compare to reservoirs, sewage treatment plants, and other humanly modified ecosystems.

Restoration of natural service is emerging as an important resource management strategy. However, restoration science is relatively recent and much remains to be learned. Some types of resource restoration are easier to accomplish than others. Rebuilding vegetative biomass for soil formation, erosion control, and sequestering greenhouse carbon will take a much shorter time than rebuilding biodiversity once it is lost. However, complete recovery of ecosystem biomass, soils, and water quality is contingent on biodiversity in ways that are now incompletely understood.

Sustainable Development in U.S. River Corridors

As indicated in the President's Council on Sustainable Development (1996), sustainable development is relevant in all aspects of natural resources management in the

United States. Nowhere, however, is sustainable development more evident than in the river corridors of the United States. Here urban, agricultural, forest, recreational, wildlife, fisheries, and water management activities converge where a disproportionate number of endangered species reside. River corridors provide ecosystem connections that support many aquatic and terrestrial species at risk of extinction. River management provides some of the most insightful trends because many different types of land and water resource use interact and the economic stakes are especially high. Improved management of river corridors, including restoration of desirable natural services, is the object of recent federal legislation involving several federal resource agencies.

Anthropogenic changes in the river corridors exacted insidious costs from environmental services, which have yet to be totaled. Degradation of certain river and floodplain services has been associated with agricultural and urban runoff, urban sprawl, forest management, and water resources management. Past environmental legislation—the Clean Water Act in particular—has improved human welfare in river corridors. While dams and levees have improved navigation, reduced flood damage and increased water supplies, they also have contributed to the fragmentation of aquatic systems and the gradual decline of numerous aquatic species (Noss and Cooperider 1994). Until recently, the biodiversity losses were overlooked or viewed as unfortunate but justifiable costs for the greater benefit derived from dams and levees.

Most recently, restoration has been initiated in some of the ecosystems where environmental degradation has been particularly costly. Although most existing restoration projects are local and small, some in progress are substantial, such as restoration of the Everglades now starting in south Florida (see Davis and Ogden 1994 for background). The Kissimmee River in central Florida, located in the upper part of the Everglades watershed, is being returned to its natural configuration, having been previously straightened by dredging and confined by levees (Toth 1995).

Additional engineering for flood control has reached a point where the costs exceed the benefits in many river systems. While removal of large dams or levee systems was at one time unthinkable, it is now being practiced, planned, or contemplated in North Carolina, Maine, Florida, Idaho, and California. Because of concern over losing salmonid diversity in the Columbia River watershed, river managers and stakeholder publics have been considering operational modification or bypassing four hydroelectric reservoirs in the Snake River (a tributary to the Columbia); perhaps ultimately removing them.

It was not until the Endangered Species Act of 1973 that the costs associated with possible species extinction were seriously considered in the mix of developmental costs and benefits. We have also learned much more about other natural service values of floodplains. In the past, small towns sometimes were moved so reservoirs could be constructed to prevent downstream areas from flooding. It has only been within the last decade that homes and small villages have been moved out of floodplains to avoid future flooding costs and to restore ecosystem integrity. Future management is likely to increasingly include breaching and setback of levees and removal of old dams as we gain improved understanding of the total costs and benefits of river corridor services. These efforts will require comprehensive management planning and integration at scales rarely attempted in the past.

Land Control and Sustainable Development

One of the practical problems associated with sustainable development is the issue of property rights. Nowhere is this more evident than in wetlands conservation. An amendment to the Clean Water Act, called the Comprehensive Wetlands Conservation and Management Act of 1995, makes it clear that wetland protection should not inhibit economic development and should protect private property rights (Hammer

1996). This includes compensation to the owner if the property value is diminished by 20% or more. Most wetlands involve private property. A national goal now guiding wetland management is no net loss of wetland function and acreage. Therefore, wetland filling for development requires compensation by creating, restoring, or protecting wetlands elsewhere. Similar issues continue to emerge with respect to declaring critical habitat for threatened and endangered species on private lands. Even more difficulty is encountered when threats originate in remote areas, as is often the case with watersheds.

How much value do wetlands, an endangered species, or other environmental services contribute to sustaining national and global wealth? Is it enough to justify payment to private property owners at rates determined by real estate values? Protection and restoration of ecosystem services may require conversion of privately owned lands in part or entirety to public ownership. The present inability to accurately value many environmental resources is of concern because of the need to provide fair monetary compensation to private owners for deprivation. Restoration of valued natural services is not likely to progress extensively until such values can be assessed accurately and private landowners are compensated fairly for the benefits they must forego in the national interest.

ECOSYSTEM MANAGEMENT

Defining Ecosystem Management

The primary means for pursuing sustainable development of natural resources is through a process now commonly known either as ecosystem management or as integrated resource management. **Ecosystem management** is the process of identifying management problems and implementing solutions in the context of communities and environmental variables in a specified area. The management space is defined by ecosystem boundaries instead of political and jurisdictional boundaries. Because ecosystems are influenced by material and energy flows originating outside their boundaries, much of ecosystem management must address interactions across boundaries, including the actions of living organisms and, most obviously, the actions of humanity. Franklin (1997) pointed out that a common theme of ecosystem management is managing toward the goal of sustainable ecosystems. Haney and Boyce (1997) and Vogt et al. (1997) emphasized that social goals are usually incorporated, including resource output goals linked closely to the concept of sustainable development.

Ecosystem management is a concept that goes back at least 30 years (Van Dyne 1969) but has only in the last decade been adopted as a philosophical approach to federal management of resources. It has only been in recent decades that understanding of ecosystem function and linkages to other ecosystems have become scientifically well enough understood that ecosystem management could begin to be put into practice. In 1995, a Memorandum of Understanding agreeing to the use of an ecosystems approach to management was signed by the federal agencies that are involved with natural resources management. One intent of ecosystem management is development of a vision of desired managed condition that is shared by all stakeholders. Because of its complexity, ecosystem management can be nothing other than an interdisciplinary and interagency team approach to public resource management. In addition to ecological attributes, Grumbine (1994) found that ecosystem management implied scientific research and monitoring, interagency cooperation, organizational change, and management goals determined by human values. Because ecosystem boundaries cross administrative boundaries, one of the most challenging aspects of ecosystem management is the need for interagency and interdisciplinary team integration. Typically a diverse array of stakeholder needs and agency functions need to be accommodated. Because detailed understanding of ecosystem process at a local level is rare, research, monitoring, and adaptive management need to be integral parts of effective ecosystem management.

An important stimulus for ecosystem management has been associated with the shortcomings of the Endangered Species Act of 1973. Because the act provides no authority for preventing losses before species become threatened or endangered, it is a last resort for biodiversity protection. In over 20 years of listing 727 species, only 16 species were delisted, 7 because of extinction (Nester 1997), and more than 3,000 are being considered for listing. Ecosystem management is viewed as a promising way to identify stresses before they result in threatened or endangered (T&E) status and, through comprehensive planning, prevent species from becoming endangered. However, much confusion remains about ecosystem management because ecosystem functions and structure are not completely understood.

One major problem faced in ecosystem management derives from the open-system nature of most ecosystems, meaning that boundaries often are difficult to determine and can be placed on ecosystems in a variety of justifiable ways depending on management objectives and perspectives. The diverse stakeholders in ecosystem management typically identify and emphasize different problems with different boundaries. Thus, flexibility is important and Vogt et al. (1997) believe that the precise definition of ecosystem management is likely to introduce too much specificity and rigidity in a concept that derives meaning from comprehensiveness and inclusiveness. On the other hand, when ecosystem management is designed too generally, as is the tendency, it loses utility as a guiding concept. Franklin (1997) emphasized the importance of choosing the right scale proportional to the problem. He cited as an example the problems associated with managing the large mammals of Yellowstone Park, which range through the "greater Yellowstone ecosystem" surrounding and including the park.

Integrating Resources Management into Ecosystems Management

The concept of integrated resource management is an approach that often parallels and links closely with the concept of ecosystem management. In some applications they are virtually indistinguishable. **Integrated resource management** is the synthesis of resource management system fragments into a functional whole. Thus, the objective of integrated resource management is to bring all of the management fragments into a planning and implementation whole that more effectively sustains resource outputs than the traditional fragmented approach to management. By its very nature, ecosystem management is a form of integrated resource management that develops a systems synthesis across the full range of resources yielded from the ecosystem.

The integrated approach to natural resource management has been advanced most clearly in watershed, coastal-zone, agricultural, and forest-resource management (see, for example, Heathcoate 1998 and Kidd and Pimentel 1992). The concept of more integrated public management of resources extends back at least to the surge of conservation laws passed in the 1930s. Many of these laws involved coordination of agencies in attaining water, soil, and wildlife resource ideals. Managing certain federal lands for multiple use goes back to 1960 legislation that requires coordinated integration of stakeholder demands and assessments of resource supply. In 1965, this concept also was incorporated into federal water resource planning legislation in river basins with the intent of improving water supply for agricultural, municipal, industrial, navigational, and recreational uses while protecting environment quality. The scope of that approach has increased through the years as social awareness has grown about the values of other watershed resources and their interaction with water supply. For agricultural and forestry resources, Kidd and Pimentel (1992, page 22) define integrated resource management in terms of "the design and operation of agricultural systems so that all of the parts of each system operate together to enhance productivity and sustainability."

Compared to integrated resource management, ecosystem management is a more inclusive concept that emphasizes optimization of the whole array of resource outputs as objects of management instead of some specific selection. Increasingly the outputs are described as ecosystem services. In Franklin's (1997) view, sustainability pertains to the potential of ecosystems for producing all goods and services for indefinitely long periods (see also Daily 1997). A key challenge for ecosystem management is to integrate all of the different resource systems associated with range, forest, farm, outdoor recreation, urban, environmental, fish, wildlife and biodiversity resources into a holistic framework. That holistic level of interaction among management agencies and stakeholder groups remains more an aspiration than a reality because of the different resource perspectives.

Agencies and nongovernment organizations come to ecosystem management with different expectations based on their different concepts of the organizational mission. It is common for those organizations and disciplines concerned with water-based resources to accept watershed boundaries for the ecosystem approach. However, for those organizations concerned with the movements of large mammals or birds, the watershed approach has little meaning. Terrestrial migrants, such as waterfowl, pay little attention to watershed boundaries. Similarly, such boundaries have little meaning to those concerned about atmospheric and oceanic resources. Thus, ecosystem management must somehow layer different ecosystem perspectives over one another and integrate the management of those layers if the optimal emphasis on diverse resources and human services is to result. A computer tool used for facilitating this approach is geographic information systems (GIS). Landscape ecology is rapidly emerging as the coordinating discipline for integrating interdisciplinary teams in economic analysis.

Some confusion exists because some philosophers equate ecosystem management with restoration and maintenance of natural ecosystems. Ecosystems, however, may be managed to emphasize goods and service provision without necessarily providing them in the wilderness state. These services include both market services, such as timber and food fishes, and nonmarket services, such as clean water supply, recreation, and sustaining genetic diversity. Whereas many services can be priced either directly through the economic market, or indirectly through benefits analysis, other service values, such as biodiversity provision, have as yet to be satisfactorily estimated in monetary terms. For agencies and stakeholders organized according to special interest advocacy this integrative concept is difficult to assimilate. However, welfare economists have wrestled for decades with developing a benefits-cost paradigm for facilitating such integration (see, for example, Loomis 1994), and progress is being made. Even so, the inability of economists to estimate the value assigned to certain services is likely to frustrate the decision process for the foreseeable future.

The ecosystem concept is so comprehensive that practitioners of ecosystem management need to have more than an introductory knowledge with ecological science and more than an intuitive sense of how to facilitate positive stakeholder input into planning and management process. Because all values cannot be reduced to interchangeable monetary units, the trade-offs between services need to be negotiated through stakeholder participation in the planning process. This requires an ability to work in partnerships and with diverse interests to identify all stakeholder problems and their interactions. It also requires framing solutions in a shared vision of desired ecosystem function while sustaining options for future generations. The concept requires managers who can interact effectively in an interdisciplinary and often contentious environment. While natural resource managers need special expertise in a particular area, they also need to be broadly enough informed so they can contribute effectively on interdisciplinary teams.

Ecosystem Health and Adaptive Management

Two concepts closely associated with ecosystem management are ecosystem health and adaptive management. All strategies of ecosystem management are increasingly being judged in terms of the ecosystem health that results. Costanza et al. (1992) indicated that self-maintenance signified ecosystem health. A degraded state is one in which functions could not be sustained indefinitely without costly management intervention. Self-sustaining states have real cost advantages over ones that must be maintained at human expense.

Whereas scientifically developed cause-and-effect understanding should underlie ecosystem-based natural resource management (see, for example, Boyce and Haney 1997, Vogt et al. 1996), controlled experimentation at ecosystems scale often is technically infeasible and prohibitively expensive. An alternative to controlled experimentation is adaptive management (Holling 1978, Walters 1986). In adaptive management, managed ecosystems are scientifically monitored to assess the effectiveness of their management and the results analyzed and published to improve management. It might be thought of as learning by doing and redoing from what was learned. Adaptive management is a long-established concept, if otherwise labeled such as "experience is the best teacher." The more formal process espoused by Holling (1978) and Walters (1986) promotes an organized approach that includes formal concept development, often in the form of a mathematical model of how management is believed to work in the ecosystem context, measurement of system outputs indicative of management effectiveness, analysis of results, sharing of results through a scientific peer review process, and modification of the model for future management action.

The institutionalization of adaptive management has been slow to occur, however. One of the primary fears is that it will divert scarce management funds into research with relatively little management application. There is concern that society and policy makers have insufficient understanding of science and scientific process to tolerate the uncertainty implied by a significant diversion of funds into investigation of results. Much can be said about that position, which identifies one of the greatest potential impediments to future resource management progress in democratic societies—the growing gap between scientific management of resources and the ability of the voting public to understand the needs, risks and benefits associated with scientific management of the human environment. Thus ecosystems management education cannot stop at the level of resource management professionals.

TECHNOLOGY

In addition to more effective global and local ecosystem-based strategies, the future effectiveness of natural resource management in sustaining ecosystem services will depend on human mastery of its technology. The history of technology reveals it has been an environmental double-edged sword, producing both problems and solutions. Many technological advances are underway that could facilitate more effective resource mangement, once they are carefully integrated into a comprehensive management planning process. Some of the technologies most relevant to future resource management are briefly discussed here.

Computers

Computers have revolutionized how all of us work, including natural resource professionals. Managers must gather, organize, and analyze large amounts of information to predict the outcome of different management practices. In the early years, data were entered and stored on paper forms and analyzed by hand. Invention of adding machines greatly facilitated development of modern data management and analytical methods. Since the first commercial computers were built following World War II computer technology has advanced at a phenomenal rate. This has allowed information processing at speeds several magnitudes greater than before World War II.

Computers "process" information input through software programs that direct computer hardware operations. Computers are unsurpassed as a means for organizing and managing data, including storage and retrieval when needed. Many specialized software programs are available to aid data analysis and management decisions. Portable computers are now routinely used in various natural resource field assessments. Data are either entered by hand or by way of a wide variety of electronic meters, including digital cameras, and then carried or relayed to a larger processor computer.

For most natural resources management, data are increasingly being organized through GIS technology. GIS data often are organized in the form of maps that may be analyzed for variable distribution and alignment, including vegetation types, wildlife species, land use, soil types, geological formations, mineral deposits, watershed processes, land ownerships and jurisdictions, and proposed development. The advantages of this type of data organization are numerous considering the potential for computer software to evaluate systematic relationships. One of the more important uses at this time is the identification of threatening land-use patterns with respect to the distribution of rare and endangered species and ecosystems.

Another advance associated with computers is information networking facilitated by e-mail and the Internet. Information that once had limited intraorganizational reach is now made instantaneously available worldwide to anyone who can access the Internet. The potential for sharing information resources may be rivaled only by the challenge of ensuring its validity.

Systems Analysis

Successful natural resource management increasingly depends on current knowledge of biological and financial outcomes of various management practices. In tomorrow's world, any successful natural resource program will depend on highly skilled personnel who can use computers to assess the risk/reward ratios of various management options. We consider a strong background in computer-mediated tool use, social skills, communication skills, social sciences, and economics to be as important to today's natural resource manager as breadth of knowledge in soil science, range science, ecology, agronomy, forestry, wildlife management, and land-use planning.

Natural resource managers usually must consider large amounts of information to predict the outcome of different management practices. These managers can and do benefit from computer data compilation and synthesis systems geared toward practical problem solving. Since the 1970s, computers have been used increasingly by the natural resource managers in both data storage and data analysis. George Van Dyne is considered the pioneer in development of computer applications to natural resource problems. During the 1960s and 1970s, he developed computerized models that integrate knowledge on various aspects of rangeland ecosystems into a framework for management decisions (Van Dyne 1966a). The approach adapted by Van Dyne and others to facilitate analysis of ecological process is referred to as **systems analysis** and is discussed in detail, as it relates to natural resources, by Van Dyne (1966b), Van Dyne (1969), and Shugart and O'Neill (1979).

Analysis of management systems is the process of defining goals and objectives for ecosystem management and identifying procedures for their efficient accomplishment. The most important attributes of systems analysis are that it provides a unifying structure for interrelating facts and observations, and that it serves as a dynamic force as new facts and observations become available.

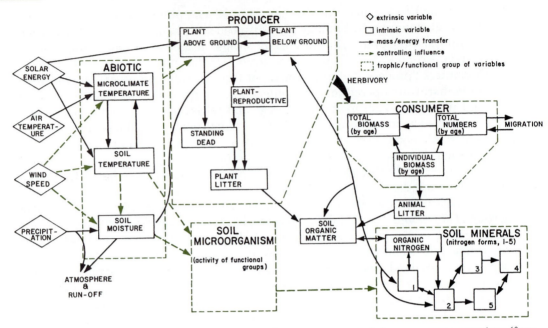

FIGURE 23.1 Diagram of a grassland ecosystem. Each arrow corresponds to one or more equations (from Bledsoe and Jameson 1969).

Model Software for Ecosystem Analysis

A mathematical model of an ecosystem is one means of organizing information about an ecosystem to test management proposals in advance of their actual implementation (Figure 23.1). The real system variables have analogous mathematical variable counterparts. It is an abstract simplification that is reduced to the minimum needed for management purposes. It does not represent the system in its entirety, but it does simulate the ecosystem in aspects regarded as essential by the modeler. No conceivable model could possibly represent all the plant and animal species present in an ecosystem, but if it is to be useful, it must represent the behavior of that portion of the system modeled, consistent with need.

Models are useful for analysis of numerous natural resources phenomena both in and outside an ecosystem context. Models may be empirical, based on statistical analysis of past behavior, or theoretical, based on explicit mathematical expressions describing natural behavior. Many models combine both empirical and theoretical attributes. Models are validated by testing the accuracy of predicted real-system behavior when the model is loaded with real-world input data. Once validated, models provide a tool for examining the effects of many manipulations of the ecosystem without doing field trials. Models also may be used to determine the sensitivity of model output results to levels of uncertainty in the model mathematical structure. Those parts of the model showing a combination of high effect on result and high uncertainty in system representativeness are identified for high-priority research attention.

Artificial Intelligence

The use of artificial intelligence in management of natural resources began with the development of expert systems for problem solving and decision making. An *expert system* is a type of artificial intelligence program that follows a few general procedures for solving problems. It uses facts (often obtained from written reports), experience, and models stored in the memory of a computer by human experts. The computer can "arrive at conclusions" in two ways. It can reason forward, going from facts to solution, or it can work backward from a hypothetical solution to seek supporting evidence. Using the two reasoning approaches to solve a particular problem, the computer suggests a set of hypotheses based on data input by the user. The system then considers each hypothesis in turn, attempting to find a specific solution. The artificial intelligence programs analyze decisions, can interpret the meanings, or can ask appropriate questions. Thus computer-operated software aids managers who often reach conclusions from partial or uncertain evidence by following possible lines of reasoning. Also, the system can add to its database as it gains experience. Maintaining a high degree of accuracy in decision-support functions requires a commitment to collect, develop, and verify the experts' knowledge.

The use of expert systems led to the development of other artificial intelligence procedures pertinent to the management of natural resources. Some of these are (1) integrated expert systems, which link management models with natural resource models, (2) intelligent GIS files, and (3) artificial intelligence modeling of ecosystem components and their interaction with the environment (Coulson et al. 1987). Expert systems have led to knowledge systems, intelligent decision systems, creativity systems, and literature research systems (Truett 1989). An expert system was used to develop a rangeland simulation model (Ritchie 1989). Although many advances have been made in artificial intelligence, their promise remains far from fully realized.

Virtual Reality

Virtual reality (VR) is a computer simulation usually experienced through headgear, goggles, and sensory gloves that lets users feel they are in another place. VR devices let you see, hear, feel, and interact with real or abstract data from computer-generated models. Perspective views may be generated by combining digital terrain data with satellite imagery, scanned aerial photographs or maps, airborne video, ground-penetrating radar, and GIS files (Truman et al. 1988, Everitt et al. 1990, Smith et al. 1991). The three-dimensional views are used to visualize and verify the results of the natural resource models. Database visualization in 3-D enables the decision makers and the general public to more readily understand complex data.

Economic Analysis

The natural resource manager often must determine the level of production that yields the greatest net returns across the variety of resources managed. New computer-mediated tools are becoming available to aid that and other economic evaluations. Computers provide an essential tool in determining the cost/return ratios of various management alternatives to find the ideal combination of short- and long-term benefits.

Models of ecosystem processes have been linked to economic and management process to aid management policy analysis (Figure 23.2). Cole et al. (1990), for example, describe a model for analyzing the sport fishery of New Mexico, the benefits of which depend greatly on integrating water management with biological management and management of an array of amenities including boat ramps, campgrounds, and sanitary facilities. Once developed, models of this type can be analyzed for the most beneficial combination of management provisions. They are, however, data demanding, and require an investment in adaptive management that has yet to be widely accepted in natural resource management. Tools are also available for assessing the best return on production for the investment costs associated with alternative management approaches.

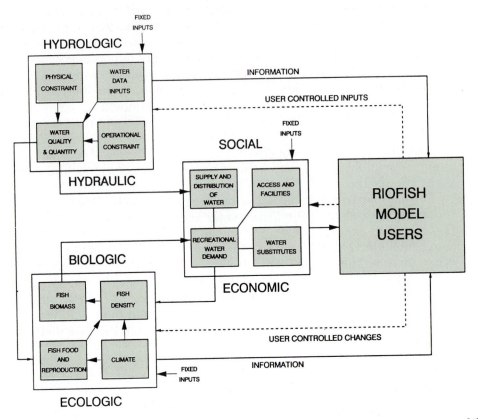

FIGURE 23.2   The integration of ecological and economic process into managment systems models can facilitate policy analysis concerned with providing efficient and sustained benefits. This diagram illustrates the main interaction among geophysical, ecological, economic, and management components in a state sport-fishery managment system (from Cole et al. 1990).

Farming is so sophisticated now that many farmers have computerized maps of soil characteristics of their individual fields. These maps are used to precisely determine what level of fertilization and irrigation will maximize yield. Modern irrigation equipment uses these computerized maps to allocate nutrients and water to each part of the field so financial returns will be maximized.

THE FUTURE

When looking to the future there is always a danger of omitting an event of great importance or including something that does not happen. However, we feel that it is very important to look at the future and discuss some possible events that may be important.

Over the past 50 years, agriculture and forestry in developed countries have been transformed from resource-based industries to science-based industries. Future improvements in human welfare will depend very heavily on maintaining or increasing the role of technological advance. Any natural resource management strategy must be developed with some idea of what the future (twenty-first century) will be like socially, politically, culturally, and economically. It is important to recognize that social demand determines the value of any natural resource product or commodity. Therefore, natural resource scientists, managers, and producers must have at least a basic understanding of consumer needs and demand in the future. This is particularly important when the long production cycles of many natural resource products are taken into account such as oil, water, timber, fish, and various types of wildlife.

Because of massive technological breakthroughs, an economic boom was predicted for the 1990s that would last far into the twenty-first century. Naisbitt and Aburdene (1990), in their book on future megatrends, made the following statement:

> The global boom of the 1990s will be free of the limits on growth we have known in the past. There will be an abundance of natural resources throughout the 1990s from agricultural products and raw materials to oil. Everything that comes out of the ground will be in oversupply for the balance of this century and probably much longer. We will need fewer raw materials, as we have been moving away from material-intensive products for decades. A prototypical example of the shift away from the material intensive products is fiber-optic cable. Just seventy pounds of fiber-optic cable can transmit as many messages as 1 ton of copper wire. Equally important, those 70 pounds of fiber-optic cable require less than 5% of the energy needed to produce one ton of copper wire. There will be no energy crisis to impede the 1990s global boom. Each year since 1979, the United States has used less energy than the year before.

So far, the preceding predictions have been mostly realized, although there are signs that sustainablility of economic development may be undergoing strain. The recent depression in growth of Asian economies is one such sign, as is the upswing in automotive fuel use in the United States. An additional concern is accelerating loss of biodiversity providing resource development options for the future. A spin-off from the new economic boom is a renewed concern and appreciation for the natural environment. On this subject, Naisbitt and Aburdene (1990) stated, "The world's preoccupation with defense and the cold war, which is receding, is being replaced by concerns about the destruction of our natural environment, now our most important common problem."

In his vision of the future beyond the year 2000, Drexler et al. (1991) stated that nanotechnology (extreme miniaturization) will permit the human race to feed itself with ordinary, naturally grown, pesticide-free foods while returning more than 90% of today's agricultural lands back to nature. If present trends continue, natural resource managers will lose credibility if they use the gloom and doom predictions foretold since the time of Malthus to justify future research and management. Some economists consider the only food problems of today's world to be those of distribution and political incompetence. However, many ecologists and environmentalists question whether the technological breakthroughs in the future can keep up with those of the past.

Nature—the environment in a natural state—is increasingly attaining monetary value as a source of wealth itself (Rolston 1989). The trend is toward cleaner and greener wealth being equated with greater wealth (Drexler et al. 1991). We believe peoples throughout the world will continue to increase the importance they place on environmental quality relative to material goods in their concept of personal and economic well-being. This does not mean that natural resource managers can ignore the importance of basic commodity production. We believe short periods of food, energy, and timber scarcity will occur in the twenty-first century, as they have in other centuries, because of climatic, political, and other types of adversity. However commodity production will increasingly be considered as only one aspect of natural resource management, particularly in the more developed countries.

A new electronic heartland was predicted for the United States by Naisbitt and Aburdene (1990). They stated, "Linked by telephones, fax machines, Federal Express, and computers, a new breed of information worker is reorganizing the landscape of America. Free to live almost anywhere, more and more individuals are deciding to live in small cities and towns and rural areas. A new electronic heartland is spreading throughout developed countries around the globe, especially in the United States. Quality-of-life rural areas are as technologically linked to urban centers as are other cities. This megatrend of the next millennium is laying the ground work for the decline of cities." More recently Swasy (1994) made a similar statement: "Today's white-collar boom towns were yesterday's cow pastures." In the future, this will only mean greater demands on natural ecosystems, with the preservation of open space turning into perhaps the biggest natural resource challenge.

Another prediction we will make is that the 1990s will be considered the beginning of the age of biological engineering. We may have the ability to create organisms by genetic engineering to produce whatever product society demands (Naisbitt and Aburdene 1990). Many scientists involved in genetic engineering are making this prediction. Boyd and Samid (1993) stated, "The amazing pace of discovery within the field of molecular biology has expanded the understanding of as well as the potential for genetic engineering in laboratory and domestic animals. The future of the planet earth and the species that inhabit it can be significantly affected by these technologies. The use of transgenic animals is limited only in the human imagination."

Nanotechnology

Breakthroughs in molecular biology could lead to a prosperity never before experienced by humankind (Drexler 1986, Drexler et al. 1991). They could completely change production processes as we know them today. These new mind-be nding technological possibilities are explored by Drexler (1986) and Drexler et al. (1991).

Molecular biology is presently causing a revolution in agriculture. The secrets of DNA are rapidly being unlocked with the aid of computers and biotechnology. The new industry of genetic engineering has developed pest-resistant strains of crops and hormones that dramatically increase meat and milk yields from livestock. These innovations and the use of computerized machinery are reducing the use of fertilizers, pesticides, and water for crop production. As the genes of food and fiber crops are redesigned, food-processing requirements could shrink. Meats, cereal grains, fruits, and various vegetables may have their flavorings programmed.

However, in 50 to 100 years, these developments may seem crude and simplistic compared to what lies ahead. Drexler (1986) points out that scientists in the United States and Japan are involved in molecular engineering efforts that could result in using the techniques of genetic coding to form microscopic computers. The potential outcome is computers that are so tiny they would fit into a human cell.

Today production processes are organized around processes that rearrange masses of atoms (Davidson and Rees-Mogg 1987). These manipulations involve heating, cooling, forging, casting, pressing, pounding, and chemical mixing. People involved in these acts work with trillions of atoms. In contrast, the new nanotechnologies (microscopic computers) will build products based on the manipulation of individual atoms. While genetic engineering is the first step in this process, the techniques could be used to make nonliving products. Keep in mind that genetic codes appear to be merely "machine tool systems" for putting together living molecules. There is no known reason once these processes are understood they cannot be used for nonliving things. Drexler (1986) made the following statement regarding the outcome of nanotechnology:

> AI (artificial intelligence) systems will bring still swifter automated engineering, evolving technological ideas at a pace set by systems a million times faster than a human brain. The rate of technological advance will then quicken to a great for-

ward leap: in a brief time many areas of technology will advance to the limits set by natural law . . . This information is a dizzying prospect. Beyond it, if we survive, lies a world with replicating assemblers, able to make whatever they are told to make, without need for human labor.

Practically any product desired by humankind could be built by molecular assemblers programmed with artificial intelligence. Imagine products such as watches, washing machines, and stoves with perfectly fitting parts assembled with almost no human participation. If this scenario should become reality, scarcity of material goods would virtually be a thing of the past.

Human life spans might be more than doubled from a new medicine practiced at the level of individual human cells. Molecular computers and machines in individual cells could monitor their performance and preserve them indefinitely. We do recognize, however, that, although nanotechnology has great possibilities, it may never come to pass as it is presently envisioned.

Optimism versus Pessimism

The darker side of the future centers around economic cycles and demographic changes (human population increase). Generally those who have predicted gloom and doom for the future, such as Malthus and Ehrlich (1968), have focused on the human population overrunning the supply of natural resources. However, throughout the history of the world, we have made the case that economic chicanery and mismanagement have had far more to do with human suffering and environmental deterioration than natural resource scarcity. The decline of ancient societies in Egypt, Greece, and Italy; the extended Dark Ages of Europe; the French Revolution; the rise and fall of Adolf Hitler in Nazi Germany; the depression of the 1930s in the United States; and the rise and fall of Communism in the Soviet Union and China all provide examples of great human hardship that resulted from a failure of government policies, economic systems, and political leaders.

Ever since the late 1960s, more pessimistic analysts have predicted that consumer and public debt expansion in the United States and several other developed countries would lead to a severe worldwide economic depression (Davidson and Rees-Mogg 1993). However, the day of reckoning has been delayed by productivity increases, inflation, a wide variety of innovative ways to expand credit, and better systems of banking. Based on history, all debt is leveled sooner or later by payback, default, and/or inflation. Many economists believe the greatest problem that confronts the United States over the next 30 years will be how to deal with its debt problem (Casey 1993, Davidson and Rees-Mogg 1993).

In closing, we believe that the interaction of technological development, information transfer, and application of improved economic and management systems will have far more to do with world human condition in the near and distant future than a scarcity of natural resources or indebtedness. We are optimistic about the future of humanity because of major changes in most developed and developing countries that reflect a trend toward democracy, market economies, protection of individual rights, racial equality, sexual equality, free trade, increased global competition, improved financial management, and zero population growth. At the same time, there is growing appreciation for natural environments and the products they contribute. We believe major advances in human thinking will continue to lead to a sustainable and improving future. This will mean finding the appropriate balance among the human population size, its material needs, and its environmental needs for the existing level of technology. While the future, as always, is uncertain, we look forward to it with cautious optimism, and have no desire to return to the past when life was shorter, more physical hardships were endured, and opportunities were far more limited.

LITERATURE CITED

Bledsoe, L. and D. A. Jameson. 1969. Model structure for a grassland ecosystem. In R. L. Dix and R. G. Beidleman, eds., *The grassland ecosystem: A preliminary synthesis.* Range Science Series No. 2. Colorado State University, Fort Collins. 417 pp.

Boyce, M. S and A. Haney, eds. 1997. *Ecosystem management: Applications for sustainable forest and wildlife resources.* New Haven, CT: Yale University Press.

Boyd, A. L. and D. Samid. 1993. Review: Molecular biology of transgenic animals. *Journal Animal Science Supplement* 3(71):1–9.

Cairns, J. Jr. 1996. Determining the balance between technological and ecosystem services, (pp. 13–30). In P. C. Schulze, ed., *Engineering within ecological constraints.* Washington, DC: National Academy of Engineering, National Academy Press.

Casey, D. 1993. *Crisis investing for the rest of the 90's.* New York: Carol Publishing Group.

Cole, R. A., F. A. Ward, T. J. Ward, and R. M. Wilson. 1990. Development of an interdisciplinary planning model for water and fishery management. *Water Resources Bulletin* 26:597–609.

Costanza, R. B., G. Norton, and B. D. Haskell, eds. 1992. *Ecosystem health.* Washington, DC: Island Press.

Costanza, R., R. D'Arge, R. De Groot, S. Farber, M. Grasso, B. Hannon, K. Limburg, S. Haeem, R. V. O'Neill, J. Paruelo, R. G. Raskin, P. Sutton, and M. van den Belt. 1997. The value of the world's ecosystem services and natural capita. *Nature* 387:253–259.

Coulson, R. N., J. L. Folse, and D. K. Loh. 1987. Artificial intelligence and natural resource management. *Science* 237:262–267.

Daily, G. C., ed. 1997. *Nature's services. Societal dependence on natural ecosystems.* Washington, DC: Island Press.

Daly, H. 1997. *Beyond growth: The economics of sustainable development.* Boston: Beacon Press.

Davidson, J. D. and W. Rees-Mogg. 1987. *Blood in the streets.* New York: Summit Books, Simon & Schuster.

Davidson, J. D. and W. Rees-Mogg. 1993. *The great reckoning.* New York: Simon & Schuster.

Davis, S. M. and J. C. Ogden, eds. 1994. *Everglades: The ecosystem and its restoration.* Delray Beach, FL: St. Lucie Press.

Drexler, K. E. 1986. *Engines of creation.* New York: Anchor Press/Doubleday Publishing Co.

Drexler, K. E., C. Peterson, and G. Pergait. 1991. *Unbounding the future: The nanotechnology revolution.* New York: William Morrow.

Ehrlich, P. R. 1968. *The population bomb.* New York: Ballantine Books.

Everitt, J. H., K. Lulla, D. E. Escobar, and A. J. Richardson. 1990. Aerospace video imagining systems for rangeland management. *Photo Engineering and Rem. Gens.* 56:343–349.

Franklin, J. F. 1997. Ecosystem management: An overview (pp. 21–53). In M. S. Boyce and A. Haney, eds., *Ecosystem management: Applications for sustainable forest and wildlife resources.* New Haven, CT: Yale University Press.

Gardner, B. D. 1984. *The role of economic analysis in public range management. Developing strategies for rangeland management.* National Research Council/ National Academy of Sciences, eds. Boulder, CO: Westview Press.

Grumbine, R. E. 1994. What is ecosystem management? *Conservation Biology* 8:27–38.

Hammer, D. A. 1996. *Creating freshwater wetlands.* 2nd ed. New York: CRC Press.

Haney, A. and M. S. Boyce 1997. Chapter 1 Introduction (pp. 1–17). In M. S. Boyce and A. Haney, eds., *Ecosystem management: Applications for sustainable forest and wildlife resources.* New Haven, CT: Yale University Press.

Heathcoate, I. W. 1998. *Integrated watershed management: Principles and practice.* New York: John Wiley & Sons.

Holling, C. S., ed. 1978. *Adaptive environmental assessment and management.* New York: John Wiley & Sons.

Kidd, C. V. and D. Pimentel. 1992. *Integrated resource management: Agroforestry for development.* New York: Academic Press.

Loomis, J. B. 1994. *Integrated public lands management: Principles and applications to national forests, wildlife refuges, and BLM lands.* New York: Columbia University Press.

Lubchenco, J. et al. 1991. The sustainable biosphere initiative: An ecological research agenda. *Ecology* 12:371–412.

Muschett, F. D. 1997. *Principles of sustainable development.* Delray Beach, FL: St. Lucy Press.

Naisbitt, J. and P. Aburdene. 1990. *Megatrends 2000: Ten new directions for the 1990s.* New York: William Morrow.

Nester, W. R. 1997. *The war for America's natural resources.* New York: St. Martin's Press.

Noss, R. F. and A. Y. Cooperider. 1994. *Saving nature's legacy: Protecting and restoring biodiversity.* Washington, DC: Island Press.

President's Council on Sustainable Development. 1996. *Sustainable America: A new consensus for prosperity, opportunity, and a healthy environment for the future.* Washington, DC: U.S. Government Printing Office, Superintendent of Documents.

Ritchie, J. R. 1989. An expert system for a rangeland simulation model. *Ecological Modeling* 46:91–105.

Rolston, H. III. 1989. *Philosophy gone wild.* Buffalo, NY: Prometheus Books.

Schiller, B. S. 1994. *The economy today.* 6th ed. New York: McGraw-Hill.

Shugart, H. H. and R. V. O'Neill, eds. 1979. *Systems ecology.* Stroudsburg, PA: Dowden, Hutchinson & Ross.

Smith, S. M., H. E. Schreier, and S. Brown. 1991. Spatial analysis of forage parameters use geographic information system and image-analysis techniques. *Grass and Forage Science* 46:183–189.

Swasy, A. 1994. America's 20 white-collar addresses. *Wall Street Journal* 129: B1, B3.

Toth, L. 1995. Principles and guidelines for restoration of river/floodplain ecosystems—Kissimmee River, Florida (pp. 49–73). In John Cairns Jr., ed., *Rehabilitating damaged ecosystems,* 2nd ed. Ann Arbor, MI: CRC Press, Lewis Publishers.

Truett, W. L. 1989. Artificial intelligence and applications across the 1990s: An applied approach. *American Laboratory* 21(4):40–47.

Truman, C. C., H. G. Perkins, L. F. Asmeissen, and H. D. Allison. 1988. Using ground-penetrating radar to investigate variability in soil properties. *Journal of Soil and Water Conservation* 43:341–345.

Van Dyne, G. M. 1966a. Application and integration of multiple linear regression and linear programming in renewable resource analyses. *Journal of Range Management* 19:356–362.

Van Dyne, G. M. 1966b. *Ecosystems, systems ecology, and systems ecologists.* ORNL–3957. Oak Ridge, TN: Oak Ridge National Laboratory.

Van Dyne, G. M., ed. 1969. *The ecosystem concept in natural resource management.* New York: Academic Press.

Vogt, K. A., J. C. Gordon, J. P Wargo, D. J. Vogt, H. Asbjornsen, P. A. Palmiotto, H. J. Clark, J. L. O'Hara, W. S. Keeton, T. Petel-Weynand, and E. Witten. 1997. *Ecosystems: Balancing science with management.* New York: Springer.

Walters, C. 1986. *Adaptive management of renewable resources.* New York: Macmillan.

West, N. E., ed. 1995. *Biodiversity on rangelands.* Logan, UT: College of Natural Resources, Utah State University.

Wilson, E. O., ed. 1988. *Biodiversity.* Washington, DC: National Academy Press.

World Commission on Environment and Development. 1987. *Our common future.* Oxford, England: Oxford University Press.

Glossary

abiotic Without life.

abyssal zone The deep, offshore ocean occupied by dark, cold waters and ocean bottom beyond the shallower depths surrounding continents and islands.

acid A chemical compound that disassociates into hydrogen ions and ions of other elements. Acid strength is in proportion to the concentration of hydrogen ions free for chemical reaction and measured by pH.

acid rain Precipitation made more acidic than environmental background levels by natural and artificial emissions that form sulfuric, nitric, and other strong acids. Fossil fuel power plants and automotive emissions are major sources.

adaptation A life-promoting adjustment to environment. Evolutionary adaptation derives from genetic change and retention of genes in a population, resulting from greater reproduction by the most environmentally fit members.

administration The process of executing policy through organizational communication and coordination into an effective management force. Also, the executive body that leads in goal development, comprehensive planning, implementation of policy, and performance evaluation.

aerobic Relying on oxygen for normal function such as aerobic organisms. *See also* **respiration.**

age structure The distribution of population membership among age categories. Stable populations form a pyramid structure with young organisms at the base.

agricultural revolution The first known transition from hunter-gatherer subsistence to crop culture and animal husbandry about 10,000 years ago in the Middle East.

air Low density material composing the atmosphere made up mostly of nitrogen, oxygen, water vapor, carbon dioxide, inert gasses, and a large number of trace gaseous and particulate (e.g., dust, pollen) materials of natural and artificial origin.

alfisols Deep, organic soils with moderately developed soil horizons associated with deciduous forests.

algae Important primary producer organisms without complex vascular tissue to conduct nutrients and metabolic gasses as in the higher mosses, ferns, and flowering plants. They vary in size from microscopic plankton to marine kelp many meters long.

alkalinity Chemical compounds containing the potential to form a hydroxide base in water with the capacity to reduce acidity and raise pH. Carbonates and bicarbonates are common forms of alkalinity in natural waters.

alluvium Unconsolidated earth of mostly gravel, sand, and clay sizes that accumulates as a consequence of erosion, transport and deposition by running water at the foot of mountains and in flood plains, river valleys, and deltas.

alpha particle Proton emitted as radiation from the nucleus of a radioisotope.

anadromous Pertaining to migration between freshwater, where reproduction occurs, and marine environments, where maturation occurs. Many salmon and sturgeon species are anadromous.

anaerobic Metabolism that relies on chemistry without oxygen to function. Anaerobic bacteria, for example, are able to decompose organic matter without oxygen present, albeit much less efficiently than aerobic decomposition.

angiosperms Those flowering plants that develop seeds in an ovary.

animal unit In range management, a standard unit of animal mass for comparing the grazing intensity of different animal sizes based on a cow and calf combination of about 1,000 pounds. Animal-unit months are used to correct for time differences in grazing use.

aquaculture The husbandry of aquatic species for production of food, sport, and other goods and services.

aquifer A geological formation that contains groundwater. It may be unconfined above an impermeable layer, appear at the surface as groundwater springs, or be confined between two impermeable layers. Water wells penetrate aquifers.

arable Can be cultivated by ploughing or tilling.

aridisols Basic desert soils with poorly developed horizons. Some form a subsurface layer of impermeable calcium carbonates (caliche).

artesian Confined aquifers and their natural springs and artificial wells in which water under pressure boils up through openings in the upper impermeable layer containing the aquifer.

aspect In climatology and ecology, refers to directional orientation of slopes, especially with respect to the sun.

assimilation In physiology and ecology, the incorporation of nutrient into living biomass where it provides for respiration and production. Also, in environmental science, the ability of an ecosystem to take up and neutralize contaminant effects.

atmosphere The circulating air enveloping the earth and made up mostly of nitrogen (78%) and oxygen (21%). The atmosphere includes the lower troposphere, where most weather occurs, and the upper stratosphere.

atomic mass The sum of neutrons and protons in the atom's nucleus.

atoms The basic components of all matter, themselves composed of a nucleus of neutrons and protons orbited by electrons. Each chemical element is composed of atoms with a unique number of protons and a unique atomic weight.

autotrophic Adjective for primary production.

baby boom A spurt of human reproduction and population growth. In the United States, the most dramatic baby boom occurred during the two decades following World War II.

bacteria A diverse group of single-celled organisms, of which some cause disease, others are primary producers, and most decompose organic matter into inorganic components.

barrier island Elongated coastal deposits, mostly of sand, eroded from continental and offshore locations. Barrier islands often protect estuarine wetlands and other coastal lowlands from the full force of storms.

base A chemical compound that disassociates into hydroxide ions and ions of other elements. Bases react with acids to form neutral salts.

bathyal   Pertaining to the cold, dimly lit ocean zone too dark for photosynthesis.

benefit-cost analysis (BCA)   An economic comparison of alternative actions to determine which one provides society with the most beneficial use of its resources.

benthic   Pertaining to the bottom of water bodies, such as a benthic habitat.

benthos   Bottom organisms and associated organic matter of aquatic ecosystems.

beta particle   Electrons emitted from the nucleus of a radioisotope.

bioamplification   Biomagnification.

biodegradable   Anything that can be reduced to smaller and simpler form by living processes including, most importantly, biological decomposition.

biodiversity   The variety of life forms. Biodiversity occurs at genetic, population, species, community, and ecosystem levels. Among many possible expressions the most frequent is the number of species (the species richness).

biofuel   Any combustible, nonfossil organic matter used as fuel (e.g., wood), especially gas and liquid fuels produced from agricultural crops.

biogas   A mixture of methane and CO_2 produced by anaerobic decomposition of wet, nonwoody organic matter.

biogeochemical   Pertaining to processes of material flow, transformation, and cycling through living and nonliving chemical forms in the biosphere. *See also* **cycle.**

biological oxygen demand (BOD)   Under controlled conditions, the respiratory uptake of oxygen during aerobic consumption and decomposition of organic matter in water.

biomagnification   Increased concentration of chemical contaminants in organisms as chemicals pass through the food web to higher trophic levels. Certain contaminants, such as DDT and mercury, accumulate in living tissue because they are excreted at a slower rate than they are assimilated.

biomanipulation   Management of ecosystem biomass, production, and resource partitioning through the food web usually by adding to or deleting from upper trophic levels, such as by stocking fish or fishing for them.

biomass   The volume or weight of living organic matter usually expressed as dry weight or carbon weight per unit area or volume, such as kilograms (kg)/hectare (ha) or grams (g)/meter cubed (m^3).

biome   Terrestrial ecosystem defined by the dominant vegetation form occupying the landscape such as forest, woodland, grassland, desert shrub, and tundra.

biosphere   Those interactive parts of the earth's lithosphere, atmosphere, and hydrosphere where life is supported and where life influences Earth's forms and processes.

biotic potential   The capacity of populations to expand in abundance and to recover abundance when more favorable environments are provided.

birth rate   Population regeneration during some specified time interval (e.g., number of eggs hatched/year and number of live births/month).

bitumen   Heavy, high-sulfur oil extracted from tar sand.

blue-green algae   The Cyanobacteria. A bacterial group of primary producers, many species of which are capable of fixing nitrogen gas directly into organic nitrogen.

bog   A wetland type typically characterized by sphagnum mosses and other acid-tolerant plants, low calcium, deep peat accumulation, and acidic water-stained tea browned by dissolved organic matter.

breeder reactor   A nuclear power reactor that, as it operates, produces fissionable plutonium-239 by irradiating nonfissionable uranium-238.

brownfields   Abandoned industrial sites characterized by persistent physical and chemical alterations of the host ecosystem.

browser   In wildlife and range management, those animals that consume the growing tips of woody plants (e.g., deer).

buffer   In water chemistry, a chemical compound that reacts with strong acids or bases to form compounds that reduce the rate of pH change.

business cycle   Fluctuation in the economy between recession or depression and recovery and expansion caused by changing supply and demand relationships.

caecum   A large digestive organ in the alimentary canal of certain herbivores, which increases digestion efficiency of fibrous plants. Horses and ruffed grouse have caecum digestion.

calorie   The energy that raises the temperature of one gram of pure water 1°C.

calorimeter   A device for measuring calories released upon material combustion.

cambium   In woody plants, a thin layer of cells actively growing between the inner wood and outer bark.

canopy   The uppermost layer of branches and leaves in a forest—the overstory.

capital   The properties in natural resources, buildings, equipment, and other materials needed to develop and distribute goods and services.

capitalism   An approach to economic development based on a market economy and the freedom to invest in private capital for production of goods and services.

carbon   The element contributing most to organic matter including living biomass, detritus, and fossil fuel. Carbon also occurs in carbonate rock, carbon dioxide, and other inorganic matter. Carbon is cycled through ecosystems by primary production, consumption, and decomposition.

carbon sinks   A natural, long-term storage location for carbon such as soil, forest, bog peat, and deep ocean.

carcinogen   Cancer-causing agent.

carnivore   Consumers of animal matter situated in the third trophic level and higher.

carrying capacity   The ability of a specified ecosystem condition to support a population, usually expressed as the maximum population number or weight that can be sustained indefinitely.

catadromous   Aquatic migration in which the species, such as the American eel, develops to maturity in fresh inland waters, descends to the oceans to reproduce, and returns to points of origin in juvenile form.

chain reaction   A sequence of nuclear fissions created as the products of each fission cause subsequent nuclear fission in the critical mass of a fissionable material.

chemosynthesis   Primary production based in the use of chemically bound energy, often in compounds of iron or sulfur, to drive the uptake of nutrient elements. Chemosynthesis can occur in darkness and is much rarer than photosynthesis.

chlorofluorocarbons (CFCs)   Widely used synthetic compounds of carbon, chlorine, and fluorine that destroy atmospheric ozone through a chemical reaction.

chlorophyll   Photosensitive pigments, at least one form of which appears to be common to all organisms that undergo photosynthesis. The chlorophyll in cells is the site of active photosynthesis and an indicator of primary production and biomass.

clear-cutting   Cutting all trees in an area and removing them.

climate   The long-term pattern of weather determining mean and extreme seasonal conditions of atmospheric radiation, temperature, precipitation, wind, humidity, and other weather phenomena.

climax community   The last stage of natural community succession toward a dynamic equilibrium with environmental conditions in terrestrial ecosystems. The climax stage usually is more structurally diverse, more productive, and more efficient in cycling nutrients than earlier stages of community succession.

climographs   A graphic representation of population distribution with respect to environmental temperature and precipitation.

coal   The most used fossil fuel. Coal is a solid composed of combustible fossil plant matter with high carbon content. A variety of forms differ in hardness, chemical composition, and energy content. The carbon, sulfur, and nitrogen content are sources of air pollution.

coal liquefaction   An artificial process of coal conversion to synthetic oil, alcohol, and other liquid fuels.

coke   The highly carbonized energy-concentrated product of heating coal to about 1,000°C in the absence of O_2.

coliform bacteria   Usually harmless bacteria used as an indicator of fecal waste contamination in water, soils, and food.

command economy   An economic system in which an administrative authority determines the production rate and kinds of goods and services and their prices. *See also* **communism; socialism.**

commerce (commercial)   Flow of market goods and services. Commercial fishing and hunting are for market trade and commercial extinction occurs when plants or animals are too scarce to harvest profitably.

commodity   Something of use and an object of trade.

common use   *See* **commons.**

commons   Property open to common use without restriction or assigned responsibility.

communism   In political ideology, a classless social system in which all members of society are the state and share equally in resources and production capacity according to need.

community   In ecology, an assemblage of plant, animal, and microbial populations interacting through nutritional webs and other means to form a generally identifiable whole wherever similar environments and ecological connections occur.

compensation point   That point at which light intensity becomes too little for photosynthesis to sustain primary production.

competition   Striving to gain use of the same scarce resources. In ecology, competition for food, space, and other resources occurs among individuals within species and between species. In economics, competition for greater purchasing power occurs among individuals offering up the same quality of goods and services for a lower price.

compost   Organic matter stored and treated to enhance partial decomposition for use as a soil conditioner and fertilizer.

comprehensive planning   An organization-wide planning process that addresses the relative contribution of programs to meeting organizational mission, appropriate emphasis on future programs, development of new programs, and the retirement of program elements that no longer justify their costs.

concentration   Amount of one material within a specified volume or weight of another material. Concentrations in liquids and gasses typically are expressed volumetrically such as milligrams (mg)/liter (l). Concentrations in solids often are expressed by weight such as kilograms (kg) /metric ton (t).

coniferous   Pertaining to trees that produce seed in cones, have narrow, needlelike leaves, and relatively soft wood. Most are evergreen, gradually losing old leaves and adding new leaves.

conservation   A philosophy and approach to resource management that seeks the greatest human benefit from resource use over the long run through efficient use and protection of renewable resource regeneration capacity.

conservation biology   The study and management of phenomena associated with sustaining biodiversity in all of its many expressions from genetic diversity to ecosystem diversity.

conservation movement   A moral movement first involving a loose coalition of nineteenth-century scientists, educators, publishers, editors, and politicians who espoused efficient natural resource development and use for the greatest long-term benefit to society.

conservation of energy   From the laws of thermodynamics. Energy is neither created nor destroyed but is simply transformed. Energy transformation efficiency cannot exceed 100% and the input energy in any transformation is degraded to a less intense form able to do less work.

conservation of matter   A natural law stating that matter is neither created nor destroyed in any physical or chemical transformation.

conservation tillage   *See* **tillage.**

consumer price index   An index to the annual amount that prices increase because of inflation.

consumers   In economics, all users of all kinds of resources. In ecology, those users of food resources that ingest organic matter and assimilate it internally.

consumptive use   Any resource use that results in lost resource utility.

continental shelf   Those parts of continents submerged by shallow ocean.

continuous grazing   In range management, the grazing of a particular pasture throughout the grazing season year after year.

contour farming   Plowing and planting that follow the land contour and create mini-terraces that slow water runoff and soil erosion.

conventional-tillage farming   Land cultivation by plowing, disking it to break up the soil clods, then smoothing it.

cooling tower   An engineered structure designed to cool thermal effluents from steam-generation power plants and other industries through evaporation or atmospheric heat exchange.

coppice method   Tree regeneration by sprouting from stumps and roots following cutting.

Coriolis force   The force from the rotation of the earth that causes a deflection of air and water movements to the right and counterclockwise in the Northern Hemisphere and to the left and clockwise in the Southern Hemisphere. Its effect shows in large atmospheric, lake, and ocean circulation patterns.

cover   In wildlife and fishery science, those attributes of habitat that provide refuge from the physical environment and predation.

critical mass   The fissionable material needed for nuclear chain reaction.

crop rotation   Alternating field planting between crops with the intent of sustaining soil structure and fertility. Nitrogen-fixing beans and alfalfa often are planted alternately with other crops, for example.

crown fire   Fire in the forest overstory.

cycle   In ecology, material cycles involve the movement—mostly by air and water transport—of elements through different forms and locations in ecosystems and their environments. Gaseous cycles include atmospheric phases such as for oxygen, carbon, nitrogen, and hydrogen. In contrast, phosphorus and other sedimentary-cycle nutrients have no gaseous phase and full cycling is very slow, depending on continental erosion and uplift.

Cynanobacteria Blue-green algae.

DDT A synthetic pesticide, dichlorodiphenyltrichloroethane, now outlawed in the United States because of environmental damage, but widely used elsewhere in the world.

death rate Mortality. The number or fraction of deaths in a population, occurring in some specified period of time; frequently one year.

deciduous Describes woody plants that shed all of their leaves at one time just before extreme cold or dry seasons. Examples include maples, oaks, poplars, alders, and many other broad-leaved trees, and a few conifers such as larches.

decomposers Organisms that externally secrete enzymes on to adjacent organic matter where it is digested to molecule sizes that can be assimilated through cell membranes. The vast majority are bacteria and fungi.

decreaser In range management, those highly palatable plants that decrease as grazing increases in intensity.

deferred-rotation A grazing management system in which livestock are rotated between two or more range areas to increase the long-term efficiency of range conversion into livestock production.

deforestation Forest loss or removal.

degradation Reduction in form, capacity, potential, or value, such as the breakdown of organic matter by decomposition, the erosion of continents, the lowering of energy available to do work, or the lowering of potential for resource sustainability.

demand In economics, the desire shown for goods and services indicated by the willingness to pay for them. *See* **supply and demand.**

demographics Statistics of human populations including age, sex, ethnicity, income, education, health, and various other descriptive attributes.

dendrology The study of trees leading to their taxonomic classification.

denitrification In the nitrogen cycle, a decomposition process that reduces solid forms of nitrogen to nitrogen gas, which can return to the atmosphere.

density Number or weight of items per standard areal unit (e.g., hectare, acre) or volume (e.g., cubic meter).

density dependent Ecological processes that depend on population density, such as prey mortality due to predation.

density independent Ecological processes that are independent of population density, such as weather effects where cover is not a limiting factor.

depletion rate (time) Reduction in material amount within a specified time; most usually applied to resource supply and concentrations.

depression In economics, a condition characterized by overproduction of goods and services for demand, sluggish trade, lowered employment, and lower average income and buying power.

desalination Salt removal by distillation or other techniques.

desert A biome characterized by low vegetation density, a mix of woody and herbaceous species, and succulent or waxy-leaved species. The natural community is adapted to spotty precipitation typically totaling to less than an annual mean of 300 mm.

desertification Transformation of nondesert ecosystems to desertlike conditions with less vegetation, less organic matter in the soil, more rapid drying of the soil, and more wind and water erosion. Desertification typically is caused by over-grazing and improper crop culture.

detritivore   A detritus-eating animal that derives part of its nutrition from decomposers.

detritus   Dead organic matter usually in some state of decomposition by associated bacteria or fungi.

deuterium   An isotope of hydrogen.

dimictic   Twice mixing; pertaining to lakes that mix each spring and fall and remain stratified in summer and under winter ice.

disclimax   Conditions of community succession different from climatic climax and maintained by chronic soil conditions or by fire, flood, storm, tidal flux, and other periodic environmental disturbance.

discounted interest rate   *See* **interest rate.**

dish engine   An external combustion engine formed from a parabolic dish, which concentrates solar energy, and a receiver that captures and transfers the heat to a chamber where heated and cooled helium gas expands and contracts, driving a piston. The dish engine can be linked to an alternator to generate electricity.

diversity   The variety of form and function in ecosystems, societies, and any other manifestation. Opposite of monotony.

drainage basin   Watershed.

dredging   A process of shoveling or pumping sediment out of aquatic environments for purposes of mining, navigation improvement, or beach replacement and aug-mentation.

drift net   A net usually made of fine monofilament synthetic material that is set adrift to entangle fish and invertebrates for commercial harvest.

drip irrigation   Precise and efficient delivery of water to plant roots through a sys-tem of slowly leaking hoses and outlet fixtures.

drought   A prolonged period of dry weather sometimes defined by a precipitation less than 75% of average.

dust bowl   The result of extreme drought of the early 1930s centered in the south-central U.S. plains. Named after huge dust storms caused in part by unwise farm-ing of shortgrass prairie.

dynamic equilibrium   Fluctuation without consistent trend around some stable condition.

ecology   The study of interactions among living organisms and their physical, chemical, and biological environment.

economic growth   Increased value of goods and services indicated by such measures as the Gross National Product.

economic system   An organized complex of processes and institutions based in hu-man preferences, motivation, and choices made in developing natural resources, capital goods (physical facilities), and labor for production, distribution, and con-sumption of goods and services. Often referred to simply as "the economy."

economics   The study of demand for, supply of, and trade in goods and services. Economics facilitates more beneficial decisions where resources are scarce.

ecosphere   Biosphere.

ecosystem   A cohesive, self-regulating, functional, and adaptive organization formed from the interactions of natural communities and their habitats (physical environments). Ecosystems are hierarchical (i.e., communities and habitats inter-act to form larger systems such as the different communities and habitats in a lake or in a terrestrial biome). The ecosphere is the largest ecosystem.

ecosystem management A holistic approach to sustainable natural resource management, which seeks resolution of resource use conflicts based on sound understanding of systems process. It identifies and works with the natural integration of ecological and social processes and within and across boundaries of natural systems such as watersheds and biomes. *See* **integrated resource management.**

ecotone A zone of gradual transition between two ecosystems showing attributes of both as well as unique attributes.

edaphic Pertaining to soils, such as the nutrient and structural environment of plant roots.

edge In ecology, a sharp transition and boundary between ecosystems.

efficiency Output divided by input. Efficiency can be physical (e.g., energy transformation from one form to another), ecological (e.g., ingested food conversion to production), and economic (e.g., benefits for the costs).

electrofishing A technique used to sample freshwater fishes by passing a battery of gasoline-generated electricity through metal curtains, seines, or, most commonly, loops at the ends of insulated handheld poles.

electromagnetic Forces that include solar radiation of different wavelengths and energy level. Radiation prominently includes, in increasing wavelength, ultraviolet, visible light, infrared, and radio waves.

electron A negatively charged particle of nearly no mass that orbits the nucleus of an atom. Electron sharing by atoms is the means by which the molecules of complex structures are formed.

electrostatic precipitator A device installed in smokestacks for removing materials from emissions.

elements In chemistry, the 92 naturally occurring atomic structures that form the basic building material of all natural abiotic and biotic structure. Some of the most important for life processes include carbon (C), hydrogen (H), oxygen (O), nitrogen (N), and phosphorus (P). Elements combine to form chemical compounds.

emigration Population departure from an area.

emissions trading A market approach to pollution control in which credits are earned whenever a polluting source reduces emissions by more than the legally required amount. The credits might be saved for later emission use or sold to others.

endangered species A designation applied when wild populations could soon become extinct in their natural range. The designation usually includes unique subspecies. In the United States, listing for protection and recovery is authorized under the federal Endangered Species Act and individual state laws.

energetics Processes involving energy transformation.

energy Force that can move or change matter and cause heat transfer between objects of different temperature; i.e., the capacity to do work. Fundamental forces in nature include electromagnetic energy (e.g., visible light), gravity, and the strong and weak forces in atoms.

entisols Young volcanic soils of the Rocky Mountains, which are little more than organically enriched regolith without horizon development.

environment The sum of material and energy conditions surrounding and permeating any object or assemblage of objects, such as organisms and ecosystems.

environmental resistance The sum of all environmental factors limiting populations and biotic communities.

environmental scanning A wide-ranging review of trends and influential events in a management environment with emphasis on opportunities for and threats to the organization's mission, goals, and objectives.

environmentalists   A political grouping of people who typically favor policies that protect natural environment and partial or full restoration of altered environments to their natural condition.

epilimnion   In lakes, a warm-water layer of lower density that floats on a colder layer of higher density (hypolimnion), creating stratification in waters too deep for wind action to effectively mix all the way to the bottom.

epipelagic   Pertaining to the uppermost layer of the open ocean.

erosion   The physical or chemical displacement of rock, soil, and organic matter. Physical erosion results from flowing water, wind, and particles in suspension. Chemical erosion results from aqueous solutions that dissolve minerals. Soil erosion occurs uniformly as sheet erosion, concentrates in the small channels of rill erosion, and gouges out deep channels in gully erosion.

estuary   A transitional ecosystem between freshwater flows and ocean characterized by tidal flux and a moving wedge of relatively dense saline water under less dense freshwater. Aquatic organisms in esturaries typically are broadly tolerant of salinity and depth changes.

euphotic zone   That part of aquatic ecosystems illuminated enough to sustain primary production.

eutrophication   Increased primary production, decomposition, and associated physical, chemical, and biological changes in aquatic ecosystems resulting from nutrient enrichment. Eutrophication may be natural or culturally caused by fertilizer, organic waste, and treated sewage, among other sources.

eutrophy   A state of high nutrient enrichment and high aquatic productivity.

evaporation   A physical process in which enough heat is gained in water or other material to vaporize it, in the process transferring heat from the material to the gas.

evapotranspiration   The joint processes of evaporation and transpiration.

even-aged   About the same age. Term used in forestry for trees.

evergreens   Trees that do not shed their leaves all at once or seasonally, including most coniferous trees and many broadleaf semitropical and tropical trees.

evolution   Organic evolution is the process by which life adapts to environment through change in genetic structure and associated anatomical, physiological, and behavioral attributes. The most adaptive traits are retained by differential reproductive success and gene transfer to successive generations.

exotic species   Organisms of nonnative and usually foreign origin.

external benefit   *See* **externality.**

external cost   *See* **externality.**

externality (ies)   In economics, either the exclusion of benefits or costs from consideration in the price of an economic transaction. For example, pollution cleanup costs incurred by society are excluded from the price of goods charged a company and people other than those benefitting from the goods subsidize the low price. In turn, the buyers of the goods are externally benefitted by society's subsidy.

extinction   Total loss of a genetically unique group of organisms at subspecies and species levels of genetic distinction. Extinction may be local or global and is assumed global when not stated otherwise.

fecundity   The number of embryos per mature adult in a population. Fecundity is a measure of population reproductive potential and can be used to estimate birthrate if embryonic mortality is known.

feedlot   An enclosure where livestock, sometimes in the thousands, are intensively husbanded.

fingerling   A postlarval immature game fish about as long as any one of the fingers on a man's hand, or typically between about 2 and 5 inches.

fish ladder   A structure designed to allow fish passage around or over a dam or other obstruction. Typically, a series of steplike pools are constructed up an incline.

fisheries   A location where, or a group of organisms from which, aquatic species (not necessarily true fishes) are actually or potentially captured by humans. Fisheries exist for commercial sale, sport, and subsistence.

fission   In atomic energy, the splitting of a fissionable isotope by a neutron or other atomic particle.

fixed costs   Capital investment, rent, and other costs that are independent of variation in goods production.

flashflood   A rapidly rising flood following a storm event typically in steep and impermeable watersheds, such as in mountainous desert.

floodplain   The relatively flat land adjacent to streams and rivers that is formed and maintained by periodic flooding.

fluidized-bed combustion (FBC)   A process for removing sulfur dioxide from emissions of burning coal. Powdered coal and limestone are mixed in a suspension maintained by forced air as the mixture is burned. The limestone reacts with the sulfur dioxide to form a calcium sulfate precipitate.

flyways   Routes repeatedly flown by birds in migration from one area to another. Four distinct flyways are recognized for waterfowl of North America: Pacific, Central, Mississippi, and Atlantic.

food chain   The predator-prey feeding sequence from plant to herbivore, herbivore to carnivore, and carnivore to top carnivore. Recognizable food chains are rarely longer than five feeding levels.

food web   The interconnection of two or more food chains. *See* **food chain.**

forage   Plants fed on by herbivores; most often applied to the foods of large grazing and browsing livestock and wildlife.

forb   A term used in range science that lumps together all of the broadleaf, branching, herbaceous vegetation of grasslands to distinguish them from the grasses, rushes, sedges, and other grasslike herbaceous plants.

forecasting   Trends and other analyses used in planning to project future conditions shaping the organizational environment and shaped by the organization.

forest   A biome classification identified by the dominance of medium to tall woody plants. Forests occur in areas of moderate to high precipitation.

forestry   The practice of forest resource management.

fossil fuel   Incompletely decomposed organic matter altered in structure and chemistry by long exposure to an anoxic, pressurized environment under the weight of sedimentary rock. Examples include petroleum, natural gas, and coal.

fragmentation   The loss of wholeness in ecosystems, habitat, law, and management that threatens ecosystem integrity and dependent functions and services; most especially biodiversity.

fry   In fisheries, the larval stage of fish following hatching and before all embryonic characteristics disappear.

function   In ecology, the rates and directions of material change that sustain generally predictable systems interactions and outputs. Energy drives function.

fungi   One of the life kingdoms. Most fungi live by decomposing organic matter while some are parasitic. Fungi penetrate detritus and hosts with threadlike hyphae. Examples are mushrooms, molds, rusts, and yeasts.

game   When applied to wildlife and fish, animals that are harvested for both sport and food.

gamma rays   High-energy electromagnetic radiation from radioisotopes. Gamma rays readily penetrate and damage living tissue.

gasohol   A fuel mixture of about one part methyl alcohol and five to ten parts gasoline.

GDP   *See* **gross domestic product.**

generalists   Broadly adapted species that can take advantage of a wide range of ecological resources, although less efficiently in each resource category than narrowly adapted specialists.

genes   The sites on deoxyribonucleic acid (DNA) molecules that are inherited through reproduction and direct the structural development and functioning of all life.

genetic diversity   The variety of genetic expression in the deoxyribonucleic acid (DNA) of chromosomes within species and all other life classifications.

genetic engineering   Alteration of genetic material with the intent of changing organism traits to improve attributes such as food production and disease prevention.

geographic information system (GIS)   A computer-mediated approach to organizing information through maps, plans, and other means of spatial referencing.

geothermal   Earth-generated heat. Useful geothermal energy is typically transferred from molten rock and extruded from deep mantle sources into solid rock, liquid water, and steam.

goal   A desired achievement typically expressed in general and holistic terms. Goals identify broad resource categories (e.g., range, forage, ecosystems), direction of change (e.g., more, fewer), and approaches (e.g., stock, restore, protect, educate) to achievement by objective and leave specific details and schedules to the objectives.

goods   In economics, material wealth that can be traded and sold; merchandise. Goods can be natural resources used as they are or in manufacturing.

gram   A unit of weight measure equivalent to 0.001 kilograms, 1,000 milligrams, 0.0022 pounds, and 0.035 ounces.

grassland   A biome (terrestrial ecosystem) dominated by grasses and forbs and found in areas of moderate precipitation.

grazer   In wildlife and range management, those animals that consume forbs and grasses such as cattle. In ecology, any animal that consumes a primary producer in the *grazing* food chain such as herbivorous zooplankton.

green revolution   The relatively dramatic increase in grain and other crop yields following World War II. The green revolution was based on more effective integration of improved soil preparation, fertilizers, irrigation, pesticides, plant breeding, and selection for desirable crop attributes. Often associated was costly soil erosion, pesticide contamination, fertilizer pollution of runoff waters, and high fossil fuel energy consumption.

greenhouse effect   Atmospheric warming as a consequence of differential transmission of electromagnetic radiation by atmospheric gasses known as greenhouse gasses.

greenhouse gasses   Atmospheric gasses that cause the greenhouse effect; examples are carbon dioxide, water vapor, chlorofluorocarbons, methane, and nitrous oxide.

greenway   Linear corridors of lawn, garden, and natural area often paralleling roads, walkways, bicycle paths, drainages, utility rights-of-way, and other natural space compatible with urban development.

gross domestic product (GDP)   The total monetary value of all goods and services priced inside a nation's borders over the course of one year. It approximates the sum of income from all sources. Average per capita GDP is a measure of average individual welfare. Real GDP is adjusted for the estimated rate of inflation.

gross national product   The gross domestic product of nations.

gross production   *See* **production.**

groundwater   Water that flows at various velocities through and accumulates in the ground. While some groundwater flows relatively rapidly to surface springs—depending on soil, slope, and geology—other water accumulates in deep aquifers where it moves very slowly.

gymnosperms   Plants with seeds exposed; usually in a cone and not in an ovary.

habitat   The environment either actually or potentially lived in (inhabited) by a population or by a biotic community.

habitat conservation plans   A process through which land containing critical habitat of a federally threatened or endangered species can be used for other purposes while assuring protection of the species despite some incidental loss of population members.

hardwoods   The wood or whole plants of any of a wide variety of broad-leaved angiosperms with dense arrangements of vessels and tracheids.

hazardous materials   Used, stored, or disposed materials that are toxic or otherwise potentially harmful to people, property, and supporting ecosystems. Legally, hazardous materials in the United States are any on a long list declared hazardous and regulated by the federal EPA and state agencies.

heartwood   Inner wood, typically darker in color, its vessels and tracheids have filled and hardened to strengthen the trunk.

heat   The kinetic energy held in matter, which flows spontaneously between objects holding different concentrations (i.e., temperatures) toward a condition of uniform concentration.

hectare   A unit of areal measure equal to 10,000 m^2 and about 2.5 acres.

herbivore   Consumers of living primary producers; examples are deer, cattle, grasshoppers, and certain zooplankton.

heterotroph   Secondary producers.

histosols   Peaty soils of very high organic matter content found primarily in lowland areas with swamps and bogs.

holistic grazing   An approach to grazing that usually involves short and intense grazing in rotation among numerous areas and is compatible with forage growing cycles and nutritional value.

homeostasis   Self-regulation.

horizon   In soil science, a distinct soil stratum with unique soil attributes that set it apart.

host   The living nutritional source for a parasite.

humidity   The concentration of water vapor in the air.

humus   A mix of decaying organic matter of generally unrecognizable origin, inorganic matter, small consumers, and decomposers that occurs in topsoil just below the plant litter. Humus typically has an open, well-aerated structure that holds water and binds nutrients well.

hydrocarbons   A general class of chemical compounds with a high proportion of carbon, hydrogen, and oxygen.

hydroelectric   Electrical power produced from the force of gravity operating on water falling through turbines connected by drive shafts to generators.

hydrologic cycle   Movement of water from earth to atmosphere through evapotranspiration and back again through precipitation.

hydrology   Water science and study.

hydropower   Any mechanical or electrical power derived from falling water.

hydrosphere   The water world, including all of the ice, liquid water, and water vapor in, on, and above the earth as well as the water included in living things.

hypolimnion   The colder and more dense layer of water found at lake bottoms during summer. Because the hypolimnion does not mix with overlying water it often develops marked differences in water chemistry and life processes.

igneous rock   Magma of diverse composition extruded from the earth's mantle and cooled into surface and subsurface formations. Granite, basalt, schist, and gneiss are common forms.

immigration   Movement of new population members, including humans, into an established population or area.

impoundment   Reservoir.

inceptisols   Slightly more developed variations of entisols found on volcanic ash primarily in the Pacific Northwest.

increaser   In range management, plants that are slightly to highly unpalatable to grazing animals and increase in abundance as grazing intensity increases.

incremental benefits   Marginal benefits.

incremental cost   *See* **marginal benefits.**

industrial revolution   The rapid transition from agricultural economies to economies based on manufacture of material goods. The harnessing of water and steam power in the seventeenth and eighteenth centuries initiated and accelerated the industrial revolution.

infiltration   In watershed science, the penetration of soil surface by water.

inflation   Rate of increase in the average price charged for all goods and services.

information age   A time, now continuing, in which information processes dominate work and lifestyle. The information age rose to prominence during the middle of the twentieth century when the economy became most influenced by information services facilitated by major advances in computing and communications technology.

inorganic   Composed without carbon.

institutions   In economics, practices or organizations for governing human behavior. Markets and government regulations are two primary examples.

integrated management   A holistic approach to planning and implementation across organizational authorities with the intent on capturing all important interactive elements of the management focus, such as integrated coastal zone management.

integrated pest management   Combining chemical, crop cultivation, predator, and other biological methods of pest control to promote the most cost-effective sustainable agriculture.

integrated resource management   A systematic approach to natural resource management, which seeks greater efficiency and sustainability through comprehensive understanding of interactions among discrete areas of resource management. *See* **ecosystem management.**

integrity   environmental law and ecological science, the completeness of ecological structure and function resulting in an adaptive, self-regulating ecosystem.

intercropping   Cultivating a mix of two or more crops in the same field, usually for improved pest and soil management.

interest rate   The price charged to borrow money for use. The real interest rate is corrected for inflation. The prime interest rate is charged by the largest banks to the best customers. It establishes a basis for comparison of all other interest rates and is monitored and managed by the Federal Reserve, which is the government lending institution of last resort. The Federal Reserve lends at a discounted interest rate a fraction lower than the prime.

internal combustion engine   Form of power conversion to the piston drive of a crankshaft by repeated explosion of combustible fuel inside a confined space.

intertidal   Pertaining to the space and ecological processes that occur between low and high tides.

intrinsic rate of increase   In population ecology, the maximum rate a population can increase under the most favorable conditions of environment, population density, and other population attributes.

invasive species   Nonnative aggressively colonizing species that displace native species. Examples include zebra mussels, purple loosestrife, Kentucky bluegrass, and starlings.

inventory   Taking stock; in management planning, the gathering and organizing of information about existing natural and human resources, facilities, finances, trends, issues, opportunities, threats, organizational strengths and weakness, and other knowledge pertaining to management effectiveness and efficiency.

inversion   *See* **thermal inversion.**

ion   An electrically charged atom, either positive or negative.

island biogeography   The study of the effects of island size and distance from the mainland on species colonization rates, extinction rates, and diversity. The principles extend to any habitat fragment isolated in a very different landscape setting (matrix) such as lakes in a forest or mountaintops in a desert.

isotope   A variant form of a chemical element with the same number of protons but a different number of neutrons in the atom nucleus.

issue   A disruption in planning continuity that begs attention and response. Issues may be either threats to or opportunities for goal accomplishment.

keystone species   Like the keystone that holds up a stone arch, a keystone species has disproportionate effect on sustaining the integrity of an ecosystem. Beaver, for example, engineer the pond ecosystems that many species depend on. Humans are the most influential keystone species on Earth.

kilo-   A word prefix that signifies 1,000 such as kilocalorie (1,000 calories), kilowatt (1,000 watts), and kilogram (1,000 grams).

kinetic energy   Energy contained in matter above a temperature of absolute zero.

labor   All forms of physical and mental work applied in an economic system.

lake   Any closed, water-filled basin with variable flow direction differing substantially from the ocean in chemical composition. Lakes are of artificial and natural origin, temporary and permanent, fresh and salty, and include small ponds and the very large "great" lakes.

lampreys   A family of fishes, many of which are external parasites of other fish species and are often lethal in effect. The marine lamprey is especially large and lethal.

land-use planning   A process by which alternative land-use proposals are evaluated and a plan is selected for the most cost-efficient provision of all services desired by the community. Land-use planning typically centers on coordination of private development with public infrastructure and environment.

landfill   A common means for disposing of solid wastes by burial in and under soil and unconsolidated regolith.

landscape ecology   Ecological study of arrangements and interactions among like and different vegetation, topographic, anthropogenic, and other geographic features, typically at a regional scale.

leaching   The dissolving and removal of materials by a solvent such as water.

levee   An artificial or natural wall of earth bordering a shoreline. Artificial levees are constructed to restrict flooding.

life expectancy   The length of life expected at any particular age based on the average length of life under the expected environmental conditions.

limiting factor   A principle in population ecology, related to tolerance, which states that too little or too much of any one factor can prevent further growth and expansion of a population regardless of how near to optimum all other factors may be.

limnetic zone   The offshore surface waters of a lake illuminated enough to sustain photosynthesis, or to about 1% of the illumination above the lake surface.

liquefied natural gas (LNG)   Liquefaction of natural gas by intense cooling.

liter   A unit measure of volume equivalent to 0.001 m^3, 1,000 milliliters, and 1.10 quarts.

lithosphere   The world of inorganic solid and molten rock and its breakdown products, including the inorganic parts of soil.

littoral zone   Waters illuminated enough to sustain photosynthesis on the bottom, or to about 1% of illumination above the water surface.

loam   A soil texture with near equal mix of sand and clay.

long-range planning   Far future assessment of trends and events that could profoundly influence organizational management effectiveness in the long run.

magma   Molten matter in the earth's mantle, and in fissures leading upward toward the earth's surface, including volcanic vents.

management   For natural resources, the planning and implementation of change with the intent of directing resource outputs toward beneficial outcomes. Usually preceded by a resource focus such as wildlife, forest, range, urban land, recreation, fishery, or, more holistically, ecosystem management.

marginal benefits   In economics, the benefits gained from a specified input, the marginal cost. Wise economic decisions are based on maximizing the marginal net benefit, which is the marginal gross benefit in excess of marginal cost.

marginal cost   *See* **marginal benefits.**

market economy   An economic system in which individual buyers and sellers price, buy, and trade goods and services freely based on individual knowledge of competitive offers. Unlike farmer's markets, most competitive markets are not physically consolidated and rely on advertising media to compare prices and qualities.

market hunting   Hunting with intent to sell wildlife products.

marketable pollution permits   A tradeable allocation of air, water, soil, or other space for energy or material disposal once government decides what can be safely assimilated. The allocation may be sold to others to the advantage of both parties.

marsh   A wetland dominated by herbaceous plants such as cattails, sedges, and grasses tolerant of wet soils.

mass   The material, or matter, in a defined volume or area.

mass transit   Buses, trains, trolleys, and other forms of transportation that carry a large mass of people.

maturation   In population biology, becoming reproductively capable. In community ecology, the process of succession to the climax stage.

maximum sustained yield   *See* **sustained yield.**

mega-   A prefix meaning very large or more specifically, 1,000,000. For example, power plant output is reported in megawatts (millions of watts).

megapolis (also megalopolis)   An outsized city typically created from suburban sprawl among once separate cities.

mensuration   The measurement of timbered area, whole forests, single trees, logs, and other pieces and units of forest products.

meristem   Actively growing plant tissue of shoots and roots.

mesic   Moist; with precipitation typically over 500 mm.

mesopelagic   A dimly lit oceanic zone too dark for photosynthesis where many species find refuge during the day before rising to the epipelagic zone to feed at night.

metamorphic rock   Rock transformed by heat and pressure from original igneous or sedimentary form typically to forms with greater density. Examples are marble from granite and slate from shale.

meter   A unit of linear measure equal to 100 centimeters, 1,000 millimeters, 39.4 inches, and 1.09 yards.

micro-   A word prefix for very small or more specifically 1/1,000,000 (0.000001). A micrometer is 1 millionth of a meter, equal to 1 micron.

migration   Repeated population movements between two or more locations. Migrations typically occur in response to seasonal temperature and wetness cues or physiological cues. They may follow the same track coming and going or follow a circle without backtracking.

milli-   A prefix meaning one thousandth (0.001), such as milligram (one thousandth of a gram).

mineral   Any naturally occurring element or compound usually crystalline and solid in structure and inorganic in composition. Although organic, coal is sometimes included.

mission   The organizational purpose. For management organizations, the organization's "reason for being" often is captured in a mission statement, which ideally guides goal and objective development.

model   A simplified abstraction of reality that intends to capture the primary attributes of real-world form and function for research and management analysis. Models may be physical, such as water-resource engineering models, or mathematical such as many geophysical, ecological, and economic process models.

molecule   Atomic union of the same or different chemical elements. By sharing electrons, two atoms of hydrogen and one atom of oxygen form a molecule of water.

mollisols   Natural grassland soils, which typically are deep, rich in organic matter, alkaline, and show moderate horizon development in a soil profile.

monoculture   One-crop cultivation, most evident in areas where a single crop dominates the landscape such as corn and wheat in parts of the midwestern United States.

monopoly   In economics, control over the supply of specific goods and services in the absence of competition.

mortality   Population death rate.

multiple use   A land management approach, promoted in certain federal land-management legislation, which advocates an overlay of compatible land uses resulting in the greatest total profit on private lands or public benefit on public lands.

municipal Pertaining to village, town, and city service organizations and regional coordinative institutions.

mutagen Anything that increases the rate of mutation from background rates.

mutation Material change, most often applied to genetic and chromosomal change in living cells leading to structural and functional change in living organisms.

nano- A word prefix for a unit of measure equal to 0.000000001 (1 billionth) and for very small things measured in billionths of a meter (e.g., nanoplankton, nanotechnology).

natality Population birthrate.

natural gas Underground deposits of gas consisting of 50 to 90% methane and small amounts of heavier gaseous hydrocarbons such as propane and butane. Natural gas is a fossil fuel like coal and petroleum and is often found associated with them.

natural law A 100% predictable natural process under specified conditions.

natural radioactivity Nuclear change in which unstable nuclei of atoms spontaneously emit mass, energy, or both at a fixed rate.

natural resources Actually or potentially useful materials and energy untransformed by manufacturing for the most part. Undeveloped land, water, air, and life, in all their diverse forms, are the most basic of natural resources.

natural selection The process by which some genes and gene combinations in a population are passed on to the next generation more often than others because the parents having the most adaptive gene-linked traits survive and reproduce more successfully.

natural system Complex natural organization with structures and functions that generate predictable material and energy outputs when provided historic environmental inputs of material and energy.

nekton The swimming inhabitants of water.

neritic zone The ocean zone closest to the intertidal zone near shore and including the relatively shallow waters of the continental shelves and similar depths around offshore islands.

net production *See* **production.**

networking Informal and formal interaction within and across administrative lines of authority, often by way of advanced communication technologies. Worldwide networks are rapidly developing through personal computers and the Internet.

neutron An uncharged particle with a relative mass of 1.0 found in the nuclei of atoms.

niche In ecology, how a species fits into an ecosystem both physically and functionally. The physical (or environmental) niche defines the tolerable environment for all life stages and successful reproduction. The functional niche is what the species does in an ecosystem, including how it competes for resources and influences its environment.

nitrogen A required and often limiting life nutrient found sparsely in inorganic form and commonly in gaseous form. Nitrogen undergoes a complex cycle through relatively small quantities of inorganic solids, organic form, and gas. Bacteria play an essential role in the cycling rate.

nitrogen fixation The transformation of gaseous nitrogen to inorganic form by lighting-caused oxidation and to organic form by bacterial (including blue-green algae) assimilation.

nonconsumptive use Resource use that does not affect resource abundance, such as certain recreational use of wilderness and water use where evaporation loss is nil.

nonpoint source A diffuse source of materials, often pollutants, such as entire watersheds or the windswept landscapes contributing to atmospheric dust. Many of the remaining pollutants in water are from nonpoint sources.

nonrenewable resources A nonregenerative resource that is irreversibly diminished with consumptive use resulting in a useless form.

nuclear energy Nuclear fission, fusion, or spontaneous release of radiation.

nuclear fission The nuclei of certain isotopes with large mass numbers split apart into two lighter nuclei when struck by a neutron. This process releases neutrons and a large amount of energy.

nuclear fusion The forcing of nuclei of two elements at very high temperature into a fusion forming a nucleus heavier than either of the initial nuclei. Much energy is released in the process.

nucleus In physics, the concentration of mass at the center of an atom, composed of positively charged protons and (except for hydrogen) uncharged neutrons. Each proton and neutron has a relative mass of 1.0. Atomic mass of elements varies as the number of protons and neutrons (atomic number) in the nucleus varies.

nutrient An element or compound needed by all living things for material growth and other life functions.

objective A measurable, highly specified intended result of planning and action, including schedules for completion (e.g., increase black duck population number by 50% within a decade). Objective specification seeks general direction from goals.

oil *See* **petroleum.**

oil shale A sedimentary rock containing kerogen, a waxy organic residue that can be vaporized upon heating and condensed to an oil.

old-growth forest Forest in a mature or climax successional stage in which some trees are approaching maximum age. In the United States, it typically has not been previously cut or intentionally burned, at least by Europeans.

oligopoly Determination of the supply of a good or service by a few suppliers; a cartel.

oligotrophic Waters of low nutrient concentration and low productivity.

omnivore Animals that can feed from both primary and secondary production sources (e.g., vegetable and meat). Examples are pigs and people.

operations planning The process of organizing specific actions and facilities to accomplish management objectives, including the assignment of budgets to specific projects and tasks.

opportunity Open access. Opportunities taken are offset by lost opportunities, or tradeoffs. Taking the opportunity to benefit from building on land, for example, foregoes the opportunity to benefit from farming it.

opportunity cost The value lost in any economic transaction resulting in foregone opportunities. For example, the cost of developing a mine in a valuable wilderness area should include the opportunity cost (i.e., the value lost with the wilderness).

optimum sustained yield *See* **sustained yield.**

ore Minerals with concentrations of extractable metal.

organic Matter produced by life processes that always contains carbon and hydrogen, typically oxygen (hydrocarbons), and other elements less consistently.

orographic effect Topography influences precipitation because air masses cool as they move upward over mountains.

overburden Layer of soil and rock overlying a mineral deposit that is removed during surface mining.

overfishing   Reducing fish supply by fishing to the point where population sustainability is threatened or is at too low a level to sustain beneficial resource use.

overgrazing   Reducing forage supply by grazing to the point where sustainable forage is threatened or is at too low a level to sustain beneficial use.

overpopulated   For organisms, including humans, too many individuals for supporting resources to sustain at some prescribed quality level.

oxisols   Highly developed soils of the wet Tropics; the surface horizon is leached by rain of most nutrients, forming a soil of low fertility and farming suitability.

oxygen   A basic element in life structure, a by-product of photosynthesis, and the final receptor for carbon elimination (forming carbon dioxide) in aerobic metabolic cycles. Because photosynthesis breaks down carbon dioxide and aerobic respiration reforms it, the oxygen and carbon cycles are closely linked as oxygen and carbon move between organic states in the biosphere and gaseous states in the atmosphere.

ozone   A compound of three oxygen atoms (O_3), which forms a layer in the stratosphere that filters out life-harming ultraviolet radiation from the sun. Ozone also is generated from by-products of internal combustion in the lower atmosphere, where it becomes a harmful pollutant in high enough concentrations.

ozone hole   A depletion of ozone in the outer atmosphere linked to emission of artificially synthesized chlorofluorocarbon gasses such as Freon.

P/B ratio   The production/biomass ratio. A measure of population or community turnover rate of biomass. Low P/B indicates low recovery rate and relatively high vulnerability to resource overuse.

P/E ratio   Climatic classification for vegetation development potential based on the ratio of precipitation to evaporation. Desert P/E is low; rain forest P/E is high.

parasite   Organisms that live on (ectoparasite) or in (endoparasite) a living host and consume from it without killing it directly—at least immediately. Parasites vary in the harm they cause hosts, some feeding only on dead tissue, with little other effect, and others feeding slowly on vital tissue until the host ultimately dies.

passive solar heating   The controlled trapping and conversion to heat of solar radiation by greenhouse or other material absorption, usually for heating building interiors.

pathogen   Disease-causing organisms.

pelagic   Pertaining to open lake and ocean waters offshore and off bottom.

per capita   For each individual, as in mean per capita income. Latin meaning "by head."

percolation   In watershed science, the downward movement of water through soil.

permafrost   Permanent layer of ice beneath ground surface in arctic and alpine regions. Permafrost is an environmental determinant of tundra ecosystem form and function.

petrochemicals   Petroleum-derived chemicals used to make many products including gasoline, kerosene, lubricant oils, plastics, fertilizers, pesticides, synthetic fibers, paints, and various medications.

petroleum   Literally (from the French), rock oil. A slippery, flammable, liquid fossil residue varying in thickness and color and containing hydrocarbon compounds with traces of sulfur, nitrogen, and other biochemical elements. Petroleum is a source of petrochemicals.

pH   A measure of acid and base strength expressed as a log-base 10 transformation of the hydrogen concentration. A solution of hydrogen ion concentration equal to 0.0001 molecular weight of water has a pH of 4.0. Pure water equally dissociates into hydrogen ions (acid) and hydroxide ions (base) resulting in a neutral pH of 7.0.

phosphorus   An element in solid form that frequently limits primary production because it occurs relatively sparsely in soil and water. It is a major component of commercial fertilizers and the first nutrient to suspect as a cause of natural or cultural eutrophication of freshwater.

photosynthesis   A process of organic synthesis in which solar radiation provides energy for combining the carbon from carbon dioxide (CO_2) and the hydrogen and oxygen of water (H_2O) to produce glucose ($C_6H_{12}O_6$) and other simple compounds with oxygen released back to the environment.

photovoltaic   Pertaining to a solid-state device, the solar cell (or photovoltaic cell), which converts sunlight directly to electricity via electron excitement in selenium or other sensitive metal.

phyto-   A word prefix meaning plant or plantlike (e.g., phytoplankton, phytobenthos).

pioneer community   First interactive association of primary producers, consumers, and decomposers and first stage of primary succession in an area previously devoid of life.

plankton   Plant, animal, bacteria, and other organisms largely adrift in lake, large river, and ocean currents. Although most members (plankters) are microscopic, many species are large enough to see, and some marine jellyfish are very large. Many plankters can control their position in the water column.

plate tectonics   *See* **tectonics.**

point source   In environmental law and technology, any consolidated and readily identifiable origin of pollutants, such as drain pipes, smokestacks, and landfill or feedlot drainage.

policy   A specific or collective course of action defined for individuals and organizations. Policy consists of the rules, regulations, codes, and other officially sanctioned guidance that directs behavior within organizations, both public and private.

politics   The art and science of guiding, enforcing, shaping, or circumventing policy through analysis, persuasion, and, sometimes, deception.

pollution   An alteration of matter or energy level that diminishes resource value, including threats to present or future human health, economic welfare, and other quality of life.

pollution tax   A tax levied on the discharge of harmful waste that is set equal to the environmental cost incurred. Such taxes stimulate pollution control by the taxed parties.

polyclimax   An alternative to the succession concept of a single, climate-controlled climax, which proposes that many variations of community composition and form may appear in equilibrium with climate, depending on circumstances.

polyculture   Mixed-crop interspersion often used in tropical areas to sustain production.

population   In ecology and genetics, a reproductively interactive group of organisms.

population dynamics   Changes in population numbers and related measures as a consequence of changed birthrate, death rate, emigration, and immigration.

porosity   The volume and other dimensions of open-space networks (interstices) typically filled with gasses or liquids, in otherwise solid material.

potential energy   Stored energy materially constrained from kinetic expression such as the constraint of a dam on water exposed to gravity or the forces contained in the structure of atoms and molecules.

precipitation   In a climatic reference, the fall of liquid or solid form (condensing fog, rain, sleet, hail, and snow) from the atmosphere. Dry precipitation is dust or other nonaqueous fallout. Chemical precipitation follows reaction in a solute that forms nonsoluble matter.

predation   In the usual sense, the killing of an animal by another animal and subsequent consumption, either whole or in parts.

prescribed burning   Intentional burning under controlled conditions to prevent wildfire and, usually, to encourage the effect of natural burning in terrestrial ecosystems where fire plays an important role in sustaining fire-adapted vegetation.

preservation   Reservation and maintenance of resources by forestalling use and destruction.

prey   In the usual sense, an animal taken alive and killed by another animal for food. Also, the act of predation (i.e., to prey on).

primary production   The conversion of inorganic nutrient elements into organic matter by solar energy via photosynthesis or chemical energy via chemosynthesis (some bacteria).

prime interest rate   *See* **interest rate.**

prior appropriation   In water law, the granting of water use priority, especially in time of water shortage, in accord with the time of first documented claim for use. The senior users get their allotment before the junior users.

private property   Land, resources, manufactured goods, and ideas owned by a fraction of the public, either an individual or group, which is authorized exclusive control over use and disposal within bounds imposed by public law.

production   Output. In economics, production is the output of goods and services, usually specified for some area and time period. In ecology, production is the output of living matter. Gross production includes that matter which is respired away in metabolism. Net production is gross production minus respiration, including all biomass consumed and decomposed by the next trophic level.

productivity   Ecological productivity is the rate at which output of living matter is generated. In economics, it is the output per unit input. *See also* **production.**

profile   Vertical cross section revealing strata and other structure in soil, water, forest, and other systems.

profit   Income in excess of costs including salaries, materials, rent, and all other costs.

profit incentive   Motivation by promise of greater profit.

profundal zone   Offshore zone of a lake too dark for photosynthesis and including both benthic environments and associated waters.

program planning   Planning organized around provision of specific customer services and associated benefits. Program planning typically is linked to specific funding sources, management goals, and projects.

progressive movement   A political movement, which, at the turn of the twentieth century, promoted resource conservation and fairer distribution of opportunity and wealth among human resources.

project planning   The process of evaluating alternative approaches to goal achievement through design that accomplishes a specific set of objectives. Projects result in a demonstrable change such as physical creation, restoration, or protection; new technologies, procedures, or basic knowledge; or other outputs serving organizational goals.

public   All citizens in a declared jurisdiction, such as a city, state, or nation.

public domain   Land and other environment ultimately owned and controlled by the public, but not assigned to a specific management authority.

public service   Aid offered with the intent of improving the general public welfare.

public trust   The authority placed in government agencies to manage properties owned collectively by the public for the public, both present and future.

purse seine   A commercial fishing net set at the surface and towed in a large circle around fish then drawn to closure under the fish and winched in. Purse seines are used to capture schooling fish such as herring, mackerel, and tuna.

pyrolysis   In biofuel technology, the thermal destruction of organic materials in the absence of oxygen to convert biomass to a mixture of gases and organic vapors.

radiation   Movement of particles or energy waves outward in all directions from an emitting source such as the sun (electromagnetic radiation) and atomic fission (both particle and energy radiation).

radiation balance   An indicator of energy level and dynamics in objects and areas determined from comparison of solar or other radiation input to radiation output.

radioactive isotope   *See* **radioactivity; isotope.**

radioactivity   The spontaneous radiation of particles and energy from unstable nuclei of a radioisotope. The major forms of radiation include energy in the form of gamma rays and alpha and beta particles.

rain shadow   The effect of topography on air movements resulting in low precipitation on the downwind side of mountains.

range   A place that may be roamed over at large; more specifically, any land area producing forage available for on-site feeding by grazing or browsing animals.

rational policy analysis   A policy approach that considers all of the relevant alternatives, identifies and evaluates all the consequences that would follow from the adoption of each alternative, and selects that alternative with the most preferable consequences.

real interest rate   *See* **interest rate.**

recharge   To restore potential use especially in groundwater hydrology or aquifer replenishment. Areas of high-infiltration soil, regolith, and fissured rock form "recharge zones" of exceptionally high groundwater contribution.

reclamation   In environmental and natural resource management, a goal and action taken to improve the services of an artificially degraded ecosystem by creating an ecosystem different from the original in form and function.

recycling   In environmental management, the collection, reprocessing, and reuse of materials in the same or modified form.

redundancy   In community ecology, the backing up of one species' functions by other species in a community.

regional planning   Geographical planning organized within natural or political boundaries and usually focused on modifying the distribution, type, and intensity of land uses for the general improvement of public welfare.

regolith   Unconsolidated inorganic material overlaying rock formations and often topped with stratified soils (with identifiable horizons). In young landscapes, such as those recently exposed by glaciers, regolith often grades upward into unstratified soils of very low organic content.

renewable energy   Biochemical energy that can be renewed through life process or abiotic sources of energy of continuous supply, such as solar radiation and gravity.

renewable resource   A resource that regenerates through biological reproduction or through biogeophysical cycles such as forest, wildlife, soil, and water resources.

rent seeking   In economics, the action of special interests who use their private resources to lobby government to pass laws that economically benefit the group at a net cost to society.

reserve   Identified resources or resource areas set aside for some future use.

reservoir   A storage place. In water resources management, a reservoir is an artificial lake, usually a stream or river backed up behind a dam in an impoundment.

resilience   Ability to bounce back to original shape or condition, including the self-restoration of a population, community, or ecosystem following temporary exposure to environmental stress.

resource partitioning   In ecology, the allocation of spatial, nutritional, and other ecological resources among the populations of a natural community.

respiration   A process of energy release from and material change in chemical compounds as they are reorganized in metabolic pathways. In aerobic respiration, atmospheric oxygen helps drive metabolic cycles by removing toxic buildup of waste carbon and hydrogen by-products in the form of carbon dioxide and water vapor.

rest-rotation grazing   In range management, a grazing system that excludes pasture use for 12 months while other pastures absorb the grazing load.

restoration   In resource management, a goal and actions taken to reestablish the original population, environment, or ecosystem following anthropogenic change.

riparian   Pertaining to stream and riverbanks or, less commonly, to lake and estuary shores. Riparian vegetation often is distinct from adjacent uplands because of groundwater and flooding influences.

riparian right   Water laws that are based on water use by riparian landowners and water return in a reusable state. Riparian law is common in humid regions where little water is lost in use. *See* **prior appropriation** *for comparison.*

river   Any water-filled, open channel in which the direction of flow is determined mostly by gravity in the absence of tidal effects. Rivers and streams are loosely differentiated according to relative size but inconsistently so among regions.

river continuum   The generally predictable continuous transport and change of materials, habitats, and associated aquatic communities from the smallest riparian-dominated headwater streams to the largest rivers.

rock   Natural solids composed of one or more minerals in the lithosphere.

rock cycle   *See* **tectonic.**

roundwood   Raw timber product; logs.

ruminants   Certain herbivorous hoofed mammals with a multichambered stomach, which allows more thorough plant digestion through swallowing, partial digestion, regurgitation, thorough chewing (of cud), swallowing again, microbial fermentation, and final digestion. Sheep and cattle are ruminants.

runoff   Intermittent and perennial surface flow of water on watershed surfaces and through stream channels to surface bodies of water, soil percolation, and groundwater aquifers.

salinity   Saltiness; usually of water or soil. In seawater and estuaries, salinity is expressed as parts per thousand based on weight. The oceans average nearly 35 parts per thousand (3.5%) salinity.

salinization   Salt accumulation, as in soil.

saltwater intrusion   Movement of saltwater into fresh groundwater usually by over pumping the groundwater, which essentially sucks saltwater into the freshwater aquifer.

sanitary landfill   A place of solid waste disposal in which waste is spread in thin layers between layers of soil or fine regolith and care is taken to prevent erosion and contamination of ground- and surface waters.

sapwood   Outer wood, typically lighter in color, that conducts water and dissolved materials.

satellite city   A city that grows at the edge of a well-established city, usually at a major intersection with a superhighway encircling the original city.

savannah   Biome condition in which grassland and open woodland mix.

seagrass   Several plant genera (e.g., *Zostera, Thalassia, Phyllospadix*) that have adapted to shallow estuaries and coastal ocean.

secondary consumers   Carnivorous consumers at the third or higher trophic level.

secondary forest   Immature forest believed to have started from secondary succession on a site disturbed by cutting, fire, storm, or other natural or artificial cause.

secondary producers   Heterotrophs (i.e., organisms that consume or decompose the organic matter that originated through primary production of autotrophs).

secondary succession   *See* **succession.**

sediment   In earth processes, settled particulate material at the bottom of a water body or left behind by receding waters, such as the sediment left after a flood.

sedimentary   Pertaining to sediment and especially to limestones, dolomites, sandstones, and shales that form following sedimentation under the pressure of accumulated sediment.

seed-tree method   Trees are left during cutting to seed the next tree generation.

selective method   Harvest of only trees specified for reasons of management or product quality such as trees of a certain species, size, maturity, or condition.

semiarid   Between xeric (arid desert) and mesic (moist to wet) conditions with between 300 and 500 mm of precipitation.

sere (seral)   The sequence of community changes that link successional stages through time from pioneer to climax.

sewage treatment   Partial to complete removal and neutralization of sewage wastes from water by a variety of means including primary (filtration and sedimentation), secondary (bacterial decomposition promoted by turbulent mixing), and advanced treatment (chemical and biological removal of inorganic nutrients).

sheet erosion   *See* **erosion.**

shelterbelt   Windbreak.

shelterwood method   A rotational selective cutting designed to leave a seed source and protective cover for forest regeneration.

short-duration grazing   In range management, a grazing system in which each pasture is grazed briefly and intensively, followed by a long period of nonuse.

shrublands   Land where the predominate vegetation is less than 3–4 meters tall and woody.

silviculture   Forest culture that manages forest growth, health, and composition, and draws heavily on ecology, soils, and dendrology.

skidding   A method of moving logs by dragging.

sludge   In sewage treatment, the sediment from treated wastewater, usually rich in organic matter, bacteria, viruses, and various natural and synthetic chemicals, many of which are toxic. Sludge often presents a recycling or disposal dilemma.

smog   Atmospheric contamination with a mixture of irritating gasses, colloids, suspended solids, and acidic droplets usually appearing as a haze of various tints. Smog from industrial sources is predominantly sulfur dioxide and suspended solids. Photochemical smog is the result of solar radiation causing chemical reaction of hydrocarbons and nitrogen oxides.

socialism   A social system in which the government owns all or most of the natural resources and economic production capacity.

softwoods   In forest management, trees with low density and easy-to-work wood; in contrast with hardwoods. So many softwoods of the United States are considered conifers that the two designations are nearly the same.

soil   Complex mixture of fine inorganic particles (mostly clay, silt, and sand), decaying organic matter, water, air, roots, bacteria, fungal filaments, and other living organisms.

solar cell   *See* **photovoltaic.**

solar collector   Device for collecting radiant energy from the sun and converting it into heat.

solar energy   Electromagnetic energy directly from the sun. Other forms of energy derive indirectly from solar energy such as wind, ocean and pond thermal gradients, and biomass.

solar heating   Heat in solar energy captured by solar collectors and stored in water, rocks, or other materials for later distribution. Solar heating is done by active means, which rely on additional energy to move the heat, and by passive means without auxiliary energy sources.

solar furnace   Creation of high temperature by artificial focus of solar radiation for electric generation or other purpose.

solar pond   Shallow waterbody managed to create strong thermal stratification and transform the energy extracted from the thermal gradient to electricity generation.

solid waste   Any unwanted or discarded material that is not a liquid or a gas.

specialists   In ecology, narrowly adapted species that take advantage of a small selection of ecological resources, although much more efficiently than broadly adapted generalists.

speciation   Evolution of new species through genetic recombination, mutation, and natural selection of adaptive characteristics by environmental conditions.

species   The most basic classification of living things, which ideally includes all organisms with potential for successful reproductive interaction under natural environmental conditions, and excludes all others.

species diversity   The variety of different species, which is determined by the richness of species numbers and the relative evenness of species abundances. Greatest diversity occurs where richness and evenness of abundance are both high.

spectrophotometry   Measurement technology based on the wavelengths of visible light.

spillover benefits   Benefits accrued to others as a by-product of economic trade, such as the bird-watching farmer who invests more in careful pesticide use and thereby benefits other bird-watchers and hunters at no expense to them. *See also* **externality.**

spillover costs   Costs passed off to others as a by-product of economic trade, such as the costs a power company is not charged for the fish killed by thermal pollution and acid rain, or the respiratory disease and climate warming caused by air pollution. *See also* **externality.**

spodosols   Soils with moderate horizon development in areas supporting coniferous forests in the northeastern United States. They have a leached mineral horizon with a distinct organic horizon and relatively low clay accumulation.

stability   In ecology, maintenance of function despite environmental stress.

stakeholder   Any individual or group significantly affected by management decisions either positively or negatively. Stakeholders include recipients of management services, competitors, partners, and anyone suffering or benefitting from anticipated side effects.

stand   In plant ecology and forest and range management, a homogeneous and close grouping of plants of the same species.

standing crop   *See* **biomass.**

stocking rate   Number of a particular kind of animal grazing on a given area of rangeland.

stomata   Small openings, especially those in plant leaves, which allow uptake of carbon dioxide for photosynthesis.

strategic planning   Planning that identifies appropriate approaches to goal accomplishment on a time horizon typically 5 to 10 years in the future.

strategy   An approach to goal accomplishment.

stratification   Layering, as in the thermal layers formed in lakes and oceans, the story levels in forests, and the horizons of soils.

stratosphere   Rarefied outer part of the atmosphere, extending from about 12 to 30 miles above the earth's surface.

strip cropping   Planting different crops in alternating rows to help reduce depletion of soil nutrients and need for pesticide applications.

strip mining   Removal (stripping) of surface material, the overburden, and digging out the underlying coal, iron, copper, stone, and other materials. Resources are mined from the surface by digging deep trenches, large open pits, or strips along mountain contours.

structure   The spatial organization of matter, including form and arrangement.

sublimation   Direct vaporization of ice and snow, bypassing the liquid form of water.

subsidence   The sinking of part of the earth's crust due to natural processes or to underground excavation, such as a coal mine, or removal of groundwater.

subsistence   Sustaining food and other basic needs for oneself and family using agricultural, wildlife, fisheries, forest, range, water, and other resources.

subspecies   Genetically discrete populations within a species typically associated with one region in the species' range.

suburban sprawl   The often environmentally destructive and socially undesirable spread of domestic housing and retail services without benefit of careful regional planning.

succession   In ecology, the more-or-less orderly and gradual replacement of ecosystem form and function through species colonization and population aging, as each successional stage alters the site in favor of the next. Primary succession starts from bare rock, regolith or open water. Secondary succession starts from some stage of less mature community development following a disturbance such as fire.

superfund   The informal abbreviation given to a fund created for hazardous waste cleanup under federal legislation enacted in 1980.

supply and demand   A basic theory ("law") of economics that states that where price is the only variable controlling consumer demand for and supplier provision of a good or service, and the price increases, the demand decreases and supply increases. Also, as price decreases demand increases and supply decreases. A market equilibrium results where supply equals demand.

surface fire   Forest fire limited to near the ground surface in litter and understory.

surface mining   Removal of soil, subsoil, and other strata and then extracting a mineral deposit found fairly close to the earth's surface. *See* **strip mining.**

sustainable agriculture movement   An approach to agriculture that substitutes polyculture over monoculture and minimizes use of fossil fuels, inorganic fertilizer, pesticide use, and irrigation water.

sustainable development   An evolving philosophy of global resource development that promotes provision for the present generation while protecting opportunities for future generations. Among the main tenets are developing more efficient resource use, maintaining a healthy environment, controlling population growth, and conserving resource options.

sustained yield   Maintenance of renewable resource harvest or other take with no indicated reduction of future yield potential. Maximum sustained yield is the greatest possible quantity of resource that may be taken over the long run. Optimum sustained yield is the amount of resource that can be harvested for maximum long-term profit or public benefit. Optimum yield is usually less than, and no more than, maximum sustained yield.

swamp   Wetland characterized by emergent woody vegetation.

SWOT analysis   A method for inventorying important influences on organizational management effectiveness in four categories: organizational strengths (S) and weaknesses (W), and the opportunities (O) and threats (T) originating outside the organization.

symbionts   Species that functionally complement and mutually benefit one another such as the reproductive gain of flowers pollinated by the consumers of their nectar.

synfuels   Synthetic fuels developed by gasification and liquefaction technology of coal, tars, and other organic matter.

systems analysis   A comprehensive and systematic approach to examining complex systems interactions either directly or through use of mathematical models. For natural resource management, analysis usually involves controlling or monitoring system inputs (usually some material, energy or information flow), monitoring system outputs, and modifying the system to evaluate the effectiveness and efficiency of alternative management approaches.

tailings   Mineral remains following the processing of ore, usually piled or washed into collecting basins. While many are no more harmful than eyesores, other abandoned tailings in numerous locations have been found to leach or radiate environmental contaminants.

tar sand   A natural mixture of fine sand, clays, and bitumen that can be heat-processed to extract and refine oil products.

tectonic(s).   In Earth science, pertaining to Earth's crustal structure and movement. Plate tectonics involves movement of the large solid plates that form the crust, float on the molten mantel, shift against one another and ride over, or slide under each other. Plate tectonics results from volcanic process and can cause earthquakes, lift mountains, rift valleys, and move continents (continental drift).

temperature   Heat concentration. At 15°C, one calorie of heat absorbed by 1 ml (1 cm^3) of water raises the temperature 1°C.

terracing   The slope is converted into an alternating series of broad, level benches and steep slopes.

tertiary sewage treatment   *See* **sewage treatment.**

thermal inversion   The layering of cooler dense air below warmer and lighter air during atmospheric calm. Inversions usually occur where cool air can drain off mountains and under the warmer air at lower elevations.

thermal pollution   Wastewater return flow or other addition of heat that reduces resource values in receiving ecosystems.

thermocline   The transition zone between warm surface water and colder deep water in lakes and oceans. The density change in the thermocline results in a stable stratification and isolation of the layers.

threatened species   One step removed from endangered species status. A legal definition declared by responsible agencies (for U.S. federal law, the Fish and Wildlife Service) when a species is vulnerable to changes that could endanger its continued existence. *See also* **endangered species.**

tide   The periodic fluctuation of oceanic water elevation along coasts created by interaction of Earth's revolution, shoreline shape, and lunar gravity.

till   In agriculture, to plow and otherwise work the soil. In geology, the mixed clay, sand, gravel, and boulders resulting from past glacial deposition.

tillage   Mechanized soil disturbance for crop cultivation by plowing, disking, and other means. Conservation tillage reduces fuel energy needs and erosion loss by minimizing or eliminating mechanized soil disturbance.

tolerance   In ecology, the range of environmental variation over which an organism or population can function. Usually tolerance is judged with respect to the limits set by a single parameter, such as temperature, and is revealed by distribution, relative abundance, and experiment. *See also* **limiting factor.**

tonne   A metric ton, which is equal to 1,000 kg and 2,200 pounds (long ton).

tracheids   Tubelike structures that conduct water and waterborne material through the wood of gymnosperms and angiosperms. In cross sections they appear as tiny pores of different dimensions.

tradeoffs   The losses in benefit incurred by actions taken for any gain in benefit.

tragedy of the commons   Coined by Garrett Hardin who explained why the sustainability of natural resources held in common by all (the commons), but free to anyone, is rapidly degraded (the tragedy). Examples include the present status of international oceanic fisheries and past game and range resource use on U.S. public domain.

transcendentalism   Frequently associated with Emerson, it is a nineteenth-century philosophical movement that was a precursor of the preservation movement led by Muir and others.

transpiration   A physiological process in which water vapor is pumped out leaf openings and water flow up through the plant is maintained.

trawl   A form of fishing net that is dragged behind a boat on or above bottom then winched aboard to remove the catch.

trend analysis   Examination of chronological changes and their interactions for insight into future conditions.

tritium (T- hydrogen-3)   A heavy isotope of hydrogen with an atomic mass number of 3.

trophic   Pertaining to food; used alone or as a suffix in other words, such as eutrophic.

trophic dynamics   Food web processes characterized by energy and material flow through trophic levels and populations in natural communities.

trophic levels   Nutritional levels in a natural community, including primary production at the first level, herbivore production at the second level, and carnivore production at higher levels. Because most energy and carbon is lost at each energy transfer, the depiction of community trophic levels in a diagram forms a pyramid typically with fewer than five levels.

troposphere   The lower atmosphere, which contains most atmospheric mass and weather phenomena. It lies below the stratosphere.

turnover rate   The number of times a population or group of populations is totally replaced through death and birth during some specified period.

ultisols   Forest and savannah soils of the southeastern United States with a leached acidic upper horizon and an often impermeable lower horizon rich in clay.

underpriced resources   Resources with prices that do not reflect all costs associated with their development. The costs incurred by acid rain, for example, are not included in the price of coal. Underpricing favors resource exploitation over conservation.

understory   In ecology and forestry, the lower strata of forest vegetation beneath the highest stratum, known as the forest canopy or overstory.

uneven-aged management   Method of forest management in which trees in a given stand are maintained at many ages and sizes to permit continuous natural regeneration.

upwelling   Locations along continental shelves where ocean currents force cold, nutrient-rich water to the surface, usually on the continent's west side.

urban areas   Concentrations of human populations in towns and cities and their suburban peripheries.

urbanization   The net movement of people into town and city concentrations. In the United States, urbanization rate is estimated by the change in percentage of the population living in officially defined urban settlements of 2,500 people or more.

vegetation   The collective structure of plant communities.

vertisols   Soils, mostly of the southcentral United States, with a high clay content that shrinks and cracks when wet.

vessels   Elongated cells important in conducting water and waterborne materials in angiosperm wood. They contribute to the appearance of pores in cross sections.

viable population   A population that is self-sustaining and not at risk of extinction.

vision   In the planning process, vision is a concept of the future desired condition for the management system of an organization, including most importantly the condition of the managed resources. Shared vision planning involves many partners and stakeholders who, in a crucial first step, develop one shared vision, held in common, for some desired future condition.

watershed   The region delivering water and its transported material to any selected point along a topographic gradient, usually a stream or river channel. Watersheds have depth as well as topographic surface, and groundwater as well as surface water is shed from both the surface and underlying rock formations.

watt   Measure of power used for electricity, equivalent to one joule-second.

weather   Day to day dynamics of the atmosphere including air circulation patterns, temperature variation, precipitation, and other measures. *See* **climate.**

weathering   Physical and chemical disintegration of solid materials by natural processes of abrasion, impact, freezing and thawing, and erosion.

wetland   Land area temporarily or permanently flooded by marine or inland waters to depths shallow enough for rooted wetland plants to occur under otherwise favorable conditions. Wetlands include marshes, bogs, swamps, and barrens, such as mudflats.

wilderness   Immense wild area with no vehicular access and only temporary human visitation. Under the Wilderness Act of 1964, wilderness is a public land-use classification for lands sufficiently remote and unmodified by use.

wildlife   In the broadest sense, all free-ranging wild organisms, including plants. More usually, the term is applied to attractive or otherwise valued wild animals in terrestrial ecosystems, especially birds and mammals.

wind farm   Wind turbines grouped in suitably windy locations to "harvest" reliable wind resources for electricity generation.

wind turbines   Wind-driven generators of electricity, consisting of blades, rotor, transmission, electrical generator, and control system, all mounted on a tower.

windbreak   Any impediment to wind, but commonly rows of dense trees or tall hedges planted to reduce wind erosion on cultivated land or to protect homes in open country.

woodland   A terrestrial ecosystem classification for any of a number of plant communities dominated by short trees and shrubs, often in a mix with grasslands and savannahs.

work   In physics, the outcome of force exerted on mass expressed in motion. The power held in energy is defined in terms of the work it can do.

xeric   Dry; annual precipitation less than 300 mm.

xeriscaping   Urban landscaping sustainable under desert climates.

yarding   In forestry, the process of moving logs from where they fell to a transport point or landing.

yellowcake   Uranium oxide. Also called uranium concentrate, yellowcake is 85% pure uranium in an isotopic mixture of 0.7 percent fissionable uranium-235 and 99.3% nonfissionable uranium-238.

zoning   Land-use regulation, planning, and management by zone. Although the term is most often used in municipal and county regulation of private land use, any designated land use of public lands also is a form of zoned use.

zoo-   A word prefix meaning animal (e.g., zooplankton, zoobenthos).

Index